nature

The Living Record of Science

《自然》学科经典系列

总顾问：李政道（Tsung-Dao Lee）

英方总主编：Sir John Maddox
Sir Philip Campbell

中方总主编：路甬祥

生命科学的进程 I
PROGRESS IN LIFE SCIENCES I

（英汉对照）

主编：许智宏

外语教学与研究出版社 · 麦克米伦教育 · 《自然》旗下期刊与服务集合

FOREIGN LANGUAGE TEACHING AND RESEARCH PRESS · MACMILLAN EDUCATION · NATURE PORTFOLIO

北京 BEIJING

图书在版编目（CIP）数据

生命科学的进程 . I：英汉对照 ／ 许智宏主编 . -- 北京 ：外语教学与研究出版社，2021.4

（《自然》学科经典系列／路甬祥等总主编）

ISBN 978-7-5213-2510-2

I．①生… II．①许… III．①生命科学－文集－英、汉 IV．①Q1-53

中国版本图书馆 CIP 数据核字（2021）第 056339 号

出 版 人	徐建忠
项目统筹	章思英
项目负责	刘晓楠　顾海成
责任编辑	刘晓楠
责任校对	夏洁媛
封面设计	孙莉明　高　蕾
版式设计	孙莉明
出版发行	外语教学与研究出版社
社　　址	北京市西三环北路 19 号（100089）
网　　址	http://www.fltrp.com
印　　刷	北京华联印刷有限公司
开　　本	787×1092　1/16
印　　张	61
版　　次	2021 年 6 月第 1 版　2021 年 6 月第 1 次印刷
书　　号	ISBN 978-7-5213-2510-2
定　　价	568.00 元

购书咨询：（010）88819926　电子邮箱：club@fltrp.com

外研书店：https://waiyants.tmall.com

凡印刷、装订质量问题，请联系我社印制部

联系电话：（010）61207896　电子邮箱：zhijian@fltrp.com

凡侵权、盗版书籍线索，请联系我社法律事务部

举报电话：（010）88817519　电子邮箱：banquan@fltrp.com

物料号：325100001

记载人类文明
沟通世界文化
www.fltrp.com

《自然》学科经典系列

（英汉对照）

总顾问：李政道（Tsung-Dao Lee）

英方总主编：Sir John Maddox
Sir Philip Campbell

中方总主编：路甬祥

英方编委：

Philip Ball

Arnout Jacobs

Magdalena Skipper

中方编委（以姓氏笔画为序）：

万立骏

朱道本

许智宏

武向平

赵忠贤

滕吉文

生命科学的进程

（英汉对照）

主编：许智宏

审稿专家 （以姓氏笔画为序）

王乃彦	王晓晨	田伟生	冯兴无	邢 松	同号文	吕 扬
刘 力	刘 武	刘 京	刘京国	江丕栋	杨 志	吴秀杰
吴新智	陈平富	林圣龙	昌增益	金 侠	金 城	周筠梅
赵见高	赵凌霞	秦志海	莫 辐	顾孝诚	徐 星	董 为
程祝宽						

翻译工作组稿人 （以姓氏笔画为序）

王耀杨	刘 明	关秀清	李 琦	何 铭	蔡 迪

翻译人员 （以姓氏笔画为序）

王耀杨	毛晨晖	田晓阳	刘 霞	刘冉冉	刘振明	刘晓辉
刘皓芳	李 飞	李世媛	吴 彦	张玉光	张立召	张锦彬
周志华	郑建全	赵凤轩	姜 薇	高如丽	韩玲俐	管 冰

校对人员 （以姓氏笔画为序）

王帅帅	王阳兰	王丽霞	王晓萌	王晓蕾	甘秋玲	田晓阳
丛 岚	乔萌萌	刘 明	刘 征	刘晓辉	齐文静	关秀清
孙 娟	孙 琳	苏 慧	李 琦	李世媛	李红菊	何 铭
何思源	张 帆	张玉光	张世馥	张竞凤	张媛媛	周玉凤
郑 琪	郑期彤	赵凤轩	夏洁媛	顾海成	唐 颖	黄小斌
崔天明	韩少卿	韩玲俐	曾红芳	游 丹	蔡 迪	管 冰

Foreword by Tsung Dao Lee

We can appreciate the significance of natural science to human life in two aspects. Materially, natural science has achieved many breakthroughs, particularly in the past hundred years or so, which have brought about revolutionary changes to human life. At the same time, the spirit of science has taken an ever-deepening root in the hearts of the people. Instead of alleging that science is omnipotent, the spirit of science emphasizes down-to-earth and scrupulous research, and critical and creative courage. More importantly, it stands for the dedication to working for the wellbeing of humankind. This is perhaps more meaningful than scientific and technological achievements themselves, which may be closely related to specific backgrounds of the times. The spirit of science, on the other hand, constitutes a most valuable and constant component of humankind's spiritual civilization.

In this sense, *Nature: The Living Record of Science* presents not only the historical paths of the various fields of natural science for almost a century and a half, but also the unremitting spirit of numerous scientists in their pursuit of truth. One of the most influential science journals in the whole world, *Nature*, reflects a general picture of different branches of science in different stages of development. It has also reported many of the most important discoveries in modern science. The collection of papers in this series includes breakthroughs such as the special theory of relativity, the maturing of quantum mechanics and the mapping of the human genome sequence. In addition, the editors have not shunned papers which were proved to be wrong after publication. Included also are the academic debates over the relevant topics. This speaks volumes of their vision and broadmindedness. Arduous is the road of science; behind any success are countless failures unknown to outsiders. But such failures have laid the foundation for success in later times and thus should not be forgotten. The comprehensive and thoughtful coverage of these volumes will enable readers to gain a better understanding of the achievements that have tremendously promoted the progress of science and technology, the evolution of key and cutting-edge issues of the relevant fields, the inspiration brought about by academic controversies, the efforts and hardships behind these achievements, and the true meaning of the spirit of science.

China now enjoys unprecedented opportunities for the development of science and technology. At the policy level, the state has created a fine environment for scientific research by formulating medium- and long-term development programs. As for science and technology, development in the past decades has built up a solid foundation of research and a rich pool of talent. Some major topics at present include how to introduce the cream of academic research from abroad, to promote Sino-foreign exchange in science and technology, to further promote the spirit of science, and to raise China's development in this respect to the advanced international level. The co-publication of *Nature: The Living Record of Science* by the Foreign Language Teaching and Research

李政道序

如何认识自然科学对人类生活的意义，可以从两个方面来分析：一是物质层面，尤其是近百年来，自然科学取得了很多跨越性的发展，给人类生活带来了许多革命性的变化；二是精神层面，科学精神日益深入人心，这种科学精神并不是认为科学万能、科学可以解决一切问题，它应该是一种老老实实、严谨缜密、又勇于批判和创造的精神，更重要的是，它具有一种坚持为人类福祉而奋斗的信念。这种科学精神可能比物质意义上的科技成就更重要，因为技术进步的影响可能与时代具体的背景有密切关系，但科学精神却永远是人类精神文明中最可宝贵的一部分。

从这个意义上，这套《〈自然〉百年科学经典》丛书的出版，不仅为读者呈现了一个多世纪以来自然科学各个领域发展的历史轨迹，更重要的是，它展现了无数科学家在追求真理的过程中艰难求索、百折不回的精神世界。《自然》作为全世界最有影响力的科学期刊之一，反映了各个学科在不同发展阶段的概貌，报道了现代科学中最重要的发现。这套丛书的可贵之处在于，它不仅汇聚了狭义相对论的提出、量子理论的成熟、人类基因组测序完成这些具有开创性和突破性的大事件、大成就，还将一些后来被证明是错误的文章囊括进来，并展现了围绕同一论题进行的学术争鸣，这是一种难得的眼光和胸怀。科学之路是艰辛的，成功背后有更多不为人知的失败，前人的失败是我们今日成功的基石，这些努力不应该被忘记。因此，《〈自然〉百年科学经典》这套丛书不但能让读者了解对人类科技进步有着巨大贡献的科学成果，以及科学中的焦点和前沿问题的演变轨迹，更能使有志于科学研究的人感受到思想激辩带来的火花和收获背后的艰苦努力，帮助他们理解科学精神的真意。

当前，中国科学技术的发展面临着历史上前所未有的机遇，国家已经制定了中长期科学和技术发展纲要，为科学研究创造了良好的制度环境，同时中国的科学技术经过多年的积累也已经具备了很好的理论和人才基础。如何进一步引进国外的学术精华，促进中外科技交流，使科学精神深入人心，使中国的科技水平迅速提升至世界前列就成为这一阶段的重要课题。因此，外语教学与研究出版社和麦克米伦出

Press, Macmillan Publishers Limited and the Nature Publishing Group will prove to be a huge contribution to the country's relevant endeavors. I sincerely wish for its success.

Science is a cause that does not have a finishing line, which is exactly the eternal charm of science and the source of inspiration for scientists to explore new frontiers. It is a cause worthy of our uttermost exertion.

T. D. Lee

Editor's note: The foreword was originally written for the ten-volume *Nature: The Living Record of Science.*

版集团合作出版这套《〈自然〉百年科学经典》丛书，对中国的科技发展可谓贡献巨大，我衷心希望这套丛书的出版获得极大成功，促进全民族的科技振兴。

科学的事业永无止境。这是科学的永恒魅力所在，也是我们砥砺自身、不断求索的动力所在。这样的事业，值得我们全力以赴。

李政道

编者注：此篇原为《〈自然〉百年科学经典》（十卷本）的序。

Foreword by Lu Yongxiang

Since the birth of modern science, and in particular throughout the 20th century, we have continuously deepened our understanding of Nature, and developed more means and methods to make use of natural resources. Technological innovation and industrial progress have become decisive factors in promoting unprecedented development of productive forces and the progress of society, and have greatly improved the mode of production and the way we live.

The 20th century witnessed many revolutions in science. The establishment and development of quantum theory and the theory of relativity have changed our concept of time and space, and have given us a unified understanding of matter and energy. They served as a theoretical foundation upon which a series of major scientific discoveries and technological inventions were made. The discovery of the structure of DNA transformed our understanding of heredity and helped to unify our vision of the biological world. As a corner-stone in biology, DNA research has exerted a far-reaching influence on modern agriculture and medicine. The development of information science has provided a theoretical basis for computer science, communication technology, intelligent manufacturing, understanding of human cognition, and even economic and social studies. The theory of continental drift and plate tectonics has had important implications for seismology, geology of ore deposits, palaeontology, and palaeoclimatology. New understandings about the cosmos have enabled us to know in general terms, and also in many details, how elementary particles and chemical elements were formed, and how this led to the formation of molecules and the appearance of life, and even the origin and evolution of the entire universe.

The 20th century also witnessed revolutions in technology. Breakthroughs in fundamental research, coupled to the stimulus of market forces, have led to unparalleled technological achievements. Energy, materials, information, aviation and aeronautics, and biological medicine have undergone dramatic changes. Specifically, new energy technologies have helped to promote social development; new materials technologies promote the growth of manufacturing and industrial prosperity; information technology has ushered in the Internet and the pervasive role of computing; aviation and aeronautical technology has broadened our vision and mobility, and has ultimately led to the exploration of the universe beyond our planet; and improvements in medical and biological technology have enabled people to live much better, healthier lives.

Outstanding achievements in science and technology made in China during its long history have contributed to the survival, development and continuation of the Chinese nation. The country remained ahead of Europe for several hundred years before the 15th century. As Joseph Needham's studies demonstrated, a great many discoveries and innovations in understanding or practical capability—from the shape of snowflakes to the art of cartography, the circulation of the blood, the invention of paper and sericulture

路甬祥序

自近代科学诞生以来，特别是 20 世纪以来，随着人类对自然的认识不断加深，随着人类利用自然资源的手段与方法不断丰富，技术创新、产业进步已成为推动生产力空前发展和人类社会进步的决定性因素，极大地改变了人类的生产与生活方式，使人类社会发生了显著的变化。

20 世纪是科学革命的世纪。量子理论和相对论的创立与发展，改变了人类的时空观和对物质与能量统一性的认识，成为了 20 世纪一系列重大科学发现和技术发明的理论基石；DNA 双螺旋结构模型的建立，标志着人类在揭示生命遗传奥秘方面迈出了具有里程碑意义的一步，奠定了生物技术的基础，对现代农业和医学的发展产生了深远影响；信息科学的发展为计算机科学、通信技术、智能制造提供了知识源泉，并为人类认知、经济学和社会学研究等提供了理论基础；大陆漂移学说和板块构造理论，对地震学、矿床学、古生物地质学、古气候学具有重要的指导作用；新的宇宙演化观念的建立为人们勾画出了基本粒子和化学元素的产生、分子的形成和生命的出现，乃至整个宇宙的起源和演化的图景。

20 世纪也是技术革命的世纪。基础研究的重大突破和市场的强劲拉动，使人类在技术领域获得了前所未有的成就，能源、材料、信息、航空航天、生物医学等领域发生了全新变化。新能源技术为人类社会发展提供了多元化的动力；新材料技术为人类生活和科技进步提供了丰富的物质材料基础，推动了制造业的发展和工业的繁荣；信息技术使人类迈入了信息和网络时代；航空航天技术拓展了人类的活动空间和视野；医学与生物技术的进展极大地提高了人类的生活质量和健康水平。

历史上，中国曾经创造出辉煌的科学技术，支撑了中华民族的生存、发展和延续。在 15 世纪之前的数百年里，中国的科技水平曾遥遥领先于欧洲。李约瑟博士曾经指出，从雪花的形状到绘图的艺术、血液循环、造纸、养蚕，包括更有名的指南针和

and, most famously, of compasses and gunpowder—were first made in China. The Four Great Inventions in ancient China have influenced the development process of the world. Ancient Chinese astronomical records are still used today by astronomers seeking to understand astrophysical phenomena. Thus Chinese as well as other long-standing civilizations in the world deserve to be credited as important sources of modern science and technology.

Scientific and technological revolutions in 17th and 18th century Europe, the First and Second Industrial Revolutions in the 18th and 19th centuries, and the spread of modern science education and knowledge sped up the modernization process of the West. During these centuries, China lagged behind.

Defeat in the Opium War (1840–1842) served as strong warning to the ancient Chinese empire. Around and after the time of the launch of *Nature* in 1869, elite intellectuals in China had come to see the importance that science and technology had towards the country's development. Many scholars went to study in Western higher education and research institutions, and some made outstanding contributions to science. Many students who had completed their studies and research in the West returned to China, and their work, together with that of home colleagues, laid the foundation for the development of modern science and technology in the country.

In the six decades since the founding of the People's Republic of China, the country has made a series of achievements in science and technology. Chinese scientists independently developed the atomic bomb, the hydrogen bomb and artificial satellite within a short period of time. The continental oil generation theory led to the discovery of the Daqing oil field in the northeast. Chinese scientists also succeeded in synthesizing bovine insulin, the first protein to be made by synthetic chemical methods. The development and popularization of hybrid rice strains have significantly increased the yields from rice cultivation, benefiting hundreds of millions of people across the world. Breakthroughs in many other fields, such as materials science, aeronautics and life science, all represent China's progress in modern science and technology.

As the Chinese economy continues to enjoy rapid growth, scientific research is also producing increasing results. Many of these important results have been published in first-class international science journals such as *Nature*. This has expanded the influence of Chinese science research, and promoted exchange and cooperation between Chinese scientists with colleagues in other countries. All these indicate that China has become a significant global force in science and technology and that greater progress is expected in the future.

Science journals, which developed alongside modern science, play an essential role in faithfully recording the path of science, as well as spreading and promoting modern science. Such journals report academic development in a timely manner, provide a platform for scientists to exchange ideas and methods, explore the future direction of science, stimulate academic debates, promote academic prosperity, and help the public

火药,都是首先由中国人发现或发明的。中国的"四大发明"影响了世界的发展进程,古代中国的天文记录至今仍为天文学家在研究天体物理现象时所使用。中华文明同其他悠久的人类文明一样,成为了近代科学技术的重要源泉。

但我们也要清醒地看到,发生在17~18世纪欧洲的科学革命、18~19世纪的第一次和第二次工业革命,以及现代科学教育与知识的传播,加快了西方现代化的进程,同时也拉大了中国与西方的差距。

鸦片战争的失败给古老的中华帝国敲响了警钟。就在《自然》创刊前后,中国的一批精英分子看到了科学技术对于国家发展的重要性,一批批中国学子到西方高校及研究机构学习,其中一些人在科学领域作出了杰出的贡献。同时,一大批留学生回国,同国内的知识分子一道,为现代科学技术在中国的发展奠定了基础。

新中国成立60年来,中国在科学技术方面取得了一系列成就。在很短的时间里,独立自主地研制出"两弹一星";在陆相生油理论指导下,发现了大庆油田;成功合成了牛胰岛素,这是世界上第一个通过化学方法人工合成的蛋白质;杂交水稻研发及其品种的普及,显著提高了水稻产量,造福了全世界几亿人。中国人在材料科学、航天、生命科学等许多领域,也取得了一批重要成果。这些都展现了中国在现代科技领域所取得的巨大进步。

当前,中国经济持续快速增长,科研产出日益增加,中国的许多重要成果已经发表在像《自然》这样的世界一流的科技期刊上,扩大了中国科学研究的影响,推动了中国科学家和国外同行的交流与合作。现在,中国已成为世界重要的科技力量。可以预见,在未来,中国将在科学和技术方面取得更大的进步。

伴随着现代科学产生的科技期刊,忠实地记录了科学发展的轨迹,在传播和促进现代科学的发展方面发挥了重要的作用。科技期刊及时地报道学术进展,交流科学思想和方法,探讨未来发展方向,以带动学术争鸣与繁荣,促进公众对科学的理解。中国在推动科技进步的同时,应更加重视科技期刊的发展,学习包括《自然》在内

to better understand science. While promoting science and technology, China should place greater emphasis on the betterment of science journals. We should draw on the philosophies and methods of leading science journals such as *Nature*, improve the standards of digital access, and enable some of our own science journals to extend their impact beyond China in the not too distant future so that they can serve as an advanced platform for the development of science and technology in our country.

In the 20th century, *Nature* published many remarkable discoveries in disciplines such as biology, geoscience, environmental science, materials science, and physics. The selection and publication of the best of the more than 100,000 articles in *Nature* over the past 150 years or so in English-Chinese bilingual format is a highly meaningful joint undertaking by the Foreign Language Teaching and Research Press, Macmillan Publishers Limited and the Nature Publishing Group. I believe that *Nature: The Living Record of Science* will help bridge cultural differences, promote international cooperation in science and technology, prove to be high-standard readings for its intended large audience, and play a positive role in improving scientific and technological research in our country. I fully endorse and support the project.

The volumes offer a picture of the course of science for nearly 150 years, from which we can explore how science develops, draw inspiration for new ideas and wisdom, and learn from the unremitting spirit of scientists in research. Reading these articles is like vicariously experiencing the great discoveries by scientific giants in the past, which will enable us to see wider, think deeper, work better, and aim higher. I believe this collection will also help interested readers from other walks of life to gain a better understanding of and care more about science, thus increasing their respect for and confidence in science.

I should like to take this opportunity to express my appreciation for the vision and joint efforts of Foreign Language Teaching and Research Press, Macmillan Publishers Limited and the Nature Publishing Group in bringing forth this monumental work, and my thanks to all the translators, reviewers and editors for their exertions in maintaining its high quality.

President of Chinese Academy of Sciences

Editor's note: The foreword was originally written for the ten-volume *Nature: The Living Record of Science*.

的世界先进科技期刊的办刊理念和方法，提高期刊的数字化水平，使中国的一些科技期刊早日具备世界影响力，为中国科学技术的发展创建高水平的平台。

20 世纪的生物学、地球科学、环境科学、材料科学和物理学等领域的许多重大发现，都被记录在《自然》上。外语教学与研究出版社、麦克米伦出版集团和自然出版集团携手合作，从《自然》创刊近一百五十年来发表过的十万余篇论文中撷取精华，并译成中文，以双语的形式呈现，纂为《〈自然〉百年科学经典》丛书。我认为这是一项很有意义的工作，并相信本套丛书的出版将跨越不同的文化，促进国际间的科技交流，向广大中国读者提供高水平的科学技术知识文献，为提升我国科学技术研发水平发挥积极的作用。我赞成并积极支持此项工作。

丛书将带领我们回顾近一百五十年来科学的发展历程，从中探索科学发展的规律，寻求思想和智慧的启迪，感受科学家们百折不挠的钻研精神。阅读这套丛书，读者可以重温科学史上一些科学巨匠作出重大科学发现的历程，拓宽视野，拓展思路，提升科研能力，提高科学道德。我相信，这套丛书一定能成为社会各界的良师益友，增强他们对科学的了解与热情，加深他们对科学的尊重与信心。

借此机会向外语教学与研究出版社、麦克米伦出版集团、自然出版集团策划出版本丛书的眼光和魄力表示赞赏，对翻译者、审校者和编辑者为保证丛书质量付出的辛苦劳动表示感谢。

是为序。

中国科学院院长

编者注：此篇原为《〈自然〉百年科学经典》（十卷本）的序。

Foreword by Xu Zhihong

In a certain sense, all human knowledge begins with the observation of natural phenomena. The assertion that "Man patterns himself on the operation of the earth; the earth patterns itself on the operation of heaven; heaven patterns itself on the operation of the Tao; the Tao patterns itself on what is natural" is found in the Chinese philosophical classic, *Tao Te Ching*. In English, the word *nature* comes to us via the Latin *natura*, which itself was influenced by the Greek *phusis* (φύσις). This term, which originally meant "growth", referred to the intrinsic relations that plants, animals, and other features of the world develop of their own accord. In these ancient concepts, human beings are merely part of nature; in modern terms, part of the Earth's biosphere. The subject matter of the life sciences, then, has always been integral to the concept of nature, and the connection between humanity and the biosphere has been understood from time immemorial.

Life sciences, or what is traditionally called biology, have become an important category of modern natural science, comprising a variety of branches and inclusive methods, from which new objects of study continually emerge. Since the naturalist Jean-Baptiste Lamarck first used the term "biology" in an 1802 publication, bioscience has experienced dramatic change. Starting from the description and generalization of biological characteristics through observation and documentation in the early years of the discipline, researchers moved on to exploring inherent causal connections by artificially changing given conditions in well-designed experiments. Furthermore, continual advances from the 19th century onward, in disciplines such as physics, chemistry, mathematics and computer science, have enabled us to focus our investigations at varying micro and macro scales. Molecular biology, biochemistry and structural biology enable us to understand the structure and function of biological macromolecules at the molecular and atomic scales. Traditional cytology has gradually evolved into cell biology and molecular cell biology, which investigates cell structure and function, cell differentiation, the formation of tissues and organs, growth and development, senescence, disease and immunity, and other life phenomena at the cellular and molecular levels. Genomics, functional genomics and epigenetics allow us to obtain a better understanding of gene function, as well as the complex mechanisms of gene expression and regulation. Evolutionary biology and ecology enlarge our horizons such that we may inspect the entire biological world and even contemplate the origin of life itself, and study the interaction between different species in a specific ecosystem and the interaction between organisms and the environment. And while the rapid advancement of life sciences has opened the door for humanity to understand the essence of life, it has perhaps also made us realize that the complexity and

许智宏序

人类对大千世界的认识，从某种意义上来说，始于对自然的观察。《道德经》中便有"人法地，地法天，天法道，道法自然"的说法。英文中 Nature 一词，其起源可以追溯到希腊文，原本指的就是植物、动物及世间其他事物演化而来的内在特征。人类也只是地球上生物圈的一个成员。如此看来，生物学与自然、人与生物圈之间的联系，实可谓源远流长。

传统上我们所说的生物学，或者当今所说的生命科学，是现代自然科学的重要门类，所涉及的研究领域纷繁芜杂，研究方法兼容并蓄，新的研究对象层出不穷。自从 1802 年博物学家拉马克最先提出"生物学"这一概念至今，生物学研究已经发生了天翻地覆的变化。从最初通过观察、记录的方法描述并归纳生物的特征，到通过设计实验，人为地改变某些条件来观测生物的变化和反应，进而探究其内在的因果联系。而 19 世纪物理学、化学、数学、计算机科学等学科日臻成熟，人们也逐渐可以将目光投向更为微观或者宏观的尺度。分子生物学、生物化学、结构生物学使人们可以从分子和原子的尺度认识生物大分子的结构和功能，传统的细胞学也逐渐演变为细胞生物学乃至分子细胞生物学，使人们可以在细胞和分子水平上研究细胞的结构和功能、细胞分化、组织和器官的形成、生长发育和衰老、疾病和免疫等等生命现象。基因组学、功能基因组学、表观遗传学则让人们更好地了解基因的功能及其表达调控的复杂机理。而演化生物学和生态学则帮助人们从时间和空间的视角审视整个生物界的演化乃至生命起源、研究特定生态系统内不同物种之间以及生物与环境的相互作用。生命科学的突飞猛进，打开了人类认识生命本质的大门，同时让我们逐渐认识到生命科学的复杂性和深刻性远超想象，而这也呼唤着科学家之间

profoundness of the life sciences may reach far beyond our imagination. This also calls for more extensive and in-depth cooperation between scientists, and the importance of scientific communication has become increasingly prominent. In this sense, today's life sciences are becoming increasingly interdisciplinary.

Nature was one of the world's earliest academic journals, and has played a pivotal role in academic exchange and the dissemination of knowledge. As a vital discipline in natural science, life sciences have been a major focus of *Nature* since it first began publication. *Progress in Life Sciences* is a collection co-published by Foreign Language Teaching and Research Press, Macmillan Education and Nature Portfolio, which includes the most influential bioscience papers published in *Nature* since its inception in 1869. While no single science journal, even the world's most prominent, can be expected to portray each and every aspect of the evolution of bioscience, or of its branches and interdisciplinary subjects, the papers published in *Nature* which are reproduced in this volume most certainly provide an overview of the history of bioscience and its seminal moments, and a glimpse of future directions.

Nature has always been an important platform for biologists to publish breakthroughs in life sciences and to report upon important biological phenomena. Many studies that are now seen to have changed the course of human history have been published in *Nature*. Watson and Crick's 1953 paper on the double helix structure of deoxyribonucleic acid (DNA) explained the nature of biological heredity, spurring the development of modern genetics, particularly molecular genetics, the cornerstone of modern genomics. Wilmut and his colleagues astonished the world with the birth of the first mammal cloned from an adult somatic cell, the cloned sheep Dolly born in 1996. In 2001, the first draft sequence of the human genome was released, symbolizing the arrival of a brand-new era in life sciences. The data were the result of a collaboration involving 20 sequencing centres from six different countries, including China, and the project became a model for international collaboration in academia.

Challenges remain for life sciences in the 21st century. Why does biological aging occur? How did the brain evolve? What is the biological basis of consciousness? Could the efficiency of photosynthesis be further improved? Exactly how and when did life on Earth originate? What was the cause of the Cambrian explosion? These problems have remained unresolved for years, and now await the tireless efforts of a new generation. With further investigation in life sciences and related disciplines, new questions and challenges will surely follow. Part of the wonder of life sciences is our own unique position: as living organisms, we study ourselves and other living beings, are able to contemplate the past and

更为广泛和深层次的合作，科学交流的重要性也日益凸显。从这个意义上说，今天的生命科学体现出越来越多的交叉学科的特征。

作为世界上最早的学术期刊之一，《自然》在传播科学知识和促进学术交流上起到了举足轻重的作用。而生命科学作为自然科学中重要的学科，在《自然》创刊以来便一直是其关注的重点。这套《〈自然〉学科经典系列——生命科学的进程》是外语教学与研究出版社、麦克米伦教育与《自然》杂志合作出版的《自然》杂志分学科论文精选集，收录了国际顶尖学术期刊《自然》自 1869 年创刊以来生命科学及其相关领域最具代表性和开创性的经典论文。尽管如此，对于生物学这门自然科学中的重要学科，发展到今天的生命科学，其研究对象之多，加上生命现象之复杂，即便是像《自然》这样具有世界级影响力的顶级学术期刊，也难言囊括了生命科学及其各个分支学科和交叉学科的历史发展全貌。但至少借助发表在《自然》上的文章，读者可以从中一瞥学科发展的基本脉络，亦可以一探其未来的发展趋势。

一直以来，《自然》都是生物学家发表生命科学研究中突破性进展及报道重要生命现象的重要平台。不少如今看来彻底改写人类历史进程的研究都发表于《自然》上。沃森和克里克 1953 年关于脱氧核糖核酸（DNA）双螺旋结构的论文解释了生物遗传的本质，同时极大地推动了当代遗传学，特别是分子遗传学的发展，也成为现代基因组学的发端。1997 年报道了世界首例由成年体细胞克隆获得的哺乳动物——克隆羊多莉诞生，威尔穆特小组的研究也将克隆技术推向了崭新的阶段。2001 年人类基因组第一个草图序列发布，更是标志着生命科学迎来了新纪元。而测序过程中来自 6 个国家（包括中国）的 20 个测序中心的通力配合，更为学术界的国际间合作提供了范本和榜样。

21 世纪，生命科学仍然面临着诸多挑战。生物衰老的机制是怎样的？人类大脑如何进化？意识的本质是什么？植物光合作用的效率能进一步提高吗？地球上的生命何时以及如何起源？寒武纪大爆发的起因是什么？这些多年以来悬而未决的难题仍然有待新一代学者们孜孜不倦的探索。而随着生命科学以及其他相关学科研究的进一步深入，新的难题也一定会接踵而至。人类作为生命体，研究自身和其他生命

the future, and, via our conscious agency, can decide upon which actions are required of us. Amidst these challenges and mysteries, various new concepts, techniques, and methods developed in life sciences have accelerated the improvement of contemporary medicine and agriculture, via gene therapy, stem cell therapy, gene editing, molecular design breeding, and so on. I hope that *Progress in Life Sciences*, with the most influential papers from *Nature*, could encourage you, enlighten you, and enrich your knowledge of bioscience, and I sincerely hope that there will be more and more young people entering research in life sciences, dedicating their youth and talent to this most spectacular of journeys.

Xu Zhihong

Chair Professor of PKU School of Life Sciences

体，并思考它们过去和未来，探讨当下应该采取的行动，这便是生命科学的魅力所在。同时，生命科学研究发展出来的各种新概念、新技术、新方法，又推动了当代医学、农业的发展，如基因治疗、干细胞治疗、基因编辑、分子设计育种，等等。愿这套承载着《自然》杂志 150 余年精华文献的《〈自然〉学科经典系列——生命科学的进程》可以为读者开启窥探生命科学奥秘之窗，拓展对生命世界的认知，亦希望有更多年轻人可以踏入生命科学研究之门，投身这一激动人心的伟大征途。

北京大学生命科学学院讲席教授

2021 年 3 月 2 日

Contents
目　　录

V

Volume I

On the Fertilisation of Winter-flowering Plants

A. W. Bennett

Editor's Note

If many plants rely on insects for pollination, how is this accomplished in those plants that habitually flower in winter, when flying insects are few? Such was the question posed in the very first research paper to be published in *Nature*. The answer, according to Alfred Bennett, a London publisher and bookseller, is that many winter-flowering plants are self-fertile, the pollen discharging from the anthers while the flower is yet a bud. This is obviously not possible in those plants in which male and female organs are carried in separate flowers. In those cases, such as the hazel (*Corylus*), pollen is shed very liberally from male flowers close to female ones, which "favours the scattering of the pollen by the least breath of wind".

THAT the stamens are the male organ of the flower, forming unitedly what the older writers called the "androecium", is a fact familiar not only to the scientific man, but to the ordinary observer. The earlier botanists formed the natural conclusion that the stamens and pistil in a flower are intended mutually to play the part of male and female organs to one another. Sprengel was the first to point out, about the year 1790, that in many plants the arrangement of the organs is such, that this mutual interchange of offices in the same flower is impossible; and more recently, Hildebrand in Germany, and Darwin in England, have investigated the very important part played by insects in the fertilisation of the pistil of one individual by the stamens of another individual of the same species. It is now generally admitted by botanists that cross-fertilisation is the rule rather than the exception. The various contrivances for ensuring it, to which Mr. Darwin has especially called the attention of botanists, are most beautiful and interesting; and the field thus opened out is one which, from its extent, importance, and interest, will amply repay the investigation of future observers. For this cross-fertilisation to take place, however, some foreign agency like that of insects is evidently necessary, for conveying the pollen from one flower to another. The question naturally occurs, How then is fertilisation accomplished in those plants which flower habitually in the winter, when the number of insects that can assist in it is at all events very small? I venture to offer the following notes as a sequel to Mr. Darwin's observations, and as illustrating a point which has not been elucidated by any investigations that have yet been recorded. I do not here refer to those flowers of which, in mild seasons, stray half-starved specimens may be found in December or January, and of which we are favoured with lists every year in the corners of newspapers, as evidence of "the extraordinary mildness of the season." I wish to call attention exclusively to those plants, of which we have a few in this country, whose normal time of flowering is almost the depth of winter, like the hazel-nut *Corylus avellana*, the butcher's broom *Ruscus aculeatus*, and the gorse *Ulex europoeus*; and to that more numerous class which flower and fructify all through the year, almost regardless of season or temperature; among which

论冬季开花型植物的受精作用

贝内特

编者按

如果说植物的传粉过程都依赖于昆虫，那么那些通常在冬季开花的植物是如何完成传粉的呢？冬季可是很少有昆虫飞行的。这正是《自然》上发表的第一篇研究论文提出的问题。根据阿尔弗雷德·贝内特（伦敦的一位出版商和书商）的观点，许多冬季开花型植物是自花受精的，在花朵还没绽放的时候花粉就从花药中释放出来了。对于那些雄性器官和雌性器官分别位于不同花朵上的植物来说，这显然是不可能的。而在诸如榛树之类的一些个例中，雄花中的花粉能够自由地释放到邻近的雌花中，这种情况"有利于花粉在最微弱风力作用下成功散播"。

雄蕊是花的雄性器官，它们聚集在一起就形成了年长的学者所称的"雄蕊群"，这是科学工作者和普通观察者都很熟悉的事实。早期的植物学家自然而然地得出了如下结论：一朵花的雄蕊和雌蕊倾向于对其他花朵发挥雄性器官和雌性器官的作用。大概在 1790 年，施普伦格尔就首先指出许多植物中器官的安排都是如此，这种职责的互换是不可能在同一朵花里实现的。最近，德国的希尔德布兰德和英国的达尔文对昆虫在同种植物不同个体的雄蕊与雌蕊间的受精中所起的重要作用进行了研究。异花受精是自然界的规律而非例外这一观点，现在已经得到了植物学家的普遍认可。植物有很多种确保异花受精的策略，达尔文认为这些策略是非常美妙有趣的，他特别呼吁植物学家能对此给予关注。于是一个新的研究领域出现了，从广度、重要性和利益等方面来看，这一领域都将给未来从事此领域研究的人员带来充分的回报。然而，一些外界媒介对异花受精的发生显然也是必要的，例如昆虫可以将花粉从一朵花传到另一朵花。于是我们便有了如下问题：对于那些总是在冬天开花的植物来说，那时能够帮助它们传粉的昆虫数量特别少，在这种情况下，这些花是如何完成受精的呢？本人斗胆给出如下摘记，作为对达尔文先生的观察研究的延续，同时也是对一个在已有记载的研究中从未被阐明过的问题进行的说明。有些植物在气候温和的年份里开的花可能会一直保持到 12 月或 1 月，但这只是些零散出现而且濒临死亡的花朵；还有些植物每年都被列在报纸的边边角角处，作为"今年气候异常温和"的证据。本人在这里并不想谈这些植物，而是想专门关注那些在本国数量不多、正常开花时间几乎在深冬的植物，例如欧洲榛、假叶树和荆豆，同时也会关注那些几乎不受季节或温度影响、一年四季都可以开花结果的植物，这样的植物种类相对更多，其中包括短柄野芝麻、紫花野芝麻、阿拉伯婆婆纳、雏菊、蒲公英、千里光、

may be mentioned the white and red dead-nettles *Lamium album* and *purpureum*, the *Veronica Buxbaumii*, the daisy, dandelion, and groundsel, the common spurge *Euphorbia peplus*, the shepherd's purse, and some others.

During the winter of 1868–1869, I had the opportunity of making some observations on this class of plants; the result being that I found that, as a general rule, fertilisation, or at all events the discharge of the pollen by the anthers, takes place in the bud before the flower is opened, thus ensuring *self-fertilisation* under the most favourable circumstances, with complete protection from the weather, assisted, no doubt, by that rise of temperature which is known to take place in certain plants at the time of flowering. The dissection of a flower of *Lamium album* (Fig. A) gathered the last week in December, showed the stamens completely curved down and brought almost into contact with the bifid stigma, the pollen being at that time freely discharged from the anthers. A more complete contrivance for self-fertilisation than is here presented would be impossible. The same phenomena were observed in *Veronica Buxbaumii*, where the anthers are almost in contact with the stigma before the opening of the flower, which occurs but seldom, *V. agrestis* and *polita*, the larger periwinkle *Vinca major*, the gorse, dandelion, groundsel, daisy, shepherd's purse, in which the four longer stamens appear to discharge their pollen in the bud, the two shorter ones not till a later period, *Lamium purpureum*, *Cardamine hirsuta*, and the chickweed *Stellaria media*, in which the flowers open only under the influence of bright sunshine. In nearly all these cases, abundance of fully-formed, seed-bearing capsules were observed in the specimens examined, all the observations being made between the 28th of December and the 20th of January.

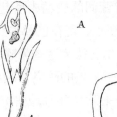
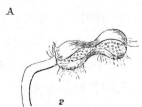

A. Lamium Album
1. Section of bud, calyx and corolla removed.
2. Stamen from bud, enlarged, discharging pollen.

In contrast with these was also examined a number of wild plants which had been tempted by the mind January to put forth a few wretched flowers at a very abnormal season, including the charlock *Sinapis arvensis*, wild thyme *Thymus serpyllum*, and fumitory *Fumaria officinalis*; in all of which instances was there not only no pollen discharged before the opening of the flower, but no seed was observed to be formed. An untimely specimen of the common garden bean *Faba vulgaris*, presented altogether different phenomena from its relative the gorse, the anthers not discharging their pollen till after the opening of the flower; and the same was observed in the case of the *Lamium Galeobdolon* or yellow archangel (Fig. B) gathered in April, notwithstanding its consanguinity to the dead-nettle.

南欧大戟、荠菜等。

在 1868~1869 年间的冬天，本人有幸对这类植物进行了一些观察。结果发现：受精作用是在开花之前发生于花蕾中的，这是一般规律，或者说在花药释放花粉的所有情况下都是如此。这样就保证了**自花受精**能在最有利的环境下进行而完全不受天气的影响，而且我们知道某些植物是在温度升高时开花的，这就更加加强了环境的有利性。从采集于 12 月最后一周的短柄野芝麻花的解剖图（图 A）可以看出，雄蕊完全弯曲下垂，几乎与二裂柱头相接触，此时花粉便从花药中自由散出。对于自花受精而言，不可能存在比这更完善的设计了。同样的现象在阿拉伯婆婆纳中也有发现，即开花前花药几乎与柱头相接触。在直立婆婆纳、双生婆婆纳、大蔓长春花、荆豆、蒲公英、千里光、雏菊和荠菜中，这种现象并不多见，而是 4 枚较长的雄蕊将花粉释放到花蕾上，2 枚较短的雄蕊则需再经过一段时间才能成熟。紫花野芝麻、碎米荠和繁缕是否开花只受日照影响。在检测过的样本中，几乎所有我们观察到的植物都产生了大量完全成形的结有种子的蒴果。上述所有观察是在 12 月 28 日到 1 月 20 日之间进行的。

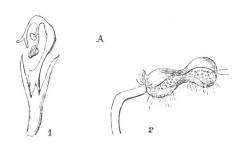

A. 短柄野芝麻

1. 去除了花萼和花冠后的花蕾部分

2. 花蕾上生出的膨大雄蕊，花粉正从中散出

作为对照，我们也研究了另外一些野生植物，这些植物在 1 月里受温暖天气的诱导反季节地开出了几朵可怜的小花，这包括野芥、铺地百里香和球果紫堇。对于所有这些植物，不仅没有发现开花前花粉的释放，而且也没有观察到种子的形成。一个不太合常理的例子是蚕豆。蚕豆的花药是在开花之后才释放花粉，这与其亲缘植物荆豆的表现完全不同。无独有偶，我们于 4 月采集的花叶野芝麻(图 B)也是这样，尽管它是野芝麻的近亲，但其花药也是在开花后才释放花粉。

B. Lamium Galeobdolon—Pistil and stamens from open flower; the latter discharging pollen.

Another beautiful contrast to this arrangement is afforded by those plants which, though natives of warmer climates, continue to flower in our gardens in the depth of winter. An example of this class is furnished by the common yellow jasmine, *Jasminium nudiflorum*, from China, which does not discharge its pollen till considerably after the opening of the flower, and which never fructifies in this country. But a more striking instance is found in the "allspice tree", the *Chimonanthus fragrans*, or *Calycanthus praecox* of gardeners, a native of Japan, which, flowering soon after Christmas, has yet the most perfect contrivance to prevent self-fertilisation (Fig. C). In a manner very similar to that which has been described in the case of *Parnassia palustris**, the stamens, at first nearly horizontal, afterwards lengthen out, and rising up perpendicularly, completely cover up the pistil, and then discharge their pollen outwardly, so that none can possibly fall on the stigma. As a necessary consequence, fruit is never produced in this country; but may we not conjecture that in its native climate the *Chimonanthus* is abundantly cross-fertilised by the agency of insects, attracted by its delicious scent, in a similar manner to our Grass of Parnassus?

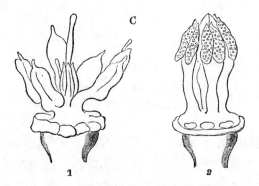

C. Chimonanthus Fragrans

1. Early stage of flower, calyx and corolla removed.

2. Later stage, stamens surrounding the pistil, and discharging their pollen outwardly.

---

* *Journal of the Linnaean Society* for 1868–1869, Botany, p. 24.

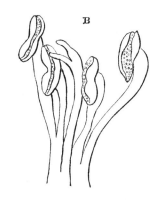

B. 花叶野芝麻——盛开花朵中的雌蕊和雄蕊；雄蕊正在释放花粉

　　另一个比较典型的对照是我们花园里的植物，尽管它们的原产地是气候较为温暖的地方，但在我们的花园里它们却一直开花到深冬。这类植物的一个例子是原产于中国的迎春花，其花粉在开花后相当长一段时间才散出，而且这种植物在我国从不结果。另一个更具代表性的例子是原产于日本的腊梅，这种植物在圣诞后不久就开花，它有最完美的阻止自花受精的机制（图C）。与梅花草采取的方式相似*，腊梅的雄蕊最初几乎是水平的，随后不断延长并垂直耸立，完全包裹住雌蕊，然后才向外散出花粉，这样就使得花粉不可能落到柱头上。一个必然的结果就是这些花在我国永远不会结出果实。但是，难道我们不能猜想腊梅在其原产地的气候下会像我国的帕纳色斯草一样，通过花朵的诱人香气来吸引昆虫作为媒介而完成异花受精吗？

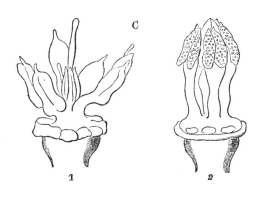

C. 腊梅
1. 去除花萼和花冠后的早期花朵
2. 晚期花朵，雄蕊包围着雌蕊并在向外散发花粉

---

* 《林奈学会会刊》，1868~1869年，植物学，第24页。

The description detailed above cannot of course apply to those winter-flowering plants in which the male and female organs are produced on different flowers; but here we find commonly another provision for ensuring fertilisation. In the case of the hazel-nut the female flowers number from two to eight or ten in a bunch, each flower containing only a single ovule destined to ripen. To each bunch of female flowers belongs at least one catkin (often two or three) of male flowers, consisting of from 90 to 120 flowers, and each flower containing from three to eight anthers. The pollen is not discharged till the stigmas are fully developed, and the number of pollen-grains must be many thousand times in excess of what would be required were each grain to take effect. The arrangement in catkins also favours the scattering of the pollen by the least breath of wind, the reason probably why so many of the timber-trees in temperate climates, many of them flowering very early in the season, have their male inflorescence in this form.

The *Euphorbias* or spurges have flowers structurally unisexual, but which, for physiological purposes, may be regarded as bisexual, a single female being enclosed along with a large number of male flowers in a common envelope of involucral glands. Two species are commonly found flowering in the winter, and producing abundance of capsules, *E. peplus* and *helioscopia*. In both these species the pistil makes its appearance above the involucral glands considerably earlier than the bulk of the stamens (Fig. D).

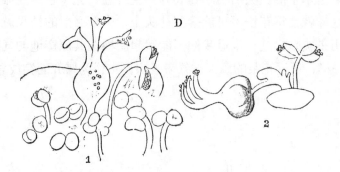

D. Euphorbia Helioscopia

1. Head of flowers opened, pistil and single stamen appearing above the involucral glands.

2. The same some what later, with the stigmas turned upwards.

A single one, however, of these latter organs was observed to protrude beyond the glands simultaneously, or nearly so, with the pistil, and to discharge its pollen freely on the stigmas, thus illustrating a kind of quasi-self-fertilisation. The remaining stamens do not discharge their pollen till a considerably later period, after the capsule belonging to the same set has attained a considerable size. In *E. helioscopia* the capsules are always entirely included within the cup-shaped bracts, and the stigmas are turned up at the extremity so as to receive the pollen freely from their own stamens. Now contrast with this the structure of *E. amygdaloides*, which does not flower before April (Fig. E). The heads of flowers which first open are entirely male, containing no female flower; in the hermaphrodite heads, which open subsequently, the stigmas are completely exposed beyond the involucral

　　上述描述当然并不适用于那些雄性器官和雌性器官分别生长于不同花朵上的冬季开花型植物。但是我们发现了另一种保证受精的常见方式。以欧洲榛为例，一簇花中雌花的数目是 2~8 或 2~10 不等，每朵花只含有一个可以成熟的胚珠。每簇雌花至少对应有一个（通常是 2 个或 3 个）雄花序，每个雄花序由 90~120 朵花（每朵花有 3~8 个花药）组成。在柱头充分发育后花粉才会散发出来。如果每个花粉粒真的都能发挥作用，那么实际花粉粒的数量就会超出所需数量数千倍。雄花序中的这种安排也有利于花粉在最微弱风力作用下成功散播，这很可能就是温带气候下有许多成材木的原因，因为其中许多在刚进入冬季时开花的植物都具有这种形式的雄花序。

　　大戟属的花从结构上来说是单性花，但是从生理学功能上来说又可以将其看作是双性的，因为一朵雌花与许多雄花被共同包裹在总苞腺体中。该属的南欧大戟和泽漆两种植物都是在冬季开花的，并且能产生大量蒴果。在这两种植物中，雌蕊出现在总苞腺体之上的时间比大部分雄蕊都早许多（图 D）。

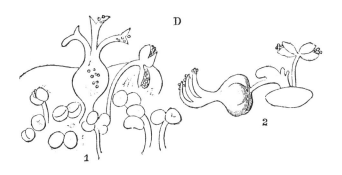

D. 泽漆
1. 盛开花朵的头部，雌蕊和单枚雄蕊位于总苞腺体之上
2. 同一朵花较晚期时的头部，这时柱头已经转弯向上

　　然而，我们观察到只有一枚雄蕊是和雌蕊同时或者接近同时显露在总苞腺体外的，这枚雄蕊会自由地将花粉散发到柱头上，这种受精方式被称为准自花受精。其余的雄蕊则是经过很长一段时间直到同组的蒴果长到一定大小的时候才会释放花粉。泽漆的蒴果自始至终都被完全包裹在杯状的苞叶之中，而柱头的末端会转而向上以便接受自身雄蕊自由释放的花粉。与此形成对比的是在 4 月之前并不开花的扁桃状大戟（图 E）的结构。最先开放的一批花朵，其头部是完全雄性的，不含任何雌花。随后开放的花朵头部则是雌雄同体的，在雄蕊从总苞腺体中伸出之前，柱头就已经完全暴露在总苞腺体之外了。因此，就会发生完全的异花受精，即首先开放的雄花

glands long before any stamens protrude from the same glands. Here, therefore, complete cross-fertilisation takes place, the pollen from the first-opened male heads no doubt fertilising the stigma from the next-opened hermaphrodite heads, and so on. In this species the bracts are not cup-shaped, but nearly flat; the stigmas hang out very much farther than in *E. helioscopia;* and the styles are perfectly straight.

E. Euphorbia Amygdaloides—Head of flower, pistil appearing above the involucral glands, all the stamens still undischarged.

The above observations are very imperfect as a series, and I can only offer them as a contribution towards an investigation of the laws which govern the cross-fertilisation or self-fertilisation of winter-flowering plants. On communicating some of them to Mr. Darwin, he suggested that the self-fertilised flowers of *Lamium album,* and other similar plants, may possibly correspond to the well-known imperfect self-fertilised flowers of *Oxalis* and *Viola*; and that the flowers produced in the summer are cross-fertilised; a suggestion which I believe will be found correct.

In conclusion, I may make two observations. The time of flowering of our common plants given in our textbooks is lamentably inexact; for the hazel, March and April for instance! and for the white dead-nettle, May and June! according to Babington. Great care also should be taken to examine the flowers the moment they are brought in-doors; as the heat of the room will often cause the anthers to discharge their pollen in an incredibly short space of time. This is especially the case with the grasses.

(**1**, 11-13; 1869)

头部释放的花粉必然会使随后开放的雌雄同体花头部的柱头受精，以此类推。这种植物的苞叶不是杯状的，而是几乎扁平的，其柱头也比泽漆的远得多，并且是完全直立的。

E. 扁桃状大戟——花朵头部，雌蕊位于总苞腺体之上，所有雄蕊尚未开始释放花粉

上述观察研究作为一个系列还很不完善，所以我只希望这些结果可以为研究冬季开花型植物的自花受精或异花受精的调控规律提供一些参考。与达尔文先生进行相关交流时，他提出短柄野芝麻和其他类似植物的自花受精方式可能与酢浆草和堇菜属植物那著名的不完全自花受精方式相似，而夏天开的花是异花受精的。我相信这种说法终将会被证明是正确的。

综上，通过观察研究我可以提出如下两点。首先，我们的教科书中给出的普通植物的开花时间是非常不准确的。例如，巴宾顿的书中竟然说榛树是在 3 月和 4 月开花，短柄野芝麻是在 5 月和 6 月开花！另外，我们也应该特别关注花被搬进室内时的反应，因为室内的热量通常会使花药在很短的时间内就释放出花粉，稻科植物尤其如此。

（刘皓芳 翻译；刘京国 审稿）

Fertilisation of Winter-flowering Plants

A. W. Bennett

Editor's Note

Alfred Bennett's article in *Nature*'s first issue on fertilisation in winter-flowering plants made much mention of work on plant fertilisation by Charles Darwin. Following the publication of that paper, Darwin drew Bennett's attention to various small errors and clarifications. Thus Bennett got a letter in *Nature*'s second issue to address Darwin's points, notably that *Vinca* (periwinkle) absolutely requires the attention of insects for pollination. Bennett went back to his notes and found that Darwin was, of course, right—apparently self-fertile periwinkles do not set fruit in the winter.

MR. Darwin has done me the honour of calling my attention to one or two points in my paper, published in your last number, "On the Fertilisation of Winter-flowering Plants". He thinks there must be some error in my including *Vinca major* among the plants of which the pollen is discharged in the bud, as he "knows from experiment that some species of *Vinca* absolutely require insect aid for fertilisation." On referring to my notes, I find them perfectly clear with respect to the time at which the pollen is discharged. My observation, however, so far agrees with Mr. Darwin's, that I find no record of any fruit being produced in January; it was, in fact, the absence of capsules on the *Vinca* which induced me to qualify the sentence on this subject, and to say "in nearly *all* these cases, abundance of fully formed seed-bearing capsules were observed." It is worthy of remark, that the *Vinca* is the only species in my list of apparently bud-fertilised plants not indigenous to this country. The second point relates to the white dead-nettle, with respect to which Mr. Darwin says, "I covered up *Lamium album* early in June, and the plants produce no seed, although surrounding plants produced plenty." This again would agree with my conjecture that it is only the flowers produced in winter that are self-fertilised. I may, however, be permitted to suggest that the test of covering up a plant with a bell-glass is not conclusive on the point of cross-fertilisation, as it is quite probable that with plants that are ordinarily self-fertilised, the mere fact of a complete stoppage of a free circulation of air may prevent the impregnation taking place. Has the experiment ever been tried with grasses, which, according to the French observer, M. Bidard are necessarily self-fertilised?

(**1**, 58; 1869)

Alfred W. Bennett: 3, Park Village East, Nov. 8, 1869.

冬季开花型植物的受精作用

贝内特

编者按

阿尔弗雷德·贝内特在《自然》的第 1 期上发表了一篇关于冬季开花型植物受精作用的论文，该论文中大量提到了查尔斯·达尔文关于植物受精作用的研究工作。在贝内特的论文发表后，达尔文提请贝内特注意文章中存在的多处小错误，并请他作一些说明。于是，在这篇发表于《自然》第 2 期的快报文章中，贝内特公布了达尔文的观点，即长春花的传粉完全需要昆虫的参与。贝内特在重新查阅了他的记录后发现，达尔文的观点确实是正确的，很显然自花受精型的长春花在冬天是不结果的。

在贵刊的上一期中有一篇我发表的论文——《论冬季开花型植物的受精作用》。我很荣幸达尔文先生给了我一些提醒，让我注意到该文中的一两个问题。我在论文中提到，大蔓长春花是一种在花蕾中就有花粉散落出来的植物，他认为这种说法肯定存在某些错误，因为他"由实验得知，长春花属的某些种只有在昆虫的辅助下才能完成受精。"于是我再次查阅了我的记录，发现关于花粉释放时间的记录是非常清楚的。不过，到目前为止，我的观察结果与达尔文先生的是一致的，因为我并没有发现长春花属的植物在 1 月结果的任何记录。事实上，正是长春花属的植物没有结出蒴果这一现象让我仔细考虑了该如何用文字来表述这个问题，后来我写下了"几乎**所有**我们观察到的植物都产生了大量完全成形的结有种子的蒴果"这句话。需要说明的是，长春花属是我观察到的花蕾受精植物中唯一一种非本土的植物。第二点是与短柄野芝麻有关的问题。对于这方面，达尔文先生说："我在 6 月初将一株短柄野芝麻罩住，当它周围的植物都硕果累累的时候，它却并没有结出任何种子。"这再一次与我提出的只有冬季开花型植物才能自花受精的推测相吻合。不过，请允许我提出如下建议：用钟形玻璃罩罩住植物以检验异花受精的方法并不完全可信，因为这样对于本来是自花受精的植物来说，很可能会由于完全中断了空气的自由流通而阻止受精作用的发生。这不就是当初法国观察者比达尔自以为是地认为肯定发生了自花受精所依据的那个用稻科植物进行的实验吗？

（刘皓芳 翻译；刘京国 审稿）

The Fertilisation of Winter-flowering Plants

C. Darwin

Editor's Note

The correspondence on fertilisation in winter-flowering plants, begun by Alfred Bennett in the journal's first issue, continues here with a letter from Charles Darwin himself, who cautions against using bell-jars to isolate flowers from their environment (and thus any external pollination agency) because such a practice "is injurious from the moisture of the contained air". Darwin recommends "what is called by ladies, 'net'", a practice which he had followed for twenty years and was able to observe fertilisation in thousands of plants. As regards Bennett's observation in *Vinca*, Darwin cautions that the observation of pollen falling on a stigma, and the formation of pollen tubes, is in itself "a most fallacious indication of self-fertilisation".

WILL you permit me to add a few words to Mr. Bennett's letter, published at p. 58 of your last number? I did not cover up the *Lamium* with a bell-glass, but with what is called by ladies, "net". During the last twenty years I have followed this plan, and have fertilised thousands of flowers thus covered up, but have never perceived that their fertility was in the least injured. I make this statement in case anyone should be induced to use a bell-glass, which I believe to be injurious from the moisture of the contained air. Nevertheless, I have occasionally placed flowers, which grew high up, within small wide-mouthed bottles, and have obtained good seed from them. With respect to the *Vinca*, I suppose that Mr. Bennett intended to express that pollen had actually fallen, without the aid of insects, on the stigmatic surface, and had emitted tubes. As far as the mere opening of the anthers in the bud is concerned, I feel convinced from repeated observations that this is a most fallacious indication of self-fertilisation. As Mr. Bennett asks about the fertilisation of Grasses, I may add that Signor Delpino, of Florence, will soon publish some novel and very curious observations on this subject, of which he has given me an account in a letter, and which I am glad to say are far from being opposed to the very general law that distinct individual plants must be occasionally crossed.

(**1**, 85; 1869)

Charles Darwin: Down, Beckenham, Kent, Nov. 13.

冬季开花型植物的受精作用

达尔文

编者按

关于冬季开花型植物受精作用这一主题的讨论，始于阿尔弗雷德·贝内特在《自然》第1期中发表的文章，随后就有了这篇来自查尔斯·达尔文的快报文章。达尔文提出不要使用钟形广口瓶将植物从其周围环境（和任何外部媒介）中隔离开来，因为这种操作"不利于保持容器内空气的湿度"。达尔文建议使用一种"被女士们称为'网'的工具"，他持续使用这种工具有20年，并观察到数千种植物的受精作用。至于贝内特对长春花的观察研究，达尔文提出，把花粉掉落到柱头上的现象以及花粉管的形成"看作是自花受精的象征是非常荒谬的"。

请允许我对贝内特先生发表于贵刊上一期第58页的文章补充几句。我并没有用钟形玻璃罩罩住野芝麻，而是用了一种被女士们称为"网"的工具。在过去的20年中，我一直用这种方法，并且已经使成数千种被这样遮罩的花成功受精，但从未观察到因为这样的处理而使它们的生殖能力受到丝毫损伤的例子。我作此声明以防有人可能会被误导而使用钟型玻璃罩，我认为钟型玻璃罩不利于保持容器内空气的湿度。不过，我曾经偶然地将一些已经长得很高的花放到一个小广口瓶中，最后这些花也结出了很好的种子。关于长春花属的植物，我猜想贝内特先生想表达的意思是：在没有昆虫帮助的情况下，花粉实际上是落到了柱头表面并萌发出花粉管。至于文中提到的在花蕾中出现的花药释放，通过反复观察我确信，把这种现象看作是自花受精的象征是非常荒谬的。此外，既然贝内特先生说到了稻科植物的受精作用，我想补充的是，佛罗伦萨的德尔皮诺先生不久就会发表一些针对此问题的观察结果，那将是非常新颖、非常奇妙的。他已经给我来信说明了这些结果，我非常高兴地告诉大家，他的观察结果与完全不同的植物个体间肯定会偶尔发生杂交这一基本规律并不矛盾。

（刘皓芳 翻译；刘京国 审稿）

The Origin of Species Controversy

A. R. Wallace

Editor's Note

The British naturalist Alfred Russel Wallace, whose theory of evolution paralleled Darwin's, here reviews a new book of scientific essays by Joseph John Murphy. Murphy had argued that evolution by natural selection could not account for the observed species if all variations were small, for the time required would be longer than earth had apparently been habitable. Wallace argued that Murphy was mistaken, suggesting that the difficulty might be resolved with further understanding of geological history. Murphy also touched on the nature of intelligence, both unconscious and conscious, the latter which he proposed must be linked to human instinct. Wallace (who developed mystical and spiritualist enthusiasms in later life) felt more favourably towards these speculations.

Habit and Intelligence, in their Connection with the Laws of Matter and Force.
A Series of Scientific Essays. By Joseph John Murphy. (Macmillan and Co., 1869)

I

THE flood of light that has been thrown on the obscurest and most recondite of the forces and forms of Nature by the researches of the last few years, has led many acute and speculative intellects to believe that the time has arrived when the hitherto insoluble problems of the origin of life and of mind may receive a possible and intelligible, if not a demonstrable, solution. The grand doctrine of the conservation of energy, the all-embracing theory of evolution, a more accurate conception of the relation of matter to force, the vast powers of spectrum analysis on one side, showing us as it does the minute anatomy of the universe, and the increased efficiency of the modern microscope on the other, which enables us to determine with confidence the structure, or absence of structure, in the minutest and lowest forms of life, furnish us with a converging battery of scientific weapons which we may well think no mystery of Nature can long withstand. Our literature accordingly teems with essays of more or less pretension on the development of living forms, the nature and origin of life, the unity of all force, physical and mental, and analogous subjects.

The work of which I now propose to give some account, is a favourable specimen of the class of essays alluded to, for although it does not seem to be in any degree founded on original research, its author has studied with great care, and has, in most cases, thoroughly understood, the best writers on the various subjects he treats of, and has brought to the task a considerable amount of original thought and ingenious criticism. He thus effectually raises the character of his book above that of a mere compilation, which, in less able hands, it might have assumed.

物种起源论战

华莱士

编者按

英国博物学家阿尔弗雷德·拉塞尔·华莱士关于进化的观点与达尔文的观点很类似，这里他评论了约瑟夫·约翰·墨菲新出的一本科学论文集。墨菲早先提出，如果所有的变异都很微小的话，那么依靠自然选择的进化将无法解释可见的物种，因为进化所需的时间将比地球适宜生物生存的时间更漫长。华莱士认为墨菲是错误的，他指出，在深入了解地质历史之后这一问题将迎刃而解。墨菲还提到了智能的本质，包括无意识的和有意识的，他认为有意识的智能一定与人类的本能有关联。对于这些猜想，华莱士（他晚年对神秘主义与唯灵论产生了兴趣）则比较赞成。

《习性与智能，及其与物质和力的法则的关系》
约瑟夫·约翰·墨菲的一本科学论文集（麦克米伦公司，1869 年）

I

通过最近几年的研究，人们已经揭示了关于自然界中力和组成结构的许多最为晦涩且极其深奥的问题，这使很多性急又有投机心理的知识分子们以为，对于到目前为止还未解决的关于生命和思维起源的问题，即使还不能得到一个可论证的答案，也应该是时候给出一个合理而又清楚的解释了。一方面，极其重要的能量守恒学说、囊括一切的进化论、关于物质和力关系的更准确的概念以及具有巨大威力的光谱分析，给我们展示了宇宙万物结构和机理的本来面目；另一方面，现代显微镜技术性能的提升，使我们可以很有把握地确定最微小和最低等的生命形式的结构或者确认它们没有结构；这些都给我们提供了强大的科学武器，使我们相信自然界没有永久的秘密。因此，这部书里的文章或多或少都对生命形式的发展、生命的起源和本质、各种力的统一、肉体和精神以及诸如此类的问题给出了自信的答案。

我将要介绍的这本书，是人们曾经提到的这类著作中的一个很好的范例。虽然这本书不是建立在原始研究的基础之上，但作者进行了细致谨慎的研究，而且在大多数情况下都能深刻地理解他所涉及的各个科目中最优秀的作者，此外，还精心引入了大量原创性的思想和独创性的批评。这样，作者就有效地提高了这本书的品质，使之超过了能力稍显逊色的人可能炮制出的纯粹的合集。

The introductory chapter treats of the characteristics of modern scientific thought, and endeavours to show, "that the chief and most distinctive intellectual characteristic of this age consists in the prominence given to historical and genetic methods of research, which have made history scientific, and science historical; whence has arisen the conviction that we cannot really understand anything unless we know its origin; and whence also we have learned a more appreciative style of criticism, a deeper distrust, dislike, and dread, of revolutionary methods, and a more intelligent and profound love of both mental and political freedom." The first six chapters are devoted to a careful sketch of the great motive powers of the universe, of the laws of motion, and of the conservation of energy. The author here suggests the introduction of a useful word, *radiance*, to express the light, radiant heat, and actinism of the sun, which are evidently modifications of the same form of energy,—and a more precise definition of the words *force* and *strength*, the former for forces which are capable of producing motion, the latter for mere resistances like cohesion.

He enumerates the primary forces of Nature as, gravity, capillary attraction, and chemical affinity, and notices as an important generalisation "that *all primary forces are attractive*; there is no such thing in Nature as a primary repulsive force" (p. 43). Now here there seem to be two errors. Cohesion, which is entirely unnoticed, is surely as much a primary force as capillary attraction, and, in fact, is probably the more general force, of which the other is only a particular case; and elasticity is the effect of a primary repulsive force. In fact, at p. 26, we find the author arguing that *all matter is perfectly elastic*, for, when two balls strike together, the lost energy due to imperfect elasticity of the mass is transferred to the molecules, and becomes heat. But this surely implies repulsion of the molecules; and Mr. Bayma has shown, in his "Molecular Mechanics", that repulsion is as necessary a property of matter as attraction.

The eighth chapter discusses the phenomena of crystallisation; and the next two, the chemistry and dynamics of life. The reality of a "vital principle" is maintained as "the unknown and undiscoverable something which the properties of mere matter will not account for, and which constitutes the differentia of living beings." Besides the formation of organic compounds, we have the functions of organisation, instinct, feeling, and thought, which could not conceivably be resultants from the ordinary properties of matter. At the same time it is admitted that conceivableness is not a test of truth, and that all questions concerning the origin of life are questions of fact, and must be solved, not by reasoning, but by observation and experiment; but it is maintained that the facts render it most probable that "life, like matter and energy, had its origin in no secondary cause, but in the direct action of creative power." Chapters X to XIV treat of organisation and development, and give a summary of the most recent views on these subjects, concluding with the following tabular statement of organic functions:—

Formative or Vegetative Functions, essentially consisting in the Transformation of Matter

Chemical	Formation of organic compounds
Structural	Formation of tissue
	Formation of organs

前言中提到了现代科学思想的特点，并力图展示"这一时期最重要也是最独具一格的思维特点是人们大多习惯于采用历史学和遗传学的研究方法，这些方法使历史科学化，也使科学历史化，由此我们可以确信：除非我们知道起源，否则我们不能够真正地认识任何事物；从这里我们也学到了带有更多赞赏眼光的批判风格，带有更深的怀疑、不赞成和担忧的创新方法，以及对精神和政治自由更理智更深刻的爱。"前 6 章主要致力于详细描述宇宙巨大的动力、运动的规律和能量守恒。作者在这里引入了**辐射**这个有用的词语，用来表示光、热辐射以及太阳的光化作用，这些实际上是同一种形式的能量的不同表现。此外，作者还更精确地定义了 *force* 和 *strength* 这两个词，前者代表能够产生运动的力，而后者只代表阻力，比如黏滞力。

作者列举了自然界中几种基本的力，如重力、毛细引力和化学亲和力，然后他总结出了一个重要的结论：**"所有基本的力都是引力，自然界中没有任何一种基本的力是排斥力"**（第 43 页）。这里似乎有两个错误。首先，和毛细引力一样，黏滞力显然也是一种基本的力，而作者则完全忽视了这种力。事实上，黏滞力可能是更普遍存在的，相比而言，毛细引力倒要算是特殊的力了。另外，弹力就是一种基本的排斥力。实际上，在该书第 26 页，我们发现作者对**所有物质都是完全弹性的**这一观点提出了质疑，他认为，当两个球撞击时，损失的能量转移给了分子并最终变成热，而这部分能量的损失应归于物质的不完全弹性。但是这显然暗示了分子排斥的存在。就像拜马先生在他的《分子力学》中所写的那样，排斥力就像引力一样是物质必需的属性。

第 8 章讨论了结晶现象，紧接着的两章讲了生命化学和生命动力学。"生命法则"存在于"未知的和未发现的事物"这些实体当中，"这些事物不能用纯粹的物质属性来解释，它们造成了生物体的差异性。"除了有机化合物的形成外，还有组织功能、本能、感觉和思想，这些都不能用普通的物质属性来解释。同时，我们也承认猜测绝不是对真理的检验，关于生命起源的所有问题都是事实的问题，必须通过观察和实验来解决，而不是只靠思考与推理。但是书中声称，事实最可能是"就像物质和能量一样，生命的起源也是第一创造力的直接作用结果，而没有任何别的原因。"第 10~14 章提到了组织和发育，总结了这些领域的最新观点，并以如下的表格形式就有机物的功能给出了最终的结论：

构形功能或营养功能，本质上来说都是通过物质转化而形成的

化学的	有机化合物的形成
结构的	组织的形成
	器官的形成

Animal Functions, consisting essentially in the Transformation of Energy

Motor	Spontaneous
	Reflex
	Consensual
	Voluntary
Sensory	Sensation
	Mind

In the fifteenth chapter we first come to one of the author's special subjects;—the Laws of Habit. He defines habit as follows: "The definition of habit and its primary law, is that all vital actions tend to repeat themselves; or, if they are not such as can repeat themselves, they tend to become easier on repetition." All habits are more or less hereditary, are somewhat changeable by circumstances, and are subject to spontaneous variations. The *prominence* of a habit depends upon its having been recently exercised; its *tenacity* on the length of time (millions of generations it may be) during which it has been exercised. The habits of the species or genus are most tenacious, those of the individual often the most prominent. The latter may be quickly lost, the former may appear to be lost, but are often latent, and are liable to reappear, as in cases of reversion. The fact that active habits are strengthened, while passive impressions are weakened, by repetition, is due in both cases to the law of habit; for, in the latter, the organism acquires the habit of not responding to the impression. As an example, two men hear the same loud bell in the morning; it calls the one to work, as he is accustomed to listen to it, and so it always wakes him; the other has to rise an hour later, he is accustomed to disregard it, and so it soon ceases to have any effect upon him. Habit has produced in these two cases exactly opposite results. Habits are capable of any amount of change, but only a slight change is possible in a short time; and in close relation with this law are the following laws of variation.

Changes of external circumstances are beneficial to organisms if they are slight; but injurious if they are great, unless made gradually.

Changes of external circumstances are agreeable when slight, but disagreeable when great.

Mixture of different races is beneficial to the vigour of the offspring if the races mixed are but slightly different; while very different races will produce either weak offspring, or infertile offspring, or none at all. Even the great law of sexuality, requiring the union of slightly different individuals to continue the race, seems to stand in close connection with the preceding laws.

The next seven chapters treat of the laws of variation, distribution, morphology, embryology, and classification, as all pointing to the origin of species by development; and we then come to the causes of development, in which the author explains his views as

动物性机能，本质上来说是通过能量转化而形成的

动作	自发性的
	反射性的
	交感性的
	自主性的
感官	感觉
	思维

在第 15 章，我们首次看到了作者一个独特的观点——习性法则。他是这样定义习性的："习性的定义及基本法则是，所有的生命活动都有自我重复的趋势，或者如果不能进行自我重复，那么它们倾向于使重复变得更加容易。"所有的习性几乎都是遗传性的，随着环境变化会有少许改变，并且受到自发变异的影响。一个习性是否有**明显的影响力**取决于它最近是否得到运用，它的**强度**取决于运用时间的长短（可能是几百万代）。种或属的习性是非常稳定和牢固的，而个体的习性经常是差异较大而十分凸显的。个体习性可能会很快消失，种属习性可能也会消失，但通常是很难察觉到的，而且种属习性还容易再次出现，例如返祖现象。通过不断重复，主动习性得到强化，而被动受到的影响会减弱，这两个事实也可以根据习性法则推断出来。因为，在后一种情况下，生物体获得了对被动受到的影响不作出反应的习性。比如，有两个人在早上听到相同的钟声。对于其中一个人，钟声会叫醒他去上班，而他也习惯于去听钟声，因此钟声总能叫醒他。而对另一个人，在钟声响起一小时后他才起床，他习惯于忽略钟声，这样不久之后钟声对他就没有任何作用了。在这两个例子中习性产生了两种完全相反的结果。习性可以发生任何程度的改变，但在短期内它只能发生微小的改变。与习性法则密切相关的是接下来要谈及的变异法则。

外界环境的轻微改变对生物体是有利的，但是，外界环境的较大改变对生物体却是有害的，除非这种改变是逐渐发生的。

当外界环境发生轻微的改变时,生物体是可以适应的,而当外界环境变化较大时,生物体就不能适应了。

如果混居在一起的各种族之间只有微小的差异，那么种族混居是有利于提高后代素质的。但是，差别较大的种族却会产生虚弱的后代或者不育的后代，甚至根本无法产生后代。伟大的性法则告诉我们，要延续种族就需要群体中有稍微不同的个体。这样看来，性法则与我们前面提到的法则之间似乎是有紧密联系的。

接下来的 7 章谈到了变异法则、分布、形态学、胚胎学以及分类法则，然后谈到了发育的原因。这样，以生物体的发育为纽带，所有问题都指向了物种的起源。

followsm:—

> These two causes, self-adaptation and natural selection, are the only *purely physical* causes that have been assigned, or that appear assignable, for the origin of organic structure and form. But I believe they will account for only part of the facts, and that no solution of the questions of the origin of organization, and the origin of organic species, can be adequate, which does not recognise an Organising Intelligence, over and above the common laws of matter…But we must begin the inquiry by considering *how much* of the facts of organic structure and vital function may be accounted for by the two laws of self-adaptation and natural selection, before we assert that any of those facts can only be accounted for by supposing an Organising Intelligence.

Again:

> Life does not suspend the action of the ordinary forces of matter, but works through them. I believe that wherever there is life there is intelligence, and the intelligence is at work in every vital process whatever, but most discernibly in the highest…Nutrition, circulation, and respiration are in a great degree to be explained as results of physical and chemical laws;—but sensation, perception, and thought cannot be so explained. They belong exclusively to life; and similarly the organs of those functions—the nerves, the brain, the eye, and the ear—can have originated, I believe, solely by the action of an Organising Intelligence.

Admitting Mr. Herbert Spencer's theory of the origin of the vascular system, and possibly of the muscular, by self-adaptation, he denies that any such merely physical theory will account for the origin of the special complexities of the visual apparatus:

> Neither the action of light on the eye, nor the actions of the eye itself, can have the slightest tendency to produce the wondrous complex histological structure of the retina; nor to form the transparent humours of the eye into lenses; nor to produce the deposit of black pigment that absorbs the stray rays that would otherwise hinder clear vision; nor to produce the iris, and endow it with its power of closing under a strong light, so as to protect the retina, and expanding again when the light is withdrawn; nor to give the iris its two nervous connections, one of which has its root in the sympathetic ganglia, and causes expansion, while the other has its root in the brain and causes contraction.

Nor will he allow that Natural Selection (which he admits may produce any simple organ, such as a bat's wing) is applicable to this case; and he makes use of two arguments which have considerable weight. One is that of Mr. Herbert Spencer, who shows that in all the higher animals natural selection must be aided by self-adaptation, because an alteration in any part of a complex organ necessitates concomitant alterations in many other parts, and these cannot be supposed to occur by spontaneous variation. But in the case of the eye he shows that self-adaptation cannot occur, whence he conceives it may be proved to be almost an infinity of chances to one against the simultaneous variations necessary to produce an eye ever having occurred. The other argument is, that well-developed eyes occur in the higher orders of the three great groups, Annulosa,

在书中，作者是这样阐释他的观点的：

对于生物体结构和形态的起源，在人们明确给出或者暗示性地提出的原因中，只有两个是**完全只与自然界有关**的原因，这两个原因是：自适应和自然选择。但是我认为这两个原因只能解释部分事实。而且，对于组织起源和物种起源这些问题，这两个原因给出的解释并不比根据普通的物质法则作出的解释更充分，并且根本没有注意到组织智能……但是，在我们断言只要加入组织智能的假定就能解释所有这些事实之前，我们还必须弄清楚，自适应和自然选择这两个法则到底能够解释**多少**有关有机体结构和功能的问题。

此外，作者还写道：

生命并没有中止物质通常的力的作用，而是通过它们来发挥作用。我相信，所有的生命都是有智能的。智能在所有的生命过程中都发挥作用，只是在最高等的生命过程中最容易被觉察到……营养、循环和呼吸在很大程度上是可以用物理和化学规律解释的。但是，感觉、感知和思考却不能这样解释。这些过程属于生命所专有。类似地，对于具有这些功能的器官，即神经、脑、眼睛和耳朵，我相信它们应该都是起源于组织智能的活动。

作者赞同赫伯特·斯宾塞先生的理论，即血管系统起源于自适应，肌肉系统也很可能起源于自适应。由此，作者认为单靠物理学理论并不能解释异常复杂的视觉器官的起源。他写道：

眼睛的结构和功能是很神奇的：视网膜的组织构造非常巧妙而复杂；眼睛里的透明体液形成了晶状体；眼睛中黑色素的沉淀能够吸收杂散光线从而使视觉清晰不受影响；眼睛中的虹膜具有在强光下关闭以保护视网膜而在光线转弱时又再次张开的能力；虹膜中有两类神经连接，一类是根部位于交感神经节中的神经连接，它能引起扩张，另一类是根部位于脑中的神经连接，它能引起收缩。所有这些结构与功能的产生，应该都不会受到光对眼睛的作用和眼睛自身活动的丝毫影响。

作者也不认为自然选择适用于这个例子（他承认，自然选择可以产生一些简单的器官，比如蝙蝠的翅膀）。他提到了两个相当有分量的观点。其中一个是来自赫伯特·斯宾塞先生的观点，即，在所有高等动物中自然选择必须借助于自适应，因为一个复杂器官任何部位的改变必然伴随着其他部位的许多改变，而这些改变是不可能通过自发变异产生的。但是作者指出，在眼睛这个例子中自适应并不会发生，由此，他觉得这个例子可能给那些认为同步变异并不是产生眼睛的必要条件的人提供了一个屡用不爽的例证（事实上这些人此前确实用过这个例子）。另一个观点是，在环节动物、软体动物和脊椎动物这三类高等群体中才出现了结构和功能发展得很好的眼睛，而在较低等的群体中，眼睛还是很初等的或者根本就没有眼睛。因此，完美

Mollusca, and Vertebrata, while the lower orders of each have rudimentary eyes or none; so that the variations requisite to produce this wonderfully complicated organ must have occurred three times over independently of each other. In the first of these objections, he assumes that many variations must occur simultaneously, and on this assumption his whole argument rests. He notices Mr. Darwin's illustration of the greyhound having been brought to its present high state of perfection by breeders selecting for one point at a time, but does not think it possible "that any apparatus, consisting of lenses, can be improved by any method whatever, unless the alterations in the density and the curvature are perfectly simultaneous." This is an entire misconception. If a lens has too short or too long a focus, it may be amended either by an alteration of curvature, or an alteration of density; if the curvature be irregular, and the rays do not converge to a point, then any increased regularity of curvature will be an improvement. So the contraction of the iris and the muscular movements of the eye are neither of them essential to vision, but only improvements which might have been added and perfected at any stage of the construction of the instrument. Thus it does not seem at all impossible for spontaneous variations to have produced all the delicate adjustments of the eye, once given the rudiments of it, in nerves exquisitely sensitive to light and colour; but it does seem certain that it could only be effected with extreme slowness; and the fact that in all three of the primary groups, Mollusca, Annulosa, and Vertebrata, species with well-developed eyes occur so early as in the Silurian period, is certainly a difficulty in view of the strict limits physicists now place to the age of the solar system.

(**1**, 105-107; 1869)

II

In his chapter on "The Rate of Variation", Mr. Murphy adopts the view (rejected after careful examination by Darwin) that in many cases species have been formed at once by considerable variations, sometimes amounting to the formation of distinct genera and he brings forward the cases of the Ancon sheep, and of remarkable forms of poppy and of *Datura tatula* appearing suddenly, and being readily propagated. He thinks this view necessary to get over the difficulty of the slow rate of change by natural selection among minute spontaneous variations; by which process such an enormous time would be required for the development of all the forms of life, as is inconsistent with the period during which the earth can have been habitable. But to get over a difficulty it will not do to introduce an untenable hypothesis; and this one of the rapid formation of species by single variations can be shown to be untenable, by arguments which Mr. Murphy will admit to be valid. The first is, that none of these considerable variations can possibly survive in nature, and so form new species, unless they are *useful* to the species. Now, such large variations are admittedly very rare compared with ordinary spontaneous variability, and as they have usually a character of "monstrosity" about them, the chances are very great against any particular variation being useful. Another consideration pointing in the same direction is, that as a species only exists in virtue of its being tolerably well adapted

的具有复杂结构的眼睛的产生，必然经过了三次以上的独立变异过程。作者对这个观点的第一条反对意见是，他认为大多数变异都是同时发生的。他的所有论述都是基于这个假设的。达尔文在关于灰色猎犬的阐述中说到，饲养者通过每次只筛选某一个特征可以使猎犬达到目前的高级状态。虽然作者也注意到了这些，但是他认为下面的论述是不可能的："除非透镜材料的密度和曲率同时得到完美的改变，否则，任何方法都不可能改进任何由透镜组成的仪器的性能。"这个观点是完全错误的。如果一个透镜的焦距太长或者太短，那么我们就可以通过改变曲率或者改变密度使透镜的性能得到改善。如果曲率不规则导致光线无法汇聚到一点，那么任何能使曲率变得规则的方法都可以改善透镜的性能。因此，对于视力来说，虹膜的收缩和眼部肌肉的运动都不重要，只有那些可能在眼睛形成过程的某些阶段已经发生而且完善化了的改进才是至关重要的。这样看来，一旦有了眼睛的雏形，那么在那些对光和色彩具有敏锐感知能力的神经中发生的多个自发变异完全有可能都使眼睛的微观结构发生有利改变。但是应该可以肯定的是，眼睛受到这种影响的速度是异常缓慢的。不过，考虑到已经被严格界定了的志留纪的确切时限，那么三个主要类群（软体动物、环节动物和脊椎动物）中具备完美进化的眼睛的物种早在志留纪就已出现这一事实无疑就很难解释了。

（郑建全 翻译；陈平富 审稿）

II

在《变异速率》一章中，墨菲先生采纳的观点（被达尔文在认真检验之后驳斥了的）是，在许多情况下，物种是通过显著的变异即刻形成的，有时可以积累变异而形成不同的属。他列举了一些案例，如安康羊，多种类型的罂粟，以及突然出现变异并迅速繁殖的紫花曼陀罗。他认为有必要凭借这一观点来克服微小的自发变异中自然选择面临变化速度太慢的难题；任何形式的生命通过这一过程发育的话，都将需要无比漫长的时间，而这与地球上生命出现的时期相矛盾。但是为了克服这种困难，引入一个站不住脚的假说也毫无意义；而通过墨菲一厢情愿地认为有效的论据来看，这个通过单一变异就能快速形成物种的论点确实是站不住脚的。首先，这些显著的变异都不可能在自然状态下幸存，更别提形成新物种了，除非他们对物种是**有用的**。现在，人们认为大变异与普通的自发变异相比少之又少，而通常这种变异的特征很"畸形"，于是出现任何有用的特定变异的机率非常小。另外，通常要思考的另一点是，由于物种只是因为具有良好的耐性而与环境相适应才生存下来，同时环境只是**缓慢地**变化，所以物种更需要小的变化而非大的变化来维持适应性。但是即使环境迅速发生了巨大的变化，比如一些新物种的侵入或者由于几英尺的下陷

to its environment, and as that environment only changes *slowly*, small rather than large changes are what are required to keep up the adaptation. But even if great changes of conditions may sometimes occur rapidly, as by the irruption of some new enemy, or by a few feet of subsidence causing a low plain to become flooded, what are the chances that among the many thousands of *possible* large variations the one exactly adapted to meet the changed conditions should occur at the right time? To meet a change of conditions this year, the right large variation *might* possibly occur a thousand years hence.

The second argument is a still stronger one. Mr. Murphy fully adopts Mr. Herbert Spencer's view, that a variation, however slight, absolutely requires, to ensure its permanence, a number of concomitant variations, which can only be produced by the slow process of self-adaptation; and he uses this argument as conclusive against the formation of complex organs by natural selection in all cases where there is no tendency for action to produce self-adaptation; *à fortiori*, therefore, must a sudden large variation in any one part require numerous concomitant variations; it is still more improbable that they can accidentally occur together; it is impossible that the slow process of self-adaptation can produce them in time to be of any use; so that we are driven to the conclusion, that any large single variation, unsupported as it must be by the necessary concomitant variations, can hardly be other than hurtful to the individuals in which it occurs, and thus lead in a state of nature to its almost immediate extinction. The question, therefore, is not, as Mr. Murphy seems to think, whether such large variations occur in a state of nature, but whether, having occurred, they could possibly maintain themselves and increase. A calculation is made by which the more rapid mode of variation is shown to be necessary. It is supposed that the greyhound has been changed from its wolf-like ancestor in 500 years; but it is argued that variation is much slower under nature than under domestication, so that with wild animals it would take ten times as long for the same amount of variation to occur. It is also said that there is ten times less chance of favourable variations being preserved, owing to the free intermixture that takes place in a wild state; so that for nature to produce a greyhound from a wolf would have required 50,000 years. Sir W. Thomson calculates that life on the earth must be limited to some such period as one hundred million years, so that only two thousand times the time required to produce a well-marked specific change has, on this theory, produced all the change from the protozoon to the elephant and man.

Although many of the data used in the above calculation are quite incorrect, the result is probably not far from the truth; for it is curious that the most recent geological researches point to a somewhat similar period as that required to change the specific form of mammalia. The question of geological time is, however, so large and important that we must leave it for a separate article.

The second volume of Mr. Murphy's work is almost wholly psychological, and can be but briefly noticed. It consists to a great extent of a summary of the teachings of Bain, Mill, Spencer, and Carpenter, combined with much freshness of thought and often submitted to acute criticism. The special novelty in the work is the theory as to the "intelligence"

导致低地平原洪水泛滥，那么在成千上万种**可能的**大变异中，恰好适应变化之后的环境的变异能在恰当的时间发生吗？为了适应今年的一种环境变化，相应的大变异**也许**需要一千年才会发生。

第二个论据更充分一些。墨菲先生全盘接受了赫伯特·斯宾塞先生的观点，即，认为一种变异无论多么微不足道都一定需要许多伴生的变异来保证它的持久性，这些伴生的变异只能通过缓慢的自我适应过程来产生；他认为这一论据无可置疑，并借此反对在没有产生自我适应的行为趋势的所有情况中复杂器官都是通过自然选择形成的；因此，更不容置疑的是突然发生在任何部位的一种大变异肯定需要大量伴生变异；这些变异偶然同时出现是非常不可能的；自我适应的缓慢过程不可能及时产生具有任何作用的伴生变异；因此我们不可避免地得到如下结论：任何单一的大变异如果缺乏必需的伴生变异，那它们除了对发生这些变异的个体产生伤害之外几乎不会起别的作用，因此在自然状态下，它们几乎可以导致物种的迅速灭绝。所以问题并不如墨菲先生臆想的那样，也就是说，问题并不是这种大变异能否在自然状态下发生，而是已经发生的变异能否维持并加强。计算结果表明必然有更快的变异模式存在。人们猜测灰狗在 500 年内从它的类狼祖先变化而来；但是有争论说，变异在自然条件下比在驯养条件下发生得更加缓慢，因此野生动物发生同样程度的变异花费的时间将是驯养动物变异时间的 10 倍。也有人说，由于野生状态下会发生自由交配，保留有利变异的机会将少 10 倍；因此在自然界中狼演化为灰狗将需要 50,000 年时间。经计算，汤姆孙爵士认为地球上的生命肯定是在一亿年这样有限的时段内产生的，因此根据这一理论，只需要 2,000 倍于产生可以明确识别的特异性变化的时间，就可以形成从原生动物到大象和人所需的特异性变化。

尽管上述计算中使用的许多数据都相当不准确，但是结果可能距离真相并不遥远；因为最近的地质学研究指出了一个与哺乳动物发生特定形式的变化所需要的时间有点类似的时期，这点很令人好奇。然而，地质年代这个问题非常庞大也非常重要，所以我们不得不在另外一篇文章里单独进行介绍。

墨菲先生著作的第 2 卷几乎全是心理学方面的内容，这里只能简要地提一下。这一卷主要概述了贝恩、米尔、斯宾塞和卡彭特所倡导的学说，并结合了很多引发激烈批评的新鲜思想。这篇著作的特别新颖之处在于提出"智能"是通过组织结构和

manifested in organisation and mental phenomena, and this is so difficult a conception that it must be presented in the author's own words:—

"I believe the unconscious intelligence that directs the formation of the bodily structures is the same intelligence that becomes conscious in the mind. The two are generally believed to be fundamentally distinct: conscious mental intelligence is believed to be human, and formative intelligence is believed to be Divine. This view, making the two to be totally unlike, leaves no room for the middle region of instinct; and hence the marvellous character with which instinct is generally invested. But if we admit that all the intelligence manifested in the organic creation is fundamentally the same, it will appear natural, and what might be expected, that there should be such a gradation as we actually find, from perfectly unconscious to perfectly conscious intelligence; the intermediate region being occupied by intelligent though unconscious motor actions—in a word, by instinct... The intelligence which forms the lenses of the eye is the same intelligence which in the mind of man understands the theory of the lens; the intelligence that hollows out the bones and the wing-feathers of the bird, in order to combine lightness with strength, and places the feathery fringes where they are needed, is the same intelligence which in the mind of the engineer has devised the construction of iron pillars hollowed out like those bones and feathers... It will probably be said that this identification of formative, instinctive, and mental intelligence is Pantheistic... I am not a Pantheist: on the contrary, I believe in a Divine Power and Wisdom, infinitely transcending all manifestations of power and intelligence that are or can be known to us in our present state of being... Energy or force is an effect of Divine power; but there is not a fresh exercise of Divine power whenever a stone falls or a fire burns. So with intelligence. All intelligence is a result of Divine Wisdom, but there is not a fresh determination of Divine thought needed for every new adaptation in organic structure, or for every new thought in the brain of man. Every Theist will admit that there is not a fresh act of creation when a new living individual is born. I go a little further, and say that I do not believe in a fresh act of creation for a new species. I believe that the Creator has not separately organised every structure, but has endowed vitalised matter with intelligence, under the guidance of which it organised itself; and I think there is no more Pantheism in this than in believing that the Creator does not separately cause every stone to fall and every fire to burn, but has endowed matter with energy, and has given energy the power of transposing itself."

I am not myself able to conceive this impersonal and unconscious intelligence coming in exactly when required to direct the forces of matter to special ends, and it is certainly quite incapable of demonstration. On the other hand, the theory that there are various grades of conscious and personal intelligences at work in nature, guiding the forces of matter and mind for their purposes as man guides them for his, is both easily conceivable and is not necessarily incapable of proof. If therefore there are in nature phenomena which, as Mr. Murphy believes, the laws of matter and of life will not suffice to explain, would it not be better to adopt the simpler and more conceivable solution, till further evidence can be obtained?

The only other portion of the work on which my space will allow me to touch, is the chapter on the Classification of the Sciences, in which a scheme is propounded of great simplicity and merit. Mr. Murphy does not appear to be acquainted with Mr. Herbert

精神现象彰显出来的这一理论。这种概念太难理解，因此这里必须奉上作者的原话：

"我相信指导身体结构形成的无意识的智能，与能在头脑中变成意识的智能是相同的。人们通常认为这两种智能具有本质的不同：有意识的精神智能只为人类所具有，而造型智能则是神所拥有的。这种观点，使得两种智能毫不相干，没有为中间领域的本能所赋予的奇异性留下任何余地。但是如果我们承认生物创造中反映的所有智能在本质上都是相同的话，我们就会自然而然地想到并作出预期，即现实中应存在一种智能的等级，如同我们也确实发现的一样，从完全无意识的智能到完全有意识的智能之间的这种等级；而这些中间区域由无意识但有智能的动机行为所占据——用一个词形容就是本能……形成眼睛晶状体的智能与人类的头脑理解透镜理论的智能是一样的；使鸟类的骨头和翅膀羽毛成为空心的以将轻巧和力量结合起来的智能以及在需要的地方长出羽毛状边缘的智能，与工程师的头脑里将建筑物的铁柱设计得像那些空心的骨头和羽毛的智能没有差别……有人可能会说，这种对造型的、本能的、精神的智能的鉴别是泛神论的……我不是一个泛神论者：相反，我相信神的支配力量和智慧，它们远远超越了我们现存生命状态中所知道的或可能知道的所有力量和智能的体现……能量或者力量是神圣支配力量的结果；但是无论何时，石头掉落或大火燃烧都不是神圣支配力量的一种冒失操练。智能也是这样。所有智能都是神的智慧的结果，但是并不存在一种神圣意志为需要它的生物结构的每种新的适应，或者人类头脑中的每种新思想给出冒失判决。每位有神论者都会承认当一个新生个体降生时，不会有创造的冒失行为，我将这点引申一下，就是说我不相信存在一种创造新物种的冒失行为。我相信创世主并没有分别组建每种结构，而是赋予有生命的物质以智能，使其在这一智能的指导下自我组建；我认为泛神论是不会相信创世主没有令每块石头分别掉落、没有令每场火分开燃烧，只是赋予物质以能量，并且给予能量变换自身的能力。"

我自己不能设想出这样的情况：非人的、无意识的智能在需要指导物质力量达到特定目标时恰好出现，而且这显然是无法论证的。另一方面，自然界中处于运作中的有意识的、个体的智能所具有的不同等级，按照它们的目的引导物质和精神的力量，就像人类为自己的目的来引导它们一样，上述理论既容易想到又必然能被证明。因此如果自然界存在如墨菲先生相信的那种物质和生命法则所不能解释的现象的话，在得到进一步的证据之前，采用简单的、更容易想到的解决方法不是更好么？

在该著作其余部分中我具有发言权的只有谈论科学的分类那一章，这部分提出了一种非常简单又有价值的方案。墨菲先生好像对赫伯特·斯宾塞先生关于这一主题的文章并不熟悉，而且值得注意的是，他得到了非常相似的结果，尽管这一结

Spencer's essay on this subject, and it is somewhat remarkable that he has arrived at so very similar a result, although less ideal and less exhaustively worked out. In one point his plan seems an improvement on all preceding ones. He arranges the sciences in two series, which we may term primary and secondary. A primary science is one which treats of a definite group of *natural laws*, and these are capable of being arranged (as Comte proposed) in a regular series, each one being more or less dependent on those which precede it, while it is altogether independent of those which follow it. A secondary science, on the other hand, is one which treats of a group of *natural phenomena*, and makes use of the primary sciences to explain those phenomena; and these can also be arranged in a series of decreasing generality and independence of those which follow them, although the series is less complete and symmetrical than in the case of the primary sciences. The two series somewhat condensed are:—

Primary Series	Secondary Series
1. Logic	1. Astronomy
2. Mathematics	2. Terrestrial Magnetism
3. Dynamics	3. Meteorology
4. Sound, Heat, Electricity, &c.	4. Geography
5. Chemistry	5. Geology
6. Physiology	6. Mineralogy
7. Psychology	7. Palaeontology
8. Sociology	8. Descriptive Biology

Taking the first in the list of secondary or compound sciences, Astronomy, we may define it as the application of the first five primary sciences to acquiring a knowledge of the heavenly bodies, and we can hardly say that any one of these sciences is more essential to it than any other. We are, perhaps, too apt to consider, as Comte did, that the application of the higher mathematics through the law of gravitation to the calculation of the planetary motions, is so much the essential feature of modern astronomy as to render every other part of it comparatively insignificant. It will be well, therefore, to consider for a moment what would be the position of the science at this day had the law of gravitation remained still undiscovered. Our vastly multiplied observations and delicate instruments would have enabled us to determine so many empirical laws of planetary motion and their secular variations, that the positions of all the planets and their satellites would have been calculable for a moderate period in advance, and with very considerable accuracy. All the great facts of size and distance in planetary and stellar astronomy, would be determined with great precision. All the knowledge derived from our modern telescopes, and from spectrum analysis, would be just as complete as it is now. Neptune, it is true, would not have been discovered except by chance; the nautical almanack would not be published four years in advance; longitude would not be determined by lunar distances, and we should not have that sense of mental power which we derive from the knowledge of Newton's grand law;—but all the marvels of the nebulae, of solar, lunar, and planetary structure, of the results of spectrum analysis, of the velocity of light, and of the vast

30

果不够完美也没有完全解决该问题，不过他的计划看上去对前人的各种研究成果有所改进。他将科学划分成两个系列，我们可以将它们称为初级的和次级的。初级科学是探讨一组确切的**自然法则**的科学领域，这些科学领域能够按一规则序列罗列下来（正如孔德提议的），每一领域都或多或少依赖于在它之前产生的那些领域，而总的来说，它又独立于随后产生的那些领域。另一方面，次级科学是探索一系列**自然现象**的科学领域，这些科学领域利用初级科学来解释这些现象；这些领域也可以通过一系列逐渐减弱的普遍性和随后产生的领域的独立性加以罗列，但是该系列与初级科学的系列相比，它们不够完整也不够系统。这两个压缩后的系列如下：

初级系列	次级系列
1. 逻辑学	1. 天文学
2. 数学	2. 地磁学
3. 动力学	3. 气象学
4. 声学、热学、电学等	4. 地理学
5. 化学	5. 地质学
6. 生理学	6. 矿物学
7. 心理学	7. 古生物学
8. 社会学	8. 描述生物学

以次级科学或者复合科学目录中的第一个（即天文学）为例来说明，我们可以将其定义为使用前5种初级科学来获取关于天体的知识的学科，我们几乎不能说这些科学中的哪一个比另外某个更重要。也许我们会有与孔德一样的倾向，认为将高等数学运用到地心引力法则来计算行星运动是现代天文学的基本特征，相比之下天文学的其他部分都无关紧要了。因此，不妨考虑一下如果地心引力法则尚未发现，那么今天这一科学会被置于何地？我们大大增加的观测资料和精密的仪器使我们能够确定许多行星运动及其长期变化的经验法则，从而可以提前一段时间精确地将所有行星及其卫星的位置计算出来。所有这些关于行星天文学和恒星天文学的大小和距离的事实都将被非常精确地确定，所有这些出自现代望远镜以及光谱分析的知识都会像现在一样完备。如若不然，那么除非靠运气，不然就不会发现海王星；航海天文年历也不会提前4年就出版；我们也不能通过月球距离确定出经度，我们也不会知晓由牛顿伟大定律的知识而获得的那种智能；但关于星云、太阳、月亮和行星结构，光谱分析结果，光速以及行星和恒星空间的广阔维度的所有奇迹都会像现在一样全部为我们所熟知，这些知识将形成一门天文学，而这门科学在尊严、庄重和强烈的趣味性方面都不会逊色于我们现在拥有的天文学。

31

dimensions of planetary and stellar spaces, would be as completely known to us as they now are, and would form a science of astronomy hardly inferior in dignity, grandeur, and intense interest, to that which we now possess.

Mr. Murphy guards us against supposing that the series of sciences he has sketched out includes all that is capable of being known by man. He professes to have kept himself in this work to what may be called positive science, but he believes equally in metaphysics and in theology, and proposes to treat of their relation to positive science in a separate work, which from the author's great originality and thoughtfulness will no doubt be well worthy of perusal.

(**1**, 132-133; 1869)

墨菲先生反对我们把他概括出来的科学系列假定为囊括所有能够为人类所知的领域。他声称已经全身心投入到这一可以被称为实证科学的工作之中了，但是他同样相信形而上学和神学，准备在另一篇著作中单独讨论它们与实证科学的关系，从作者的伟大原创性和思虑的慎重性方面考虑，这本著作无疑具备精读的价值。

（刘皓芳 翻译；陈平富 审稿）

On the Dinosauria of the Trias, with Observations on the Classification of the Dinosauria

Editor's Note

As a news magazine, *Nature* reported on the activities of various scientific societies. This report of a meeting of the Geological Society held on 24 November 1869 describes the reading of a paper by the President of the Society, Thomas Henry Huxley, entitled "On the Dinosaurs of the Trias, with Observations on the Classification of the Dinosauria". The anonymous third-person account notes the discussion, in which Harry Seeley (an expert on Pterosaurs) and others disputed various points with the President. Huxley's paper was followed by one on biogeography as elucidated by fossil corals. This and many other detailed accounts give a flavour of the daily scientific life of London: from today's perspective, they read like the business of titans.

GEOLOGICAL Society, November 24.—"On the Dinosauria of the Trias, with observations on the Classification of the Dinosauria", by Prof. T. H. Huxley, LL.D., F.R.S., President. The author commenced by referring to the bibliographical history of the Dinosauria, which were first recognised as a distinct group by Hermann von Meyer in 1830. He then indicated the general characters of the group, which he proposed to divide into three families, viz.: —

I. The Megalosauridae, with the genera *Teratosaurus*, *Palaeosaurus*, *Megalosaurus*, *Poikilopleuron*, *Laelaps*, and probably *Euskelosaurus;*

II. The Scelidosauridae, with the genera *Thecodontosaurus*, *Hylaeosaurus*, *Pholacanthus*, and *Acanthopholis;* and

III. The Iguanodontidae, with the genera *Cetiosaurus*, *Iguanodon*, *Hypsilophodon*, *Hadrosaurus*, and probably *Stenopelys.*

Compsognathus was said to have many points of affinity with the Dinosauria, especially in the ornithic character of its hind limbs, but at the same time to differ from them in several important particulars. Hence the author proposed to regard *Compsognathus* as the representative of a group (*Compsognatha*) equivalent to the true Dinosauria, and forming, with them, an order to which he gave the name of Ornithoscelida. The author then treated of the relations of the Ornithoscelida to other Reptiles. He indicated certain peculiarities in the structure of the vertebrae which serve to characterise four great groups of Reptiles, and showed that his Ornithoscelida belong to a group in which, as in existing Crocodiles, the thoracic vertebrae have distinct capitular and tubercular processes springing from the arch of the vertebra. This group was said to include also the Crocodilia, the Anomodontia, and the Pterosauria, to the second of which the author was inclined to approximate the Ornithoscelida. As a near ally of these reptiles, the author cited the Permian *Parasaurus*, the structure of which he discussed, and stated that it seemed to be a terrestrial reptile, leading back to some older and less specialised reptilian form.

关于三叠纪恐龙以及恐龙分类的研究

编者按

作为新闻杂志，《自然》一直在报道各个科学学会的活动。这篇关于 1869 年 11 月 24 日地质学会举行的一次会议的通讯，记述了学会主席托马斯·亨利·赫胥黎在会上所作的报告，报告的题目是《关于三叠纪恐龙以及恐龙分类的研究》。匿名的第三方提到了哈里·西利（翼龙方面的专家）等人与赫胥黎主席在许多观点上的争论。在赫胥黎的这篇文章之后，是一篇生物地理学方面的阐述珊瑚化石的文章。从这篇文章以及其他许多详细的报道，可以看到当时伦敦日常学术活动的情况。以现在的眼光来看，他们作学术报告就像是现在的商界巨亨们发表演说一样。

11 月 24 日，在地质学会的会议上宣读了《关于三叠纪恐龙以及恐龙分类的研究》，报告人是法学博士、英国皇家学会会员、地质学会主席赫胥黎教授。作者在文章开篇首先介绍了有关恐龙的文献历史，然后又指出了这一类群的基本特征。1830 年，赫尔曼·冯·迈尔首次将恐龙划定为一个独立的类群。作者提议将恐龙这个类群分为 3 个科，即：

I. 巨齿龙科，包括怪晰龙属、远古龙属、巨齿龙属、杂肋龙属、暴风龙属，此外很可能还包括优肢龙属。

II. 腿龙科，包括槽齿龙属、林龙属、多刺甲龙属和棘甲龙属。

III. 禽龙科，包括鲸龙属、禽龙属、棱齿龙属、鸭嘴龙属和狭盘龙属。

有人曾指出，有很多特征可以说明美颌龙与恐龙具有较近的亲缘关系，特别是它的后肢具有鸟类的特征，但同时，二者在一些重要细节上又存在差异。因此作者提出将以美颌龙为代表的一个类（美颌龙类）与真正的恐龙类视为同一等级，并共同构成一个目，命名为鸟臀目。之后作者论述了鸟臀目与其他爬行动物的关系。他列举了爬行动物 4 个主要类群的脊椎结构特征，并证明鸟臀目属于其中的一类。这一类群的胸椎有一个独特的从椎骨弓形区发出的小头突，与现存的鳄鱼相似。这一类群还包括鳄目，缺齿亚目和翼龙目，而作者认为缺齿亚目与鸟臀目亲缘关系更近。由于这些爬行动物关系较近，作者引述了二叠纪的鸡冠龙属，讨论了它的结构特征，认为它属于陆生爬行动物，是较古老的、非特化的爬行动物类群。关于鸟臀目与鸟类的关系问题，作者说，他所知的所有特征都表明，鸟类在结构上和爬行动物并没有区别，而爬行动物的结构在鸟臀目中已有先兆。他还简要讨论了翼手龙和鸟类

With regard to the relation of the Ornithoscelida to birds, the author stated that he knew of no character by which the structure of birds as a class differs from that of reptiles which is not foreshadowed in the Ornithoscelida, and he briefly discussed the question of the relationship of Pterodactyles to birds. He did not consider that the majority of the Dinosauria stood so habitually upon their hind feet as to account for the resemblance of their hind limbs to those of birds, by simple similarity of function. The author then proceeded to notice the Dinosauria of the Trias, commencing with an historical account of our knowledge of the occurrence of such reptilian forms in beds of that age. He identified the following Triassic reptilian-forms as belonging to the Dinosauria:— *Teratosaurus, Plateosaurus,* and *Zanclodon* from the German trias; *Thecodontosaurus* and *Palaeosaurus* from the Bristol conglomerate (the second of these genera he restricted to *P. cylindrodon* of Riley and Stutchbury, their *P. platyodon* being referred to *Thecodontosaurus*); *Cladyodon* from Warwickshire; *Deuterosaurus* from the Ural; *Ankistrodon* from Central India; *Clepsysaurus* and *Bathygnathus* from North America; and probably the South African *Pristerosaurus.*—Sir Roderick Murchison, who had taken the chair, inquired as to the lowest formation in which the bird-like character of Dinosaurians was apparent, and was informed that it was to be recognised as low as the Trias, if not lower.—Mr. Seeley insisted on the necessity for defining the common plan both of the Reptilia and of the ordinal groups before they could be treated of in classification. He had come to somewhat different conclusions as to the grouping and classification of Saurians from those adopted by the President. This would be evident, in so far as concerned Pterodactyles, from a work on Ornithosauria which he had just completed, and which would be published in a few days.—Mr. Etheridge stated that the dolomitic conglomerate, in which the Thecodont remains occurred near Bristol, was distinctly at the base of the Keuper of the Bristol area, being beneath the sandstones and marls which underlie the Rhaetic series. There were no Permian beds in the area. He regarded the conglomerates as probably equivalent to the Muschelkalk. It was only at one point near Clifton that the Thecodont remains had been found.—Prof. Huxley was pleased to find that there was such a diversity of opinion between Mr. Seeley and himself, as it was by discussion of opposite views that the truth was to be attained. He accepted Mr. Etheridge's statement as to the age of the Bristol beds.

(**1**, 146-147; 1869)

的关系。作者认为，虽然大部分恐龙类也习惯后脚站立，但不能仅凭这种功能上的简单相似就认为恐龙类的后肢和鸟类的后肢相似。之后，作者开始关注三叠纪恐龙，首先阐述了那个时期出现的爬行类相关的历史发展知识。他鉴别出以下几个属于恐龙目的三叠纪爬行类：德国的三叠纪巨齿龙属、远古龙属和镰齿龙属；英国布里斯托尔砾岩中的槽齿龙属和远古龙属（其中远古龙属的代表是由赖利和斯塔奇伯里命名的柱齿龙，而槽齿龙属的代表则是板齿龙）；英格兰沃里克郡的独巨齿龙属；乌拉尔地区的亚次龙属；印度中部的钩齿龙属；北美的伏龙属和深颚龙属；还可能包括来自南非的原始龙属。曾任学会主席的罗德里克·麦奇生爵士就恐龙类最早是在哪一个地层出现了与鸟类相似的特征提出问题，他得到了如下的答案：如果在更早期的地层中没有发现的话，那就是出现在三叠纪。西利先生坚持认为，无论是对于爬行类，还是对于有序排列的各个类群，在进行分类之前都必须先明确基本的排序方法。对于蜥臀类的分组和分类，他得到了一些与学会主席所持观点不太相同的结论。根据他的一部关于鸟蜥亚纲的著作，就翼手龙而言这一点是非常明显的。西利先生的这部著作刚刚完成，不久后将会发表。埃瑟里奇先生则提到，在布里斯托尔附近，发现过槽齿类遗骸的白云石砾岩地层明显是处在考依波地层的底部，其上方是瑞替期的砂岩和泥灰岩。这个地区没有二叠纪的地层。他认为白云石砾岩地层和壳灰岩地层很可能是等同的。在克利夫顿附近，只在一处发现有槽齿类的遗骸。赫胥黎教授表示，很高兴听到西利先生提出不同的观点，因为真理是从不同意见的碰撞和冲突中产生的。对于埃瑟里奇先生关于布里斯托尔地区地层年代的论述，他则表示认可。

（刘冉冉 翻译；徐星 审稿）

Darwinism and National Life

H.

Editor's Note

This comment by a writer identified only as "H." (perhaps Thomas Huxley?) shows how Darwinian evolutionary theory was from its inception linked to social engineering and eugenics. H. implies that natural selection should be able to explain all national characteristics (which here are really just stereotypes), such as the "self-reliance" of the inhabitants of the United States. Little account is taken of the tremendous timescales that Darwin assumed to be necessary for significant evolutionary change (although Darwin himself echoed many of the assumptions and presumptions made here in "The Descent of Man" (1877)). Nonetheless, the idea that evolutionary forces can explain why one culture thrives and another fails is still current, thanks to researchers who are probing the links between environment and history.

THE Darwinian theory has a practical side of infinite importance, which has not, I think, been sufficiently considered. The process of natural selection among wild animals is of necessity extremely slow. Starting with the assumption (now no longer a mere assumption) that the creature best adapted to its local conditions must prevail over others in the struggle for existence, the final establishment of the superior type is dependent at each step upon three accidents—first, the accident of an individual sort or variety better adapted to the surrounding conditions than the then prevailing type; secondly, the accident that this superior animal escapes destruction before it has had time to transmit its qualities; and, thirdly, the accident that it breeds with another specimen good enough not to neutralise the superior qualities of its mate. In the case of domesticated animals the progress is incomparably more rapid, because it is practicable, first, to modify the conditions of life, so as to encourage the appearance of an improved specimen; next, to cherish and protect it against disaster; and, lastly, to give it a consort not altogether unworthy of the honour of reproducing its qualities. The case of man is intermediate in rapidity of progress to the other two. The development of improved qualities cannot be insured by judicious mating, because as a rule human beings are capricious enough to marry without first laying a case for opinion before Mr. Darwin. Neither would it be easy, nor, perhaps, even allowable, to extend any special protection by law or custom to those who may be physically and intellectually the finest examples of our race. Still, two things may be done: we may vary the circumstance of life by judicious legislation, and still more easily by judicious non-legislation, so as to multiply the conditions favourable to the development of a higher type; and by the same means we may also encourage, or at least abstain from discouraging, the perpetuation of the species by the most exalted individuals for the time being to be found. Parliament, being an assembly about as devoid of any scientific insight as a body of educated men could possibly be, has not as yet consciously legislated with a view to the improvement of the English type of character. Without knowing it, however, the legislature has sometimes stumbled on the right course, though

达尔文学说与国民生活

通过这篇作者仅署名为"H."（也许是托马斯·赫胥黎？）的评论文章我们可以看到，从一开始达尔文的进化论就与社会工程学和优生学紧密相连。作者暗示，应该可以用自然选择来解释所有民族的特征（文章中论述的实际上只是些模式化的见解），例如美国土著居民"自力更生"的特点。文章几乎没有考虑漫长的时间尺度这一在达尔文看来对明显的进化改变来说十分必要的因素（尽管达尔文本人在文中反复提到了1877年的《人类的由来》一书中的多项假设）。不过，根据一直在探索环境与历史之间关联的研究人员的工作，进化动力的概念可以解释为什么一种文化繁盛而另一种文化衰落的现象持续存在。

　　达尔文的进化理论有极其重要的实际应用，我认为人们并没有充分认识到这一点。在野生动物中，自然选择的过程必然是十分缓慢的。我们从一个假设（现在已经不仅仅是一个假设了）开始论述这个问题，假设最适应当地环境的生物必定在生存斗争中胜过其他生物，那么，一种高等生物最终确立的每一步都是依赖于以下三个事件：第一，出现了一个新种类的个体或变异体，比当时占优势的物种更适应周围环境；第二，这个更高等的个体在死亡之前有足够的时间能够产生后代以传递自己的特性；第三，这个个体与其他足够优良的个体交配来繁殖后代，以保证它的优等生物性状不会被中和掉。对于家养动物来说，进化的速度无比之快，因为在实际应用中为了促进改良品种的产生首先会改变动物的生活环境；其次，会照料并保护出现的改良品种使其免受灾害；最后，还会给它指定配偶以保证其优良性状的延续。人类的进化速度则介于两者之间。人类中出现的优良性状不能通过明智的交配保证其延续，因为在达尔文提出进化论的观点之前，人类本能地在婚配问题上表现得十分任性，通常不会提出任何婚配的理由。企图通过法律或者惯例给那些在体能和智力两方面都是我们种族的精英的个体提供特别的保护既不容易，甚至也不被允许。尽管如此，我们仍然可以做这样两件事情：我们也许可以通过明智的立法来改变生活的环境，不过可能通过明智的非立法手段会更容易些，这样一来就能提供更多适合于更高等个体出现和发展所需要的环境；我们也可以通过同样的办法来促进，或者至少不去妨碍由那些在当时能找到的最好的个体所组成的物种的绵延不绝。国会尚未有意识地从提高英国人素质的角度来进行立法，因为国会就是这样的一个群体，它缺乏一批有素养的人可能具备的科学的洞察力。然而，由于不理解这一观点，立法机构有时会在正确的问题上犹豫，而更经常的则是犯下愚蠢的错误。我们的自由

it has more often blundered into the wrong. Our free trade policy has furnished special scope and special advantages to the energetic enterprising character, and so far has tended to perpetuate and intensify the type which has given to little England her wonderful prominence in the world. On the other hand, the steady refusal to make a career for scientific men has drained away most of our highest intellect from its proper field, and has subjected the rest to an amount of discouragement by no means favourable to increase and improvement. Our laws and customs practically check the growth of the scientific mind as much as they tend to develop the speculative and energetic commercial character.

We do not expect for a long time to hear an orator in the House of Commons commence his speech by announcing, (as a distinguished member of the Austrian Reichsrath recently did, in a debate on the relation of the different nationalities in the empire), that the whole question is whether we are prepared to accept and act upon the Darwinian theory. But even an average English M.P. may be brought to see that it may be possible, indirectly, to influence the character and prosperity of our descendants by present legislation, and none will deny that, if this is practicable, a higher duty could not be cast upon those who guide the destinies of a nation.

A glance at the operation of Darwinism in the past, will best show how potent it may be made in the future. Look at English progress and English character, and consider from this point of view to what we owe it. There were originally some natural conditions favourable to the growth of our commercial and manufacturing energy. We had an extensive coast and numerous harbours. We had also abundance of iron-stone in convenient proximity to workable coal. Other nations either wanted these advantages or were ignorant that they possessed them. These favourable conditions developed in many individuals a special adaptability to commercial pursuits. The type was rapidly reproduced and continually improved until England stood, in the field of commerce, almost alone among the nations of the world. And what is there now to sustain our pre-eminence? Nothing, or next to nothing, except the type of national character, which has been thus produced. Steam, by land and sea, has largely diminished the superiority which we derived from the nature of our coast; and coal and iron are now found and worked in a multitude of countries other than our own. Our strength in commerce, like our weakness in art, now rests almost exclusively on the national character which our history has evolved.

Take another example of the character of a people produced partly by natural conditions of existence, but far more by the artificial conditions to which evil legislation has exposed it. What has made the typical Irishman what he now is? The Darwinian theory supplies the answer. Ireland is mainly an agricultural country, with supplies of mineral wealth altogether inferior to those of England, though by no means contemptible if they were but developed. This is her one natural disadvantage, and it is trifling compared with those which we in our perversity created. For a long period we ruled Ireland on the principles of persecution and bigotry, and left only two great forces at work to form the character of the people. All that there was of meanness and selfishness and falsehood was tempted to servility and apostacy, and flourished and perpetuated itself accordingly. All that there was

贸易政策为那些富有强烈进取心的人们提供了特殊的机会和特别有利的条件，迄今为止，此政策也使得已经让不大的英格兰在世界范围内获得荣耀的那种特质类型得以存续和加强。另一方面，由于从事科学研究的人们在成就一番事业上还存在很大的阻力，使得我们这些智力最高的人中的大多数无法在合适的领域从事相关的工作，而其余的人也由于没有良好的发展环境而备受挫折。我们的法律和习俗在很大程度上发展了富有活力的商业投机特质，但反过来实际上也同样多地限制了科学思维的发展。

长期以来，我们并不期望听见一个演说家在下议院开始他的演讲时就宣称（就像奥地利议会的一位著名成员最近在关于帝国中不同民族之间关系的辩论中所做的那样），所有的问题就是我们是否准备接受并且按照达尔文的理论行事。但是，即使是一名普通的英国下议院议员也应该明白，通过现有的立法是有可能间接地对我们后代的特质和富足产生影响的，而且如果这是切实可行的话，也不会有人否认这才是赋予那些引导民族发展方向的人的最为重大的责任。

回顾一下进化论在过去所起的作用会让我们对它在未来将有多大的潜力看得更加清楚。看看英国人取得的进步和英国人的特质，并从这个角度来考虑一下我们应该把这一切归因于什么。这里本来就有适合我们的商业和制造业发展的一些自然条件。我们拥有广阔的海岸线和大量的港口。我们也拥有丰富的与可经营煤矿相邻的铁矿。其他国家要么没有这样的便利，要么还不知道他们拥有这样的便利。这些有利条件造就了许多个体在商业活动方面的特殊适应能力。这些特质迅速地扩散并且持续提高，直至英格兰在商业领域几乎是独一无二地屹立于世界民族之林。如今，又要靠什么来维持我们的这种卓越呢？除了这种历史造就的民族特质，我们没有或者说几乎没有别的任何东西可以依赖。无论是在内陆还是沿海，蒸汽的应用都已经大大地降低了我们从海岸线得到的优越性，煤和钢铁如今在其他许多国家也被大量发现和使用。我们在商业领域的优势，就好比我们在艺术领域的弱势一样，现在几乎只能完全依赖于我们从历史演化中得来的民族特质了。

再举另外一个例子，其中一部分人们的特质是由于自然条件造成的，而更多的则是遭受了那些有害的立法的影响。是什么造就了现在典型的爱尔兰人呢？达尔文的进化理论提供了答案。爱尔兰很大程度上是一个农业国家，矿藏资源总量比英格兰少，即便他们真的发展起来也还是如此，这绝不是轻视他们的看法。这对他们来说是一个天然的不足，而这个不足与我们自以为是地创造出的东西相比就显得微不足道了。长期以来，我们带着迫害和偏见的原则统治着爱尔兰，结果只留下了两种巨大的力量影响着这里人民特质的形成。一方面，吝啬和自私到了如此地步，为了达到使他们接受奴役并放弃信仰的目的，谎言得到鼓励和使用，以至于谎言自身竟然因

of nobleness and heroic determination was drawn into a separate circle, where the only qualities that throve and grew were irreconcilable hatred of the oppressor and resolute but not contented endurance. The two types rapidly reproduced themselves, and as long as the external conditions remained unaltered, they absorbed year by year more and more of the people's life; as, if Darwinism is true, they could not but do. And what is the result now? A great part of a century has elapsed since we abandoned the wretched penal laws, and yet none can fail to see in Ireland the two prevailing types of character which our ancestors artificially produced, the only change being that the two types have become, to a certain extent, amalgamated in a cross which reflects the peculiarities of each. Whether future legislation may so far modify the conditions of Irish existence as to work a gradual change in the national character, is a question of much interest, but too large to be discussed just now. In any case we can scarcely expect the results of centuries upon a national type to be reversed in less than a succession of generations.

Still confining myself to the past, let me point again to the very marked qualities which the conditions of their existence have produced in the people of the United States. They started with a large element of English energy already ingrained into them; they have been reinforced by millions of emigrants presumably of more than the average energy of the various races which have contributed to swell the tide. Added to this, the Americans have enjoyed the natural stimulus of a practically unlimited field for colonisation. Only the resolute, self-reliant settler could hope to prosper in the early days of their national existence; and self-reliance approaching to audacity is the special type of character which on the Darwinian hypothesis we should expect to see developed, transmitted, and increased. How far this accords with actual experience, no one can be at a loss to say. There is probably not a nation in the world whose peculiarities might not be traced with equal ease to the operation of the same universal principle. And the moral of the investigation is this: Whenever a law is sufficiently ascertained to supply a full explanation of all past phenomena falling within its scope, it may be safely used to forecast the future; and if so, then to guide our present action with a view to the interest and well-being of our immediate and remote descendants. Read by the light of Darwinism, our past history ought to solve a multitude of perplexing questions as to the probable supremacy of this or that nation in times to come in the field of commerce, as to the effects of emigration and immigration on the ultimate type likely to be developed in the country that loses and in that which gains the new element of national life, and many another problem of no less interest to ourselves and to humanity.

The subject I have thus slightly indicated seems to me to deserve a closer investigation than it has yet received: and, strange as it will sound to the ears of politicians, I cannot doubt that, in this and other ways, statesmen, if they could open their eyes, might derive abundant aid from the investigations of science, which they almost uniformly neglect and despise.

(**1**, 183-184; 1869)

此而大行其道。另一方面，高尚和英勇却被引入另一个不同的循环，在这里唯一得到繁荣和发展的特性就是对于压迫者无可调和的憎恨以及坚定而无尽的忍耐。这两种类型的特质迅速地自我复制着，并且只要外部环境保持不变，它们就会年复一年地同化越来越多的人；如果达尔文的进化论是正确的，那么他们本不该如此，但却确实如此。那么，现在的结果是什么呢？从我们废除了那些肮脏的刑律算起，已经过去大半个世纪了，然而在爱尔兰，没有人不会发现由我们的祖先人为导致的这两种主流特质。唯一的变化是，这两种特质在一定程度上混杂在一起，交叉体现着各自的特性。是否未来的立法能在一定程度上改变爱尔兰的生存环境，从而使得这个民族的特质慢慢有所变化，这个问题是十分有意义的，但它太大了，不是现在就能讨论的。不管怎样，我们都难以期待短短几代人的时间就能使几个世纪以来形成的民族特质得到改变。

我还是把眼光放在过去吧，让我再来说一下由美国人自身的生存环境造就的美国人的显著的特质。一开始，他们身上就携带着大量的英国人所具有的根深蒂固的特质，而数百万超出各个种族平均水平的移民可能使这种特质得到了进一步加强，促进了这种趋势的增强。另外，因殖民而占据的几乎无限的土地极大地激励着美国人。也只有坚定的、自力更生的移民者有希望能在他们民族生存的早期获得成功；自力更生以至无畏进取，这种特殊的特质，正是我们根据达尔文的理论预期会得到发展、传承和提高的。这与实际的经历究竟有多一致，大概没有人会说不出来。可能对于世界上任何一个民族的特质，都能够按照相同的普遍原则轻易地追溯其来源。研究的意义应当是这样的：如果一个法则能充分确定地为过去在这个范围内的现象提供充分解释的话，那么用它来预测未来应该就是可行的。如果确实如此，那么为我们子孙后代的利益和安康着想，就应该用它来引导我们目前的行为。从达尔文的进化论来看，通过研究我们过去的历史应该可以解决很多复杂的问题，如关于将来哪个民族在商业领域可能占据至高无上的地位的问题，关于出入境移民致使民族特质元素丧失或新增，从而影响民族特质最终发展方向的问题，以及其他许多对我们自身和人类的利益来说都相当重要的问题。

我在这里极其简略地提出的这个主题，在我看来实在是值得进行比以往更深入的调查研究，这对政治家来说可能还是一个陌生的东西，但我不会怀疑，无论通过什么样的方法，如果政治家们打开视野，他们将能够从科学研究中得到充分的帮助，而这些他们几乎都无一例外地忽略和轻视了。

（刘晓辉 翻译；江丕栋 审稿）

A Deduction from Darwin's Theory

W. S. Jevons

Editor's Note

This comment on Darwin's theory of evolution, from British economist W. Stanley Jevons, is typical of the racial chauvinism of its time, from which Darwin himself was by no mean immune. It argues that temperate climates, like those of Jevons' own country, are best suited to nurturing the "highest forms of civilization". Jevons takes it for granted that Europeans have a "superior degree of energy and intellect" than black Africans. Despite Darwin's humanitarian principles and opposition to slavery, he and most of his supporters shared this notion of racial hierarchy, which is very evident in Darwin's "The Descent of Man", published two years after this letter.

THERE is one important consequence deducible from Darwin's profound theory which has not yet been noticed so far as I am aware. The theory is capable under certain reasonable conditions of accounting for the fact that the highest forms of civilisation have appeared in temperate climates.

Although some apparent exceptions might be adduced, it is no doubt true that man displays his utmost vigour and perfection, both of mind and body, in the regions intermediate between extreme heat and extreme cold, allowance being made for the reduced temperature of elevated mountain districts. The explanations hitherto given of this fact are of a purely hypothetical and shallow character. It is said, for instance, that the prolific character of the tropical climate too easily furnishes man with subsistence, so that his powers are never properly called into action. On the other hand in the Arctic regions nature is too sterile and no exertions can lead to the accumulation of much wealth. This explanation obviously involves the gratuitous hypothesis that man has been created with powers exactly suited to be called forth by just that degree of difficulty experienced in a temperate climate. There are those even who maintain our peculiar British climate to be the very best possible, because it taxes our powers of endurance to the last point which they can bear, and thus calls forth the greatest amount of energy. But here again is the assumption that the British people and the British climate were specially created to suit each other.

The theory of natural selection, on the other hand, represents that great method by which infinitely numerous adaptations will always be produced throughout time. Whatever happens in this material world must happen in consequence of the properties originally impressed upon matter, and our notions of the wisdom embodied in the Creation must be infinitely raised when we understand, however imperfectly, its true method. The continual resort to special inventions and adaptations must surely be below the greatness of a Power which could so design and create matter from the first that it must go on thenceforth

44

达尔文理论的一个推论

杰文斯

编者按

这篇文章是英国经济学家斯坦利·杰文斯对达尔文进化论的评论，可以说颇具代表性地反映了当时的种族沙文主义，达尔文本人也不是完全没有受到影响。文章声称，包括杰文斯自己的祖国在内的许多地方所具有的温带气候最适宜于孕育"最高级别的文明"。杰文斯据此就想当然地认为，与非洲黑人相比，欧洲人拥有"更高等的体力和智力"。与达尔文提倡人道主义精神并反对奴隶制不同，杰文斯和他的许多支持者都持这种种族等级观念，这一点从本文发表两年后出版的达尔文的著作《人类的由来》一书中可以很明显地看到。

据我所知，至今还没有人注意到，达尔文理论有一个重要的推论，即在一定的理想条件下，根据达尔文理论可以证明，最高等形式的文明出现在温带气候中。

尽管也可能存在一些明显的例外，但无疑，生活在介于极寒冷和极炎热地区之间的中间气候带中的人类，无论是身体方面还是心智方面都是最富有活力、最完美的，这里也考虑到了因海拔升高而温度降低的山区。至今为止，对这一现象的解释还都只是一些单纯的假说或粗浅的说明。例如，有些人认为热带地区自然物产十分丰富，可以很容易地满足人类的物质需要，因此在热带地区居住的人类的能力永远无需全面施展从而也难以发展。另一方面，极地地区的自然条件则太恶劣，以至于任何努力都不会带来大量财富的积累。这种说法中显然包含着一个毫无根据的假设：人类被创造出来时具备的能力恰好与温带气候下会遭遇的困难程度相一致。一些人坚持认为我们不列颠特殊的气候是最佳的可能，因为这样的气候使我们的忍耐力达到我们所能承受的极限程度，因此最大限度地挖掘出了我们的潜力。这里又是一个假设：不列颠的人和气候是特别创造出来的，彼此相适应。

另一个方面，自然选择的理论表明：伴随着时间的推移，会有多种多样的适应现象不断产生。物质世界发生的任何事情必然反馈到原本强加于物质的特性上，尽管这看起来不那么完美，但当我们真正理解了这一点时，可以肯定，我们对蕴涵在创造过程中的智慧的认知也必将不断得到提升。持续不断的创造和适应一定是受某种伟大力量支配的，就好像从一开始就是这种力量按照某种始终如一的法则设计并

inventing and adapting forms of life without apparent limit, in pursuance of one uniform principle.

I conceive it to be the essential consequence of Darwin's views that no form of life is to be regarded as a fixed form; but that all living beings, including man, are in a continual process of adjustment to the conditions in which they live. If this be so, it will of necessity follow that the longer any race dwells in given circumstances, the more perfectly will it become adapted to those circumstances. A migratory race, on the contrary, will always be liable to enter climates unsuited to it, and less favourable to the development of the greatest amount of energy. Negroes can bear a tropical heat simply because the race has grown more accustomed to it than Europeans, who bring with them indeed a superior degree of energy and intellect, but soon sicken and fail to reproduce themselves in equal perfection.

The intellect of man renders him far more migratory than most other animals, and when we look over long periods of time we must regard him as in a constant state of oscillation between the equator and the borders of perpetual snow. It will of necessity follow that the race, as a whole, will be better adapted to a medium than to an extreme climate. Not only may the same race have passed alternately through colder and hotter climates, but it is obvious that the tribes which intermix and intermarry in temperate regions will have come, some from a hotter and some from a colder region. The amalgamated race will therefore be precisely adapted to a medium climate. The inhabitants of the Arctic regions, on the contrary, must have come entirely from a warmer climate, and those of a tropical region from a colder climate, so that ages must pass before either re-adapts itself perfectly to its new circumstances.

It is hardly to be expected that history can afford complete corroboration of this theory; but I do not think that historical facts can be adduced in serious opposition to it. The progress of archaeological and linguistic inquiry shows more and more clearly that the civilised parts of the earth have been inhabited by a succession of different races. A really aboriginal and indigenous people, growing upon a single island or spot of ground without kinship with other races, is not known to exist; and it is almost certain that all races have descended from a few stocks, if not from a single one. The evidences of extensive and frequent migrations are thus most complete, even if we had not distinct historical facts concerning the rapid and extensive movements of the Goths, Huns, Moors, Scandinavians, and many other races.

If the historical evidence disagrees with the theory in any point, it is that the migrations from temperate to extreme climates greatly over-balance any opposite movement. It would hardly, perhaps, be too much to represent the temperate regions of the Old World as the birthplace of successive races, which have diverged and died away more or less rapidly in distant and extreme climates. But if such be the conclusion from historical periods, it would only indicate that the human race had already acquired, in prehistoric times, a

创造了各种物质，之后又不受明显约束地继续创造了各种生命形式并使它们不断适应。

我认为达尔文观点的一个必然结论是，没有任何一种生命形式是固定不变的；所有的生物，包括人类，都是处于不断调整改变自己以适应生存环境的过程之中。如果的确如此，那么一个必然的结论是，任何一个种族在给定的环境中生活的时间越长，他们就越适应这种环境。相反，一个种族迁移后，很有可能迁入他们不适应的气候环境，因而不利于他们最大能力的发展。例如，黑人能忍受热带的高温，只不过是因为他们在热带生活了更长时间而更加习惯热带的环境，相反，欧洲人虽然在身体和智力上的确都更有优势，但他们并不习惯热带的环境，因而很快就会生病，并且无法繁衍和他们一样优良的后代。

人类的智慧使他们比其他大部分动物都更具迁移性，当我们纵观历史时，我们可以看出人类一直在赤道和终年积雪的区域之间不断徘徊着。这样就可以得出一个必然的结论，即人类作为一个整体应该更适应于温和的气候而不是极端的气候。不但同一个种族可能交替地经历较冷和较热的气候，而且很明显温带地区将出现混居或通婚的部落，一部分是来自较热地区，而另一部分可能来自较寒冷地区。融合后的种族因此也正好可以适应温和的气候。相反，如果北极地区的居民一定是全部来自较温暖气候的，而热带地区的居民全部是来自较寒冷气候带，那么许多年后他们才能重新适应他们的新环境。

我们很难期待历史能提供事实来完全证实这一理论，但我也相信不会有与这一理论强烈相悖的历史事实。考古学和语言学的研究成果越来越清晰地告诉我们，地球上出现过文明的地方都曾相继居住过不同的种族。至今还没有发现那种生活在一个孤岛或某一片孤立的土地上，和其他外族没有任何亲缘关系的纯粹的土著种族；另外也几乎可以确定，任何一个种族如果不是一个血统的话则肯定是来源于几个血统。就算没发现像哥特人、匈奴人、摩尔人、斯堪的纳维亚人以及其他一些种族的快速而广泛的迁徙这样清楚的历史记录，我们关于广泛而频繁迁移的证据也是充分的。

如果说有历史证据在某些方面与该理论不相吻合，那便是人类从温带地区向极端气候区的迁移要比相反方向的迁移多很多。可能我们可以毫不夸张地认为，旧大陆温带地区就是那些在偏远且气候极端的地区中很快分散并渐渐消失了的后继种族的发祥地。但是，如果通过历史学的研究发现果真如此的话，那只能认为在温带地区的人类在史前就已经获得一种非凡的体质，从而展示出他们强大的生命力。无疑，

constitution displaying its greatest vitality in temperate regions. There can be no doubt that, were the rest of the world uninhabited by man, a very inferior race, such as the negroes of tropical Africa, would gradually re-people it; but they cannot do so in the present state of things, because they come into conflict with races of superior intellect and energy.

I would add in conclusion that the utmost result of speculations of this kind, supposing them to be valid, would consist in establishing a *general tendency*, so that the probabilities will be in favour of a great display of civilisation occurring in temperate climates rather than elsewhere. I do not for a moment suppose that any common physical cause, such as soil, climate, mineral wealth, or geographical position, or any combination of such causes, can alone account for the rise and growth of civilisation in Assyria, Egypt, Greece, Italy, or England. Material resources are nothing without the mind which knows how to use them. No physiology of protoplasm, no science that yet has a name, or perhaps ever will have a name, can account for the evolution of intellect in all its endless developments. The vanity of the Comtists leads them to suppose that their philosophy can compass the bounds of existence and account for the evolution of history; but the scientific man remembers that however complicated the facts which he reduces under the grasp of his laws, yet beyond all doubt there remain other groups of facts of surpassing complication. Science may ever advance, but, like an improved telescope in the hands of an astronomer, it only discloses the unsuspected extent and difficulty of the phenomena yet unreduced to law.

(**1**, 231-232; 1869)

世界上可能还有一些无人居住的地方，那么某些劣等的种族，比如非洲热带地区的黑人，将逐渐移居进去；但是在目前的状况下，他们就无法做到了，因为他们正在与更具智慧和力量的优等种族发生冲突。

最后，我还要再补充一点，假设前面的论述都是正确的话，那么最可能的推论结果将体现为一种**普遍趋势**的建立，那就是来自温带而不是其他地区的文明将有更大的机会得到展示。我一直认为，单靠各种常见的物质因素，如土壤、气候、矿藏或地理位置，或是这些因素的各种组合，是不能解释亚述、埃及、希腊、意大利或英格兰等地区文明的出现和发展的。如果没有利用物质资源的意识，物质本身就是完全无用的。如果没有原生质基础上的生理机能，现在或将来都不会有一门能够对智力进化的整个无穷的发展过程作出说明的科学学科。实证主义哲学家们自夸地认为他们的哲学可以解释世间万物，可以解释历史的演变。但是从事科学的人们清醒地认识到，尽管复杂的事物可以简化到已有的科学规律中，但毋庸置疑，总有更复杂的事物没法简化处理。科学会不断进步，但正如天文学家手中不断改进的望远镜一样，它揭示的只是那些未知的问题和尚未简化成科学规律的疑难现象。

（刘冉冉 翻译；陈平富 审稿）

The Velocity of Thought

M. Foster

Editor's Note

The author here is Michael Foster, a pioneer of physiology at Cambridge. He describes experiments reminiscent of Luigi Galvani's famous studies of muscle action in a dead frog's leg induced by electrical currents. Foster aims to measure how fast nerve signals travel when inducing muscle contraction. He uses an ingenious apparatus of levers, rotating cylinders and tuning forks to measure the very short time intervals involved, and finds a resulting speed—about 28 metres per second—similar to that measured by the German physiologist Hermann von Helmholtz. But he points out that mental processes involve some processing time too. His measurements of reaction times lead him to assert that it takes about 1/26 of a second "to think".

"As quick as thought" is a common proverb, and probably not a few persons feel inclined to regard the speed of mental operations as beyond our powers of measurement. Apart, however, from those minds which take their owners so long in making up because they are so great, rough experience clearly shows that ordinary thinking does take time; and as soon as mental processes were brought to work in connection with delicate instruments and exact calculations, it became obvious that the time they consumed was a matter for serious consideration. A well-known instance of this is the "personal equation" of the astronomers. When a person watching the movement of a star, makes a signal the instant he sees it, or the instant it seems to him to cross a certain line, it is found that a definite fraction of a second always elapses between the actual falling of the image of the star on the observer's eye, and the making of the signal—a fraction, moreover, varying somewhat with different observers, and with the same observer under differing mental conditions. Of late years considerable progress has been made towards an accurate knowledge of this mental time.

A typical bodily action, involving mental effort, may be regarded as made up of three terms; of sensations travelling towards the brain, of processes thereby set up within the brain, and of resultant motor impulses travelling from the brain towards the muscles which are about to be used. Our first task is to ascertain how much time is consumed in each of these terms; we may afterwards try to measure the velocity of the various stages and parts into which each term may be further subdivided.

The velocity of motor impulses is by far the simplest case of the three, and has already been made out pretty satisfactorily. We can assert, for instance, that in frogs a motor impulse, the message of the will to the muscle, travels at about the rate of 28 metres a second, while in man it moves at about 33 metres. The method by which this result is obtained may be described in its simplest form somewhat as follows: —

50

思考的速度

福斯特

编者按

这篇文章的作者迈克尔·福斯特是剑桥生理学领域的先驱。他描述的实验很容易让我们想到路易吉·伽伐尼用电流引起死蛙腿肌肉运动的著名研究。福斯特的目标是测量神经信号在诱导肌肉收缩时传递得有多快。他使用了一个精巧的杠杆装置，利用该装置可以通过旋转圆筒和调整叉子来测量过程中所涉及的极短的时间间隔。通过测量他得到的速度是大约每秒28米，这与德国生理学家赫尔曼·冯·亥姆霍兹测得的结果很接近。不过，福斯特指出精神活动同样需要一些处理时间。通过对反应时间的测量，福斯特提出人的"思考"大约需要1/26秒。

人们常说"像思考一样快"，而且可能不少人都倾向于认为思考的速度已经快到无法测量的程度。然而，人们除了作重大决定时需要花很长时间思考以外，粗略的经验清楚地告诉我们，普通的思考也是需要花费时间的。而且，一旦将思考过程与精密仪器和精确计算联系起来以后，思考所花的时间就变得不容忽视了。天文学家的"测者误差"就是一个著名的例子。当观测者在观察星体运动时，会在看到星体或其特定的运动轨迹后立即记录，人们发现观测者作出记录的时间总是比星体实际在他眼中成像的时间滞后零点几秒，并且这个滞后时间的长短因人而异，就算是同一个观测者，其精神状态不同，滞后时间也不同。近年来，人们在对这一思考时间的准确认识方面已经取得了相当大的进展。

一个典型的身体动作需要大脑的帮助，可能包括三个步骤：首先感觉信息向大脑传输，然后在大脑中建立动作过程，最后大脑再将产生的运动脉冲传递给相关的肌肉。我们的首要任务是确定每个步骤花费的时间，然后测定每个步骤可能再被细分成的不同阶段和部分的速率。

目前，运动脉冲的传输速率是三个速率中最容易测定的，现在已经得到了满意的结果。例如，我们可以确定，运动脉冲（即由大脑向肌肉传输的信息）的传输速率对于蛙来说是每秒28米，而对于人则是每秒33米。现将获得上述速率的测定方法以最简单的方式描述如下：

The muscle which in the frog corresponds to the calf of the leg, may be prepared with about two inches of its proper nerve still attached to it. If a galvanic current be brought to bear on the nerve close to the muscle, a motor impulse is set up in the nerve, and a contraction of the muscle follows. Between the exact moment when the current breaks into the nerve, and the exact moment when the muscle begins to contract, a certain time elapses. This time is measured in this way: —A blackened glass cylinder, made to revolve very rapidly, is fitted with two delicate levers, the points of which just touch the blackened surface at some little distance apart from each other. So long as the levers remain perfectly motionless, they trace on the revolving cylinder two parallel, horizontal, unbroken lines; and any movement of either is indicated at once by an upward (or downward) deviation from the horizontal line. These levers further are so arranged (as may readily be done) that the one lever is moved by the entrance of the very galvanic current which gives rise to the motor impulse in the nerve, and thus marks the beginning of that motor impulse; while the other is moved by the muscle directly this begins to contract, and thus marks the beginning of the muscular contraction. Taking note of the direction in which the cylinder is revolving, it is found that the mark of the setting-up of the motor impulse is always some little distance ahead of the mark of the muscular contraction; it only remains to be ascertained to what interval of time that distance of space on the cylinder corresponds. Did we know the actual rate at which the cylinder revolves this might be calculated, but an easier method is to bring a vibrating tuning-fork, of known pitch, to bear very lightly sideways on the cylinder, above or between the two levers. As the cylinder revolves, and the tuning fork vibrates, the latter will mark on the former a horizontal line, made up of minute, uniform waves corresponding to the vibrations. In any given distance, as for instance in the distance between the two marks made by the levers, we may count the number of waves. These will give us the number of vibrations made by the tuning-fork in the interval; and knowing how many vibrations the tuning-fork makes in a second, we can easily tell to what fraction of a second the number of vibrations counted corresponds. Thus, if the turning-fork vibrates 100 times a second, and in the interval between the marks of the two levers we count ten waves, we can tell that the time between the two marks, *i.e.* the time between the setting-up of the motor impulse and the beginning of the muscular contraction, was 1/10 of a second.

Having ascertained this, the next step is to repeat the experiment exactly in the same way, except that the galvanic current is brought to bear upon the nerve, not close to the muscle, but as far off as possible at the furthest point of the two inches of nerve. The motor impulse has then to travel along the two inches of nerve before it reaches the point at which, in the former experiment, it was first set up.

On examination, it is found that the interval of time elapsing between the setting up of the motor impulse and the commencement of the muscular contraction is greater in this case than in the preceding. Suppose it is 2/10 of a second—we infer from this that it took the motor impulse 1/10 of a second to travel along the two inches of nerve: that is to say, the rate at which it travelled was one inch in 1/20 of a second.

将蛙小腿的肌肉制成带有大约 2 英寸长固有神经的样品。如果给临近肌肉的神经通电，神经内部就会产生运动脉冲，进而导致随后的肌肉收缩。从给神经通电到肌肉开始收缩需要经过一段时间。而这段时间是这样测定的：将两根细杆连到一个快速旋转的涂黑的玻璃圆柱上，让它们的末端彼此分开一段微小的距离，并分别与圆柱黑色表面接触。只要两根细杆保持完全静止，那么当圆柱体旋转时，它们就会在圆柱黑色表面上画出两条平行的水平实线。而如果任何一根细杆有一丁点的运动，水平线就会立即发生向上（或向下）的偏移。进一步对这些细杆作出如下安排（实施起来很容易）：一根细杆的运动由在神经中引起运动脉冲的特定电流控制，以此来标记运动脉冲的产生；另一根细杆的运动由产生收缩的肌肉控制，以此来标记肌肉收缩的开始。记录下玻璃圆柱的旋转方向，我们发现运动脉冲产生的标记总是比肌肉收缩开始的标记靠前一小段距离。现在我们只需要确定圆柱上这段距离所对应的时间间隔。如果知道圆柱旋转的准确速度，那么就可以计算出这段时间的间隔，然而我们还有更简单的方法：在两根细杆之间或上面放置一个已知音调的振动音叉，轻轻贴在圆柱两侧。随着圆柱的旋转和音叉的振动，音叉会在圆柱表面标记出一条水平线，这条水平线由与振动相对应的微小而又均一的波组成。对于任何给定的距离，比如细杆画出的两个标记之间的距离，我们都能数出其中的波数。这就给出了在上述时间间隔内音叉的振动次数。我们知道了一秒内音叉振动的次数，就能很容易地算出相应音叉振动次数对应的时间间隔。因此，假设音叉在一秒内振动 100 次，在两个细杆画出的标记之间我们数出了 10 个波，那么我们就可以算出这两个标记之间的时间间隔，也就是说，从产生运动脉冲到肌肉开始收缩的时间间隔是 1/10 秒。

确定了上述时间间隔，下一步就是以几乎完全相同的方式重复上述实验，唯一不同的是，不是给临近肌肉的神经通电，而是在 2 英寸长的神经上尽可能远离肌肉的末端通电。在到达临近肌肉的末端之前，运动脉冲沿着这 2 英寸神经传递，而在之前的实验中，运动脉冲则是在这一临近肌肉的末端产生的。

检测后发现，在这种实验条件下，从产生运动脉冲到肌肉开始收缩的时间间隔比之前的实验中的时间间隔更长。假设从产生运动脉冲到肌肉开始收缩的时间间隔是 2/10 秒，那么我们就可以由此推测出运动脉冲在 2 英寸神经中传递的时间是 1/10 秒，也就是说，运动脉冲的传递速度是每 1/20 秒 1 英寸。

By observations of this kind it has been firmly established that motor impulses travel along the nerves of a frog at the rate of 28 metres a second, and by a very ingenious application of the same method to the arm of a living man, Helmholtz and Baxt have ascertained that the velocity of our own motor impulses is about 33 metres a second.* Speaking roughly this may be put down as about 100 feet in a second, a speed which is surpassed by many birds on the wing, which is nearly reached by the running of fleet quadrupeds, and even by man in the movements of his arm, and which is infinitely slower than the passage of a galvanic current. This is what we might expect from what we know of the complex nature of nervous action. When a nervous impulse, set up by the act of volition, or by any other means, travels along a nerve, at each step there are many molecular changes, not only electrical, but chemical, and the analogy of the transit is not so much with that of a simple galvanic current, as with that of a telegraphic message carried along a line almost made up of repeating stations. It has been found, moreover, that the velocity of the impulse depends, to some extent, on its intensity. Weak impulses, set up by slight causes of excitement, travel more slowly than strong ones.

The contraction of a muscle offers us an excellent objective sign of the motor impulse having arrived at its destination; and, all muscles behaving pretty much the same towards their exciting motor impulses, the results obtained by different observers show a remarkable agreement. With regard to the velocity of sensations or sensory impulses, the case is very different; here we have no objective sign of the sensation having reached the brain, and are consequently driven to roundabout methods of research. We may attack the problem in this way. Suppose that, say by a galvanic shock, an impression is made on the skin of the brow, and the person feeling it at once makes a signal by making or breaking a galvanic current. It is very easy to bring both currents into connection with a revolving cylinder and levers, so that we can estimate by means of a tuning-fork, as before, the time which elapses between the shock being given to the brow and the making of the signal. We shall then get the whole "physiological time", as it is called (a very bad name), taken up by the passage of the sensation from the brow to the brain, by the resulting cerebral action, including the starting of a volitional impulse, and by the passage of the impulse along the nerve of the arm and hand, together with the muscular contractions which make the signal. We may then repeat exactly as before, with the exception that the shock is applied to the foot, for instance, instead of the brow. When this is done, it is found that the whole physiological time is greater in the second case than in the first; but the chief difference to account for the longer time is, that in the first case the sensation of the shock travels along a short tract of nerve (from the brow to the brain), and in the second case through a longer tract (from the foot to the brain). We may conclude, then, that the excess of time is taken up by the transit of the sensation through the distance by which the sensory nerves of the foot exceed in length those of the brow. And from this we can calculate the rate at which the sensation moves.

---

* Quite recently M. Place has determined the rate to be 53 metres per second. This discordance is too great to be allowed to remain long unexplained, and we are very glad to hear that Helmholtz has repeated his experiments, employing a new method of experiment, the results of which we hope will soon be published.

通过这样的研究已经确定了蛙神经中运动脉冲传递的速率是每秒 28 米。亥姆霍兹和巴克斯特将该方法非常巧妙地应用在活人的胳膊上，测定出我们人类神经中运动脉冲的传递速度是每秒 33 米。*粗略来说这个速度大约为每秒 100 英尺，这超过了许多鸟的飞行速度，几乎接近四足动物的奔跑速度，甚至接近人挥动手臂的速度，但远比电流的传输速度慢得多。这可能就是神经行为的复杂性所在。在由动作意识或其他方式激发的神经脉冲沿神经传递的过程中，每一步都发生许多分子变化，不仅有电学的还有化学的变化。神经脉冲的传递，与其说是类似于简单的电流传递，不如说是类似于含有多个重复信号站的电话线中电信号的传递。此外，研究发现，神经脉冲的传递速度，在某种程度上取决于其强度。轻度兴奋导致的弱神经脉冲比强神经脉冲传递得要慢一些。

肌肉收缩是运动脉冲到达目的地的客观标志。所有肌肉对兴奋性运动脉冲的反应完全相同，而且由不同观测者得到的观测结果相当一致。感觉或感觉脉冲的传递速度，却有着不同的情况。我们没有找到感觉到达大脑的客观指标，因此只能采取间接的研究方法。我们或许可以这样做：假设当一个人的额头皮肤受到电击后，人在感知的同时立即通过形成或切断电流而产生一个信号。我们可以很容易地通过旋转的圆柱和细杆将这两种电流联系起来，这样我们就能用上文中描述的音叉法测定出从额头遭受电击到产生信号的时间间隔。这样我们就可以获得整个的“生理时间”，正如它的名字（一个很糟糕的名字）所表示的一样，在这段时间里感觉从额头传递到大脑，然后大脑皮层作出反应，形成意识脉冲，意识脉冲沿着胳膊和手的神经传递，到达目的地后引起肌肉收缩从而产生信号。然后我们又进行了类似的测量，不过这次电击的是脚部而不是额头。结果发现脚部的整个生理时间比额头的长，主要原因是刺激感觉从脚部到大脑的神经传递路径比从额头到大脑长得多。由此，我们可以得出结论，多出的时间花费在感觉的传递上了，传递的距离就是从脚部到大脑的神经长度超过从额头到大脑的神经长的那段距离。这样我们就可以算出感觉的传递速度了。

---

* 最近，普莱斯测定出这一速度是每秒 53 米。这么大的差异，不能任其长期得不到解释。我们很高兴地听说，亥姆霍兹已经用一种新的实验方法重复了他的实验，我们期望能尽快看到该结果的公布。

Unfortunately, however, the results obtained by this method are by no means accordant; they vary as much as from 26 to 94 metres per second. Upon reflection, this is not to be wondered at. The skin is not equally sentient in all places, and the same shock might produce a weak shock (travelling more slowly) in one place, and a stronger one (travelling more quickly) in another.

Then, again, the mental actions involved in the making the signal may take place more readily in connection with sensations from certain parts of the body than from others. In fact, there are so many variables in the data for calculation that though the observations hitherto made seem to show that sensory impressions travel more rapidly than motor impulses (44 metres per second), we shall not greatly err if we consider the matter as yet undecided.

By a similar method of observation certain other conclusions have been arrived at, though the analysis of the particulars is not yet within our reach. Thus nearly all observers are agreed about the comparative amount of physiological time required for the sensations of sight, hearing, and touch. If, for instance, the impression to be signalled be an object seen, a sound heard, or a galvanic shock felt on the brow, while the same signal is made in all three cases, it is found that the physiological time is longest in the case of sight, shorter in the case of hearing, shortest of all in the case of touch. Between the appearance of the object seen (for instance, an electric spark) and the making of the signal, about 1/6; between the sound and the signal, 1/5; between the touch and signal, 1/7 of a second, is found to intervene.

This general fact seems quite clear and settled; but if we ask ourselves the question, why is it so? Where, in the case of light, for instance, does the delay take place? We meet at once with difficulties. The differences certainly cannot be accounted for by differences in length between the optic, auditory, and brow nerves. The retardation in the case of sight as compared with touch may take place in the retina during the conversion of the waves of light into visual impressions, or may be due to a specifically lower rate of conduction in the optic nerve, or may arise in the nervous centre itself through the sensations of light being imperfectly connected with the volitional mechanism in the brain put to work in the making of the signal. One observer (Wittich) has attempted to settle the first of these questions by stimulating the optic nerve, not by light, but directly by a galvanic current, and has found that the physiological time was thereby decidedly lessened; while conversely, by substituting a prick or pressure on the skin for a galvanic shock, the physiological time of touch was lengthened. But there is one element, that of intensity (which we have every reason to think makes itself felt in sensory impressions, and especially in cerebral actions even more than in motor impulses), that disturbs all these calculations, and thus causes the matter to be left in considerable uncertainty. How can we, for instance, compare the intensity of vision with that either of hearing or of touch?

The sensory term, therefore, of a complete mental action is far less clearly understood than the motor term; and we may naturally conclude that the middle cerebral term is

然而，不幸的是，用这种方法得出的结果彼此之间很不一致，传输速度从每秒26 米到每秒 94 米不等。细想之下，有这样的结果也并不奇怪。因为各个部分皮肤的敏感度是不一样的，同样的刺激在一个地方引起较弱的脉冲（传递得较慢），而在另一个地方则引起较强的脉冲（传递得较快）。

同样地，人体某些部位的感觉比其他部位的感觉更容易引起这种产生信号的神经行为。事实上，这些用于计算的数据变化太大，以至于尽管已有许多测定结果表明感觉的传递快于运动脉冲的传递（每秒 44 米），我们却无法精确计算。如果我们能将未知因素考虑在内，就不会犯太大的错误。

对那些特例我们鞭长莫及，不过用类似的方法我们还是获得了另外一些结果。几乎所有的观测者在视觉、听觉及触觉的生理时间比较上都获得了一致的结论。例如，研究发现，如果信号的获得方式是看到一个物体，听到一个声音，或者额头感觉到电击，尽管这三种情况都产生同样的信号，但是视觉的生理时间最长，其次是听觉，而触觉的生理时间最短。不过也出现了不同于上述结论的研究结果：从看到物体（如电火花）到产生信号的时间是 1/6 秒；从听到声音到产生信号的时间是 1/5 秒；而从接触物体到产生信号的时间是 1/7 秒，发现这一结果造成了干扰。

看起来大体上事实问题似乎已经很清楚并已经得到了确定，不过如果我们自问，为什么会是这样呢？比如光，是在哪一步发生了延迟呢？我们立刻遇到了问题。很显然，生理时间长短的不同不是由视神经、听觉神经和额头的神经在长度上的差异引起的。与触觉相比，视觉延迟可能发生在视网膜将光波转化为视觉印象的过程中，也可能与视神经中特殊的低传导速度有关，或者还可能发生在中枢神经中，对光的感觉并没有与产生信号的大脑意识机制很好地协调起来。一位观测者（威蒂克）已经尝试解决上述问题中的首要问题，他用电替代光去刺激视神经，发现生理时间明显减少了；然而相反地，用尖刺或按压替代电击去刺激皮肤，结果触觉的生理时间延长了。不过有一个因素——强度（我们有理由相信，我们可以感受到感觉印象的强度，而与感受运动脉冲的强度相比，我们能更强烈地感受到大脑行为的强度），扰乱了所有这些计算结果，使得问题具有相当大的不确定性。例如，我们如何对视觉的强度与听觉或触觉的强度进行比较呢？

因此，对于感觉这样一种完全精神性的活动，人们对它的理解远没有对运动的理解那么清晰。于是我们可能会很自然地认为，人们对处于中间地位的大脑仍然

still less known. Nevertheless, here too it is possible to arrive at general results. We can, for instance, estimate the time required for the mental operation of deciding between two or more events, and of willing to act in accordance with the decision. Thus, if a galvanic shock be given to one foot, and the signal be made with the hand of the same side, a certain physiological time is consumed in the act. But if the apparatus be so arranged that the shock may be given to either foot, and it be required that the person experimenting, not knowing beforehand to which foot the shock is coming, must give the signal with the hand of the same side as the foot which receives the shock, a distinctly longer physiological time is found to be necessary. The difference between the two cases, which, according to Donders, amounts to 66/1,000, or about 1/15 of a second, gives the time taken up in the mental act of recognising the side affected and choosing the side for the signal.

A similar method may be employed in reference to light. Thus we know the physiological time required for any one to make a signal on seeing a light. But Donders found that when matters were arranged so that a red light was to be signalled with the left hand and a white with the right, the observer not knowing which colour was about to be shown, an extension of the physiological time by 154/1,000 of a second was required for the additional mental labour. This of course was after a correction (amounting to 9/1,000 of a second) had been made for the greater facility in using the right hand.

The time thus taken up in recognising and willing, was reduced in some further observations of Donders, by the use of a more appropriate signal. The object looked for was a letter illuminated suddenly by an electric spark, and the observer had to call out the name of the letter, his cry being registered by a phonautograph, the revolving cylinder of which was also marked by the current giving rise to the electric spark.

When the observer had to choose between two letters, the physiological time was rather shorter than when the signal was made by the hand; but when a choice of five letters was presented, the time was lengthened, the duration of the mental act amounting in this case to 170/1,000 of a second.

When the exciting cause was a sound answered by a sound, the increase of the physiological time was much shortened. Thus, the choice between two sounds and the determination to answer required about 50/1,000 of a second; while, when the choice lay between five different sounds, 88/1,000 of a second was required. In these observations two persons sat before the phonautograph, one answering the other, while the voices of both were registered on the same revolving cylinder.

These observations may be regarded as the beginnings of a new line of inquiry, and it is obvious that by a proper combination of changes various mental factors may be eliminated and their duration ascertained. For instance, when one person utters a sound, the nature of which has been previously arranged, the time elapsing before the answer is given corresponds to the time required for simple recognition and volition. When, however, the first person has leave to utter any one, say of five, given sounds, and the second person

了解甚少。然而，事实并非如此。例如，我们能估计出大脑在两三件事之间作出抉择以及将决定付诸行动所需的时间。因此，如果一只脚遭到电击后，用同一侧的手作出反应，这一动作过程需要一定的生理时间。如果受试者事先并不知道哪只脚会遭到电击，并被要求必须用与被电击的那只脚同侧的手作出反应的话，那么需要的生理时间会明显变长。东德斯得出，这两种情况的生理时间差是 66/1,000 秒，大约 1/15 秒，这就是精神活动用于识别哪一侧受刺激（识别哪只脚被电击）和决定用哪一侧作出反应（决定用哪只手作出反应）所需的时间。

类似的方法也可以用于光刺激的实验中，这样我们就可以知道人看见一束光后作出反应所需的生理时间。不过东德斯发现，当受试者被要求看到红光举左手，看到白光举右手，并且事先同样不知道将会出现什么颜色的光时，生理时间延长了 154/1,000 秒，这是额外的精神活动需要的时间。当然，这是针对人们使用右手更为熟练这种情况进行校正（9/1,000 秒）后的结果。

东德斯的进一步研究发现，如果使用更合适的信号，大脑进行识别和作出决定所需的时间将会缩短。受试者需要去寻找突然被电火花照亮的字母，看到后要大喊出字母名称，喊声用声波记振仪记录，记振仪的旋转圆柱也用产生电火花的电流进行了标记。

受试者在 2 个字母之间进行选择需要的生理时间比前述的用手作出反应所需的生理时间短很多，而当被选择的字母增加到 5 个时，生理时间变长，这一过程中思考的时间是 170/1,000 秒。

当用声音回应声音刺激时，延长的生理时间将大大缩短。在 2 种声音之间作出选择并决定回应所需的时间大约是 50/1,000 秒。然而，当在 5 种声音之间作出选择时，所需的生理时间为 88/1,000 秒。在上述观测研究中，2 名受试者坐在声波记振仪前，他们的声音被记录在同一个旋转圆柱上。

这些观测结果也许可以看作是新一轮研究的起点，很明显，适当地综合运用多种方法，可以减少精神因素的影响，并确定它们的生理时间。例如，人发出声音的过程是预先安排好的，在给出回答之前消耗的时间与简单的识别和决定所需的时间相对应。然而，当第一个人可以发出 5 种音中的任一种，而第二个人需在听到后发出同样的声音回应前者时，这一思维过程要复杂很多。在这一过程中，首先要对声

to make answer by the same sound to any and every one of the five which he thus may hear, the mental process is much more complex. There is in this case first the perception and recognition of sound, then the bare volition towards an answer, and finally the choice and combination of certain motor impulses which are to be set going, in order that the appropriate sound may be made in answer. All this latter part of the cerebral labour may, however, be reduced to a minimum by arranging that though any one of five sounds may be given out, answer shall be made to a particular one only. The respondent then puts certain parts of his brain in communication with the origin of certain out going nerves; he assumes the attitude, physical and mental, of one about to utter the expected sound. To use a metaphor, all the trains are laid, and there is only need for the match to be applied. When he hears any of the four sounds other than the one he has to answer, he has only to remain quiet. The mental labour actually employed when the sound at last is heard is limited almost to a recognition of the sound, and the rise of what we may venture to call a bare volitional impulse. When this is done, the time is very considerably shortened. In this way Donders found, as a mean of numerous observations that the second of these cases required 75/1,000 of a second, and the third only 39/1,000 over and above the first. That is to say, while the complex act of recognition, rise of volitional impulse, and inauguration of an actual volition, with the setting free of coordinated motor impulses, took 75/1,000 of a second, the simple recognition and rise of volitional impulse took 39/1,000 only. We infer, therefore, that the full inauguration of the volition took $(75 - 39)/1,000 = 36/1,000$. In rough language, it took 1/20 of a second to think, and rather less to will.

We may fairly expect interesting and curious results from a continuation of these researches. Two sources of error have, however, to be guarded against. One, and that most readily appreciated and cared for, refers to exactitude in the instruments employed; the other, far more dangerous and less readily borne in mind, is the danger of getting wrong in drawing averages from number of exceedingly small and variable differences.

(**2**, 2-4; 1870)

音进行感知和识别，然后形成初步的回答意识，最后对某些运动脉冲进行选择和组合，以便选用合适的声音进行回答。然而，所有这些后来的脑力活动是可以通过如下安排而减到最少的：尽管发出声音时可以选择 5 种声音中的任何一种，但是回答时只能选择特定的一种。然后，应答者将大脑中的某些部分与某些外周神经联系起来，身体上和精神上都采取一种准备发出预期声音的姿态。打个比方，所有的火车都停在那，只能开那辆最合适的。当应答者听到其他 4 种声音中的一种，而不是那个他必须回答的声音时，他就得保持沉默。他最后听到声音时消耗的脑力劳动几乎只是识别声音并唤起意识脉冲。完成了这些脑力劳动，生理时间就大大缩短了。用这样的方式，东德斯发现，综合大量观察结果，第二种情况所花的平均时间是 75/1,000 秒，而第三种情况只需 39/1,000 秒，优于第一种情况。也就是说，识别、唤起意识脉冲和作出决定，在不产生与之相协调的运动脉冲的情况下，需要的时间是 75/1,000 秒，而识别和唤起意识冲动的时间只有 39/1,000 秒。因此，我们推出，作出决定的时间是 (75 − 39)/1,000 = 36/1,000 秒。粗略地说，思考只需要 1/20 秒，作出决定的时间则更短。

我们很期待后续的实验研究能进一步获得有趣而奇妙的结果。不过，应该注意避免两种错误来源：首先，对于所使用仪器的准确性要多加注意，研究者在这一点上是最小心谨慎的；其次，对只有细微差别的数据取平均值时，要避免出现错误，研究者经常在这个问题上犯错并且很容易忽视它。

（高如丽 翻译；刘力 审稿）

Pasteur's Researches on the Diseases of Silkworms

J. Tyndall

Editor's Note

This is an account of French chemist Louis Pasteur's first proof that an infectious disease is caused by the physical transmission of minute entities or "germs" (bacteria). It is written by John Tyndall of the Royal Institution in London, one of the leading British scientists of his time. Pasteur had been asked by the French chemist Jean-Baptiste Dumas to investigate the cause of a disease ravaging France's silkworms. Pasteur focused on the microscopic "corpuscles" seen by others in silkworm blood but previously thought to be native to the creatures. Pasteur's initial presentation in France of his theory on infectious germs met with strong criticism, but his famously systematic experiments eventually made the case irrefutable.

I have recently received from M. Pasteur a copy of his new work, "Sur la Maladie des vers à soie", a notice of which, however brief and incomplete, will, I am persuaded, interest a large class of the readers of *Nature*. The book is the record of a very remarkable piece of scientific work, which has been attended with very remarkable practical results. For fifteen years a plague had raged among the silkworms of France. They had sickened and died in multitudes, while those that succeeded in spinning their cocoons furnished only a fraction of the normal quantity of silk. In 1853 the silk culture of France produced a revenue of one hundred and thirty millions of francs. During the twenty previous years the revenue had doubled itself, and no doubt was entertained as to its future augmentation. "Unhappily, at the moment when the plantations were most flourishing, the prosperity was annihilated by a terrible scourge." The weight of the cocoons produced in France in 1853 was twenty-six millions of kilogrammes; in 1865 it had fallen to four millions, the fall entailing in the single year last mentioned a loss of one hundred millions of francs.

The country chiefly smitten by this calamity happened to be that of the celebrated chemist Dumas, now perpetual secretary of the French Academy of Sciences. He turned to his friend, colleague, and pupil, Pasteur, and besought him with an earnestness which the circumstances rendered almost personal, to undertake the investigation of the malady. Pasteur at this time had never seen a silkworm, and he urged his inexperience in reply to his friend. But Dumas knew too well the qualities needed for such an inquiry to accept Pasteur's reason for declining it. "Je met," said he, "un prix extrême à voir votre attention fixée sur la question qui interesse mon pauvre pays; la misère surpasse tout ce que vous pouvez imaginer." Pamphlets about the plague had been showered upon the public, the monotony of waste paper being broken at rare intervals by a more or less useful publication. "The Pharmacopoeia of the Silkworm," wrote M. Cornalia in 1860,

巴斯德对家蚕疾病的研究

廷德尔

编者按

法国化学家路易·巴斯德提出，感染性疾病是由很小的实体或者说"病菌"（细菌）的物理传播造成的，这篇文章报道的正是巴斯德关于此项研究的首例证据。文章作者是伦敦皇家研究院的约翰·廷德尔，他是当时英国第一流的科学家之一。早先，法国化学家让-巴蒂斯特·杜马邀请巴斯德研究当时给法国家蚕带来极大破坏的一种疾病的起因。巴斯德把重点放在了一种微观"粒子"上。早先就有人在家蚕的体液中发现了这种粒子，但先前人们认为这种粒子是家蚕本身就有的。巴斯德的感染性病菌理论一经在法国公布便遭到了强烈的批判，不过，通过出色的系统实验，他最终无可辩驳地证明了其理论的正确。

最近我收到了一本巴斯德先生寄来的他的最新著作，书名是《关于家蚕疫病》，这里将对这本书作个介绍，我想尽管这个介绍非常简要甚至不太完整，但一定有很多《自然》的读者们对此很感兴趣。这本书中记录了一段不平凡的科学研究，并附有一些极具价值的应用成果。在法国，一种家蚕疫病肆虐了15年之久，导致大量家蚕生病死亡，而那些成功结茧的家蚕提供的丝只有正常产量的一小部分。1853年，法国丝绸业创收1.3亿法郎，较20年前翻了一番，毫无疑问丝绸业将继续保持上升势头。"可很不幸的是，就在种桑养蚕最为兴旺的时候，一场严重的灾难突然降临，完全破坏了整个行业繁荣的局面。"1853年，法国的蚕茧产量是2,600万千克，到了1865年则下降到了400万千克，仅1865年这一年，蚕茧数量减少导致的丝绸业损失就高达1亿法郎。

这一灾难使法国受到重创，时任法国科学院常务秘书长的著名化学家杜马向他的朋友、同事兼学生巴斯德求助，他几乎是以个人的名义真诚地恳求巴斯德来研究这种家蚕疾病的。当时的巴斯德根本就没见过家蚕，他极力强调自己的经验不足，并以此答复了杜马。不过杜马很清楚巴斯德具备承担这项研究所需的能力，因而没有接受巴斯德经验不足的拒绝理由。他说："我们国家的状况可谓惨不忍睹，我愿出巨资请您来研究并解决这个问题。"当时，关于这种家蚕疾病的小册子大量地出现，除了偶尔有那么一两本还算有些用处外，其他大部分都是单调空洞的废纸。1860年，科尔纳利阿先生曾写道："眼下，治疗家蚕疾病的药就像人吃的药一样复杂，有气体的，有液体的，也有固体的。从盐酸到硫酸，从硝酸到朗姆酒，从糖类到硫酸奎宁，

"is now as complicated as that of man. Gases, liquids, and solids have been laid under contribution. From chlorine to sulphurous acid, from nitric acid to rum, from sugar to sulphate of quinine, —all has been invoked in behalf of this unhappy insect." The helpless cultivators, moreover, welcomed with ready trustfulness every new remedy, if only pressed upon them with sufficient hardihood. It seemed impossible to diminish their blind confidence in their blind guides. In 1863 the French Minister of Agriculture himself signed an agreement to pay 500,000 francs for the use of a remedy which its promoter declared to be infallible. It was tried in twelve different departments of France and found perfectly useless. In no single instance was it successful. It was under these circumstances that M. Pasteur, yielding to the entreaties of his friend, betook himself to Alais in the beginning of June 1865. As regards silk husbandry, this was the most important department in France, and it was also that which had been most sorely smitten by the epidemic.

The silkworm had been previously attacked by *muscardine*; a disease proved by Bassi to be caused by a vegetable parasite. Muscardine, though not hereditary, was propagated annually by the parasitic spores, which, wafted by winds, often sowed the disease in places far removed from the centre of infection. According to Pasteur, muscardine is now very rare; but for the last fifteen or twenty years a deadlier malady has taken its place. A frequent outward sign of this disease are the black spots which cover the silkworms, hence the name pébrine, first applied to the plague by M. de Quatrefages, and adopted by Pasteur. Pébrine also declares itself in the stunted and unequal growth of the worms, in the languor of their movements, in their fastidiousness as regards food, and in their premature death. The discovery of the inner workings of the epidemic may be thus traced. In 1849 Guerin Méneville noticed in the blood of certain silkworms vibratory corpuscles which he supposed to be endowed with independent life, and to which he gave a distinctive name. As regards the motion of the particles, Filippi proved him wrong; their motion was the well-known Brownian motion. But Filippi himself committed the error of supposing the corpuscles to be normal to the life of the insect. They are really the cause of its mortality—the form and substance of its disease. This was studied and well described by Cornalia; while Lebert and Frey subsequently found the corpuscles not only in the blood, but in all the tissues of the silkworm. Osimo, in 1857, discovered the corpuscles in the eggs, and on this observation Vittadiani founded, in 1859, a practical method of distinguishing healthy from diseased eggs. The test often proved fallacious, and it was never extensively applied.

The number of these corpuscles is sometimes enormous. They take possession of the intestinal canal, and spread thence throughout the body of the worm. They fill the silk cavities, the stricken insect often going through the motions of spinning without any material to answer to the act. Its organs, instead of being filled with the clear viscous liquid of the silk, are packed to distension by these corpuscles. On this feature of the plague Pasteur fixed his attention. He pursued it with the skill which appertains to his genius, and with the thoroughness that belongs to his character. The cycle of the silkworm's life is briefly this:—From the fertile egg comes the little worm, which grows, and after some time casts its skin. This process of moulting is repeated two or three times at subsequent

所有这些药品都被用来处理患病的家蚕。"无助的家蚕养殖者们大胆地尝试着各种新的治疗方法，尽管盲目却信心十足。1863 年，法国农业部部长亲自签署了一项协议，花费 50 万法郎推行一个据称是绝对有效的新治疗方案，该方案在法国 12 个不同的地区试行，可惜的是，试行结果表明该方案完全无效。当时所有试图治疗家蚕疾病的尝试没有一例成功。在这种情况下，巴斯德只好答应朋友的请求，并于 1865 年 6 月初在艾雷斯开展研究。艾雷斯是法国最重要的生产丝绸的地区，同时也是疫情最严重的地区。

之前，家蚕曾受到**白僵病**的攻击，巴锡的研究表明，这是一种由蔬菜寄生虫引起的疫病。白僵病虽然不具有遗传性，但是通过随风飘浮的寄生虫孢子，它也会传播到远离疫情中心的地区。巴斯德说，白僵病现在已经非常罕见了，但最近的 15~20 年间出现了一种更严重的疾病。这种疫病的一个鲜明标志是患病的蚕身上通常有许多黑点，根据这一特征，德卡特勒法热先生最先将这种病命名为微粒子病，巴斯德沿用了这一名称。患微粒子病的家蚕生长受到阻碍且体型发育不均，蠕动缓慢无力，挑食且易早死。关于传染性疾病内在机制的研究从此揭开了序幕。1849 年，介朗·莫奈维勒发现一些蚕的体液中存在以振动形式运动的微粒，他认为这些微粒拥有独立的生命并给它们取了一个特别的名字。关于微粒的运动方式，后来菲利皮证明介朗·莫奈维勒是错误的，这些微粒的运动方式正是著名的布朗运动。不过，菲利皮本人错误地认为这些微粒的出现对于蚕来说没什么不正常。其实，正是这些微粒导致了疫情中蚕的不正常死亡——这种疾病的表现和本质。科尔纳利阿仔细地研究并描述了这种微粒，而莱贝特和弗赖最终发现，不仅蚕的体液中存在这些微粒，而且其他所有组织中都有。1857 年，奥希姆在蚕卵中也发现了这种运动微粒，基于这一发现维塔迪尼在 1859 年建立了区分健康蚕卵与患病蚕卵的操作方法。不过，这个区分方法时常会出错，所以从来没有得到广泛应用。

有时候，蚕体中这种微粒的数量大得惊人。它们占据肠道并且向蚕体其他部位蔓延，有时甚至填满蚕的腹腔，使得蚕在整个吐丝过程中动弹不得并且吐不出丝。蚕体内不是充满了洁净蚕丝黏液，而是胀满了这些微粒。巴斯德把注意力集中在疾病的这一特征上。他靠那种似乎只有他才具有的天赋抓住了这一重点，而他深入钻研的科学作风又使他能对此进行彻底的研究。家蚕的生命周期大概如下：从受精卵长成幼虫，幼虫逐渐长大，随后开始蜕皮。在家蚕的整个生命中，蜕皮过程会重复 2~3 次。最后一次蜕皮结束后，家蚕会爬到提前放好的荆棘丛中吐丝织茧。这样家

intervals during the life of the insect. After the last moulting the worm climbs the brambles placed to receive it, and spins among them its cocoon. It passes thus into a chrysalis; the chrysalis becomes a moth, and the moth when liberated lays the eggs which form the starting-point of a new cycle. Now Pasteur proved that the plague-corpuscles might be incipient in the egg, and escape detection; they might also be germinal in the worm, and still baffle the microscope. But as the worm grows, the corpuscles grow also, becoming larger and more defined. In the aged chrysalis they are more pronounced than in the worm; while in the moth, if either the egg or the worm from which it comes should have been at all stricken, the corpuscles infallibly appear, offering no difficulty of detection. This was the first great point made out in 1865 by Pasteur. The Italian naturalists, as aforesaid, recommended the examination of the eggs before risking their incubation. Pasteur showed that both eggs and worms might be smitten and still pass muster, the culture of such eggs or such worms being sure to entail disaster. He made the moth his starting-point in seeking to regenerate the race.

And here is to be noted a point of immense practical importance. The worms issuing from the eggs of perfectly healthy moths may afterwards become themselves infected through contact with diseased worms, or through germs mixed with the dust of the rooms in which the worms are fed. But though the moths derived from the worms thus infected may be so charged with corpuscles as to be totally unable to produce eggs fit for incubation, still Pasteur shows that the worms themselves, in which the disease is not hereditary, never perish before spinning their cocoons. This, as I have said, is a point of capital importance; because it shows that the moth-test, if acted upon, even though the worms during their "education" should contract infection, secures, at all events, the next subsequent crop.

Pasteur made his first communication on this subject to the Academy of Sciences in September 1865. It raised a cloud of criticism. Here forsooth was a chemist rashly quitting his proper *métier* and presuming to lay down the law for the physician and biologist on a subject which was eminently theirs. "On trouva étrange que je fusse si peu au courant de la question; on m' opposa des travaux qui avaient paru depuis longtemps en Italie, dont les resultats montraient l'inutilité de mes efforts, et l'impossibilité d'arriver à un resultat pratique dans la direction que je m'étais engagé. Que mon ignorance fut grande au sujet des recherches sans nombre qui avaient paru depuis quinze années." Pasteur heard the buzz, but he continued his work. In choosing the eggs intended for incubation, the cultivators selected those produced in the successful "educations" of the year. But they could not understand the frequent and often disastrous failures of their selected eggs; for they did not know, and nobody prior to Pasteur was competent to tell them, that the finest cocoons may envelop doomed corpusculous moths. It was not, however, easy to make the cultivators accept new guidance. To strike their imagination and if possible determine their practice, Pasteur hit upon the expedient of prophecy. In 1866 he inspected at St. Hippolyte-du-Fort fourteen different parcels of eggs intended for incubation. Having examined a sufficient number of the moths which produced these eggs, he wrote out the prediction of what would occur in 1867, and placed the prophecy as a sealed letter in the hands of the Mayor of St. Hippolyte.

蚕就成了被蚕丝包裹的蚕蛹，随后蚕蛹变成蛾子，最后蛾子破茧而出并产下蚕卵，这样就又开始新一轮的生命周期。现在巴斯德证明蚕卵中的微粒还处在繁殖的最初阶段，所以无法检测到，在幼虫体内也只是萌芽期，所以通过显微观察还很难确认。而随着幼虫逐渐长大，这些微粒也开始变大从而更容易被确认。在蚕蛹期，这些颗粒比在幼虫期时更加清晰可辨。如果一开始的蚕卵或幼虫就受到感染，那么在相应的蛾子体内就可以很容易地准确检测到颗粒的存在。这是巴斯德在 1865 年提出的关于这一疫病的第一个重要观点。前面已经提过，一些意大利的博物学家建议在孵化前先对蚕卵进行检测。而巴斯德则认为，这些蚕卵和幼虫都可能是染病的但仍能通过检测，这些通过患病的蚕卵和幼虫发育而来的家蚕显然还是染病的。为了筛选出不染病的家蚕品种，他把蛾子作为检测控制的起点。

还有一点在实际应用中非常重要。由完全健康的蛾子产的卵孵化而来的健康家蚕随后也会受到感染，比如通过与病蚕的接触，或者通过饲养蚕的房间内混在灰尘中的病菌而感染。巴斯德通过研究表明，尽管家蚕在养殖过程中可能会染病，从而使蛾子体内充满微粒而完全不能产生适于孵化的健康卵，但这些家蚕本身在吐丝之前从来不会死亡，同时这种病在家蚕中并不会发生遗传。正如前文已经说过的，这一点是相当重要的，它表明即便家蚕在"成长"过程中被感染，但只要对蛾子进行切实的检测控制，至少接下来的蚕丝收成是有保障的。

1865 年 9 月，巴斯德就此课题与法国科学院的成员进行了首次交流，结果招来一片指责。在这里巴斯德被认为是一位**不务正业**的化学家，对从事本专业研究的医生和生物学家指手画脚。"认为我不了解眼下的问题，这是个奇怪的想法；他们认为：我很早之前在意大利发表的研究结果毫无意义，我在自己的研究领域内没有得到过实际有效的成果，而过去 15 年来我表现出来的愚昧无知更是数不胜数。"巴斯德当然听到了那些非议，不过他还是继续他的研究。在选择适于养殖的蚕卵方面，养殖者们一般选择那些当年正常"成长"的蚕产的卵。不过他们不明白为什么精心挑选出来的蚕卵时常会导致惨痛的损失，因为在巴斯德之前还没有人能告诉他们为什么看起来最为完好的蚕茧却可能包裹着带病的蛾子。不过，要想说服家蚕养殖者们接受新的方法并非易事，为了给养殖者们直观的印象并尽可能改变他们的做法，巴斯德想到了通过对孵育结果进行预测来证实自己理论的正确性。1866 年，巴斯德在圣希波利特堡检查了 14 份装有适于孵育的蚕卵的包裹。在检测完产生这些蚕卵的蛾子后，他写下了对 1867 年孵育结果的预测，并将预测结果密封后交给了圣希波利特市的市长。

In 1867 the cultivators communicated to the mayor their results. The letter of Pasteur was then opened and read, and it was found that in twelve out of fourteen cases, there was absolute conformity between his prediction and the observed facts. Many of the educations had perished totally; the others had perished almost totally; and this was the prediction of Pasteur. In two out of the fourteen cases, instead of the prophesied destruction, half an average crop was obtained. Now, the parcels of eggs here referred to were considered healthy by their owners. They had been hatched and tended in the firm hope that the labour expended on them would prove remunerative. The application of the moth-test for a few minutes in 1866 would have saved the labour and averted the disappointment. Two additional parcels of eggs were at the same time submitted to Pasteur. He pronounced them healthy; and his words were verified by the production of an excellent crop. Other cases of prophecy still more remarkable, because more circumstantial, are recorded in the work before us.

These deadly corpuscles were found by Leydig in other insects than the silkworm moth. He considers them to belong to the class of psososperms founded by J. Müller. "This," says Pasteur, "is to regard the corpuscular organism as a kind of parasite, which propagates itself after the manner of parasites of its class." Pasteur subjected the development of the corpuscles to a searching examination. With admirable skill and completeness he also examined the various modes by which the plague is propagated. He obtained perfectly healthy worms from moths perfectly free from corpuscles, and selecting from them 10, 20, 30, 50, as the case might be, he introduced into the worms the corpusculous matter. It was first permitted to accompany the food. Let us take a single example out of many. Rubbing up a small corpusculous worm in water, he smeared the mixture over the mulberry leaves. Assuring himself that the leaves had been eaten, he watched the consequences from day to day. Side by side with the infected worms he reared their fellows, keeping them as much as possible out of the way of infection. These constituted his "lot temoign", his standard of comparison. On the 16th of April, 1868, he thus infected thirty worms. Up to the 23rd they remained quite well. On the 25th they seemed well, but on that day corpuscles were found in the intestines of two of the worms subjected to microscopic examination. The corpuscles begin to be formed in the tunic of the intestine. On the 27th, or eleven days after the infected repast, two fresh worms were examined, and not only was the intestinal canal found in each case invaded, but the silk organ itself was found charged with the corpuscles. On the 28th the twenty-six remaining worms were covered by the black spots of pébrine. On the 30th the difference of size between the infected and non-infected worms was very striking, the sick worms being not more than two-thirds of the size of the healthy ones. On the 2nd of May a worm which had just finished its fourth moulting was examined. Its whole body was so filled with corpuscles as to excite astonishment that it could live. The disease advanced, the worms died and were examined, and on the 11th of May only six out of the thirty remained. They were the strongest of the lot, but on being searched they also were found charged with corpuscles. Not one of the thirty worms had escaped; a single corpusculous meal had poisoned them all. The standard lot, on the contrary, spun their fine cocoons, and two only of their moths were found to contain any trace of corpuscles. These had doubtless been introduced during the rearing of the worms.

到了 1867 年，家蚕养殖者们先将实际的养殖结果报告给市长。然后当巴斯德的预测结果被打开并宣读之后，人们发现 14 份样本中有 12 份的养殖结果和巴斯德的预测完全相同。其中大部分在养殖时彻底被疾病摧毁了，其余的也几乎完全被疾病摧毁。这恰恰是巴斯德的预测结果。14 份样本中的另外 2 份没有被疾病毁掉，最终大概有将近一半的收成。而当初这些蚕卵的主人都认为他们的蚕卵是健康的。他们悉心地孵化，仔细地照料幼蚕，殷切地期望着付出的劳动能得到回报。如果当初他们在 1866 年进行几分钟的蛾子检测，就能够节省劳动并避免失望一场了。当时，另外还有 2 份样本也寄来让巴斯德检测。他检测后宣布这 2 份是健康的，来年丰硕的收成证明了他的预言完全正确。当然，除此之外还有一些间接的但同样具有重要意义的预测此前已经报道过了。

在此之前，莱迪希在其他昆虫体内也发现过类似的致命性微粒。他认为这些微粒属于米勒发现的寄生病菌家族。巴斯德指出，"这等于是将微粒看作是寄生虫，它们在完成寄生行为的同时还繁殖后代。"他通过实验研究了微粒的发育过程，熟练而全面地检查了疾病的各种传播途径。他还筛选出了由完全不携带病原微粒的蛾子产的卵孵化的完全健康的幼蚕，然后根据实验需要，他又从这些完全健康的幼蚕中选出 10 条、20 条、30 条或 50 条，向其体内引入那些微粒物质。一开始这些微粒是混在食物中喂给家蚕的。我们就说说其中一个简单的实例吧。巴斯德先把携带微粒的蚕体在水中磨碎，然后将所得的混合物涂在桑叶上，再让家蚕吃下这些混有微粒的桑叶，然后每天追踪观察结果。另外，他还在这些受感染的蚕的附近同时饲养健康蚕，并尽量让它们保持距离不相互感染。这就组成了他"大量检验"的对比标准。就是通过这样的方法，他在 1868 年 4 月 16 日得到了 30 条受感染的家蚕。到 23 日它们状态仍然良好。到 25 日它们看起来也还可以，不过通过显微镜在 2 条家蚕的小肠内观察到了微粒，可以发现在小肠膜上微粒已经形成。到 27 日，也就是健康蚕吃桑叶而发生感染的 11 天后，检查发现有 2 条家蚕不仅在它们的肠道内而且在吐丝器官上出现了微粒。到 28 日，剩下的 26 条家蚕全身布满了微粒子病特有的黑点。到 30 日时，受感染的蚕的蚕体大小已经和没受感染的蚕有了明显的差异，病蚕蚕体不超过健康蚕蚕体的 2/3。到 5 月 2 日，经检查发现，一条刚完成第四次蜕皮的成蚕的体内充满了大量病原微粒，这样的蚕居然还能存活真是令人惊奇！随着病情的发展，病蚕开始死亡，巴斯德对它们都进行了检查。一开始的 30 条家蚕到 5 月 11 日只有 6 条还活着，它们是这 30 条蚕中最强壮的，不过观察发现它们体内同样充满了病原微粒。最终，30 条家蚕都死掉了，无一幸免。只因为吃了一次带病原微粒的桑叶，它们就统统丢了性命。相比之下，没受感染的对照组则顺利地吐丝结茧，随后只在其中 2 个蛾子中发现有少量的病原微粒。毫无疑问，这是在饲养幼虫时不小心引入的。

As his acquaintance with the subject increased, Pasteur's desire for precision augmented, and he finally gives the growing number of corpuscles seen in the field of his microscope from day to day. After a contagious repast the number of worms containing the parasite gradually augmented until finally it became cent per cent. The number of corpuscles would at the same time rise from 0 to 1, to 10, to 100, and sometimes even to 1,000 or 1,500 for a single field of his microscope. He then varied the mode of infection. He inoculated healthy worms with the corpusculous matter, and watched the consequent growth of the disease. He showed how the worms inoculate each other by the infliction of visible wounds with their "crochets". In various cases he washed the "crochets", and found corpuscles in the water. He demonstrated the spread of infection by the simple association of healthy and diseased worms. In fact, the diseased worms sullied the leaves by their dejections, they also used their crochets, and spread infection in both ways. It was no hypothetical infected medium that killed the worms, but a definitely-organised and isolated thing. He examined the question of contagion at a distance, and demonstrated its existence. In fact, as might be expected from Pasteur's antecedents, the investigation was exhaustive, the skill and beauty of his manipulation finding fitting correlatives in the strength and clearness of his thought.

Pébrine was an enigma prior to the experiments of Pasteur. "Place," he says, "the most skilful educator, even the most expert microscopist, in presence of large educations which present the symptoms described in our experiments; his judgment will necessarily be erroneous if he confines himself to the knowledge which preceded my researches. The worms will not present to him the slightest spot of pébrine; the microscope will not reveal the existence of corpuscles; the mortality of the worms will be null or insignificant; and the cocoons leave nothing to be desired. Our observer would, therefore, conclude without hesitation that the eggs produced will be good for incubation. The truth is, on the contrary, that all the worms of these fine crops have been poisoned; that from the beginning they carried in them the germ of the malady; ready to multiply itself beyond measure in the chrysalides and the moths, thence to pass into the eggs and smite with sterility the next generation. And what is the first cause of the evil concealed under so deceitful an exterior? In our experiments we can, so to speak, touch it with our fingers. It is entirely the effect of a single corpusculous repast; an effect more or less prompt according to the epoch of life of the worm that has eaten the poisoned food."

It was work like this that I had in view when, in a lecture which has brought me much well-meant chastisement from a certain class of medical men, and much gratifying encouragement from a different class, I dwelt on the necessity of experiments of physical exactitude in testing medical theories. It is work like this which might be offered as a model to the physicians of England, many indeed of whom are pursuing with characteristic skill and energy the course marked out for them by this distinguished master. Prior to Pasteur, the most diverse and contradictory opinions were entertained as to the contagious character of pébrine; some stoutly affirmed it, others as stoutly denied it. But on one point all were agreed. "They believed in the existence of a deleterious medium, rendered epidemic by some occult and mysterious influence, to which was attributed the cause of

随着巴斯德对这一领域的不断熟悉，他逐渐期望能提高实验的精确性，后来他开始记录每天用显微镜观察时视野中不断增加的微粒数目。给家蚕喂食一顿带病原微粒的食物之后，体内出现病原微粒的家蚕数量就开始逐渐增加，直到最终所有家蚕体内都出现病原微粒。与此同时，在显微镜单个视野中观察到的病原微粒数目也从 0 增加到 1、10 以至 100，有时甚至高达 1,000 或者 1,500。接着，巴斯德又改用另一种方式感染家蚕。他直接把微粒物质注入健康家蚕体内，然后观察病情的变化。他发现，家蚕会通过自己的"腹足趾钩"对在一起生活的其他家蚕造成外伤，从而使病原颗粒在家蚕之间互相传染。在多次实验中，他用水清洗这些"腹足趾钩"后在水中都发现了微粒物质。因此，他提出只是把健康蚕和病蚕放在一起饲养就会造成疾病在家蚕之间传染。事实上，疫病在家蚕之间传染有两条途径，一个是病蚕的粪便污染桑叶，另一个是通过腹足趾钩。杀死家蚕的并不是假想的什么传染介质，而是确实存在并能分离出来的物质。另外巴斯德还研究了远距离传染的问题，并证实远距离传染确实可能发生。实际上，由巴斯德之前的研究过程可以想到，这项研究是非常繁杂的，巴斯德高超的实验技能和艺术性的实验操作反映了他思考的深度和清晰的思路。

在巴斯德做这些实验前，人们对微粒子病的了解是一团迷雾。巴斯德说道："即使是让那些训练有素的专业人员甚至是非常杰出的显微镜专家来分析我们的实验，如果他们局限在我们实验之前的那些知识而并不了解我们在实验中发现的关于这种疾病的许多信息，那他们的判断也将必错无疑。他们不会看到得微粒子病的蚕的躯体上很小的斑点，通过显微镜观察也根本看不到微粒，而病蚕的死亡率接近零或者非常低，至于蚕茧就更看不出有什么问题了。那么观察者可能就会毫不犹豫地认为，这样的蚕卵是适合孵育的。但是事实正好相反，所有这些正常吐丝的蚕都被喂食了带病原微粒的食物，从一开始它们产的卵就携带着死亡的种子，这些病原微粒在蚕蛹和蛾体内准备进行疯狂的自身繁殖，然后传入蚕卵并彻底杀死下一代，导致下一代无法产卵。那么在这种欺骗性的表象之下导致这种灾难的首要原因是什么呢？在我们的实验中，我们是亲手触摸般地发现这一原因的。其实这完全就是吃入带病原微粒的食物引起的，结果的严重程度一定程度上和家蚕吃这些带病食物时所处的生命周期有关。"

像这样的研究我曾经也想到过，那是在一次报告上，一部分医生对我提出了并无恶意的责备，而另一部分医生则充分鼓励我，当时我就想到，实验的物理精度对于医学理论的检验是很必要的。我们应该把巴斯德这样的研究工作作为范例提供给英国的医生们，实际上许多既有专业技能又有热情的医生正期望着能接受大师为他们定制的课程指导。在巴斯德之前，关于微粒子病传播特征的观点形形色色、相互矛盾，有些人轻率固执地肯定它，也有些人同样轻率固执地否定它。不过有一点当时是达成共识的，"他们相信必定存在某种有毒的媒介物质以未知的神秘的方式传播疫情，从而导致了疾病灾难。"我想，任何一个头脑清晰的人，都会在这种观念和巴斯德的研究结果之间毫不犹豫地作出正确的选择。

the malady." Between such notions and the work of Pasteur, no physically-minded man will, I apprehend, hesitate in his choice.

Pasteur describes in detail his method of securing healthy eggs, which is nothing less than a mode of restoring to France her ancient prosperity in silk husbandry. And the justification of his work is to be found in the reports which reached him of the application, and the unparalleled success of his method, at the time he was putting his researches together for final publication. In France and Italy his method has been pursued with the most surprising results. It was an up-hill fight which led to this triumph, but it is consoling to think that even the stupidities of men may be converted into elements of growth and progress. Opposition stimulated Pasteur, and thus, without meaning it, did good service. "Ever," he says, "since the commencement of these researches, I have been exposed to the most obstinate and unjust contradictions; but I have made it a duty to leave no trace of these contests in this book." I have met with only a single allusion to the question of spontaneous generation in M. Pasteur's work. In reference to the advantage of rearing worms in an isolated island like Corsica, he says:—"Rien ne serait plus facile que d'éloigner, pour ainsi dire, d'une manière absolue la maladie des corpuscles. Il est au pouvoir de l'homme de faire disparaître de la surface du globe les maladies parasitaires, si, comme c'est ma conviction, la doctrine des générations spontanées est une chimère." It is much to be desired that some really competent person in England should rescue the public mind from the confusion now prevalent regarding this question.

M. Pasteur has investigated a second disease, called in France *flacherie*, which has co-existed with pébrine, but which is quite distinct from it. Enough, I trust, has been said to send the reader interested in these questions to the original volumes for further information. I report with deep regret the serious illness of M. Pasteur; an illness brought on by the labours of which I have tried to give some account. The letter which accompanied his volumes ends thus:—"Permettez-moi de terminer ces quelques lignes que je dois dicter, vaincu que je suis par la maladie, en vous faisant observer que vous rendiez service aux Colonies de la Grande Bretagne en repandant la connaissance de ce livre, et des principes que j'établis touchant la maladie des vers à soie. Beaucoup de ces colonies pourraient cultiver le mûrier avec succès, et en jetant les yeux sur mon ouvrage vous vous convaincrez aisement qu'il est facile aujourdhui, non seulement d'éloigner la maladie régnante, mais en outre de donner aux récoltes de la soie une prospérité qu'elles n'ont jamais eue."

(**2**, 181-183; 1870)

John Tyndall: Royal Institution, 30th June.

巴斯德详细地叙述了他保证蚕卵处于健康状态的方法，用这种方法完全能够恢复法国丝绸业以往的繁荣。描述巴斯德工作的应用并介绍其方法所获得的空前成功的报告合理地解释了他的研究工作，当时巴斯德还正在整理他的研究工作以供最终出版。巴斯德的方法在法国和意大利取得了惊人的成果。这是持续不断的努力换来的最终胜利，令人欣慰的是，即便是人类的那些愚蠢意见后来也变成了发展进步大道上的铺路石。那些批驳的意见激励着巴斯德，无意中起到了推动作用。巴斯德说："从我开始这项研究起，我就一直受到极为顽固而不公的批判，不过我在书中不会提起这些事情。"的确，在巴斯德的著作中我只发现有一处暗示性地提了一下。当谈到在类似科西嘉岛这样的孤立岛屿上养殖家蚕的好处时，他说："可以说，没有什么比完全清除致病微粒更容易解决这个问题了。如果如我所坚信的那样，自然发生学说只是不切实际的空想，那么就只有通过人类的力量才能消灭地球上的各种寄生性疾病。"而现在在英国，还真需要一批有识之士向民众澄清关于这些问题的种种流言。

巴斯德先生已经开始研究另一种疾病了。这种病在法国叫做**软腐病**，它与微粒子病共存，不过与微粒子病又完全不同。我相信，我所写的这些内容已经足以吸引那些对这一系列问题感兴趣的读者去阅读巴斯德先生的原著以获得更多的信息。这里我很遗憾地告诉读者，巴斯德先生现在得了重病，他是在进行前面详细介绍了的实验中染病的。巴斯德先生在随著作一起给我寄来的信的末尾写道："由于我身患重病，这封短信是我口述完成的。在信的末尾，我想说，你那些来自英国的桑树种子不仅有利于这本书中知识的传播，而且有利于我在对影响家蚕的疾病的研究中得到相关的原理。那些种子大部分都顺利地长成了桑树，从我的工作中你可以很容易地看到，现在可以很容易地消灭这种流行的疾病，并恢复以往丝绸业的繁荣了。"

（高如丽 翻译；刘力 审稿）

Spontaneous Generation

J. A. Wanklyn

Editor's Note

The interest of this comment from English chemist James Alfred Wanklyn on the topic of spontaneous combustion, discussed recently by H. Christian Bastian, is that it shows how tangled up the issue was becoming with the germ theory of disease. Wanklyn implies that the "vitalists" now invoked invisible airborne "germs" to explain spontaneous generation in water. A specialist in the analysis of water quality, Wanklyn complains that there is simply not enough nitrogen-rich organic matter in air to make that plausible. But of course microscopic airborne germs were precisely what Louis Pasteur was then correctly advancing as the cause of many diseases. Bastian himself was later to have an acerbic exchange in *Nature* with Thomas Huxley over spontaneous generation.

DR. H. C. Bastian, who has recently called attention to the nature of the evidence before scientific men in favour of the theory of so-called spontaneous generation, has supplemented it by fresh experiments of his own. The dilemma in which the opponents of this doctrine are now placed is that they must either admit it, or else allow that a temperature of 150 °C maintained for four hours, and applied by means of liquid, is incapable of killing the germs of infusoria. Many, doubtless, of these opponents will courageously mount this horn of the dilemma, and make the requisite enlargement of their ideas on the subject of vital resistance to change. There are, however, other difficulties in the way. For instance, great difficulties are involved in the assumption that the atmosphere constitutes a storehouse of germs of all kinds ready to burst out into life on the occurrence of suitable conditions.

However small these germs may be, still they must weigh something. And there must be very many of them, seeing that there must be an immense number of kinds of germs, if a volume of air is to supply to any given infusion precisely the right kinds of germs suitable to the conditions provided by the infusion.

Now chemists are in possession of data showing that the possible amount of organic nitrogenous matter in common clear water and common good air is remarkably small—so small, indeed, that the question may fairly be asked—Is it large enough to admit of the requisite number of germs, the existence of which the vitalists assume in water and air?

By the employment of our ammonia method, Chapman, Smith, and myself have shown that the organic ammonia from a kilogramme of good filtered water often falls as low as 0.05 milligramme, and Dr. Angus Smith has shown that a kilogramme of good air sometimes contains as little as 0.085 milligramme of organic ammonia.

A gramme of air—that is about 700 cubic centimetres—contains only 0.000085

自然发生学说

万克林

编者按

英国化学家詹姆斯·阿尔弗雷德·万克林对最近克里斯蒂安·巴斯蒂安讨论的自然发生学说作了评论，这篇评论的意义在于，它表明了该学说是怎样与疾病的细菌理论纠缠起来的。万克林暗示，"活力论者"现在要通过引入一个看不见的在空气中传播的"细菌"来解释水中的自然发生现象。作为一名水质分析专家，万克林抗议说，空气中根本没有足量的含氮有机物，因此该观点是不合理的。但恰恰是这些只有在显微镜下才可以看到的在空气中传播的细菌，被路易·巴斯德正确地判定为造成很多疾病的根源。后来巴斯蒂安本人在《自然》上围绕自然发生学说与托马斯·赫胥黎展开了一场激烈的争辩。

巴斯蒂安博士最近呼吁科学家们关注那些呈现在他们眼前的支持所谓自然发生学说的证据的本质，现在他通过自己刚做的实验对这个问题进行了补充。这一学说的反对者们目前所处的困境是：他们必须承认这一学说，或者承认将某种液体的温度维持在 150℃ 并持续 4 小时也不能杀死其中的纤毛虫类细菌。无疑，许多反对者将勇敢地面对这一困境，并且对他们的有关对外界变化的重要抗性的观点进行必要的扩展。然而，他们还面临着许多其他的困难。例如，关于大气是由各种各样随时准备好在适宜条件下萌发生命的细菌组成的大仓库这一假说就面临着许多巨大的困难。

不管这些细菌有多么小，它们肯定是有重量的。细菌的种类也一定非常多，因为，如果大量空气能让其中任意的内含物中都含有细菌，而这些细菌又适于在该内含物所提供的条件下生长的话，那么空气中就一定有大量不同种类的细菌存在。

现在化学家们所拥有的确切数据显示，有机含氮物质在普通清洁水和普通新鲜空气中的含量非常少，少得以至于人们很可能要问：其含量是否足够大到能容许活力论者所假定的在水和空气中存在的细菌的必需数目？

通过使用我们的测氨法，查普曼、史密斯和我本人已经发现，从 1 千克优质过滤水中得到的有机氨经常少到只有 0.05 毫克。安格斯·史密斯博士也发现 1 千克新鲜空气中有时仅含有少至 0.085 毫克的有机氨。

1 克空气大约是 700 立方厘米，其中仅含有 0.000085 毫克有机氨。将有机氨的

milligramme of organic ammonia. Expressing the organic ammonia in its equivalent of dry albumen we have in 700 cubic centimetres of air 0.00085 milligramme of dry albumen. Translated into volume this 0.00085 milligramme of dry albumen will fall short of a cube, the face of which is 1/10 millimetre in diameter.

Expressed in English measures, the result is, that rather more than one pint of average atmospheric air does not contain so much organic nitrogenous matter as corresponds to a cube of dry albumen of the 1/250 part of an inch in diameter.

Now is this quantity adequate to admit of the existence of the immense multitudes of germs, the existence of which in atmospheric air is assumed by the vitalists?

(**2**, 234-235; 1870)

含量用与其等量的干燥蛋白来表示的话，就是 700 立方厘米空气中有 0.00085 毫克干燥蛋白。换算成体积的话，这 0.00085 毫克干燥蛋白还装不满一个表面直径只有 1/10 毫米的立方体。

以英制单位表示的话，结果是，多于 1 品脱的普通大气中含有的有机含氮物质对应的干燥蛋白还不到直径为 1/250 英寸的一个立方体那么多。

现在这一数量结果能否足以让我们承认，大气中确实存在如活力论者所假定的那样大量的细菌吗？

（刘皓芳 翻译；刘力 审稿）

Spontaneous Generation

C. Ekin

Editor's Note

Earlier in *Nature*, pugnacious chemist James Alfred Wanklyn had questioned whether the low amounts of nitrogenous matter in the air were really enough to account for the multitude of invisible microorganisms being proposed as the cause of many diseases. Instead, he favoured spontaneous generation, the idea that living forms could emerge from purely inorganic building blocks. Wanklyn's letter had needled one reader to suggest his calculations supported rather than undermined the germ theory of disease. Here Charles Ekin expresses surprise that there is anything to argue about, citing results published by Louis Pasteur almost a decade earlier. In spite of the growing evidence in support of germ theory, some, like Wanklyn, were to dispute it for several decades to come.

IF there is one thing more curious than another in the "Spontaneous Generation" theory, it is the way in which so-called matters of fact, as proved by careful experiment, are brought forward by the one side to be disproved by the other, one need only instance Pasteur's famous flask experiments, which were thought to be so overwhelming at the time, but which were afterwards refuted, I think by Frémy and others.

I notice with surprise the letters of Prof. Wanklyn and Dr. Lionel Beale in *Nature*, with regard to the presence of germs in the air; there is an experiment of Pasteur's, given in his "Memoirs upon the organised Corpuscles which exist in the Atmosphere, 1862", and which I have never seen disproved, and if not disproved it must surely settle at least this part of the question. He passed a quantity of air, taking various precautions to eliminate error, which I need not here detail, by means of an aspirator through a plug of gun-cotton; he then dissolved the gun-cotton in ether, and on examining the sediment which subsided in the course of an hour or two, he found abundant evidence of the presence of organised corpuscles.

(**2**, 296; 1870)

Charles Ekin: Bath.

自然发生学说

伊金

编者按

好辩的化学家詹姆斯·阿尔弗雷德·万克林曾在《自然》上发表文章提出，鉴于空气中含氮物质的含量很低，是否真的可能存在那么多被认为足以导致疾病的看不见的细菌？而他更倾向于自然发生说，即生命体可以从纯无机物中生成。一位读者读了万克林的快报文章后提出，万克林的计算实际上是对疾病的细菌学说的支持而不是反对。在这篇文章中查尔斯·伊金表示对还有人在争论以上两种理论感到吃惊，因为路易·巴斯德早在近十年前就已经提出了这些观点。尽管有越来越多的证据支持了疾病的细菌学说，但有些人，如万克林在后来的几十年中还在不断地对这个理论提出质疑。

如果在"自然发生"学说中存在最令人好奇的事情的话，那就是由某一方提出的所谓经过精细实验证实的事实证据却被另一方证伪的过程了，这里只需要举出巴斯德著名的烧瓶实验一例即可，这一实验当时被认为是非常具有说服力的，但是，我记得后来却被弗雷米等人驳倒了。

我很惊奇地注意到，在万克林教授和莱昂内尔·比尔博士发表于《自然》上的关于空气中存在细菌这一主题的快报文章中提到了巴斯德的一个实验，该实验在巴斯德于 1862 年发表的《对存在于大气中的有序微粒子的研究》中介绍过。我到现在还没看到有报道对该实验表示异议，如果没有人表达过不同意的意见，那么可以肯定这一实验至少解决了相应的问题。巴斯德借助抽气机使得一定量的空气通过一块火棉，同时他采取了各种预防措施以排除误差，具体的我想就不需要在这里详细介绍了。然后，他将火棉溶解在乙醚中，检测在一两个小时这么长的时间里沉积到火棉上的沉淀物，通过这一实验他得到了大量的能够证明有序微粒子存在的证据。

（刘皓芳 翻译；刘力 审稿）

Address of Thomas Henry Huxley

T. H. Huxley

Editor's Note

In 1870, Thomas Henry Huxley was president of the British Association for the Advancement of Science, and used the occasion for an eloquent (and long) paper arguing that life is never generated spontaneously but only from pre-existing living things. This appears to have been a direct reply to a paper published in instalments in *Nature* on 30 June, 7 July and 14 July 1870 by H. Charlton Bastian describing experiments to observe the minute organisms that appeared in apparently sterile liquids apparently left to stand for long periods of time and from which air was not excluded. Bastian's paper had been submitted to the Royal Society but withdrawn by its author after a long delay and, apparently, accepted for publication in *Nature*. The paper occupied 26 pages of the journal.

MY Lords, Ladies, and Gentlemen,—It has long been the custom for the newly installed President of the British Association for the Advancement of Science to take advantage of the elevation of the position in which the suffrages of his colleagues had, for the time, placed him, and, casting his eyes around the horizon of the scientific world, to report to them what could be seen from his watch-tower; in what directions the multitudinous divisions of the noble army of the improvers of natural knowledge were marching; what important strongholds of the great enemy of us all, ignorance, had been recently captured; and, also, with due impartiality, to mark where the advanced posts of science had been driven in, or a long-continued siege had made no progress.

I propose to endeavour to follow this ancient precedent, in a manner suited to the limitations of my knowledge and of my capacity. I shall not presume to attempt a panoramic survey of the world of science, nor even to give a sketch of what is doing in the one great province of biology, with some portions of which my ordinary occupations render me familiar. But I shall endeavour to put before you the history of the rise and progress of a single biological doctrine; and I shall try to give some notion of the fruits, both intellectual and practical, which we owe, directly or indirectly, to the working out, by seven generations of patient and laborious investigators, of the thought which arose, more than two centuries ago, in the mind of a sagacious and observant Italian naturalist.

It is a matter of everyday experience that it is difficult to prevent many articles of food from becoming covered with mould; that fruit, sound enough to all appearance, often contains grubs at the core; that meat, left to itself in the air, is apt to putrefy and swarm with maggots. Even ordinary water, if allowed to stand in an open vessel, sooner or later becomes turbid and full of living matter.

托马斯·亨利·赫胥黎的致词

赫胥黎

编者按

1870 年，时任英国科学促进会主席的托马斯·亨利·赫胥黎借致词的机会宣读了一篇非常雄辩有力（而且很长）的论文。在文章中他指出，生命体绝不可能自发产生，而只能由先前已经存在的生物体得到。此前，查尔顿·巴斯蒂安在 1870 年 6 月 30 日、7 月 7 日以及 7 月 14 日的《自然》上分期发表了一篇描述实验结果的论文，声称在长时间放置的与空气隔绝的无菌液体中发现了很小的生物体。对于巴斯蒂安的系列论文来说，赫胥黎的这篇致词可以说是针锋相对的回应。早先，巴斯蒂安曾将其论文提交给皇家学会，不过在被搁置了很长一段时间后作者自己就将提交的论文撤销了。后来《自然》接受了巴斯蒂安的论文，该论文占据了《自然》的 26 个页面。

尊敬的各位来宾、女士们、先生们——长久以来，对于新上任的英国科学促进会主席，有一个传统是：他要利用职位的升迁这次机会，向那些为使其此次能够处于主席之位而投票的同事们作报告，报告人应放眼于科学世界，告诉大家从他的高度可以看到些什么：例如，为了丰富自然知识的贵族军队，其大部队是朝什么方向前进的；我们的大敌（愚昧无知）中有哪些重要据点最近被攻克；此外，本着公平公正的原则，他要标出何处是科学的前沿，或者在长久持续的围攻下都没有取得进展的领域。

我打算以一种适合于我有限的知识和能力的方式来尽力遵循这一古老的惯例。我不会冒昧地试图去统览整个科学世界的全貌，甚至不会去概括生物学某一重大领域里大家都在研究些什么，尽管我平时的工作使我熟悉其中的部分内容。但是我会尽力将一个生物学法则的产生与发展历史呈现在你们面前，并给出一些关于智力成果和实践成果的看法，这些都直接或间接地归功于七代孜孜不倦的研究者对两个多世纪前一位聪明睿智且具有敏锐观察力的意大利博物学家的思想的研究与解读。

通过日常经验我们知道，很难阻止许多食物表面上长出霉菌；外观十分完整的水果却经常在果核里生了虫；放在空气中的肉很容易腐烂、生满蛆虫。即使是普通的水，如果将其放在敞口的容器里，迟早会变得浑浊，并充满各种生物。

The philosophers of antiquity, interrogated as to the cause of these phenomena, were provided with a ready and a plausible answer. It did not enter their minds even to doubt that these low forms of life were generated in the matters in which they made their appearance. Lucretius, who had drunk deeper of the scientific spirit than any poet of ancient or modern times except Goethe, intends to speak as a philosopher, rather than as a poet, when he writes that "with good reason the earth has gotten the name of mother, since all things are produced out of the earth. And many living creatures, even now, spring out of the earth, taking form by the rains and the heat of the sun." The axiom of ancient science, "that the corruption of one thing is the birth of another," had its popular embodiment in the notion that a seed dies before the young plant springs from it; a belief so wide spread and so fixed, that Saint Paul appeals to it in one of the most splendid outbursts of his fervid eloquence:—

"Thou fool, that which thou sowest is not quickened, except it die."

The proposition that life may, and does, proceed from that which has no life, then, was held alike by the philosophers, the poets, and the people, of the most enlightened nations, eighteen hundred years ago; and it remained the accepted doctrine of learned and unlearned Europe, through the middle ages, down even to the seventeenth century.

It is commonly counted among the many merits of our great countryman, Harvey, that he was the first to declare the opposition of fact to venerable authority in this, as in other matters; but I can discover no justification for this wide spread notion. After careful search through the "Exercitationes de Generatione", the most that appears clear to me is, that Harvey believed all animals and plants to spring from what he terms a *"primordium vegetale"*, a phrase which may nowadays be rendered "a vegetative germ;" and this, he says, is "oviforme", or "egg-like"; not, he is careful to add, that it necessarily has the shape of an egg, but because it has the constitution and nature of one. That this *"primordium oviforme"* must needs, in all cases, proceed from a living parent is nowhere expressly maintained by Harvey, though such an opinion may be thought to be implied in one or two passages; while, on the other hand, he does, more than once, use language which is consistent only with a full belief in spontaneous or equivocal generation. In fact, the main concern of Harvey's wonderful little treatise is not with generation, in the physiological sense, at all, but with development; and his great object is the establishment of the doctrine of epigenesis.

The first distinct enunciation of the hypothesis that all living matter has sprung from pre-existing living matter, came from a contemporary, though a junior, of Harvey, a native of that country, fertile in men great in all departments of human activity, which was to intellectual Europe, in the sixteenth and seventeenth centuries, what Germany is in the nineteenth. It was in Italy, and from Italian teachers, that Harvey received the most important part of his scientific education. And it was a student trained in the same schools, Francesco Redi—a man of the widest knowledge and most versatile abilities, distinguished alike as scholar, poet, physician, and naturalist—who, just two hundred and

82

古代的哲学家们询问起这些现象的成因时，得到的是经过精心准备的、貌似合理的答案。他们甚至不想去质疑这些低级的生物是否是从那些它们所出现的物质中产生的。卢克莱修是一位比其他任何古代以及现代诗人（除了歌德）都更充分地领会了科学精神的人。当他写下"万物都是由地球而生，所以地球具有充分的理由获得母亲这一称呼。即便现在，也有许多生物从地球涌出，然后在雨水的滋润和阳光的照耀下成长"的诗句时，他是试图以一位哲学家而非诗人的口吻说话。"一种事物的灭亡意味着另一种事物的诞生"，这一古代科学的公理自有其能够为大家广泛接受的具体化身，即种子会在幼苗萌发之前死掉。这一观念传播得如此广泛并且根深蒂固，以至于圣保罗在一篇极具爆发力的狂热雄辩中呐喊道：

"无知的人啊！除非你播种的植物死掉，否则它不会长出新的生命。"

生命可能并且确实由无生命之物产生，这一命题与 1,800 年前那些最开明国家的哲学家、诗人以及平民所持的观点相同；并且，整个中世纪，甚至直到 17 世纪，这一主张依然是整个欧洲不管是博学的人还是平民都接受的信条。

在我们的伟大同胞哈维的众多功绩之中，经常被提到的是，正如在其他一些观点上一样，他是第一个就此观点向德高望重的权威们提出相反事实的人；但是我发现这一广泛传播的观念并没有任何合理性可言。在仔细查阅了《论生物的发生》之后，我最清楚的一点就是：哈维相信所有动植物都起源于他称为 **"植物原基"** 的东西，这一短语现在可以被表述成"植物性胚芽"；他说这种植物性胚芽是"卵形的"；并谨慎地补充说，它不一定具有卵的形状，这样称呼只是因为它具有卵的组成和性质。无论在什么情况下，这种 **"卵形原基"** 肯定需要由活着的亲本来产生，这一观点尽管可能被认为曾在一两个段落里有所暗示，但是哈维并没有明确地这样宣称过；另一方面，他不止一次地表达出完全相信自然发生或者不明确发生的想法。事实上，哈维精彩小论文的主要关注点根本不是生理意义上的发生，而是发育；他的伟大目标是建立渐成论。

所有生命物质都是从已存在的生命物质中产生而来的，第一次清楚阐明这一假说的是与哈维同时代的一个年轻人。这个年轻人所在的国家在人类活动的各个方面都涌现出了非常伟大的人物，16 世纪和 17 世纪时这个国家对于开明的欧洲而言，就像是 19 世纪的德国。这个国家就是意大利，哈维就是在意大利师从意大利的教师，接受了他科学教育培养中最重要的部分。弗朗切斯科·雷迪也是同一所学校培育出来的学生。他是一个具有非常渊博的知识和极其多样的才能的人，以学者、诗人、内科

two years ago, published his "Esperienze intorno alla Generazione degl' Insetti", and gave to the world the idea, the growth of which it is my purpose to trace. Redi's book went through five editions in twenty years; and the extreme simplicity of his experiments, and the clearness of his arguments, gained for his views, and for their consequences, almost universal acceptance.

Redi did not trouble himself much with speculative considerations, but attacked particular cases of what was supposed to be "spontaneous generation" experimentally. Here are dead animals, or pieces of meat, says he; I expose them to the air in hot weather, and in a few days they swarm with maggots. You tell me that these are generated in the dead flesh; but if I put similar bodies, while quite fresh, into a jar, and tie some fine gauze over the top of the jar, not a maggot makes its appearance, while the dead substances, nevertheless, putrefy just in the same way as before. It is obvious, therefore, that the maggots are not generated by the corruption of the meat; and that the cause of their formation must be a something which is kept away by gauze. But gauze will not keep away aëriform bodies, or fluids. This something must, therefore, exist in the form of solid particles too big to get through the gauze. Nor is one long left in doubt what these solid particles are; for the blowflies, attracted by the odour of the meat, swarm round the vessel, and, urged by a powerful but in this case misleading instinct, lay eggs out of which maggots are immediately hatched upon the gauze. The conclusion, therefore, is unavoidable; the maggots are not generated by the meat, but the eggs which give rise to them are brought through the air by the flies.

These experiments seem almost childishly simple, and one wonders how it was that no one ever thought of them before. Simple as they are, however, they are worthy of the most careful study, for every piece of experimental work since done, in regard to this subject, has been shaped upon the model furnished by the Italian philosopher. As the results of his experiments were the same, however varied the nature of the materials he used, it is not wonderful that there arose in Redi's mind a presumption, that in all such cases of the seeming production of life from dead matter, the real explanation was the introduction of living germs from without into that dead matter. And thus the hypothesis that living matter always arises by the agency of pre-existing living matter, took definite shape; and had, henceforward, a right to be considered and a claim to be refuted, in each particular case, before the production of living matter in any other way could be admitted by careful reasoners. It will be necessary for me to refer to this hypothesis so frequently, that, to save circumlocution, I shall call it the hypothesis of *Biogenesis*; and I shall term the contrary doctrine—that living matter may be produced by not living matter—the hypothesis of *Abiogenesis*.

In the seventeenth century, as I have said, the latter was the dominant view, sanctioned alike by antiquity and by authority; and it is interesting to observe that Redi did not escape the customary tax upon a discoverer of having to defend himself against the charge of impugning the authority of the Scriptures; for his adversaries declared that the generation of bees from the carcase of a dead lion is affirmed, in the Book of Judges, to have been the origin of the famous riddle with which Samson perplexed the Philistines:—

84

医师和博物学家著称。他恰于 202 年前出版了《关于昆虫世代的实验》一书，向世人展示了他的观点，该观点的发展历程也是我决定要探寻的。雷迪的书在 20 年间出了 5 版；他的实验极其简单易懂，并且论证清晰，这为他的观点以及实验结果几乎赢得了普遍的认同。

雷迪并没有使自己过于被推理性因素羁绊，而是通过实验攻克了那些被认为是"自然发生"的几个特殊的案例。他说，这些案例中有死亡的动物或者碎肉块；在炎热的天气里，我把它们暴露在空气中，几天后它们就会长满蛆虫。你们告诉我这些蛆虫是在死肉中产生的；但是如果我把类似的只是非常新鲜的死肉放进一个坛子里，然后将坛子的顶部用干净的纱布扎起来，那么就不会有任何蛆虫出现，尽管死肉仍会以一种与暴露在空气中时同样的方式腐烂掉。因此很明显，蛆虫不是由肉的腐烂产生的；它们产生的原因肯定是某种被纱布隔离掉的东西。但是，纱布不能隔离掉气体和液体。因此，这里的某种东西肯定是以固体颗粒形式存在的，并且由于太大而不能通过纱布。那么，这些固体颗粒是什么？这个长久以来一直困扰着我们的问题，已经得到了解决。答案是：绿头苍蝇被肉的臭味吸引，云集在容器周围，受一种强大的但在本例中却是产生误导的本能驱使，将卵产在纱布之外，于是这些卵立刻在纱布之上被孵化成蛆虫。因此，必然的结论就是，蛆虫并非由肉产生，而是那些苍蝇通过空气带来的卵产生的。

这些实验好像简单得近乎幼稚，有人会好奇为什么以前没有人想到过这些。然而，尽管这些实验很简单，但是却值得对它们进行最认真的研究，因为从那时开始针对这一问题所做的每项实验工作，都被这位意大利哲学家设计的模型定型了。不管他使用的材料的性质如何变化，他的实验结果都一样，因此，雷迪的脑海中出现如下设想并不奇怪，他认为在所有这些好像是从死物质产生了生命的例子中，其真正的原因是将有生命的微生物引入了原本没有这些微生物的死物质当中。因此生命物质总是借助于已存在的生命物质而产生这一假说就明确成形了；于是，从此以后，在每一个特殊具体的例子中，在通过谨慎的推理能够确认有任何其他方式产生生命物质之前，大家都有权进行考虑并拒绝接受某种假说。我必须频繁提到以上的假说，为了节省迂回累赘的陈述，我称之为**生源论**假说；并且将与此相反的学说——即认为生命物质可以由非生命物质产生，称为**无生源论**假说。

正如我已说过的，无生源论假说是 17 世纪的主流观点，被旧势力和权威们认可；有趣的是我们发现雷迪并没有逃脱作为一个因抨击经典权威、受到控诉而进行申辩的发现者通常需要承受的压力；因为他的对手声称，死亡狮子的尸体能够产生蜜蜂，这已经在《士师记》一书中被证实是萨姆森用来难住腓力斯人的著名谜语的起源：

> "Out of the eater came forth meat,
> And out of the strong came forth sweetness."

Against all odds, however, Redi, strong with the strength of demonstrable fact, did splendid battle for Biogenesis; but it is remarkable that he held the doctrine in a sense which, if he had lived in these times, would have infallibly caused him to be classed among the defenders of "spontaneous generation." "Omne vivum ex vivo", "no life without antecedent life", aphoristically sums up Redi's doctrine; but he went no further. It is most remarkable evidence of the philosophic caution and impartiality of his mind, that although he had speculatively anticipated the manner in which grubs really are deposited in fruits and in the galls of plants, he deliberately admits that the evidence is insufficient to bear him out; and he therefore prefers the supposition that they are generated by a modification of the living substance of the plants themselves. Indeed, he regards these vegetable growths as organs, by means of which the plant gives rise to an animal, and looks upon this production of specific animals as the final cause of the galls and of at any rate some fruits. And he proposes to explain the occurrence of parasites within the animal body in the same way.

It is of great importance to apprehend Redi's position rightly; for the lines of thought he laid down for us are those upon which naturalists have been working ever since. Clearly, he held *Biogenesis* as against *Abiogenesis*; and I shall immediately proceed, in the first place, to inquire how far subsequent investigation has borne him out in so doing.

But Redi also thought that there were two modes of Biogenesis. By the one method, which is that of common and ordinary occurrence, the living parent gives rise to offspring which passes through the same cycle of changes as itself—like gives rise to like; and this has been termed *Homogenesis*. By the other mode the living parent was supposed to give rise to offspring which passed through a totally different series of states from those exhibited by the parent, and did not return into the cycle of the parent; this is what ought to be called *Heterogenesis*, the offspring being altogether, and permanently unlike the parent. The term Heterogenesis, however, has unfortunately been used in a different sense, and M. Milne-Edwards has therefore substituted for it *Xenogenesis*, which means the generation of something foreign. After discussing Redi's hypothesis of universal Biogenesis, then, I shall go on to ask how far the growth of science justifies his other hypothesis of Xenogenesis.

The progress of the hypothesis of Biogenesis was triumphant and unchecked for nearly a century. The application of the microscope to anatomy in the hands of Grew, Lecuwenhoek, Swammerdam, Lyonet, Vallisnieri, Réaumur, and other illustrious investigators of nature of that day, displayed such a complexity of organisation in the lowest and minutest forms, and everywhere revealed such a prodigality of provision for their multiplication by germs of one sort or another, that the hypothesis of Abiogenesis began to appear not only untrue, but absurd; and, in the middle of the eighteenth century, when Needham and Buffon took up the question, it was almost universally discredited.

"吃的从吃者出来，

甜的从强者出来。"

尽管如此，雷迪还是凭借可论证的事实的力量，为生源论进行了精彩的斗争；但是值得注意的是，如果他生活在这种时代，那么从某种意义上说，他坚持的学说将肯定会使他被归为"自然发生学说"的拥护者。"生命源于生命"，"没有先前的生命就不会有新的生命产生"，这些以警句的形式概括了雷迪的学说；但是他没有更进一步。以下证据最显著地证明了他头脑中对待自然科学的谨慎和公正：尽管他曾经推测性地预见了存在于水果以及植物虫瘿中的幼虫产生的方式，但是他谨慎地承认那些证据不足以证实自己的观点，因此他更倾向于推测这些幼虫是通过对植物自身的生命物质加以修饰而产生的。实际上，他将这些植物的生长视为器官的生长，植物借助于器官产生动物，并且将这样产生的特殊动物看成是瘤而且至少是某些果实产生的最终根源。他提议以同样方式解释动物体内寄生虫的出现。

正确理解雷迪的观点具有重要的意义，因为从那以后博物学家一直依据他为我们呈现的思想脉络进行研究工作。很明显，他坚持**生源论**，反对**无生源论**。首先，我会立即着手考查后续的研究能在多大程度上证实他的想法。

但是雷迪也认为有两种生源论模式。一种模式是普通平常的发生方法，活着的亲本产生后代，后代经历同亲本一样的变化周期，即同类型产生同类型，这被称为**同型生殖**。另一种模式认为活着的亲本产生的后代经历与亲本表现出的完全不同的一系列状态，并不会回到亲本经历的周期中去，其后代全然、永远地与亲本不同，这应该被称为**异型生殖**。然而，不幸的是，异型生殖一词已经在一个不同的意义下被使用了，因此米尔恩·爱德华兹用**异源发生**来代替异型生殖，其意思是外源事物的产生。在讨论了雷迪的普遍生源论假说之后，我想再问一个问题，那就是科学的发展能在多大程度上证明其另一个假说——异源发生假说的合理性？

生源论假说在近一个世纪里的发展是顺利的，没有受到任何遏制。格鲁、列文虎克、斯瓦默丹、莱尔尼特、瓦利斯涅里、雷奥米尔及其他同时代的著名自然科学研究者将手中的显微镜应用到解剖学，展示了最低级、最微小形式的生物体组织结构竟是如此复杂，而且在所有例子中都揭示了通过某种微生物进行繁殖的规律，以致无生源论假说开始变得不仅仅不正确，而且很荒谬。18世纪中期，当尼达姆和布丰着手研究这一问题时，无生源论假说几乎已经被普遍放弃了。

But the skill of the microscope-makers of the eighteenth century soon reached its limit. A microscope magnifying 400 diameters was a *chef d'oeuvre* of the opticians of that day; and at the same time, by no means trustworthy. But a magnifying power of 400 diameters, even when definition reaches the exquisite perfection of our modern achromatic lenses, hardly suffices for the mere discernment of the smallest forms of life. A speck, only 1/25 of an inch in diameter, has, at 10 inches from the eye, the same apparent size as an object 1/10,000 of an inch in diameter, when magnified 400 times; but forms of living matter abound, the diameter of which is not more than 1/40,000 of an inch. A filtered infusion of hay, allowed to stand for two days, will swarm with living things, among which, any which reaches the diameter of a human red blood-corpuscle, or about 1/3,200 of an inch, is a giant. It is only by bearing these facts in mind, that we can deal fairly with the remarkable statements and speculations put forward by Buffon and Needham in the middle of the eighteenth century.

When a portion of any animal or vegetable body is infused in water, it gradually softens and disintegrates; and, as it does so, the water is found to swarm with minute active creatures, the so-called Infusorial Animalcules, none of which can be seen, except by the aid of the microscope; while a large proportion belong to the category of smallest things of which I have spoken, and which must have all looked like mere dots and lines under the ordinary microscopes of the eighteenth century.

Led by various theoretical considerations which I cannot now discuss, but which looked promising enough in the lights of that day, Buffon and Needham doubted the applicability of Redi's hypothesis to the infusorial animalcules, and Needham very properly endeavoured to put the question to an experimental test. He said to himself, if these infusorial animalcules come from germs, their germs must exist either in the substance infused, or in the water with which the infusion is made, or in the superjacent air. Now the vitality of all germs is destroyed by heat. Therefore, if I boil the infusion, cork it up carefully, cementing the cork over with mastic, and then heat the whole vessel by heaping hot ashes over it, I must needs kill whatever germs are present. Consequently, if Redi's hypothesis hold good, when the infusion is taken away and allowed to cool, no animalcules ought to be developed in it; whereas, if the animalcules are not dependent on pre-existing germs, but are generated from the infused substance, they ought, by-and-by, to make their appearance. Needham found that, under the circumstances in which he made his experiments, animalcules always did arise in the infusions, when a sufficient time had elapsed to allow for their development.

In much of his work Needham was associated with Buffon, and the results of their experiments fitted in admirably with the great French naturalist's hypothesis of "organic molecules", according to which, life is the indefeasible property of certain indestructible molecules of matter, which exist in all living things, and have inherent activities by which they are distinguished from not living matter. Each individual living organism is formed by their temporary combination. They stand to it in the relation of the particles of water to a cascade, or a whirlpool; or to a mould, into which the water is poured. The form of the

但是 18 世纪时，显微镜制造技术达到了极限。一台能够放大 400 倍的显微镜就算是当时光学仪器商的杰作了；与此同时，绝对不可以完全信赖这种仪器。但是即使显微镜的分辨率达到了我们现代的消色差透镜的敏锐精细程度，放大 400 倍的能力几乎连仅仅想辨别最微小的生命形式的要求都满足不了。一个直径仅仅 1/25 英寸的灰尘，在距离观察者的眼睛 10 英寸时，与将一个直径是 1/10,000 英寸的物体放大 400 倍，具有相同的表观尺寸；但是生物的形式丰富多样，很多生物的直径都不足 1/40,000 英寸。放置了两天的干草过滤浸液会长满有生命的物质，其中任何达到人类血红细胞直径大小或者大约 1/3,200 英寸大小的物质都算得上是庞然大物了。只有记住这些事实，我们才能公平地对待 18 世纪中期由布丰和尼达姆提出的引人注目的言论和推测。

将任何动物体或植物体的一部分浸泡在水中时，这部分就会逐渐软化、分解；此时，会发现水中充满微小的活动着的生物，即所谓的藻类微生物，除非借助于显微镜，否则仅仅依靠肉眼是看不到它们的；它们大部分属于我已经讲过的最小生物一类，用 18 世纪的普通显微镜来观察的话，它们看起来肯定都不过是些小点和线状物。

在一些这里我不会讨论但是当时看起来非常有前景的各种理论思考的引领下，布丰和尼达姆对雷迪的假说是否适用于藻类微生物表示怀疑，并且尼达姆尽力用一个非常恰当的实验来检验这一问题。他对自己说，如果这些藻类微生物来源于细菌，那么这些细菌肯定存在于浸泡的物质当中，或者存在于制取浸液用的水中，或者存在于上方的空气当中。而所有细菌的生命力都可以用热来摧毁。因此，如果我把浸液煮沸，将它仔细地用软木塞塞住，并用树脂封好软木塞，然后通过在容器上堆积热灰来加热整个容器，势必会杀死所有存在的细菌。于是，如果雷迪的假说是正确的话，那么当取走浸液、冷却下来之后，浸液中应该不会再出现新的微生物；反过来，如果微生物并不依赖于已存在的细菌，而是从浸泡的物质中产生的话，那么不久之后它们就应该会出现。尼达姆发现，在他的实验环境下，当经过足够满足微生物的发育要求的时间后，浸液中确实总会出现微生物。

尼达姆的许多工作都与布丰有关，他们的实验结果与法国著名博物学家的"有机分子"假说非常吻合。根据该假说，生命是某些不可毁灭的分子物质的固有属性，存在于所有生命体当中，具有内在的区别于非生命物质的活动。每个有生命的生物个体都由临时的组合构成，如同水滴与小瀑布、水滴与漩涡或者水与注入了水的模子的关系，生物体坚守着自己的临时组合。因此，生物的形式是由外界条件和构成它的有机分子的内在活动之间的相互作用共同决定的。如同漩涡的中断只是一

organism is thus determined by the reaction between external conditions and the inherent activities of the organic molecules of which it is composed; and, as the stoppage of a whirlpool destroys nothing but a form, and leaves the molecules of the water, with all their inherent activities intact, so what we call the death and putrefaction of an animal, or of a plant, is merely the breaking up of the form, or manner of association, of its constituent organic molecules, which are then set free as infusorial animalcules.

It will be perceived that this doctrine is by no means identical with *Abiogenesis*, with which it is often confounded. On this hypothesis, a piece of beef, or a handful of hay, is dead only in a limited sense. The beef is dead ox, and the hay is dead grass; but the "organic molecules" of the beef or the hay are not dead, but are ready to manifest their vitality as soon as the bovine or herbaceous shrouds in which they are imprisoned are rent by the macerating action of water. The hypothesis therefore must be classified under Xenogenesis, rather than under Abiogenesis. Such as it was, I think it will appear, to those who will be just enough to remember that it was propounded before the birth of modern chemistry, and of the modern optical arts, to be a most ingenious and suggestive speculation.

But the great tragedy of Science—the slaying of a beautiful hypothesis by an ugly fact—which is so constantly being enacted under the eyes of philosophers, was played, almost immediately, for the benefit of Buffon and Needham.

Once more, an Italian, the Abbé Spallanzani, a worthy successor and representative of Redi in his acuteness, his ingenuity, and his learning, subjected the experiments and the conclusions of Needham to a searching criticism. It might be true that Needham's experiments yielded results such as he had described, but did they bear out his arguments? Was it not possible, in the first place, that he had not completely excluded the air by his corks and mastic? And was it not possible, in the second place, that he had not sufficiently heated his infusions and the superjacent air? Spallanzani joined issue with the English naturalist on both these pleas, and he showed that if, in the first place, the glass vessels in which the infusions were contained were hermetically sealed by fusing their necks, and if, in the second place, they were exposed to the temperature of boiling water for three-quarters of an hour, no animalcules ever made their appearance within them. It must be admitted that the experiments and arguments of Spallanzani furnish a complete and a crushing reply to those of Needham. But we all too often forget that it is one thing to refute a proposition, and another to prove the truth of a doctrine which, implicitly or explicitly, contradicts that proposition, and the advance of science soon showed that though Needham might be quite wrong, it did not follow that Spallanzani was quite right.

Modern chemistry, the birth of the latter half of the eighteenth century, grew apace, and soon found herself face to face with the great problems which biology had vainly tried to attack without her help. The discovery of oxygen led to the laying of the foundations of a scientific theory of respiration, and to an examination of the marvellous interactions of organic substances with oxygen. The presence of free oxygen appeared to be one of

种形式的中止，并不会损坏任何东西，水分子中所有的内在活动都是完好的，我们所说的动物或植物的死亡和腐烂只是构成它们的有机分子形式的分解或者结合方式的破裂，它们随后都会以藻类微生物的形式释放出来。

经常会有人将这一学说与**无生源论**混淆，实际上两者是截然不同的。根据这一假说，一块牛肉或者一把干草只是在有限的意义上是死亡的。牛肉是死亡的牛，而干草是死亡的草；但是牛肉或干草中的"有机分子"并没有死，而是随时准备好，一旦其受到的束缚被水的浸离作用破坏掉，它们就会展现自己的生命力。因此该假说肯定应该被归入异源发生，而非无生源论。尽管这一假说可能不太准确，但对于那些刚好能记住它是在现代化学及现代光学技术诞生之前就被提出来了的人而言，我认为这会是一个非常具有独创性和启示性的推测。

但是为了布丰和尼达姆的利益，科学的巨大灾难——使用一个丑陋的事实来残杀一个美丽的假说——几乎立即上演，这种灾难非常频繁地在哲学家的眼皮底下上演。

再一次地，一个意大利人，阿贝·斯帕兰扎尼，一个值得尊敬的雷迪的继承者和代表者，以其独有的敏锐性、独创性和学识，将尼达姆的实验和结论置于一个有待于评论的境地。也许尼达姆的实验结果真如他自己描述的那样，但是它们证实他的论点了吗？首先，有没有可能通过软木塞和树脂他并没有完全排除空气呢？其次，有没有可能他没有对浸液和上面的空气进行足够的加热呢？斯帕兰扎尼在这两个问题上与英国的博物学家存在争议。他指出，如果首先将盛有浸液的玻璃器皿通过热熔其颈部来达到密封的目的，然后再将它们暴露于沸水中达 45 分钟的话，那么浸液中就不会再有微生物出现。必须承认斯帕兰扎尼的实验和论证对尼达姆的实验和论证给予了彻底的、粉碎性的反击。但是我们也经常忘记驳斥一个命题是一件事，含蓄或明确地证明一个与此命题相矛盾的另一学说的正确性又是另一回事。科学的进步不久就表明，尽管尼达姆可能是非常错误的，但并不表示斯帕兰扎尼就是非常正确的。

现代化学于 18 世纪下半叶诞生，并且发展迅速，它很快就发现自己面对着巨大的难题，这些难题都是生物学家曾经试图在没有化学的帮助下进行攻克却徒劳无获的。氧气的发现奠定了呼吸这一科学理论的基础，引发了对有机物质与氧气之间奇妙的相互作用的研究。自由氧的存在似乎是生命存在的条件之一，也是有机物发生

the conditions of the existence of life, and of those singular changes in organic matters which are known as fermentation and putrefaction. The question of the generation of the infusory animalcules thus passed into a new phase. For what might not have happened to the organic matter of the infusions, or to the oxygen of the air, in Spallanzani's experiments? What security was there that the development of life which ought to have taken place had not been checked or prevented by these changes?

The battle had to be fought again. It was needful to repeat the experiments under conditions which would make sure that neither the oxygen of the air, nor the composition of the organic matter, was altered in such a matter as to interfere with the existence of life.

Schulze and Schwann took up the question from this point of view in 1836 and 1837. The passage of air through red-hot glass tubes, or through strong sulphuric acid, does not alter the proportion of its oxygen, while it must needs arrest or destroy any organic matter which may be contained in the air. These experimenters, therefore, contrived arrangements by which the only air which should come into contact with a boiled infusion should be such as had either passed through red-hot tubes or through strong sulphuric acid. The result which they obtained was that an infusion so treated developed no living things, while if the same infusion was afterwards exposed to the air such things appeared rapidly and abundantly. The accuracy of these experiments has been alternately denied and affirmed. Supposing them to be accepted, however, all that they really proved was that the treatment to which the air was subjected destroyed *something* that was essential to the development of life in the infusion. This "something" might be gaseous, fluid, or solid; that it consisted of germs remained only an hypothesis of greater or less probability.

Contemporaneously with these investigations a remarkable discovery was made by Cagniard de la Tour. He found that common yeast is composed of a vast accumulation of minute plants. The fermentation of must or of wort in the fabrication of wine and of beer is always accompanied by the rapid growth and multiplication of these *Torulae*. Thus fermentation, in so far as it was accompanied by the development of microscopical organisms in enormous numbers, became assimilated to the decomposition of an infusion of ordinary animal or vegetable matter; and it was an obvious suggestion that the organisms were, in some way or other, the causes both of fermentation and of putrefaction. The chemists, with Berzelius and Liebig at their head, at first laughed this idea to scorn; but in 1843, a man then very young, who has since performed the unexampled feat of attaining to high eminence alike in Mathematics, Physics, and Physiology—I speak of the illustrious Helmholtz—reduced the matter to the test of experiment by a method alike elegant and conclusive. Helmholtz separated a putrefying or a fermenting liquid from one which was simply putrescible or fermentable by a membrane which allowed the fluids to pass through and become intermixed, but stopped the passage of solids. The result was, that while the putrescible or the fermentable liquids became impregnated with the results of the putrescence or fermentation which was going on on the other side of the membrane, they neither putrefied (in the ordinary way) nor fermented; nor were any of the organisms which abounded in the fermenting or putrefying liquid generated in them.

发酵和腐烂等异常变化的条件之一。因此藻类微生物如何发生的问题进入了一个新的阶段。在斯帕兰扎尼的实验中，为什么浸液中的有机物质或者空气中的氧气似乎并没有发生什么变化呢？斯帕兰扎尼实验中的变化提供了什么安全措施使得本应发生的生命发育过程没有被检测到或者被阻止了呢？

这场战斗不得不再打一次。有必要再在确保空气中的氧气和有机物质的构成成分不改变，并且没有干扰生命存在的物质的情况下，再次重复这些实验。

舒尔策和施旺在 1836 年和 1837 年开始从这一角度出发进行研究。通过灼热的玻璃管或通过浓硫酸之后，空气中所含氧气的比例都不会改变，但是这样肯定会抑制或者破坏空气中可能含有的任何有机物质。因此，这些实验被设计成，只有那些通过了炽热的玻璃管或者浓硫酸的空气才会与沸腾的浸液相接触。他们得到的结果是，这样处理的浸液并没有产生任何有生命的物质，而如果过后将相同的浸液暴露在空气中，那么很快就会产生大量的有生命的物质。这些实验的准确性时而被否决，时而被证实。然而，假设被接受了，那么这些实验真正证明的就是，对空气的处理毁掉了对于浸液中生命发育必不可少的**某种物质**。这里所说的“某种物质”可能是气态的、液态的或者固态的；认为其中含有微生物还只是一个有一定可能性的假说。

在得到这些研究结果的同时，卡尼亚尔·德拉图尔取得了一项引人注目的发现。他发现普通酵母菌由大量微小的植物积累构成。在酿造葡萄酒和啤酒的过程中，葡萄汁或麦芽汁的发酵总是伴随着这些**圆酵母**的快速生长和增殖。因此，伴随着大量微生物的发育过程的发酵作用，类似于普通动物性物质或植物性物质的浸液的腐烂过程；而且在某种意义上，这明确提示了是微生物引起了发酵和腐烂。以伯齐利厄斯和李比希为首的化学家们首先嘲笑了这个想法并表示不屑；但是在 1843 年，一个当时还非常年轻的人通过精巧而又令人信服的方法来减少检测实验中的物质。我说的这个年轻人就是著名的亥姆霍兹，他曾在数学、物理学和生理学领域都作出了极其卓越的无可比拟的业绩。亥姆霍兹使用了一种薄膜将正在腐烂的或者正在发酵的液体与仅仅可能会腐烂或可能会发酵的液体分开，该薄膜允许液体通过并发生混合，但是可以阻止固体通过。结果是，当膜的另一侧的正在腐烂或者发酵的产物渗入到可能会腐烂或者可能会发酵的液体中时，它们既不以通常的方式腐烂也不发酵；并且它们中也不会产生任何充满于发酵液或者腐烂液中的微生物。因此这些微生物发育的原因肯定在于某种不能通过膜的物质；由于亥姆霍兹的研究比格雷姆对胶体进行的研究早很多，所以他自然而然地得出结论，认为被拦截的媒介肯定是固体物质。

Therefore the cause of the development of these organisms must lie in something which cannot pass through membranes; and as Helmholtz's investigations were long antecedent to Graham's researches upon colloids, his natural conclusion was that the agent thus intercepted must be a solid material. In point of fact, Helmholtz's experiments narrowed the issue to this: that which excites fermentation and putrefaction, and at the same time gives rise to living forms in a fermentable or putrescible fluid, is not a gas and is not a diffusible fluid; therefore it is either a colloid, or it is matter divided into very minute solid particles.

The researches of Schroeder and Dusch in 1854, and of Schroeder alone, in 1859, cleared up this point by experiments which are simply refinements upon those of Redi. A lump of cotton-wool is, physically speaking, a pile of many thicknesses of a very fine gauze, the fineness of the meshes of which depends upon the closeness of the compression of the wool. Now, Schroeder and Dusch found, that, in the case of all the putrefiable materials which they used (except milk and yolk of egg), an infusion boiled, and then allowed to come into contact with no air but such as had been filtered through cotton-wool, neither putrefied nor fermented, nor developed living forms. It is hard to imagine what the fine sieve formed by the cotton-wool could have stopped except minute solid particles. Still the evidence was incomplete until it had been positively shown, first, that ordinary air does contain such particles; and, secondly, that filtration through cotton-wool arrests these particles and allows only physically pure air to pass. This demonstration has been furnished within the last year by the remarkable experiments of Professor Tyndall. It has been a common objection of Abiogenists that, if the doctrine of Biogeny is true, the air must be thick with germs; and they regard this as the height of absurdity. But Nature occasionally is exceedingly unreasonable, and Professor Tyndall has proved that this particular absurdity may nevertheless be a reality. He has demonstrated that ordinary air is no better than a sort of stirabout of excessively minute solid particles; that these particles are almost wholly destructible by heat; and that they are strained off, and the air rendered optically pure by being passed through cotton-wool.

But it remains yet in the order of logic, though not of history, to show that among these solid destructible particles there really do exist germs capable of giving rise to the development of living forms in suitable menstrua. This piece of work was done by M. Pasteur in those beautiful researches which will ever render his name famous; and which, in spite of all attacks upon them, appear to me now, as they did seven years ago, to be models of accurate experimentation and logical reasoning. He strained air through cotton-wool, and found, as Schroeder and Dusch had done, that it contained nothing competent to give rise to the development of life in fluids highly fitted for that purpose. But the important further links in the chain of evidence added by Pasteur are three. In the first place he subjected to microscopic examination the cotton-wool which had served as strainer, and found that sundry bodies clearly recognizable as germs, were among the solid particles strained off. Secondly, he proved that these germs were competent to give rise to living forms by simply sowing them in a solution fitted for their development. And, thirdly, he showed that the incapacity of air strained through cotton-wool to give rise

事实上，亥姆霍兹的实验将问题缩小到了下述范围：引起发酵和腐烂作用的、同时在可发酵或可腐烂液体中产生生命的既不是气体也不是扩散性的液体；因此这种物质要么是胶体，要么是一种被分成非常微小的固体颗粒的物质。

施罗德和杜施在 1854 年进行的研究以及施罗德自己在 1859 年进行的研究都通过实验澄清了这个问题，这些实验都只是在雷迪实验的基础上进行了简单的改良。从物理学角度而言，一块脱脂棉就是一堆不同厚度的纱布，纱布的网眼细度取决于毛料压缩的密实度。现在，施罗德和杜施发现，在他们使用的所有会腐烂的材料中（除了牛奶和蛋黄），如果煮沸后的浸液不与空气接触，而且用脱脂棉进行过滤，那么也是既不会腐烂，也不会发酵，且没有任何生命形式发育出来。很难想象脱脂棉形成的精细筛子除了阻止微小的固体颗粒外还阻止了别的什么东西。证据要得以完整，需要确定以下两点：第一，普通空气中确实含有这种颗粒；第二，脱脂棉的过滤作用阻止了这些颗粒的通过，而只允许物理意义上纯净的空气通过。廷德尔教授已于去年通过自己的著名实验为此提供了实证。这招来了无生源论者的普遍异议，即如果生源论法则是正确的，那么空气中肯定充满了细菌；他们认为这荒唐透顶。但是自然界有时就是非常不合常理，廷德尔教授已经证明了这个特别的谬论确实是真实的。他证明了普通空气几乎跟一种含有大量微小固体颗粒的稀饭一样；通过加热，可以使这些颗粒几乎完全被破坏掉；也可以通过过滤除掉它们，所以使空气通过脱脂棉后，从视觉上看它就变得纯净了。

尽管根据历史记录细菌已经被除掉了，但是依逻辑学次序考虑，仍然表明在这些固态可破坏颗粒中，确实存在能够在适当溶剂下引发生命发育的细菌。巴斯德通过一些完美的研究完成了这一工作，并使他的名字为众人所知。尽管存在各种各样的攻击，但是正如七年前这些研究给我的感觉那样，我现在仍觉得它们是准确的实验和逻辑推理的典范。巴斯德使空气通过脱脂棉从而被过滤，结果跟施罗德和杜施的发现一样，即这样处理过的空气不含有那些特别的液体中所含有的任何能够产生生命发育的物质。但是巴斯德进一步补充了这一证据链中的 3 个重要环节。首先他借助显微镜检测了作为过滤器的脱脂棉，发现被过滤掉的固体颗粒中，有各种可以被明确识别为细菌的物体。其次，他证明了把这些微生物散播到适合它们发育的溶液中能够产生生命。再次，他证明了通过脱脂棉过滤的空气不能产生生命，并不是由于脱脂棉本身影响空气组分发生了任何超自然变化。如果完全不用脱脂棉，那么

to life, was not due to any occult change effected in constituents of the air by the wool, by proving that the cotton-wool might be dispensed with altogether, and perfectly free access left between the exterior air and that in the experimental flask. If the neck of the flask is drawn out into a tube and bent downwards; and if, after the contained fluid has been carefully boiled, the tube is heated sufficiently to destroy any germs which may be present in the air which enters as the fluid cools, the apparatus may be left to itself for any time and no life will appear in the fluid. The reason is plain. Although there is free communication between the atmosphere laden with germs and the germless air in the flask, contact between the two takes place only in the tube; and as the germs cannot fall upwards, and there are no currents, they never reach the interior of the flask. But if the tube be broken short off where it proceeds from the flask, and free access be thus given to germs falling vertically out of the air, the fluid which has remained clear and desert for months, becomes, in a few days turbid and full of life.

These experiments have been repeated over and over again by independent observers with entire success; and there is one very simple mode of seeing the facts for oneself, which I may as well describe.

Prepare a solution (much used by M. Pasteur, and often called "Pasteur's solution") composed of water with tartrate of ammonia, sugar, and yeast-ash dissolved therein. Divide it into three portions in as many flasks; boil all three for a quarter of an hour; and, while the steam is passing out, stop the neck of one with a large plug of cotton-wool, so that this also may be thoroughly steamed. Now set the flasks aside to cool, and when their contents are cold, add to one of the open ones a drop of filtered infusion of hay which has stood for twenty-four hours, and is consequently full of the active and excessively minute organisms known as *Bacteria*. In a couple of days of ordinary warm weather the contents of this flask will be milky from the enormous multiplication of *Bacteria*. The other flask, open and exposed to the air, will, sooner or later, become milky with *Bacteria*, and patches of mould may appear in it; while the liquid in the flask, the neck of which is plugged with cotton-wool, will remain clear for an indefinite time. I have sought in vain for any explanation of these facts, except the obvious one, that the air contains germs competent to give rise to *Bacteria*, such as those with which the first solution has been knowingly and purposely inoculated, and to the mould-*Fungi*. And I have not yet been able to meet with any advocate of Abiogenesis who seriously maintains that the atoms of sugar, tartrate of ammonia, yeast-ash, and water, under no influence but that of free access of air and the ordinary temperature, rearrange themselves and give rise to the protoplasm of *Bacterium*. But the alternative is to admit that these *Bacteria* arise from germs in the air; and if they are thus propagated, the burden of proof that other like forms are generated in a different manner, must rest with the assertor of that proposition.

To sum up the effect of this long chain of evidence: —

It is demonstrable that a fluid eminently fit for the development of the lowest forms of life, but which contains neither germs, nor any protein compound, gives rise to living things

就在外界空气与实验中所用的长颈瓶中的空气之间形成了完全自由的通道。如果将长颈瓶的颈部拉长成一个管状并向下弯曲，然后将其中盛入的液体小心煮沸，之后如果将管子也充分加热以破坏当液体冷却时进入管子的空气中的所有细菌的话，那么无论将装置放置多久，其中的液体都不会产生任何生命。原因很简单。尽管充满着微生物的大气和长颈瓶中的无菌空气间存在自由的流通，但是这两个地方的空气只能在管子里发生接触。因为细菌不能朝上落，也没有气流，所以它们永远都到达不了长颈瓶的内部。但是如果从长颈瓶延伸处将管子截短，这样外界空气中的细菌就可以自由地垂直下降，那么原本放置数月仍然保持清澈且无生命的液体在几天内就会变浑并且充满生命。

这些实验已经由不同的观察者独立重复了一遍又一遍，并且都取得了成功；有一套非常简单的，可以亲自观察这些事实的模式，这里我不妨也描述一下。

准备一份溶有酒石酸铵、糖和酵母粉的水溶液。（由于这是巴斯德先生经常采用的溶液，所以通常被称为"巴斯德溶液"。）将该溶液分成三等份分别盛入 3 个长颈瓶中；将 3 瓶溶液都煮沸一刻钟；当看见有蒸汽冒出时，将其中一个长颈瓶的颈部用一大块脱脂棉堵住，这样这一瓶就可以被彻底蒸透了。现在将旁边的长颈瓶放置至冷却，当其中的溶液冷却后，向开口的长颈瓶中加入一滴已放置了 24 小时的过滤了的干草浸液，因而这瓶溶液中会充满大量活动的微小生物，即通常所说的细菌。在经历几天普通的温暖天气后，这只长颈瓶中的内含物将会因细菌的大量繁殖而变成乳状。另一个开口并暴露于空气中的长颈瓶迟早也会长满细菌而变成乳状，也可能出现多片霉菌；而颈部被脱脂棉塞住的长颈瓶中的液体无论放置多长时间都是澄清的。我曾经试图解释这些现象，但是除了空气中含有能够产生细菌（比如故意向第一种溶液中接种的那些）和霉菌的微生物这个明显的原因外，别无所获。无生源论的支持者坚持认为，在没有受到任何影响、只是在空气的自由流通和普通温度下，糖、酒石酸铵、酵母粉和水的原子发生了重排，从而产生了细菌的原生质。对于他们的观点，我到目前都不能苟同。但是另一种观点承认这些细菌是由空气中的微生物产生的；如果它们是这样繁殖的，那么举证责任，即举出其他类似的形式是以不同方式产生的，就必须得靠那一命题的主张者了。

这一长串证据的影响可以概括如下：

可以证实：一种不含有微生物和任何蛋白质复合物的，特别适合于最低级形式的生命发育的液体，如果暴露于普通空气中就可以产生大量的生物；而如果通过机

in great abundance if it is exposed to ordinary air, while no such development takes place if the air with which it is in contact is mechanically freed from the solid particles which ordinarily float in it and which may be made visible by appropriate means.

It is demonstrable that the great majority of these particles are destructible by heat, and that some of them are germs or living particles capable of giving rise to the same forms of life as those which appear when the fluid is exposed to unpurified air.

It is demonstrable that inoculation of the experimental fluid with a drop of liquid known to contain living particles gives rise to the same phenomena as exposure to unpurified air.

And it is further certain that these living particles are so minute that the assumption of their suspension in ordinary air presents not the slightest difficulty. On the contrary, considering their lightness and the wide diffusion of the organisms which produce them, it is impossible to conceive that they should not be suspended in the atmosphere in myriads.

Thus the evidence, direct and indirect, in favour of *Biogenesis* for all known forms of life must, I think, be admitted to be of great weight.

On the other side the sole assertions worthy of attention are that hermetically sealed fluids, which have been exposed to great and long-continued heat, have sometimes exhibited living forms of low organization when they have been opened.

The first reply that suggests itself is the probability that there must be some error about these experiments, because they are performed on an enormous scale every day with quite contrary results. Meat, fruits, vegetables, the very materials of the most fermentable and putrescible infusions are preserved to the extent, I suppose I may say, of thousands of tons every year, by a method which is a mere application of Spallanzani's experiment. The matters to be preserved are well boiled in a tin case provided with a small hole, and this hole is soldered up when all the air in the case has been replaced by steam. By this method they may be kept for years without putrefying, fermenting, or getting mouldy. Now this is not because oxygen is excluded, inasmuch as it is now proved that free oxygen is not necessary for either fermentation or putrefaction. It is not because the tins are exhausted of air, for *Vibriones* and *Bacteria* live, as Pasteur has shown, without air or free oxygen. It is not because the boiled meats or vegetables are not putrescible or fermentable, as those who have had the misfortune to be in a ship supplied with unskilfully closed tins well know. What is it, therefore, but the exclusion of germs? I think that Abiogenists are bound to answer this question before they ask us to consider new experiments of precisely the same order.

And in the next place, if the results of the experiments I refer to are really trustworthy, it by no means follows that Abiogenesis has taken place. The resistance of living matter to heat is known to vary within considerable limits, and to depend, to some extent, upon

械方法除去与液体接触的空气中的固体颗粒（这些固体颗粒通常悬浮于空气中，并且可以通过适当的方式使其可见）的话，那么生物便不会产生。

这些颗粒中的绝大部分可以通过加热破坏掉，其中有些颗粒是微生物或者有生命的颗粒，它们能够产生的生命形式与液体暴露于未净化的空气中时产生的生命形式相同，这一点也是可以证实的。

将已知含有有生命颗粒的一滴液体接种到实验液体中，能够产生与暴露于未净化空气中同样的现象，这一点同样是可以证实的。

此外，可以进一步肯定的是，这些有生命的颗粒太小了，因此假设它们悬浮于普通空气中时，不存在任何困难。相反，考虑到它们的轻微以及产生它们的有机体的广泛扩散，必然能想到它们会以极大的数量悬浮于大气之中。

因此我认为必须承认，那些支持对于所有已知生命形式都成立的**生源论**的直接和间接证据，是非常重要的。

另一方面，值得注意的独特论断是，对密封液体进行长时间、充分的加热后，有时也会产生低级的生命形式。

对于此论断，第一个回复是这些实验本身可能有某种误差，因为这些实验每天都在很大的范围内进行着，并且得到了相当矛盾的结果。我估计，每年光是运用斯帕兰扎尼的实验方法腌制的肉、水果和蔬菜，以及最易发酵和腐烂的浸液材料就有成千上万吨。将要腌制的物质放在一个有小孔的铁罐里充分煮沸，当罐里的所有空气都被蒸汽替换的时候，将这个孔焊住。通过这种办法，这些材料可以保存多年而不会腐烂、发酵或发霉。这不是因为排除了氧气，而是因为氧气并不是发酵或腐烂的必要条件，这一点现在已经得到了证明。正如巴斯德表明的，这并不是由于罐子里的空气被排除了，因为弧菌和细菌的生存并不需要空气或氧气。这也不是因为煮熟的肉或者蔬菜不会腐烂或者发酵，那些在船上不幸食用了没有密闭好的罐子中的肉和蔬菜的人们都清楚地知道这一点。那么到底是什么排除了微生物呢？我认为无生源论者一定会在要求我们想出同样精准、有条理的新实验之前就对这一问题作出回答。

第二，如果我引用的实验结果真的值得信任，那么得出的观点绝对不会与无生源论的观点相同。我们知道生命物质对热的耐受在一定范围内变化，并在某种程度

the chemical and physical qualities of the surrounding medium. But if, in the present state of science, the alternative is offered us, either germs can stand a greater heat than has been supposed, or the molecules of dead matter, for no valid or intelligible reason that is assigned, are able to rearrange themselves into living bodies, exactly such as can be demonstrated to be frequently produced in another way, I cannot understand how choice can be, even for a moment, doubtful.

But though I cannot express this conviction of mine too strongly, I must carefully guard myself against the supposition that I intend to suggest that no such thing as Abiogenesis ever has taken place in the past or ever will take place in the future. With organic chemistry, molecular physics, and physiology yet in their infancy, and every day making prodigious strides, I think it would be the height of presumption for any man to say that the conditions under which matter assumes the properties we call "vital" may not, some day, be artificially brought together. All I feel justified in affirming is that I see no reason for believing that the feat has been performed yet.

And looking back through the prodigious vista of the past, I find no record of the commencement of life, and therefore I am devoid of any means of forming a definite conclusion as to the conditions of its appearance. Belief, in the scientific sense of the word, is a serious matter, and needs strong foundations. To say, therefore, in the admitted absence of evidence, that I have any belief as to the mode in which the existing forms of life have originated, would be using words in a wrong sense. But expectation is permissible where belief is not; and if it were given me to look beyond the abyss of geologically recorded time to the still more remote period when the earth was passing through physical and chemical conditions, which it can no more see again than a man can recall his infancy, I should expect to be a witness of the evolution of living protoplasm from not living matter. I should expect to see it appear under forms of great simplicity, endowed, like existing fungi, with the power of determining the formation of new protoplasm from such matters as ammonium carbonates, oxalates and tartrates, alkaline and earthy phosphates, and water, without the aid of light. That is the expectation to which analogical reasoning leads me; but I beg you once more to recollect that I have no right to call my opinion anything but an act of philosophical faith.

So much for the history of the progress of Redi's great doctrine of Biogenesis, which appears to me, with the limitations I have expressed, to be victorious along the whole line at the present day.

As regards the second problem offered to us by Redi, whether Xenogenesis obtains, side by side with Homogenesis; whether, that is, there exist not only the ordinary living things, giving rise to offspring which run through the same cycle as themselves, but also others, producing offspring which are of a totally different character from themselves, the researches of two centuries have led to a different result. That the grubs found in galls are no product of the plants on which the galls grow, but are the result of the introduction

上依赖于周围介质的化学和物理性质。但是在目前的科学水平下，我们面对的是二选一，不是微生物可以承受超过我们预期的更充分的加热，就是死亡物质的分子能够将自己重排成新的有生命的物质（这一点不能用任何有效的或可理解的已知原因来解释），但是事实证明有生命的物质经常是以其他方式产生的，于是我不明白在二者中进行选择怎么会令人生疑，哪怕只是一时的怀疑。

尽管我不能将自己的这种信念表达得过于强烈，但我必须谨慎地使自己不去猜想——我觉得无生源论这种事情根本就没有发生过，将来也不会发生。有机化学、分子物理学和生理学尚处于初期阶段，每天都在取得巨大的进步，我认为任何人都可能会极为自以为是地认为物质表现出来的我们称为"有生命的"属性的条件终有一天不再被人为地整合到一起。我有依据能确定迄今尚无证据表明这一假设已经奏效。

回顾过去的奇异景象，我发现没有关于生命开端的记录，因此对于生命出现的条件，我找不到任何依据来形成确定的结论。从词语的科学意义上说，信念是一个严肃的事情，它需要坚实的基础。因此，如果要在公认缺乏证据的情况下让我说出我相信生命形式通过哪种方式产生而来，那么我只能说我无从相信任何一种对现存生命形式的产生模式的假设，否则这种信念就是对词语意义的误用。但是在信念不存在的地方却是允许期望存在的。如果让我向地质记录年代更加久远的深渊望去，甚至是地球正在经历物理化学变化的那些更遥远的时期——当然这些情景如同一个人要回想起他的婴儿时期一样都是不可能再看到的事情，那么我希望自己能够见证生命原生质如何从没有生命的物质进化而来。我希望看到生命物质以非常简单的形式出现，就像现存的真菌，它们被赋予了决定如何在不依靠光的情况下，从碳酸铵、草酸盐和酒石酸盐、碱土金属磷酸盐和水等物质中形成新的原生质的力量。这就是类推推理引导我想到的期望。但是我再次请求你们记住我只能将我的观点称为哲学信仰的结果。

对雷迪的伟大生源论法则的进展历史，我就介绍到此。尽管它存在我前面已经提到的局限性，但是即使在今天，在我看来它也取得了彻底的胜利。

雷迪为我们提出的第二个问题是异源发生是否伴随同型生殖一同存在，或者说是否不仅存在普通的生物，其产生的后代与自身经历同样的生命周期，而且存在其他生物，能够产生具有完全不同于自己特征的后代。至于这一问题，两个世纪以来的研究已经产生了不同的结果。虫瘿中发现的幼虫不是长了虫瘿的植物产生的，而是昆虫的卵进入植物体内的结果。这一现象是由瓦利斯涅里、雷奥米尔等人于

of the eggs of insects into the substance of these plants, was made out by Vallisnieri, Reaumur, and others, before the end of the first half of the eighteenth century. The tapeworms, bladderworms, and flukes continued to be a stronghold of the advocates of Xenogenesis for a much longer period. Indeed, it is only within the last thirty years that the splendid patience of Von Siebold, Van Beneden, Leuckart, Küchenmeister, and other helminthologists, has succeeded in tracing every such parasite, often through the strangest wanderings and metamorphoses, to an egg derived from a parent, actually or potentially like itself; and the tendency of inquiries elsewhere has all been in the same direction. A plant may throw off bulbs, but these, sooner or later, give rise to seeds or spores, which develop into the original form. A polype may give rise to Medusae, or a pluteus to an Echinoderm, but the Medusa and the Echinoderm give rise to eggs which produce polypes or plutei, and they are therefore only stages in the cycle of life of the species.

But if we turn to pathology it offers us some remarkable approximations to true Xenogenesis.

As I have already mentioned, it has been known since the time of Vallisnieri and of Reaumur, that galls in plants, and tumours in cattle, are caused by insects, which lay their eggs in those parts of the animal or vegetable frame of which these morbid structures are outgrowths. Again, it is a matter of familiar experience to everybody that mere pressure on the skin will give rise to a corn. Now the gall, the tumour, and the corn are parts of the living body, which have become, to a certain degree, independent and distinct organisms. Under the influence of certain external conditions, elements of the body, which should have developed in due subordination to its general plan, set up for themselves and apply the nourishment which they receive to their own purposes.

From such innocent productions as corns and warts, there are all gradations to the serious tumours which, by their mere size and the mechanical obstruction they cause, destroy the organism out of which they are developed; while, finally, in those terrible structures known as cancers, the abnormal growth has acquired powers of reproduction and multiplication, and is only morphologically distinguishable from the parasite worm, the life of which is neither more nor less closely bound up with that of the infested organism.

If there were a kind of diseased structure, the histological elements of which were capable of maintaining a separate and independent existence out of the body, it seems to me that the shadowy boundary between morbid growth and Xenogenesis would be effaced. And I am inclined to think that the progress of discovery has almost brought us to this point already. I have been favoured by Mr. Simon with an early copy of the last published of the valuable "Reports on the Public Health", which, in his capacity of their medical officer, he annually presents to the Lords of the Privy Council. The appendix to this report contains an introductory essay "On the Intimate Pathology of Contagion", by Dr. Burdon Sanderson, which is one of the clearest, most comprehensive, and well-reasoned discussions of a great question which has come under my notice for a long time. I refer you to it for details and for the authorities for the statements I am about to make.

18世纪上半叶末之前搞清楚的。绦虫、囊尾幼虫和吸虫类在很长一段时期内都一直是异源发生论的拥护者捍卫的要塞。实际上，就是在过去的短短30年间，冯西博尔德、范贝内登、洛伊卡特、库彻梅斯特和其他蠕虫学家通过不懈的努力才成功地将上述每一种寄生虫追溯到与其自身存在实际的或者是潜在的相似性的亲本的卵，这些卵通常经过了最奇特的游走和变态。其他方面的研究趋势与此一致。一株植物可能会脱掉球茎，但是这些球茎迟早会产生种子或孢子，它们可以发育成原来的形式。水螅体可以产生水母，长腕幼虫可以长成棘皮动物，而水母和棘皮动物产生的卵又可以长成水螅体或长腕幼虫，因此它们（水螅体和长腕幼虫）只是这些物种生命周期中经历的阶段而已。

但是当我们转向病理学时就会发现它给我们提供了一些与真正的异源发生非常接近的情形。

正如我已提到的，从瓦利斯涅里和雷奥米尔时代起人们就已经知道植物的虫瘿和牛的瘤都是由昆虫引起的。昆虫将自己的卵产在动物或植物的某些部位，于是这些部位就会长成病态结构。另外，对每个人来说都很熟悉的一种经历是，皮肤受到压力后会角质化。植物的虫瘿、动物的瘤及角质化结构都是生命体的一部分，但在某种程度上，它们已经成为了独立的、截然不同的生命体。在一定外界条件的影响下，本应按照总体规划按部就班发育的身体的某些部分，会自行调整，并且按照自己的目的来利用吸收到的营养。

诸如角质化结构和疣等这些无害的产物都有可能渐变成严重的肿瘤，这些肿瘤会由于其大小和引起的机械性阻塞而破坏培育出它们的机体；最终，在那些可怕的结构（即众所周知的肿瘤）中，不正常的生长已经获得了复制和增殖的能力，因此肿瘤只是在形态学上不同于寄生虫，而其生命与被感染的生物体的生命恰好是密切相关的。

如果说存在一种病态结构，即能够脱离机体而独立存在的组织结构，那么对我而言似乎病态生长和异源发生之间的模糊界限就不存在了。我倾向于认为发现的进展差不多已经将我们带到了这一步。西蒙先生在他最后出版的极具价值的《公共卫生报告》的早期版本中对我的观点进行了有力的支持。作为一名卫生官员，他有责任每年向枢密院的高级官员呈交一份这样的报告。这份报告的附录中包含一篇入门短文，由伯登·桑德森撰写，题目是《探秘传染病的病理学》，这篇小文章是长久以来最清晰、最易理解、最合理地讨论了我关注的这一重大问题的作品之一。为了便于我进行接下来的陈述，在此特向你们提一下其中的细节及权威性的引文。

You are familiar with what happens in vaccination. A minute cut is made in the skin, and an infinitesimal quantity of vaccine matter is inserted into the wound. Within a certain time a vesicle appears in the place of the wound, and the fluid which distends this vesicle is vaccine matter, in quantity a hundred or a thousandfold that which was originally inserted. Now what has taken place in the course of this operation? Has the vaccine matter, by its irritative property, produced a mere blister, the fluid of which has the same irritative property? Or does the vaccine matter contain living particles, which have grown and multiplied where they have been planted? The observations of M. Chauveau, extended and confirmed by Dr. Sanderson himself, appear to leave no doubt upon this head. Experiments, similar in principle to those of Helmholtz on fermentation and putrefaction, have proved that the active element in the vaccine lymph is non-diffusible, and consists of minute particles not exceeding 1/20,000 of an inch in diameter, which are made visible in the lymph by the microscope. Similar experiments have proved that two of the most destructive of epizootic diseases, sheep-pox and glanders, are also dependent for their existence and their propagation upon extremely small living solid particles, to which the title of *microzymes* is applied. An animal suffering under either of these terrible diseases is a source of infection and contagion to others, for precisely the same reason as a tub of fermenting beer is capable of propagating its fermentation by "infection", or "contagion", to fresh wort. In both cases it is the solid living particles which are efficient; the liquid in which they float, and at the expense of which they live, being altogether passive.

Now arises the question, are these microzymes the results of *Homogenesis*, or of *Xenogenesis*; are they capable, like the *Torulae* of yeast, of arising only by the development of pre-existing germs; or may they be, like the constituents of a nutgall, the results of a modification and individualisation of the tissues of the body in which they are found, resulting from the operation of certain conditions? Are they parasites in the zoological sense, or are they merely what Virchow has called "heterologous growths"? It is obvious that this question has the most profound importance, whether we look at it from a practical or from a theoretical point of view. A parasite may be stamped out by destroying its germs, but a pathological product can only be annihilated by removing the conditions which give rise to it.

It appears to me that this great problem will have to be solved for each zymotic disease separately, for analogy cuts two ways. I have dwelt upon the analogy of pathological modification, which is in favour of the xenogenetic origin of microzymes; but I must now speak of the equally strong analogies in favour of the origin of such pestiferous particles by the ordinary process of the generation of like from like.

It is, at present, a well-established fact that certain diseases, both of plants and of animals, which have all the characters of contagious and infectious epidemics, are caused by minute organisms. The smut of wheat is a well-known instance of such a disease, and it cannot be doubted that the grape-disease and the potato-disease fall under the same category. Among animals, insects are wonderfully liable to the ravages of contagious and infectious diseases caused by microscopic *Fungi*.

接种疫苗过程中发生的情况你们肯定都很熟悉。在皮肤上开一个微小的口，然后将极微量的疫苗物质注入伤口。一定时间内，在伤口处会出现一个小水泡，其中充满的液体就是疫苗物质，数量上相当于最初注入的一百倍或一千倍。那么在这一过程中发生了什么？是具有刺激性的疫苗物质导致了水泡的产生吗？水泡中的液体是否具有同疫苗物质相同的刺激性？或者疫苗物质中含有有生命的颗粒吗？这些颗粒会在它们被接种的地方生长并繁殖吗？肖沃通过观察确定无疑地回答了这些问题，桑德森博士自己又对此观察结果进行了扩充和证实。与亥姆霍兹进行的发酵和腐烂实验依据的原理相似，这些实验证明了痘浆中的活性成分是不可扩散的，由直径不超过 1/20,000 英寸的微小颗粒构成，可以在显微镜下观察到。相似的实验证明，最具破坏性的两种动物流行病，即绵羊痘和马鼻疽，其存在和传播也依赖于极小的有生命的固体颗粒，这些小的固体颗粒被称为**酵母菌**。动物如果染上了这两种可怕疾病中的任何一种，对其他动物来说都会成为间接传染源和接触传染源，原因与一桶正在发酵的啤酒能够通过"间接传染"或"接触传染"使其他新鲜的麦芽汁也发酵一样。这两个例子中，发生作用的都是固体生命颗粒；这些颗粒漂浮于其中的液体只是提供颗粒生存的环境，除了颗粒以外的物质都是不起作用的。

于是就有了如下的问题：这些酵母菌是**同型生殖**的结果还是**异源发生**的结果？这些酵母菌能够像**圆酵母**一样，只有通过已存在的微生物才可以发育出来吗？或者它就像五倍子的组分一样，是通过对特定条件的操作，由它们所在的有机体的组织经过修饰和个性化作用而产生的结果？这些寄生菌是动物学意义上的，还是仅仅是菲尔绍所谓的"异源发生"的产物？很明显，无论我们从实践的角度还是理论的角度来思考这一问题，它都具有至关重要的意义。通过破坏宿主微生物可以消灭某种寄生菌，但是病理学的产物则只能通过破坏其产生条件来消除。

我觉得这个大问题需要通过分别解决每种发酵病来各个击破，根据相似性可以分为两种方式。我已经详细叙述了病理学修饰方式，它支持酵母菌的异源发生起源；但是我现在必须要提及与此同等重要的方式，即支持这样的传染病颗粒是通过同类型产生同类型的一般过程产生的。

现在非常确定的事实是植物和动物的某些具有间接传染性和接触传染性的流行病学特征的疾病都是由微小的有机体引起的。小麦的黑粉病就是这种病的著名例子，毋庸置疑的是葡萄短节病和马铃薯腐烂病也属于这一类。在动物界，昆虫很容易受到微观真菌引起的间接传染性疾病和接触传染性疾病的攻击。

In autumn, it is not uncommon to see flies, motionless upon a window-pane, with a sort of magic circle, in white, drawn round them. On microscopic examination, the magic circle is found to consist of innumerable spores, which have been thrown off in all directions by a minute fungus called *Empusa muscae*, the spore-forming filaments of which stand out like a pile of velvet from the body of the fly. These spore-forming filaments are connected with others which fill the interior of the fly's body like so much fine wool, having eaten away and destroyed the creature's viscera. This is the full-grown condition of the *Empusa*. If traced back to its earlier stages, in flies which are still active, and to all appearance healthy, it is found to exist in the form of minute corpuscles which float in the blood of the fly. These multiply and lengthen into filaments, at the expense of the fly's substance; and when they have at last killed the patient, they grow out of its body and give off spores. Healthy flies shut up with diseased ones catch this mortal disease and perish like the others. A most competent observer, M. Cohn, who studied the development of the *Empusa* in the fly very carefully, was utterly unable to discover in what manner the smallest germs of the *Empusa* got into the fly. The spores could not be made to give rise to such germs by cultivation; nor were such germs discoverable in the air, or in the food of the fly. It looked exceedingly like a case of Abiogenesis, or, at any rate, of Xenogenesis; and it is only quite recently that the real course of events has been made out. It has been ascertained, that when one of the spores falls upon the body of a fly, it begins to germinate and sends out a process which bores its way through the fly's skin; this, having reached the interior cavities of its body, gives off the minute floating corpuscles which are the earliest stage of the *Empusa*. The disease is "contagious," because a healthy fly coming in contact with a diseased one, from which the spore-bearing filaments protrude, is pretty sure to carry off a spore or two. It is "infections" because the spores become scattered about all sorts of matter in the neighbourhood of the slain flies.

The silkworm has long been known to be subject to a very fatal and infectious disease called the *Muscardine*. Audouin transmitted it by inoculation. This disease is entirely due to the development of a fungus, *Botrytis Bassiana*, in the body of the caterpillar; and its contagiousness and infectiousness are accounted for in the same way as those of the fly-disease. But of late years a still more serious epizootic has appeared among the silkworms; and I may mention a few facts which will give you some conception of the gravity of the injury which it has inflicted on France alone.

The production of silk has been for centuries an important branch of industry in Southern France, and in the year 1853 it had attained such a magnitude that the annual produce of the French sericulture was estimated to amount to a tenth of that of the whole world, and represented a money-value of 117,000,000 of francs, or nearly five millions sterling. What may be the sum which would represent the money-value of all the industries connected with the working up of the raw silk thus produced is more than I can pretend to estimate. Suffice it to say that the city of Lyons is built upon French silk as much as Manchester was upon American cotton before the civil war.

秋天时，苍蝇落在玻璃窗上一动不动，同时还有一种白色的幻圈围绕着它们，这种现象很常见。镜检时，发现该幻圈是由无数的孢子构成的，它们是由被称为蝇单枝虫霉的一种微小的真菌向四面八方散发出来的，这种霉菌的孢子形成的长纤丝就像从苍蝇身体上长出的一堆天鹅绒一样竖立着。这些孢子形成的长纤丝互相连接在一起，就像很多细羊绒一样充满了苍蝇的身体内部，吞食并破坏着苍蝇的内脏。这是蝇疫霉菌发育完成时的状态。如果追溯到其早期生长阶段，在尚且活跃、外观看起来完全健康的苍蝇体内，可以看到它们以微粒子形式漂浮在苍蝇的血液中。这些微粒子靠消耗苍蝇的体质成分来繁殖并延伸成长纤丝；当最终杀死宿主后，它们从宿主的身体中长出来并释放出孢子。健康的苍蝇如果患上这一致命疾病就会像其他有病的苍蝇一样死亡。一位非常杰出的观察者——科恩非常认真地研究了苍蝇中的蝇疫霉菌，却全然没有发现蝇疫霉菌的最小菌体是以何种方式进入到苍蝇体内的。通过培养孢子并不能产生这种细菌，在空气以及苍蝇的食物中也都没有发现这种细菌。这看起来太像是无生源论的例子了，即使不是无生源论，至少也算是异源发生的例子；直到最近才弄清楚这些事件的真正过程。已经确认的是当一个孢子落到苍蝇的身体上时，它就开始萌芽，并在苍蝇皮肤上钻孔而进入到苍蝇体内；到达苍蝇内腔后，排出微小的悬浮粒子，这些粒子就是蝇疫霉菌的最早期阶段。这种疾病是"接触传染性的"，这是因为如果健康的苍蝇与一只伸出含有孢子的长纤丝的患病苍蝇接触的话，那么这只健康的苍蝇肯定也会带走一两个孢子。这种疾病也是"间接传染性的"，这是因为孢子会在被杀死的苍蝇附近散布得到处都是。

我们早就知道，蚕一直遭受着一种非常致命且具有传染性的疾病的困扰，这种疾病被称为**白僵病**。奥杜安通过接种完成了这种疾病的传染。该病完全是由于毛虫体内感染了一种被称为白僵菌的真菌；这种疾病也具有接触传染性和间接传染性，原因与上述苍蝇疾病的相同。但是近年来，在蚕中出现了一种更严重的家畜流行病。我会提到一些事例以使大家了解这种伤害对法国带来了多大的痛苦。

在法国南部，丝绸的生产作为工业体系的一个重要分支已经有几个世纪了。1853 年，法国该产业的年产量已经达到了占据整个世界该产业年产量的 1/10 的程度，这代表着 117,000,000 法郎的货币价值，或者说将近 500 万英镑的货币价值。所有产业中与生丝的生产相关的行业创造的货币价值的总额，我想比我能够估计出来的还要多得多。可以这么说，里昂是靠法国的蚕丝建立起来的，就像内战前的曼彻斯特是靠美国的棉花建立起来的一样。

Silkworms are liable to many diseases; and even before 1853 a peculiar epizootic, frequently accompanied by the appearance of dark spots upon the skin (whence the name of "Pébrine" which it has received), has been noted for its mortality. But in the years following 1853 this malady broke out with such extreme violence, that, in 1858, the silk-crop was reduced to a third of the amount which it had reached in 1853; and, up till within the last year or two, it has never attained half the yield of 1853. This means not only that the great number of people engaged in silk growing are some thirty millions sterling poorer than they might have been; it means not only that high prices have had to be paid for imported silkworm eggs, and that, after investing his money in them, in paying for mulberry-leaves and for attendance, the cultivator has constantly seen his silkworms perish and himself plunged in ruin; but it means that the looms of Lyons have lacked employment, and that for years enforced idleness and misery have been the portion of a vast population which, in former days, was industrious and well to do.

In 1858 the gravity of the situation caused the French Academy of Sciences to appoint Commissioners, of whom a distinguished naturalist, M. de Quatrefages, was one, to inquire into the nature of this disease, and, if possible, to devise some means of staying the plague. In reading the Report made by M. de Quatrefages in 1859, it is exceedingly interesting to observe that his elaborate study of the Pébrine forced the conviction upon his mind that, in its mode of occurrence and propagation, the disease of the silkworm is, in every respect, comparable to the cholera among mankind. But it differs from the cholera, and so far is a more formidable disease, in being hereditary, and in being, under some circumstances, contagious as well as infectious.

The Italian naturalist, Filippi, discovered in the blood of the silkworms affected by this strange disease a multitude of cylindrical corpuscles, each about 1/6,000 of an inch long. These have been carefully studied by Lebert, and named by him *Panhistophyton*; for the reason that in subjects in which the disease is strongly developed, the corpuscles swarm in every tissue and organ of the body, and even pass into the undeveloped eggs of the female moth. But are these corpuscles causes, or mere concomitants, of the disease? Some naturalists took one view and some another; and it was not until the French Government, alarmed by the continued ravages of the malady, and the inefficiency of the remedies which had been suggested, dispatched M. Pasteur to study it, that the question received its final settlement; at a great sacrifice, not only of the time and peace of mind of that eminent philosopher, but, I regret to have to add, of his health.

But the sacrifice has not been in vain. It is now certain that this devastating, cholera-like Pébrine is the effect of the growth and multiplication of the *Panhistophyton* in the silkworm. It is contagious and infectious because the corpuscles of the *Panhistophyton* pass away from the bodies of the diseased caterpillars, directly or indirectly, to the alimentary canal of healthy silkworms in their neighbourhood; it is hereditary, because the corpuscles enter into the eggs while they are being formed, and consequently are carried within them when they are laid; and for this reason, also, it presents the very singular peculiarity of being inherited only on the mother's side. There is not a single one of all the apparently

蚕容易得很多种疾病。有一种经常伴随着皮肤上出现暗色斑点（因此该病被称为"微粒子病"）的特殊家畜流行病，甚至在 1853 年以前，这种病就以其高致死率而著称。在 1853 年之后的几年中，这种疾病的爆发带来了极大的破坏力。1858 年，蚕作物的产量减少到了 1853 年年产量的 1/3；直到过去的一两年，其产量都没有再达到过 1853 年年产量的一半。这意味着不只是为数众多的从事养蚕的人的总收入比过去减少了约 3,000 万英镑；也不仅仅意味着人们不得不花高价进口蚕种，而且在投入这些钱之后，还要付钱买桑叶以及付费进行照料，但是养殖者却总是看到他的蚕死掉了，自己也破产了；而且还意味着里昂的织造业缺少就业机会，数年来，无奈的失业和痛苦曾一度成为那些昔日勤劳而又富裕的人们生活的一部分。

1858 年，形势的严峻性迫使法国科学院指派专员去调查这种疾病的原因，并且如果有可能的话，就设计一些方法来控制疫情。著名的博物学家德卡特勒法热就是专员之一。读到德卡特勒法热先生于 1859 年递交的报告时，我发现了非常有趣的一点，即他对微粒子病的精细研究使他坚信：蚕的这种疾病的发生和传播模式，从任何一方面看，都与人类中出现的霍乱有类似之处。但它又不同于霍乱，迄今为止，它是一种更难对付的疾病，因为它是可遗传的，并且在某些情况下，它是可以接触传染以及间接传染的。

意大利自然学家菲利皮发现，受此奇怪疾病影响的蚕的血液中出现了大量柱状小体，每个小体长约 1/6,000 英寸。莱贝特曾经仔细研究了这些小体，并且将其命名为微粒子。这种疾病在机体内猛烈地发作，这些小体会在体内的每个组织和器官中大量出现，甚至会进入雌蛾尚未发育完全的卵子中。但是，究竟是这些小体还是其附随物引起了疾病？自然学家们各持己见。直到法国政府因持续的疾病攻击而感到惊慌并且所有建议的补救措施都无济于事，于是委托巴斯德来研究这种疾病时，这一问题才最终得以解决；为此也付出了巨大的代价，牺牲了卓越的哲学家的时间和内心的平静，我很遗憾地补充一句，而且也牺牲了他的健康。

但是这些牺牲并没有白费。现在已经确定，这种破坏性类似霍乱的微粒子病就是微粒子在蚕体内生长和繁殖造成的。这种病是可以接触传染及间接传染的，因为造成微粒子病的粒子通过得病蚕的身体直接或间接传递到邻近的健康蚕的消化管中；这种病也是可遗传的，因为这种粒子可以进入到正在形成的卵中，于是在产卵时就被卵子携带；也因为这个原因，这种疾病表现出母系遗传的独特性质。这并不是微粒子病表现出的所有多变、难以解释的现象中的唯一一个，但是这一问题已经通过

capricious and unaccountable phenomena presented by the Pébrine, but has received its explanation from the fact that the disease is the result of the presence of the microscopic organism, *Panhistophyton*.

Such being the facts with respect to the Pébrine, what are the indications as to the method of preventing it? It is obvious that this depends upon the way in which the *Panhistophyton* is generated. If it may be generated by Abiogenesis, or by Xenogenesis, within the silkworm or its moth, the extirpation of the disease must depend upon the prevention of the occurrence of the conditions under which this generation takes place. But if, on the other hand, the *Panhistophyton* is an independent organism, which is no more generated by the silkworm than the mistletoe is generated by the oak or the apple-tree on which it grows, though it may need the silkworm for its development in the same way as the mistletoe needs the tree, then the indications are totally different. The sole thing to be done is to get rid of and keep away the germs of the *Panhistophyton*. As might be imagined, from the course of his previous investigations, M. Pasteur was led to believe that the latter was the right theory; and, guided by that theory, he has devised a method of extirpating the disease, which has proved to be completely successful wherever it has been properly carried out.

There can be no reason, then, for doubting that, among insects, contagious and infectious diseases, of great malignity, are caused by minute organisms which are produced from preexisting germs, or by homogenesis; and there is no reason, that I know of, for believing that what happens in insects may not take place in the highest animals. Indeed, there is already strong evidence that some diseases of an extremely malignant and fatal character to which man is subject, are as much the work of minute organisms as is the Pébrine. I refer for this evidence to the very striking facts adduced by Professor Lister in his various well-known publications on the antiseptic method of treatment. It seems to me impossible to rise from the perusal of those publications without a strong conviction that the lamentable mortality which so frequently dogs the footsteps of the most skilful operator, and those deadly consequences of wounds and injuries which seem to haunt the very walls of great hospitals, and are, even now, destroying more men than die of bullet or bayonet, are due to the importation of minute organisms into wounds, and their increase and multiplication; and that the surgeon who saves most lives will be he who best works out the practical consequences of the hypothesis of Redi.

I commenced this Address by asking you to follow me in an attempt to trace the path which has been followed by a scientific idea, in its long and slow progress from the position of a probable hypothesis to that of an established law of nature. Our survey has not taken us into very attractive regions; it has lain, chiefly, in a land flowing with the abominable, and peopled with mere grubs and mouldiness. And it may be imagined with what smiles and shrugs, practical and serious contemporaries of Redi and of Spallanzani may have commented on the waste of their high abilities in toiling at the solution of problems which, though curious enough in themselves, could be of no conceivable utility to mankind.

如下事实得到了解释，即这种疾病是微生物——微粒子的存在造成的结果。

这就是微粒子病的情况，那么这些情况对于预防该病的方法有什么启示呢？很明显这依赖于微粒子产生的方式。如果它是通过无生源方式产生的或者是通过异源发生方式产生的，那么要想消灭蚕或蛾的这种疾病就一定要阻止能够产生这种疾病的条件的出现。但是另一方面，如果微粒子是一种独立的生物，它不是由蚕产生的，而是像槲寄生是由其所寄生的橡树或苹果树产生的一样，那么尽管它的发育可能与槲寄生需要树的方式相同，它们也会需要蚕，但是所给出的启示就完全不同了。这时，唯一要做的事情就是除去微粒子这种微生物并远离它们。正如由巴斯德以前的研究过程可以想象到的，巴斯德相信后者才是正确的理论；于是，在此理论的指导下，他发明了一种消灭这种疾病的方法，事实已经证明，无论在什么地方，只要准确操作，这种方法都能取得圆满的成功。

没有理由怀疑，昆虫界极具毒害性的接触传染性疾病和间接传染性疾病都是由先前存在的微生物产生的微小生命体引起的，或者由同源生殖引起的；我也知道没有理由去相信昆虫中发生的这些疾病并不会发生在高等动物中。事实上，已经有有力的证据表明人类所遭受的极具危害性和致命性的一些疾病正如微粒子病一样，也是由微生物的作用引起的。我这里提到这个证据是想引出利斯特教授在其论述杀菌处理方法的几篇知名的著作中援引的一些非常令人震惊的事例。我研读这些著作后深信，经常与技术最好的外科医生形影相随的令人惋惜的死亡，以及那些萦绕在各大医院的围墙之内的创伤和损伤造成的致命后果，即使现在，都在摧毁着很多人，其杀伤力超过子弹和刺刀，这些都是由于微小的生物进入了创口，而后不断增加、繁殖引起的；拯救了众生的外科医生将是最能够搞清楚雷迪假说的实践结果的人。

在从一个可能的假说发展成一条确定的自然法则的漫长而缓慢的过程中，科学的思想有其发展所遵循的路径，我在这篇致词一开始就邀请你们随我一起尝试追踪这条路径。我们的研究并没有将我们带入非常引人入胜的领域；简单地说，这是一片充满着令人憎恶的蛆虫和霉菌的土地。可以想象一下，与雷迪以及斯帕兰扎尼同时代的那些实际而严肃的人可能如何微笑着耸耸肩，去评论他们将自己的优秀能力浪费在苦苦寻求这些问题的解决办法上，尽管这些问题本身非常令人好奇，但是对人类而言，它们当时没有任何可以想象到的实用性。

Nevertheless you will have observed that before we had travelled very far upon our road there appeared, on the right hand and on the left, fields laden with a harvest of golden grain, immediately convertible into those things which the most sordidly practical of men will admit to have value—viz., money and life.

The direct loss to France caused by the Pébrine in seventeen years cannot be estimated at less than fifty millions sterling; and if we add to this what Redi's idea, in Pasteur's hands, has done for the wine-grower and for the vinegar-maker, and try to capitalise its value, we shall find that it will go a long way towards repairing the money losses caused by the frightful and calamitous war of this autumn. And as to the equivalent of Redi's thought in life, how can we over-estimate the value of that knowledge of the nature of epidemic and epizootic diseases, and consequently of the means of checking, or eradicating, them, the dawn of which has assuredly commenced?

Looking back no further than ten years, it is possible to select three (1863, 1864, and 1869) in which the total number of deaths from scarlet-fever alone amounted to ninety thousand. That is the return of killed, the maimed and disabled being left out of sight. Why, it is to be hoped that the list of killed in the present bloodiest of all wars will not amount to more than this! But the facts which I have placed before you must leave the least sanguine without a doubt that the nature and the causes of this scourge will, one day, be as well understood as those of the Pébrine are now; and that the long-suffered massacre of our innocents will come to an end.

And thus mankind will have one more admonition that "the people perish for lack of knowledge;" and that the alleviation of the miseries, and the promotion of the welfare, of men must be sought, by those who will not lose their pains, in that diligent, patient, loving study of all the multitudinous aspects of Nature, the results of which constitute exact knowledge, or Science. It is the justification and the glory of this great meeting that it is gathered together for no other object than the advancement of the moiety of science which deals with those phenomena of nature which we call physical. May its endeavours be crowned with a full measure of success!

(**2**, 400-406; 1870)

然而，你会发现我们在自己的路上长途跋涉以前，道路的左右两边都出现了长满金黄稻谷的田地，但是它们会立刻变成最肮脏的只讲实际的人类才承认有价值的那些东西——就是金钱和生命。

17 年来微粒子病对法国造成的直接损失估计至少有 5,000 万英镑；如果我们把巴斯德掌握的雷迪的想法对葡萄酒栽培者和醋酿造者已造成的影响加上去，并且试着计算它的价值时，我们会发现要想弥补发生在这个秋天的由可怕的灾难性战争引起的金钱损失，恐怕将有很长的路要走。至于雷迪对生命的想法的等价物，我们怎么会高估对传染病和家畜流行病的本质的了解，以及检测或者消灭它们的方法的价值呢？可以肯定在这些方面我们已经迎来了曙光。

回顾最近 10 年之内，我们可以选出 3 年（1863 年、1864 年和 1869 年），这 3 年中仅仅由猩红热导致死亡的总共就有 9 万人。曾经远离我们视线的那些被杀死的、受重伤的和变残疾的人又重新出现在我们的面前。我们怎么可以期望，在人类与疾病的历次斗争中，伤亡最惨重的这次斗争中死亡的人数将不会超过以上的数据！你们可能幻想着，就像现在对微粒子病的了解一样，总有一天我们会充分了解这一灾害的本质和原因，我们无辜的患者长期遭受的大屠杀也将结束，然而毫无疑问，我这里向你们陈述的事实一定会令人们失去所有乐观的希望。

因此人类还有一个训诫，即"人类因知识的贫乏而灭亡"；人类痛苦的减轻和福利的提高都要由那些愿意不辞辛劳的人来努力达成，因为他们勤奋、耐心且忠诚地研究大自然各个方面而得到的结果构成了真正的知识或者叫科学。我们聚集在这里，没有其他目的，只是希望能推动研究自然界现象的科学的发展，这是本次大会召开的理由，也是本次大会的光荣。祝愿本次大会圆满成功！

（刘皓芳 翻译；刘力 审稿）

Dr. Bastian and Spontaneous Generation

T. H. Huxley

Editor's Note

Thomas Huxley's address to the British Association in 1870, published earlier in *Nature*, was a critique of the assertion by H. Charlton Bastian (itself made in a long paper in *Nature*) that micro-organisms could be spontaneously generated in sterile water. Bastian had responded to Huxley at length, and here Huxley engages in the next round of battle. His letter, magisterially dismissive of Bastian's claims, is a fine example of Victorian polemic. Huxley was quite correct to be sceptical, but the debate shows that spontaneous generation was still entertained at that time in serious scientific circles. It was eventually ruled out, thanks in part to careful experiments by Louis Pasteur.

I find that the "Address" which it was my duty to deliver at Liverpool, fills thirteen columns of *Nature*. The "Reply" with which Dr. Bastian has favoured you occupies fifteen columns, and yet professes to deal with only the first portion of the "Address". Between us, therefore, I should imagine that both you and your readers must have had enough of the subject; and, so far as my own feeling is concerned, I should be disposed to leave both Dr. Bastian and his reply to the benign and Lethean influences of Time.

But I am credibly informed that there are persons upon whom Dr. Bastian's really wonderful effluence of words weighs as much as if it were charged with solid statements and accurate reasonings; and I am further told that it is my duty to the public to state why such distinguished special pleading makes not the least impression on my mind. With your permission, therefore, I will do so in the briefest possible manner.

The first half of Dr. Bastian's "Reply" occupies seven columns of your number for the 22nd of September. In all this wilderness of words there is but one paragraph which appears to me to be worth serious notice. It is this: —

"In the first place, he does not attempt to deny—he does not even allude to the fact—that *living things may and do arise as minutest visible specks, in solutions in which, but a few hours before, no such specks were to be seen.* And this is in itself a very remarkable omission. The statement must be true or false—and if true, as I and others affirm, the question which Professor Huxley has set himself to discuss is no longer one of such a simple nature as he represents it to be. It is henceforth settled that as far as *visible* germs are concerned, living beings can come into being without them."

If I did not allude to the assertion which Dr. Bastian has put in italics—it is because it bears absurdity written upon its face to any one who has seriously considered the

巴斯蒂安博士与自然发生学说

赫胥黎

编者按

早些时候,《自然》发表过托马斯·赫胥黎于 1870 年向英国科学促进会作的报告。该报告批评了查尔顿·巴斯蒂安主张的无菌水中可能会自发产生微生物的观点(这是在一篇发表于《自然》上的长篇论文中提出的)。巴斯蒂安详尽地回复了赫胥黎的意见,于是赫胥黎又发起了新一轮辩论。赫胥黎的文章权威地驳斥了巴斯蒂安的观点,可以说是维多利亚时代学术争辩的典范。赫胥黎的观点是非常正确无可怀疑的,不过,通过他们之间的辩论可以看到,当时在严肃的学术圈内也还有人坚持自然发生学说的观点。一定程度上来说,是因为路易·巴斯德的一系列精细实验,自然发生学说才最终被人们抛弃。

我发现,我因个人职责而在利物浦发表的"致词"居然占据了《自然》13 栏的版面。巴斯蒂安博士支持贵杂志而作出的"回复"又占据了 15 栏的版面,但却声称只针对我的"致词"中的第一部分进行了讨论。因此,我应该想象得出《自然》及其读者对于我们俩之间讨论的话题已经有了充分的了解;另外,就我个人情感而言,我愿意让巴斯蒂安博士和他的回复给时代留下一个友好的、能够遗忘过去的影响。

但是我据可靠消息得知,有一些人认为巴斯蒂安博士的精彩论断就像充满了可靠的论述和精准的推理一样有分量;我还被进一步告知,我有责任对公众说明为什么这么卓越非凡的抗辩却没有给我留下一丝一毫的印象。因此,如果您许可,我将尽可能以最简洁的方式在此做一说明。

在贵杂志 9 月 22 日那一期上,巴斯蒂安博士的"回复"的前半部分就占了 7 栏。在我看来,这些长篇大论中只有一段值得认真注意一下。该段原文如下:

"首先,他并没有试图否认,甚至没有提及如下事实——**溶液中可能并且确实出现了生物,那是一些极小的可见斑点,而在几个小时前这些溶液中还看不到这种斑点。**这本身是一个非常值得注意的遗漏,因为该陈述肯定非对即错。如果是对的,那么就像我和其他人断言的那样,赫胥黎教授自己着力讨论的问题并不像他描绘出来的那样简单。从今以后我们可以确定,生物可以在不存在**可见细菌**的情况下产生。"

假如我没有提及巴斯蒂安博士那些突出表示的论断——那是因为对于任何一个认真考虑了显微观察条件的人来说,这种论断都相当于在自己脸上写下荒谬的印记。

conditions of microscopic observation. I have tried over and over again to obtain a drop of a solution which should be optically pure, or absolutely free from distinguishable solid particles, when viewed under a power of 1,200 diameters in the ordinary way. I have never succeeded; and, considering the conditions of observation, I never expect to succeed. And though I hesitate to speak with the air of confident authority which sits so well on Dr. Bastian, I venture to doubt whether he ever has prepared, or ever will prepare, a solution, in a drop of which no "minutest visible specks" are to be seen by a careful searcher. Suppose that the drop, reduced to a thin film by the cover-glass, occupies an area 1/3 of an inch in diameter; to search this area with a microscope in such a way as to make sure that it does not contain a germ 1/40,000 of an inch in diameter, is comparable to the endeavour to ascertain with the unassisted eye whether the water of a pond, a hundred feet in diameter is or is not absolutely free from a particle of duckweed. But if it is impossible to be sure that there is no germ 1/40,000 of an inch in diameter in a given fluid, what becomes of the proposition so valuable to Dr. Bastian that he has made your printer waste special type upon it?

I now pass to the second part of the "Reply", which, though longer than the first, is really more condensed, inasmuch as it contains two important statements instead of only one.

The first is, that Dr. Bastian has found *Bacterium* and *Leptothrix* in some specimens of preserved meats. I should have been very much surprised if he had not. If Dr. Bastian will boil some hay for an hour or so, and then examine the decoction, he will find it to be full of *Bacteria* in active motion. But the motion is a modification of the well-known Brownian movement, and has not the slightest resemblance to the very rapid motion of translation of active living *Bacteria*. The *Bacteria* are just as dead as those which Dr. Bastian has seen in the preserved meats and vegetables; and which were, I doubt not, as much put in with the meat, as they are with the hay, in the experiment to which I invite his attention.

The second important statement in the second part of the "Reply" is: —

"Professor Huxley is inclined to believe that there has been some error about the experiments recorded by myself and others."

In this I cordially concur. But I do not know why Dr. Bastian should have expressed this my conviction so tenderly and gently as regards his own experiments; inasmuch as I thought it my duty to let him know both orally and by letter, in the plainest terms, six months ago, not only that I conceived him to be altogether in the wrong, but why I thought so.

Any time these six months Dr. Bastian has known perfectly well that I believe that the organisms which he has got out of his tubes are exactly those which he has put into them; that I believe that he has used impure materials, and that what he imagines to have been

我曾经用普通的可以放大 1,200 倍的显微镜进行观察，反复尝试以期得到一滴视觉上纯净的溶液，或完全不含可辨别固体颗粒的溶液。但都失败了；考虑到这种观察条件，我也没指望过会成功。尽管我不愿像巴斯蒂安博士那样以自信的权威口吻讲话，但我还是冒昧地对其表示怀疑——他是否已经制备好了，或者将会制备出一种溶液，即使是很仔细的观察者也看不到这样的一滴溶液中有"极小的可见斑点"。假如通过盖玻片将这滴溶液压成薄膜后，只占据直径为 1/3 英寸的区域，那么以这种方法用显微镜搜索这一区域，以确定该区域是否含有直径为 1/40,000 英寸的细菌，就如同仅靠肉眼来确定直径为 100 英尺的一方池塘的水中是否绝对不含一片浮萍一样。但如果不可能确定给定的液体中不含直径为 1/40,000 英寸的细菌的话，那么巴斯蒂安博士的哪些主张如此有价值以至你们的印刷机肯为其浪费呢？

现在我要开始讨论"回复"的第二部分，尽管这部分比第一部分长，但是其实这部分内容已经更加精简了，因为它包含的是两个重要的陈述而不是一个。

第一个重要陈述是，巴斯蒂安博士已经在某些腌肉样本中发现了杆菌和纤毛菌。如果他没有发现的话，那就太令我惊讶了。如果巴斯蒂安博士将一些干草煮沸一小时左右，然后再观察煮出来的草汁，那么他将发现草汁中充满了活跃运动着的杆菌。但是这种运动只是著名的布朗运动的一种变体，与活跃的、有生命的杆菌的快速运动方式没有一点相像之处。这些杆菌与巴斯蒂安博士看到的腌肉和咸菜中的杆菌一样都是死的。并且，我相信这些细菌是在实验中同肉一起放进去的，就像它们随干草一起被放进去一样，对于这一点，我希望能够引起他的注意。

"回复"第二部分的第二个重要陈述是：

"赫胥黎教授倾向于认为由我自己和其他人记录的实验中存在错误。"

我由衷地赞成这一点。但是我不知道巴斯蒂安博士为什么将我对他的实验的这种指责表达得如此温和婉转；因为我觉得通过口头和书面形式让他知道这一点是我的职责，所以 6 个月前，我就以最简明的措辞让他不仅知道他是完全错误的，而且告诉了他我为什么这么认为。

这 6 个月里，无论何时，巴斯蒂安博士都已经清楚地知道，我认为他试管中的生物其实正是被他放进试管中去的那些；我还认为他使用了不纯的物质，而他所想

the gradual development of life and organisation in his solutions, is the very simple result of the settling together of the solid impurities, which he was not sufficiently careful to see, in their scattered condition when the solutions were made.

Any time these six months Dr. Bastian has known why I hold this opinion. He will recollect that he wrote to me asking permission to bring for my examination certain preparations of organic structures, which he declared he had clear and positive evidence to prove to have been developed in his closed and digested tubes. Dr. Bastian will remember that when the first of these wonderful specimens was put under my microscope, I told him at once that it was nothing but a fragment of the leaf of the common Bog Moss (*Sphagnum*); he will recollect that I had to fetch Schacht's book "Die Pflanzenzelle", and show him a figure which fitted very well what we had under the microscope, before I could get him to listen to my suggestion; and that only actual comparison with *Sphagnum*, after he had left my house, forced him to admit the astounding blunder which he had made.

To any person of critical mind, versed in the preliminary studies necessary for dealing with the difficult problem which Dr. Bastian has rashly approached—the appearance of a scarlet geranium, or of a snuff-box, would have appeared to be hardly more startling than this fragment of a leaf, which no one even moderately instructed in vegetable histology could possibly have mistaken for anything but what it was; but to Dr. Bastian, agape with speculative expectation, this miracle was no wonder whatever. Nor does Dr. Bastian's chemical criticality seem to be of a more susceptible kind. He sees no difficulty in the appearance of living things in potash-alum, until Dr. Sharpey puts the not unimportant question, whence did they get their nitrogen? And then it occurs to him to have the alum analysed and he finds ammonia in it.[*]

And as to the elementary principles of physics—in his last communication to you, Dr. Bastian shows, that he is of opinion that water in a vessel with a hole in it, from which the steam freely issues, may be kept at a temperature of "230° to 235° F for more than an hour and a half."[†] I hope that Professor Tyndall, whom Dr. Bastian scolds as authoritatively and as unsparingly as he does me, will take note of this revolutionary thermotic discovery, in the next edition of his work on Heat.

It is no fault of mine if I am compelled to write thus of Dr. Bastian's labours. I have been blamed by some of my friends for remaining silent as long as I have done concerning them. But when, because I have preserved a silence, which was the best kindness I could show to Dr. Bastian, he presumes to accuse me publicly of unfairness, and to tell your readers that my Address "is calculated to mislead" them, I have no alternative left but to give them the means of judging of the competency of my assailant.

(**2**, 473; 1870)

T. H. Huxley: Jermyn Street, Oct. 10.

---

[*] See *Nature*, No. 36, p. 198.

[†] Ibid, No. 48, p. 433.

象的那些在他的溶液中经历了生命和组织结构的逐步发育的东西，其实就是配制溶液时以分散状态一起放入试管中的固体杂质，他只是没有足够认真地去观察而已。

这 6 个月里，无论何时，巴斯蒂安博士都应该已经清楚我为什么会持有这种观点。他应该记得，他曾写信请我对他的某些具有有机结构的制品进行检查，并声称有清楚确凿的证据证明这些有机物是在他的密闭消化管中得到的。他也应该记得，当这些极好的样品中的第一个被放在我的显微镜下时，我立刻告诉他这只不过是普通泥炭藓的叶子碎片而已；他还应该记得，当我不能让他信服我的意见时，我不得不找出沙赫特的《植物细胞》一书，给他指出其中的一幅图画，那与我们在显微镜下所看到的完全吻合；离开我家后，他又将自己的实验结果与真实的泥炭藓进行了比较，才被迫承认自己所犯的惊人错误。

对于任何具有批判思维的人，在对待巴斯蒂安博士鲁莽提出的难题时，都会谨慎地进行必要的初步研究——出现一株绯红色的天竺葵或者一个鼻烟盒恐怕都不会比这片叶子碎片更令人吃惊，并且，即使是仅仅受过中等植物组织学教育的人，也不可能将它误认为其他东西；但是对于这位对根据推理而得出的预期都会瞠目结舌的巴斯蒂安博士来说，这样的奇事也就不足为奇了。巴斯蒂安博士对化学制品的知识好像也不是那么敏感。他不费吹灰之力就在明矾中发现了生物，直到沙比博士提出了如下重要问题，即，它们是从哪里得到所需的氮气的？这时候他才开始分析明矾，并且又从中发现了氨。*

至于物理学的基本原理——在与贵刊的最后一次通信中，巴斯蒂安博士表明，他的观点是对于有孔容器中的水，由于水蒸气可以从小孔处自由排出，所以水温可以保持在"华氏 230~235 度之间达一个半小时以上。"†巴斯蒂安博士曾像斥责我一样以一种权威的口吻毫不宽容地斥责过廷德尔教授，我希望廷德尔教授能够在他下一项关于热量的研究工作中留意这一革命性的热学发现。

我不得不将巴斯蒂安博士的工作以这种方式写出来，这并非我的过错。我的一些朋友已经责备过我，埋怨我自从关注到这些工作以来就一直在保持沉默。我已经保持了沉默，这是我对巴斯蒂安博士表现出的最友好的态度，但是当他不正当地妄自公开控诉我，并且告诉《自然》的读者我的致词"旨在欺骗误导"他们的时候，除了给读者们判断攻击我的人的能力素质的方法之外，我别无选择。

（刘皓芳 翻译；刘力 审稿）

---

* 《自然》，第 36 期，第 198 页。

† 同上，第 48 期，第 433 页。

The Evolution of Life: Professor Huxley's Address at Liverpool

H. C. Bastian

Editor's Note

This is the fifth part of an exchange between H. Charlton Bastian and Thomas Huxley, president of the British Association, on the topic of the spontaneous generation of life in water. Bastian supported the idea; Huxley did not. Here Bastian continues to assert his case, with perhaps more rhetoric than real evidence. Huxley was eventually proved correct; but a point of interest in Bastian's reply here is his argument over whether tiny particles seen in the water are self-propelled and thus living, or just moving by Brownian motion caused by random impacts of water molecules. Brownian motion was soon to emerge as central to the molecular theory of matter.

BELIEVING that readers of *Nature* can feel no interest in the extended personalities with which Prof. Huxley almost fills his letter this week, and believing also that such matters are little worthy of occupying your columns, I shall only allude to that part of his letter which contains statements having a scientific bearing.

The distinct issue raised in my experiments was, were *living* things to be found in the fluids of my flasks? If so, such living things must either have braved a higher degree of heat than had been hitherto thought possible, or else they had been evolved *de novo*.

The effect of the very high temperature upon pre-existing living things, which were purposely exposed thereto, was shown by their complete disorganisation in an experiment which is recorded in *Nature*, No. 37, p. 219, and to this I would especially direct Prof. Huxley's attention.

Prof. Huxley advances an explanation of the mode of origin of the distinct fungi, bearing masses of fructification (*Nature*, No. 36, figs. 12, 14, and 17) and of the inextricably tangled coils of spiral fibres (figs. 13 and 15) found in my flasks after exposure to temperatures at and beyond Pasteur's standard of destructive heat; his theory is entirely novel, apparently extemporised for the occasion, and is very startling. He says, and in justice to Prof. Huxley I quote the passage in full, "Any time these six months Dr. Bastian has known perfectly well that I believe that the organisms which he got out of his tubes are exactly those which he has put into them; that I believe that he has used impure materials, and that what he imagines to have been the gradual development of life and organisation in his solution is *the very simple result of the settling together of the solid impurities; which he was not sufficiently careful to see when in their scattered condition when the solutions were made.*"

生命的进化：赫胥黎教授在利物浦的演说

巴斯蒂安

编者按

这是查尔顿·巴斯蒂安和英国科学促进会的主席托马斯·赫胥黎就生命是否能在水中自然生成这一论题进行的一次激烈争论的第五部分。巴斯蒂安赞同这一论点，而赫胥黎则持反对意见。在这篇文章中，巴斯蒂安仍然坚持自己的观点，但更多的是卖弄言辞技巧，而不是提出真凭实据。人们最终认识到赫胥黎的看法是正确的；但令人感兴趣的是巴斯蒂安在回复中提到的问题：在水中看到的小颗粒的运动究竟是有生命物体的自发运动，还是由水分子的无规则碰撞导致的布朗运动？此后不久布朗运动就成了物质分子理论的中心话题。

赫胥黎教授本周的信件中几乎都是些无关的人身攻击，我相信《自然》的读者们对此毫无兴趣，我也相信，这种事不值得占用贵刊的版面，因此我将只提及他信中具有科学依据的部分言论。

我的实验引发的一个很清楚的问题是：在我的长颈瓶中的液体里会不会发现**有生命**的物质？如果会的话，那么这种生物要么一定能够承受高于我们目前所认为生物可能承受的高温，要么它们就是在实验过程中新进化出来的。

将溶液中已存在的生物故意暴露在高温下以研究其影响的实验已经被报道过了（《自然》，第 37 期，第 219 页），结果表明生物组织被完全破坏了。我特别想请赫胥黎教授注意这一点。

赫胥黎教授提出了一种对不同真菌的起源模式的解释，当我按照巴斯德的破坏性热量标准以及高于这个标准的温度处理真菌时，我发现它们在长颈瓶中产生了大量实体（《自然》，第 36 期，图 12、14 和 17）及纠缠在一起的螺旋纤维团（图 13 和 15）；而赫胥黎教授的理论完全令人耳目一新，很明显那是当场的即兴演说，相当令人吃惊。为了对赫胥黎教授公正起见，我将完整引用该段，他说"这 6 个月里，无论何时，巴斯蒂安博士都已经清楚地知道，我认为他试管中的生物其实正是被他放进试管中去的那些；我还认为他使用了不纯的物质，而他所想象的那些在他的溶液中经历了生命和组织结构的逐步发育的东西，**其实就是配制溶液时以分散状态一起放入试管中的固体杂质，他只是没有足够认真地去观察而已。**"

Now, although it was quite true that minute portions of *Sphagnum* leaf were found in two unpublished experiments, it seems very marvellous that on this slender foundation Prof. Huxley should hazard such a purely imaginative and unprecedented hypothesis as to the mode of production of fungi.

I have, moreover, not been able to see why the occurrence of the incident to which he refers should make him repudiate a number of experiments in which *unmistakably living things* were found in fluids from hermetically sealed flasks after these and their contents had been exposed to temperatures higher than those which living things are known to be capable of resisting.

Following a precept more honoured in dialectics than in science, Prof. Huxley has attacked his opponent rather than the arguments which he affects to destroy. He objects to only one passage in my "Reply", and this he thinks was not worthy of the special type in which it was printed; and yet, notwithstanding its special type, I can only conclude from his reply that Prof. Huxley has failed to appreciate its meaning. My words were: "Living things may and do arise as minutest visible specks, in solutions in which, but a few hours before, no *such* specks were to be seen." The word which now alone stands in italics was ignored by Prof. Huxley. I had no wish to tell him that certain refractive particles, or foreign bodies, might not be visible in the thin film of fluid to which I referred. I alluded to the gradual and equable development of living specks throughout a fluid containing no apparent germs. His retort that some unobserved visible germs might have become centres of development is a *contre-sens*. It does not apply to the gradual appearance of myriads of *equally diffused* motionless particles in a motionless film of fluid.

The very authoritative tone which Prof. Huxley has lately assumed in his remarks concerning Brownian movements and those of living organisms, fails to impress me very much. His knowledge about these movements, as I have good reason to know, is of quite recent growth. Movements which, in the month of March of the present year, Prof. Huxley did not regard as Brownian, he now does believe to have been of this nature. If he is now right, what value is to be set upon his knowledge of Brownian movements six months ago; and what guarantee have we that in another six months Prof. Huxley may not again take a different view?

Let me assure Prof. Huxley, however, that the duty which he is "credibly informed" he owes to the public remains still undischarged. I protested against his "Address" on scientific grounds which are fully stated, and those who have read my protest will see that Prof. Huxley cannot dispose of the question really at issue by recounting any mistakes of mine, whether real or imaginary. If, as I believe, he has failed to give any worthy or serious view of the question, this could have been in no way necessitated by a disbelief, however strong, in my experiments. The labours of Profs. Wyman, Mantegazza, and Cantoni had already taken the question into regions never attained by M. Pasteur, and therefore they demanded a fair consideration. Is Prof. Huxley, in his capacity of President of the British Association, warranted in ignoring their labours, and therefore in misrepresenting the

尽管现在我已经在两个尚未发表的实验中发现了泥炭藓叶子的一小部分，这确实是事实，但是不可思议的是，赫胥黎教授居然凭借不实依据而对真菌产生模式得出一个纯粹想象的和史无前例的假说。

此外，我并不明白他提到的那些事件的发生为什么会令他否认许多如下情况的实验，在那些实验中，当密封的长颈瓶及其内部容纳的物质被加热到高于已知的生物能够承受的温度时，仍然可以在其中的液体里发现**生物，这是确定无疑的**。

赫胥黎教授更推崇的是辩证法而非科学原则，即他攻击的是他的反对者而非他佯装想要驳倒的论点。他只反对我的"回复"中的一个段落，而且他认为这部分不值得出现在用来发表它的特殊版面上；然而，且不论是不是特殊版面，我从他的回复中得到的唯一结论是：赫胥黎教授根本不理解它的意义。我的话如下："溶液中可能并且确实出现了生物，那是一些极小的可见斑点，而在几个小时前这些溶液中还看不到**这种**斑点。"此句中突出强调的词被赫胥黎教授忽视了。我原本不想告诉他在我所用的液体薄膜上可能看不到某些折光颗粒或者外来物质。我说的是不含有明显的微生物的液体中有生命的颗粒可以逐渐、温和地进行发育。他在反驳中提到的某些未观察到的可见微生物可能变成了发育的主要物质，是与本来的意义相反的。这并不适用于一张静止的液体薄膜上逐渐出现大量**同等分散**的静止颗粒的情况。

赫胥黎教授在他最近发表的关于布朗运动和生物的布朗运动的言论中所使用的权威腔调并没有给我留下太深的印象。他只是最近才对这些运动有所了解，对于这一点我很清楚。今年3月赫胥黎教授还不认为是布朗运动的那些运动，他现在却相信它们具有这样的属性。如果他现在是正确的，那么对于6个月之前他对布朗运动的了解，又该赋予什么样的价值呢？另外，怎么能保证再过6个月赫胥黎教授不会又有不同的观点呢？

不过，让我来向赫胥黎教授保证，他依然承担着"据可靠消息得知"的对公众所负有的职责。我对他的"致词"提出的抗议是站在科学立场上的，这已经表述得很充分了，那些读过我的抗议的人会看到，就算赫胥黎教授重提我的那些真实的或者想象中的错误，也不能逃避那些存在争议的问题。因为我相信，如果他无法给出关于此问题的任何有价值的或严肃的观点，那么就无法强迫别人不相信我的实验。怀曼教授、曼泰加扎教授和坎托尼教授的工作已经将该问题归入到巴斯德先生从未涉足过的领域之内，他们也需要公平的对待。难道在英国科学促进会主席的能力范围内，赫胥黎教授有权忽视他们的劳动，乃至可以对这一问题在科学领域的当前状

present state of science on the subject, because, owing to two errors among my many experiments, he declares himself to have altogether lost faith in my skill or capacity as an investigator? The answer cannot be doubtful. Is it, again, consistent with his high responsibilities that he should pervert the real issues, and should do a grave injustice to others, in order that he might preserve a "silence" which should be his "best kindness" to me? Let me tell Prof. Huxley that I repudiate such "kindness", as any honest man would who is simply seeking after truth, and relegate it to the same regions as I would that indescribable air of restrained omniscience whereby he endeavours to crush arguments and facts, to which he altogether fails to reply.

(**2**, 492; 1870)

H. Charlton Bastian: University College, Oct. 17.

态加以曲解吗？他有权可以因为我众多实验中存在的两个错误就声称自己对我作为一位研究者的技术和能力完全失去信任了吗？答案是不容置疑的。那么他曲解真正的争议、对其他人严重不公，保持"沉默"来作为他"最友好的态度"，这些难道与他的高度责任心是相符的吗？让我来告诉赫胥黎教授，我与任何只追求真理的诚实的人一样，拒绝这样的"友好"，我认为他是试图凭借他有限的全知全能来造成莫可名状的氛围，借以不遗余力地打压争论和事实，因为他根本无法对这些作出回应。

（刘皓芳 翻译；刘力 审稿）

The Descent of Man

P. H. Pye-Smith

Editor's Note

"If Mr. Darwin had closed his rich series of contributions to Science by the publication of the 'Origin of Species', he would have made an epoch in Natural History like that which Socrates made in philosophy, or Harvey in Medicine." Such extravagant plaudits opened physiologist Philip Henry Pye-Smith's two-part essay on Darwin's two-volume work in which he presented evidence for the biological origin of humanity, attempted to trace our evolution from lower forms of life, and then advanced his pioneering ideas on sexual selection as a potent driver of evolutionary change: "But though in the lists of Love the battle is often to the strong," Pye-Smith writes, "even more frequently it is to the beautiful."

The Descent of Man, and Selection in relation to Sex.
By Charles Darwin, M. A., F. R. S., &c.
In two volumes, pp. 428, 475. (Murray, 1871)

I

IF Mr. Darwin had closed his rich series of contributions to Science by the publication of the "Origin of Species", he would have made an epoch in Natural History like that which Socrates made in philosophy, or Harvey in medicine. The theory identified with his name has stimulated ethnological and anatomical inquiries in every direction; it has been largely adopted and followed out by naturalists in this country and America, but most of all in the great work-room of modern science, whence a complete literature on "Darwinismus" has sprung up, and there disciples have appeared who stand in the same relation to their master as Muntzer and the Anabaptists did to Luther. Like most great advances in knowledge, the theory of Evolution found everything ripe for it. This is shown by the well-known fact that Mr. Wallace arrived at the same conclusion as to the origin of species while working in the Eastern Archipelago, and scarcely less so by the manner in which the theory has been worked out by men so distinguished as Mr. Herbert Spencer and Prof. Haeckel. But it was known when the "Origin of Species" was published, that instead of being the mere brilliant hypothesis of a man of genius, of which the proofs were to be furnished and the fruits gathered in by his successors, it was really only a summary of opinions based upon the most extensive and long-continued researches. Its author did not simply open a new province for future travellers to explore, he had already surveyed it himself, and the present volumes show him still at the head of his followers. They are written in a more popular style than those on "Animals and Plants under Domestication", as they deal with subjects of more general interest; but all the great

人类的由来

编者按

生理学家菲利普·亨利·派伊-史密斯发表了一篇由两部分组成的文章，对达尔文的两卷本著作《人类的由来及性选择》给予了极高的称赞。史密斯在开篇写道，"如果《物种起源》为达尔文先生丰富的科学著述画上句号的话，那么他在自然史上的里程碑作用就如同苏格拉底在哲学领域、哈维在医学领域一样。"达尔文在《人类的由来及性选择》中展示了关于人类的生物学起源的证据，试图寻找人类从低等生命形式进化而来的足迹，并开创性地提出了性选择是人类进化的驱动力的观点。派伊-史密斯认为，"尽管在爱的竞技场上，胜利经常属于强者，但更多时候是美丽者获胜。"

《人类的由来及性选择》
作者：查尔斯·达尔文，文学硕士，皇家学会会员。
共两卷，428 页，475 页。（默里，1871 年）

I

如果《物种起源》为达尔文先生丰富的科学著述画上句号的话，那么他在自然史上的里程碑作用就如同苏格拉底在哲学领域、哈维在医学领域一样。这项用他的名字命名的理论激发了学者对人种学和解剖学各方面问题的讨论，得到了我国和美国的自然学家的广泛接受和推崇。但最主要的是在现代科学的大工作室里诞生了关于"达尔文主义"的一套完整学说，达尔文主义的信奉者们与达尔文的关系，就像闵采尔和再浸礼论者与路德之间的那种复杂关系。正如知识方面的许多巨大进展一样，进化论在各方面都具备了成熟的条件。可证明这一点的著名事例有，华莱士先生在东部群岛工作的时候对物种起源问题得到了同样的结论；著名的赫伯特·斯宾塞先生和海克尔教授也提出了十分相似的理论。但《物种起源》出版时，它还只是各种基于广泛且长期的研究而得出的观点的综述，并非某个天才的杰出理论假说；该假说的证据将由后来的追随者提供，最终成果也由他们获得。实际上，其作者不只是简单地为未来的探索者开辟了一个新的研究领域，他本人已经亲自进行了调查，而且这两卷著作表明他至今仍然先于他的追随者。该书的书写风格比《动物和植物在家养下的变异》更通俗，因为讨论了更多人们普遍感兴趣的话题。但是，在研究的勤勉与准确、构建假说的能力以及判断的公正性这些重要方面，则一如达尔文先

127

qualities of industry and accuracy in research, of fertility in framing hypotheses, and of impartiality in judgment, are as apparent in this as in Mr. Darwin's previous works. To one who bears in mind the too frequent tone of the controversies these works have excited, the turgid rhetoric and ignorant presumption of those "who are not of his school—or any school", and the still more lamentable bad taste which mars the writings of Vogt and even occasionally of Haeckel, it is very admirable to see the calmness and moderation (for which philosophical would be too low an epithet) with which the author handles his subject. If prejudice can be conciliated, it will surely be by a book like this.

It consists of two parts. The first treats of the origin of man, his affinities to other animals, and the formation of the races (or sub-species) of the human family. Besides the obvious interest to all Mr. Darwin's readers of a discussion on the subject of their "proper knowledge", naturalists will find the detailed application of the laws of natural selection to a single common and well-known species an excellent test of their truth and illustration of their difficulties. It is in dealing with the latter, which are never extenuated or passed by, that the author introduces the subject of sexual selection. This is dealt with in the second part, which forms more than two-thirds of the work, and that not only as it affects man, but in its entire range. Reserving this division of the book for a future article, we will endeavour here to give a summary of the course of argument in the earlier portion.

The author, justly assuming that the general principles of natural selection are admitted by all who have examined the evidence on the subject, with the exception of many of "the older and honoured chiefs in natural science", proceeds at once to discuss the proofs of the origin of man considered apart from those affecting all animals in common. The first group of facts adduced to show his kinship with other forms of animal life, relate to the strict correspondence of his bodily parts with those of other mammalia. To say that these structures are the same because they have the same uses, is untrue, for many of them have no use in the sense of active function, and we constantly find the same structures in animals turned to different uses, and the same uses subserved by different structures. To say that the bodies of men and animals are alike because they are formed on the same plan, or because they are the realisation of the same idea in the Creator, is true enough, but is beside the mark; for natural science inquires how or by what steps these things have become so, not why and from what first cause. If one sees two men very much alike, one naturally supposes that they are brothers; if they are rather less so, they may be cousins; if only agreeing in general characters, we recognise them as at least belonging to the same race or nation; and so, when the facts to be accounted for are once ascertained, nothing but prejudice or repugnance to acknowledge our true relations, can explain why it was so long before naturalists admitted the hypothesis of community of origin between men and other animals. What is called the Darwinian theory accounts for the way in which diversities have arisen, and thus has converted an apparently obvious hypothesis into a well-grounded theory. But in expounding the likeness between men and animals, the author does not confine himself to anatomical structure, but shows how the same resemblance extends to the laws of disease, the distribution of parasites, and other minute particulars.

生先前的著作。如果一个人牢记这些著作所引发的太过频繁的争议，"非达尔文学派或其他任何学派的人"的浮夸修辞和愚昧假设，以及更令人遗憾的损害沃格特作品，甚至有时是海克尔作品的低品位论调，那么作者在处理书中问题时所表现出的镇定与谦和（对此使用冷静这个词远远不够）就值得人们敬仰了。如果偏见可以被纠正，那肯定是像这样的书才能够做到。

这本书包括两部分。第一部分讲述人类的起源、人类与其他动物的亲缘关系以及人科的种（亚种）的形成。除了讨论他们自己的"专业知识"以引起所有达尔文先生的读者的兴趣之外，自然学家也将通过对真实性的精确检验和对难点问题的图解说明来找到自然选择法则在一种简单而又为人所熟知的物种上的具体应用。在处理从未被藐视或忽视的自然选择的具体应用时，作者引入了性选择的观点。这在文章的第二部分中谈到，内容超过整个著作的 2/3，它不仅谈论人类，还涉及其生活环境。该书的这部分内容我们会在以后的文章中谈及，这里我们将尽可能对前一部分提到的争论进行总结。

作者理所当然地认为，自然选择的总法则得到了除一些"长者和受人尊敬的领导们"之外的所有分析过自然选择论据的人的承认，因此，他只讨论人类起源的直接证据，而没有谈论对所有动物都有影响的共同因素。第一组用来说明人类与其他动物关系的亲缘证据与人类和其他哺乳动物在身体结构上的严格对应有关，认为功能相同从而结构也相同，这种看法是不正确的，因为这些结构中有很多并没有实际的功能，而且我们也经常能在动物中发现，相同的结构实现不同的功能，而相同的功能又可以由不同的结构来完成。人类的身体结构与动物相似是因为他们是由同一方案设计形成的，或者说是按照造物主的同一个想法实现出来的，这种说法有足够的理由，但无法得到检验，因为自然科学要弄清楚的是生物如何或者说通过什么步骤变成现在的样子，而不是为何这样以及最初的起因。如果看到两个人长得非常相像，很自然就会认为他们是兄弟；如果不是特别相像，则可能是堂（表）兄弟；如果只是一般特征相似，则认为他们至少属于同一种族或民族。因此，一旦要说明的事物得到证实，只有在我们承认真正起源时出现的偏见或嫌恶才能解释为何过了那么久自然学家才承认人类与其他动物具有共同起源的假说。达尔文理论解释了多样性是如何产生的，并将显而易见的假说转化为证据充分的理论。在详细说明人类与动物的相似性时，作者并没有将自己局限于解剖结构方面，而是充分阐述了在疾病、寄生虫的分布及其他细节方面同样具有的相似之处。

The next argument brought forward is the equally familiar one drawn from the likeness of the human embryo to that of other vertebrata. Then follows an account of the rudimentary organs in man, which in all other species are justly held among the most important indications of affinities. One such rudiment is mentioned which is, we believe, hitherto unrecorded. It is a slight projection of the rim of the helix of the auricle, which would correspond when unfolded to the point of an erect ear. (See illustration.) This occasional abnormity may, perhaps, be recognised by future anatomists as the *Angulus Woolnerii* after its first observer.

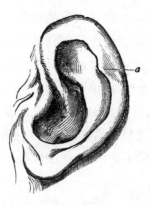

Fig. 1. Human Ear, Modelled and Drawn by Mr. Woolner.
a. The projecting point.

In the second chapter Mr. Darwin shows that a consideration of the mental faculties of man, including the use of language, which has been held the greatest difficulty to admitting his kinship to other animals, may rather strengthen than weaken the arguments derived from his bodily structure. Memory and curiosity, jealousy and friendship, and even the power of correct reasoning, and of communication by sounds, are shown to belong to many of the lower animals, while the faculty of reflection and self-consciousness, and "the ennobling belief in the existence of an Omnipotent God", cannot be ascribed to the lowest tribes of the human family. At the same time it is argued that the use of articulate language, the power of forming abstract ideas, and even the sense of right and wrong, may have been gradually acquired by steps which here and there it is not impossible to trace. The question of the origin of the moral sense leads to the proposition of the following theory. Some natural emotions are of great intensity but short endurance, and their force is not easily recalled by memory; others, though less powerful at certain times, exert a constant influence, or one which is only interrupted by being overpowered for a time by the former. Accordingly, during the greater part of life, and always when there is leisure for reflection, the gratification connected with the more violent passions, such as hunger, sexual desire, and revenge, appears small, whereas the social instincts of sympathy and the pleasures of benevolence exert their full power. Hence we find social virtues, as courage, fidelity, obedience, among savages and even animals, long before the "self-regarding" virtues begin to appear. This theory is analogous to that by which Mr. Bain explains the higher character of the pleasures of sight compared with those of smell; they can be more easily recalled; and corresponds to the distinction drawn by the same writer between the acute and the more "massive" and permanent pleasures.

随后提出的论据同样是大家都熟悉的，即人类胚胎与其他脊椎动物胚胎的相似性。接着又对人类的退化器官进行了说明，这些也存在于其他物种中的器官是判断亲缘关系的最重要的指示之一。作者提到了一个这样的痕迹器官，我们相信，至今尚无记载。这个痕迹器官是外耳耳轮边缘的一个微小突起（结节），它显示时表现为一竖直耳朵的耳尖（见图示）。这一偶然的不规则形状有可能被今后的解剖学家根据其第一个发现者的名字而命名为**伍尔纳角**。

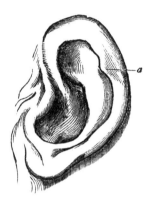

图 1. 该图是由伍尔纳先生构思并绘制的人耳
a. 结节

在第二章中达尔文先生指出，一项关于人类智力水平的研究可能会加深而不是削弱由身体结构引发的争论。该研究包括了语言的使用，而这被认为是承认达尔文提出的人类与其他动物的亲缘关系的最大困难。现已发现，记忆和好奇，妒嫉和友谊，甚至正确推理的能力和用声音沟通的能力在许多低等动物中都存在，但是思考的能力和主动意识，甚至"坚信存在万能上帝的高尚信念"却并不为人类家族中最低级的类群所拥有。与此同时，使用有音节的语言、形成抽象的概念，乃至形成是非观都是循序渐进地获得的，而且是可以追溯其起源的。道德观念的起源问题引出了下述理论。自然情绪中有一些来势凶猛但持续时间短，它们的影响力不容易通过记忆回想起来；而其他情绪，尽管在某些时间不如前者那么强烈，但能产生持久的影响力，或者产生一种只会因无法忍受前一种情绪的影响而被打断的影响力。相应地，生命的大部分时间通常是有闲暇进行思考的，这时，与非常强烈的感情如饥饿、性欲和报复等有关的个人满足感就显得微不足道，而同情这种社会本能和行善的快乐一直都在发挥它们全部的力量。因此我们发现，野人甚至动物中的社会道德，如勇气、忠诚和顺从，在"利己主义"品德开始出现以前已存在很久。这一看法与贝恩先生的视觉愉悦比嗅觉愉悦更高级的说法相似，视觉愉悦更易于被记起，这与该作者所说的短暂快乐与更"大众化的"、更永久性的快乐之间的区别是相对应的。

In the fourth chapter Mr. Darwin discusses the manner in which man was developed. It is shown that the broad facts on which the theory of Natural Selection rests apply to him. He is prolific enough to share in the struggle for existence. In him, as in all organic forms, there is a constant tendency to growth, which being checked and modified by external influences, proceeds in the direction of least resistance, and so produces the variations which are often ascribed to an assumed inherent tendency. Among the various forms produced, those will survive which are best fitted for the surrounding conditions, and they will transmit their character to their descendants, still subject to the same liability to vary. Next the author argues that the mental endowments of man, including language, his social habits, his upright position, and perfect hands, are of direct advantage to him in the struggle with other animals and with his fellows. It has always appeared that the difficult point in the development of man by Natural Selection is at the period when he was more defenceless than an anthropoid ape and less intelligent than the lowest savage; but Mr. Darwin thinks that the transition may have been safely made in some large tropical island where there was abundance of forest and of fruit. That man, once developed, can maintain himself, is obvious from his present existence. The arguments in favour of civilised man being the descendant of savages, which have been so admirably developed by Sir John Lubbock and Mr. Tylor, are of course brought forward in support of the author's view, and the important question is discussed how far we may hope for future improvement in the race by means of continued Natural Selection. Thus, while admitting that the process undergoes many checks and complications among human beings, the author does not assent to the arguments urged by Mr. Wallace that it would cease to operate as soon as the moral faculties came into play.[*] One human peculiarity which is apparently inexplicable by Natural Selection, the nakedness of the body and presence of a beard, is referred by Mr. Darwin to the operation of Sexual Selection. To this same agency is attributed the origin of the so-called Races of Man, which is discussed with admirable clearness and impartiality in the last chapter, and this leads to the complete exposition of the theory of Sexual Selection which occupies the second part of this work, and must be considered in a future article.

It only remains here to add a word on the account of the affinities and genealogy of man contained in the sixth chapter. As a kind of retribution for the attempt to raise Cuvier's order *Bimana* into a sub-class, not only have most naturalists now reverted to a modified definition of the *Primates* of Linnaeus, but Mr. Darwin shows reasons for refusing to the genus *Homo* even the rank of a family in this order, which Prof. Huxley admits, and regards it simply as an aberrant member of the Catarrhine division of the *Simiadae*. This conclusion, which seems to us to be a just one, will only be distasteful to those who so little appreciate the true characters of man as a spiritual being, that they could feel self-complacency in the brevet-rank of a sub class.

---

* In reviewing in these columns the contributions of the latter eminent writer, we took occasion to quote the estimate he expresses of Mr. Darwin's claims. Should anyone be disposed to overlook the original value of Mr. Wallace's work, he will be corrected by a somewhat similar passage in the present volume. See pp. 137, note, and 416.

在第四章中，达尔文先生探讨了人类的发育模式。文中表明，自然选择理论所依赖的大量事实也是达尔文先生本人的生活写照。他在为生存而斗争方面积累了大量的生活经验。他就像所有的生物体一样，始终有生长的趋势，这一生长受到外界影响的检验和修饰，朝着阻力最小的方向发展，并最终产生了变异，通常假设这种变异是由内在的趋势引起的。在产生的各种变异类型中，只有最能适应周围环境的类型才能存活下来，并将自己的性状传递给后代，再由后代进行新的变异。随后作者又探讨了人类的心智天赋，包括语言、社会习惯、直立行走和完美的双手，这些都是人类与其他动物或同类争斗的有利条件。在人类的发展过程中，自然选择最难解释的阶段就是当他们的防御能力不及类人猿、聪明程度不及最低级野人的阶段。不过，达尔文先生认为这一过渡时期可以在有丰富森林和水果的大型热带岛屿上安全度过。人类一旦发展成熟，便能够立足于世，这从人类的现存状态就能明显地看出来。书中引用了约翰·卢伯克爵士和泰勒先生提出的文明人是野人的后代的观点，以支持作者的看法，同时还讨论了一个重要问题，即通过不断的自然选择，我们可以预期人类种族能发展到什么程度。因此，尽管作者承认人类发展过程经受了许多考验和复杂情况，但并不赞成华莱士先生提出的如下论断：一旦道德职能开始起作用，自然选择就会失灵。* 人类的某些特性，比如躯体裸露和长有胡子，很难用自然选择来解释，达尔文先生将其解释为性选择的作用。最后一章清楚而公正地讨论了所谓的人类种族的起源问题。这就引出了对该工作的第二部分内容——性选择理论进行完整说明的必要性，这将在下一篇中阐述。

第六章对人类的亲缘关系和系谱关系作了进一步的阐述。为了反对居维叶试图将他命名的二手目提升为一个亚纲，不仅大多数自然学家重新采用林耐对灵长目的修正定义，而且达尔文先生也给出了拒绝使用该目中人属甚至是科级阶元名称的理由，并得到了赫胥黎教授的认可，赫胥黎认为它只是旧大陆猿中狭鼻猿的一个异常成员而已。我们可能觉得这一结论是公正的，但是对于那些没有认识到人类作为智慧生物的真实特性的人来说，却是令他们不悦的，这些人只会自我满足于亚纲地位提升的虚衔。

---

* 在查阅后者这位著名作家的专栏文章时，我们趁机引用了他对达尔文先生的观点的评价。没有人存心忽略华莱士先生的工作的原创价值，本卷书中一段有点类似的内容纠正了他的观点。参见第 137 页和第 416 页。

Mr. Darwin mentions Africa as the possible seat of the Catarrhine progenitors of man, but shows the futility of speculations on this point, until we know more of the recent changes of the earth, the records of palaeontology, and the laws affecting the rapidity of animal modifications. He does not advert to Prof. Haeckel's hypothesis of a "Lemuria" in the Indian Ocean, but agrees with him in next tracing the phylum of man to the *Prosimiae*. These again were developed from "forms standing very low in the deciduate mammalian series" (possibly, as Prof. Huxley suggests, most nearly allied to the existing *Insectivora*), and thus, through the Marsupials and Monotremes from the Reptilian stock, and thence through the *Dipnoi* and Ganoids from the *Urtyhus* of the vertebrate series, represented by the Lancelet alone. Nor does Mr. Darwin stop here, but adds the weight of his judgment to the theory based on the observations of Kowalewsky and Kuppfer, which deduces the primeval *Vertebrata* from a form resembling a Tunicate larva. Perhaps the most brilliant of the many new suggestions in these volumes is one thrown out incidentally in a note to p. 212, and based upon this supposed relation of man to the Ascidians. Beyond the organic world Mr. Darwin does not attempt to trace the genealogy of man. Considering how essential this extension of the theory of evolution is held by men so distinguished as Haeckel, and how keenly the question of Abiogenesis has recently been discussed, the reticence shown in avoiding allusion to the subject is perhaps the most remarkable among the many remarkable characters of this great work.

(**3**, 442-444; 1871)

II

That selection in relation to sex has been an important factor in the formation of the present breeds of animals was more than indicated in the "Origin of Species", and the theory has since been especially worked out by Professor Haeckel. It includes two distinct hypotheses. One is that in contests between males, the weakest would go to the wall, and thus either be killed outright, or at least debarred more or less completely from transmitting their characters to another generation. This may be regarded as a particular case of Natural Selection, and may be compared with the theory of protection by mimicry, suggested by Mr. Bates, and carried out by him and by Mr. Wallace. But though in the lists of Love the battle is often to the strong, even more frequently it is to the beautiful. This introduces a new process, of which the effects are not nearly so obvious as those of Natural Selection, either in its simplest form or in the more complicated cases of mimicry, and of sexual selection by battle. Many circumstances must combine in order that the most successful wooers shall have a larger and more vigorous progeny than the rest. In the first place, all hermaphrodite and all sessile animals may be excluded, and also those cases in which sexual differences depend on different habits of life. Mr. Darwin then shows that secondary sexual characters are eminently variable, and that males vary more than females from the standard of the species, a standard determined by the young, by allied forms, and sometimes by the character of the male himself when his peculiar functions are only periodical, or when they have been artificially prevented. Moreover it is the males who

达尔文先生提到非洲可能是狭鼻猿祖先的起源地，但他也表示目前还无法证实这一推测，这需要对地球近期的变化、古生物学记录以及诱发动物变异的速度的规律有更多的了解。他没有提及海克尔教授的"利莫里亚"大陆沉入印度洋的假说，而是对其另一个观点表示认同，即将人类追溯到原猴亚目。原猴亚目由"低等蜕膜哺乳动物"（据赫胥黎教授提示，最可能与现存的食虫目同属一类）发育而来，即从爬行类经过有袋类和单孔类发育而来，而爬行类则从以文昌鱼为代表的脊椎动物头索动物亚门经过肺鱼和硬鳞鱼发育而来。达尔文先生并没有停止于此，他还评价了柯瓦莱夫斯基和库普费尔通过观察建立的理论，这一理论推断原始的脊椎动物是从一种类似于被囊动物幼虫的生物形式进化而来的。可能在众多新建议中最杰出的就是基于人类与海鞘类的这种猜测性联系而在第 212 页所作的一条附带说明。达尔文先生追溯人类谱系自始至终都没有超出有机世界。考虑到像海克尔这么著名的人物对人类进化理论所作的扩充的重要性以及最近围绕无生源论的讨论的激烈程度，那么，作者谨慎避免提及这个话题可能是这一伟大著作众多特点中最重要的一点。

II

性选择是现代动物养殖的重要依据，这一理论早在《物种起源》中就提到过，之后海克尔教授对其进行了详细研究并作为理论提出。该理论包括两个不同的假说。假说一是雄性间的竞争。失败的弱者将被直接杀死，或者至少得不到交配机会，从而不能将性状传递给下一代。性选择可以看作自然选择的一个特例，可以与拟态保护理论相比，后者是由贝茨提出并与华莱士共同完成的。尽管在爱的竞技场上，胜利经常属于强者，但更多时候是美丽者获胜。这里涉及一个新过程，拟态无论是简单形式还是复杂形式，其影响都不如自然选择和通过争斗进行的性选择明显。许多情况必须结合起来以便最成功的求爱者拥有更多更强壮的后代。首先，所有的雌雄同体动物和固着动物可以被排除，性别差异受不同生活习性影响的动物也被排除在外。达尔文先生随后指出第二性征的变化是非常明显的，从物种的标准来看，雄性比雌性发生的变化更多，这种标准由幼体和联姻形式决定，当雄性功能是周期性的或者被人为阻止时，则由雄性的自身特点决定。此外，在交配中主动的一方通常是雄性，它们不仅要进行战斗来争夺配偶，还要展示它们的颜色、声音或者其他有独特魅力的方面来吸引异性。这一规律得到了食火鸡和一些其他物种的实例的证实：

take the active part in pairing, and who not only fight for the possession of their mates, but display their colours, their voice, of whatever be their peculiar attractions, in order to gain the same end. This rule is confirmed by the exceptional case of the cassowary and a few other species in which the hens court the male birds, fight together in rivalry, and accordingly assume the brighter colours and more attractive shape usually worn by the male. Not only the parental and incubating instincts, but the usual moral qualities of the two sexes are in these cases reversed: "the females being savage, quarrelsome, and noisy, the males gentle and good." But it is further necessary to show that the females exert a choice among the males, and that the latter are polygamous, or arrive earlier at the place of pairing, as is the case with some birds, or else exceed in numbers, at least when both sexes are mature. On this point a series of observations is recorded relating chiefly to man, to domesticated mammals, and to insects. The rule as to transmission of male characters to both sexes appears to be that when variations appear late in life they are usually developed in the same sex only of the next generation, although they are, of course, transmitted in a latent condition through both; while, on the other hand, the differences which appear before maturity in the parent are equally developed in both sexes when transmitted to the offspring. The numerous apparent exceptions to these laws of inheritance and of sexual selection are examined with wonderful fairness and fertility in resource. I may particularly refer to the discussion of the ways in which the young and adults of both sexes differ among birds. The extreme intricacy of some of the questions considered is best shown by a postscript in which, with characteristic candour, the author corrects "a serious and unfortunate error" in the eighth chapter.

The remainder of the first and the greater portion of second volume are occupied by a survey of sexual variations throughout the animal kingdom. Passing rapidly over the other invertebrate classes, the author devotes two chapters to the secondary sexual characters of insects. The weapons, the ornaments, and the sounds peculiar to the males of this vast group of animals are briefly described, and the remarkable analogy between insects and birds which is seen in so many other particulars is traced here also. The brilliant colours of many caterpillars, which, of course, cannot be due to sexual selection, offer one of the many difficulties which are faced, and this is explained by the aid of what the author terms Mr. Wallace's "innate genius for solving difficulties", as being due to natural selection. The bright colours warn the enemies of the caterpillars that they are unfit for food, and so benefit the latter, "on nearly the same principle that certain poisons are coloured by druggists for the good of man." Many cases are probably further complicated by mimicry, savoury caterpillars assuming the colours of distasteful ones so as to share in their immunity, in the same way that a druggist might label his bottles of sweetmeats "poison," to keep them from the shop boy.

In the frigid classes of the lower Vertebrata one would think that sexual selection would have little play; yet Mr. Darwin gives several instances among fishes, amphibians, and reptiles in which weapons or ornaments, peculiar to the males, appear to have been acquired by this means. (See Fig. 2.) But it is in the great class of birds that the most complete series of examples is found, and our advanced knowledge of the habits of this

雌性食火鸡互相打斗以追求雄鸟，因此雌鸟也拥有更鲜艳的色彩和更迷人的体型，而这些通常是雄性的特征。在这些例子中，不仅双亲本能和孵育本能在两性中是相反的，就连一般的品行也是相反的，"雌性野蛮、爱争吵、聒噪，而雄性则绅士而友好。"但是有必要进一步说明的是，雌性对雄性有选择权，而它们实行一妻多夫制；或者像一些鸟类那样，雄性先到达交配地点，或者至少在两性成熟时数量更多。在这一点上，一系列与人类、家养哺乳动物及昆虫有关的观察已经被记录下来。雄性特征向两种性别的后代传递的法则应该是，出现在生命后期的变异通常只会传递给下一代中的同性，尽管这些变异肯定是在某种潜在的条件下传递给两种性别的子代；另一方面，父母亲成熟前产生的变异可以平等地传递给两种性别的子代。在资源丰富而又公平分配的条件下，也观察到了大量并不符合遗传和性选择法则的明显例外。我要特别提到关于鸟类中两种性别的幼鸟和成鸟之间不同点的讨论。其中有些非常复杂的问题，作者在附言中对其进行了很好的阐述，同时特别坦诚地对第八章中"一系列严重的和令人遗憾的错误"进行了纠正。

第一卷其余部分及第二卷的大部分内容是关于整个动物王国中普遍存在的性别变异的调查。作者用两章内容来阐述昆虫的第二性征，而对其他无脊椎动物则快速带过。该部分对雄性昆虫特有的武器、装饰结构和声音作了简要描述，同时还描绘了昆虫与鸟类在很多细节方面的相似之处。许多蝴蝶幼虫具有明丽的颜色，当然用性选择理论是无法解释的，这成为该研究面临的困难之一，而作者凭借华莱士先生所称的"先天就具有的解决困难的天赋"对该现象作出了解释，认为这是自然选择的结果。蝴蝶幼虫明丽的颜色意在警告天敌它们不适合作为食物，从而对后者也是有益的，"与此同理，药剂师将某些毒药染上颜色以利于人们识别。"伪装使许多情况变得复杂，具有伪装本领的、美味可口的毛毛虫把自己伪装成看起来味道很差的样子从而避免被吃掉，同样，药剂师可能在盛有甜食的瓶子上贴上"毒药"的标签，以免被店里的男孩吃掉。

人们可能会认为，性选择在脊椎动物的几个低等冷血的纲中几乎没有作用，然而达尔文先生给出了鱼、两栖动物和爬行动物的例子，它们的雄性特有的武器或装饰结构都是通过性选择作用获得的（见图2）。最完整的例子是在鸟类中发现的，由于我们对鸟的习性了解得比较多，使得鸟类最可能成为揭示整个理论的领域。当在

class renders it the best possible field for the exposition of the whole theory. Again and again our author forestalls the evidence adduced in the chapters on sexual selection among birds, when tracing its first obscure operation among lower classes, and falls back on the same stronghold when explaining its less obvious working in the mammalia.

Fig. 2. *Chamaeleon Owenii.*
Upper figure, male; lower figure, female.

Among birds the rivalry of beauty has led to far more striking results than has the rivalry of strength. Foremost of these is the power of song, which, in accordance with the law of the least waste, is usually confined to birds of inconspicuous colours, while the combination of the harsh note with the magnificent plumage of the peacock is a familiar converse example. The object of the adornment of birds is conclusively proved by its being, as a rule, confined to males, and often to them only during the breeding season, as well as by the pains they take to exhibit their beauties to the hens. The difficulty is to show the precise way in which the results have been attained by gradual selection. In two remarkable instances, the wings of the Argus pheasant and the train of the peacock, Mr. Darwin succeeds in tracing the gradations in the same bird or the same family by which these wonderful and elaborate ornaments have been brought to their present perfection. The woodcut which illustrate these gradations are unfortunately too numerous to be reproduced here; they are admirably drawn, and convey the impression of the feathers as nearly as is possible by the means employed. Indeed, we may here remark that throughout these volumes the original cuts generally of details of structure, contrast very favourably with the figures of species taken from Brehm's "Thierleben", which are feebly drawn and ill-engraved.

Sexual selection has, of course, been continually checked and modified by the never-ceasing influence of natural selection, sometimes, as in the case of the horns of stags, being only somewhat diverted, but often directly opposed, as when it produces dangerously conspicuous colours, and dangerously cumbersome ornaments. In the case of birds, Mr. Darwin holds that the usual tendency of sexual selection being to produce variation in males, its transmission to hen birds has been checked by natural selection. Mr. Wallace, on the other hand, believes that both tendencies have generally operated together, in opposite

低级动物中探索这一理论的运作机制时，作者屡屡避开在讨论鸟类性选择时所采用的举证环节，在解释哺乳动物更不明显的性选择作用时也采用了同样的方式。

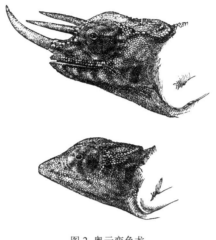

图 2. 奥云变色龙
上图：雄性；下图：雌性

　　鸟类的美丽之争比力量之争产生的结果更具影响力。其中最重要的是歌声的力量，那些没有亮丽颜色的鸟儿往往具备美妙的歌声，这也与最少消耗原则相吻合；而与之相反的如孔雀拥有美丽的翅膀但声音却很刺耳也是我们非常熟悉的例子。鸟类装饰结构的作用通常完全可由其表现证实为，只出现在雄性身体上，而且只出现于繁殖期间雄鸟不辞辛苦地向雌鸟展示美丽的时候。研究的难点在于弄清楚通过逐步选择从而获得最后结果的具体方式。两个具有代表性的例子是阿耳戈斯雉的翅膀和孔雀的长尾，达尔文先生成功地追踪到同种鸟或同一家族鸟的演变过程，现在的装饰结构已变得更加华丽而精致。很遗憾，描绘这一渐变过程的木刻画太复杂而无法在此复制。这些图画尽可能真实地表达出了各个阶段的羽毛结构。的确，这里我们可以注意到，达尔文的卷册里的原创细节结构图与布雷姆《动物的生命》中苍白有误的物种形象形成了鲜明的对比。

　　当然，性选择一直以来就受到自然选择的影响而不断被检验和被修正；有时只是略微修改，如雄鹿的角的例子，而通常则是直接向相反方向转变，例如出现危险性的醒目颜色和笨重装饰的例子。达尔文先生坚持认为，鸟类性选择的一般趋势是雄性发生变异，这些变异向雌鸟的传递受到自然选择的控制。另一方面，华莱士先生则相信，两种趋势通常是从不同方向共同起作用的，以使雄性子代具有比亲代更

directions, so as to make successive generations of males more and more conspicuous than the primitive type, and those of females less so. The fact that, as a rule, young birds resemble hens in their plumage, is a strong argument for the former opinion since most naturalists admit that early characters are the most trustworthy guide to natural alliances, *i.e.*, to true genealogy. To explain the transmission in some cases of brilliant colours (acquired probably by sexual selection, and therefore properly a male character) to both sexes indiscriminately, Mr. Wallace has framed the ingenious hypothesis, that the females have been protected from the dull uniformity threatened by natural selection, by their very general habit of building covered nests. Our author looks at the facts in a reversed way, and supposes that in most cases these hen birds, having inherited bright colours from the males, were led to the habit of building covered nests for the sake of protection.

Among mammals sexual selection has chiefly operated by increasing the size and strength of the males, and furnishing them with weapons of offence;[*] but besides allurements to the senses of smell and hearing, this class offers not a few instances, especially among the Quadrumana, of brilliant colouring being developed as a secondary sexual character. Here also we have the most striking instances of the production of defensive organs by the same process, as in the manes of lions, the cheekpads of some of the *Suidae*, and possibly the upper tusks of that ancient enigma, the barbirusa. Lastly, it is in the class of mammals that we meet with cases of what may be called primary sexual ornament, as in *Cercopithecus cynosurus*, which make one wonder, with a thankful wonder, why such apparently obvious results are not more common. We must, however, admit that such adornment is not more disgusting, nor that of which we copy a figure more ludicrous, than the personal decorations of savages. Sir Joshua Reynolds says that if a European in full dress and pigtail were to meet a Red Indian in his warpaint, the one who showed surprise or a disposition to laugh would be the barbarian.[†] But who could stand this test when meeting *Semnopithecus rubicundus* or *Pithecia satanas*?

Fig. 3. Head of *Semnopithecus rubicundus*.
This figure (from Prof. Gervais) is given to show the odd arrangement and development of the hair on the head.

---

[*] The very general transmission of such weapons to both sexes may, perhaps, be explained by the need females have of means to defend their young.

[†] Discourse delivered at the Royal Academy, December 10, 1776.

加显著的性状，而雌性子代则更不明显。但是作为一般规则，幼鸟的羽毛都与雌鸟相像，这和前面的观点了形成了强烈的冲突，因为大多数自然学家都认为早期性状是研究自然联姻也就是真正的谱系的最可靠向导。为了解释在一些例子中，亮丽的颜色（可能通过性选择获得，因此是雄性性状）可以不加区别地传递给两种性别的后代，华莱士先生提出了一个独特的假说，即雌性可通过建造有盖巢穴的普遍习性来保护自己不受自然选择的威胁。该书作者则从相反的角度认为，大多数继承了雄性的明亮颜色的雌鸟都会为了保护自己而获得建造有盖巢穴的习性。

哺乳动物的性选择主要通过增加雄性个体的大小和力量以及装备抵御的武器来起作用。* 除了在嗅觉和听觉的诱惑特征，此纲中有许多的例子表明，发育形成的亮丽色彩成了第二性征，尤其是在灵长类动物中。这里介绍了一些引人注目的通过性选择产生防御器官的例子，如狮子的鬃毛、猪科动物的颊垫以及远古之谜巨獠猪的上牙。最后，在哺乳动物纲（如长尾猴）中，我们发现了被称之为原始性装饰结构的例子，这一发现令我们好奇而欣慰，好奇的是为什么这种明显的装饰作用不具有普遍性。不过，我们必须承认，这样的装饰以及我们复制的图示与野人的装饰一样，不但滑稽可笑，而且令人难以接受。乔舒亚·雷诺兹爵士说，如果一个身着正式礼服、扎着辫子的欧洲人要会见一个涂了涂料的印第安人，那么那个表现出惊讶或者大笑的人就是没有教养的人。† 但是在看到婆罗州长尾猴或者黑色狐尾猴时，谁又能忍住不笑呢？

图 3. 婆罗州长尾猴
此图（引自热尔韦教授）用于显示头上毛发的奇怪布局和生长情况

---

* 这些抵御武器可传递给两种性别的后代，可以解释为，雌性需要用它们来保护幼体。
† 参见 1776 年 12 月 10 日皇家学会会议上所作的报告。

We must admit, notwithstanding such anomalies, that, on the whole, birds and other animals admire the same forms and colours which we admire, and this, perhaps, may be admitted as an additional argument in favour of their kinship with us. Some of the ugliest creatures (like the hippopotamus) appear to have been quite uninfluenced by sexual selection, while the magnificent plumes of pheasants and birds of paradise are undoubtedly due to its operation. That it has occasionally led to unpleasing results in birds and monkeys of aberrant taste, is no more strange than that all savages do not carve and colour as well as the New Zealanders, or that most Englishmen admire ugly buildings and vulgar pictures. The prevailing aspect of nature is beauty, and the prevailing taste of man is for beauty also. The *means* by which natural beauty has been attained are various. Natural selection is one, by which the healthiest, and therefore the most symmetrical forms survive the rest. Protective mimicry is another, by which fishes have assumed the bright colours of a coral garden and butterflies the delicate venation of leaves. Flowers again have in many cases obtained their gay petals and fantastic shapes from the advantage thus gained for fertilisation by insects. The successive steps which have led to the graceful forms and brilliant tints of shells, to the intricate symmetry of an echinus-spine or a nummulite, these are as yet untraced even in imagination.

But that many of the most striking ornaments of the higher animals, and almost all those which are peculiar to one sex, have been developed by means of sexual selection, is a conclusion which can no longer be distrusted. There remain doubtless many exceptions to be accounted for, many modifying influences to be discovered; but the existence of a new principle has been established which has helped to guide the organic world to its present condition. Side by side with the struggle for existence has gone on a rivalry for reproduction, and the survival of the fittest has been tempered by the success of the most attractive.

(**3**, 463-465; 1871)

尽管有些反常情况，但我们必须承认，总体上鸟类和其他动物喜好的形式和颜色也是我们人类所欣赏的，这或许可以作为支持它们与我们具有亲缘关系的佐证。有些丑陋的生物（如河马）好像并未受到性选择的影响，但是雉和极乐鸟具有的华丽羽毛，肯定是性选择的作用。有时品味怪异的鸟类和猴子也会令人不快，但这并不比并非所有的野人都像新西兰土著居民那样擅长雕刻粉饰、许多英国人喜欢丑陋的建筑和粗俗的图画更奇怪。自然的主流是美丽，人类的主流品位也是追求美丽。达到自然美的**途径**是多种多样的。自然选择就是其中之一，它使得最健康也是最对称的类型更易于存活。保护性的拟态是另外一种途径，它使得鱼儿呈现出珊瑚丛般明亮的色彩，使得蝴蝶拥有树叶般精美的翅脉。在许多情况下，花朵都因其鲜艳的花瓣和漂亮的外形而优先获得昆虫的帮助从而完成受精。贝壳具有优雅的外形和鲜艳的色彩，货币虫具有复杂而对称的海胆棘，它们的发生过程让人难以琢磨，甚至无法想象。

但是，高等动物中许多醒目的装饰结构，以及仅为一种性别所特有的结构，都是通过性选择的方式发育而来，这已经成为毋庸置疑的结论。诚然，还有许多例外的情况尚待解释，许多修饰作用的影响尚待发现，但一个新法则已经建立起来，在它的帮助指导下，有机世界才成为现在的样子。伴随生存竞争的是繁衍竞争，最具吸引力者的成功繁衍则调和了适者生存的法则。

（刘皓芳 翻译；冯兴无 审稿）

Pangenesis

C. Darwin

Editor's Note

Despite receiving much praise, Charles Darwin's evolutionary theory was at the same time under attack by scientific colleagues. Francis Galton, from University College London, had carried out experiments to test Darwin's theory of heredity, called pangenesis. In this scheme, Darwin had supposed that the germ cells which unite in the fertilisation of the ovum of an animal or plant are physically made from entities called "gemmules" derived from their organs or structures. _Nature_ published Darwin's comments on Galton's experiments. He almost admits in his closing sentence that his theory is indefensible.

IN a paper, read March 30, 1871, before the Royal Society, and just published in the Proceedings, Mr. Galton gives the results of his interesting experiments on the inter-transfusion of the blood of distinct varieties of rabbits. These experiments were undertaken to test whether there was any truth in my provisional hypothesis of Pangenesis. Mr. Galton, in recapitulating "the cardinal points", says that the gemmules are supposed "to swarm in the blood". He enlarges on this head, and remarks, "Under Mr. Darwin's theory, the gemmules in each individual must, therefore, be looked upon as entozoa of his blood", &c. Now, in the chapter on Pangenesis in my "Variation of Animals and Plants under Domestication", I have not said one word about the blood, or about any fluid proper to any circulating system. It is, indeed, obvious that the presence of gemmules in the blood can form no necessary part of my hypothesis; for I refer in illustration of it to the lowest animals, such as the Protozoa, which do not possess blood or any vessels; and I refer to plants in which the fluid, when present in the vessels, cannot be considered as true blood. The fundamental laws of growth, reproduction, inheritance, &c., are so closely similar throughout the whole organic kingdom, that the means by which the gemmules (assuming for the moment their existence) are diffused through the body, would probably be the same in all beings; therefore the means can hardly be diffusion through the blood. Nevertheless, when I first heard of Mr. Galton's experiments, I did not sufficiently reflect on the subject, and saw not the difficulty of believing in the presence of gemmules in the blood. I have said (Variation, &c., vol. II, p. 379) that "the gemmules in each organism must be thoroughly diffused; nor does this seem improbable, considering their minuteness, and the steady circulation of fluids throughout the body." But when I used these latter words and other similar ones, I presume that I was thinking of the diffusion of the gemmules through the tissues, or from cell to cell, independently of the presence of vessels,—as in the remarkable experiments by Dr. Bence Jones, in which chemical elements absorbed by the stomach were detected in the course of some minutes in the crystalline lens of the eye; or again as in the repeated loss of colour and its recovery after a few days by the hair,

泛生论

达尔文

编者按

当时，尽管获得了许多称赞，查尔斯·达尔文的进化论还是受到了科学界同行的攻击。伦敦大学学院的弗朗西斯·高尔顿进行了相关实验来检验被达尔文称为泛生论的遗传学理论。达尔文在这一理论中假设生殖细胞（动物或植物的卵子受精后的结合）是由来源于生物器官或组织的被称为"微芽"的实体组成的。《自然》发表了达尔文对高尔顿的实验的评论。达尔文在其评论的结尾几乎承认了自己的理论还很脆弱。

在 1871 年 3 月 30 日的皇家学会会议上，高尔顿先生公布了一篇论文，该论文刚被发表在《皇家学会学报》上。在这篇论文中，高尔顿先生给出了他的实验结果，是关于不同种兔子间输血的有趣实验。这些实验是用来检验我提出的关于泛生论的暂定假说是否可信。高尔顿先生在概括"最重要的几点"时说，微芽被认为是"充满在血液中的"。他对此进行了详细阐述，并评论说，"因此，根据达尔文先生的理论，每个个体的微芽必须被看作是其血液内的寄生物。"现在我要说，在我的《动物和植物在家养下的变异》一书介绍泛生论的章节中，我没有提到关于血液或任何循环系统中的任何液体的内容。很明显，血液中微芽的存在实际上并不是我的假说的必要组成部分。因为我在说明该假说时提到了最低级的动物，例如没有血液或任何脉管的原生动物，另外我也提到了植物，在它们的脉管中的液体并不能被当作是真正的血液。整个有机王国普遍存在的生长、繁殖、遗传等基本法则是非常相似的，因此微芽（暂时假定是存在的）在体内扩散的方式在所有生物中可能都是一样的，那么就不大可能是通过血液扩散的。不过，当我第一次听说高尔顿先生的实验时，我并没有充分思考这个问题，也没有想到要相信血液中存在微芽有什么困难。我已经说过（见《动物和植物在家养下的变异》，第 2 卷，第 379 页），"每种生物的微芽一定是完全分散的，考虑到微芽的微小及液体在全身的稳定循环，这好像不无可能。"但当我说这些话和其他类似的话时，我认为我考虑的是微芽在细胞间的扩散或穿过组织的扩散，这与有无脉管无关——正如本斯·琼斯博士所做的了不起的实验，即被胃吸收的化学物质经过一段时间后可以在眼睛的晶状体中观察到，或者又如佩吉特先生记录的一个患有神经痛的妇女的特例，其中头发会发生重复性掉色并且几天后又可以自行恢复。另外，也不能否认微芽不能通过组织或细胞壁，因为每种花粉

in the singular case of a neuralgic lady recorded by Mr. Paget. Nor can it be objected that the gemmules could not pass through tissues or cell-walls, for the contents of each pollen-grain have to pass through the coats, both of the pollen-tube and embryonic sack. I may add, with respect to the passage of fluids through membrane, that they pass from cell to cell in the absorbing hairs of the roots of living plants at a rate, as I have myself observed under the microscope, which is truly surprising.

When, therefore, Mr. Galton concludes from the fact that rabbits of one variety, with a large proportion of the blood of another variety in their veins, do not produce mongrelised offspring, that the hypothesis of Pangenesis is false, it seems to me that his conclusion is a little hasty. His words are, "I have now made experiments of transfusion and cross circulation on a large scale in rabbits, and have arrived at definite results, negativing, in my opinion, beyond all doubt the truth of the doctrine of Pangenesis." If Mr. Galton could have proved that the reproductive elements were contained in the blood of the higher animals, and were merely separated or collected by the reproductive glands, he would have made a most important physiological discovery. As it is, I think every one will admit that his experiments are extremely curious, and that he deserves the highest credit for his ingenuity and perseverance. But it does not appear to me that Pangenesis has, as yet, received its death blow; though, from presenting so many vulnerable points, its life is always in jeopardy; and this is my excuse for having said a few words in its defence.

(**3**, 502-503; 1871)

粒的内含物都必须穿过花粉管和胚囊共同构成的外被。对于液体穿越细胞膜，我想补充的是，正如我自己在显微镜下观察到的那样，它们以一定的速率在活体植物的吸收根毛的细胞间穿越，这是非常令人惊讶的。

因此，当高尔顿先生根据"静脉血管中含有大量的另一种兔子血液的兔子并没有产生混血后代"这一事实而断定泛生论假说是错误的时候，我觉得这个结论下得有点轻率。他的原话如下，"我目前已经用兔子进行了大规模的输血和交叉循环实验，并且已经得到了确定的阴性结果。在我看来，这些结果毫无疑问地否定了泛生论学说的真实性。"如果高尔顿先生可以证实高等动物的生殖物质包含在血液里，仅由生殖腺隔开或聚集，那么他可就是取得了一个非常重大的生理学发现。如果真是这样，我觉得每个人都会承认他的实验是非常奇妙的，他也将值得凭借其独到的实验设计和锲而不舍的毅力获得最高荣誉。但到目前为止，尽管泛生论暴露出了许多弱点，就此而言，其生存一直处于危险之中，但我认为泛生论并没有受到致命的打击，而且我也有理由为它辩护。

（刘皓芳 翻译；刘力 审稿）

Pangenesis

F. Galton

Editor's Note

In response to comments from Charles Darwin on his recent experiments on heredity, his cousin Francis Galton offered for publication the letter below, in which he acknowledges that his work with rabbits did not amount to a disproof of Darwin's notion of pangenesis, which, he suggested, had now been redefined by Darwin.

IT appears from Mr. Darwin's letter to you in last week's *Nature**, that the views contradicted by my experiments, published in the recent number of the "Proceedings of the Royal Society", differ from those he entertained. Nevertheless, I think they are what his published account of Pangenesis (Animals, &c., under Domestication, II, 374, 379) are most likely to convey to the mind of a reader. The ambiguity is due to an inappropriate use of three separate words in the only two sentences which imply (for there are none which tell us anything definite about) the *habitat* of the Pangenetic gemmules; the words are "circulate", "freely", and "diffused". The proper meaning of circulation is evident enough—it is a re-entering movement. Nothing can justly be said to circulate which does not return, after a while, to a former position. In a circulating library, books return and are re-issued. Coin is said to circulate, because it comes back into the same hands in the interchange of business. A story circulates, when a person hears it repeated over and over again in society. Blood has an undoubted claim to be called a circulating fluid, and when that phrase is used, blood is always meant. I understood Mr. Darwin to speak of blood when he used the phrases "circulating freely", and "the steady circulation of fluids", especially as the other words "freely" and "diffusion" encouraged the idea. But it now seems that by circulation he meant "dispersion", which is a totally different conception. Probably he used the word with some allusion to the fact of the dispersion having been carried on by eddying, not necessarily circulating, currents. Next, as to the word "freely". Mr. Darwin says in his letter that he supposes the gemmules to pass through the solid walls of the tissues and cells; this is incompatible with the phrase "circulate freely". Freely means "without retardation"; as we might say that small fish can swim freely through the larger meshes of a net; now, it is impossible to suppose gemmules to pass through solid tissue without *any* retardation. "Freely" would be strictly applicable to gemmules drifting along with the stream of the blood, and it was in that sense I interpreted it. Lastly, I find fault with the use of the word "diffused", which applies to movement in or with fluids, and is inappropriate to the action I have just described of solid boring its way through solid. If Mr. Darwin had given in his work an additional paragraph or two to a description of the whereabouts of the gemmules which, I must remark, is a cardinal point of his theory, my

---

* *Nature* vol. III, p. 502.

泛生论

高尔顿

编者按

查尔斯·达尔文解释了他最近关于遗传学方面的实验，作为回应，他的表弟弗朗西斯·高尔顿寄来如下的回信要求发表。在这封回信中他提出，他用兔子进行研究的结果并不是对达尔文已经重新定义过的泛生论观点的反驳。

从达尔文先生上周写给《自然》的信来看*，他持有的观点与我发表在最近一期《皇家学会学报》上的实验结果所反驳的观点似乎并不相同。我觉得，我的结果反驳的不过是那些他发表的关于泛生论的说明（见《动物和植物在家养下的变异》，第2卷，第374、379页）中很可能会传递给读者的观点。这一歧义是由两个句子中三个使用不当的词引起的，这两个句子暗示了（因为并没有明确地告诉我们）泛生微芽的**聚集地**。这三个词是"循环"，"自由地"和"分散的"。循环的恰当含义是非常明显的——它是一种再次进入的运动。任何东西如果没有在后来返回到原先的位置就不能称为循环。在一个循环的图书馆里，书籍能够返回并被再利用。钱币被说成是循环流通的，是因为在商业交换中它能返回到同一个人手中。故事是循环传播的，是因为当某个人听到它时，该故事在社会上已经被重复叙述过许多遍了。毫无疑问，血液可以被称作是一种循环的液体，而且当人们说到循环液体的时候，通常就是指血液。因此当达尔文先生使用"自由地循环""液体的稳定循环"这些词时，我认为他就是在说血液，尤其是"自由地"和"分散的"这两个词更坚定了我的理解。但现在看来，好像他是在用循环来表示"散布"的意思，但这是个完全不同的概念。可能他使用这个词时想的是散布是由涡流造成的，而未必是环流。接下来是第二个词"自由地"。达尔文先生在他的信中说，他认为微芽可以穿过坚固的组织壁和细胞壁。这与"自由地循环"这一表达是不相符的。"自由地"就意味着"没有阻碍"，例如我们可以说小鱼能"自由地"游过渔网上的大网眼，但我们现在不可能认为微芽可以没有**任何**阻碍地穿越坚固的组织。严格来说，"自由地"适用于微芽随着血流漂移的情况，我就是从这个意义上来理解这个词的。我要挑的最后一个毛病是"分散的"这个词的使用。这个词适用于液体中的或与液体一起进行的运动，而不适用于我刚刚描述过的固体穿越固体的运动。如果达尔文先生在其著作中再用一两个段落来描述微芽的所在之处，那么我就不会像之前那样没有太

---

* 参见《自然》，第3卷，第502页。

misapprehension of his meaning could hardly have occurred without more hesitancy than I experienced, but I certainly felt and endeavoured to express in my memoir some shade of doubt; as in the phrase, p. 404, "that the doctrine of Pangenesis, pure and simple, as I have interpreted it, is incorrect."

As I now understand Mr. Darwin's meaning, the first passage (II, 374), which misled me, and which stands: "...minute granules...which circulate freely throughout the system" should be understood as "minute granules...which are dispersed thoroughly and are in continual movement throughout the system"; and the second passage (II, 379), which now stands: "The gemmules in each organism must be thoroughly diffused; nor does this seem improbable, considering...the steady circulation of fluids throughout the body", should be understood as follows: "The gemmules in each organism must be dispersed all over it, in thorough intermixture; nor does this seem improbable, considering...the steady circulation of the blood, the continuous movement, and the ready diffusion of other fluids, and the fact that the contents of each pollen grain have to pass through the coats, both of the pollen tube and of the embryonic sack." (I extract these latter *addenda* from Mr. Darwin's letter.)

I do not much complain of having been sent on a false quest by ambiguous language, for I know how conscientious Mr. Darwin is in all he writes, how difficult it is to put thoughts into accurate speech, and, again, how words have conveyed false impressions on the simplest matters from the earliest times. Nay, even in that idyllic scene which Mr. Darwin has sketched of the first invention of language, awkward blunders must of necessity have often occurred. I refer to the passage in which he supposes some unusually wise, ape-like animal to have first thought of imitating the growl of a beast of prey so as to indicate to his fellow monkeys the nature of expected danger. For my part, I feel as if I had just been assisting at such a scene. As if, having heard my trusted leader utter a cry, not particularly well articulated, but to my ears more like that of a hyena than any other animal, and seeing none of my companions stir a step, I had, like a loyal member of the flock, dashed down a path of which I had happily caught sight, into the plain below, followed by the approving nods and kindly grunts of my wise and most-respected chief. And I now feel, after returning from my hard expedition, full of information that the suspected danger was a mistake, for there was no sign of a hyena anywhere in the neighbourhood. I am given to understand for the first time that my leader's cry had no reference to a hyena down in the plain, but to a leopard somewhere up in the trees; his throat had been a little out of order—that was all. Well, my labour has not been in vain; it is something to have established the fact that there are no hyenas in the plain, and I think I see my way to a good position for a look out for leopards among the branches of the trees. In the meantime, *Vive* Pangenesis.

(**4**, 5-6; 1871)

过犹豫就误解了他的意思，另外我必须指出，微芽的所在之处是其理论的一个关键点。我确实有一些怀疑并且尽力在我的文章里表达出来，例如第 404 页写道的那句"如我所解释的单纯而简单的泛生论是不正确的。"

因为我现在理解了达尔文先生的意思，所以曾误导过我的表述为"……小颗粒……在整个系统中自由地循环"的第一段（第 2 卷，第 374 页）应当理解成"……小颗粒……在整个系统中完全散布并不停地移动"。误导我的第二段（第 2 卷，第 379 页），原描述为"每种生物的微芽一定是完全分散的；考虑到……液体在全身的稳定循环，这好像不无可能，"应理解为"每种生物的微芽一定是以充分混合的状态散布在全身各处的；考虑到……血液稳定的循环，其他液体连续的移动和充分的分散，以及每种花粉粒的内含物都必须穿过花粉管和胚囊共同构成的外被，这好像不无可能。"（后半部分**附加的**内容摘自达尔文先生的信件。）

我并不是要过多地抱怨模棱两可的语言使我产生了误解，因为很早之前我就知道达尔文先生在写出所有这些话时是多么认真尽责，我也知道要将自己的思想准确地付诸语言是多么困难，以及文字曾经怎样在最简单的事情上传递了错误的观念。不但如此，就算是在达尔文先生曾经描绘过的，语言刚被发明出来的田园时代，一些笨拙的错误也一定不可避免地经常发生。在某段文章中，达尔文先生指出有些非常聪明的猿类动物能够一下子想到通过模仿野兽捕食猎物时的咆哮声来提示他的猴子同伴们他所预料到的危险的性质，在这里我要提一下这段内容。对我来说，我觉得我好像也一直处在这样的情境中。就好像是听到我信任的领袖发出一声呼喊，尽管这种声音的意义不是特别明确，但对我的耳朵来说这种声音最像土狼的叫声，当我发现我的同伴都没有动身的意向时，我已经像一个群体中的一名忠实的成员一样，沿着我先前很幸运地发现的一条通往下方平原的小路飞奔出了一段距离，我非常敬重的英明领袖给了我赞许的点头和友好的喷喷声。当我从自己的艰难探险中回来之后，我才发现我错误地判断了危险，因为周围任何地方都没有土狼出现的迹象。我第一次明白，我的领袖的喊声并不是提示平原上有土狼，而是树上某处有豹。他的嗓音有点失控了——这就是全部。总之，我的劳动并非徒然，因为我的劳动确定了平原上没有土狼的事实，另外我觉得我也知道了用以留神藏于树枝间的豹的好方法。同时，祝泛生论万岁。

（刘皓芳 翻译；刘力 审稿）

A New View of Darwinism

Editor's Note

While bowing to Darwin's authority, correspondent Henry Howorth here advanced what he saw as a problem with Darwin's idea of evolution by natural (and sexual) selection. If survival goes to the strongest such that they reproduce, how is it that plantsmen and animal-breeders conspire to induce reproduction in their charges by pruning them and even actively reducing their condition? The next issue of *Nature* contained a brief and courteous response from Darwin referring to his refutation of the idea in his published work, followed by a longer and altogether far less courteous letter from Alfred Russel Wallace.

I have noticed that *Nature* is very catholic in its sympathies, and allows all views which are not palpably absurd to be discussed in its pages, and I therefore venture to ask for some space in which to present a few of the difficulties which have been suggested by Mr. Darwin's theory of Natural Selection, and which have not, so far as I know, been as yet discussed. I have not the taste for the language nor the arguments which were used by a *Times* reviewer, and I have much too great a reverence for one of the most fearless, original, and accurate investigators of modern times, to speak of Mr. Darwin and his theory in the terms used by that very ignorant person. Approaching the subject in this spirit, and knowing how very small a section of biologists are now opposed to Mr. Darwin, I may be very rash, but hardly impertinent, in stating my difficulties.

I cannot dispute the validity and completeness of many of Mr. Darwin's proofs to account for individual cases of variation and isolated changes of form. Within the limits of these proofs it is impossible to deny his position. But when he heaves these individual and often highly artificial cases, and deduces a general law from them, it is quite competent for me to quote examples of a much wider and more general occurrence that tell the other way. In this communication I shall confine myself to Mr. Darwin's theory, and shall not trespass upon the doctrine of evolution, with which it is not to be confounded.

The theory of Natural Selection has been expressively epitomised as "the Persistence of the Stronger", "the Survival of the Stronger". Sexual selection, which Mr. Darwin adduces in his last work as the cause of many ornamental and other appendages whose use in the struggle for existence is not very obvious, is only a by-path of the main conclusion. Unless by the theory of the struggle for existence is meant the purely identical expression that those forms of life survive which are best adapted to survive, I take it that it means in five words the Persistence of the Stronger.

对达尔文学说的新看法

编者按

通讯员亨利·霍沃思在认可达尔文的权威性的同时，也在此对达尔文关于自然选择（性选择）的进化论观点提出了一点质疑。如果物种延续都是靠强者的繁殖，那为什么栽培植物的园丁和养殖动物的饲养员要通过修剪枝干甚至主动限制被养生物的生活空间的方法来促使它们繁殖后代呢？在后一期的《自然》杂志上，达尔文简短而客气地回应了霍沃思对其著作的驳斥，紧随其后的是阿尔弗雷德·拉塞尔·华莱士写的一封毫不客气的长信。

我发现《自然》对所有不是特别荒唐的观点都有广博的宽容心，兼收并蓄，在自己的版面上允许出现对这些观点的探讨，因此我在此冒昧地借贵刊对达尔文先生的自然选择学说提出几点疑问。据我所知，到目前为止，这些疑问还未曾被讨论过。我不太喜欢《泰晤士报》的评论员在谈及达尔文先生及其学说时所使用的语言和持有的观点，觉得他所使用的那些用语很无知，而我对这位现代最无畏、最具独创性、最准确无误的研究者充满着无限敬意。我知道现在只有很少的生物学家反对达尔文先生，但我是抱着这样一种精神来讨论这一话题的，因此虽然我在陈述自己的疑问时可能有些急躁，但并非鲁莽。

我不想争论达尔文先生用来说明变异和隔离变化形式的个案时所用的许多证据的有效性和完整性。仅仅用这些证据是不足以否认达尔文学说的地位的。但是当他举出这些个别的而且往往在很大程度上是人为造成的事件，并从中推导出一条普遍规律时，我就有权引用更普遍、更广泛发生的事件来得出其他的观点。在这封信里，我会将问题局限在达尔文先生的学说范围内，不会涉及关于进化的学说，以防混淆。

自然选择学说已经被概括表述为"强者的延续""强者生存"。达尔文先生在其最近的一部著作中提出，性选择是许多装饰性的以及其他的附属部分的成因，这些附属部分在生存斗争中的用处并不是很明显，性选择在这里只是主要结论的一个分支而已。生存斗争理论除了意味着幸存的生命形式是最适于生存的这条完全一致的表述之外，我对它的理解就是那五个字：强者的延续。

Among the questions which stand at the very threshold of the whole inquiry, and which I have overlooked in Mr. Darwin's books if it is to be found there, is a discussion of the causes which produce sterility and those which favour fertility in races. He no doubt discusses with ingenuity the problem of the sterility of mules and of crosses between different races, but I have nowhere met with the deeper and more important discussion of the general causes that induce or check the increase of races. The facts upon which I rely are very common-place, and are furnished by the smallest plot of garden or the narrowest experience in breeding domestic animals. The gardener who wants his plants to blossom and fruit takes care that they shall avoid a vigorous growth. He knows that this will inevitably make them sterile; that either his trees will only bear distorted flowers, that they will have no seed, or bear no blossoms at all. In order to induce flowers and fruit, the gardener checks the growth and vigour of the plant by pruning its roots or its branches, depriving it of food, &c., and if he have a stubborn pear or peach tree which has long refused to bear fruit, he adopts the hazardous, but often most successful, plan of ringing its bark. The large fleshy melons or oranges have few seeds in them. The shrivelled starvelings that grow on decaying branches are full of seed. And the rule is universally recognised among gardeners as applying to all kinds of cultivated plants, that to make them fruitful it is necessary to check their growth and to weaken them. The law is no less general among plants in a state of nature, where the individuals growing in rich soil, and which are well-conditioned and growing vigorously, have no flowers, while the starved and dying on the sandy sterile soil are scattering seed everywhere.

On turning to the animal kingdom, we find the law no less true. "Fat hens won't lay", is an old fragment of philosophy. The breeder of sheep and pigs and cattle knows very well that if his ewes and sows and cows are not kept lean they will not breed; and as a startling example I am told that to induce Alderney cows, which are bad breeders, to be fertile they are actually bled, and so reduced in condition. Mr. Doubleday, who wrote an admirable work in answer to Malthus, to which I am very much indebted, has adduced overwhelming evidence to show that what is commonly known to be true of plants and animals is especially true of man. He has shown how individuals are affected by generous diet and good living, and also how classes are so affected. For the first time, so far as I know, he showed why population is thin and the increase small in countries where flesh and strong food is the ordinary diet, and large and increasing rapidly where fish or vegetable or other weak food is in use; that everywhere the rich, luxurious, and well-fed classes are rather diminishing in numbers or stationary; while the poor, under-fed, and hard-worked are very fertile. The facts are exceedingly numerous in support of this view, and shall be quoted in your pages if the result is disputed. This was the cause of the decay of the luxurious power of Rome, and of the cities of Mesopotamia. These powers succumbed not to the exceptional vigour of the barbarians, but to the fact that their populations had diminished, and were rapidly being extinguished from internal causes, of which the chief was the growing sterility of their inhabitants.

在提出所有这些正处于全部调查出发点的问题，以及我所忽略但可以从达尔文先生的书中发现的问题之前，我们要先讨论是什么原因导致了不育以及有哪些因素可以促进物种的繁殖。毫无疑问，达尔文先生对骡子的不育和不同物种间杂交问题的探讨很有独创性，但是我没有在该书其他部分读到更深刻、更重要的有关导致或阻碍种群数目增加的普遍原因的讨论。我所信赖的事实其实是老生常谈的问题，这些事实可以通过花园里最细微的情节或者仅通过饲养家畜的极为有限的经历来提供。一名园丁如果想要让自己种植的植物开花结果，就会注意避免他的植物旺盛生长。他知道旺盛生长肯定会导致植物不育；一旦生长过于旺盛，那么他的树要么只开出畸形的不会结种子的花朵，要么根本就不开花。为了诱导出花和果实，园丁会通过修剪树木的根部和枝干、减少施肥等措施来阻止植物的生长和活力。假如他有一棵很久都没有结过果实的顽固的梨树或桃树，他会采用冒险但通常会很成功的环状剥皮措施。园丁界流传着一条公认的规律，即认为大个的甜瓜或桔子几乎没有种子，而缺少养分只能依靠腐坏枝叶生长的干枯植株上面的果实通常结满了种子，园丁们认为这一法则适用于所有的栽培植物，并且认为要使各种栽培植物产生多汁的果实，就需要阻碍或者削弱他们的生长。这一法则对处于自然状态下的植物也一样适用，那些生长在肥沃土壤里、状态良好、生长旺盛的植物都不开花，而那些生长在贫瘠的沙地中营养不良的垂死植物却会将它们的种子散播到各处。

再来看看动物界，我们发现这一规律同样存在。"肥鸡不下蛋"，是一句古老的哲语。羊、猪、牛的饲养者都清楚地知道，如果不让母羊、母猪和母牛保持瘦弱的话，它们就不会繁殖；别人告诉过我一个惊人的例子，即为了使难以繁殖的奥尼德尼母牛能够多产，它们竟然被放血来使其身体状况变差。作为对马尔萨斯的回答，道布尔迪先生写了一部令人敬佩的著作，这部著作也使我受益匪浅，其中援引了大量证据来表明对动植物普遍适用的法则对人类也同样适用。他向我们说明了，个体是如何受丰富的饮食和舒适的生活条件影响的，以及社会阶层是如何受此影响的。据我所知，他首次解释了为什么在以肉和高能量食物为日常饮食的国家里人口数量少并且增长幅度小，而在以鱼或蔬菜或其他低能量食物为日常饮食的国家里，人口数量却很大并且增长迅速；无论在何处，富有的、奢侈的、营养好的种群的数量都是处于减少的状态或保持稳定不变；而贫穷的、营养不良的和辛苦工作的种群则往往大量繁殖。支持这种观点的事实不计其数，如果有人对这一观点还存在异议的话，我可以在贵刊上引述这些事实。这就是使罗马的强大力量衰退的原因，也是令美索不达米亚城没落的原因。这些力量不是被野蛮人的异常体力打垮的，而是被自己群体不断缩小的现实击败的，他们是由于内因而迅速灭亡，而其中主要的内因就是其居民的不育情况在不断增加。

The same cause operated to extinguish the Tasmanians and other savage tribes which have decayed and died out, when brought into contact with the luxuries of civilisation, notwithstanding every effort having been made to preserve them. In a few cases only have the weak tribes been supplanted by the strong, or weaker individuals by stronger; the decay has been internal, and of remoter origin. It has been luxury and not want; too much vigour and not too little, that has eviscerated and destroyed the race. If this law then be universal both in the vegetable and animal kingdoms, a law too, which does not operate on individuals and in isolated cases only, but universally, it is surely incumbent upon the supporters of the doctrine of Natural Selection, as propounded by Mr. Darwin, to meet and to explain it, for it seems to me to cut very deeply into the foundations of their system. If it be true that, far from the strong surviving the weak, the tendency among the strong, the well fed, and highly favoured, is to decay, become sterile, and die out, while the weak, the under-fed, and the sickly are increasing at a proportionate rate, and that the fight is going on everywhere among the individuals of every race, it seems to me that the theory of Natural Selection, that is, of the persistence of the stronger, is false, as a general law, and true only of very limited and exceptional cases. This paper deals with one difficulty only, others may follow if this is acceptable.

Henry H. Howorth

(**4**, 161-162; 1871)

*   *   *

I am much obliged to Mr. Howorth for his courteous expressions towards me in the letter in your last number. If he will be so good as to look at p. 111 and p. 148, vol. II of my "Variation of Animals and Plants under Domestication", he will find a good many facts and a discussion on the fertility and sterility of organisms from increased food and other causes. He will see my reasons for disagreeing with Mr. Doubleday, whose work I carefully read many years ago.

Charles Darwin

(**4**, 180-181; 1871)

*   *   *

The very ingenious manner in which Mr. Howorth first misrepresents Darwinism, and then uses an argument which is not even founded on his own misrepresentation, but on a quite distinct fallacy, may puzzle some of your readers. I therefore ask space for a few lines of criticism.

Mr. Howorth first "takes it" that the struggle for existence "means, in five words, the persistence of the stronger". This is a pure misrepresentation. Darwin says nothing of the kind. "Strength" is only one out of the many and varied powers and faculties that lead

　　同样的原因使塔斯马尼亚和其他的野蛮部落灭绝了，进入奢侈的文化状态时，尽管他们做了一切努力来保护自己，但还是衰退并最终消亡了。只有在少数的几个例子中，弱小的部落才被强壮的部落取代，或者较弱的个体被较强的取代；衰退起源于远古时代传承下来的内因。种族元气大伤或毁灭是由于奢侈而非欲望，是由于他们的精力太充沛而非太少。如果这一规律在植物界和动物界都是普遍的，它们不是只对个体起作用或者只存在于极个别的案例中，而是普遍存在的，那么这就是支持达尔文先生提出来的自然选择学说的人应该解决的问题了，因为这似乎已经触及到了自然选择学说体系的基础。如果这是真的，那么远非强者生存弱者淘汰，而是强壮的、营养充足的、非常有利的群体趋向于衰退、不育，直至灭绝，而瘦弱的、营养不良的、多病的群体却以适当的比率在逐渐增加，并且各个地方的每个种族内个体间的战斗一直在进行着，我觉得作为一般规律而言，自然选择理论所认为的强者的延续这一观点是错误的，这种现象只在几个有限的、个别的例子中才存在。这篇文章旨在讨论这一疑问，如果您接受该观点，可以继续讨论下去。

亨利·霍沃思

*　　*　　*

　　对于霍沃思先生在贵刊上一期刊登的信中对我的恭维，我很感激。我想恳请他看一下我的《动物和植物在家养下的变异》第 2 卷，第 111、148 页，他就会发现许多事实，以及对食物增加和其他原因导致的生物繁殖和不育性进行的讨论。关于道布尔迪先生，我在许多年前就仔细读过了他的著作，在上述两页中，霍沃思先生也会看到我不同意道布尔迪先生的观点的原因所在。

查尔斯·达尔文

*　　*　　*

　　霍沃思先生首次以非常独创的方式对达尔文学说进行了歪曲，他使用了一个论据，该论据甚至不是建立在他自己曲解的基础上，而是建立在一个非常明显的谬论的基础上，这些有可能会误导贵刊的读者。因此我请求贵刊允许我在此占用一点空间作几句评论。

　　霍沃思先生首先将生存斗争"理解"为"五个字：强者的延续"。这是一个纯粹的曲解。达尔文没有说过任何这种意思的话。"强"只是许多不同的有助于在生存斗

to success in the battle for life. Minute size, obscure colours, swiftness, armour, cunning, prolificness, nauseousness, or bad odour, have any one of them as much right to be put forward as the cause of "persistence". The error is so gross that it seems wonderful that any reader of Darwin could have made it, or, having made it, could put it forward deliberately as a fair foundation for a criticism. He says, moreover, that the theory of Natural Selection "has been expressively epitomised" as "the persistence of the stronger", "the survival of the stronger". By whom? I should like to know. I never saw the terms so applied in print by any Darwinian. The most curious and even ludicrous thing, however, is that, having thus laid down his premisses. Mr. Howorth makes no more use of them, but runs off to something quite different, namely, that *fatness* is prejudicial to fertility. "Fat hens won't lay", "overgrown melons have few seeds", "overfed men have small families",—these are the *facts* by which he seeks to prove that the *strongest* will not survive and leave offspring! But what does nature tell us? That the strongest and most vigorous plants *do* produce the most flowers and seed, not the weak and sickly. That the strongest and most healthy and best fed wild animals *do* propagate more rapidly than the starved and sickly. That the strong and thoroughly well-fed backwoodsmen of America increase more rapidly than any half-starved race of Indians upon earth. No *fact*, therefore, has been adduced to show that even "the persistence of the stronger" is not true; although, if this had been done, it would not touch Natural Selection, which is the "survival of the fittest".

Alfred R. Wallace

(**4**, 181; 1871)

Henry H. Howorth: Derby House, Eccles.
Charles Darwin: Down, Beckenham, Kent, July 1.

争中取得成功的力量和能力中的一种。微小的体型、晦暗的颜色、敏捷、防护具、狡猾、大量繁殖、腐臭或者不好的气味，拥有这些特点中的任何一点都可以被当作是"延续"下来的原因。达尔文学说的读者中竟然有人会犯这种显而易见的错误，或者已经犯了，并且还故意地将这种错误作为一种公正的批评根据提出来，这实在太让人吃惊了。此外，他还说自然选择理论"已经被概括表述"为"强者的延续"和"强者生存"。是谁这样概括的？我倒是很想知道。我从没有看到任何达尔文主义者在任何出版物中使用过这样的词语。然而，最令人好奇和最为荒谬的事情是，霍沃思先生在已经作出这样的谬论铺垫之后却再没有使用这些前提，而是转到了别的毫不相关的问题上，也就是，**肥胖**有损于生育能力。"肥鸡不下蛋""生长过度的甜瓜几乎没有种子""吃得太多的人往往只有很小的家庭"——这些就是他想用来证明**最强者**将不能幸存并且不能留下后代的**事实**！但是自然界告诉我们的又是什么？产出了最多的花朵和种子的**正是**那些最强壮、最有活力的植物，而不是那些虚弱多病的植物。同样最强壮、最健康、吃得最好的野生动物**确实**比那些挨饿、多病的动物繁殖得更快。那些住在美洲大陆森林地区中的强壮且营养非常充足的人也比那片土地上总是处于半饥饿状态的印第安人增加得更快。因此没有任何**事实**可以用来说明"强者的延续"是不真实的，退一步说，即使存在这样的例子，那也撼动不了自然选择学说所认为的"适者生存"的法则。

阿尔弗雷德·华莱士

（刘皓芳 翻译；刘力 审稿）

Australopithecus africanus: the Man-ape of South Africa

Editor's Note

Raymond Dart's discovery of the face and brain cast of a juvenile ape-like creature in South Africa, reported here, can be marked as the beginning of the modern era of the study of fossil man. Until that date, all members of the fossil human family were either definitely apes, such as Dryopithecus, or clearly close to humans, such as Neanderthal Man or Pithecanthropus (nowadays *Homo erectus*). Having something so clearly transitional raised challenging questions about the course of human evolution. One was the very human-like teeth associated with a small, ape-like brain, at complete variance with the then-current dogma that human ancestors evolved bigger brains before human-like teeth—amplified by Piltdown Man, now known to have been a hoax.

TOWARDS the close of 1924, Miss Josephine Salmons, student demonstrator of anatomy in the University of the Witwatersrand, brought to me the fossilised skull of a cercopithecid monkey which, through her instrumentality, was very generously loaned to the Department for description by its owner, Mr. E. G. Izod, of the Rand Mines Limited. I learned that this valuable fossil had been blasted out of the limestone cliff formation—at a vertical depth of 50 feet and a horizontal depth of 200 feet—at Taungs, which lies 80 miles north of Kimberley on the main line to Rhodesia, in Bechuanaland, by operatives of the Northern Lime Company. Important stratigraphical evidence has been forthcoming recently from this district concerning the succession of stone ages in South Africa (Neville Jones, *Jour. Roy. Anthrop. Inst.*, 1920), and the feeling was entertained that this lime deposit, like that of Broken Hill in Rhodesia, might contain fossil remains of primitive man.

I immediately consulted Dr. R. B. Young, professor of geology in the University of the Witwatersrand, about the discovery, and he, by a fortunate coincidence, was called down to Taungs almost synchronously to investigate geologically the lime deposits of an adjacent farm. During his visit to Taungs, Prof. Young was enabled, through the courtesy of Mr. A. F. Campbell, general manager of the Northern Lime Company, to inspect the site of the discovery and to select further samples of fossil material for me from the same formation. These included a natural cercopithecid endocranial cast, a second and larger cast, and some rock fragments disclosing portions of bone. Finally, Dr. Cordon D. Laing, senior lecturer in anatomy, obtained news, through his friend Mr. Ridley Hendry, of another primate skull from the same cliff. This cercopithecid skull, the possession of Mr. De Wet, of the Langlaagte Deep Mine, has also been liberally entrusted by him to the Department for scientific investigation.

南方古猿非洲种：南非的人猿

达特

编者按

这篇文章报道的是雷蒙德·达特在南非发现了一件幼年类人猿的头骨，这被认为是现代人类化石研究的开端。在此之前，所有的人类化石要么确定无疑地属于猿类，比如森林古猿，要么非常接近人类，比如尼安德特人和爪哇猿人（现在称为直立人）。这块化石具有非常明确的过渡性特征，这对人类的进化历程提出了质疑。这块化石中的牙齿与人类的牙齿非常相似，脑比较小，类似于古猿，这与当时的主流观点是完全相悖的。由于皮尔当人的发现，当时人们广泛接受的主流观点认为，人类祖先脑量的增大先于类似现代人牙齿的出现，不过现在看来皮尔当人不过是场骗人的闹剧。

在 1924 年岁末年终之际，威特沃特斯兰德大学解剖学专业的学生助教约瑟芬·萨蒙斯小姐给我带来了一件猴的头骨化石。因为她的关系，这件化石的主人兰德矿业有限公司的伊佐德先生才非常慷慨地把头骨化石借给学院用于描述研究。我得知这块珍贵的化石是从垂直高 50 英尺，水平宽 200 英尺的一个石灰岩悬崖中炸出来的，地点是在汤恩。汤恩位于金伯利北部 80 英里，在通往贝专纳兰的罗得西亚市的主干线上，为北方石灰公司所有。最近出现的重要的地层学证据表明，这一地区与南非岩层年代的连续性有关（内维尔·琼斯，《皇家人类学研究院院刊》，1920 年），我想这个石灰岩沉积层可能像罗得西亚的布罗肯希尔山一样，可能包含原始人类的化石遗迹。

我立即与威特沃特斯兰德大学的地质学教授扬博士讨论了相关发现。非常巧合的是，几乎同时，他被派遣到汤恩附近的一个农场去调查石灰岩沉积层的地质情况。在汤恩，经北方石灰公司总经理坎贝尔先生的首肯，扬教授获准探查化石发现地，并从同一形成层中给我挑选了更多的化石标本。这些标本包括一个天然的猴颅内模，另一个更大的颅内模，和一些漏出部分骨头的岩石碎块。后来，一位年长的解剖学教师科登·莱恩博士通过他的朋友里德利·亨德里先生获知，同一悬崖中又发现了一件灵长类头骨。这块头骨来自兰拉格特深矿，已经由他的拥有者德威特先生委托给学院作科研之用。

Fig. 1. Norma facialis of *Australopithecus africanus* aligned on the Frankfort horizontal

The cercopithecid remains placed at our disposal certainly represent more than one species of catarrhine ape. The discovery of Cercopithecidae in this area is not novel, for I have been informed that Mr. S. Haughton has in the press a paper discussing at least one species of baboon from this same spot (Royal Society of South Africa). It is of importance that, outside of the famous Fayüm area, primate deposits have been found on the African mainland at Oldaway (Hans Reck, *Silsungsbericht der Gesellsch. Naturforsch. Freunde*, 1914), on the shores of Victoria Nyanza (C. W. Andrews, *Ann. Mag. Nat. Hist.*, 1916), and in Bechuanaland, for these discoveries lend promise to the expectation that a tolerably complete story of higher primate evolution in Africa will yet be wrested from our rocks.

In manipulating the pieces of rock brought back by Prof. Young, I found that the larger natural endocranial cast articulated exactly by its fractured frontal extremity with another piece of rock in which the broken lower and posterior margin of the left side of a mandible was visible. After cleaning the rock mass, the outline of the hinder and lower part of the facial skeleton came into view. Careful development of the solid limestone in which it was embedded finally revealed the almost entire face depicted in the accompanying photographs.

It was apparent when the larger endocranial cast was first observed that it was specially important, for its size and sulcal pattern revealed sufficient similarity with those of the chimpanzee and gorilla to demonstrate that one was handling in this instance an anthropoid and not a cercopithecid ape. Fossil anthropoids have not hitherto been recorded south of the Fayüm in Egypt, and living anthropoids have not been discovered in recent times south of Lake Kivu region in Belgian Congo, nearly 2,000 miles to the north, as the crow flies.

All fossil anthropoids found hitherto have been known only from mandibular or maxillary fragments, so far as crania are concerned, and so the general appearance of the types they

图 1. 南方古猿非洲种的前面观（已在眼耳平面上对齐）

我们手里有的这些猴化石绝不仅仅只代表狭鼻猴的一个种。在这一地区发现猴类物种并不是什么新闻，就我所知，霍顿先生的一篇已投稿的文章中讨论了至少一种来自这一地区的狒狒（南非皇家学会）。更重要的是，在著名的法尤姆地区之外，非洲大陆的奥德威（汉斯·雷克，《研究者协会会刊》，1914 年），维多利亚–尼亚萨湖岸（安德鲁斯，《自然史年鉴》，1916 年），以及贝专纳兰都有灵长类化石发现，这些发现使我有希望从我们这些化石研究中获得一个关于非洲高等灵长类动物进化的相对完整的故事，当然这需要费一番辛苦。

在处理扬教授带回的这些化石岩块时，我发现较大的那块颅内模在额骨前端破裂了，但它正好与另一块相连，其中可以清楚地看到下颌骨左侧的后下缘。清理完这些岩块，面部骨骼的后下部分轮廓就呈现出来了。然后，经过小心处理包埋着头骨的坚硬的石灰石，最终一张几乎完整的面部出现了（如图所示）。

很明显，这个大的颅内模的首次发现相当重要，因为它的尺寸和沟回形状与黑猩猩和大猩猩非常相似，这表明我们手中的并非猴类化石，而是一个类人猿。到目前为止在埃及法尤姆以南至今也没有发现类人猿化石，比属刚果的基伍湖（按照直线距离计算距北部将近 2,000 英里）以南的地区也没有关于现存类人猿的记录。

到目前为止，所有关于类人猿化石的知识，就头骨而言，仅仅来源于上颌骨或者下颌骨，这些化石所代表的类型也都不清楚。因此事实上，长满牙齿的完整面部

represented has been unknown; consequently, a condition of affairs where virtually the whole face and lower jaw, replete with teeth, together with the major portion of the brain pattern, have been preserved, constitutes a specimen of unusual value in fossil anthropoid discovery. Here, as in *Homo rhodesiensis*, Southern Africa has provided documents of higher primate evolution that are amongst the most complete extant.

Apart from this evidential completeness, the specimen is of importance because it exhibits an extinct race of apes *intermediate between living anthropoids and man.*

In the first place, the whole cranium displays *humanoid* rather than anthropoid lineaments. It is markedly dolichocephalic and leptoprosopic, and manifests in a striking degree the *harmonious relation* of calvaria to face emphasised by Pruner-Bey. As Topinard says, "A cranium elongated from before backwards, and at the same time elevated, is already in harmony by itself; but if the face, on the other hand, is elongated from above downwards, and narrows, the harmony is complete." I have assessed roughly the difference in the relationship of the glabella-gnathion facial length to the glabella-inion calvarial length in recent African anthropoids of an age comparable with that of this specimen (depicted in Duckworth's "Anthropology and Morphology", second edition, vol. I), and find that, if the glabella-inion length be regarded in all three as 100, then the glabella-gnathion length in the young chimpanzee is approximately 88, in the young gorilla 80, and in this fossil 70, which proportion suitably demonstrates the enhanced relationship of cerebral length to facial length in the fossil (Fig. 2).

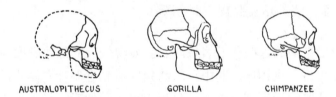

AUSTRALOPITHECUS   GORILLA   CHIMPANZEE

Fig. 2. Cranial form in living anthropoids of similar age (after Duckworth) and in the new fossil. For this comparison, the fossil is regarded as having the same calvarial length as the gorilla.

The glabella is tolerably pronounced, but any traces of the salient supra-orbital ridges, which are present even in immature living anthropoids, are here entirely absent. Thus the relatively increased glabella-inion measurement is due to brain and not to bone. Allowing 4 mm for the bone thickness in the inion region, that measurement in the fossil is 127 mm; *i.e.* 4 mm less than the same measurement in an adult chimpanzee in the Anatomy Museum at the University of the Witwatersrand. The orbits are not in any sense detached from the forehead, which rises steadily from their margins in a fashion amazingly human. The interorbital width is very small (13 mm) and the ethmoids are not blown out laterally as in modern African anthropoids. This lack of ethmoidal expansion causes the lacrimal fossae to face posteriorly and to lie relatively far back in the orbits, as in man. The orbits, instead of being subquadrate as in anthropoids, are almost circular, furnishing an orbital index of 100, which is well within the range of human variation (Topinard,

和下颌，连同脑结构的主要部分，被一起保留了下来，这构成了类人猿发现史上一件不寻常的宝贵的标本。在这里，如同罗德西亚人，南非提供了有关高等灵长类动物进化的现存最完整的资料。

抛开证据的完整性不谈，这个标本的重要性在于它揭示了一种已经灭绝的猿类，**介于现存类人猿与人类之间**的中间类型。

首先，整个头骨轮廓显示的是**人类的**特征，而非类人猿的。其显著特征：颅长，面窄，这些都非常符合普瑞纳贝（人类学家）强调的颅顶与脸的**和谐关系**。如托皮纳尔所说："头盖骨从前向后延伸，同时抬高，本身已经是和谐的；但另一方面如果面部 也从下向上延伸，并且变窄，就和谐完整了。"在粗略估计了这个标本和与之年龄相近的近代非洲类人猿的面长（眉间至颌下点）与颅长（眉间至枕骨隆突）比例的差异后（详见迪克沃斯的《人类学与形态学》，第2版，第1卷），我发现，如果三个物种的颅长都为100，那么年轻黑猩猩的面长为88，年轻大猩猩的面长为80，这个化石标本的面长则为70。这一比例很好地证明了化石中颅长与面长之比不断增加的关系（图2）。

AUSTRALOPITHECUS　　　GORILLA　　　CHIMPANZEE

图2. 新发现化石及年岁接近的现存类人猿（根据迪克沃斯的著作）的颅骨形状通过这个比较可以看到，新发现化石的颅盖骨与大猩猩的一样长

在这个化石标本中，眉间还算突出，然而突出的眶上脊却完全缺失，这一特征即使在现存的幼年类人猿中也是存在的。这就是说相对增长的眉间至枕外隆突之间的距离是由于脑量增长而并非骨骼生长所致。留出4毫米作为枕外隆突区域的骨骼厚度，这个化石的颅长为127毫米，也只比威特沃特斯兰德大学解剖学博物馆中成年黑猩猩的颅长少4毫米。其眼眶没有任何从前额分离的迹象，而是从边缘平稳突出，与人类的眼眶模式极其相似。眼间宽度很小（13毫米），而筛骨也没有像现代非洲类人猿一样从侧面鼓起。由于筛骨没有膨大，使得其泪腺沟朝后，像人一样位于眼眶相对较后的位置。眼眶几乎是圆形的，不像类人猿的方形。设定眼眶指数为100，则它完全位于人类变化范围内（托皮纳尔，《人类学》）。颧骨，颧弓，上颌骨和下颌骨，所有这些显示的是精巧的人类的特征。面部突颌度相对轻微，弗劳尔颌指数

"Anthropology"). The malars, zygomatic arches, maxillae, and mandible all betray a delicate and humanoid character. The facial prognathism is relatively slight, the gnathic index of Flower giving a value of 109, which is scarcely greater than that of certain Bushmen (Strandloopers) examined by Shrubsall. The nasal bones are not prolonged below the level of the lower orbital margins, as in anthropoids, but end above these, as in man, and are incompletely fused together in their lower half. Their maximum length (17 mm) is not so great as that of the nasals in *Eoanthropus dawsoni*. They are depressed in the median line, as in the chimpanzee, in their lower half, but it seems probable that this depression has occurred post-mortem, for the upper half of each bone is arched forwards (Fig. 1). The nasal aperture is small and is just wider than it is high (17 mm × 16 mm). There is no nasal spine, the floor of the nasal cavity being continuous with the anterior aspect of the alveolar portions of the maxillae, after the fashion of the chimpanzee and of certain New Caledonians and negroes (Topinard, *loc. cit.*).

In the second place, the dentition is *humanoid* rather than anthropoid. The specimen is juvenile, for the first permanent molar tooth only has erupted in both jaws on both sides of the face; *i.e.* it corresponds anatomically with a human child of six years of age. Observations upon the milk dentition of living primates are few, and only one molar tooth of the deciduous dentition in one fossil anthropoid is known (Gregory, "The Origin and Evolution of the Human Dentition", 1920). Hence the data for the necessary comparisons are meagre, but certain striking features of the milk dentition of this creature may be mentioned. The tips of the canine teeth transgress very slightly (0.5–0.75 mm) the general margin of the teeth in each jaw, *i.e.* very little more than does the human milk canine. There is no diastema whatever between the premolars and canines on either side of the lower jaw, such as is present in the deciduous dentition of living anthropoids; but the canines in this jaw come, as in the human jaw, into alignment with the incisors (Gregory, *loc. cit.*). There is a diastema (2 mm on the right side, and 3 mm on the left side) between the canines and lateral incisors of the upper jaw; but seeing, first, that the incisors are narrow, and, secondly, that diastemata (1 mm–1.5 mm) occur between the central incisors of the upper jaw and between the medial and lateral incisors of both sides in the lower jaw, and, thirdly, that some separation of the milk teeth takes place even in mankind (Tomes, "Dental Anatomy", seventh edition) during the establishment of the permanent dentition, it is evident that the diastemata which occur in the upper jaw are small. The lower canines, nevertheless, show wearing facets both for the upper canines and for the upper lateral incisors.

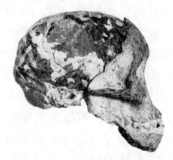

Fig. 3. Norma lateralis of *Australopithecus africanus* aligned on the Frankfort horizontal.

为 109，这一指数几乎不大于舒本萨尔检测过的布希曼人。鼻骨向下延伸不低于眶下沿水平，这点像类人猿，而止于眶下沿之上，这点又像人类，并且下半部分不完全地融合在一起。鼻骨的最大长度为 17 毫米，不像道森曙人的那么大。鼻骨下半部分从中线部位下陷，这点像黑猩猩，然而这种下陷也可能是发生在死后，因为其上半部分的每块骨头都向前拱起（图 1）。鼻孔小，其宽度刚好大于其高度（17 毫米 × 16 毫米）。没有鼻棘，鼻腔底面与上颌齿槽部的前面相连，这与黑猩猩和一些新苏格兰人及黑人的样式相仿（如前面托皮纳尔所述）。

其次，其齿系是**人类的**而非类人猿的。从两侧第一恒臼齿刚刚萌出可以判断这一标本还是幼年，从解剖学上说大约相当于人类的 6 岁儿童。对现存灵长类乳齿系的研究还很少，而化石类人猿中也只有一个乳臼齿的记录（格雷戈里，《人类齿系的起源与进化》，1920 年）。因此对乳齿作必要比较的数据太少，但是这个物种乳齿系显著的特征还是值得一提的。在两颌中，犬齿尖端略微超出整个齿列（0.5 ~ 0.75 毫米），例如，只比人类的犬齿突出一点点。下颌两侧的前臼齿和犬齿之间没有齿隙，现存类人猿的乳齿也是如此；然而其犬齿与门齿排列在一起，这点与人类一样（格雷戈里，见上述引文）。上颌犬齿与侧门齿之间有齿隙（右侧为 2 毫米，左侧为 3 毫米），但应注意到：（1）门齿狭窄；（2）上颌中门齿之间以及下颌两侧中门齿与侧门齿之间都有间隙裂（1~1.5 毫米）；（3）在恒齿形成过程中，即使在人类中，乳齿分开的现象也有发生（托姆斯，《牙齿解剖学》，第 7 版）。显然上颌的牙间隙比较窄。然而，下犬齿有相对于上犬齿以及上侧门齿的磨损面。

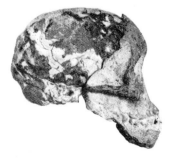

图 3. 南方古猿非洲种的侧面观（已在眼耳平面上对齐）

The incisors as a group are irregular in size, tend to overlap one another, and are almost vertical, as in man; they are not symmetrical and well spaced, and do not project forwards markedly, as in anthropoids. The upper lateral incisors do project forwards to some extent and perhaps also do the upper central incisors very slightly, but the lateral lower incisors betray no evidence of forward projection, and the central lower incisors are not even vertical as in most races of mankind, but are directed slightly backwards, as *sometimes* occurs in man. Owing to these remarkably human characters displayed by the deciduous dentition, when contour tracings of the upper jaw are made, it is found that the jaw and the teeth, as a whole, take up a parabolic arrangement comparable only with that presented by mankind amongst the higher primates. These facts, together with the more minute anatomy of the teeth, will be illustrated and discussed in the memoir which is in the process of elaboration concerning the fossil remains.

In the third place, the mandible itself is *humanoid* rather than anthropoid. Its ramus is, on the whole, short and slender as compared with that of anthropoids, but the bone itself is more massive than that of a human being of the same age. Its symphyseal region is virtually complete and reveals anteriorly a more vertical outline than is found in anthropoids or even in the jaw of Piltdown man. The anterior symphyseal surface is scarcely less vertical than that of Heidelberg man. The posterior symphyseal surface in living anthropoids differs from that of modern man in possessing a pronounced posterior prolongation of the lower border, which joins together the two halves of the mandible, and so forms the well-known *simian shelf* and above it a deep genial impression for the attachment of the tongue musculature. In this character, *Eoanthropus dawsoni* scarcely differs from the anthropoids, especially the chimpanzee; but this new fossil betrays no evidence of such a shelf, the lower border of the mandible having been massive and rounded after the fashion of the mandible of *Homo heidelbergensis*.

Fig. 4. Norma basalis of *Australopithecus africanus* aligned on the Frankfort horizontal.

其门齿大小不规则，倾向于彼此重叠，几乎垂直，这点像人类；它们呈不对称分布，空间分布也不甚合理，没有显著向前突出，这点像类人猿。上侧门齿的确有一定程度的向前突出，上中门齿似乎也略为有点向前突出；然而下侧门齿丝毫没有向前突出的迹象，并且下中门齿不是像大多数人种中那样整齐地垂直向上，而是像人类中**有时**发生的那样略为向后倾斜。基于乳齿系所显示的这些显著的人类特征，当绘出上颌的轮廓线之后，发现颌与牙齿在整体上呈抛物线的排列方式，仅能与高等灵长类动物中的人类相比拟。这些事实，以及更多的牙齿微细解剖特征将在即将发表的有关化石的论文中阐述和讨论。

第三，下颌骨本身是**人类的**而非类人猿的。从整体上看，下颌支短而纤细，与类人猿相仿，但骨头本身比同龄的人类的大。其联合区近乎完整，并且从前面显示比类人猿甚至比皮尔当人更为垂直的轮廓。其前端联合面几乎与海德堡人的一样垂直。现存类人猿下颌骨后联合面与现代人类的区别在于其下沿明显地向后延伸，将下颌骨后部连接在一起形成著名的**猿板**结构，这一结构上的深印迹在于舌头肌肉组织的附着处。在这一特征上，道森曙人与类人猿，尤其是黑猩猩几乎没有差别，但在这一新化石标本中没有找到这一结构，其下颌骨下沿粗壮而圆隆，类似于海德堡人的下颌骨。

图 4. 南方古猿非洲种的底面观（已在眼耳平面上对齐）

169

That hominid characters were not restricted to the face in this extinct primate group is borne out by the relatively forward situation of the foramen magnum. The position of the basion can be assessed within a few millimetres of error, because a portion of the right exoccipital is present alongside the cast of the basal aspect of the cerebellum. Its position is such that the basi-prosthion measurement is 89 mm, while the basi-inion measurement is at least 54 mm. This relationship may be expressed in the form of a "head-balancing" index of 60.7. The same index in a baboon provides a value of 41.3, in an adult chimpanzee 50.7, in Rhodesian man 83.7, in a dolichocephalic European 90.9, and in a brachycephalic European 105.8. It is significant that this index, which indicates in a measure the poise of the skull upon the vertebral column, points to the assumption by this fossil group of an attitude appreciably more erect than that of modern anthropoids. The improved poise of the head, and the better posture of the whole body framework which accompanied this alteration in the angle at which its dominant member was supported, is of great significance. It means that a greater reliance was being placed by this group upon the feet as organs of progression, and that the hands were being freed from their more primitive function of accessory organs of locomotion. Bipedal animals, their hands were assuming a higher evolutionary role not only as delicate tactual, examining organs which were adding copiously to the animal's knowledge of its physical environment, but also as instruments of the growing intelligence in carrying out more elaborate, purposeful, and skilled movements, and as organs of offence and defence. The latter is rendered the more probable, in view, first, of their failure to develop massive canines and hideous features, and, secondly, of the fact that even living baboons and anthropoid apes can and do use sticks and stones as implements and as weapons of offence ("Descent of Man", p. 81 *et seq.*).

Lastly, there remains a consideration of the endocranial cast which was responsible for the discovery of the face. The cast comprises the right cerebral and cerebellar hemispheres (both of which fortunately meet the median line throughout their entire dorsal length) and the anterior portion of the left cerebral hemisphere. The remainder of the cranial cavity seems to have been empty, for the left face of the cast is clothed with a picturesque lime crystal deposit; the vacuity in the left half of the cranial cavity was probably responsible for the fragmentation of the specimen during the blasting. The cranial capacity of the specimen may best be appreciated by the statement that the length of the cavity could not have been less than 114 mm, which is 3 mm greater than that of an adult chimpanzee in the Museum of the Anatomy Department in the University of the Witwatersrand, and only 14 mm less than the greatest length of the cast of the endocranium of a gorilla chosen for casting on account of its great size. Few data are available concerning the expansion of brain matter which takes place in the living anthropoid brain between the time of eruption of the first permanent molars and the time of their becoming adult. So far as man is concerned, Owen ("Anatomy of Vertebrates", vol. III) tells us that "The brain has advanced to near its term of size at about ten years, but it does not usually obtain its full development till between twenty and thirty years of age." R. Boyd (1860) discovered an increase in weight of nearly 250 grams in the brains of male human beings after they had reached the age of seven years. It is therefore reasonable to believe that the adult forms typified by our present specimen possessed brains which were larger than

在这一灭绝的灵长类中这样的人类特征不仅仅局限于面部，其枕骨大孔处于相对朝前的位置也是一个很好的例证。由于沿小脑基部轮廓的外侧处的一部分右枕骨保存了下来，对颅底点的测量可以控制在几毫米的误差之内。它的位置可以参考两个位置点：即颅底点距上颌齿槽前缘点 89 毫米，颅底点至枕外隆突至少 54 毫米。这可以用头"平衡指数"的方式表达为 60.7。狒狒这一指数为 41.3，成年黑猩猩为 50.7，罗得西亚人为 83.7，长头欧洲人为 90.9，短头欧洲人为 105.8。很重要的一点是，这一指数反映的是头骨在脊柱上的姿态，从这一指数可以推论，化石标本所代表的类群表现出比现代类人猿更直立的一种姿态。头部姿势的提升，以及伴随这一角度改变而来的整个身体框架的姿态的改进，对于生物体本身来说太重要了。这意味着这一类群的生物更多地依赖于脚作为身体行进的器官，而手则被解放出来，其功能不再只是原始的移动器官的附属品。两足动物的手被认为是一种高等的进化，因为手已经不仅仅只是一个精妙的触觉上的感知器官，使动物获得更丰富的物理环境知识，更重要的是手已经成为提高智力的一种工具，能够承担更精细的，更有目的性的，更有技巧性的运动，并作为防御以及进攻的器官。而后者的可能性居多，这些体现在，首先，没有形成巨大的犬齿和丑陋的面貌特征，其次，一个不容忽略的事实是，即使现存的狒狒和类人猿也能够并确实在使用树枝和石头作为工具以及攻击的武器（《人类的由来》，第 81 页）。

最后，除了考虑面部特征外，对于颅内模型还有一些新的考虑。这个标本包含右大脑半球和小脑半球（很幸运两者都贯穿整个背侧长度达到中线位置）以及左大脑半球的前面部分。化石标本的左面部包裹了一层别致的石灰石晶体沉积物，由此推断颅腔的其余部分可能是空的；而颅腔左半部分的中空可能是因为爆破中标本碎裂了。标本的颅容量可以通过腔体长度计算，这一标本的颅腔长度不小于 114 毫米，这比威特沃特斯兰德大学解剖学博物馆中成年黑猩猩的颅腔长度长 3 毫米，而比一只由于其体形巨大被挑选出来用于制作颅内模型的大猩猩的颅腔最大长度仅仅少 14 毫米。目前，在长出第一恒臼齿到成为成体这段时间，有关现存类人猿脑量扩张的数据还基本没有。就人类而言，欧文（《脊椎动物解剖学》，第 3 卷）指出，"在大约 10 岁时脑能够发育到接近成年的体积大小，而通常要到 20~30 岁才能发育完全。"博伊德（1860 年）发现 7 岁以后人类男性个体的脑重量将增长近 250 克。因此有理由相信，目前我们手头的这个化石标本代表的成年脑量，应该比这个幼年标本的大，如果不超过的话，也应该等于一个发育完全的成年大猩猩的脑量。

that of this juvenile specimen, and equalled, if they did not actually supersede, that of the gorilla in absolute size.

Whatever the total dimensions of the adult brain may have been, there are not lacking evidences that the brain in this group of fossil forms was distinctive in type and was an instrument of greater intelligence that that of living anthropoids. The face of the endocranial cast is scarred unfortunately in several places (cross-hatched in the dioptographic tracing—see Fig. 5). It is evident that the relative proportion of cerebral to cerebellar matter in this brain was greater than in the gorilla's. The brain does not show that general pre- and post-Rolandic flattening characteristic of the living anthropoids, but presents a rounded and well-filled-out contour, which points to a symmetrical and balanced development of the faculties of associative memory and intelligent activity. The pithecoid type of parallel sulcus is preserved, but the sulcus lunatus has been thrust backwards towards the occipital pole by a pronounced general bulging of the parieto-temporo-occipital association areas.

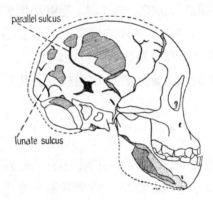

Fig. 5. Dioptographic tracing of *Australopithecus africanus* (right side), $\times \frac{1}{3}$.

To emphasise this matter, I have reproduced (Fig. 6) superimposed coronal contour tracings taken at the widest part of the parietal region in the gorilla endocranial cast and in this fossil. Nothing could illustrate better the mental gap that exists between living anthropoid apes and the group of creatures which the fossil represents than the flattened atrophic appearance of the parietal region of the brain (which lies between the visual field on one hand, and the tactile and auditory fields on the other) in the former and its surgent vertical and dorso-lateral expansion in the latter. The expansion in this area of the brain is the more significant in that it explains the posterior *humanoid* situation of the sulcus lunatus. It indicates (together with the narrow interorbital interval and human characters of the orbit) the fact that this group of beings, having acquired the faculty of stereoscopic vision, had profited beyond living anthropoids by setting aside a relatively much larger area of the cerebral cortex to serve as a storehouse of information concerning their objective environment as its details were simultaneously revealed to the senses of vision and touch, and also of hearing. They possessed to a degree unappreciated by living anthropoids the use of their hands and ears and the consequent faculty of associating with the colour,

172

无论成体脑的大小如何，我们都不难发现形成这种化石的种群的脑的类型，与现存类人猿相比不仅是明显不同的，而且是更加高级的。不幸的是颅腔标本表面有多处破损（图5，用交叉平行线画出的阴影部分）。显然，这个颅腔中大脑与小脑的相对比例高于大猩猩。这个脑没有表现出一般现存类人猿的罗蓝氏区前后扁平的模式特征，而是显示出一个变圆的、更充盈的轮廓，表明与记忆和智力活动相关的官能得到了对称的、平衡的发展。类人猿类型的平行沟得以保存下来，然而由于顶骨–颞骨–枕骨联合区域显著的整体突出使得月状沟被向后推向枕侧。

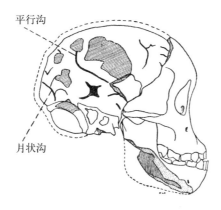

图 5. 南方古猿非洲种透视素描图（右侧），$\times \frac{1}{3}$

为了强调这一点，我绘制了大猩猩和这个化石标本的颅内模顶区最宽处的轮廓叠加线（图6）。大猩猩脑顶区（处于视觉区与触觉区及听觉区之间）扁平、萎缩，而化石标本相应区域陡立并向背部膨大，最好地说明了现存类人猿与化石标本代表的生物类群之间的智力差异。更重要的是这一脑区的膨大揭示了月状沟靠后这种**人类化**情况的原因。这一特征，加上窄的眶间隔，具有人类特征的眼眶，表明这一类群的生物已经获得了立体视觉的能力，这一超越现存类人猿的能力得益于将一块相对较大的大脑皮层区域设置为同时向视觉、触觉以及听觉传递有关客观环境细节的信息储藏库。一定程度上它们拥有类人猿不能相比的支配它们手、耳的使用的能力，并随之而来获知有关颜色、形状和物体的总体面貌、重量、质地、弹性和柔韧性，以及物体发出的声音所代表的意义的能力。换言之，与近代的猿类相应的器官相比，更有意识性和目的性地用它们的眼睛看，用它们的耳朵听，用它们的手进行操作。

173

form, and general appearance of objects, their weight, texture, resilience, and flexibility, as well as the significance of sounds emitted by them. In other words, their eyes saw, their ears heard, and their hands handled objects with greater meaning and to fuller purpose than the corresponding organs in recent apes. They had laid down the foundations of that discriminative knowledge of the appearance, feeling, and sound of things that was a necessary milestone in the acquisition of articulate speech.

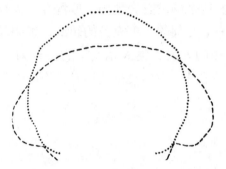

Fig. 6. Contour tracings of coronal sections through the widest part of the parietal region of the endocranial casts in Australopithecus... and in a gorilla...

There is, therefore, an ultra-simian quality of the brain depicted in this immature endocranial cast which harmonises with the ultra-simian features revealed by the entire cranial topography and corroborates the various inferences drawn therefrom. The two thousand miles of territory which separate this creature from its nearest living anthropoid cousins is indirect testimony to its increased intelligence and mastery of its environment. It is manifest that we are in the presence here of a pre-human stock, neither chimpanzee nor gorilla, which possesses a series of differential characters not encountered hitherto in any anthropoid stock. This complex of characters exhibited is such that it cannot be interpreted as belonging to a form ancestral to any living anthropoid. For this reason, we may be equally confident that there can be no question here of a primitive anthropoid stock such as has been recovered from the Egyptian Fayüm. Fossil anthropoids, varieties of Dryopithecus, have been retrieved in many parts of Europe, Northern Africa, and Northern India, but the present specimen, despite its youth, cannot be confused with anthropoids having the dryopithecid dentition. Other fossil anthropoids from the Siwalik hills in India (Miocene and Pliocene) are known which, according to certain observers, may be ancestral to modern anthropoids and even to man.

Whether our present fossil is to be correlated with the discoveries made in India is not yet apparent; that question can only be solved by a careful comparison of the permanent molar teeth from both localities. It is obvious, meanwhile, that it represents a fossil group distinctly advanced beyond living anthropoids in those two dominantly human characters of facial and dental recession on one hand, and improved quality of the brain on the other. Unlike Pithecanthropus, it does not represent an ape-like man, a caricature of precocious hominid failure, but a creature well advanced beyond modern anthropoids in just those characters, facial and cerebral, which are to be anticipated in an extinct

它们已经具有了辨别事物外貌、触感以及声音的基础，这对于语言能力（能够清晰发音）的获得是一个重要的里程碑。

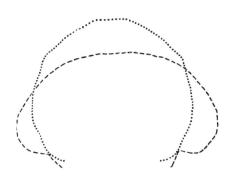

图 6. 南方古猿非洲种和大猩猩颅内模顶区最宽处的冠状面轮廓图

因此，这个未成熟的颅内模描绘的是一个具有超类人猿性质的脑，这与整个头骨解剖形态显示的超类人猿特征是一致的，并进一步证实了以此而来的各种推论。这个生物和它最近的现存类人猿兄弟们相隔了 2,000 英里的区域，这也间接证实了其智力的提高以及对环境的掌握。显然，摆在我们面前的是一个前人类类群，既不是黑猩猩也不是大猩猩，它已经拥有了一系列至今为止任何类人猿都不具有的特征。这些特征的复杂性表明，它不能被认为是任何现存类人猿的祖先。因此，我们同样可以确信，如同在埃及法尤姆所发现的，这里毫无疑问有一个原始类人猿类群的存在。在欧洲、南非、北印度的多个地区曾发现多种森林古猿的化石类人猿，然而我们的标本，尽管年轻，也不可能与具有森林古猿齿系的类人猿相混淆。根据其他人的观察，在印度（中新世和上新世）西瓦利克山脉发现的一些其他化石类人猿可能是现代类人猿甚至人类的祖先。

我们目前的化石是否与印度的发现相互关联还不清楚，这一问题只有通过仔细比较来自两地的恒白齿才能解决。然而，很显然的是它所代表的化石类群毫无疑问超越了现存类人猿，主要表现在，一方面已经具有显著的人类特征的面部和齿系，另一方面脑质的提高。它不像猿人属代表的是与猿相像的人类，一种早熟的人类的失败类型，而是代表一种从面部及脑部特征上已经远远超过当代类人猿的生物，这些面部及脑部特征被认为可能是人类及其猿类祖先的已经灭绝的连接类型中应有的

link between man and his simian ancestor. At the same time, it is equally evident that a creature with anthropoid brain capacity, and lacking the distinctive, localised temporal expansions which appear to be concomitant with and necessary to articulate man, is no true man. It is therefore logically regarded as a man-like ape. I propose tentatively, then, that a new family of *Homo-simiadae* be created for the reception of the group of individuals which it represents, and that the first known species of the group be designated *Australopithecus africanus*, in commemoration, first, of the extreme southern and unexpected horizon of its discovery, and secondly, of the continent in which so many new and important discoveries connected with the early history of man have recently been made, thus vindicating the Darwinian claim that Africa would prove to be the cradle of mankind.

It will appear to many a remarkable fact that an ultra-simian and pre-human stock should be discovered, in the first place, at this extreme southern point in Africa, and, secondly, in Bechuanaland, for one does not associate with the present climatic conditions obtaining on the eastern fringe of the Kalahari desert an environment favourable to higher primate life. It is generally believed by geologists (*vide* A. W. Rogers, "Post-Cretaceous Climates of South Africa", *South African Journal of Science*, vol. XIX, 1922) that the climate has fluctuated within exceedingly narrow limits in this country since Cretaceous times. We must therefore conclude that it was only the enhanced cerebral powers possessed by this group which made their existence possible in this untoward environment.

In anticipating the discovery of the true links between the apes and man in tropical countries, there has been a tendency to overlook the fact that, in the luxuriant forests of the tropical belts, Nature was supplying with profligate and lavish hand an easy and sluggish solution, by adaptive specialisation, of the problem of existence in creatures so well equipped mentally as living anthropoids are. For the production of man a different apprenticeship was needed to sharpen the wits and quicken the higher manifestations of intellect—a more open veldt country where competition was keener between swiftness and stealth, and where adroitness of thinking and movement played a preponderating role in the preservation of the species. Darwin has said, "no country in the world abounds in a greater degree with dangerous beasts than Southern Africa", and, in my opinion, Southern Africa, by providing a vast open country with occasional wooded belts and a relative scarcity of water, together with a fierce and bitter mammalian competition, furnished a laboratory such as was essential to this penultimate phase of human evolution.

In Southern Africa, where climatic conditions appear to have fluctuated little since Cretaceous times, and where ample dolomitic formations have provided innumerable refuges during life, and burial-places after death, for our troglodytic forefathers, we may confidently anticipate many complementary discoveries concerning this period in our evolution.

In conclusion, I desire to place on record my indebtedness to Miss Salmons, Prof. Young, and Mr. Campbell, without whose aid the discovery would not have been made; to

特征。同时，同样可以确信，一个生物只具有与类人猿同样大小的脑量并且缺少局部的颞区扩张（而这些是成为具有语言能力的人所必需的），它还不是真正意义上的人。因此，从逻辑上讲它应当被称为像人的猿。这样我试探性地建议，创立一个名为人猿科的新科来接纳标本代表的生物类群，并且将这一类群第一个已知的种命名为南方古猿非洲种，以纪念：第一，其极南的发现地和出人意料的地层，第二，其所在的大陆。在这一大陆上近来有非常多的重要发现与人类的早期历史相关，这将为达尔文所主张的非洲是人类的摇篮这一提议提供依据。

鉴于卡拉哈里沙漠东缘现在的气候条件无法与适宜高等灵长类生存的环境条件相联系，将会发生的一个毋庸置疑的事实是，超级猿类和前人类类群的发现应当在，第一，非洲大陆的极南端，第二，贝专纳兰。地质学家普遍承认（参见罗杰斯的文章《后白垩纪南非的气候》，刊载于《南非科学杂志》1922 年第 19 卷），从白垩纪以来这片大陆的气候只是在一个非常小的范围内波动。因此我们必然能推出，只有这个大脑能力已经提高的类群才能在这种不利的环境下存在。

在预期位于热带地区的国家将会发现连接猿类与人类真正的"接环"时，存在一种忽略如下事实的倾向，那就是在热带繁茂的森林中，大自然用其多产而慷慨的手，通过适应性的特化，为具有类人猿智力程度的生物们的生存问题提供了一个简单但却迟缓的解决方案。为了人类的产生，需要一个不同的学徒期以磨砺它们的智力，提高它们的理解力。然而正是一片更为广阔的草原地带为之提供了必要的条件，在这里迅捷与隐秘之间的竞争更加尖锐，在这里敏捷的思维和运动在物种生存中扮演了极为重要的角色。达尔文说过，"世界上没有任何国家比南非拥有更多的危险野兽"，以我的观点，南非，由于拥有广阔的空旷地带和稀少的林带，水源相对匮乏，加上哺乳动物之间的残忍严酷的竞争，提供了对于人类进化史上这倒数第二个阶段来说至关重要的实验室。

南非，这里的气候从白垩纪以来波动就很小，而丰富的白云岩层则为我们的穴居人祖先提供了无数的生活居所和死亡墓地，我们可以自信地预期，还将会继续发现有关这一时期的更多的人类进化的补充证据。

最后，我要感谢萨蒙斯小姐、扬教授和坎贝尔先生，没有他们的帮助就没有这些发现；感谢莱恩·理查森先生提供的照片；感谢莱恩博士和我实验室的同事们的热

Mr. Len Richardson for providing the photographs; to Dr. Laing and my laboratory staff for their willing assistance; and particularly to Mr. H. Le Helloco, student demonstrator in the Anatomy Department, who has prepared the illustrations for this preliminary statement.

(**115**, 195-199; 1925)

Prof. Raymond A. Dart: University of the Witwatersrand, Johannesburg, South Africa.

情帮助；特别要感谢解剖学学院的学生管理员勒埃洛克先生为这个初步报告准备了插图。

（刘晓辉 翻译；赵凌霞 审稿）

The Fossil Anthropoid Ape from Taungs

Editor's Note

Raymond Dart's announcement of *Australopithecus africanus* in *Nature* the previous week, and his assertion that it was intermediate between apes and humans, was greeted with a chorus of very faint praise in this quartet of letters from the anthropological establishment. All welcomed the discovery, but preferred to consider Australopithecus as very definitely an ape whose human-like features could be attributed to the fact that the fossil was of a child, whose adult form could not yet be discerned. The tone of all letters was courteous—except for Smith Woodward's criticism of the "barbarous" merger of Latin and Greek to create "Australopithecus".

THE discovery of fossil remains of a "man ape" in South Africa raises many points of great interest for those who are studying the evolution of man and of man-like apes. No doubt when Prof. Dart publishes his full monograph of his discovery, he will settle many points which are now left open, but from the facts he has given us, and particularly from the accurate drawing of the endocranial cast and skull in profile, it is even now possible for an onlooker to assess the importance of his discovery. I found it easy to enlarge the profile drawing just mentioned to natural size and to compare it with corresponding drawings of the skulls of children and of young apes. When this is done, the peculiarities of Australopithecus become very manifest.

Prof. Dart regrets he has not access to literature which gives the data for gauging the age of young anthropoids. In the specimen he has discovered and described, the first permanent molar teeth are coming into use. Data which I collected 25 years ago show that these teeth reach this stage near the end of the 4th year, two years earlier than is the rule in man and two years later than is the rule in the higher monkeys. In evolution towards a human form there is a tendency to prolong the periods of growth. Man and the gorilla have approximately the same size of brain at birth; the rapid growth of man's brain continues to the end of the 4th year; in the gorilla rapid growth ceases soon after birth.

Prof. Dart recognises the many points of similarity which link Australopithecus to the great anthropoid apes—particularly to the chimpanzee and gorilla. Those who are familiar with the facial characters of the immature gorilla and of the chimpanzee will recognise a blend of the two in the face of Australopithecus, and yet in certain points it differs from both, particularly in the small size of its jaws.

In size of brain this new form is not human but anthropoid. In the 4th year a child has reached 81 percent of the total size of its brain; at the same period a young gorilla has obtained 85 percent of its full size, a chimpanzee 87 percent. From Prof. Dart's accurate diagrams one estimates the brain length to have been 118 mm—a dimension common in

汤恩发现的类人猿化石

编者按

一周前，雷蒙德·达特在《自然》上宣布了南方古猿非洲种的发现，他断言这是介于古猿与人类之间的中间类型。在下述4封信中，人类学研究领域的重要人物一致对此给出了名褒实贬的评价。所有人都表示欢迎这个发现，但同时也提出更倾向于认为南方古猿非常明显就是古猿，其与人类相像的特征可能是因为这块化石属于幼年的个体，其成年的形态还不能被辨认出来。史密斯·伍德沃德批评道，造出"南方古猿"这个名字属于"野蛮"拼合拉丁语和希腊语，除此之外，这几封信的语气都还算客气。

南非发现的"人猿"化石激起了那些正在研究人类和类人猿进化的人们的巨大兴趣。当达特教授发表关于其发现的详尽专著时，他无疑将解决现在尚待进一步讨论的许多问题，但是从他向我们展示的事实来看，尤其是从他对颅腔模型和头骨剖面精确的绘图来看，即使是现在，旁观者也可以掂量出其发现的重要性。我发现很容易将刚刚提到的剖面图放大到真实大小，也很容易将其与相应的人类的孩子和幼年猿类的头骨绘图进行比较。当这样做的时候，南方古猿的特性就会变得十分明显了。

达特教授遗憾的是自己没能使用那些提供鉴定幼年猿类年龄资料的文献。在由他发现并进行描述的这个标本中，第一恒臼齿正开始被使用。我在25年前搜集的数据表明这些牙齿达到这一阶段应该是接近4岁末的时候，比人类早了两年，比高等猴子则晚了两年。在向人类形式进化的过程中，存在发育期延长的趋势。人类和大猩猩在出生时具有大约相同大小的脑；人脑的快速生长一直持续到4岁末；而大猩猩的脑的快速生长在出生后不久就停止了。

达特教授识别出了将南方古猿和大型类人猿联系起来的许多相似点——尤其是将南方古猿与黑猩猩和大猩猩联系起来的相似之处。对未成年大猩猩和黑猩猩的面部特征熟悉的人会在南方古猿的脸上看到二者的混合体，然而有些方面南方古猿与这两者都不同，尤其是其颌骨较小。

在脑尺寸方面，这个新类型与人类不同，是类人猿式的。儿童的脑在4岁时达到了其脑总尺寸的81%；而同样年龄的幼年大猩猩已经达到了全部尺寸的85%，黑猩猩则达到了87%。根据达特教授精确的图表可以估计出脑的长度已经达到了

the brains of adult and also juvenile gorillas. The height of the brain above the ear-holes also corresponds in both Australopithecus and the gorilla—about 70 mm. But in width, as Prof. Dart has noted, the gorilla greatly exceeds the new anthropoid; in the gorilla the width of brain is usually about 100 mm; in Australopithecus the width is estimated at 84 mm. The average volume of the interior of gorillas' skulls (males and females) is 470 c.c., but occasional individuals run up to 620 c.c. One may safely infer that the volume of the brain in the juvenile Australopithecus described by Prof. Dart must be less than 450 c.c., and if we allow a 15 percent increase for the remaining stages of growth, the size of the adult brain will not exceed 520 c.c. At the utmost the volume of brain in this new anthropoid falls short of the gorilla maximum. Even if it be admitted, however, that Australopithecus is an anthropoid ape, it is a very remarkable one. It is a true long-headed or dolichocephalic anthropoid—the first so far known. In all living anthropoids the width of the brain is 82 percent or more of its length; they are round-brained or brachycephalic; but in Australopithecus the width is only 71 percent of the length. Here, then, we find amongst anthropoid apes, as among human races, a tendency to roundness of brain in some and to length in others. On this remarkable quality of Australopithecus Prof. Dart has laid due emphasis.

This side-to-side compression of the head taken in conjunction with the small size of jaws throw a side light on the essential features of Australopithecus. The jaws are considerably smaller than those of a chimpanzee of a corresponding age, and much smaller than those of a young gorilla. There is a tendency to preserve infantile characters, a tendency which has had much to do with the shaping of man from an anthropoid stage. The relatively high vault of the skull of Australopithecus and its narrow base may also be interpreted as infantile characters. It is not clearly enough recognised that the anthropoid and human skulls undergo remarkable growth changes leading to a great widening of the base and a lowering or flattening of the roof of the skull. In Australopithecus there is a tendency to preserve the foetal form.

When Prof. Dart produces his evidence in full he may convert those who, like myself, doubt the advisability of creating a new family for the reception of this new form. It may be that Australopithecus does turn out to be "intermediate between living anthropoids and man", but on the evidence now produced one is inclined to place Australopithecus in the same group or sub-family as the chimpanzee and gorilla. It is an allied genus. It seems to be near akin to both, differing from them in shape of head and brain and in a tendency to the retention of infantile characters. The geological evidence will help to settle its relationships. One must suppose we are dealing with fossil remains which have become embedded in the stalagmite of a filled-up cave or fissure of the limestone cliff.

May I, in conclusion, thank Prof. Dart for his full and clear description, and particularly for his accurate drawings. One wishes that discoverers of such precious relics would follow his example, and, in place of reproducing crude tracings and photographs, give the same kind of drawings as an engineer or an architect prepares when describing a new engine or a new building.

Arthur Keith

118 毫米——这个尺寸在成年和幼年大猩猩的脑中也是很常见的。耳孔之上的脑高度在南方古猿和大猩猩中是相当的——大约 70 毫米。但是宽度方面，正如达特教授注明的那样，大猩猩远远超过了新发现的类人猿；大猩猩中，脑的宽度通常是 100 毫米左右；而南方古猿的脑的宽度估计在 84 毫米左右。大猩猩头骨（雄性和雌性）的平均内容量是 470 毫升，但是个别个体达到了 620 毫升。我们可以有把握地推测达特教授描述的幼年南方古猿的脑量肯定不足 450 毫升，如果我们容许其在剩余的生长阶段还有 15% 的增长空间的话，那么成年脑将不会超过 520 毫升。这种新型类人猿的最大脑量比大猩猩的最大脑量小。然而，即使承认南方古猿是一种类人猿，它们也是一种非常特别的类型。它们是一种真正的长头型或长颅型的类人猿——这是目前为止知道的第一种。现存的所有类人猿的脑宽度都是其长度的 82% 以上；它们的脑都是圆形的或短颅型的；而南方古猿的脑宽度只有其长度的 71%。因此我们发现在类人猿中，就像各人种之间那样，有些具有圆头型趋势，而另一些则具有长头型趋势。达特教授突出强调了南方古猿的这一显著特征。

这种与颌骨较小有关的对头颅两侧的挤压从侧面为阐明南方古猿的本质特征提供了线索。其颌骨比相应年龄的黑猩猩小得多，比幼年大猩猩的小得更多。在进化过程中有着保留婴儿特征的趋势，这种趋势对于从类人猿阶段逐渐形成人类具有紧密关联。南方古猿相对较高的头骨顶盖以及狭窄的颅底也可以理解成是婴儿时期的特征。现在还不能十分清楚地认识到类人猿和人类头骨经历了导致颅底显著变宽以及颅顶变低或变平的显著的生长变化。南方古猿具有保留胎儿形式的趋势。

当达特教授将其证据悉数列出时，他可能转变了那些像我一样怀疑过创建一个新科来接纳这种新类型是否明智的人的观点。可能结果南方古猿确实是"介于现存的类人猿和人类之间的一种中间类型"，但是依据现在所列出的证据，人们倾向于将南方古猿放到与黑猩猩和大猩猩同样的群体或亚科中。南方古猿是一个同源的属，它似乎与黑猩猩和大猩猩都具有很近的亲缘关系，但是在头部和脑形状以及保留婴儿特征的趋势等方面有所不同。地质学证据将有助于解决其关系问题。必须想到我们正在研究的是那些埋藏在被填满了的山洞或者石灰岩悬崖裂缝的石笋中的化石。

最后，请允许我感谢达特教授给出的详实而清晰的描述，尤其是他那精确的绘图。人们希望这样一类珍贵化石的发现者会遵循他的榜样，像工程师或者建筑师在描述一种新发动机或新大楼时绘制的图画那样提供同样水平的绘图，而非只是复制粗糙的描图和照片。

阿瑟·基思

*   *   *

It is a great tribute to Prof. Dart's energy and insight to have discovered the only fossilised anthropoid ape so far obtained from Africa, excepting only the jaw of the diminutive Oligocene Propliopithecus from the Egyptian Fayum. Whether or not the interpretation of the wider significance he has claimed for the fossil should be corroborated in the light of further information and investigation, the fact remains that his discovery is of peculiar interest and importance.

The simian infant discovered by him is an unmistakable anthropoid ape that seems to be much on the same grade of development as the gorilla and the chimpanzee without being identical with either. So far Prof. Dart does not seem to have "developed" the specimen far enough to expose the crowns of the teeth and so obtain the kind of evidence which in the past has provided most of our information for the identification of the extinct anthropoids. Until this has been done and critical comparisons have been made with the remains of Dryopithecus and Sivapithecus, the two extinct anthropoids, that approach nearest to the line of man's ancestry, it would be rash to push the claim in support of the South African anthropoid's nearer kinship with man. Prof. Dart is probably justified in creating a new species and even a new genus for his interesting fossil: for if such wide divergences between the newly discovered anthropoid and the living African anthropoids are recognisable in an infant, probably not more than four years of age, the differences in the adults would surely be of a magnitude to warrant the institution of a generic distinction.

Many of the features cited by Prof. Dart as evidence of human affinities, especially the features of the jaw and teeth mentioned by him, are not unknown in the young of the giant anthropoids and even in the adult gibbon.

The most interesting, and perhaps significant, distinctive features are presented by the natural endocranial cast. They may possibly justify the claim that Australopithecus has really advanced a stage further in the direction of the human status than any other ape. But until Prof. Dart provides us with fuller information and full-size photographs revealing the details of the object, one is not justified in drawing any final conclusions as to the significance of the evidence.

The size of the brain affords very definite evidence that the fossil is an anthropoid on much the same plane as the gorilla and the chimpanzee. But while its brain is not so large as the big gorilla-cast used for comparison by Prof. Dart, it is obvious that it is bigger than a chimpanzee's brain and probably well above the average for the gorilla. But the fossil is an imperfectly developed child, whose brain would probably have increased in volume to the extent of a fifth had it attained the adult status. Hence it is probable the brain would have exceeded in bulk the biggest recorded cranial capacity for an anthropoid ape, about 650 c.c. As the most ancient and primitive human brain case, that of Pithecanthropus, is at least 900 c.c. in capacity, one might regard even a small advance on 650 c.c. as a definite approach to the human status. The most suggestive feature (in Prof. Dart's Fig. 5,

*   *   *

除了在埃及法尤姆发现的小型渐新世原上猿的唯一颌骨之外，达特教授的发现是迄今为止在非洲得到的唯一一个类人猿化石，他的精力和洞察力令人万分敬佩。无论他声明的这具化石的重要意义是否应该根据更多的信息与研究加以确认，他的发现具有特殊意义和重要性这一事实都不会改变。

达特教授发现的猿孩肯定是一只类人猿，该类人猿似乎处于与大猩猩和黑猩猩同样的发育阶段，但是与两者又都不相同。迄今为止，达特教授似乎并没有对该标本进行足够的"开发"以揭示其牙冠状况，所以得到的都是过去为我们鉴定已灭绝类人猿提供了绝大部分信息的那一类证据。除非完成了揭示牙冠状况的工作，并且对另两种与人类祖先世系最接近的已灭绝类人猿（森林古猿和西瓦古猿）的化石进行比较研究之后，我们才能有理由支持南非类人猿与人类的亲缘关系更近，否则这种说法就太轻率了。达特教授在为其感兴趣的化石创建一个新物种甚至一个新属方面可能是有道理的：因为，如果新发现的类人猿与现存的非洲类人猿在可能不超过4岁的婴儿期阶段就存在如此大的差异的话，那么它们的成年个体之间肯定差异更大，能确保形成属一级的区别。

达特教授引用了许多特征作为与人类具有亲缘关系的证据，尤其是他提到的颌骨和牙齿的特征，我们并不是不知道这些特征在巨型类人猿的幼年期是什么样子，甚至成年长臂猿的也很清楚。

最有趣的，可能也是最重要的独特特征是由天然的颅腔模型呈现出来的。这些颅腔模型可能可以证明如下说法的合理性，即南方古猿在向人类状态进化的方向上确实比其他猿类都迈进了更大一步。但是除非达特教授能够提供更充分的信息以及揭示我们研究对象形态细节的真实尺寸的照片，否则我们不能信服于对证据重要性所作的任何定论。

脑的大小提供了非常明确的证据，证明了该化石是与大猩猩和黑猩猩处于大致同一水平的一种类人猿。但是它的脑没有达特教授用来进行比较的大型大猩猩的模型大，不过它肯定比黑猩猩的脑大，并且可能远远超过了大猩猩的平均尺寸。但是该化石是一个尚未发育完全的孩子的，在其达到成年状态之前其脑量可能还有1/5的增长幅度。因此它的脑可能远远超过了现有大量记录的类人猿的约650毫升最大颅腔容量。作为最古老、最原始的人类脑壳，爪哇猿人的颅容量至少有900毫升，有人可能认为甚至还需要从650毫升经过小幅增长才能达到确定的接近于人类的状

p. 197) is the position of the sulcus lunatus and the extent of the parietal expansion that has pushed asunder the lunate and parallel sulci—a very characteristic human feature.

When fuller information regarding the brain is forthcoming—and no one is more competent than Prof. Dart to observe the evidence and interpret it—I for one shall be quite prepared to admit that an ape has been found the brain of which points the way to the emergence of the distinctive brain and mind of mankind. Africa will then have purveyed one more surprise—but only a real surprise to those who do not know their Charles Darwin. But what above all we want Prof. Dart to tell us is the geological evidence of age, the exact conditions under which the fossil was found, and the exact form of the teeth.

G. Elliot Smith

*   *   *

The new fossil from Taungs is of special interest as being the first-discovered skull of an extinct anthropoid ape, and Prof. Dart is to be congratulated on his lucid and suggestive preliminary description of the specimen. As usual, however, there are serious defects in the material for discussion, and before the published first impressions can be confirmed, more examples of the same skull are needed.

First, as Prof. Dart remarks, the fossil belongs to an immature individual with the milk-dentition, and, so far as can be judged from the photograph, I see nothing in the orbits, nasal bones, and canine teeth definitely nearer to the human condition than the corresponding parts of the skull of a modern young chimpanzee. The face seems to be relatively short, but the lower jaw of the Miocene Dryopithecus has already shown that this must have been one of the characters of the ancestral apes. The symphysis of the lower jaw may owe its shape and the absence of the "simian shelf" merely to immaturity; but it may be noted that a nearly similar symphysis has been described in an adult Dryopithecus, of which it may also be said that "the anterior symphyseal surface is scarcely less vertical than that of Heidelberg man" (see diagrams in *Quart. Journ. Geol. Soc.*, vol. 70, 1914, pp. 317, 319).

Secondly, the Taungs skull lacks the bones of the brain-case, so that the amount and direction of distortion of the specimen cannot be determined. I should therefore hesitate to attach much importance to rounding or flattening of any part of the brain-cast, and would even doubt whether the relative dimensions of the cast of the cerebellum can be relied on. Confirmatory evidence is needed of the reality of appearances in such a fossil.

In the absence of knowledge of the skulls of the fossil anthropoid apes represented by teeth and fragmentary jaws in the Tertiary formations of India, it is premature to express any opinion as to whether the direct ancestors of man are to be sought in Asia or in Africa. The new fossil from South Africa certainly has little bearing on the question.

态。最具提示性的特征（达特教授的图 5，第 197 页）是月状沟的位置和顶骨的扩展范围，后者将月状沟和平行沟推离开了——这是一种人类特有的特征。

当将来出现更全面的关于脑的信息时——没有人会比达特教授更应该在这个化石上观察证据并且给出解释——就我个人而言，将充分做好准备：承认已经发现一种猿类，它的脑指向人类特有的脑和智能出现的道路。那时非洲将提供又一个惊喜——但是只是对于那些不知道他们的查尔斯·达尔文为何许人的人来说，才是一个真正的惊喜。但是我们最希望达特教授告诉我们的是关于年代的地质学证据、发现化石的地点的准确状况以及牙齿的精确形式。

埃利奥特·史密斯

*　　*　　*

在汤恩发现的新化石特别有意思，因为这是第一次发现的一种已灭绝类人猿的头骨，在此祝贺达特教授对标本进行了清晰而具有提示性的初步描述。然而，与通常一样，在讨论部分还是存在严重的缺陷，所以在发表的第一印象能被确认之前，还需要更多同样头骨的例子。

首先，正如达特教授论述的，该化石属于一只具有乳牙齿系的未成年个体，就能从照片判断出来的信息而言，我看不到其具有比现代幼年黑猩猩头骨的相应部分明显更接近于人类状况的眼眶、鼻骨和犬齿。它的面部似乎相对较短，但是中新世森林古猿的下颌骨已经显示出这肯定是猿类祖先的特征之一。下颌联合部位的形状及没有"猿板"仅是因为该个体尚未成年；但是大家也许注意到一只成年森林古猿曾经被描述为具有一个很相似的下颌联合的情况，这也可以说是"下颌联合的前表面简直与海德堡人的一样垂直"（见刊载于《地质学会季刊》1914 年第 70 卷第 317、319 页的图）。

其次，汤恩头骨缺少脑壳骨骼，所以标本扭曲变形的程度和方向都无从确定。因此我对于认为脑模型任何部分的变圆和变得扁平有重要意义的观点都深表怀疑，甚至怀疑小脑模型的相对尺寸是否可信。对于这样一个化石，要想确定其真实的外观，还需要进一步证据来确认。

由于缺乏对在印度的第三纪地层发现的牙齿和颌骨断片所代表的类人猿头骨化石的了解，所以想要表达应该在亚洲还是非洲寻找人类直接祖先的任何观点还为时尚早。南非发现的这个新化石对于这一问题的解答毫无帮助。

Palaeontologists will await with interest Prof. Dart's detailed account of the new anthropoid, but cannot fail to regret that he has chosen for it so barbarous (Latin-Greek) a name as Australopithecus.

Arthur Smith Woodward

*   *   *

Prof. Dart's description of the fossil skull found at Taungs in Bechuanaland shows that this specimen possesses exceptional interest and importance. Should the claims made on its behalf prove good, then its discovery will in effect be comparable to those of the Pithecanthropus remains, of the Mauer mandible and the Piltdown fragments. In the following paragraphs I venture to make some comments based upon perusal of the article published in *Nature* of February 7.

First of all, the fact that the fragments came immediately under notice of so competent an anatomist as Prof. Dart establishes confidence in the thoroughness of the scrutiny to which they have been subjected. That the history of the specimen should be known precisely from the time of its release from the limestone matrix, provides another cause for satisfaction.

The specimen itself at once raises a number of questions, and, as Prof. Dart evidently realises, these fall into at least two categories. The first question arising out of the discovery is the status of the individual represented by these remains. But the answer to that question, and the presence of such a creature in South Africa, affect other problems. The latter include inquiry into the probable locality of origin of the simian and human types, and the search for evidence of dispersion from a centre, or along a line of successive migrations.

In dealing with the first problem, Prof. Dart has surveyed a considerable number of structural details, and he concludes that the specimen represents an extinct race of apes intermediate between living anthropoid apes and mankind. The specimen comprises the greater part of a skull with the lower jaw still in place (or nearly so). The number and characters of the teeth testify to the immaturity of the individual. The evidence on the last-mentioned point is quite definite, and interest thus comes to be centred in the status assigned to the specimen; namely, that of a form intermediate between the living anthropoid apes and man himself.

Prof. Dart places the specimen on the side of the living anthropoid apes in relation to the interval separating these from man. At the same time, it is claimed that this new form of ape is more man-like than any of the existing varieties of anthropoid apes; and so it comes about that the decision turns on the claims made for the superiority of the new ape to these other forms.

古生物学家会继续怀着兴趣等待达特教授对该新型类人猿作出更详细的说明，但是对于他为其选择了南方古猿这样一个如此野蛮的(拉丁–希腊语)名字不得不感到遗憾。

阿瑟·史密斯·伍德沃德

*   *   *

达特教授对发现于贝专纳兰的汤恩的头骨化石的描述表明该标本具有特别的影响和重要性。如果就这件化石发表的这个主张被证明是对的，那么实际上这件头骨的发现就堪比爪哇猿人化石、毛尔下颌骨化石和皮尔当颅骨破片化石的发现了。在接下来的段落中，我冒昧地根据自己对《自然》2 月 7 日发表的文章的精读进行评论。

首先，以上那些骨骼破片出现后很快就引起了像达特教授这样杰出的解剖学家的注意，这个事实使我们相信对这件标本的研究会是彻底的。应该由这件标本从石灰岩基质挖掘出来的时间开始精确地了解它的历史，这将提供另一个使人们对此发现感到满意的理由。

标本本身马上就引出了许多问题，正如达特教授明显意识到的，问题至少可以分为两类。从这个发现引出的第一个问题是这些化石所代表的个体的身份。但是这个问题的答案以及这样一种生物存于南非影响了其他问题。后者包括对猿猴和人类诸多类型起源的可能地点的探究，以及对从中心扩散开来或沿着相继迁徙的路线向外扩散的证据的寻找。

在研究第一个问题时，达特教授调查了大量结构细节，他得出的结论是，标本代表的是一种已灭绝的猿类，介于现存的类人猿和人类之间的中间类型。该标本由头骨的大部分组成，该头骨上还连有仍然处于原位（或接近原位）的下颌骨。牙齿的数目和特征说明该个体是未成年的。最后提到的这个论点的证据是非常确定的，因此兴趣就集中在了这件化石应该属于什么身份；也就是，一种介于现存类人猿和人类自身的中间类型。

达特教授将该标本置于把现存的类人猿与人类分开的间隔当中的类人猿这一边。同时，他主张这种新类型的猿比任何现存种类的类人猿都更像人类；所以就决定转而主张这种新型猿比其他类型的猿更加优越。

189

The report shows that (as noted above) many structural details have been scrutinised, and that all accessible parts of the specimen have been examined. The observations relate not only to the external parts of the skull and lower jaw, but also to the endocranial parts exposed to view by the partial shattering of the brain-case. The claims advanced on behalf of the higher status of the specimen are based, therefore, upon a number and variety of such details. Should Prof. Dart succeed in justifying these claims, the status he proposes for the new ape-form should be conceded. Much will depend on the interpretation of the features exhibited by the surface of the brain, as also upon that of all the characters connected therewith; and since Prof. Dart is so well equipped for that aspect of the inquiry, his conclusions must needs carry special weight there. In regard to the brain and its characters, I find the tracing of the contour of an endocranial cast in a gorilla-skull shown in Fig. 6 rather surprisingly flattened, and almost suggestive of the influence of age.

Among the anatomical characters enumerated in the article, some appear to me to possess a higher value in evidence than others. As good points in favour of the claims, there may be cited, in addition to the cerebral features to which reference has just been made, the level of the lower border of the nasal bones in relation to the lower orbital margins, the (small) length of the nasal bones, the lack of brow-ridges (even though the first permanent tooth has appeared fully), the steeply-rising forehead, and the relatively short canine teeth.

On the other hand, I feel fairly certain that some of the other characters mentioned are related preponderantly to the youthfulness of the specimen. Fully to appreciate the latter, demands not only the handling of it, but also thorough survey of a collection of immature (anthropoid ape) crania. The development of the "shelf" at the back of the symphysis of the lower jaw may almost certainly be delayed in some individuals (gorillas). Even the level of the lower border of the nasal bones is subject to some variation, and in young gorillas before the first permanent tooth has emerged fully, that level may be (as in man) above the level of the orbital margin. Generally, the elimination and detachment of features influenced largely by the factor of age demand special attention.

If, however, the good points can be justified, then these characters of youth will not gravely affect the final decision.

However these discussions may end, the record remains of the occurrence of an anthropoid ape some two thousand miles to the south of the nearest region providing a record of their presence. So far as the illustrations allow one to judge, the new form resembles the gorilla rather than the chimpanzee, that is, an African, not an Asiatic form of anthropoid ape. In this respect the new ape does not introduce an obviously disturbing factor. Disturbance, and the recasting of disturbed views, might nevertheless be caused in two other directions. Thus, the determination of the geological antiquity of the embedding of the fossil remains might have such an effect, were the estimate such as to carry that event very far back in time. Again, a comparison of the new ape with the fossil forms from India (Siwaliks) remains to be made, and it may be productive of results bearing on the relation of the African and the Asiatic groups. In any case, opinion must needs conform to the situation created by this discovery.

该报告表明（正如上文所述），许多结构性细节都已经被仔细观察过了，标本所有可及的部分也都已经被查看过了。这些观察不仅涉及头骨和下颌骨的外部，还涉及脑壳一部分毁损后暴露出来的颅腔部分。因此为了表示该标本具有比较高级的身份而提出的主张是建立在许多种这类细节的基础上的。如果达特教授成功地证明这些说法是合理的，他建议的新型猿类的身份应当得到承认。更多信息将依赖于对脑表面所展示出的特征的解释，还依赖于对与此相关的所有特征的解释；而且由于达特教授对此方面的调查进行了非常充分的准备，所以他的结论肯定具有格外重要的分量。至于脑及其特征，我觉得图 6 所展示的大猩猩头骨的颅腔模型轮廓线的扁平程度堪称惊人，也暗示了年龄的影响。

在文章中列举的解剖学特征中，对于我来说，就证据价值而言，有些特征的价值比其他特征的更高。除了刚刚已经提到的大脑特征以外，可能被引用的作为支持这些说法的很好的证据还有与眼眶下缘相关的鼻骨下缘位置的水平、鼻骨的长度（短）、缺乏眉脊（尽管第一恒牙已经完全出现）、陡峭上升的前额以及相对短小的犬牙。

另一方面，我非常肯定地认为，提到的其他特征中有些肯定也是与标本的年轻性有关。为了充分理解后者，不仅需要对标本进行处理，还需要对收藏的未成年（类人猿）头盖骨进行彻底的调查。几乎可以肯定的是，下颌联合部位背面的"猿板"的发育在某些个体（大猩猩）中被延迟了。甚至鼻骨下缘的位置水平趋向于发生一定的变化，幼年大猩猩在第一颗恒牙完全出现之前，其位置水平可能（与人类一样）位于眶缘水平之上。通常来说，需要特别注意那些很大程度上受年龄因素影响而造成的特征的消失和分离。

然而，如果这些有利的证据可以被证实的话，那么这些幼年个体的特征将不会严重影响最终的决定。

无论这些讨论将会如何结束，这块化石记录了一只类人猿的存在，它距离现存最近的类人猿发现区域还要偏南 2,000 英里。就允许我们进行判断的现有说明而言，这种新类型与大猩猩的相像程度大于与黑猩猩的，也就是说，这是一种非洲类型的类人猿，而不是亚洲类型的。在这方面，这种新型猿并没有带来明显的干扰因素。干扰和对受到干扰的观点的重新认识可能来自另外两个方向。确定化石的埋藏发生在哪个地质学时期可能具有这样一种作用，这样的估计可能会将这一事件带回到非常远古的时期。再者，现在还没有将这种新型猿类与印度西瓦利克发现的化石类型进行比较，这种比较可能会得到关于非洲和亚洲群体之间关系的丰富结果。无论如何，意见必须与这一发现所产生的局面相符。

If in these notes there have been passed over those observations and reflections wherewith Prof. Dart has illustrated and supported his views, such omissions are not due to want of appreciation, but to lack of capacity and space for their adequate treatment.

W. L. H. Duckworth

(**115**, 234-236; 1925)

如果在这些说明中出现了忽略达特教授描述过并用来支持自己观点的观察和思考,那么这种忽略不是由于想索要赞赏,而是缺乏对它们进行充分处理的能力和空间。

迪克沃斯

(刘皓芳 翻译; 吴新智 审稿)

Some Notes on the Taungs Skull

R. Broom

Editor's Note

While anthropologists in distant London debated the significance of Raymond Dart's new "Man-ape", *Australopithecus africanus*, from South Africa, local palaeontologist Robert Broom went to see the actual specimen. This note looks at the geological setting, suggesting that the skull was of Pleistocene to Recent date; and also the cranial and dental anatomy, asserting the intermediate status of the creature. However, Broom presciently notes that if specimens of adults were to be discovered, "the light thrown on human evolution would be very great". A decade was to elapse before such finds came to light: and the discoverer would be Broom himself.

A few days ago I visited Johannesburg to have a look at the remarkable new skull discovered by Prof. Dart, and named by him *Australopithecus africanus*. Prof. Dart not only allowed me every facility for examining the skull, but also gave me with almost unexampled generosity full permission to publish any observations I made on it, and suggested further that I might send to *Nature* any notes that might amplify the account he had already given. As the skull is one of extreme importance, a full account with measurements and very detailed figures will in due course be published by Prof. Dart, but the world already realises the unique character of the discovery and is anxious for more immediate information.

From the cablegrams received in South Africa, it is manifest that the first demand is for further light on the geological age of the being, and unfortunately complete information on this point cannot now be given, and will possibly never be available. Though I have not myself visited the Taungs locality, I am fairly familiar with many similar deposits farther south along the Kaap escarpment. This escarpment runs for more than 150 miles along the west side of the Harts River and lower Vaal River valleys from a little south of Vryburg to 20 miles south of Douglas. The escarpment is formed for the most part of huge cliffs of dolomitic limestone of the Campbell Rand series, in most places some hundreds of feet thick. The wide valley has an interesting geological history. Originally it was carved out in Upper Carboniferous or Lower Permian times by the Dwyka glaciers. For millions of years it was steadily refilled by Dwyka, Ecca, and Beaufort beds until the whole valley was perhaps buried by more than 2,000 feet of Permian and Triassic shales. Then conditions changed and the valley was re-excavated, by denudation, until today we find it not unlike what it must have been when originally carved out by the Dwyka glaciers.

The dolomite escarpment forms the most striking feature of the landscape in this part of the world. All along the west of the Harts-Vaal valley lies the high dead-level Kaap

汤恩头骨的几点说明

布鲁姆

编者按

当伦敦的人类学家们正在为雷蒙德·达特在南非发现的新"人猿"（即南方古猿非洲种）的意义争论不休的时候，身在南非的古生物学家罗伯特·布鲁姆去看了真实的化石标本。这篇文章着眼于地质学背景，提出该头骨介于更新世到现代之间，同时也考虑了头盖骨和牙齿的解剖学特征，主张该生物处于中间的位置。不过，布鲁姆很有预见性地指出，如果能发现这种生物的成年个体，那么"其对于人类进化带来的启示将是非常重要的"。10 年之后迎来了这样的发现，而发现者正是布鲁姆本人。

几天前我访问了约翰内斯堡，参观了达特教授发现并将其命名为南方古猿非洲种的著名新头骨。达特教授不仅允许我使用查看该头骨的所有设备，而且完全许可我发表自己对该头骨进行的任何观察，这种慷慨几乎是史无前例的，他还建议我可以向《自然》投递可以扩充他发表过的报告内容的任何稿件。因为该头骨是极具重要性的标本之一，所以一份兼具测量尺寸和详细图像的完整报告将会由达特教授在适当的时间发表，但是世人已经意识到了这一发现的独特特征，并且急于得到更多的即时信息。

从南非收到的海底电报来看，很显然第一个要求就是希望对这种生物的地质学年代作更进一步的说明，不幸的是，对于这一点，目前还不能给出完整的信息，而且可能永远都无法得到。尽管我并没有亲自参观过汤恩遗址，但是我非常熟悉远在开普断崖沿线南部的许多相似的堆积物。这一断崖从弗雷堡稍南部沿着哈茨河和瓦尔河下游河谷西侧绵延 150 多英里直至道格拉斯南部 20 英里处。该断崖大部分由坎贝尔-兰德系列的巨大白云灰岩山崖组成，很多地方都有几百英尺厚。这里宽阔的河谷有一段很有趣的地质学史。最初它是在上石炭纪或早二叠纪时期由德怀卡冰川开拓出来的。数百万年间它不断地被德维卡、埃卡和博福特河床回填，直到整条河谷多半被 2,000 多英尺的二叠纪和三叠纪页岩掩埋掉为止。后来环境发生了改变，河谷受到剥蚀作用而被重新挖掘出来了，直到现在，我们发现这条河谷与最初被德维卡冰川开拓出来时的样子几乎一样。

白云石断崖在世界的这个部分形成了最具独特特征的景观。沿着哈茨-瓦尔河谷西部的所有部分都位于高而平坦的开普高原上，从 20 英里外眺望时，整个断崖看

plateau, and when viewed from 20 miles away the escarpment looks like a high black wall bounding the lower plain of the valley. Every five or ten miles along the black wall are to be seen large light-coloured patches which on examination prove to be great masses of calc-sinter formed by calcareous springs. These, of course, must have been formed after the dolomite cliffs had been denuded of their covering Dwyka shales, and may in some cases be of considerable age—perhaps even dating from moderately early Tertiary times. Other masses of this secondary limestone may be of comparatively recent date. In places the great masses of calc-sinter have been excavated by underground water and moderately large caves are formed.

At Taungs the mass of secondary limestone is some hundreds of feet thick and about 70 feet high where it is being worked. Already 250 feet have been quarried away. On the face about 50 feet below the top of the mass, an old cave is cut across which is filled up with sand partly cemented together with lime, and it is in this old cave that the skull of Australopithecus has been found. The only other bones that I have seen or heard of are skulls and bones of a baboon, a jaw of a hyrax, and remains of a tortoise. I have not seen the hyrax jaw, so cannot say if it belongs to one of the living species. The baboon has been examined by Dr. Haughton, who regards it as an extinct species and has named it *Papio capensis*. I have seen a number of imperfect skulls of this baboon, and while they belong to a different species from the living local *Papio porcarius*, the difference between them is not so very striking.

I think it can be safely asserted that the Taungs skull is thus not likely to be geologically of great antiquity—probably not older than Pleistocene, and perhaps even as recent as the *Homo rhodesiensis* skull. When later or other associated mammalian bones are discovered, it may be possible to give the age with greater definiteness. At present all we can say is that the skull is not likely to be older than what we regard as the human period. But the age of the specimen in no way interferes with its being a true "missing link", and the most important hitherto discovered.

Prof. Dart in his photographs has given the general features of the skull and the brain, but there are a number of important characters in the skull and dentition to which I should like to direct attention.

Though the parietals and occipital are almost completely lost from the brain cast, most of the sutures can be clearly made out, and are as I indicate in Fig. 1. The sutures in the temporal region can also be clearly seen. The suture between the temporal bone and the parietal is fairly horizontal as in the anthropoid apes, but in the upward development of the squamous portion we have a character which is human and not met with in the gorilla, the chimpanzee, the orang, or the gibbon.

起来就像一堵以河谷的下流平原为界的高大黑墙。沿着黑墙，每 5~10 英里就可以看到巨大的浅颜色的点缀，经检验证实它们是由石灰泉形成的大片钙华。当然，这些肯定是在白云石山崖被剥蚀掉了覆盖着的德维卡页岩之后才形成的，而且在有些情况下，它们可能有着相当长的年代——甚至可能一直追溯到第三纪的早期。这种次生石灰石的其他块体的年代可能比较晚近。有些地方，地下水挖掘了大片钙华，形成了中等的大型山洞。

在汤恩，次生石灰石有几百英尺厚，70 英尺高。采石场已经挖掉了 250 英尺。在大块石灰石顶部之下约 50 英尺处的表面上，挖掘中穿过了一个旧山洞，洞中填满了与石灰一起形成水泥的沙子，就在这个老山洞里发现了南方古猿头骨。我见过或听说过的骨骼只有狒狒的头骨和骨骼、岩狸的颌骨和龟的残骸。我没有见到岩狸的颌骨，所以不能说它是否属于现存物种之一。狒狒的遗骸已经由霍顿博士查看过了，他认为它是一种已灭绝的物种，并将其命名为狒狒开普种。我见过这只狒狒的许多不完整的头骨，尽管它们属于一种与现存的本地浅灰狒狒不同的物种，但二者之间的差异不是十分明显。

我认为可以有把握地说，从地质学角度而言，汤恩头骨不可能是非常古老的——可能不早于更新世，甚至可能与罗德西亚人头骨一样晚。在发现比较晚的或其他与之共生的哺乳动物骨骼时，就有可能给出更确定的年代了。现在我们能说的就是这个头骨不可能早于我们所认为的人类时期。但是该标本的年代绝不妨碍它作为一个真正的"缺失环节"和迄今为止发现的最重要的标本。

达特教授在他发表的照片中给出了头骨和脑的一般特征，但是我想关注的事情是，头骨和齿系还有许多重要特征。

尽管在脑的模型上几乎找不到顶骨和枕骨的影子，但是可以清楚地辨认出大部分骨缝，这些骨缝我在图 1 中都指明了。颞区的骨缝也可以清楚看到。颞骨和顶骨之间的骨缝与类人猿中的一样非常水平，但是它的鳞部向上发育，这是一个属于人类的特征，在大猩猩、黑猩猩、猩猩和长臂猿中都没有遇见过。

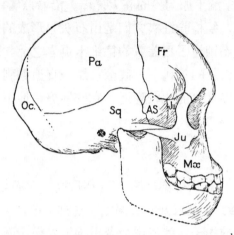

Fig. 1. Side view of skull of *Australopithecus africanus*, Dart. About $\frac{1}{3}$ natural size.

The arrangement of the sutures in the temporal region is also remarkably interesting. The upper part of the sphenoid articulates with both the parietal and the frontal. In the gorilla and chimpanzee in all the drawings I can find, the temporal bone meets the frontal and prevents the meeting of the sphenoid and the parietal. In the orang the condition varies, and I have in my possession a skull which has on the right side a spheno-parietal suture and on the left a fronto-temporal. In the baboon there is a large fronto-temporal suture, and in Cercopithecus a spheno-parietal suture. In the gibbon there is also a spheno-parietal suture. While the arrangement of the sutures in this region may not be of very great fundamental importance, it is interesting to note that Australopithecus agrees with man, the gibbon, and Cercopithecus, but differs from the gorilla, the chimpanzee, and the baboon.

The jugal or malar arch is interesting in that there is a long articulation between the jugal and squamosal. In this Australopithecus agrees rather with the anthropoids than with man.

On the face there are one or two striking characters, and of these perhaps the most important is the fusion of the premaxilla with the maxilla. On the palate the suture between these bones is seen almost as in the human child, the suture running out about two-thirds of the way towards the diastema between the second incisor and the canine. On the face there is no trace of any suture in the dental region, but on the left side of the nasal opening there is what is probably the upper part of the original premaxilla-maxillary suture. On the right side there is a faint indication of a suture just inside the nostril. In the chimpanzee the suture becomes obliterated in the dental region early, as apparently is the case in Australopithecus. In the orang and gorilla the suture remains distinct until a much later stage. In man, as is well known, all trace of the suture is obliterated from the face long before birth.

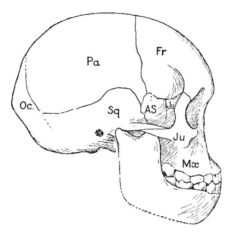

图 1. 南方古猿非洲种头骨的侧面观，达特。约相当于真实尺寸的 $\frac{1}{3}$。

颞区骨缝的分布也非常有意思。蝶骨的上部与顶骨和额骨都连接着。在我能找到的所有大猩猩和黑猩猩的图画中，颞骨都与额骨连接，而且颞骨阻止了蝶骨和顶骨相遇。猩猩则不是这种情况，我拥有的一个头骨的右侧有一条蝶顶缝，左侧有一条额颞缝。狒狒有一条大型的额颞缝，长尾猴有一条蝶顶缝。长臂猿也有一条蝶顶缝。尽管这一区域的骨缝的分布不具有重大的关键意义，但有趣的是，我们注意到南方古猿与人类、长臂猿和长尾猴都是一致的，而与大猩猩、黑猩猩和狒狒则是不同的。

颧弓的有趣之处在于，颧骨和颞鳞之间有一条长的关节。在这点上南方古猿与类人猿的一致性要多过与人类的一致性。

面部有一两点显著的特征，这些特征中最重要的可能是前颌骨和上颌骨之间的融合。可以看出硬腭的这些骨骼间的骨缝几乎与人类孩子的一样，骨缝向第二门齿和犬齿间的齿隙延伸约 2/3 的距离后消失。在面部，牙齿区域没有任何骨缝的痕迹，但是在鼻腔开口的左侧有一条骨缝，可能是原先的前颌骨–上颌骨骨缝的上半部分。右侧在鼻孔里有一条微弱的骨缝迹象。黑猩猩这条骨缝早早地就在牙齿区域消失了，南方古猿中表面上也是这样。猩猩和大猩猩中这条骨缝直到很晚的阶段依然明显。众所周知，人类的所有骨缝的痕迹在出生之前就已经从脸上消失了。

Australopithecus agrees with man and the chimpanzee in having a single foramen for the superior maxillary nerve. In the orang, gibbon, and other apes there are usually two or more foramina. In the gorilla sometimes there is one foramen; sometimes two.

In the shortness of the nasal bones and the high position of the nasal opening the Taungs skull agrees more with the chimpanzee than with the gorilla.

The dentition is beautifully preserved, and the teeth have been cleared of matrix by Prof. Dart with the greatest care. Though, owing to the lower jaw being in position, a full view of the crowns of the teeth could only be obtained by detaching the lower jaw, a sufficiently satisfactory view can be obtained to give us practically all we require of the structure.

The whole deciduous denture is present in practically perfect condition. The incisors, which are small, have been much worn down by use, and most of the crowns of the median ones have been worn off. Prof. Dart has directed attention to the vertical position of the teeth, which is a human character and differs considerably from the conditions found in the chimpanzee and gorilla. The small size of the incisors is also a human character.

The relatively small size of the canine is a character in which Australopithecus agrees with both the chimpanzee and man, and lies practically between the two.

The deciduous molars agree more closely with those of man than with those of any of the apes.

The first permanent molars of both upper and lower jaws are perfectly preserved and singularly interesting.

The first molar of the upper jaw (Fig. 2) has four large cusps arranged as in man and the anthropoid apes.

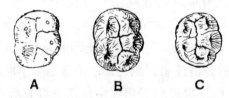

Fig. 2. First right upper molars: A, orang (after Röse); B, *Australopithecus africanus*, Dart, unworn; C, Bushman child, unworn. All natural size.

在上颌神经具有单一小孔这方面，南方古猿与人类和黑猩猩一致。而猩猩、长臂猿和其他猿类通常有两个或更多个小孔。大猩猩有时有一个小孔，有时有两个。

在鼻骨短小和鼻腔开口位置偏高这方面，汤恩头骨与黑猩猩的一致性比与大猩猩的高。

齿系的保存状况很好，而且达特教授已经十分慎重地清除掉了附着在牙齿上的基质。由于下颌骨仍保留在原位，所以要想对牙冠进行全面观察，只有将下颌骨分离下来才可以，这样就能得到一幅足以满足我们对全部结构的需要的视图。

全部乳牙的保存状况都相当完美。门齿小，由于使用而有了很大程度的磨损，大部分内侧门齿的牙冠都被磨损掉了。达特教授注意到牙齿的方位是垂直的，这是一种人类特征，与黑猩猩和大猩猩中观察到的情况差异很大。门齿尺寸较小也是一种人类特征。

相对小尺寸的犬齿是一种南方古猿与黑猩猩和人类都一致的特征，实际上南方古猿的犬齿尺寸介于这二者之间。

与所有猿类的乳臼齿相比，南方古猿的乳臼齿与人类的更加相符。

上颌骨和下颌骨的第一恒臼齿的保存状况都很好，它们都非常有意思。

上颌骨的第一臼齿（图2）有4个大的牙尖，其排列情况与人类和类人猿的一样。

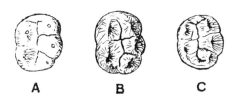

A **B** **C**

图 2. 第一右上臼齿：A,猩猩（依照罗斯）；B,南方古猿非洲种,达特,未磨损；C,儿童布希曼人,未磨损。全部都是真实尺寸。

The first lower molars (Fig. 3) has three well-developed sub-equal cusps on the outer side and two on the inner. Though in its great length and in the large development of the third outer cusp or hypoconulid the tooth differs considerably from the typical first lower molar of man, teeth of this pattern not infrequently occur in man. In general structure, however, the tooth more closely resembles that of the chimpanzee. It is interesting to compare this tooth with the corresponding tooth in Eoanthropus.

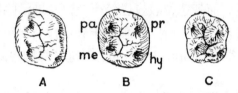

Fig. 3. First right lower molars: A, old chimpanzee, worn (after Miller); B, *Australopithecus africanus*, Dart; C, Bushman child. All natural size.

The arrangement of the furrows on the crown of the molar of Australopithecus is almost exactly similar to that in both the orang and the Bushman. In the chimpanzee and gorilla, there is usually a well-marked ridge passing from the protocone to the metacone, of which there is an indication in the Bushman tooth.

It will be seen that in *Australopithecus africanus* we have a large anthropoid ape resembling the chimpanzee in many characters, but approaching man in others. We can assert with considerable confidence that it could not have been a forest-living animal, and that almost certainly it lived among the rocks and on the plains, as does the baboon of today. Prof. Dart has shown that it must have walked more upright than the chimpanzee or gorilla, and it must thus have approached man more nearly than any other anthropoid hitherto discovered.

Eoanthropus has a human brain with still the chimpanzee jaw. In Australopithecus we have a being also with a chimpanzee-like jaw, but with a sub-human brain. We seem justified in concluding that in this new form discovered by Prof. Dart we have a connecting link between the higher apes and one of the lowest human types.

The accompanying table (Fig. 4) shows what I believe to be the relationships of Australopithecus. If an attempt be made to reconstruct the adult skull (Fig. 5), it is surprising how near it appears to come to *Pithecanthropus erectus*—differing only in the somewhat smaller brain, and less erect attitude.

While nearer to the anthropoid apes than man, it seems to be the forerunner of such a type as Eoanthropus, which may be regarded as the earliest human variety, the other probably branching off in different directions.

第一下臼齿（图3）外侧有3个发育完好的几乎相等的牙尖，内侧有2个。尽管在牙尖的巨大长度以及第三外侧牙尖或下次小尖的发达程度上，这颗牙与典型的人类第一下臼齿非常不同，但是人类中这种形式的牙齿并不罕见。然而，就总体结构而言，这颗牙与黑猩猩的更像。将这颗牙与曙人相应的牙齿进行比较会得到很有趣的发现。

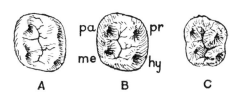

图 3. 第一右下白齿：A,年老黑猩猩,有磨损（依照米勒）；B,南方古猿非洲种,达特；C,儿童布希曼人。全部都是真实尺寸。

南方古猿白齿牙冠上沟的排列几乎与猩猩和布希曼人的完全一样。黑猩猩和大猩猩中，通常有一条从上原尖到后尖的明显的脊，布希曼人牙齿中也有这种迹象。

我们将会看到，在南方古猿非洲种中出现了一只许多特征与黑猩猩相像、但其他方面又与人类相像的大型类人猿。我们可以充满信心地断言它不是一只生活在森林中的动物，并且几乎可以肯定它是生活在岩石间和平原上的动物，就像现在的狒狒一样。达特教授已经说明了，它行走的姿势肯定比黑猩猩和大猩猩的更加直立，因此它肯定比迄今发现的其他类人猿更接近于人类。

曙人具有人类的脑、黑猩猩的颌骨。南方古猿中我们也有一只黑猩猩式的颌骨和次人的脑。我们似乎有理由作出如下结论：这种由达特教授发现的新类型使得我们拥有了一个连接高等猿类和一种人类的最低级类型的环节。

附表（图4）展示了我相信的南方古猿的亲缘关系图。如果试图对成年头骨进行复原（图5），那么就会惊奇地发现它与爪哇猿人是非常接近的——只是在脑稍小、姿势欠直立方面有所不同。

尽管与人类相比它更接近于类人猿，但是似乎仍旧可以将其认为是像曙人这样的类型的先祖，并将其当作是最早期的人类物种，即可能是另一个与人类发生了分歧而向不同方向进化的物种。

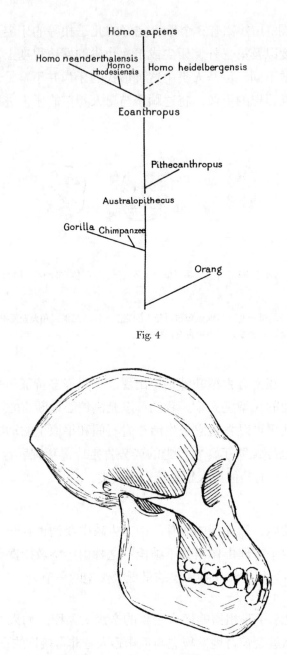

Homo sapiens

Homo neanderthalensis
Homo rhodesiensis
Homo heidelbergensis

Eoanthropus

Pithecanthropus

Australopithecus

Gorilla Chimpanzee

Orang

Fig. 4

Fig. 5. Attempted reconstruction of adult skull of *Australopithecus africanus*, Dart. About $\frac{1}{3}$ natural size.

There seems considerable probability that adult specimens will yet be secured, and if the skeleton as well as the skull is preserved, the light thrown on human evolution will be very great.

(**115**, 569-571; 1925)

R. Broom: Douglas, South Africa.

204

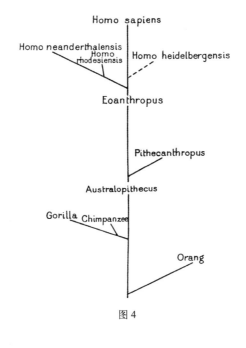

图 4

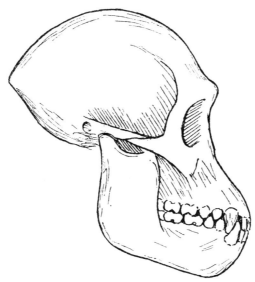

图 5. 成年南方古猿非洲种尝试性的复原头骨，达特。约相当于真实尺寸的 $\frac{1}{3}$。

似乎有足够的可能性认为成年标本会使我们更安心，如果骨架和头骨都被保存下来的话，那么其对于人类进化带来的启示将是非常重要的。

（刘皓芳 翻译；吴新智 审稿）

205

The Taungs Skull

A. Keith

Editor's Note

Months after the discovery of *Australopithecus africanus*, London anthropologists remained frustrated that they could not see the specimen—not even a cast. This may explain the rage of this letter from the eminent Scottish anatomist and anthropologist Arthur Keith, that the closest he had got to the Taungs skull was to go to an exhibition and "peer at [the casts] in a glass case" along with ordinary members of the public. Keith's conviction that Australopithecus was a fossil ape rather than a transitional form between apes and humans hardened.

THE account which Prof. Dart published of the Taungs skull (*Nature*, Feb. 7, p. 195) left many of us in doubt as to the true status of the animal of which it had formed part, and we preferred, before coming to a decision, to await an examination of the fossil remains, or failing such an opportunity, to study exact casts of them. For some reason, which has not been made clear, students of fossil man have not been given an opportunity of purchasing these casts; if they wish to study them they must visit Wembley and peer at them in a glass case which has been given a place in the South African pavilion.

The chief point which awaited decision relates to the position which must be assigned in the animal kingdom to this newly discovered form of primate. Prof. Dart, in writing of it, has used the name of anthropoid ape; he has described it as representing "an extinct race of apes intermediate between living anthropoids and man"—which is tantamount to saying that at Taungs there has been discovered the form of being usually spoken of as the "missing link". That this is his real decision is evident from the fact that he speaks of it as "ultrasimian and prehuman" and proposes the creation of a new family for its reception.

An examination of the casts exhibited at Wembley will satisfy zoologists that this claim is preposterous. The skull is that of a young anthropoid ape—one which was in the fourth year of growth—a child—and showing so many points of affinity with the two living African anthropoids—the gorilla and chimpanzee—that there cannot be a moment's hesitation in placing the fossil form in this living group. At the most it represents a genus in the Gorilla-Chimpanzee group. It is true that it shows in the development of its jaws and face a refinement which is not met with in young gorillas and chimpanzees at a corresponding age. In these respects it does show human-like traits. It is true that it is markedly narrow-headed while the other African anthropoids are broad-headed—but we find the same kind of difference in human beings of closely allied races. Prof. Dart claimed that the brain showed certain definite human traits. This depends upon whether or not he had correctly identified the position of a certain fissure of the brain—the parallel fissure. In the show-case at Wembley a drawing is placed side by side with the "brain cast"; but when we examine the brain cast at the site where the fissure is shown on the drawing, we find only a broken surface where identification becomes a matter of guess-work.

汤恩头骨

基思

编者按

在达特发现南方古猿非洲种几个月之后，伦敦的人类学家们依旧很沮丧，因为他们看不到那个标本，甚至连模型也看不到。这大概可以解释苏格兰杰出解剖学家、人类学家阿瑟·基思在这封来信中表达的愤怒，他说自己距离那件汤恩头骨最近的时候就是随着普通观众参观头骨展览时"盯着看玻璃柜里的[模型]"。基思坚信，南方古猿更像是古猿的化石，而不是古猿与人类之间的过渡类型。

达特教授就汤恩头骨发表的报道（《自然》，2月7日，第195页）中关于这种动物（这件化石就是它的一个部分）的真实身份给我们留下了许多可疑之处，在得到结论之前，我们更倾向于等待对化石进行查看，如果没有这样的机会，研究其精确的模型也行。没有人清楚是出于什么原因，研究人类化石的学者们还没有得到购买这些模型的机会；如果他们希望研究这些模型的话，他们就必须去温布利参观南非展览馆，然后盯着看玻璃柜里的模型。

有待确定的最主要的一点与将这种新发现的灵长类类型归属到动物界的哪个位置有关。达特教授写这部分的时候，使用了类人猿的名称，他将其描述为代表着"一种已经灭绝的猿类，介于现存的类人猿和人类之间的中间类型"——这种说法相当于是说汤恩发现了通常被称作是"缺失环节"的生物类型。显而易见这是他的真实判断，从他将其称为"超猿和前人类"的说法以及提议创建一个新科来接纳这种动物的事实就可以看出来。

对温布利陈列的模型进行观察将会使动物学家相信这种主张是荒谬可笑的。这具头骨是一只幼年类人猿的——该个体处于生长阶段的第4年——即一个孩子——它显示出了许多与两种现存的非洲类人猿——大猩猩和黑猩猩——的亲缘关系，于是将该化石类型列入这种现存群体中就变得不容置疑了。它最多代表了大猩猩-黑猩猩群体中的一个属。它确实表明颌骨和面部在发育过程中发生了细微的变化，这在相应年龄的幼年大猩猩和黑猩猩中都是不曾见到过的。这些方面它确实显示出了与人类相像的特点。它具有非常狭窄的头部也是真实的，而其他非洲类人猿都是阔头型的——但是我们发现具有密切亲缘关系的人类种族也存在同样的差异。达特教授声称该标本的脑显示了某些确定的人类特征。这依赖于他是否正确地辨认了某一大脑裂缝——平行裂缝的位置。在温布利的陈列柜里，一幅画与"脑模型"一起陈列在那里；但是当我们观察脑模型，查看图画上展示的裂缝所在位置时，我们发现只有一处裂开的表面，在这里，鉴定变成了一种猜测。

In every essential respect the Taungs skull is that of a young anthropoid ape, possessing a brain which, in point of size, is actually smaller than that of a gorilla of a corresponding age. Only in the lesser development of teeth, jaws, and bony structures connected with mastication can it claim a greater degree of humanity than the gorilla. Its first permanent molar teeth which have just cut are only slightly smaller than those of the gorilla, while the preparations which are being made in the face for the upper permanent canines show that these teeth were to be of the large anthropoid kind.

The other point on which we awaited information related to the geological age of the Taungs skull. Fortunately, Dr. Robert Broom (*Nature*, April 18, p. 569) has thrown a welcome light on this matter. The skull was blasted out of a cave which had become filled up by sand washed in from the Kalahari. The fossil baboons found in neighbouring caves differ in only minor structural details from baboons still living in South Africa. In Dr. Broom's opinion the Taungs skull is of recent geological date; it is not older than the Pleistocene; he thinks it probable that it may not be older than the fossil human skull found in a limestone cave at Broken Hills, Rhodesia. It is quite possible—nay, even probable—that the Taungs anthropoid and Rhodesian man were contemporaries. Students of man's evolution have sufficient evidence to justify them in supposing that the phylum of man had separated from that of anthropoid apes early in the Miocene period. The Taungs ape is much too late in the scale of time to have any place in man's ancestry.

In a large diagram, placed in the show-case at Wembley, Prof. Dart gives his final conception of the place occupied by the Taungs ape in the scale of man's evolution. He makes it the foundation stone of the human family tree. From the "African Ape Ancestors, typified by the Taungs Infant", Pithecanthropus, Piltdown man, Rhodesian man, and African races radiate off. A genealogist would make an identical mistake were he to claim a modern Sussex peasant as the ancestor of William the Conqueror.

In the show-case at Wembley plastic reconstructions are exhibited in order that visitors may form some conception of what the young Taungs Ape looked like in life. Although the skull is anthropoid it has been marked by a "make-up" into which there have been incorporated many human characters. It is true the ears are those of the chimpanzee, but the forehead is smooth and rounded, the hair of the scalp is sleek and parted; the bushy eyebrows are those of a man at fifty-five or sixty; the neck is fat, thick, and full—extending from chin to occiput. In modelling the nose, gorilla lines have been followed, whereas the nasal part of the skull imitates closely chimpanzee characters. The mouth is wide, with a smile at each corner.

Prof. Dart has made a discovery of great importance, and the last thing I want to do is to detract from it. He has shown that anthropoid apes had extended, during the Pleistocene period, right into South Africa—into a land where anthropoid apes could not gain a livelihood today. He has found an extinct relative of the chimpanzee and gorilla but one with more man-like features than are possessed by either of these. His discovery throws light on the history of anthropoid apes but not on that of man. Java-man (Pithecanthropus) still remains the only known link between man and ape, and this extinct type lies on the human side of the gap.

(**116**, 11; 1925)

汤恩头骨在各个关键方面，都显示出一只幼年类人猿的特点，它的脑在尺寸上确实比相应年龄的大猩猩的小。只在较欠发育的牙齿、颌骨和与咀嚼相关的骨质结构方面，可以将它称为一种比大猩猩更高等的似人动物。其刚刚萌出的第一恒臼齿只比大猩猩的略小，而面部针对上恒犬齿所作的准备则表明这些牙齿是属于大型类人猿那一类的。

我们等待的另外一个方面的资料与汤恩头骨的地质学年代有关。幸运的是，罗伯特·布鲁姆博士（《自然》，4月18日，第569页）已经对这个问题进行了阐述，并且他的说法被很多人接受。这个头骨是从一个填满了来自卡拉哈里沙漠的沙子的山洞里炸出来的。在旁边山洞发现的狒狒化石与南非现存的狒狒只在微小的结构细节上存在差异。布鲁姆博士认为，汤恩头骨的地质学年代较近，应该不早于更新世；他认为一种可能的情况是，这具头骨的年代并不比在罗德西亚布罗肯希尔山的石灰石山洞中发现的人类头骨化石的时期早。非常可能——不，几乎可以肯定的是——汤恩类人猿和罗德西亚人是同时代的。研究人类进化的学者们有足够的证据证实他们所假设的人类这一支系是在中新世早期从类人猿中分离出来的。汤恩猿在时间尺度上太晚了，所以根本不可能是人类的祖先。

在温布利的陈列柜里摆放的一张大图表中，达特教授给出了他对汤恩猿在人类进化上的位置的最终想法。他把汤恩猿当作奠定人类家族树的基石。从"汤恩婴儿代表的非洲猿祖先"开始，爪哇猿人、皮尔当人、罗德西亚人和非洲人种呈辐射状散出。如果一位系谱专家宣称一位现代萨塞克斯的农民是征服者威廉的祖先，那他就是犯了同样的错误。

在温布利的陈列柜里，陈列着复原像的造型以便参观者可以对幼年汤恩猿在生活状态下看起来是什么样子有一定的概念。尽管这具头骨是类人猿的，但是许多人类特征被"捏造"出来表现在这件复原像上。确实其耳朵是黑猩猩的，但是前额平滑而圆润，头皮上的头发光滑而疏散；浓密的眉毛与55~60岁的人类很像；脖子粗、厚而丰满——从下巴一直延伸到枕部。在制作鼻子的模型时，参照了大猩猩的线条，然而这具头骨的鼻子部分模仿了与黑猩猩非常接近的特征。嘴部宽阔，嘴角带笑。

达特教授取得了非常重要的发现，但是我想做的最后一件事就是贬低它。他向我们展示了类人猿在更新世时期曾经一直扩展到南非——这是一片现在的类人猿无法生存的地域。他发现了黑猩猩和大猩猩的一种已灭绝的亲属，但是它又具有比起二者来更像人类的特征。他的发现对于类人猿历史有一定的昭示作用，但是对人类历史并不尽然。爪哇猿人仍然是唯一已知的人和猿之间的环节，这种已灭绝的类型处于缺口中人类这一边。

（刘皓芳 翻译；吴新智 审稿）

The Taungs Skull

Editor's Note

Here Raymond Dart responds to Arthur Keith's assertive criticism of Dart's claim to have identified an intermediate form between man and ape, called *Australopithecus africanus*. Dart refutes Keith's charges robustly at every point except one, which relates to the question of whether a cast of the skull should be made available to other anthropologists.

IN *Nature* of July 4, 1925, p. 11, Sir Arthur Keith has attempted to show first that I called the Taungs skull a "missing link", and secondly, that it is not a "missing link".

As a matter of fact, although I undoubtedly regard the description as an adequate one, I have not used the term "missing link." On the other hand, Sir Arthur Keith in an article entitled "The New Missing Link" in the *British Medical Journal* (February 14, 1925) pointed out that "it is not only a missing link but a very complete and important one". After stating his views so definitely in February, it seems strange that, in July, he should state that "this claim is preposterous".

Despite this reversal of opinion, Sir Arthur tells us that the skull "does show human-like traits in the refinement of its jaws and face which is not met with in young gorillas and chimpanzees at a corresponding age." He appears to have overlooked the fact that in addition to these and other facts brought forward by myself, the temporal bone, sutures, and deciduous and permanent teeth (according to Dr. Robert Broom) also show human-like traits. Moreover, as Prof. Sollas has so ably shown, the whole profile of the skull is entirely different from that in living anthropoids, thus indirectly confirming my discovery that the brain inside the skull-dome which caused this profound difference was very different from the brains inside the skulls of modern apes.

The fact that Sir Arthur was unable to find the *parallel sulcus* depression in the replica cast sent to Wembley illustrates how unsatisfactory the study of the replica can be in the absence of the original.

With reference to the question of endocranial volume, I would state with Prof. Sollas that this "is a matter of only secondary importance". Nothing could exemplify this matter better than the condition of affairs in the Boskop race, where the endocranial volume was in the vicinity of 1,950 c.c. (The average European's endocranial volume is 1,400–1,500 c.c.) Indeed, the world's record in human endocranial volume (2,000 c.c.) was discovered in a "boskopoid" skull by Prof. Drennan in a dissecting room subject at Capetown this year. It is well known that the elephant and the whale have brains much larger than those of

汤恩头骨

编者按

雷蒙德·达特此前声称发现并鉴定了一种介于人与古猿之间，被命名为南方古猿非洲种的中间类型，阿瑟·基思对此提出了批评。在这里，雷蒙德·达特对阿瑟·基思的批评作出了回应。除了是否应该将这具头骨的模型对其他人类学家开放的问题，达特对基思的每一条意见都进行了反驳。

在 1925 年 7 月 4 日的《自然》第 11 页，阿瑟·基思爵士首先试图表明我把汤恩头骨称为一个"缺失的环节"，其次，试图表明汤恩头骨不是一个"缺失的环节"。

事实上，尽管我毫不怀疑地认为这种描述是适当的，但是我并没有使用过"缺失的环节"这个短语。另一方面，阿瑟·基思爵士在发表于《英国医学杂志》（1925 年 2 月 14 日）上的一篇题为《新的缺失环节》的文章里指出"它不仅是一个缺失的环节，而且是一个非常完整而重要的环节"。在他 2 月份如此确定地陈述了自己的观点之后，在 7 月他又说"这种说法是荒谬可笑的"，这似乎有点奇怪。

尽管有这样矛盾的观点，阿瑟爵士还是告诉我们这个头骨"在颌骨和面部的细节上确实显示出了与人类相像的特征，这种特征在相应年龄的幼年大猩猩和黑猩猩中都是不曾见到过的。"看起来他似乎忽略了一个事实，那就是除了我本人提出来的这些和其他事实之外，颞骨、骨缝、乳齿和恒牙（根据罗伯特·布鲁姆）也显示出了与人类相像的特征。另外，索拉斯教授也非常巧妙地说明了该头骨的整个剖面与现存的类人猿完全不同，因此间接地证实了我的发现，即引起这种深刻差异的位于头骨之内的脑与现代猿的头骨内的脑差异很大。

阿瑟爵士未能发现送到温布利的复制品模型上的**平行沟**凹陷，这个事实显示了在缺乏原型时只是对复制品进行研究会多么令人不满。

关于颅腔容量的问题，我已经与索拉斯教授共同声明过，即这"只是第二重要的事情"。除了颅腔容量接近 1,950 毫升（欧洲人的颅腔容量平均值为 1,400 ~ 1,500 毫升）的博斯科普人（南非石器时代中期的一个人种）的情况之外，可能没有什么更好的可以用来证明这个问题的例子了。实际上，人类颅腔容量的世界记录（2,000 毫升）是德雷南教授今年在开普敦的一间解剖室里发现的一具"类博斯科普人的"头骨的容量。众所周知，大象和鲸的脑比人类的大得多，但是没有人会从这一点得出它们

human beings, but no one has inferred from that that their intelligence is greater. It is fairly certain that size of brain has some relation to size of body, as Dubois has shown. It is highly probable that the australopithecid man-apes were relatively small as compared with the gorilla. It is not the quantity so much as the quality of the brain that is significant.

Sir Arthur is harrowed unduly lest the skull *may* be Pleistocene. It is significant in this connexion that Dr. Broom, who first directed attention to this possibility (of which I was aware before my original paper was sent away), regarded it nevertheless as "the forerunner of such a type as Eoanthropus". It should not need explanation that the Taungs infant, being an infant, was ancestral to nothing, but the family that he typified are the nearest to the prehuman ancestral type that we have.

In view of these facts, there is little justification for the attempted witticism that in making the "African ancestors typified by the Taungs infant" the "foundation stone of the human family tree"—whatever that may be—I am making "a mistake identical with that of claiming a Sussex peasant as the ancestor of William the Conqueror". This is merely a case of mistaken identity on the part of Sir Arthur. I have but translated into everyday English the genealogical table suggested by Dr. Robert Broom (*Nature*, April 18, 1925), with which I agree almost entirely. I take it, however, as a mark of his personal favour that Sir Arthur should have attacked my utterance and spared Dr. Broom's.

Sir Arthur need have no qualms lest his remarks detract from the importance of the Taungs discovery—criticism generally enhances rather than detracts. Three decades ago Huxley refused to accept Pithecanthropus as a link. Today Sir Arthur Keith regards Pithecanthropus as the only known link. There is no record that Huxley first accepted it, then retracted it, but history sometimes repeats itself.

Raymond A. Dart

*　　*　　*

Prof. Dart is under a misapprehension in supposing that I have in any way or at any time altered my opinion regarding the fossil ape discovered at Taungs. From the description and illustrations given by him (*Nature*, Feb. 7, 1925, p. 195) the conclusion was forced on me that Australopithecus was a member of "the same group or sub-family as the chimpanzee and gorilla" (*Nature*, Feb. 14, 1925, p. 234). In the same issue of *Nature*, Prof. G. Elliot Smith expressed a similar opinion, describing Australopithecus as "an unmistakable anthropoid ape that seems to be much on the same grade of development as the gorilla and chimpanzee without being identical with either."

All the information which has come home since Prof. Dart made his original announcement in *Nature* has gone to support the close affinity of the Taungs ape to the gorilla and to the chimpanzee—it is a member of that group. Prof. Bolk, of Amsterdam,

的智慧也比人类高的推论。正如杜波伊斯指出的，脑的尺寸肯定与身体的尺寸有一定的关系。有一种很大的可能性是，与大猩猩相比，南方古猿属的人猿相对较小。显然脑的量不如质那样具有重大意义。

阿瑟爵士因为唯恐该头骨**可能**是更新世时期的而过度苦恼了。在这个关系上，值得注意的是，首先将注意力放到这种可能性（我意识到这种可能性是在我的初稿送出去之前）上的人是布鲁姆博士，他认为其不过是"像曙人一样类型的先祖"。需要解释的不是汤恩婴儿作为一个婴儿能不能作为任何生物的祖先，而是他所代表的科与我们拥有的人类之前的祖先型的亲缘关系是否是最近的。

鉴于这些事实，企图抓住将"汤恩婴儿代表的非洲祖先"当作"人类家族树的基石"——无论可能是什么——这些话，说我正在犯着"与宣称一位萨塞克斯农民是征服者威廉的祖先同样的错误"这种风凉话，几乎是没有道理的。这只不过是阿瑟爵士自己搞错罢了。我只是将罗伯特·布鲁姆建议的家族树（《自然》，1925 年 4 月 18 日）翻译成日常用的英语，对于该家族树，我几乎完全赞成。阿瑟爵士攻击我的意见，却没有批评布鲁姆博士的意见，我将这当作是他个人偏爱的一个标志。

阿瑟爵士不必担心他的言论是否会贬低汤恩发现的重要性——批评通常会提高影响而不会贬低影响。大约 30 年前，赫胥黎拒绝接受爪哇猿人作为一个环节而存在。今天阿瑟·基思爵士又认为爪哇猿人是唯一一个已知的环节。虽然没有关于赫胥黎首先接受它、而后又摒弃它的记载，但是历史有时是会重演的。

雷蒙德·达特

*　　*　　*

达特教授认为我随随便便改变对汤恩发现的猿化石的观点，他的这种想法是一种误解。从他给出的描述和说明（《自然》，1925 年 2 月 7 日，第 195 页）来看，他是将如下结论强加于我，即南方古猿是"与黑猩猩和大猩猩一样的群体或亚科"的成员之一（《自然》，1925 年 2 月 14 日，第 234 页）。在同期的《自然》中，埃利奥特·史密斯教授表达了相似的观点，将南方古猿描述为"肯定是一只类人猿，该类人猿似乎处于与大猩猩和黑猩猩同样的发育阶段，但是与两者又都不相同。"

自从达特教授在《自然》上最初公布结果之后，这里的所有信息都趋向于支持汤恩猿与大猩猩和黑猩猩存在亲密的亲缘关系——认为它是那一群体的成员之一。阿姆斯特丹的博尔克教授和俄亥俄州克利夫兰的温盖特·托德教授都注意到了一个事

and Prof. Wingate Todd, of Cleveland, Ohio, have directed attention to the fact that the skulls of occasional gorillas show the same kind of narrowing and lengthening as has been observed in that of the Taungs ape. Prof. Arthur Robinson has shown that there is a wide variation in the size of jaws of young chimpanzees of approximately the same age, the smaller of the jaws approaching in size and shape to the development seen in the Taungs ape. The dimensions of the erupting first permanent molar of the Taungs ape and the form of its cusps point to the same conclusion—that Australopithecus must be classified with the chimpanzee and gorilla. It is, therefore, "preposterous" that Prof. Dart should propose to create "a new family of Homo-simiadae for the reception of the group of individuals which it (Australopithecus) represents". It is preposterous because the group to which this fossil ape belongs has been known and named since the time of Sir Richard Owen.

The position which Prof. Dart assigns to the Taungs ape in the genealogical tree of man and ape has no foundation in fact. A large diagram in the exhibition in Wembley, prepared by Prof. Dart, informs visitors that the Taungs ape represents the ancestor of all forms of mankind, ancient and modern. Before making such a claim one would have expected that due inquiry would first be made as to whether or not the geological evidence can justify such a claim. From his letter one infers that Prof. Dart does not set much store by geological evidence. Yet it has been customary, and I think necessary, to take the time element into account in constructing pedigrees of every kind. Dr. Robert Broom and, later, Prof. Dart's colleague, Prof. R. B. Young, have reviewed the evidence relating to the geological antiquity of the Taungs fossil skull, and on data supplied by them one can be certain that early and true forms of men were already in existence before the ape's skull described by Prof. Dart was entombed in a cave at Taungs. To make a claim for the Taungs ape as a human ancestor is therefore "preposterous".

Finally, Prof. Dart reminds me that whales and elephants have massive brains and that many large-headed men and women show no outstanding mental ability. Still the fact remains that every human being whose brain fails to reach 850 grams in weight has been found to be an idiot. Size as well as convolutionary pattern of brain have to be taken into account in fixing the position of every fossil type of being that has any claim to be in the line of human evolution—the Taungs brain cast at Wembley possesses no feature which lifts it above an anthropoid status.

Arthur Keith

(**116**, 462-463; 1925)

Raymond A. Dart: University of the Witwatersrand, Johannesburg.
Arthur Keith: Royal College of Surgeons of England, Lincoln's Inn Fields, London, W.C., September 5.

实，即个别的大猩猩头骨显示出与在汤恩猿中观察到的同样形式的变窄、拉长的现象。阿瑟·鲁滨逊教授也表明大约同样年龄的幼年黑猩猩的颌骨大小存在很广泛的变异范围，其中稍小的颌骨就与汤恩猿中见到的大小和形状的发育程度接近。汤恩猿正在萌出的第一颗恒臼齿的尺寸和其牙尖的形式指向同样的结论——南方古猿肯定与黑猩猩和大猩猩属于同类。因此达特教授建议创建"一个名为人猿科的新科来接纳标本（南方古猿）代表的生物类群"的想法是"荒谬可笑的"。因为这例化石猿所属的群体是已知的，早在理查德·欧文先生时期就已经被命名过了，所以说他是荒谬可笑的。

达特教授给汤恩猿在人类和猿类家族树上指定的位置事实上没有任何基础。陈列在温布利的一幅达特教授制定的大图表告诉参观者，汤恩猿代表了古代和现代所有人类类型的祖先。一个人在提出这样的主张之前，应该会想到先通过适当的调查，看看地质学证据是否支持这种主张。从达特教授的信中可以推断出他并没有对地质学证据投入太多精力。但是通常习惯性的做法是，在构建每一类谱系时将时间因素考虑在内，我觉得这是必须的。罗伯特·布鲁姆博士审视过与汤恩头骨化石的地质学年代相关的证据，后来达特教授的同事扬教授也对此作过调查，根据他们所提供的数据，可以肯定早期的真正的人类类型早在达特教授描述的猿类头骨埋入汤恩洞穴之前就已经存在了。因此声称汤恩猿是人类祖先是"荒谬可笑的"。

最后，达特教授提醒我，鲸和大象具有巨大的脑，并且许多大脑袋的男人和女人并没有过人的智力。但是事实仍然是，研究发现每个脑重量不足 850 克的人都是白痴。在确定每种与人类进化世系有所关联的生物化石类型的地位时，脑的大小和脑回式样都是必须考虑的因素——温布利的汤恩脑模型不具备将其提升到类人猿以上地位的任何特征。

阿瑟·基思

（刘皓芳 翻译；吴新智 审稿）

Tertiary Man in Asia: the Chou Kou Tien Discovery*

D. Black

Editor's Note

The fossil-bearing cave site of Chou Kou Tien (Zhoukoudian in China) had been discovered in 1921, and soon yielded bones of various fossil mammals from horses to bats. This note is a secondary account of discoveries made to date that included, notably, two teeth attributable to *"Homo ? sp."*, the first human remains known to science in mainland Asia. The reporter, Canadian-born Davidson Black, Anatomy Professor at Peking Union Medical College, would go on to make his name at Chou Kou Tien with discoveries of other remains that he would call Sinanthropus, now regarded as *Homo erectus*. Black died in his office of heart problems in 1934, the remains of "Peking Man" close by. He was 49.

A rich fossiliferous deposit at Chou Kou Tien, 70 li [about 40 kilometres] to the southwest of Peking, was first discovered in the summer of 1921 by Dr. J. G. Andersson and later surveyed and partially excavated by Dr. O. Zdansky. A preliminary report on the site was published by Dr. Andersson in March 1923 (*Mem. Geol. Surv. China*, Ser. A, No. 5, pp. 83–89), followed in October of that year by a brief description of his survey by Dr. Zdansky (*Bull. Geo. Surv. China*, No. 5, pp. 83–89). The material recovered from the Chou Kou Tien cave deposit has been prepared in Prof. Wiman's laboratory in Upsala and afterwards studied there by Dr. Zdansky. As a result of this research, Dr. Andersson has now announced that in addition to the mammalian groups already known from this site, there have also been identified representatives of the Cheiroptera, one cynopithecid, and finally two specimens of extraordinary interest, namely, one premolar and one molar tooth of a species which cannot otherwise be named than *Homo ? sp.*

Judging from the presence of a true horse and the absence of Hipparion, Dr. Andersson in his preliminary report considered that the Chou Kou Tien fauna was possibly of Upper Pliocene age, an opinion also expressed by Dr. Zdansky. It is possible, however, in the light of recent research, that the horizon represented by this site may be of Lower Pleistocene age. Whether it be of late Tertiary or of early Quaternary age, the outstanding fact remains that, for the first time on the Asiatic continent north of the Himalayas, archaic hominid fossil material has been recovered, accompanied by complete and certain geological data. The actual presence of early man in eastern Asia is therefore now no longer a matter of conjecture.

---

* Announcement of the Chou Kou Tien discovery was first made by Dr. J. G. Andersson on the occasion of a joint scientific meeting of the Geological Society of China, the Peking Natural History Society and the Peking Union Medical College held in Peking on October 22, 1926, in honour of H. R. H. the Crown Prince of Sweden.

亚洲的第三纪人：周口店的发现[*]

步达生

编者按

1921年人们在中国发现了藏有化石的周口店洞穴遗址，很快地，从中发掘出了从马到蝙蝠等多种哺乳动物的化石标本。这篇文章间接地介绍了之前在周口店遗址取得的各种发现，其中包括两件著名的来自"人属（种名不确定）"的牙骨，以及对于科学界来说亚洲大陆的首例人类化石。文章作者是出生于加拿大的戴维森·步达生，当时他是北京协和医学院的解剖学教授，后来他因为在周口店发现了一些其他的化石（他称之为中国猿人，现在被称为直立人）而闻名于世。1934年，49岁的步达生因心脏病死于办公室中，当时"北京人"的化石就在他身旁。

1921年夏天，安特生博士在北京西南70里 [大约40公里] 的周口店首次发现了一套富含化石的沉积层，日贾尔斯基博士随后进行了调查及部分发掘工作。1923年3月，有关这个遗址的初步报告由安特生博士发表（《中国地质专报》，A辑，第5期，第83~89页）。同年10月，日贾尔斯基博士发表了他的初步调查结果（《中国地质学会会志》，第5期，第83~89页）。从周口店洞穴沉积中发现的材料在乌普萨拉的威曼教授的实验室进行修整，之后日贾尔斯基博士在那里对这些材料进行了研究。安特生博士发布了这项研究的一个结果：在这个化石点，除了已知的哺乳动物类群，还发现了一些翼手目的代表性物种，一个猕猴的标本，以及两个非常有趣的标本，即一个前臼齿和一个臼齿，对应的物种好像只能称为"人属（种名不确定）"。

根据存在真正的马以及没有发现三趾马来判断，安特生博士在他初期的报道中认为周口店动物群属于上新世早期，日贾尔斯基博士也表达了同样的观点。最近的研究表明，这一化石点代表的层位更可能属于上新世晚期。无论它属于第三纪晚期还是第四纪早期，最引人注目的事实是，这是第一次在喜马拉雅山北部的亚洲大陆发现原始人类的化石材料，并且具有完整的相关地质学资料。从此，关于早期人类事实上存在于东亚的说法不再只是一种猜测。

---

[*] 1926年10月22日，为欢迎瑞典王储，中国地质学会、北京自然历史学会和北京协和医学院联合举办了科学会议，会上安特生博士首次宣布了周口店的发现。

While a complete description of these very important specimens may shortly be expected in *Palaeontologia Sinica*, the following brief notes may be of interest here. One of the teeth recovered is a right upper molar, probably the third, the relatively unworn crown of which presents characters appearing from the photographs to be essentially human. The posterior moiety of the crown is narrow and the roots appear to be fused. The other tooth is probably a lower anterior premolar, of which the crown only is preserved. The latter also is practically unworn, and appears in the photograph to be essentially bicuspid in character, a condition usually to be correlated with a reduction of the upper canine.

The Chou Kou Tien molar tooth, though unworn, would seem to resemble in general features the specimen purchased by Haberer in a Peking native drug shop and afterwards described in 1903 by Schlosser. The latter tooth was a left upper third molar having a very much worn crown, extensively fused lateral roots, and from the nature of its fossilisation considered by Schlosser to be in all probability Tertiary in age. It was provisionally designated as *Homo? Anthropoide?* It is of more than passing interest to recall that Schlosser, in concluding his description of the tooth, pointed out that future investigators might expect to find in China a new fossil anthropoid, Tertiary man or ancient Pleistocene man. The Chou Kou Tien discovery thus constitutes a striking confirmation of that prediction.

It is now evident that at the close of Tertiary or the beginning of Quaternary time man or a very closely related anthropoid actually did exist in eastern Asia. This knowledge is of fundamental importance in the field of prehistoric anthropology; for about this time also there lived in Java, Pithecanthropus; at Piltdown, Eoanthropus; and, but very shortly after, at Mauer, the man of Heidelberg. All these forms were thus practically contemporaneous with one another and occupied regions equally far removed respectively to the east, to the south-east, and to the west from the central Asiatic plateau which, it has been shown elsewhere, most probably coincides with their common dispersal centre. The Chou Kou Tien discovery therefore furnishes one more link in the already strong chain of evidence supporting the hypothesis of the central Asiatic origin of the Hominidae.

(**118**, 733-734; 1926)

Davidson Black: Department of Anatomy, Peking Union Medical College, Peking, China.

关于这些非常重要的标本的完整描述在《中国古生物志》上发表之前，以下这些简要的描述可能是有价值的。发现的牙齿中有一个是右上臼齿，可能是第三个，它的照片显示，牙冠几乎没有磨损，所显示出的基本上是人类的特征。其牙冠后半部分狭窄，根部融合。另一个牙齿可能是一个前部的下前臼齿，只有牙冠保存了下来。这一个也几乎没有磨损，它的照片显示出基本的二尖齿特征，这种情况通常与上犬齿退化有关。

周口店臼齿，尽管没有磨损，其总体特征却和哈贝雷尔在北京当地的一个药店买到的标本（1903 年，施洛瑟描述了这一标本的特征）似乎相似。这一来自药店的牙齿是一颗左上第三臼齿，其牙冠磨蚀严重，外侧两齿根大部分合并。从它石化的性质考虑，施洛瑟认为它很可能是第三纪的。它被临时归属于人属？或者猿属？回想起来，非常有趣的是，当施洛瑟总结他对那颗牙齿所作的描述时曾指出，将来的调查者将有希望在中国发现一种新的类似人的化石第三纪人或古老更新世人。周口店的发现是对那个预测的一个惹人注目的确认。

现在很明显了，在第三纪结束或第四纪开始之际，在东亚的确存在人类或者说一种与人类关系密切的类人猿。这在史前人类学领域是一个重要的基本发现。在大约同一个时期，爪哇存在爪哇直立猿人，皮尔当存在曙人，之后不久在毛尔地区出现海德堡人。所有这些生物类型都是属于同一时期的，并且分别在东部、东南部和西部各自占据了距中亚平原相同距离的区域，而其他的证据表明，中亚平原极有可能是它们共同的扩散中心。周口店的发现，在已有的有关人类中亚起源假说的有力证据链中又增添了一环。

（刘晓辉 翻译；徐星 审稿）

Sterilisation as a Practical Eugenic Policy

E. W. MacBride

Editor's Note

Perhaps the most chilling aspect of this discussion of eugenics by zoologist Ernest William MacBride, reviewing of a book on eugenic policies in the United States, is its ignorance and prejudice several decades after Darwin's evolutionary theory sparked calls for controls on breeding in human society. MacBride does not even respect that theory: he was a notorious late advocate of Lamarckian inheritance of acquired characteristics, evident here in how he thinks people who acquire mental defects might pass them on to offspring. There is also an elision from sterilization of "mental defectives" to the control of fertility in people whom MacBride deems merely "stupid" or feckless apparently, many of the poor. It is a stark reminder of how ideas about genetic heredity were confused and abused.

Sterilisation for Human Betterment: a Summary of Results of 6,000 Operations in California, 1909–1929. By E. S. Gosney and Dr. Paul Popenoe (A Publication of the Human Betterment Foundation.) pp. xviii+202. (New York: The Macmillan Co., 1929.) 8*s*. 6*d*. net.

THIS little book is a storehouse of information on the efforts which have been made in the United States to improve the human stock by sterilising the feeble-minded and the insane. It appears that although *more Americano* laws have been passed in about twenty states of the Union providing for the legal sterilisation of sexual perverts, and imbecile and insane patients in public institutions, these laws have been put into practical operation only in the State of California, so that in the book discussion is mainly concerned with the results obtained in that State.

The justification for these attempts to aid Nature in eliminating the unfit is set forth in the introduction. Amongst our unsentimental forefathers, no efforts were made to keep alive weakly and diseased children, and hence the race was propagated only from its most vigorous members; but nowadays, when unreflecting humanitarian sentiment is in fashion, all babies are kept alive so far as medical science can avail, and this science is paid for by levying tribute on the thrifty and self-supporting. The result is that this section of society limits its offspring, and future generations are likely to be recruited not from the fit but from the unfit.

How drastically and efficiently natural selection operated amongst the young in England during the eighteenth century may be gathered from figures given by Miss Buer in her book, "Health, Wealth, and Population in the Early Days of the Industrial Revolution". In 1730, out of all babies born in London, 74 percent died before they were five years of age; in 1750, 63 percent died; in 1770 the percentage was 50, and it did not sink to 30 until

作为一项实用优生学政策的绝育术

麦克布赖德

编者按

达尔文在进化论中曾提出应该对人类的生育有所控制，几十年后，动物学家欧内斯特·威廉·麦克布赖德仍然对此缺乏理解并存有偏见，他对一本关于美国优生政策的书所作的评论令人心寒。麦克布赖德甚至并不推崇达尔文的进化论。他作为拉马克获得性遗传假说的支持者之一而广为人知，从这篇文章中可以很明显地看出，他认为精神病患者可能会把他们的疾病基因传给后代。其实，不管是对"精神病患者"的绝育，还是对在麦克布赖德看来只是略显"愚钝"的无能之辈的生育控制，都是没有必要的，后者显然是指穷人。这篇文章是提醒人们关注遗传学的观点是如何被混淆和滥用的绝好实例。

《用于人类改良的绝育术：1909~1929 年加利福尼亚 6,000 例手术结果的概要》

戈斯尼和保罗·波普诺博士著（人类改良基金会的一部著作）

xviii + 202 页。（纽约：麦克米伦公司，1929 年。）8 先令 6 便士

这本小册子是一座知识宝库，蕴藏着美国通过使低能者和精神病患者绝育的方式来提高人口质量的过程中取得的成就。尽管美国联邦约 20 个州已经通过了**越来越多的美式法律**，为性异常者、智能低下者及精神病患者在公共机构进行绝育手术提供合法依据，但是看起来只有加利福尼亚州真正实施了这些法律，因此本书中主要是关于加利福尼亚州取得的成果的讨论。

为了帮助造物主减少不适于生存者而采取这些绝育措施的合理性，在本书的引言里已经有所陈述。我们那些无情的祖先，没有为使虚弱和得病的孩子能够活下来而进行过任何努力，因此种系都是从其最强壮的成员中繁衍而来的；但是如今，在这个浅薄的人道主义情感泛滥的时代，医学在其力所能及的范围内力求使所有婴儿都能活下来，并且通过向节俭的自食其力者征税来支付这些费用。这样的结果就是，社会上的这部分自食其力者限制自己的后代个数，未来的孩子们更可能是从那些不适者，而非适者中繁衍而来。

自然选择在 18 世纪英国的年轻人中所起的作用是多么明显有效，这可以从比埃小姐书中给出的数字推断出来，那本书的书名是《工业革命早期的健康、财富和人口》。1730 年伦敦出生的所有婴儿中有 74% 死于 5 岁之前，1750 年这一比例为 63%，1770 年为 50%，直到 1833 年这一比例才降到 30%。这一比例在该国的其他

1833. The percentage was probably even higher in other parts of the country. The help given by hospitals, and later by the State, to indigent mothers has all grown up in the last century, so that the argument that because we have maintained a vigorous, enterprising, fighting race in these islands for eight hundred years since the Norman Conquest, we shall continue to do so, is not one for which there is any sound basis.

It is, however, not practical politics to suggest a return to the old plan of *laissez-faire*. How then shall the elimination of the unfit be promoted? The authors of this book suggest "by legalised sterilisation". The method of sterilisation advocated is cutting the ducts (vasa deferentia in the male, and Fallopian tubes in the female) which convey the germ cells to the exterior. The authors point out that more than six thousand operations of this sort have been already performed in California, and that only seven failures are recorded (three in males, four in females). The operation does not interfere with sexual desire or the performance of the sexual act. The genital organs in man, as in Vertebrata generally, have two functions, namely: (1) to produce the germ cells; (2) to produce a hormone which diffuses through the system and maintains youth and vigour. In a man the spermatozoa forms a minor part of the sexual discharge, the main portion of which is constituted by the prostatic secretion, and some authorities hold that this secretion when absorbed by the female has an invigorating effect on the constitution. As to a woman, when it becomes necessary on account of tumours to remove the uterus, if a portion of one of the ovaries is preserved and sewn to the abdominal wall, this will prevent the premature onset of the menopause and maintain in the patient all the qualities of a young woman.

But are insanity and mental defect hereditary? Some British authorities hold that in many cases they are not. So far as insanity is concerned, however, there is general agreement, as our authors point out, that the condition known as "dementia praecox" is the result of an inborn weakened constitution, and that it is a mere question of time when it will manifest itself in the life of the unfortunate individual who has inherited this constitution. As to mental defect, the argument that it is sometimes not of hereditary origin, overlooks the consideration that all "mutations", of which mental defect is one, must ultimately have been produced by some external cause, and there is nothing to show that an "accidental" mental defective will not propagate mentally defective children. In any event, even if a defective should produce healthy children, such a person would make the worst possible parent to carry out the duty of caring for and training the children; and it is a little too much to ask the State to allow a defective to go on having children on the chance of some of them being normal, if the State has to support them all.

Our authors urge that sterilisation should not be regarded as a punishment but as a hygienic measure; that defectives confined in asylums might be allowed out on condition of their consenting to this operation. But whilst we agree that this argument is good so far as it goes, a little reflection will show that it only touches the fringe of the problem. The defectives most dangerous to society are those who are never confined in institutions at all! The high-grade defectives are just able to support themselves in the lowest paid and most unskilled occupations, and no civilised government would take the responsibility of

地方可能更高。上个世纪，医院给予这些贫困母亲的帮助（后来国家也给予了帮助）已经有所增加。因此有一种观点认为，诺曼征服以来的 800 年间，这些岛上的人们一直保持着一个精力充沛的、有进取心的、好战的民族应有的素质，所以应该继续保持这种做法，这一论点是缺乏可靠根据的。

然而，重新起用**自由放任**的旧政策并不切实可行。那么怎样才能促进不适于生存者数量的减少呢？本书的作者建议采取"合法绝育"的手段。这里提倡的绝育方法，是指切断向外输送生殖细胞的管道（男性是输精管，女性是输卵管）。作者指出加利福尼亚已经实施了六千多例此类手术，根据记录，只有 7 例失败（3 例男性绝育术，4 例女性绝育术）。该手术不会影响性欲和性行为的进行。与脊椎动物门其他成员一样，男性生殖器有两个功能，即：（1）产生生殖细胞；（2）产生散布于全身系统并维持机体青春与活力的激素。在男性分泌物中精子仅仅占很小部分，大部分是前列腺分泌液。一些权威人士认为，这些分泌液在被女性吸收后，会起到令女性的体质充满生气与活力的作用。对于女性，当她们因为肿瘤而必须摘除子宫时，如果保留一个卵巢的一部分，并将其与腹壁缝合在一起，就可以延缓更年期的到来并使病人保有年轻女人的所有特质。

但是精神病和心智缺陷是可遗传的吗？英国的一些权威人士认为，许多情况下它们并不遗传。然而就精神病而言，人们普遍认可的观点正如本书作者指出的那样，通常所说的"早发性痴呆"就是先天体质虚弱造成的，遗传了这一体质的不幸个体，发病只是时间问题。对于心智缺陷，有观点认为，它有时不是由于遗传。这一看法忽略了一点，那就是包括心智缺陷在内的所有"突变"最终都是由某一外因引起的，而没有任何证据表明"偶然的"心智缺陷者不会生育出心智缺陷的孩子。无论如何，即使一个有缺陷的个体可以生出健康的孩子，作为负有照顾和培养孩子的责任的父母来说，他们也是最不该成为父母的人；如果国家不得不供养所有有缺陷的个体的话，那么怀着"有缺陷的人也可能生出正常孩子"的投机心理而请求国家允许他们生育，这未免太过分了。

本书作者认为不应该将绝育视为一种惩罚，而应该看作一项卫生措施；对于那些被关在收容所的有缺陷的个体，在他们同意接受该手术的情况下，可以允许他们出去。就目前状况来看，我们承认这一想法很好，但是稍加考虑就会发现，这种做法只是刚刚触及到这一问题的边缘而已。对社会最具危险性的有缺陷的个体是那些从来没有受到公共机构限制的人！高级的有缺陷的个体能够凭最低收入和最不需技能的工作来养活自己，并且任何文明社会的政府都不会承担约束他们的责任，所以

confining them, and so they go on propagating large families as stupid as themselves. As Mr. Lidbetter has shown[*], it is from the ranks of just these classes that in the last hundred years the majority of paupers and criminals of London has been recruited.

It seems to us that in the last resort compulsory sterilisation will have to be inflicted as a penalty for the economic sin of producing more children than the parents can support. Whether a man has a large or a small family is—given a healthy wife—a matter of taste, so long as he provides for his own children; but when he comes to the State and demands that it—that is to say, his neighbours—should support these children, then the State can say, "Very well—we shall help you with the family which you have, but if after this you have any more children you shall be sterilised."

Before, however, such an alternative is presented to any citizen, he may justly claim that he should receive instruction from the State in the means of birth-control. It is obviously unfair that such knowledge should be denied to the poor whilst it is easily accessible to the rich. It is often said, and with justice, that the great objection to birth-control is that the wrong people practise it. But this knowledge once attained cannot be taken away; the middle classes possess it and cannot be prevented from putting it into practice. If, however, the knowledge and practice of birth-control were widely spread among the working-class, there would be created such a resentment against the reckless production of children that the movement to establish compulsory sterilisation of the unfit would prove irresistible.

(**125**, 40-42; 1930)

---

[*] "Pauperism and Heredity", by E. J. Lidbetter, *The Eugenics Review*, vol. 14, p. 152; 1923.

他们就会继续繁衍出和他们自己一样有缺陷的大家族。正如李德贝特先生所说 *，最近的 100 年间，伦敦的大部分乞丐和罪犯正是从这类人中产生的。

对我们而言，强制性的绝育术似乎是对那些生育了很多孩子而无力供养的父母们所犯经济罪进行惩罚的最后手段。如果一个男人拥有一个健康的妻子，那么他选择拥有大家庭还是小家庭完全是个人喜好的问题，只要他能养得起自己的孩子；但是如果他要求国家（其实也就是他的左邻右舍们）帮助抚养他的孩子们，那么国家就可以说："没问题，我们会帮你养家，但是从现在起如果你再生孩子的话，那么你将会被绝育。"

不过，在向市民公布这种可供选择的方案之前，人们可能会理直气壮地声称自己有权利了解国家的生育控制政策。富人们可以很容易地获悉这些信息，而穷人们的知情权可能被剥夺了，这显然是非常不公平的。人们常常说，生育控制的最大问题在于是错误的人在实施政策，这么说也是公平的。但是，这些知识一旦为人们获得就不可能再拿走；中产阶级们知悉了这些，就不可避免会将其付诸实践。然而，假如节育的知识和实践在工人阶级中广泛传播，那么可能会出现一种对不计后果进行生育的怨恨，到那时对不适于生存者实施强制性绝育术的行动可能就无法遏止了。

（刘皓芳 翻译；刘京国 审稿）

---

* 《贫困与遗传》，作者李德贝特，《优生学评论》，第 14 卷，第 152 页，1923 年。

Eugenic Sterilisation

Editor's Note

In the 1930s the potential benefits of eugenic sterilisation, first advocated by Darwin's cousin Francis Galton, were being considered in many European nations and in the United States. This editorial reacts to a proposal by the Eugenics Society of London for a legal change allowing sterilisation of the mentally impaired on their own consent, or that of a parent or legal guardian. The society had argued this would achieve a 17% reduction in mentally deficient individuals in one generation. The editorial counters that this figure is unreasonably optimistic, as it assumed permissions could be obtained for some 300,000 certifiable mental defectives in England and Wales. Yet like many scientists of the times, this editorial supports the principle of sterilisation.

SOME of the young people of Germany would have us believe that much of the time that can be spared from their more materially fruitful exploits is given over to singing a song which they call "Deutsche Jugend, heraus". Its language, borrowed from historical romanticism, permits, if it does not foster, a certain diversity of interpretation, and some lines with a frankly Christian significance may even be omitted at the discretion of the singer. Claim to popularity is thus made more catholic.

> Wollt Ihr ein neues bauen
> mit Händen stark und rein,
> in gläubigem Vertrauen
> lasst dies die Losung sein:
> Den Feind in eigner Mitte
> gefällt in ernstem Strauss....

Moralists, it is easy to see, may use these lines to assist them in focusing attention upon that enemy in their midst distinguished as the *beam*, while the nationalist may recognise more immediately its particular referability to the communistic *mote*.

We are assured, however, that the resiliency of this *credo* unites rather than divides, and such demonstrations as we have enjoyed tend to reinforce the assurance audibly. But we cannot help wondering what will happen in the world when the youth of one country or another not only present accessible enemies in their patriotic songs but also define them with scientific precision.

The real enemies of mankind are made, yearly, more and more accessible to attack by science, and if it were not for the protective screens, intangible and often fantastic, thrown up by the unscientific for whom nakedness, even the enemy's, still seems to possess terrible powers, mankind might subjugate very speedily its worst foes. But if, as Sir Walter Fletcher

优生绝育

编者按

20世纪30年代，美国和很多欧洲国家都开始认识到，由达尔文表弟弗朗西斯·高尔顿最先提出的优生绝育将对社会产生积极的影响。这篇评论回应了伦敦优生学会的一项倡议，该倡议要求修改法律，以便在取得本人或父母中的一方或法定监护人同意的条件下，允许对心智缺陷者实施绝育。优生学会认为，这样做将使下一代中心智缺陷者的数量减少17%。这篇评论认为这个数字过于乐观，因为它假设了英格兰和威尔士的300,000名确诊的心智缺陷者都会同意进行绝育。不过和当时的许多科学家一样，这篇评论的作者也是支持绝育原则的。

富有成效的物质文明建设使德国的年轻人有了更多的闲暇时间，但他们中的一些人把大部分多出来的时间都花费在吟唱一首被他们称为《德国青年》的歌曲上了。其带有历史浪漫主义色彩的歌词给整首歌赋予了多种解释，甚至是一些本来并不包含的意思。歌词中有些颇具虔诚基督教意义的部分可能会被歌手酌情删掉，从而可以更广泛地向大众普及。

> 你愿意建立一个新世界吗?
> 用强壮的双手建立一个纯净的世界，
> 坚定这一信念
> 并使它成为口号:
> 让处于中间阶层的敌人
> 就像虔诚的施特劳斯信徒一样……

很容易就能想到，道德家们可能会借用这些歌词来帮助他们将注意力集中在那些混杂于人民中间并以**国家栋梁**而著称的敌人，而民族主义者们通过这些歌词可能会更直接地意识到，相对于共产主义的**瑕疵**而言他们的主张具有的特殊借鉴意义。

尽管如此，我们还是确信这一**信条**的弹性有利于团结而非分裂，而且这种我们一直很喜欢的表述方式显然很有利于增强其可信性。但我们不禁要问，当一个国家或另一个国家的年轻人不仅在自己的爱国歌曲里提到触手可及的敌人，而且用科学精确的语言定义这些敌人的时候，这个世界将会发生什么呢?

一年一年过去，人类真正的敌人变得越来越容易被科学击倒，如果不是因为无知者甚至是反对者们非科学地抛出的一些无形的、通常是幻想出来的保护屏障看上去似乎拥有可怕力量的话，人类可能很快就能征服他们最大的敌人了。但是正如沃尔特·弗莱彻爵士最近指出的，如果对于一种纯粹的疾病（例如癌症），只有在破

has lately pointed out, a mere ailment, like cancer, has only been made accessible to scientific study through the lifting of foolish and superstitious taboos, how can we expect the direr social maladies to be approached courageously? A protective hedge of errors and superstitions hems them in on every side, so rank and poisonous that it seems that even science is infected and intimidated while it attacks.

How else is it possible to explain the demand just put forward[*] by a committee of the Eugenics Society for permissive legislation which would take a whole generation to achieve a reduction in the incidence of mental defect not of a hundred, not of fifty, not of twenty-five, but, problematically, of seventeen percent? Between our people and the realisation of this slender benefit stands "an ambiguity of the law" which the Society proposes to remove. A person may, with consent, be sterilised in the interests of his *own* health. In the interest of the public health, present or future, he may not be sterilised. By a curious legal inversion, the "willing mind" of the individual cannot take away the offence against the public even should he be prepared to save it from all possibility of contamination by his own progeny. The offence consists in a "maim" which deprives the individual, or so it may be contended, of martial courage, and the State of a vessel, however unsuitable otherwise, for this same virtue. To contentions of this sort, surely the monosyllabic genius of Mr. H. G. Wells's latest novel has supplied the only effective answer.

To meet the practical situation, the committee proposes a Bill legalising eugenic sterilisation. This would authorise the mental deficiency authority or superintendent of an institution to sterilise a mental defective, subject to the consent of the parent or guardian and of the Board of Control, and of the spouse if the defective is married. In the case of defectives deemed capable of giving consent, sterilisation would not be performed otherwise than with this consent. It would authorise the voluntary sterilisation of a person about to be discharged from a mental hospital for the insane as recovered, again with the added consent of Board and of parent, guardian, or spouse; and it would legalise voluntary sterilisation for the sole purpose of preventing the transmission of hereditary defect seriously impairing physical or mental health or efficiency.

Five members of the committee and another contributed to the *Lancet* for July 19 a letter defending this policy. The defence combats the assertion that if every certifiable mental defective had been sterilised twenty or thirty years ago it would have made little appreciable difference to the number of mental defectives existing today. It repeats a sentence of the committee's report urging that "if all the defectives in the community could be prevented from having children the effect would be even on the most unfavourable genetic assumptions with regard to defectiveness, to reduce the incidence of mental defect by as much as 17 percent in one generation".

---

* Committee for Legalising Eugenic Sterilisation. Eugenics Society, London, 1930.

除愚蠢迷信的忌讳后才能对其进行科学研究的话，那么我们又怎么能期望研究人员会大胆地去研究更加可怕的社会弊病呢？错误和迷信的保护罩将各种问题和弊病团团包围，当这些讨厌而又恶毒的保护罩发起进攻时，看起来似乎连科学都受到了影响和威胁。

优生学会的一个委员会最近提出，希望一项提案能获得立法通过[*]。他们认为提案的措施将会使下一代中心智缺陷的发生率下降，这种降低不是减少 100 个、50 个或者 25 个缺陷个体，而是使缺陷的发生率整体上降低 17%。对他们的此项要求，还有什么别的可能的解释吗？在这一微薄利益的实现和我们的民众意识之间，还存在着"法律上的模糊地带"，这正是该学会主张消除的。为了**自身**健康，一个人可能会同意做绝育手术。但如果是为了公众健康，那么无论是现在还是将来，他可能都不会去做绝育手术的。通过一种奇妙的立法转换，个人的"意愿"就不能再对公众利益有所冒犯，甚至他必须做好准备以免因为自己后代可能的缺陷而使公众利益受到损害。这里所说的冒犯包含在一种"伤害"中，这种伤害可能存在争议，它剥夺了个人的战斗勇气，剥夺了国家的命脉，然而同样的特点并不适合于其他情况。对于这种争论，单音节天才威尔斯先生的最后一部小说肯定已经提供了唯一有效的答案。

为了符合实际情况，委员会提出了一份使优生绝育合法化的议案。这将在心智缺陷者的父母或监护人和管理委员会及缺陷者的配偶（如果该缺陷者已婚）同意的情况下，赋予心智缺陷相关的权威机构或者机构管理人对心智缺陷者进行绝育的权利。如果缺陷者被认为具有作出决定的能力，则只有在本人同意的情况下才能对其进行绝育。自愿同意进行绝育的患者在康复后可以允许其离开精神病院，当然这需要来自委员会以及父母、监护人或配偶的同意；这将使把阻止遗传缺陷发生传递以防止其对生理或精神健康或功效造成严重损害作为唯一目的的自愿绝育合法化。

该委员会的 5 位委员和另外一人于 7 月 19 日向《柳叶刀》投稿捍卫这项政策。他们在辩护中反对了如下的断言：即使在二三十年前就对每一个确认具有心智缺陷的人进行了绝育，那也几乎不会对现在的心智缺陷者的数量带来多么显著的影响。文中重复了一个委员会报告中的句子，极力主张"如果能够使社会上所有有缺陷的个体不生育孩子的话，那么，即使是在最坏的遗传假设下，其影响也将使下一代中心智缺陷的发生率降低 17% 左右"。

---

[*] 优生绝育合法化委员会。优生学会，伦敦，1930 年。

Obviously a 17 percent reduction in the incidence of mental deficiency is more desirable than a 17 percent increase. But do the committee's proposals ensure this reduction? Clearly, no. The words quoted promise at least that reduction if the fertility of *all* living mental defectives is prevented. The committee's proposals, with their emphasis upon the voluntary principle, by no means ensure that the 300,000 certifiable mental defectives in England and Wales would be sterilised. Who must consent? (1) The patient, if he is capable. (2) The parent or guardian. (3) The spouse if the patient is married. (4) The Board of Control. The calculation, it is true, is based on two assumptions "highly unfavourable" to the effectiveness of the proposals—that the genetic factor responsible for defectiveness (primary amentia) would be much "carried" and would only rarely produce manifest defectives, and that defectiveness is uniformly distributed throughout the community. (The fertility of defectives also is assumed to be that of the average of the population.)

How unfavourable, on the other hand, are the chances of permission? Nothing is gained by attempts to write off opposing assumptions. A figure is a figure, right or wrong.

Again, is a 17 percent reduction all that eugenic science can promise? Disregarding altogether those so-intelligent defectives who will strive to serve the country by seeking this minor mutilation, is it the institutional class that constitutes the chief danger to society? Prof. MacBride (*Nature*, Jan. 11, 1930, p. 40) says emphatically that this but touches the fringe of the problem. "The defectives most dangerous to society are those who are never confined in institutions at all! The high-grade defectives are just able to support themselves in the lowest paid and most unskilled occupations, and no civilised government would take the responsibility of confining them, and so they go on propagating large families as stupid as themselves." His idea of penal sterilisation, a punishment "for the economic sin of producing more children than the parents can support", is one which becomes more and more difficult to apply as more and more ways are devised by the State for screening the individual from biological estimation.

Is there not a real danger that the advocates of such legislation as here may mistake the assent of the political machine for victory? If assent were gained, would it not be much more accurately determined as the hall-mark of failure? It is not the assent of the State, but the initiative and creative power of the State, that is needed to secure essential progress, and that will not exist until our legislators of all parties or of any party derive their inspiration from the cultivation of natural knowledge.

(**126**, 301-302; 1930)

很显然，心智缺陷的发生率下降 17% 比增加 17% 更有利。但是该委员会的提案是否可以保证这一减少量的实现呢？当然不能。此处引用的文字认为，如果使**所有**活着的心智缺陷者不生育的话，那么至少可以保证实现这一减少量。按照委员会所强调的自愿原则，他们的提议根本不能保证英格兰和威尔士的 300,000 例确定具有心智缺陷的个体都被绝育。谁必须同意呢？（1）病人，如果他有自主能力的话。（2）父母或监护人。（3）配偶，如果病人已婚的话。（4）管理委员会。该计算是无误的，但它建立在两个"非常不支持"该提案有效性的假设之上。一个假设是：导致缺陷（先天痴愚）的遗传因子会在大量的后代中"被携带"，但仅仅在很少情况下才会产生明显的缺陷；另一个假设是：这种缺陷在整个社会中是均匀分布的。（缺陷者的生育能力也被假定为相当于整个人群的平均生育能力）。

另一方面，允许缺陷个体生育的话情况会有多么不利呢？通过试图取消对立的假设并没有取得任何成果。无论正误，数字就是数字。

再者，17% 的减少是否是优生学能够承诺的全部效果？如果完全不考虑那些非常聪明并尽力为国家作贡献的缺陷个体来判断最小损害的话，那么是不是公共机构就成了对社会的主要威胁呢？麦克布赖德教授（《自然》，1930 年 1 月 11 日，第 40 页）强调说这仅仅触碰到问题的边缘。"对社会最具危险性的有缺陷的个体是那些从来没有受到公共机构限制的人！高级的有缺陷的个体能够凭最低收入和最不需技能的工作来养活自己，并且任何文明社会的政府都不会承担约束他们的责任，所以他们就会继续繁衍出和他们自己一样有缺陷的大家族。"当国家设计出越来越多的方法通过生物评估来筛选缺陷个体时，他那将绝育术视为"对那些生育了很多孩子而无力供养的父母们所犯经济罪"的惩罚措施的观点就变得越来越难应用于实际了。

像这里提到的支持如此立法的倡导者们可能会将政治机构的同意误解为自己取得了胜利，这难道不是真正的危险吗？一旦得到同意，那么更准确地说，这更应该被视作失败的标志。因为这不是国家的同意，而是国家自发的创造性力量的同意，这种力量是用来保障最基本的发展的，除非所有政党的立法者从自然知识的熏陶中获得启发，否则这种力量将不会存在。

（刘皓芳 翻译；刘京国 审稿）

Eugenic Sterilisation

J. S. Huxley

Editor's Note

Eugenics—the attempted elimination of "bad genes" in a population—was widely held to be important for maintaining a healthy society for long after Charles Darwin published his evolutionary theory. It was advocated by Darwin's cousin Francis Galton, and Darwin himself assented. So did Julian Huxley, grandson of Darwin's staunch advocate Thomas Henry Huxley, who served in the British Eugenics Society until the 1960s. Here he writes to defend the society's recommendations of enforced sterilization of "mental defectives" against a criticism in a *Nature* editorial. Tellingly, that criticism was of the proposed mechanism, not the principle—*Nature* fully supported eugenic arguments in the 1930s. Only later did they become seen as not just morally but scientifically flawed.

AS a member of the Committee of the Eugenics Society for Legalising Eugenic Sterilisation, I should like to be allowed to say a few words concerning the leading article in *Nature* of Aug. 30 on our proposals. It is stated there: "Is there not a real danger that the advocates of such legislation as here may mistake the assent of the political machine for victory? If assent were gained, would it not be much more accurately determined as the hall-mark of failure? It is not the assent of the State, but the initiative and creative power of the State, that is needed to secure essential progress..."

With the last sentence I entirely agree; but I fail to perceive how a step in the right direction can be regarded as the hall-mark of failure—unless, indeed, the Committee should be so stupid as to believe that the taking of this one step had brought us to our final goal, which is certainly not the case. The article opens with references to the difficulties in the way of progress which are created by timid and ignorant public opinion, and continues, "if, as Sir Walter Fletcher has lately pointed out, a mere ailment, like cancer, has only been made accessible to scientific study through the lifting of foolish and superstitious taboos, how can we expect the direr social maladies to be approached courageously?" I think I can speak for the Committee in saying that we realise to the full the extent of these intangible difficulties, and that it is precisely for that reason that we have concentrated on a small but tangible and urgent beginning. Somehow or other the public has to be made race-conscious, has to be imbued with the eugenic idea as a basic political and ethical ideal. We believe that a campaign of the kind we have launched, directing attention to a gross racial defect, will be the best possible way of turning their thoughts in the desired direction.

优生绝育

编者按

在查尔斯·达尔文提出进化论很长时间以后，优生学这种试图消除人类"不良基因"的学说被大家公认为是确保社会成员健康的重要方法。优生学是由达尔文的表弟弗朗西斯·高尔顿倡导的，并且得到了达尔文本人的支持。本文的作者朱利安·赫胥黎也是优生学的支持者，他是托马斯·亨利·赫胥黎的孙子。20世纪60年代以前，他一直在英国优生学会工作。他写这篇文章的目的是，为该学会倡导的"精神病患者"应该被强制进行绝育作辩护，以反驳《自然》上持反对意见的一篇评论。事实上，那篇评论批驳的是提案中所说的运作机制，而不是批驳基本原则。早在20世纪30年代，《自然》就完全赞成优生学的观点。只是后来这些观点变得不仅不人道，而且出现了科学上的谬误。

作为优生学会优生绝育合法化委员会的一名成员，我想就《自然》8月30日发表的那篇针对我们的提议的重要文章说几句。文中说到："像这里提到的支持如此立法的倡导者们可能会将政治机构的同意误解为自己取得了胜利，这难道不是真正的危险吗？一旦得到同意，那么更准确地说，这更应该被视作失败的标志。因为这不是国家的同意，而只是国家自发的创造性力量的同意，这种力量是用来保障最基本的发展的……"

我完全同意最后一句话，但是我却不理解方向正确的措施怎么会被当作失败的标志呢？除非委员会愚蠢到相信仅仅通过实施这一措施就可以实现我们的终极目标，但事实上他们肯定没有这么认为。那篇文章一开篇就提出，怯懦而又无知的公众意识对优生绝育的实施造成了困难，紧接着写到"正如沃尔特·弗莱彻爵士最近指出的，如果对于一种纯粹的疾病（例如癌症），只有在破除愚蠢迷信的忌讳后才能对其进行科学研究的话，那么我们又怎么能期望研究人员会大胆地去研究更加可怕的社会弊病呢？"我想我可以代表委员会说，我们已经认识到了所有这些无形的困难，也正是因为这个原因，我们才集中精力从一个比较小但很明确很紧迫的问题入手。不管通过什么方法，都必须使公众具有种族意识，必须让他们把优生思想当作一项基本的政治道德理想。我们相信，我们发起的这种引导公众关注整个种族缺陷的运动，是把公众思想扭转到我们预期方向上的最可能有效的途径。

Comment is also made on the fact that the prevention of reproduction by all defectives would only lower the incidence of mental defect by about 17 percent in one generation. The article fails to remind readers that the process is cumulative, and also does not point to any other way in which it could be reduced more rapidly. Finally, the most relevant fact of all is omitted, namely, that one of the greatest obstacles to securing assent to the sterilisation of defectives has been and is the widespread belief that, since two normal persons may have a defective child, therefore preventing defectives from reproducing will have no effect on the proportion of defectives in later generations. Dr. R. A. Fisher has gone carefully into the matter, and has shown that, even when the most unfavourable assumptions are made, prevention of reproduction by all defectives would result in a reduction of some 17 percent—which to me at least seems considerable, as it would mean that there would be above 50,000 less defectives in Great Britain after the lapse of the, biologically speaking, trivial span of one generation.

I am glad that *Nature* has directed attention to the gravity of the problem, and look forward with interest to further discussion of the problem in its columns.

(**126**, 503; 1930)

J. S. Huxley: King's College, London, W. C. 2.

该文章还评论了如下事实：一代人中所有的缺陷个体都不生育，也将只能使下一代中心智缺陷的发生率减少大约17%。该文章并没有提醒读者注意这一过程是累积的，也没有提出任何别的能使心智缺陷的发生率减少得更加迅速的方法。最后，该文章还遗漏了与减少心智缺陷的发生率关系最密切的事实，即，确保对缺陷个体实施绝育得到人们赞同的最大障碍之一，曾经是并且现在依然是广为流传的一种看法，那就是即使两个正常人也可能会生出一个有缺陷的孩子，因此阻止缺陷个体的生育这一做法对于后代中缺陷个体的出现比例不会有任何影响。费希尔博士已经对这一问题进行了仔细深入的研究，结果表明即使是在最坏的假设下，通过阻止所有缺陷个体的生育也会使下一代中的缺陷个体减少17%左右，这个数字至少对我来说算是相当可观了，因为从生物学意义上来说，这意味着仅仅是经过一代人的生育之后，英国心智缺陷者的数量的减少就会超过50,000。

我很高兴《自然》能够关注这一问题的重要性，我也热切期待其专栏里能够出现对此问题的深入讨论。

（刘皓芳 翻译；刘京国 审稿）

The Problem of Epigenesis

E. W. MacBride

Editor's Note

Ernest William MacBride is perhaps in retrospect not the ideal author for this discourse on embryogenesis, being a late supporter of the Lamarckian view of inheritance. Yet he was considered an expert on embryology, and here he anticipates some of the key themes in modern biology: how is the development of an embryo related to its evolutionary heritage (the topic now dubbed "evo-devo"), and how does the organism, "considered as a machine", function? The former question had received much attention in Germany (MacBride's article is a review of three recent German books), especially in Ernst Haeckel's notion that embryogenesis recapitulates evolutionary history. The main issue debated here, however, is the origin of the force that organizes an undifferentiated egg into a structured body, and how that depends on the emerging concept of genes.

(1) *Grundriss der Entwicklungsmechanik.* Von Prof. Dr. Bernhard Dürken. Pp. vii + 208. (Berlin: Gebrüder Borntraeger, 1929.) 12.50 gold marks.

(2) *Die Determination der Primitiventwicklung: eine zusammenfassende Darstellung der Ergebnisse über das Determinationsgeschehen in den ersten Entwicklungsstadien der Tiere.* Von Prof. Dr. Waldemar Schleip. Pp. xii + 914. (Leipzig: Akademische Verlagsgesellschaft m.b.H., 1929.) 85 gold marks.

(3) *Experimentelle Zoologie: eine Zusammenfassung der durch Versuche ermittelten Gesetzmässigkeiten tierischer Formen und Verrichtungen.* Von Prof. Dr. Hans Przibram. Band 6 : *Zoonomie; eine Zusammenfassung der durch Versuche ermittelten Gesetzmässigkeiten tierischer Formbildung (Experimentelle, theoretische und literarische Übersicht bis einschliesslich 1928).* Von Prof. Dr. Hans Przibram. Pp. viii + 431 + 16 Tafeln. (Leipzig und Wien: Franz Deuticke, 1929.) 40 gold marks.

THE question of epigenesis may be justly said to constitute one of the two root problems of zoology. For if we think it out there are two main things to be discovered about an animal, namely: (1) How does it fulfil its functions?—in a word, considered as a machine, how does it work? and (2) How does it come into being?—that is, how did it develop and grow? A subsidiary question to the last is: If there be such a thing as evolution, how and why did the powers of growth change from generation to generation? For, as the late Dr. Bateson reminded us so long ago as 1894, the conception of evolution as the remoulding of the adult structures of an animal as we could alter the features of a wax doll by melting the wax and remodelling it, is an entire illusion, since the members of the parent species and of that to which it gives rise both begin as tiny formless germs and what is changed is *the powers of growth*. Now when we begin to analyse growth, we can either directly observe its successive phases—and this is the scope of descriptive embryology; or by operating on the germ by chemical and physical agencies we can seek to discover the

关于渐成论的问题

麦克布赖德

编者按

回想起来，由欧内斯特·威廉·麦克布赖德来作这个胚胎发生学方面的报告可能不是很理想，因为他是拉马克遗传学说的晚期支持者之一。不过，他毕竟是胚胎学方面的专家，而且在这篇报告中他也预见到了现代生物学的一些关键性课题：胚胎的发育过程与其种系的进化过程有什么关系（这一研究课题现在被称为"进化发育生物学"）？"被视为机器"的有机体如何发挥功能？实际上，在此之前德国科学家就对前一个问题给予了极大的关注（麦克布赖德的这篇文章就是对最近相关的 3 本德国科学家的著作所作的评论），特别是恩斯特·海克尔提出了胚胎发生重演了进化过程的观点。不过，这篇文章讨论的主要问题是，由未分化的受精卵形成有结构的机体的原始动力是什么，以及这一动力与人们最近提出的基因这一概念有何关系。

(1)《发育机制概论》。伯恩哈德·迪肯博士，教授。vii + 208 页。（柏林：施普林格兄弟，1929 年）。12.50 金马克。

(2)《原始发育的决定性：关于动物早期发育中决定因素的结果总结》。瓦尔德马·施莱普博士，教授。xii + 914 页。（莱比锡：学术出版有限责任集团公司，1929 年）。85 金马克。

(3)《实验动物学：关于实验动物挑选的法律规范和步骤概要》。汉斯博士，教授。第 6 卷：动物学；有关实验动物的法律的总结概述（理论和文学概述，以及试验，1928 年）。汉斯·普西布兰博士，教授，xiii + 431 页 + 16 表格。（莱比锡和维也纳：弗朗茨·多伊蒂克，1929 年）。40 金马克。

公正地说，渐成论的问题可能是动物学的两大根本问题之一。因为我们在研究一个动物时会思考两个主要问题：(1) 它是怎样实现它的功能的？简言之，如果把动物看作一台机器的话，它是如何工作的？ (2) 它是怎样产生的？即，它是如何发育和生长的？伴随第二个问题又会出现另一个问题：如果真的存在进化的话，那么生长的动力是如何一代代发生变化的，以及为什么会发生变化呢？正如已故的贝特森博士早在 1894 年就提醒我们的，把进化看成是对成体动物结构的重塑，就如同我们通过熔化蜡并进行重建来改变蜡人的造型特点一样，这是完全错误的概念，因为亲代物种的成员以及他们产生的子代都是从无定形的微小生殖细胞开始生长而来的，而**生长的动力**在变化。现在当我们开始分析生长的时候，我们可以直接观察到其连续进行的各个阶段——这属于描述胚胎学的范畴，我们也可以通过我们能够找到的化学和物理手段对生殖细胞进行实验来观察每一个可见因素在成年个体的生长

part which each visible element plays in the upbuilding of the adult individual—and this is the object of experimental embryology.

How this science has grown since its first beginnings with His in 1874 ("Unser Körperform und die physiologische Problem ihrer Entstehung") is witnessed by the three splendid works which are the subject of this review. Each of the three is worthy of unstinted praise: though we may differ from the authors in some of the conclusions reached by them, yet in each case the collection and setting forth of the matter is worthy of our sincere admiration. We hope that too long a time may not elapse before all are translated into English.

As an introduction to the subject Dürken's manual is to be preferred, because it is concise, well illustrated, and includes only typical cases which serve to exemplify the main principles of the subject, so that a beginner can get a good grasp of these principles without being overwhelmed by too much detail. Schleip's large and well-illustrated volume attempts to give a more or less complete account of the present state of our knowledge of the subject, and it will for a long time constitute a classic work of reference. Przibram's work—thorough and excellent as all his work is—is even more ambitious in its scope than that of Schleip, for it includes not only the facts of experimental embryology in the narrower sense, but also a considerable amount of the results of Mendelian experiments. It is, however, extremely condensed and, not being adequately illustrated, somewhat difficult to follow: it seems to us that its chief value will reside in its being a manual in which references to all the important papers on the subject can be easily looked up.

It must be obvious to the reader that, within the limits of the longest review for which space can be found in *Nature*, it would be impossible to refer to a tithe of the new matter contained in these volumes, and so we must limit ourselves to a discussion of the main problems involved and to the attitude of the three authors towards them. In fairness, however, it should be added that this new matter is almost entirely confined to an elaboration of subjects dealt with by the older authors such as Roux, Hertwig, Driesch, Herbst, Boveri, Conklin, and Wilson, and does not consist to consist to any considerable extent of discoveries in newer fields. The number of animals the eggs of which can conveniently be handled and which are tolerant of experiments is limited, and the same familiar figures crop up in successive text-books of experimental embryology. After all, as Driesch has wisely remarked, the biological experimenter cannot produce life at will—he must wait until he finds it, and he is therefore in the same position as a physicist would be if he could only study fire when he found it in the crater of a volcano.

When we approach the analysis of the development of the egg, the first question we encounter is whether the organs of the adult exist in the egg preformed in miniature and development consists essentially in an unfolding and growing bigger of these rudiments, or whether the egg is at first undifferentiated material which from unknown causes afterwards becomes more and more complicated and development is consequently an "epigenesis". This problem is *the* problem of experimental embryology; in varied forms it reappears in every experiment on development which has been made.

238

过程中都起着什么作用——这是实验胚胎学的目标。

这里即将评论的 3 部出色的著作描述了这一领域从 1874 年西斯首次发表著作（《我们的身体形态、构造和生理问题》）以来的发展过程。这 3 部作品中的任何一部都是值得高度赞扬的，尽管我们可能在某些结论上和作者有不同观点，但是每部作品中对事件的收集及详尽阐述都值得我们致以由衷的敬意。我们希望不久之后，所有这些著作都可以被翻译成英语。

迪肯的这本指南是该学科的首选入门书籍，因为该书简洁明了、插图丰富，只包含解释该学科主要原理的经典事例，所以初学者可以很好地掌握这些原理而不会被铺天盖地的细节吓倒。施莱普的那本插图丰富的大部头试图尽可能全面地向我们阐明该学科的知识现状，该书将在很长时间内成为这一领域的经典参考著作。普西布兰的这部著作和他的其他所有著作一样全面而出色，就其视野来说比施莱普的那本更具远见，因为在这本书里，不仅包括狭义实验胚胎学的内容，也包括相当一部分孟德尔实验的结果。不过，这本书极度浓缩，描述不够详细，因而想要读懂会有些难度。对于我们而言，它的主要价值似乎在于，这是一本可以从中很容易地找到该学科所有重要文章引用的参考文献的手册。

读者们都知道，《自然》上发表的评论是有字数限制的，受此所限，即使只是想谈论这些书中包含的一小部分新内容也是不可能的，因此我们只好仅限于讨论涉及的主要问题以及 3 位作者对这些问题的看法。不过，应该补充说明的是，公平地讲，这些新内容几乎完全局限于对老一辈作者所讨论主题的细化，这些作者包括鲁、赫特维希、德里施、赫布斯特、博韦里、康克林和威尔逊，而并没有涉及较新领域的任何重要发现。能够方便地对其卵子进行操作并保证卵子可以耐受实验条件的受试动物的数量是有限的，那些熟悉的图片后来连续出现在实验胚胎学的教科书中。毕竟，正如德里施曾经明智地提出过的，生物学实验者不能随意制造生命——他只有在找到受试的生命体后才能对其进行实验，因此从这个角度来看，生物学实验者的处境就如同只有在火山爆发时发现了火之后才能对火进行研究的物理学家的处境。

当我们对卵的发育进行分析时，我们面临的第一个问题就是：是不是卵子中就存在预先成形了的成体器官的缩微版，而发育过程实质上是这些器官雏形逐渐展开显露并变大的过程？又或者卵子最初只是未分化的物质，后来由于未知的原因而变得越来越复杂，因而发育是一个"渐成的"过程？这**正是**实验胚胎学要解决的问题。在研究发育的实验中，这一问题以各种各样的形式重复出现。

The answer to this question given by the earlier experiments of Driesch was that some eggs, such as those of starfish and sea-urchins, consist of undifferentiated material; but others, like those of Ctenophores, show a specialisation into parts destined to form particular organs of the adult. The experiments of Wilson, Conklin, and Crampton proved that the eggs of Annelida and Mollusca belong also to this latter category. To eggs of the first kind Driesch gave the name of "equipotential systems", since when the egg had divided into eight cells any one of these was capable of forming a tiny larva perfect in all details, and, moreover, when the egg had developed into a hollow sphere or blastula, any considerable piece of this blastula would round itself off and form a perfect blastula of reduced size, which would give rise to a correspondingly reduced larva. On these results, which were a complete surprise to him, Driesch founded his theory of vitalism, arguing that if the organism were to be regarded as a physico-chemical machine, such things could not happen, for no conceivable machine could be divided into parts, each of which would function as a similar machine of reduced size. He inferred that there must be in every egg a non-material force or "entelechy" which was capable of controlling the physical and chemical changes taking place in the germ, so as to direct them towards a definite end. This power of direction was named by Driesch "regulation". This revolutionary idea of Driesch, transcending the bounds of materialistic explanation, evoked the fiercest opposition amongst those biologists by whom life was regarded as nothing more than complicated chemistry. Yet the arguments of Driesch have never been successfully met. The utmost that can be urged against them is the assertion that, although we cannot explain life by physics and chemistry now, some day in the distant future, when we have made further discoveries, we may possibly be able to do so.

Of the authors reviewed in this article, Dürken is inclined to favour Driesch whilst Schleip and Przibram oppose him, but the alternative explanations of the two latter authors when examined in detail resolve themselves into saying the same things that Driesch said, in different phrases. All three authors agree in showing that between equipotential and specialised eggs every conceivable grade of intermediate exists, and that even the eggs of *Echinus* itself are not quite so equipotential as Driesch imagined. Schleip quotes the work of Hörstatius as proving that when the upper half of a blastula is cut off, though it will round itself off so as to form a reduced blastula, yet this will never form endoderm or proceed any further in development. The vegetative half, however, when severed will produce a completely viable gastrula. By a triumph of manipulative skill, Hörstatius succeeded in separating the vegetative pole of a blastula and grafting it in various positions on another blastula in which an appropriate defect had been produced. He thus proved that in all cases development begins in the graft, and that this graft can change cells that would otherwise produce ectoderm into endoderm, in other words, act as an "organiser" of development.

Driesch attributed specialisation in eggs to a "premature stiffening of the cytoplasm" which prevented the "entelechy" from moulding the fragment of the egg into a reduced whole. Przibram in other language comes to exactly the same conclusion. He says that the formation of definite organs is in all cases due to a *solidifying* of a portion of the

德里施早期的实验对于这一问题给出的答案是，像海星和海胆这一类的生物的卵是由未分化的物质组成的；而其他生物，如栉水母类，它们的卵则特化为几个不同的部分，每一个部分都特定发育为成体的特定器官。威尔逊、康克林和克兰普顿的实验证明，环节动物和软体动物的卵也属于第二类。对于第一类卵，德里施称它们是"等潜能系统"，因为当这类卵分裂成 8 个细胞时，其中任何一个都具有成长为所有细节部分都完整的小幼虫的能力。此外，当卵发育成中空的球体或囊胚时，该囊胚的任何一部分有一定大小的片段都可以完善自身，并形成一个体积相对较小的完好的囊胚，相应地这一囊胚可以产生一个体积较小的幼虫。这些结果使德里施感到无比惊讶，基于此，他建立了自己的活力论。该学说认为，如果把生物当作一台物理化学机器的话，这些情况就不会发生，因为想象不出任何机器可以在被分成几部分后，各个部分仍然能够像一台只是尺寸有所减小的类似机器一样正常运转。他推断每个卵中一定都存在一种非物质的驱动力或者"生机"，它具有控制生殖细胞中发生的物理和化学变化的能力，因而可以指导这些变化朝着特定的方向发展。德里施将这种指导能力称为"调控"。德里施这一革命性的想法超越了唯物论解释的范畴，激起了那些认为生命仅仅是一些复杂化学变化的生物学家们最强烈的反对。然而没有任何人在与德里施的观点的交战中取胜。这些极力反对的观点中，分量最重的也只不过是如下的论断：尽管我们现在不能用物理化学变化来解释生命，但是在遥远未来的某一天，当我们取得了进一步发现的时候，我们可能就有能力对其进行解释了。

在本文所评论的这些作者中，迪肯倾向于支持德里施的观点，而施莱普和普西布兰则反对德里施的观点，但是经过仔细推敲后可以看到，后两位作者提出的另外的解释其实与德里施的观点是一样的，只是说法不同而已。所有这 3 位作者都同意，每一种可以想象到的介于等潜能的卵与特化的卵之间的中间状态都是存在的，甚至连海胆本身的卵也并不像德里施想象的那样处于完全等潜能的状态。施莱普引用了赫斯塔提乌斯的工作，以此为证据来证明，当囊胚的上半部分被分离下来时，尽管它可以完善自己而形成一个体积减小了的囊胚，然而却并不能形成内胚层，也不会继续发育。植物极的那一半在被切下来后则可以生成一个完全能存活下去的原肠胚。由于操作技术上的巨大突破，赫斯塔提乌斯成功地分离了囊胚的植物极并将其植入到另一囊胚（此囊胚事先已经进行了相应的切除）的多个位置上。他由此证明了：在所有这些例子中，发育都是从植入物上开始的，这种植入物可以改变细胞，使本来会产生外胚层的细胞转而形成内胚层，换言之，植入物扮演了发育的"组织原"的角色。

德里施将卵的特化归因于"细胞质的过早硬化"，这阻止了"生机"将卵的片段塑造成尺寸有所减小的完整卵。普西布兰用不同的语言书写了几乎完全相同的结论。他说，在所有情况下特定器官的形成都是由部分细胞质的**硬化**引起的，从而形成了

cytoplasm, forming what he calls an "apoplasm" which, if we understand him right, he does not regard as fully alive. In proportion as "apoplasms" are deposited the potentialities of the germ are successively limited, and the reason why the higher animals approximate in their working to mechanisms is the large number of "apoplasms" included in their make-up. Only fluid cytoplasm is completely living and possesses all the potentialities of the race, and Przibram is driven to conclude that these potentialities, so far as embodied at all, must be contained in the molecules of the cytoplasm, and that, therefore, these molecules constitute the real entelechy. Schleip similarly concludes that there must be an ultra-microscopic structure in the cytoplasm which, like a crystal, tends to assume a definite form and to complete itself when a fragment is severed.

In making these admissions, however, it seems to us that both Schleip and Przibram deliver themselves into the hands of Driesch. For in the crystallisation of an inorganic substance from a solution, the crystal assumes a definite form because its molecules have definite corresponding shapes, as Sir William Bragg has taught us. But what kind of structure, whether molecular or super-molecular, are we to envisage in cytoplasm? When the limb of a young newt is cut off and the stump proceeds to regenerate a new limb, are the molecules in the stump in the form of infinitesimal fingers and toes? Moreover, when the stump is cut at different levels and only the missing piece is regenerated, are we to assume that at each level in the limb before amputation the molecules are miniatures of the part distal to them? If we are able to swallow these fantastic assumptions, what are we to say of the experiment recorded by Dürken in which the tail bud of one newt embryo was grafted into the body of another near its forelimb and developed into a new limb? Presumably the cytoplasm of the tail bud was "organised" so as to produce the tissues of an adult tail. How then was this organisation so completely changed as to produce a limb instead? No wonder that Dürken says that in cases like this, physical and chemical explanations leave us completely in the lurch, and we must have recourse to the conception of the "biological field", an influence not in the living matter itself, but in the space, presumably the ether, around it.

Schleip seeks to disprove Driesch's theory by pointing out that the supposititious entelechy sometimes does foolish things, as in the case of the eggs of Nematoda subjected to centrifugal force each of which produces two partial embryos instead of one whole one. But in this objection lurks the childish conception that the entelechy, if it exists, must be the embodiment of Divine Wisdom. the entelechy is not all-seeing—it is a rudimentary "striving" which reacts to its immediate environment, in this case the "apoplasm" or ball of dead matter ejected from the egg by centrifugal force.

The term "organiser" we owe, of course, to Spemann, who wisely abstains from giving any chemical explanation of it. In the course of his marvellous experiments on the newt, Spemann showed that a piece of the dorsal lip of the blastopore of one newt gastrula grafted on the flank of another would change the fate of all the cells in its neighbourhood and force them to develop into a supplementary nerve-cord and underlying notochord. The reviewer might humbly plead that exactly the same conception was reached by him

他称为"质外体"的结构。如果我们没有理解错的话，德里施并不认为"质外体"完全具有生命的结构。根据所形成的"质外体"的多少，生殖细胞的潜能也会成比例地受到相应的限制。高等动物具有相似的生长机理正是由于它们的组成物质中包含大量的"质外体"。只有流动态的细胞质具有完全生命力并拥有该种系的全部潜能。普西布兰认为，所有个体发育中都包含的这些潜能一定存在于细胞质的分子中，因此这些分子构成了真正的生机。与此相似，施莱普认为细胞质中一定存在一种超微观结构，这种结构像晶体一样倾向于形成一种固定的形式，并且当其中一部分被切下来时，它可以再完善自己。

不过，如果认可这些说法，那在我们看来施莱普和普西布兰似乎都成了德里施的支持者。一种无机物从溶液中结晶析出时，正如威廉·布拉格爵士告诉我们的，析出的晶体会呈现出一定的形状，因为这种物质的分子具有相应的确定形状。但是我们应该弄清楚的是，细胞质中的结构到底是什么样子的？构成细胞质的是分子还是超分子？当蝾螈幼体的四肢被切除后，其残肢可以继续再生出新的四肢，那么残肢的分子是不是以极小的手指和脚趾的形式存在？此外，当上述残肢被不同程度地切除时，只有缺失的部分能够再生出来，那么我们是否可以认为，在以不同程度切除四肢之前，四肢末端的微缩版就已经存在于四肢的分子之中？如果我们轻信这些荒谬的设想的话，那么对于迪肯记录的现象，即，在实验中将一只蝾螈胚胎的尾芽移植到另一只蝾螈前肢附近的躯体上，结果发育出一只新的前肢，我们又如何解释呢？大概是尾芽的细胞质被"组织化"而产生了成体尾巴的组织。那么这一组织是如何完全改变发育方向而产生前肢的呢？难怪迪肯说，像这种情况，物理和化学解释根本无能为力，我们必须依靠"生物学领域"的概念，这种影响力不在生命物质上，而可能是在其周围的空间中，有可能是以太。

施莱普不同意德里施的理论。他指出，生机假设有时是很愚蠢的，例如当线虫类的卵受到离心力作用时，每个卵并不是形成一个完整的胚胎，而是产生两个不完整的胚胎。但是这一反对观点中潜藏着一个幼稚的概念，即，如果生机真的存在的话，那么它一定是神性智慧的化身。生机并不是全能的——它是对周围环境作出反应的最基本的"努力"，在上述例子中，它就是"质外体"或者被离心力驱逐出卵的无机球状物。

当然，我们认为"组织原"一词的发明者是施佩曼，他很明智地拒绝给这个词赋予任何化学解释。在其令人称奇的蝾螈实验中，他向我们展示了，将一只蝾螈的原肠胚中的胚孔背唇片断移植到另一只蝾螈的原肠胚的侧面，就会改变附近所有细胞的命运，促使它们发育成一条辅助的神经索和深层脊索。本人在此谦恭地为自己辩护，早在 1918 年我就提出来了同样的概念，并且发表在一篇题为《具有两套水管

and published in a paper which appeared in 1918 entitled "The artificial production of Echinoderm larvae with two water-vascular systems and also of larvae devoid of a water-vascular system" (*Proc. Roy. Soc.*, B, vol. 90). In this paper he showed that when under the stimulus to hypertonic sea-water a second hydrocoele bud was produced in the pluteus, it completely altered the fate of all the tissues near it. It unfortunately did not occur to him to invent the term "organiser."

Of what nature is the influence emitted from the "organiser"? Here again all physical and chemical analogies fail to help us. If the influence were merely a physical or chemical force it would *combine* with the growth-forces of the organised tissue, and what we should observe would be the *resultant* of the two forces. The complete domination of one part by another is not a physical but a vital phenomenon and an instance of Driesch's "regulation".

It would be a fair conclusion to draw from all that has been discovered in the field of embryology to say that in broad outline there are three stages in development, namely: (1) Division of the egg into cells—that is, segmentation; (2) differentiation of these cells so as to form the three primary layers—ectoderm, endoderm, and mesoderm; (3) The action of portions of one layer on the neighbouring parts of other layers so as to form definite organs—that is, the action of organisers.

The ultimate question, however, whence the original organisation of the cytoplasm of the egg is derived, must now be faced. The only answer possible is the nucleus. It is true that, as we have seen, many eggs when ready for fertilisation have an already differentiated cytoplasm. But the cytoplasm of these eggs *when young* is undifferentiated, and during ripening their nuclei are engaged in emissions into the cytoplasm. In particular the nucleolus has been repeatedly observed to become broken into fragments which pass through the nuclear membrane and become dissolved in the cytoplasm. If we take such a specialised egg as that of the Nematode *Ascaris*, Boveri has shown that if it is subjected to centrifugal force *when young*, large portions of the cytoplasm can be shorn away and yet the reduced egg will give rise to a typical embryo. To this conclusion Schleip and Przibram also consent. But it seems to us that a further conclusion follows which they have not clearly envisaged. When differentiation of the cells of the blastula takes place, this must be due to further emissions from the nuclei. But the nuclei in these early stages of development are all alike, and by means of pressure experiments, these nuclei, as Hertwig has put it, may be juggled about like a heap of marbles without altering the result. Moreover, so far as can be judged by the most minute cytological examination, they remain unchanged in their essential make-up throughout the whole of development. So we reach the conception of an *intermittent action of the nuclei on the cytoplasm* giving rise to successive differentiations, that is, stages of development; and as it is by means of these stages that development is directed towards a definite end, if there be an entelechy, we may conclude that the mode of its action is by nuclear emissions. These emissions are the physical correlates of what Uexküll in his "Theoretische Biologie" (1927) calls the "Impulse" to development and the distinguishing of which, he avers, constitutes the utmost limit to which biological analysis can go.

系统的棘皮动物及其缺少水管系统的幼虫的人工培育》（《皇家学会学报》，B 辑，第 90 卷）的论文中。我在该论文中提到，当受到高渗海水的刺激时，长腕幼虫就可以长出第二个水系腔芽体，它的出现会完全改变附近所有组织的命运。但是很遗憾，并不是我发明了"组织原"这个词。

那么"组织原"产生的影响的本质是什么呢？对于这个问题的解释，物理的和化学的推理再一次无能为力。如果这一影响仅仅是一种物理的或化学的力量，那么它就会与有序组织的生长力**结合**起来，我们观察到的就应该是这两种力量作用的**合成**。一部分相对于另一部分来说成为完全主导，这不属于物理现象而是生命现象，这是德里施的"调控"的一个实例。

根据胚胎学领域现已观察到的所有现象，可以很客观地得到如下结论，概括地说，发育可以分为 3 个阶段，即：（1）卵分裂为细胞——即卵裂；（2）这些细胞分化形成 3 个主要的胚层——外胚层、内胚层和中胚层；（3）一个胚层的某些部分作用于邻近的属于其他胚层的部分从而形成特定的器官——即组织原的作用。

然而，最根本的问题是，卵细胞质的最初组成物质来源于何处？这是我们现在必须面对的问题。唯一可能的答案是细胞核。正如我们看到的，事实上许多将要受精的卵子都拥有已分化了的细胞质。但是这些细胞质在卵子**未成熟时**并没有分化，而是在卵子逐渐成熟的过程中，它们的细胞核向细胞质中释放了物质。特别是，在许多研究中都重复观察到核仁分裂成碎片，然后这些碎片穿过核膜，融合到细胞质中。就拿线虫类蛔虫的卵子这样一个特化的卵子来说，博韦里的实验已向我们表明，如果这些卵子**未成熟时**受到离心力的作用，那么大部分细胞质都会丢掉，但是减小了的卵子仍能发育成一个典型的胚胎。施莱普和普西布兰也赞成这一结论。不过我们觉得似乎可以进一步得到他们还没有想清楚的某些结论。囊胚细胞的分化一定是由于细胞核又释放出了某些物质。但是处于这些早期发育阶段的细胞核都是很相似的。赫特维希通过压力实验发现，这些细胞核可能就像被拨弄的一堆大理石一样并没有发生任何变化。此外，从大多数细微的细胞学检查结果可以判断，在整个发育过程中，这些细胞核的基本组成都没有发生变化。所以我们想到，**细胞核对细胞质的间歇性作用**引起了细胞的连续分化，即各个发育阶段。发育过程正是通过这些阶段逐渐走向一个确定的结果，所以如果存在生机的话，我们可以断定它是通过细胞核释放出的物质来发挥作用的。这些释放物就是于克斯屈尔在其《理论生物学》（1927 年）中称为发育"推动力"的相关物质。于克斯屈尔断言，对这些释放物的区分是生物分析所能达到的最大极限。

Comparative embryology, however, can go further, and Schleip rightly insists that experimental embryology ought to be comparative. These embryonic stages are soon discovered to be merely smudged and simplified forms of larval stages which in allied forms lead a free life in the open, seeking their own food and combating their own enemies. These larval forms in turn are seen to be nothing but modified and simplified editions of adult forms in the past history of the race. Therefore, in the last resort, development is found to be due to the successive coming to the surface of a series of racial memories, and the entelechy might be defined as a "bundle" of such memories.

The so-called Mendelian "genes", however, constitute a problem for the embryologist; for the conception of the hereditary make-up which they induce in the minds of geneticists is totally at variance with that which the embryologist draws from the study of development. Schleip and Przibram struggle valiantly to reconcile the two conceptions and fail. Dürken alone boldly questions the validity of the whole conception of the genes and points out how much it is purely arbitrary and theoretical. If the results of a crossing experiment agree with expectation based on the ordinary Mendelian rules, then it proves the reality of genes; if the results do not agree, the geneticist denies that it disproves them, because he immediately postulates the action of an undiscovered "gene" which complicates the result. The real answer to the conundrum was given by Johannsen, when, in his latest publication, deploring the damage and confusion of thought caused by the invention of the word "gene", he states that it represents a mere superficial disturbance of the chromosomes and gives no insight into the real nature of heredity. Even Przibram points out that X-rays will produce "unzählige" mutations, and that there is no correlation between the rays and the nature of the mutation. With these remarks we thoroughly agree.

(**126**, 639-643;1930)

　　不过，比较胚胎学还可以走得更远，施莱普始终坚持实验胚胎学应该引入比较的方法。人们很快发现这些胚胎发育的阶段仅仅是幼体阶段被混杂和简化后的形式，各个幼体阶段的联合作用最终形成了可以在野外生存的自由生命，它们可以自己寻找食物并与敌人战斗。反过来，这些幼虫形式又仅仅被看作是该种系在过去的发育史中成年形式的修饰简化版本。因此，最终我们会得出发育是一系列种族记忆的逐次苏醒，而生机可以被定义成这些记忆中的"一束"。

　　但是所谓的孟德尔式"基因"给胚胎学家带来了一个难题，因为遗传学家们推导出的遗传组成的概念与胚胎学家从发育研究中得出的完全不同。施莱普和普西布兰大胆地尝试去调和这两种概念，但最终失败了。迪肯独自勇敢地质疑了基因整个概念的有效性，并且指出这一概念是非常主观和理论性的。如果有交叉实验结果与基于普通孟德尔法则预期的结果一致的话，那么就能证实基因的真实性；如果不一致，那么遗传学家就不会承认该结果能成为他们理论的反证，因为他可以马上假定出一个作用是使该结果变得更加复杂的尚未被发现的"基因"。约翰森给出了这一谜底的真正答案，在他的最后一部著作中，他谴责了"基因"一词的发明给人们造成的思维混乱和困惑，他指出"基因"这个词只不过是给染色体的概念带来了浅薄的干扰，而对于洞察遗传的真正本质并无帮助。甚至普西布兰也指出，X 射线能产生"无数"突变，而射线和突变的本质之间没有任何相关性。我们完全同意这些观点。

（刘皓芳 翻译；刘京国 审稿）

Natural Selection Intensity as a Function of Mortality Rate

J. B. S. Haldane

Editor's Note

J. B. S. Haldane was a prime mover behind the "modern synthesis" in evolutionary biology. This fused Darwin's ideas about selection and Mendel's insights into how traits pass from parents to offspring into a mathematical description of the genetic makeup of populations and how it changes. At the time, some believed these ideas to be antithetical to one another. *Nature* previously published a commentary on Haldane's note by E. W. MacBride, who concluded that he was "convinced that Mendelism has nothing to do with evolution." Here Haldane gives a specific example of his advocacy of mathematical analysis in evolutionary biology, and its superiority over verbal arguments, disproving the notion that selection is limited to the stage of highest mortality.

IN *Nature* of May 31, Prof. Salisbury points out that most of the mortality among higher plants occurs at the seedling stage, and concludes that natural selection is mainly confined to this stage. I believe, however, that this apparently obvious conclusion is fallacious, for the following reason:

Consider two pure lines A and B originally present in equal numbers, and with a common measurable character, normally distributed according to Gauss's law in each group. Let the standard deviations of the character be equal in each group, but its mean value in group A slightly larger than that in group B. Johanssen's beans furnish examples of this type of distribution. Now let selection act so as to kill off all individuals in which the character falls below a certain value. I think that this type of artificial selection furnishes a fair parallel to natural selection, in which chance commonly plays a larger part than heritable differences. Let x be the proportion of individuals eliminated to survivors, and $1+y$ the proportion of A to B among the survivors, so that x measures the intensity of competition, y that of selection.

Then when x is small y is roughly proportional to it. Thus when x increases from 10^{-4} to 10^{-1}, y increases 200 times. But when x is large y becomes proportional to $\sqrt{\log x}$. In consequence y only increases 9 times when x increases from 1 to 10^{12}, and is only doubled when x increases from 1 to 1,800. In other words, when more than 50 percent of the population is eliminated by natural selection, the additional number eliminated makes little difference to the intensity of selection. The theory, which I hope to publish shortly, has been extended to cover cases where the standard deviations differ, and also where populations consist of many genotypes. In general y changes its sign with x, but when x is large y never increases more rapidly than $\log x$.

自然选择强度与死亡率的关系

霍尔丹

编者按

在进化生物学领域，霍尔丹是"现代综合论"的先驱。现代综合论融合了达尔文的自然选择学说和孟德尔关于性状如何从亲代传递到子代的观点，并用数学方法描述了人类的基因构成和基因如何变化。当时，一些人认为达尔文的理论和孟德尔的观点是相互对立的。《自然》早先曾发表了麦克布赖德对霍尔丹的一篇论文的评论，麦克布赖德在评论中声称，他"对孟德尔的遗传学说对于进化毫无意义这一点深信不疑"。在这篇文章中，霍尔丹为了反驳自然选择只局限于最高死亡率阶段的观点，他给出了一个具体的例子。这个例子表明，他支持在进化生物学中采用数学分析的方法，并认为这种方法比文字说明更好。

在 5 月 31 日的《自然》中，索尔兹伯里教授指出高等植物中绝大多数的死亡发生在幼苗时期，由此他断定自然选择主要发生在这个阶段。我认为，这个看似很显然的结论是不正确的，主要原因如下：

假定纯系 A 和纯系 B 在初始状态时具有相同的个体数并都具有某一相同的可测性状，并且这一可测性状在每组群体中都服从高斯分布。假定每组群体中性状分布具有相同的标准差，而 A 系群体的均值比 B 系群体的均值略高。约翰森豆正是能够满足这种分布类型的实例。现在让选择起作用，杀死那些性状低于某一给定值的所有个体。我认为这样的人为选择是非常类似于自然选择的，在自然选择中通常是偶然事件所起的作用比遗传差异更大。用 x 表示死亡个体与存活个体的比例，$1 + y$ 表示纯系 A 的存活个体与纯系 B 的存活个体的比例。那么 x 表示的就是竞争的强度，而 y 则表示选择的强度。

这样，当 x 较小时 y 与 x 大致是成比例的。当 x 从 10^{-4} 增加到 10^{-1} 时，y 增加了 200 倍。但是当 x 较大时，y 则变成与 $\sqrt{\log x}$ 成比例。其结果是当 x 从 1 增加到 10^{12} 时，y 仅增加了 9 倍，当 x 从 1 增加到 1,800 时，y 仅翻了一番。换言之，当超过 50% 的个体在自然选择中被淘汰时，再淘汰更多个体对应的自然选择强度的改变是很小的。这一理论已经扩展到纯系间性状分布的标准差不同的情况，以及群体中包含多种基因型的情况。我希望能很快发表这一理论。总体上讲，y 的数值随 x 的改变而改变，但是当 x 较大时 y 增加的速度不会比 $\log x$ 更快。

Careful mathematical analysis seems to disclose the extraordinary subtlety of the natural selection principle, and merely verbal arguments concerning it are likely to conceal serious fallacies.

(**126**, 883; 1930)

J. B. S. Haldane: John Innes Horticultural Institution, Merton Park, London, S.W. 19, Nov. 1.

看起来，细致的数学分析才能够解析自然选择原理的精细之处，而文字说明中很可能隐藏着严重的谬误。

（刘晓辉 翻译；刘京国 审稿）

The X-ray Interpretation of the Structure and Elastic Properties of Hair Keratin

W. T. Astbury and H. J. Woods

Editor's Note

William Astbury was an X-ray crystallographer based in a university department supported by funds from the wool and leather industries of the county of Yorkshire in northern England. He was one of the first to use X-ray analysis for the study of the structure of complicated polymer molecules; during his career in Leeds, he also studied the structure of deoxyribonucleic acid (DNA) and even arrived at the correct spacing between successive units in the polymer. His biggest handicap was that he made enemies easily. Here he describes the X-ray analysis of a fibrous "structural" protein, keratin, the main component of hair. Such proteins were Astbury's forte.

RECENT experiments,[1] carried out for the most part on human hair and various types of sheep's wool, have shown that animal hairs can give rise to two X-ray "fibre photographs" according as the hairs are unstretched or stretched, and that the change from one photograph to the other corresponds to a reversible transformation between two forms of the keratin complex. Hair rapidly recovers its original length on wetting after removal of the stretching force, and either of the two possible photographs may be produced at will an indefinite number of times. Both are typical "fibre photographs" in the sense that they arise from crystallites or pseudo-crystallites of which the average length along the fibre axis is much larger than the average thickness, and which are almost certainly built up in a rather imperfect manner of molecular chains—what Meyer and Mark[2] have called *Hauptvalenzketten*—running roughly parallel to the fibre axis.

Hair photographs are much poorer in reflections than are those of vegetable fibres, but it is clear that the α-keratin, that is, the unstretched form, is characterised by a very marked periodicity of 5.15 Å along the fibre axis and two chief side-spacings of 9.8 Å and 27 Å (? mean value), respectively; while the β-keratin, the stretched form, shows a strong periodicity of 3.4 Å along the fibre axis in combination with side-spacings of 9.8 Å and 4.65 Å, of which the latter is at least a second-order reflection. The β-form becomes apparent in the photographs at extensions of about 25 percent and continues to increase, while the α-form fades, up to the breaking extension in cold water, which is rarely above 70 percent. Under the action of steam, hair may be stretched perhaps still another 30 percent, but no other fundamentally new X-ray photograph is produced. The question is thus immediately raised as to what is the significance of a crystallographically measurable transformation interpolated between two regions of similar extent where no change of a comparable order, so far as X-ray photographs show, can be detected.

X射线衍射法解析毛发角蛋白的结构与弹性

阿斯特伯里，伍兹

编者按

威廉·阿斯特伯里是一位在大学工作的 X 射线晶体学家，他所在的系受到了北英格兰约克郡羊毛和皮革业多家企业提供的研究基金的支持。他是利用 X 射线解析复杂多聚体分子结构的先驱之一。在利兹工作期间，他还研究了脱氧核糖核酸 (DNA) 的结构，甚至得出了这一聚合物中相邻单体分子之间的正确距离。对他来说最不利的是他太容易树敌。在这篇文章中，他描述了对角蛋白这种纤维状"构造的"蛋白质的 X 射线衍射分析。角蛋白是毛发的主要组成部分。阿斯特伯里非常擅长于研究这种蛋白。

最近的一些主要针对人类毛发和各种类型羊毛的实验 [1] 表明，动物毛发在拉伸状态和非拉伸状态下可以产生两种不同的 X 射线"纤维衍射图"，而且从一种衍射图向另一种衍射图的转变对应于角蛋白复合体的两种形式之间的可逆转变。在撤去外界拉力后，浸湿的毛发会很快恢复到原来的长度，因此可以无限次地重复得到这两种衍射图中的任意一种。这两种衍射图都来自晶体或伪晶体，从这一意义上来说，这两种衍射图都是典型的"纤维衍射图"。在相应的晶体和伪晶体中，沿中心轴方向的平均长度比平均厚度大得多，这些晶体和伪晶体极有可能是由一种非常不完整的、几乎平行于中心轴的分子链（被迈尔和马克 [2] 称为**主分子链**）搭建起来的。

与植物纤维的衍射图相比，毛发衍射图中的反射线非常少，但很清楚的是，非拉伸状态的 α 角蛋白明显表现出了沿中心轴 5.15Å 的周期性以及宽度分别为 9.8Å 和 27Å（均值？）的两个主要的侧向间隔。处于伸展状态的 β 角蛋白则表现出沿中心轴 3.4Å 的很强的周期性以及宽度分别为 9.8Å 和 4.65Å 的两个侧向间隔，其中后一个侧向间隔至少是二次反射。当毛发在冷水中被拉伸 25% 时，衍射图中开始出现 β 类型，随着拉伸幅度的增大 β 类型越来越多，与此同时，α 类型逐渐消失，直到毛发被拉断（在冷水中毛发的拉伸幅度很少超过 70%）。在蒸汽的作用下，毛发也许还能再被拉伸 30%，但即便这样也不会出现本质上全新的 X 射线衍射图。这样立刻就提出了一个问题，就 X 射线图而言，在拉伸程度接近、没有发生相对次序改变的两个区域之间，晶体学上可测量的相互转变的意义是什么呢？

The elastic properties of hair present a complex problem in molecular mechanics which up to the present has resisted all efforts at a satisfactory explanation, either qualitative or quantitative. Space forbids a detailed discussion here of the almost bewildering series of changes that have been observed, and we shall merely state what now, after a close examination of the X-ray and general physical and chemical data, appear to be the most fundamental.

(1) Hair in cold water may be stretched about twice as far, and hair in steam about three times as far, as hair which is perfectly dry. (2) On the average, hair may be stretched (in steam) to about twice its original length without rupture. (3) By suitable treatment with steam the discontinuities in the load/extension curve may be permanently smoothed out, the original zero is lost, so that the hair may be even contracted by as much as one-third of its original length, and elasticity of form may be demonstrated *in cold water* over a range of extensions from −30 percent to +100 percent. (4) The elastic behaviour in steam is complicated by "temporary setting" of the elastic chain and ultimately by a "permanent setting" of that part which gives rise to the fibre photograph. (5) That part of the elastic chain which is revealed by X-rays acts *in series* with the preceding and subsequent changes.

On the basis of these properties and the X-ray data, it is now possible to put forward a "skeleton" of the keratin complex which gives a quantitative interpretation of the fundamentals, and may later lead to a correct solution of the details. The skeleton model is shown in Fig. 1. It is simply a peptide chain folded into a series of hexagons, with the precise nature of the side links as yet undetermined. Its most important features may be summarised as follows:—(1) It explains why the main periodicity (5.15 Å) in unstretched hair corresponds so closely with that which has already been observed in cellulose, chitin, etc., in which the hexagonal glucose residues are linked together by oxygens. (2) When once the side links are freed, it permits an extension from 5.15 Å to a simple zigzag chain

Fig. 1

254

毛发弹性的分子机制是个复杂的问题，无论是定性的还是定量的，至今为止还没有得到任何令人满意的结论。篇幅所限，我们这里将不再详细讨论曾观测到的纷繁复杂的一系列变化，只介绍对 X 射线结果和基本的物理和化学数据进行详细考察后得到的一些看起来最重要的结论。

（1）和完全干燥的毛发相比，在冷水中毛发可以被拉伸到大约原长的两倍，而在蒸汽中可以被拉伸到大约原长的三倍。（2）平均而言，毛发（在蒸汽中）可以被拉伸到原长的两倍而不断裂。（3）适当的蒸汽处理可以永久地消除毛发载荷–伸长曲线的不连续性，原来的零点就没有了，这样毛发甚至可以收缩到其原长的 2/3，而**在冷水中**其所表现的弹性介于收缩 30% 到伸长 100% 之间。（4）在蒸汽中毛发的弹性行为由于存在弹性链的"临时形态"而变得复杂，但最后变成"固定形态"，从而产生纤维衍射图。（5）X 射线衍射图显示，部分弹性链**连续地**进行先前和随后的变化。

在这些特性以及 X 射线衍射数据的基础上，我们现在可以提出一种角蛋白复合体的"骨架"，这可以定量地解释一些基本问题，将来或许还能引导对细节问题的正确解答。骨架模型如图 1 所示。该骨架模型只是简单地展示了折叠成几个六边形的一条多肽链，其中侧链的准确性质目前还没有被测定。该模型最重要的特征可以总结如下：（1）它解释了为什么非拉伸状态下毛发中的主要周期（5.15Å）与在纤维素、几丁质等物质中观察到的周期十分相近，这是因为在这些物质中六边形的葡萄糖残基可以通过氧原子相互连接起来。（2）当侧向的连接解开后，模型中长为

图1

of length 3×3.4 Å, that is, 98 percent, and also allows for possible contraction below the original length, without altering the inter-atomic distances and the angles between the bonds. (3) It explains why natural silk does not show the long-range elasticity of hair, since it is for the most part already in the extended state,[3] with a chief periodicity of 3.5 Å. We may now hope to understand why it is that the photographs of β-hair and silk are so much alike. (4) It gives a first picture of the "lubricating action" of water and steam on the chain, since X-rays show that the direction of attack is perpendicular to the hexagons and that this spacing remains unchanged on stretching. Furthermore, it now seems clear that the new spacing, 4.65 Å, is related to the old by the equation $27/(3 \times 4.65) = 3 \times 3.4/5.15$ (very nearly), that is, the transformation elongation takes place directly at the expense of the larger of the two side-spacings. In the particular arrangement of the hexagons shown in the model, the side chains occur in pairs on each face, and it may well be that the action of water is the opening-up of an internal anhydride between such adjacent side chains. (5) The chain being built up of a succession of ring systems stabilised and linked together in some way by side chains of the various amino-acids, we have here an explanation of the well-known resistance of the keratins to solvents and enzyme action. In addition, each hexagon is effectively a diketo-piperazine ring, an interesting point in view of the evidence which has been brought forward by Abderhalden and Komm[4] that such groups pre-exist in the protein molecule. It may also throw light on the stimulating researches of Troensegaard.[5] (6) There are three principal ways of constructing the model, according to which group lies at the apex of a hexagon. It thus affords an explanation of the apportioning of a transformation involving a 100 percent elongation into three approximately equal regions which may be opened up in turn under the influence of water and temperature and other reagents. The modification shown in the model must be ascribed to the crystalline phase, since it would, alone of the three, be expected to give rise to a strong reflection at 5.15 Å, as in the α-photograph.

A detailed account of the above work will be published shortly.

(126, 913-914; 1930)

W. T. Astbury, H. J. Woods: Textile Physics Laboratory, The University, Leeds, Nov. 15.

References:

1. W. T. Astbury, *J. Soc. Chem. Ind.*, **49**, 441; 1930.

2. Meyer and Mark, "Der Aufbau der hochpolymeren organischen Naturstoffe".

3. Meyer and Mark, *Berichte*, **61**, 1932; 1928.

4. Abderhalden and Komm, *Z. Physiol. Chem.*, **139**, 181; 1924.

5. Troensegaard, *Z. physiol. Chem.*, **127**, 137; 1923.

5.15 Å 的主链单元就会伸展成 3×3.4 Å 的简单"之"字形长链，也就是伸展了 98%，同样也可能会因为收缩而使长链短于初始长度，但不论伸展还是收缩，原子间距和键–键之间的角度都不会发生改变。（3）它解释了为什么自然丝的伸缩性没有毛发那么大，因为大多数情况下自然状态的丝已经处于主周期为 3.5 Å 的伸展状态。[3] 这样我们就能够理解为什么 β 型毛发的衍射图与丝蛋白的衍射图十分相似。（4）这一模型首次给出了水和蒸汽对主链的"润滑作用"的图片。X 射线衍射结果显示，水分子攻击主链的方向是垂直于六边形平面的，而且在主链拉伸过程中六边形中各原子之间的空间间隔没有发生改变。另外，比较明确的是，拉伸后新的侧向间隔（4.65 Å）与原来的侧向间距之间的关系符合方程 27/(3×4.65) = 3×3.4/5.15（非常近似），这就是说，主链的伸长是以两个侧向间隔的减小为直接代价的。在模型中显示的六边形的某些特殊排列中，在六边形平面任何一边的侧链总是成对出现，水分子的作用很可能就是打开这些相邻侧链内部的酐键。（5）由连续的环形系统构成的主链是通过各种不同的氨基酸的侧链以某种方式保持稳定并连接在一起的，这样我们就可以解释众所周知的角蛋白对溶剂作用和酶作用的抗性。另外，每个六边形都是一个有效的二酮哌嗪环，这一点很有趣，考虑到阿布德哈尔登和科姆 [4] 曾提出证据说这种基团结构预先就存在于蛋白分子中，那么这个有趣的特点也许会对特森加德的研究工作有所启示。[5]（6）根据位于六边形顶点上的基团的不同，有 3 种主要的模型构建方法。这样就可以把长链 100% 的伸长分配到 3 个近似相等的区域上。在水、温度或其他试剂的影响下，这些区域会依次伸展开。对模型的这些修正要归因于晶相结构的特点，因为可以预期在 3 种结构中只有晶相结构会给出如 α 衍射图所示的在 5.15 Å 处出现的强反射。

关于以上工作的详细说明将在不久之后发表。

（高如丽 翻译；刘京国 审稿）

Embryology and Evolution

Editor's Note

In 1930 Irish-born zoologist Ernest William MacBride had exposed his Lamarckian leanings in a review of several books on experimental embryology, arguing that the genes-eye view of the world was "totally at variance with that which the embryologist draws from the study of development." Here, embryologist G. L. Purser conjures up a nice analogy that he thinks illuminates the role that genes do (and do not) play in embryonic development or "epigenesis". This letter contains some prescient reflections about the mutual interdependence of genes and their immediate environment. However, Purser goes too far in claiming a new proof for the inheritance of acquired characteristics when he suggests "that the environment is in some way responsible for the appearance of the gene".

I have read with much interest Prof. MacBride's review entitled "The Problem of Epigenesis", and I should like to make a few remarks upon what he says at the end. First of all, I wonder if the following analogy will help him, as it has helped me, to reconcile the conceptions of the geneticist with those of the embryologist. In a modern motor works the cars, so I understand, move along a track past a series of workmen, each of whom has one particular job to do, which is related to what has already been done and also to what is going to be done afterwards. Now if we imagine that all the parts and materials which are going to make up the finished car represent the substances in the developing embryo and that the workmen are the genes, we have an analogy which can be carried surprisingly far. Not only will it give us a picture of normal development, but we can see, by altering one of the parts, how a variation may occur; by altering a workman, how "sports" may arise; and, by adding a new workman with a new job, how progressive evolution may take place.

There is no need for me to occupy space in working the analogy out, for anyone can do it for himself: what is more important is to point out where the analogy fails. A motor-car is adapted for life on the road, and, until it is completed, it has, for all practical purposes, no environment at all comparable with that which bears upon an embryo throughout its development. So whereas a feature of a car is simply due to the action of the workman on the materials, a feature of an animal is the result of the combined action of the genes and of the environment upon the materials of the embryo. Genes without the appropriate materials can produce nothing; genes with the appropriate materials can only produce a partially developed structure; but genes with the appropriate materials and environment can produce the fully developed functional character. Hence it is that in the development of the frog, for example, the gill-clefts, etc., are full developed, whereas in the Amniota, with the radical change in the environment of the early stages, such structures are only

胚胎学与进化

编者按

1930 年，出生于爱尔兰的动物学家欧内斯特·威廉·麦克布赖德在评论几本关于实验胚胎学的著作时，表示了自己对拉马克学说的认同。他认为从基因的角度看到的世界"与胚胎学家从发育研究中得出的完全不同。"在这里，胚胎学家珀泽想出了一个很好的类比，他认为这样就可以解释基因在胚胎发育或"渐成论"中是否起作用。这篇文章在基因与周围环境的相互依赖关系这个问题上提出了一些很有远见的观点。但是珀泽在阐述获得性遗传假说的新证据时说，"环境在某种程度上决定了基因的表现形式"，这就过于偏激了。

我怀着极大的兴趣读完了麦克布赖德教授那篇名为《关于渐成论的问题》的书评，看完后我想对他的书评结尾处的内容发表一下我的看法。首先，我想知道如下的类比是否会像其帮助过我一样，也能够帮助他化解遗传学家的观念与胚胎学家的观念之间的分歧。我是这样理解的，现代汽车工业中，小汽车是沿着一条由一系列工人组成的生产线移动的，生产线上的每个人都有自己特定的工作要做，每个人的工作都与已经完成的工作以及之后将要进行的工作有关。这时如果我们想象组成成品车辆的所有零件和材料代表正在发育的胚胎里的物质，工人代表基因，那么我们就有了一个蕴含着深远意义的类比。这一类比不仅可以给我们描绘出一幅正常发育的图像，而且从中我们可以看到：变异是如何通过改变众多零件中的一个而实现的；"变种"是如何通过更换一名工人而产生的；以及渐进演化是如何通过增加一名从事新工作的新工人而发生的。

我没有必要浪费版面在这里推演这一类比，因为任何人都可以独自完成，但更重要的是指出这种类比在哪些地方不适用。一辆汽车是适于在路面上奔跑的，并且在生产出来后它就已经适于各种实用目的，而在胚胎生长发育的整个过程中都没有任何环境可以与此相类比。因此汽车的特征仅仅取决于工人对原材料进行的处理，而动物的特征则是基因和环境对胚胎物质共同作用的结果。如果没有适当的胚胎物质，那么基因什么也产生不了；如果只有适当的胚胎物质，那么基因只能产生部分发育的结构；只有既有适当的胚胎物质，又有环境时，基因才能产生充分发育的有功能的结构。因此，如腮裂等结构，在青蛙的发育过程中是充分发育的，在羊膜动物中则因为在发育早期遭遇到剧烈的环境变化而只是部分发育的。引用麦克布赖德

partially developed and the stages, to quote Prof. MacBride, are smudged.

Looked at from this point of view, two other conclusions of great importance are unavoidable. The first is that the recapitulation of an ancestral stage of the evolution of an animal, as distinct from the repetition of an ancestral character, will only occur when the early stage of development is passed in the same environment as that of the ancestor, which environment is different from that of the present-day adult. Only under such conditions will the genes responsible for the adult ancestral characters give rise to them all together without any great admixture of other features; though it must always be borne in mind that such stages in the life history, being larvae, may evolve on their own account and, therefore, may have features which the ancestor never had. In parenthesis, I should just like to add here that, so far as I know, a larva has never been properly defined: such a definition would be "A free-living stage in an animal's life history which fends for itself and possesses certain characters which it has to lose before it can become a young adult": the possession of *positive* characters distinguishes a larva, not its lack of adult ones.

The other conclusion is reached thus. The appearance of a functional feature is dependent, as we have seen, upon the interaction of three things: the materials of the embryo, the genes, and the environment. Now the facts of Mendelian inheritance give clear evidence that there need be no change in the materials of an embryo for a new gene to modify the form, so, in discussing the origin of a new feature, there is no need to consider a change in the materials as one of the essential factors. The fortuitous appearance of a gene without the appropriate environment would produce a partially developed character, but, in actual experience, we do not find features in a partially developed condition which *have never been functional* at any period in the history of the race. So the genes must, in actual fact, only arise after the suitable environment is present; and the only conclusion to be drawn from that is that there is a causal relation between the two; that is, that the environment is in some way responsible for the appearance of the gene, which is surely nothing more or less than the basis of a new proof of the inheritance of acquired characters.

G. L. Purser

*   *   *

I have read with interest Mr. Purser's thoughtful letter on the subject of my review. If he will substitute the term "race-memory" for "gene", we shall not be far apart. But the gene of the Mendelian stands out as something that is never functional. "No one," said the late Sir Archdall Reid, "ever heard of a useful gene." When one takes into consideration the fact that the Mendelian genes in *Drosophila* have been shown to increase in their damaging effect on the viability of the organism in proportion to the structural change which they involve, and when further it is discovered that genes can be artificially produced by irradiating insect eggs with X-rays—a process which kills most of the eggs—one is driven

教授的话说，这些阶段都是遗留下的痕迹。

从这个观点来看，将会不可避免地得出另外两个非常重要的结论。第一个是，只有当胚胎在发育早期阶段所处的环境与该种生物的祖先所处环境一样时，才会出现对其祖先进化过程的重演。这种环境与现在的成体所处的环境是不同的，这种重演与重现祖先特征是完全不同的概念。只有在这种情况下，那些决定祖先成体特征的基因才完全产生效果，而不与其他特征发生大规模的混合；但是我们必须时刻牢记，生命过程中的这些阶段（即幼虫）都是自行进化的，因此它们可能具有祖先从不具有的特征。另外，我想在此补充说明的是，据我所知，幼虫这一概念从未被恰当地定义过。如下描述可能是一个比较恰当的定义，"在动物生命史中的一个自由生活的阶段，在此阶段中它们自己照料自己并具有一定的特征，但一旦它们成长为年轻的成体，这些特征就会消失。"幼虫拥有**初级的**特征而成虫没有，根据这一点能够将它们区别开来。

第二个结论是按照如下所述得出的。正如我们看到的一样，功能性特征的出现依赖于 3 个条件的相互作用：胚胎物质、基因和环境。现在，孟德尔的遗传结果已经给出了明确的证据，表明一个新基因形式发生改变时胚胎物质并没有发生改变，所以在讨论新特征的起源时，就没有必要将胚胎物质的变化作为基本因素之一来考虑。偶尔会出现某个没有适当环境的基因，这时该基因会产生部分发育的结构，然而实际上在种系史的任何阶段我们都没发现部分发育的**无功能**结构的出现。因此，事实上基因肯定是在适当环境出现之后才产生的。从这一点，我们可以得到的唯一结论是：基因与环境之间存在因果关系，即，从某种意义来说，环境在某种程度上决定了基因的表现形式，这正好可以作为获得性性状遗传的新证据。

珀泽

*　　*　　*

我已经饶有兴趣地读完了珀泽先生就我的评论所写的颇具思想性的快报。如果他用"种族记忆"一词来代替"基因"，那么我们的分歧并不大。但是，孟德尔式基因正是因其根本不具有功能而格外引人注目。已故的阿奇德尔·里德爵士说："没有人曾听说过一个有用的基因。"考虑到果蝇的孟德尔式基因对该物种生存能力的损伤有所增加，且损伤增加的程度与相关基因发生的结构变化成比例这一事实，再加上人们又发现可以通过 X 射线辐射昆虫卵子来人为地产生基因（这一辐射过程能够杀

to the conclusion that a gene is germ damage of which the outward manifestation is a mutation. The only effect that natural selection would have on such aberrations would be to wipe them out. In my opinion, mutations and adaptations have nothing to do with one another and only adaptations are recapitulated in ontogeny.

E. W. MacBride

(**126**, 918-919; 1930)

G. L. Purser: The University, Aberdeen, Oct. 29.

死大部分卵子），我们就会被引向如下结论：基因是外在表现为突变的配子损伤。自然选择对这种异常的唯一反应就是清除它们。在我看来，突变和适应没有任何关系，只有适应性能够在个体发育中重演。

麦克布赖德

（刘皓芳 翻译；刘京国 审稿）

Embryology and Evolution

J. B. S. Haldane

Editor's Note

Renowned biologist J. B. S. Haldane was one of the key figures in the development of population genetics, a field underpinned by a Mendelian view of inheritance. Here Haldane has no time for the views of the neo-Lamarckian zoologist Ernest William MacBride, one of a dwindling number of scientists still prepared to dismiss the notion of a gene. With characteristic rhetorical flair, Haldane ridicules MacBride for his outdated views. Publicly aired disagreements like these played an important part in turning the scientific community towards a modern evolutionary synthesis, the idea that a Mendelian mechanism of inheritance could result in the sort of gradual natural selection that Charles Darwin envisaged.

FOUR of Prof. MacBride's statements, in *Nature* of Dec. 6, call for comment. "...no one has ever seen 'genes' in a chromosome." Genes cannot generally be seen, because in most organisms they are too small. In *Drosophila* more than 100, probably more than 1,000, are contained in a chromosome about $1\,\mu$ in length. They are therefore invisible for exactly the same reasons as molecules. But the evidence for their existence is, to many minds, as cogent. Where the chromosomes are larger, as in monocotyledons, competent microscopists—for example, Belling, in *Nature* of Jan. 11, 1930—claim to have seen genes. In a case where I (among others) postulated the absence of a gene in certain races of *Matthiola*, my friend Mr. Philp has since detected the absence of a trabant, which is normally present, from a certain chromosome. I shall be glad to show this visible gene to Prof. MacBride.

"...if Prof. Gates were a zoologist instead of being a botanist, he would know that the assumption that 'genes' have anything to do with evolution leads to results...that can only be described as farcical." I should like to direct Prof. MacBride's attention to the droll fact that in a good many interspecific crosses various characters behave in a Mendelian manner, that is, are due to genes. This is so, for example, with the coat colour of *Cavia rufescens*, which, on crossing with the domestic guinea-pig, behaves as a recessive to the normal coat colour, but a dominant to the black. Hence there has been a change in a gene concerned in its production during the course of evolution. Scores of similar cases could be cited.

"All known chemical actions are inhibited by the accumulation of the products of the reaction. An 'autocatalytic' reaction, in which the products of the reaction accelerated it, must surely be a vitalistic one!" Autocatalytic reactions are common both in ordinary physical chemistry and in that of enzymes. Thus the acid produced by the hydrolysis of an

胚胎学与进化

霍尔丹

编者按

在以孟德尔遗传学说为基础的群体遗传学的发展中，著名生物学家霍尔丹算是一个重要的人物。在这篇文章中，霍尔丹没有理会支持新拉马克主义的动物学家欧内斯特·威廉·麦克布赖德（后者仍然拒绝接受基因的概念，而持这种观点的科学家已经越来越少了）的观点，而是以他特有的文字才能嘲弄了麦克布赖德的过时思想。在公开场合进行这样的争论，对于学术界最终转向现代进化综合论起到了重要的作用。在现代进化综合论中，可以由孟德尔的遗传机制推出查尔斯·达尔文设想的自然选择学说。

我要对麦克布赖德教授发表于 12 月 6 日的《自然》上的 4 项陈述稍作评论。他说："……没有人看见过染色体上的'基因'。"确实，基因一般是看不见的，因为大多数生物的基因都太小了。果蝇的一条长约 1μ 的染色体上可能就有一百多个甚至一千多个基因。因此，就像分子是不可见的一样，基因也是不可见的。不过许多人都认为用来证明基因确实存在的证据是令人信服的。另外，一些有能力的显微镜专家曾公布过他们在具有较大染色体的单子叶植物中看到了基因，例如贝林就在 1930 年 1 月 11 日的《自然》上宣称看到了基因。我（和其他人一起）曾经在一项研究中推测紫罗兰属的某些种是没有基因的，后来我的朋友菲尔普先生检测到了特定染色体上随体的缺失，而在正常情况下这一染色体上是存在随体的。我很乐意向麦克布赖德教授展示这个可见的基因。

"……如果盖茨教授不是植物学家而是动物学家的话，他将了解'基因'与进化具有某些关联，这样的假设产生的结果……只能用滑稽可笑来形容。"我期望麦克布赖德教授能注意到下面这个很有趣的事情：即在许多种间杂交的实例中，很多性状都表现出孟德尔式遗传的特点，也就是说它们是由基因决定的。例如，有一种野生巴西豚鼠，当它与家养豚鼠杂交时，其毛色对于正常毛色表现为隐性，但对于黑色却表现为显性。因此，在进化过程中，与产生该性状相关的基因一定发生了某种变化。与此类似的例子还有很多。

"所有已知的化学反应都会被反应产物的积累所抑制。而在'自催化'反应中，反应产物却可以加快反应的进行，因而这种反应显然就是活力论的一个实例！"其实，自催化反应在普通物理化学和酶化学中都是很常见的。酯水解产生的酸可以促

ester may accelerate its further hydrolysis. As an example of an enzyme action, which for quite simple physico-chemical reasons proceeds with increasing velocity up to 75 percent completion, I would refer Prof. MacBride to Table 7 of Bamann and Schmeller's [1] paper on liver lipase.

In view of such facts, Prof. MacBride's statement that "The term 'autocatalysis' is a piece of bluff invented by the late Prof. Loeb to cover up a hole in the argument in his book" would seem to be a wholly unfounded attack on a great man who can no longer defend himself. If Prof. MacBride would acquaint himself with the facts of chemistry and genetics, he might be somewhat more careful in his criticism of those who attempt to analyse the phenomena of life. He might also cease to ask the question propounded by him in *Nature* of Oct. 25, "whether the organs of the adult exist in the egg preformed in miniature and development consists essentially in an unfolding and growing bigger of these rudiments, or whether the egg is at first undifferentiated material which from unknown causes afterwards becomes more and more complicated and development is consequently an 'epigenesis'. " The formation of bone in the embryo chick was shown by Fell and Robison[2] to be due to the action of the enzyme phosphatase, which is neither a miniature bone nor an unknown cause. But so long as he does not take cognisance of recent developments in science, Prof. MacBride will no doubt remain a convinced vitalist.

(**126**, 956; 1930)

J. B. S. Haldane: Biochemical Laboratory, Cambridge University, Dec. 8.

References:
1. *Zeit. Physiol. Chem.*, **188**, p. 167.
2. *Biochem. Jour.*, **23**, p. 766.

进酯的进一步水解。酶反应中也有因为非常简单的物理化学原因而使得反应速率提高 75% 的例子，我想请麦克布赖德教授看一看巴曼和施梅勒关于肝脂酶的那篇文章[1]中的表 7。

麦克布赖德教授还认为，"'自催化'是已故的洛布教授为掩盖其书中观点的漏洞而发明的欺骗性词语。"而从前述的事实来看，麦克布赖德教授的这一观点似乎完全是对一位不可能再为自己辩解的伟人的毫无根据的攻击。如果麦克布赖德教授了解化学和遗传学事实的话，那他在批评那些试图分析生命现象的人时可能会更加谨慎些。他可能也就不会提出自己在 10 月 25 日的《自然》上提到的问题："是不是卵子中就存在预先成形了的成体器官的缩微版，而发育过程实质上是这些器官雏形逐渐展开显露并变大的过程？又或者卵子起初只是未分化的物质，后来由于未知的原因而变得越来越复杂，因而发育是一个'渐成的'过程？"费尔和罗比森的研究[2]显示，鸡胚中骨骼的形成是由于磷酸酶的作用，因而既不存在微型骨骼，也不是由于未知的原因。但是，在麦克布赖德教授认识到科学领域的最新进展之前，他无疑将依旧是一个固执的活力论者。

（刘皓芳 翻译；刘京国 审稿）

Embryology and Evolution

E. W. MacBride

Editor's Note

The pages of *Nature* were alive with a debate about whether Mendelian genetics could account for the complexities of embryonic development and inheritance. With genetics still in its infancy, there remained many uncertainties and several scientists felt more comfortable with the Lamarckian view of inheritance: the idea that characteristics acquired by an organism during its lifetime can be passed on to its offspring. In his frequent contributions to *Nature*, Ernest William MacBride was particularly outspoken in his critique of Mendelian genetics, a stance that drew an aggressive response from those working in this expanding field. Here MacBride responds to the latest views of genetics advocate J. B. S. Haldane.

I should like to comment on two letters by Mr. Haldane: one on "Natural Selection Intensity as a Function of Mortality Rate", in *Nature* of Dec. 6, and the other on "Embryology and Evolution", in the issue of Dec. 20. In the first, Mr. Haldane criticises as "fallacious" Prof. Salisbury's argument that mortality amongst plants is mainly confined to the seedling stage and that at this period natural selection mainly works. He goes on to consider a case where two races vary as to a single character! Now, this is a travesty of what occurs in Nature. Two allied races do not differ from one another in a single character: they differ in a multitude of minute points, and it is quite impossible to say whether one or another of these points determines their survival. The "characters", in fact, are mere abstractions. The organism is a whole, and the characters are the expression of its constitution; in a word, of the vigour of its reaction to its surroundings. The whole point of Prof. Salisbury's argument was that natural selection chooses the most vigorous, not that which possesses some special character, and this argument I believe to be perfectly sound.

In his second letter Mr. Haldane objects to four of the statements in my reply to Prof. Gates. I shall deal with these seriatim.

(1) Mr. Haldane claims that some microscopists have seen "genes". What they have seen are segregations of material in the stained and fixed chromosomes which they have identified as genes—a purely hypothetical conclusion. He further says that the presence or absence of a "trabant", that is, not a gene but a small chromosome, makes a difference in the constitution of the plant *Matthiola*. This is quite possible, and I shall be glad to have it demonstrated. Prof. Gates was, I think, the first to show that an extra chromosome made a difference to the appearance of the mutant.

胚胎学与进化

欧内斯特·威廉·麦克布赖德

编者按

大家正在《自然》杂志的专栏中激烈地讨论孟德尔遗传学是否能解释胚胎发育和遗传的复杂性。由于遗传学刚刚兴起，还存在着许多不确定的因素，所以一些科学家还是更愿意接受拉马克的遗传理论，即认为有机体后天获得的特征可以传给后代。欧内斯特·威廉·麦克布赖德经常在《自然》杂志上投稿，毫不避讳地公开批评孟德尔的遗传学说，因而他的立场引起了这一迅速发展领域中的研究人员的反击。在本文中，麦克布赖德回应了孟德尔遗传学支持者霍尔丹最近提出的意见。

我想对霍尔丹先生的两篇快报发表一下自己的看法：一篇是在 12 月 6 日的《自然》杂志上刊登的《与死亡率有关的自然选择强度》，另一篇是 12 月 20 日刊登的《胚胎学与进化》。在第一篇快报里，霍尔丹先生批评索尔兹伯里教授的观点是"错误的"，索尔兹伯里教授认为植物的死亡主要发生在苗期，并且自然选择主要在此阶段发生作用。霍尔丹先生竟然继而举证两个物种区分于某一单一性状的例子！这是对大自然中发生的事实的歪曲。两个近缘种在单一性状上不能区分彼此：它们在许多微小的方面都有所不同，不可能说出这些方面中的哪一个方面能够决定它们的生存。事实上，"性状"只是抽象的概念。生物体是一个整体，性状是对其构成的表达；简言之，性状是生物体对周围环境的反应活力的表达。索尔兹伯里教授的中心论点是自然选择选出的是最有活力的生物，而非拥有某些特别性状的生物，我认为这一观点是相当合理的。

在第二篇快报中，霍尔丹先生对我在回应盖茨教授时所作的四点说明进行了反驳。下面我将逐一回复。

（1）霍尔丹先生声称有些显微镜学家已经看到了"基因"。 其实，他们看到的被认定为基因的物质只不过是经染色固定的染色体上的分离物——这是一个纯粹假想性的结论。他还说"随体"的存在或缺失会对紫罗兰属植物的构造产生影响，其中，随体并不是基因，而是一小染色体。这是完全可能的，我很高兴这一点已经得到了实证。我认为盖茨教授是第一个表明额外染色体能影响突变体外观的人。

(2) Mr. Haldane asserts that scores of cases are known where in interspecific crosses characters behave in a Mendelian manner, that is, are due to genes. All I know on this subject is that my friends who are systematists, and have devoted their lives to the study of species and races, deny that such is the case. Of course, a mutant such as the domesticated race almost always "mendelises" when crossed with the wild type; that is just what distinguishes a mutant from a racial character, and the case quoted by Mr. Haldane is such a cross.

(3) Mr. Haldane states that autocatalytic reactions are common in physical chemistry. By this is meant reactions in aqueous solutions which are accelerated by the products of the reaction. I put this question to three first-class chemists, all of them fellows of the Royal Society and one of them a bio-chemist, and as they were all unaware of any such case, I prefer to accept their testimony.

(4) Mr. Haldane objects to my posing the alternative of the organs being preformed in miniature in the embryo or being due to an "unknown cause". He says that bone is formed by an enzyme "phosphatase". This is a mere quibble. Enzymes are *means* employed by the embryo to develop its powers, and their orderly appearance is just as much a mystery as the appearance of the organs themselves.

Mr. Haldane's remarks about my refusing to take cognisance of the recent advances of science and his invitation to acquaint myself with the "facts" of genetics and chemistry I prefer to disregard. I have quoted the authorities on whom I rely in chemical matters. As to genetics, I have served for seventeen years on the Council of the Institution to which Mr. Haldane is attached as statistician, and I have watched all the work going on there, and the more I see of it the more I am convinced that Mendelism has nothing to do with evolution.

(**127**, 55-56; 1931)

E. W. MacBride: 43 Elm Park Gardens, Chelsea, S. W. 10, Dec. 23, 1930.

（2）霍尔丹先生断言有大量为人们所知的例子表明：在种间杂交中，性状以孟德尔式遗传，也就是说是由于基因。对于这一问题，我所知道的是，我的那些一生都致力于物种和种族研究的分类学家朋友们都否认情况是这样的。当然，当一个突变体，如家养种，与野生型杂交时，几乎总是按"孟德尔式"遗传的；这恰恰是区分突变性状和种族性状的要点，霍尔丹先生引用的例子就是这样一种杂交类型。

（3）霍尔丹先生认为自催化反应在物理化学界很普遍。这指的是发生在水溶液中的反应，其生成的产物可以加速该反应。我向三位一流的化学家咨询了这一问题。他们都是英国皇家学会的会员，有一位还是生化学家。由于他们都不知道有任何此类的例子，所以我更倾向于接受他们的证词。

（4）霍尔丹先生反对我提出的以下这个二者择一的说法：器官或者是在胚胎中以缩影形式预先存在或者是由于"不明原因"而发生。他说骨骼是通过一种被称为"磷酸酶"的酶形成的，这纯粹是诡辩。酶是胚胎用于发育的**工具**，酶的有序出现与器官本身的出现一样神秘。

霍尔丹先生评论说我拒绝认知最新的科学进展，并且要请我去了解一下遗传学和化学中的"事实"，对于这个提议，我宁愿漠视不理。在化学方面，我已经指出了一些我可以依靠的权威人士。至于遗传学，我已经在理事会机构工作了 17 年，而霍尔丹先生只是该机构的一名统计员。我则监视那里进行的所有工作，并且我看到的越多，就越相信孟德尔遗传学说与进化毫无关系。

（刘皓芳 翻译；陈平富 审稿）

Embryology and Evolution

C. O. Bartrum

Editor's Note

Following a long-running tit-for-tat correspondence in the pages of *Nature* between supporters and opponents of a synthesis of Darwinian natural selection and Mendelian genetics, C. O. Bartrum attempts to bring some philosophical clarity to the dispute. In an earlier issue, botanist Ruggles Gates and neo-Lamarckian zoologist Ernest William MacBride had gone head to head. Here Bartrum backs Gates and gently scolds MacBride for straying onto philosophical territory. Scientists should restrict themselves to discovering facts and marshalling them into testable hypotheses, he says.

THE discussion between Prof. R. Ruggles Gates and Prof. E. W. MacBride, in *Nature* of Dec. 6, bears in an important way upon the philosophy of science. May one without authority in biology offer what he hopes may be a useful contribution from the philosophical point of view?

It is the function of the scientific man to discover facts, to endeavour to co-ordinate them, and by generalisation to build up a useful scheme of hypotheses. Such a scheme must be a deterministic scheme or it cannot be useful, that is, it cannot be used to forecast further facts. When Prof. MacBride writes of mechanical hypotheses, he refers, presumably, to such a deterministic scheme. Whether the resulting scheme represents the truth is not the business of the scientific man as such, but of the philosopher.

As a philosopher Prof. Gates may believe himself to be a "mere mechanism" or a Drieschian entelechian organism. For science this is beside the question. A scientific man must continue to have faith in "so-called mechanical hypotheses", or, as Prof. Gates says, "there would be no further incentive to experimental embryology", and his function would cease. As a philosopher he may doubt whether such deterministic schemes will ultimately prevail, but as a scientific man he must carry on.

(**127**, 56; 1931)

C. O. Bartrum: 32 Willoughby Road, Hampstead, London, Dec. 10.

胚胎学与进化

巴特拉姆

编者按

对于综合达尔文自然选择学说和孟德尔遗传学的新理论，支持者和反对者一直在《自然》杂志的专栏中针锋相对地互相抨击，巴特拉姆想用哲学观点来解决争端。植物学家拉各尔斯·盖茨和新拉马克主义者动物学家欧内斯特·威廉·麦克布赖德曾在《自然》杂志上激烈地争辩过。巴特拉姆支持盖茨的观点，他温和地指责麦克布赖德偏离到了哲学领域。他说：科学家的本职工作应该是发现事实并基于这些事实形成可以被检验的假说。

拉各尔斯·盖茨教授和麦克布赖德教授之间的讨论与科学哲学有很重要的关联，讨论的内容发表于《自然》杂志 12 月 6 日版。尽管我并非生物学的权威，但我是否可以从哲学观点出发提供一份我所期望的可能对此讨论有用的帮助呢？

科学工作者的职责是发现事实、努力整合事实，并归纳建立一套有用的假说体系。这样的体系必须是具有决定性的，否则它们丝毫用处也没有，也就是说，不能用它们来预测更多的事实。当麦克布赖德教授在文中提到机械假说时，他理当指的就是这样一种具有决定性的体系。产生的假说体系是否代表了事实就不是科学工作者本身的事情，而是哲学家的事情了。

盖茨教授作为一位哲学家，可以认为自己是一台"纯粹的机械"，或者是一种德里施（译者注：德国生物学家与哲学家）式的蕴含生机本源的生物。对于科学而言，这与本文要讨论的问题无关。科学工作者必须继续怀着对"所谓的机械假说"的信仰，否则就如盖茨教授所说的"将不会有进一步研究实验胚胎学的动力了"，那时科学工作者的作用也就停止了。作为一位哲学家，他可以怀疑这样的决定性体系能否最终得到普及；而作为一位科学工作者，他却必须继续进行下去。

（刘皓芳 翻译；陈平富 审稿）

Vitamin B

Editor's Note

There are eight separate chemical compounds in the family of B vitamins, but all were once thought to be a single substance, deemed essential for the growth of organisms. This anonymous report describes how that former picture began to change. Following work that suggested vitamin B had two forms that relieve pain and a concatenation of ailments called pellagra, it seemed by this stage that there were at least four B vitamins. Vitamin B_1 (thiamine), the focus here, was the best characterised, its chemical composition and properties being already sketchily known. Vitamin chemistry became an important strand of biochemistry in the prewar era, although it was some time before their molecular structures and physiological roles were understood.

Assay and Vitamin B_1

THE separation of vitamin B into two factors, antineuritic and antipellagrous, a few years ago, led to considerable attention being devoted to the properties of this vitamin, with the result that it is now possible to distinguish at least four B factors, quite apart from any grouped under the name "Bios", which may be necessary for the growth of lower organisms. The factors are distinguished by differences in their chemical properties and physiological effects: their differentiation has necessitated a revision of the methods of assay, since it is possible that a failure to respond to an addition to the diet is an indication of the absence of a factor other than that for which the test was designed. In this type of research a preventive test is less delicate than a curative, whilst the growth test may be considered still cruder: a single factor should cure the specific symptoms due to its absence, preventive tests may test for more than one, whilst it is clear that a positive growth response can only be obtained when every factor is adequately supplied; and our knowledge of all the factors required for growth is still incomplete, as the recent work on vitamin B has shown.

H. W. Kinnersley, R. A. Peters, and V. Reader[1] have analysed the pigeon curative test for vitamin B_1, or the antineuritic factor. By adherence to certain principles, the test can be made reasonably accurate and has been successfully used in following the vitamin in its concentration from a yeast extract. The birds should be in the laboratory for a month on a mixed grain diet before being placed on the diet of polished rice, and only those developing symptoms within 30 days should be used. As soon as signs of head retraction appear, the bird should be transferred to a warm room for 2 hours and given 50 mgm. glucose in water by stomach tube: this procedure eliminates birds showing false cures. The dose of extract must be given within 6–12 hours of the onset of symptoms and, provided the cure lasts more than 1 and less than 10 days, the amount of active principle

维生素 B

编者按

在 B 族维生素中共有 8 种不同的化合物，但是，以前人们曾把它们看作是单一的物质，有机体的成长离不开这些物质。这份匿名报告告诉我们：以前的观念正在发生变化。研究表明：有两种形式的维生素 B 可以减轻人类的痛苦并能缓解一系列被称为糙皮病的病症，这使维生素 B 家族中的成员数达到至少 4 个。本文详细介绍了维生素 B₁（硫胺素）的特征，当时人们对它的化学组成和性质已经有所了解。虽然维生素的分子结构和生理作用是在一段时间以后才为人们所熟知的，但维生素化学在战前时代就已经成为了生物化学领域的重要分支。

检验与维生素B₁

不久之前，维生素 B 被分离为两个因子：抗神经炎因子和抗糙皮病因子，因此，人们开始对这种维生素的性质给予充分的重视，结果发现它可能至少可以被分为四种截然不同的因子，与基于"生长素"名称的分类结果相去甚远，这种"生长素"对于低等生物的生长可能是必需的。这些因子可以由化学性质和生理学活性的差异加以区别；而这种差异性要求我们对检测方法进行修订，因为当一种检测方法不能检测出食品中的相应添加物质时，人们有可能会认为此种因子并不存在，而并非实验设计有问题。在此类研究中，预防性检测不如治疗性检测那么灵敏；同时，生长检测被认为更加粗糙：一种单独的因子应该可以治愈由于其缺乏而引起的特定病症；预防性检测可以检测多种因子的存在，但同时非常清楚的是，只有当每一种因子都充分给予的时候，才能够获得阳性生长的结果；而我们对于生长所需的所有因子的了解还不全面，最近在维生素 B 方面的研究工作中也体现出了这一点。

金纳斯利、彼得斯和里德[1]分析研究了维生素 B₁ 或抗神经炎因子对鸽子的治疗效果。遵照一定的原则，这个实验可以得到相当精确的结果，而且已经被成功地应用于随后从酵母提取液中得到的维生素浓缩液实验中。这些禽类首先需要在实验室中用混合谷物饲料喂养一个月，然后用精制大米喂养，只有那些在 30 天时间内产生相应病症的个体被用于随后的实验。一旦头部萎缩现象出现，就应把这些鸟禽转移到温室中搁置 2 小时，同时通过胃管给予 50 mg 的葡萄糖水溶液。这个处理过程消除了鸟禽表现出假阳性治愈的可能性。酵母提取物必须在病理症状出现后的 6~12 小时之内喂食，如果治疗时间超过 1 天，少于 10 天，则可以认为出现的有效

present can be considered as directly proportional to the length of the cure. After the test is over, the bird is given marmite and kept warm for a few days. It is then placed on the stock diet again for about a month, when it is ready for another period of polished rice feeding. Individual birds show a remarkable constancy in the time symptoms appear after commencement of the experimental diet, but there is no correlation between this interval and the duration of the subsequent cure, or between it and the colour or weight of the bird.

H. Chick and M. H. Roscoe[2] have used the growth of young rats as a criterion for the presence of vitamin B_1. It is difficult to carry out a curative test with this animal, since there is only a very short interval between the onset of acute symptoms and death: Reader has, however, been successful and has found that the adult rat requires about one pigeon day dose each day (quoted by Peters, the Harben Lectures, 1929). Chick and Roscoe used synthetic diets free from vitamin B_1: vitamin B_2 was supplied as autoclaved yeast or as fresh egg-white. After 2–3 weeks the animals began to lose weight: growth was resumed if Peters' antineuritic concentrate was then administered. The egg-white diet, however, did not maintain growth to maturity. B. C. Guha and J. C. Drummond[3] have used both the pigeon curative and the rat growth tests: in the latter, vitamin B_2 was supplied as marmite autoclaved at an alkaline reaction.

Chick and Roscoe[4] have used a similar method for the assay of vitamin B_2, young rats being placed on a diet complete except for this vitamin, and the B_1 factor being supplied as Peters' concentrate. It was found that the caseinogen used contained traces of vitamin B_2 unless it was reprecipitated with acetic acid and thoroughly extracted with alcohol before being heated at 120°. Animals on this diet fail to grow but respond to a supplement containing vitamin B_2. If the supplement is not given, after about six weeks a generalised dermatitis appears, which can be cured by administration of the vitamin.

B. C. P. Jansen and W. F. Donath[5] obtained highly active preparations of vitamin B_1 from rice polishings by a process involving extraction with acid water, adsorption on fuller's earth, eleution with baryta, and fractionation of the extract with silver sulphate and baryta. The activity was precipitated with phosphotungstic acid, the precipitate decomposed with baryta, and after removal of barium the concentrated solution was treated with platinic chloride, which precipitated the vitamin. Further purification was effected by acetone precipitation from alcoholic solution and by treatment with picrolonic acid or gold chloride. 0.012 mgm. of the final fraction a day was sufficient to maintain pigeons in health over six weeks: C. Eykman[6] confirmed the activity with both pigeons and cocks. The final product was obtained in crystalline form, as a hydrochloride, a picrolonate, or a double salt with gold chloride.

Kinnersley and Peters[7] have continued their work on antineuritic yeast concentrates[8]. It is not yet certain whether the curative substance is the same as that obtained from rice polishings by Jansen and Donath: the activity of the final product does not appear to be quite so great and its properties are not quite the same. In all work on the concentration

成分的数量与治疗时间的长度成正比。当实验结束的时候，再给这些鸟禽喂食含酵母的食物，并在几天内保持温房喂养环境。然后再用普通食物喂养大概一个月的时间，以便准备开始下一个阶段的精制大米喂养。当开始使用实验用食物喂养后，个别鸟禽表现出了显著的周期性发病，但周期的长短与随后的治愈时间、鸟禽的颜色以及体重都没有关系。

　　奇克和罗斯科 [2] 用幼年大鼠的生长作为判断维生素 B_1 存在的标准。用这种动物很难进行治愈性实验研究，原因在于这种动物从急症发作到死亡只间隔很短的时间：然而，里德获得了成功，他发现成年个体老鼠每天需要差不多一只鸽子一天的使用量（引自彼得斯，哈本演讲稿，1929 年）。奇克和罗斯科采用不含维生素 B_1 的合成饲料进行喂养，并以喂食高压灭菌酵母或新鲜蛋清的方式补充维生素 B_2。经过两到三周之后，这些动物的体重开始减轻；如果在食物中添加彼得斯抗神经炎浓缩物质，身体会再次开始生长。然而，添加蛋清的食物并不能维持生长到成熟阶段。古哈和德拉蒙德 [3] 同时进行了鸽子治愈实验和老鼠生长实验；在后一个实验中，补充维生素 B_2 的方式是用碱性反应条件下的高压灭菌酵母。

　　奇克和罗斯科 [4] 采用了一种类似的方法来检验维生素 B_2，他们用完全缺失这种维生素的食物来喂养幼鼠，通过彼得斯抗神经炎浓缩物质提供维生素 B_1 因子。结果发现所使用的酪蛋白原含有微量的维生素 B_2；除非它在被加热到 120℃ 之前曾用乙酸沉淀并用乙醇进行彻底的萃取。采用这种类型食物喂养的动物不能生长，但在补充维生素 B_2 后会有效果。如果没有补充维生素 B_2，大概六个星期之后就会出现普遍性的皮炎，这种病症可以通过补充维生素而得到治愈。

　　詹森和多纳特 [5] 通过以下处理过程从米糠中获得了高活性的维生素 B_1 制备物。这个过程包括酸水萃取、漂白土吸附、氧化钡洗脱，以及用硫酸银和氧化钡对提取物的分馏。得到的活性物质采用磷钨酸进行沉淀，沉淀物采用氧化钡进行分解；去除钡元素之后，浓缩溶液采用能使维生素沉淀的氯化铂处理。进一步的纯化过程还包括采用丙酮从醇溶液中沉淀，以及采用苦酮酸或者氯化金进行处理。每天喂食 0.012 mg 的这种最终提取物就足以保证鸽子在六个星期的时间内处于健康状态。艾克曼 [6] 利用鸽子和公鸡进一步确定了这种生物学活性。最终产物可以以盐酸盐、苦酮酸盐或者氯化金复盐的晶体形式获得。

　　金纳斯利和彼得斯 [7] 继续了他们在抗神经炎酵母浓缩物方面的研究 [8]。然而还无法确认这种治疗性物质是否与詹森和多纳特从米糠中得到的物质完全一样，因为最终产物的活性并不是表现得非常明显，它的性质也不是十分一致。在所有有关维

of vitamin B_1, it has been found that the properties of the active fractions vary according to the nature of the accompanying impurities, so that methods developed for use with an extract of rice polishings may not be applicable without modification to an extract of yeast. The extract from the charcoal adsorption, after removal of metals, can be fractionated by successive additions of alcohol, the vitamin passing into the portion soluble in 99 percent ethyl alcohol. The authors failed to get consistently successful results with a silver fractionation, but were more successful with the use of phosphotungstic acid and platinic chloride. The most active preparations contained a day dose in 0.027 mgm., but more lately some have been obtained with a curative activity of 0.01 mgm. a day dose.

Guha and Drummond (*loc. cit.*) prepared active concentrates from wheat embryo. After extraction by means of acid alcohol, two different methods of concentration were employed: in the first, impurities were precipitated by lead acetate, and the activity adsorbed on norite charcoal at pH 4.5 and eleuted with acid alcohol: it was then precipitated by phosphotungstic acid, adsorbed on silver oxide, and the product fractionated with alcohol. Picrolonic acid then precipitated impurities from the material, which was soluble in alcohol. The first product had a pigeon day dose of 0.043 mgm. In the second method, Jansen and Donath's process was followed, namely, adsorption on fuller's earth at pH 4.5 and eleution with baryta, and fractionation with silver nitrate and baryta followed by precipitation with phosphotungstic acid. The product was then submitted to precipitation with platinic chloride, followed by gold chloride; at the last stage most of the activity passed into the precipitate, but it was observed that smaller doses of both precipitate and filtrate together restored growth in the rat or cured the pigeon than of either when given separately, suggesting that vitamin B_1 may itself be composed of more than one factor. The smallest pigeon day dose was 0.0025 mgm., and 0.015 mgm. promoted good growth in rats. These figures indicate that the preparations were more active than the crystals obtained by Jansen and Donath.

Although formulae have been assigned to vitamin B_1 preparations, it does not appear that a pure substance has yet been isolated. A certain amount is, however, known about its properties. It appears to be a tertiary base: it is soluble in water and alcohol, but is unstable in the latter solvent when highly purified: it is insoluble in the other common organic solvents. It is destroyed by alkali, but is stable to oxidising and reducing reagents and to nitrous acid. Cruder preparations give a definite Pauly reaction, but as purification proceeds the reaction becomes very weak. Sulphur is absent, and the purer preparations do not give the xanthoproteic, purine, or Millon's reactions. In extracts from rice polishings, after treatment with lead acetate and concentration of the filtrate, vitamin B_1 is destroyed by fermentation and by heating to 95°, and is removed by filtration through a Berkefeld filter[9], although it will dialyse through cellophane.

The isolation from concentrates of supposedly pure substances and the fact that false positives may be given by the pigeon test have led to claims that different pure compounds are the vitamin. J. M. Gulland and Peters[10] have examined the claims that certain

生素 B₁ 浓缩物的研究中，人们发现活性组分的性质与所含的杂质有关，因此，以使用米糠萃取物为基础的方法也许不能原样照搬到酵母提取物上。从木炭吸附得到的提取物，经过去除金属离子的操作后，可以通过连续注入乙醇的方法进行分馏处理，维生素会转移进入 99% 纯度的乙醇溶液中。研究者采用银分馏法未能得到可重复的成功结果，但是在采用磷钨酸和氯化铂时取得了更好的效果。活性最高的制备物能使一天的剂量为 0.027 mg；但是最近的一些研究已经取得了一天给药剂量只需要 0.01 mg，就可以达到治疗活性的效果。

古哈和德拉蒙德（在上述引文中）从麦芽中制备得到了活性提取物。通过酸性乙醇溶液萃取处理后，采用了两种不同的浓缩方法。在第一种方法中，杂质采用乙酸铅进行沉淀处理，活性物质在 pH 值为 4.5 的条件下吸附到苏长岩木炭上，然后用酸性乙醇溶液洗脱。接着用磷钨酸进行沉淀，用氧化银吸附；再用乙醇分馏产物。随后，采用苦酮酸将杂质从可溶于乙醇的物质中沉淀出来。第一种方法制得的产物要求鸽子每天的摄取量为 0.043 mg。第二种方法采用的是詹森和多纳特的处理过程，即首先在 pH 值为 4.5 时，采用漂白土进行吸附处理，然后用氧化钡进行洗脱；接着选用硝酸银和氧化钡进行分馏，再用磷钨酸进行沉淀处理。产物通过氯化铂转入到沉淀中，然后用氯化金进行处理。在最后阶段，绝大多数的活性物质都转移到了沉淀中；但发现同时使用沉淀和上清物质时，在恢复大鼠生长或鸽子治疗实验上的有效使用剂量都要比它们单独使用时的剂量要小，这说明维生素 B₁ 可能含有不止一种因子。鸽子每天所需的最小有效剂量为 0.0025 mg，而促进大鼠健康成长的有效剂量为 0.015 mg。这些结果表明制备物的活性要高于詹森和多纳特得到的晶体。

尽管分子式被写成维生素 B₁ 制备产物，但看上去依然没有分离到一种纯的物质。然而，现在对于它的性质已经有了一定的了解。它看上去应该是一种叔碱：它可以溶解在水和乙醇中，但纯度高时在乙醇中不稳定；它不能在其他的常用有机溶剂中溶解。它可以被碱破坏，但是对氧化和还原试剂以及硝酸都是稳定的。粗提物可以发生明显的波利反应，但是随着纯化过程的进行，这个反应会变得很弱。在没有硫黄存在的情况下，较纯的制备物质不能发生黄蛋白、嘌呤或者米隆反应。从米糠得到的萃取物质，在经过乙酸铅的处理和过滤浓缩后，维生素 B₁ 在发酵和加热到 95℃ 的过程中被破坏，可以采用贝克菲尔德滤器进行过滤去除 [9]，尽管它可以通过玻璃纸的透析。

从浓缩物中分离得到假定的纯物质以及基于鸽子实验可能给出假阳性结果的事实，使我们有理由相信这些不同的纯化合物就是维生素。格兰德和彼得斯 [10] 经仔细

279

quinoline and glyoxaline derivatives have curative properties. Without exception all those examined, including 4 (or 5) glyoxaline methylethyl carbinol hydrochloride and 2:6-dihydroxyquinoline, were quite inactive when tested on pigeons by Peters' technique.

(**127**, 95-96; 1931)

References:

1. Kinnersley, H. W., Peters, R. A., and Reader, V., *Biochem. Jour.*, **22**, 276 (1928).

2. Chick, H., and Roscoe, M. H., *Biochem. Jour.*, **23**, 498 (1929).

3. Guha, B. C., and Drummond, J. C., *Biochem. Jour.*, **23**, 880 (1929).

4. Chick, H., and Roscoe, M. H., *Biochem. Jour.*, **22**, 790 (1928).

5. Jansen, B. C. P., and Donath, W. F., *Mededeelingen van den Dienst der Vdksgezondheid in Ned.-Indië*, Part 1 (Anno 1927).

6. Eykman, C., *Kon. Akad. van Wetensch. Amsterdam*, **30**, 376 (1927).

7. Kinnersley, H. W., and Peters, R. A., *Biochem. Jour.*, **22**, 419 (1928).

8. Kinnersley, H.W., and Peters, R. A., *Nature*, **121**, 516 (1928.)

9. Rosedale, J. L., and Oliveiro, C. J., *Biochem. Jour.*, **22**, 1362 (1928).

10. Gulland, J. M., and Peters, R. A., *Biochem. Jour.*, **23**, 1122 (1929).

研究发现：某些喹啉和咪唑的衍生物具备治疗的性质。毫无例外的是，在采用彼得斯的方法对鸽子进行实验时，所有这些被研究的物质，包括4（或者5）咪唑甲基乙基甲醇盐酸盐和2:6－二羟基喹啉，都根本没有活性。

（刘振明 翻译；刘京国 审稿）

Vitamin B

Editor's Note

Following on from an earlier article on the B vitamins, this paper examines what was known about vitamins B_2 and B_3, mostly from extracts of yeast. That these extracts were rich in such nutrients contributed to the popularity in the UK of the "marmite" mentioned here—a dietary supplement whose attractions have met a mixed reception elsewhere in the world. The article serves as a testament to the difficulties of making progress in biochemistry while extraction, purification and identification techniques were still rather rudimentary.

Vitamins B_2 and B_3: Bios

THE possibility of obtaining vitamin B_1 in a relatively pure condition has facilitated the differentiation of the other factors which, with B_1, make up the vitamin B complex. Chick and Roscoe used Peters' concentrate to demonstrate that the rat required two factors, the second being known as B_2 or the antipellagrous vitamin. More recently, they have published papers dealing with the chemical properties of this factor[1].

Yeast extracts contain vitamin B_2, but the final antineuritic concentrate none: examination of the by-products of the concentration showed that about half the B_2 was precipitated by lead acetate, in the treatment of the extract with this reagent (at pH 4.5) and another third in the treatment with baryta and sulphuric acid, the remainder being precipitated during the treatment with acid mercuric sulphate and the subsequent passage of hydrogen sulphide through the filtrate. The lead acetate precipitate was the most convenient source for obtaining a concentrated preparation: examination of this stage in detail showed that all the vitamin was carried down when the precipitation was carried out at a neutral reaction, but less than half at pH 2.6. The vitamin was recovered by decomposing the precipitate with hydrogen sulphide: to ensure precipitation of the lead sulphide at an acid reaction, at which the vitamin is not adsorbed on the precipitate, it was necessary first to hydrolyse the yeast gum in the extract with hydrochloric acid. Unfortunately, the lead precipitate also carries down some vitamin B_1, and it was not found possible to obtain a preparation of B_2 free from B_1 by dialysis, by making use of their different solubilities in alcohol or their different rates of destruction by ultra-violet light. The concentrate was active in a dose of 0.03 gm., equivalent to 0.5 gm. dried yeast daily. It is possible that yeast extract is an unsuitable medium for effecting a separation. Rosedale[2] precipitated from rice polishings extract (by means of lead acetate) a factor which was not B_1, although enabling pigeons to grow and maintain health on a diet of polished rice. It cannot yet be said with certainty, however, that this factor is vitamin B_2.

维生素 B

编者按

与前面那篇介绍维生素 B 的文章不同，这篇文章分析了来自酵母提取物中的维生素 B_2 和 B_3 的已知性质。这些提取物中含有大量的维生素 B_2 和 B_3，这也是"酸制酵母"在英国如此流行的原因之一——这种膳食补充剂也已经被其他国家的人所接受。文中也谈到了生物化学领域在发展中遇到的困难，因为在当时，抽取、提纯和鉴定技术都十分不完善。

维生素B_2和B_3：生长素

在人们有可能得到较高纯度的维生素 B_1 以后就开始区分与 B_1 一起构成 B 族维生素复合体系的其他因子了。奇克和罗斯科采用彼得斯浓缩法证实大鼠需要两种因子，第二种因子被称为维生素 B_2，或者抗糙皮病的维生素。就在最近，他们又发表了一些与这种因子的化学性质相关的文章[1]。

酵母提取物包含维生素 B_2，但是在最终的抗神经炎浓缩物中却不含这种物质。对这种浓缩物产生过程中的副产物的检验结果显示，在用乙酸铅处理这种提取物的过程中（在 pH 值为 4.5 的条件下），大概一半的维生素 B_2 都被沉淀掉了；另外三分之一在用氧化钡和硫酸处理时被沉淀，剩下的在用酸性硫酸汞处理以及随后使用硫化氢过滤的过程中被沉淀了。乙酸铅沉淀法是最便于获得维生素 B_2 浓缩制备产物的方法；对这个过程的仔细研究发现，在中性反应条件下，所有的维生素都会沉淀下来；但当 pH 值为 2.6 时，沉淀的维生素量不到上述情况时的一半。使用硫化氢分解这些沉淀时，又可重新回收维生素：为了保证在酸性条件下产生硫化铅的沉淀，且维生素不会被沉淀吸附，就需要首先用盐酸对提取液中的酵母胶进行水解。不幸的是，铅沉淀中还是吸附了一些维生素 B_1；目前人们仍无法获得不含维生素 B_1 的维生素 B_2 制备物，无论是采取透析，利用它们在乙醇中的不同溶解性，还是根据它们在紫外光下被破坏的程度不同等都无法实现。这种浓缩物在剂量为 0.03 g 时就有活性，相当于每天给予 0.5 g 的干酵母。也许酵母提取物并不是最合适的分离媒介。罗斯黛尔[2] 从米糠的提取物中沉淀（采用乙酸铅的方法）出一种因子，这种因子不是维生素 B_1；尽管它可以使得采用精制米喂养的鸽子继续生长和保持健康，但不能完全确定这种因子就是维生素 B_2。

B. T. Narayanan and J. C. Drummond have also carried out experiments on the concentration of vitamin B_2[3]. Yeast was extracted with dilute alcohol and the extract concentrated; lead acetate was then added, sometimes following a preliminary hydrolysis of the extract with baryta. The lead precipitate was decomposed with sulphuric acid, vitamin B_2 passing into the filtrate. It could be adsorbed on fuller's earth in strongly acid solution, but no satisfactory method of eleuting it again was found. Norite charcoal was not efficient as an adsorbing agent, and its use led to disappearance of the activity.

A certain amount is known about the properties of vitamin B_2: it is soluble in dilute, but insoluble in strong ethyl alcohol, and exposure to the latter results in its destruction. It is stable to hydrogen peroxide and nitrous acid and, to a certain extent, to heat, provided the reaction is acid. Autoclaving or even boiling at an alkaline reaction brings about rapid destruction. It is more easily destroyed by exposure to ultra-violet light than vitamin B_1.

Narayanan and Drummond examined a number of chemical compounds, including nucleic acid, purines, nicotinic acid, betaine and inositol, for vitamin B_2 activity, but all were, without exception, inactive.

V. Reader has adduced evidence that a third factor, tentatively called vitamin B_3, is necessary for the nutrition of the rat[4]. Animals kept on an apparently complete synthetic diet, in which Peters' concentrate supplied vitamin B_1 and alkaline autoclaved yeast B_2, failed to grow after some weeks; substitution of yeast extract for the two supplements led to an immediate resumption of growth. The failure did not appear to be due to lack of either vitamin B_1 or B_2, since increasing the amounts given did not improve growth. More convincing evidence of the existence of the third factor rests in the fact that it was found to be precipitated by the mercuric sulphate used in the preparation of the B_1 concentrate, and was recoverable, to the extent of 75 percent of that present in the original extract, from the precipitate. Vitamin B_3 is even more easily destroyed by heat than B_1; under certain conditions it is soluble in ether. Rats fed on an ordinary diet carry a larger store of vitamin B_3 than of B_2; it is, therefore, possible that in short-time growth experiments, such as were used by Chick and Roscoe in the assay of B_1 or B_2, lack of B_3 does not play a part; again, it is also possible that antineuritic concentrates, unless highly purified, or autoclaved yeast, may be contaminated with traces of the third factor.

Chick and Roscoe found that vitamin B_2 was alone present in egg-white, but that rats on a synthetic diet with this material as source of the vitamin instead of autoclaved yeast, showed subnormal growth after a few weeks. They suggest that yeast contains a third B factor, which is, however, thermostable in contradistinction to Reader's B_3. M. A. Boas, some years ago, found that a diet containing dried egg-white produced skin lesions and nervous symptoms in rats, whereas fresh egg-white had no such effect[5]. A number of foodstuffs contained a factor which counteracted the ill-effects of the ingestion of dried egg-white; it could not be identified with either vitamin B_1 or B_2.

284

纳拉亚南和德拉蒙德也进行了浓缩维生素 B_2 的相关实验 [3]。先用稀释的醇对酵母进行提取，再对提取液进行浓缩处理；然后加入乙酸铅；有时候，还需要采用氧化钡对提取物进行水解预处理。铅沉淀物采用硫酸进行分解处理，维生素 B_2 转移到滤液中。在强酸溶液中它可以被漂白土吸附，但是还没有发现一种令人满意的方法可以将它从漂白土上洗脱下来。苏长岩木炭不是一种有效的吸附试剂，使用它会导致维生素丧失活性。

我们对于维生素 B_2 的性质已经有所了解；它溶解于稀醇溶液，但在高浓度乙醇中却无法溶解，接触后者会导致它被破坏。它与过氧化氢和亚硝酸不发生反应，当反应条件为酸性时，它在某种程度上对受热也是稳定的。在碱性反应条件下，使用高压蒸煮甚至只是沸腾状态就会导致它的迅速破坏。当暴露在紫外光下面时，它比维生素 B_1 更容易被破坏。

纳拉亚南和德拉蒙德检查了很多种化合物，包括核酸、嘌呤、尼克酸、甜菜碱和肌醇，希望能够找到具有维生素 B_2 的生物活性的物质；但是毫无例外，上述化合物都没有这种活性。

里德提出证据证明有第三种因子存在，暂时可以称作维生素 B_3，它是大鼠生长不可缺少的营养物 [4]。完全用合成饲料喂养的动物在几个星期后就会停止生长，其中彼得斯浓缩物提供维生素 B_1，碱性条件下高温高压处理后的酵母粉提供维生素 B_2。采用酵母提取物替代上述两种物质在食物中进行补充，动物很快就恢复了生长。由上述情况看来，停止生长不是由于缺少维生素 B_1 或者维生素 B_2，因为增加给予的量并不会改善动物的生长状况。能够证明第三种因子存在的更为令人信服的证据是它可以被维生素 B_1 制备过程中所使用的硫酸汞沉淀出来，通过处理沉淀物可重新获得；含量可达最初提取物中的 75%。维生素 B_3 比维生素 B_1 更容易因受热而被破坏；在特定条件下，它可以溶解在乙醚中。采用普通食物喂养的大鼠，相对于维生素 B_2 而言，可以吸收更多的维生素 B_3。因此，在短时间的生长实验中，例如奇克和罗斯科用于检测维生素 B_1 或者维生素 B_2 的实验，缺乏维生素 B_3 并没有造成影响。另一方面，还有可能是因为抗神经炎浓缩物没有经过高度纯化，或是在高压灭活的酵母粉中混有痕量的第三种因子。

奇克和罗斯科研究发现，蛋清中只存在维生素 B_2；但是对于用合成食物喂养的大鼠，如果只采用这种物质作为维生素的来源而没有食用高压灭菌的酵母粉，那么经过数周之后大鼠就会表现出生长缓慢的迹象。他们认为酵母中含有第三种因子，然而，它却比里德的维生素 B_3 更为耐热稳定。博厄斯在几年前发现，食用含有干蛋清的食物会导致大鼠的皮肤受损及引发神经系统病症，而选用新鲜的蛋清却不会出现这种情况 [5]。很多食物中都含有一种可以清除因摄取干蛋清而造成的不良影响的因子；它既不是维生素 B_1，也不是维生素 B_2。

The above account by no means exhausts the work which has been done on the chemistry of vitamin B. That three or four factors are included under this term appears certain, but it is not easy to relate the work of different investigators, especially when the claim for a new factor is based on the supplementary effects of different foodstuffs. Peters, in his Harben Lectures, reviews some of these investigations. Williams and Waterman found that pigeons maintained their weight on a polished rice diet when supplemented with marmite; transference to wheat, however, produced growth. Hence there is a factor in wheat which is absent from marmite and cannot be B_3. Hunt found that the residue of autolysed yeast, after thorough extraction with water, contained a factor which supplemented two obtainable from the extract; again this third factor does not appear to be B_3, which is soluble in water. Peters has also found that pigeons require for growth, in addition to B_1 and the factor of Williams and Waterman, a thermolabile factor which is present in yeast extract, and may be considered provisionally as identical with B_3: pigeons apparently do not require B_2. It appears, therefore, that the rat requires B_1, B_2, and B_3 with possibly Hunt's factor, whilst the pigeon requires B_1, B_3, and the factor of Williams and Waterman.

It may be of value to refer also to some recent work upon a substance which has been related by many observers to vitamin B, the yeast-growth stimulant or "Bios". Confusion was caused by the fact that it was not realised that different yeasts behave differently on various media, that their requirements for bios vary, and that the requirements of other micro-organisms need not necessarily be the same as those of yeast. Peters and his colleagues have investigated the growth factors required by *Streptothrix corallinus*[6]. When the organism was grown on a synthetic medium, a growth-stimulating factor was found in tryptic beef broth, yeast, rabbit muscle, serum, and wheat embryo; it is organic, soluble in water but not in ether, dialysable, and not precipitated by lead acetate. It accompanies vitamin B_1 in the preparation of the antineuritic concentrate but is stable to alkali in this concentrate (although unstable when crude) and is not, therefore, the same factor as the vitamin. It is synthesised by the meningococcus. It is not vitamin B_2. By parallel tests on the organism and on pigeons, it was shown that the factor and vitamin B_1 fractionate quantitatively together through all stages into the final concentrate.

The pitfalls encountered in such work are disclosed by certain anomalous results obtained with this organism, impure preparations of the factor sometimes appearing to contain relatively more growth-promoting activity when used in high concentrations than when used in lesser amounts. Reader has found that this effect can also be produced by adding mannitol to the purer extracts[7]. Mannose itself and related alcohols cannot replace mannitol; it appears that the organism uses the alcohol as a specific source of food supply.

Working on the relation of bios to yeast, A. M. Copping[8] has found that the necessity for bios depends on the yeast and the medium used. Those which grow in a synthetic medium without the addition of a factor such as is supplied by an autoclaved extract of yeast, produce a stimulant for other yeasts. Bios is required by yeasts which only ferment and do not respire in its absence; added bios then stimulates both respiration and fermentation.

上面的论述并没有把人们在维生素 B 化学性质方面的研究工作全部列举出来。这个专业术语下面必然包含着三到四种因子，但要在不同研究工作之间建立联系并非易事，特别是当采用不同食物喂养产生的附加效应来证实一种新的因子的存在时。彼得斯在哈本演讲中回顾了一些研究进展。威廉斯和沃特曼发现，鸽子在使用精制米作为食物喂养时，辅助加入酵母可以保持它们的体重；转而使用小麦时，会促进其生长。因此，小麦中有一种因子是酵母中所没有的，并且不可能是维生素 B_3。亨特发现在用水对自溶酵母进行彻底提取之后，剩余残渣中含有一种因子，与提取物中的另外两种因子同时存在；这种因子也不是维生素 B_3，因为后者可以溶解在水中；彼得斯还发现，为了满足鸽子的生长需要，除了维生素 B_1 和威廉斯和沃特曼发现的因子之外，还需要一种在酵母提取液中存在的、不耐热的因子，这种因子可以暂时被认为是维生素 B_3；鸽子显然不需要维生素 B_2。由此看来，大鼠需要的是维生素 B_1、维生素 B_2、维生素 B_3，还有可能需要亨特因子；鸽子需要的是维生素 B_1、维生素 B_3，以及威廉斯和沃特曼发现的因子。

许多研究者认为，酵母生长刺激因子或称"生长素"与维生素 B 有关，也许有必要谈及最近的一些基于该物质的研究工作。混乱的来源事实是，我们没有意识到不同种类的酵母在不同的介质中行为各异，它们对于生长素的需求也是变化各异的；其他微生物的需求也不一定和酵母完全一样。彼得斯和他的同事研究了珊瑚红诺卡氏菌所需的生长因子 [6]。当微生物体在合成培养基上生长时，人们发现有一种生长刺激因子存在于胰蛋白酶消化的牛肉膏、酵母、野兔肌肉、血清以及小麦胚芽中；这是一种有机物，可以溶解在水中但不溶解于醚，可以通过透析装置，不会被乙酸铅沉淀。它与维生素 B_1 一起共存于抗神经炎浓缩物的制备液中，但在浓缩物中，它对碱性环境是稳定的（尽管在粗提物状态下是不稳定的）；因此，它和维生素不属于同一个因子。它由脑膜炎球菌合成，它并不是维生素 B_2。通过在微生物体和鸽子上的平行试验，结果显示这种因子可以和维生素 B_1 一起在整个阶段都定量地分馏到最终的浓缩物中。

在这种微生物体上所获得的某些反常结果暴露出此类研究工作存在不少缺陷。当选用高浓度浓缩物时，该因子的不纯制备物有时可以显现出相对更高的生长促进活性。里德发现这种效应也可以通过在更纯净的提取物中加甘露糖醇获得 [7]。甘露糖本身以及相关的醇不能够代替甘露糖醇；似乎这种生物体将醇作为了一种特殊的食物来源。

在研究生长素与酵母之间的关系时，科平 [8] 发现是否需要生长素取决于酵母以及选用的培养介质。在不添加任何因子（例如高压灭菌后的酵母提取液）的情况下，生长在合成培养基中的酵母可以为其他酵母制造一种生长刺激因子。对于那些当生长素缺失时只发酵而不呼吸的酵母而言，生长素才是必需的；当加入生长素后可以同时促进呼吸和发酵过程。

Narayanan has separated a yeast bios from vitamin B_2 in yeast extract, since the former is not precipitated by lead acetate in the hydrolysed extract[9]. The bios is not adsorbed on norite charcoal, but is soluble in strong alcohol; it is precipitated by phosphotungstic acid but not by silver or platinic chloride. By a series of fractionations a highly active concentrate was obtained. The material contained nitrogen but no phosphorus: most tests for nitrogenous compounds were negative, but a positive Pauly reaction was obtained. It is stable to nitrous acid but destroyed by hydrogen peroxide. Narayanan also tested a large number of pure substances, such as nucleic acid, purines, nicotinic acid, betaine, lipoids, and various bases and amino acids for bios activity, but all were found to be inactive.

(**127**, 131-133; 1931)

References:

1. Chick, H., and Roscoe, M. H., *Biochem. Jour.*, **23**, 504 and 514 (1929): **24**, 105 (1930).

2. Rosedale, J. L., *Biochem. Jour.*, **21**, 1266 (1927).

3. Narayanan, B. T., and Drummond, J. C., *Biochem. Jour.*, **24**, 19 (1930).

4. Reader, V., *Biochem. Jour.*, **23**, 689 (1929): **24**, 77 (1930).

5. Boas, M. A., *Biochem. Jour.*, **21**, 712 (1927).

6. Reader, V., *Biochem. Jour.*, **22**, 434 (1928); Orr-Ewing, J., *Biochem. Jour.*, **22**, 440 and 443 (1928); Peters, R. A., and Kinnersley, H. W., *Biochem. Jour.*, **22**, 445 (1928).

7. Reader, V., *Biochem. Jour.*, **23**, 61 (1929).

8. Copping, A. M., *Biochem. Jour.*, **23**, 1050 (1929).

9. Narayanan, B. T., *Biochem. Jour.*, **24**, 6 (1930).

　　纳拉亚南已经从酵母提取得到的维生素 B_2 中分离出了一种酵母生长素，因为这种生长素在水解提取液中不被乙酸铅沉淀 [9]。这种生长素不能被苏长岩木炭所吸附，但是可以溶解在高浓度的醇溶液中；它可以被磷钨酸沉淀，但不能被氯化银或者氯化铂沉淀。在经过一系列的分馏之后可以得到一种高活性的浓缩物。这种物质含氮不含磷：针对含氮化合物的绝大多数检测结果都是阴性的，但是波利反应的结果是阳性的。该物质对于亚硝酸是稳定的，但是可以被过氧化氢破坏。纳拉亚南还检测了大量的纯净物质，例如核苷酸、嘌呤类、尼克酸、甜菜碱、类脂以及各种各样的碱和氨基酸，希望能发现生长素的活性，但研究结果表明所有这些物质都不具备这种活性。

（刘振明 翻译；刘京国 审稿）

Vitamin B

Editor's Note

The third in a series of articles on the current understanding of B vitamins, this paper looks at what was known about the distribution and roles of these substances in the body. Animals such as pigeons and rodents deficient in "vitamin B" show a range of dysfunctions, from dermatitis to weight loss and paralysis. It was a challenge to explain this breadth of effects, but the studies point to the emerging appreciation of how complex dietary factors can be in health.

Distribution and Physiology

THE general distribution of vitamin B is now fairly well defined, but the adequacy of different food substances in this respect for different species and the distribution of the various factors in the B complex are still subjects for investigation. R. H. A. Plimmer, with W. H. Raymond, J. Lowndes, and J. L. Rosedale, has examined the comparative vitamin B value of cereals, pulses, and nuts[1]. The preventive method was employed, using pigeons, and the criterion was maintenance for at least 26 weeks. All the vitamins required by the pigeon were therefore included in the estimation; symptoms of deficiency were paralysis and loss of weight. The diets used contained 5 percent fish meal, white flour or white rice, and the substance under test in varying proportions. Dried yeast contained most vitamin: of the other foodstuffs, wheat germ was about half as good as the yeast, whole wheat, bran, and middlings contained about a tenth of the amount present in yeast, and other cereals about a twentieth. The majority of the pulses and nuts examined contained between a fifth and a tenth of the quantity present in yeast. More vitamin B is required for hatching and rearing young than for maintenance. Chickens require half as much again as pigeons, rats only about half; the requirements of human beings may be intermediate between those of the pigeon and the rat.

A. L. Bacharach and E. Allchorne[2] found that the vitamin B content of malted flour was the same as that of the original unmalted flour, but that the malt extract appeared to contain more: the experiments were carried out on rats and the effect is attributed to the improvement in appetite brought about by the extract.

The content of vitamin B in seeds has been shown to be markedly influenced by the manure applied to the plant, by M. J. Rowlands and B. Wilkinson[3]. Two similar plots of grass and clover were manured with an artificial manure and pigs' dung respectively: the pigs were fed on barley meal, middlings, and a small amount of a mixture of meat meal, rye and wheat embryo, bone meal, and cod liver oil. The manured patch produced a heavier crop, containing more clover, but the growth on the dunged patch was bigger. By preventive and curative growth tests on rats, it was shown that the vitamin B content

维生素 B

编者按

这是有关B族维生素的系列文章中的第三篇，作者在文中考查了B族维生素在人体中的分布以及对人体的作用。有些动物，如鸽子和啮齿类动物因为缺乏"维生素B"而表现出一定程度的功能失调，从皮炎到体重减轻甚至瘫痪。要解释维生素B这么多的作用对我们来说是一种挑战，但是这些研究使人们认识到饮食因素对于健康的影响是多么复杂。

分布与生理学

对于维生素 B 的一般性分布规律，现在已经了解得非常清楚了；但是不同物种对各种食物的需求量到底为多少才合适以及不同因子在 B 族维生素中的分布仍然是我们研究的课题。普利玛与雷蒙德、朗兹以及罗斯黛尔一道研究和比较了谷物、豆类以及坚果中维生素 B 的相对含量 [1]。研究采用了预防性的方法，以鸽子作为实验对象，所使用的标准至少在 26 周内保持不变。鸽子所需的所有维生素都被事先计算在内；缺乏这些物质的症状是神经麻痹以及体重减轻。所选用的食物包含 5% 的鱼肉、白面或白米以及不同比例的测试用物质。干酵母中含有最多的维生素：在其他食物中，麦芽中维生素的含量只有酵母的一半，全麦、麸糠和小麦粗粉包含的维生素大概是酵母中含量的 1/10，其他谷物中的含量大概是酵母的 1/20。绝大多数用于测试的豆类和坚果中所含有的维生素量介于酵母中含量的 1/5~1/10 之间。孵化过程以及幼体的喂养过程需要更多的维生素B。小鸡对于维生素 B 的需求量比鸽子多一半，大鼠只需要鸽子需求量的一半；人体对维生素 B 的需求量也许介于鸽子和大鼠之间。

巴卡拉克和奥科恩 [2] 发现，发芽的麦粒中维生素 B 的含量与最初没有发芽的麦粒含量相同；但是，麦芽提取物中似乎含量更高一些；实验是在大鼠上进行的，产生效果的原因是提取物改善了大鼠的食欲。

罗兰兹和威尔金森 [3] 发现，种子中的维生素 B 含量明显受到对植物所施的肥料的影响。两块相似的苜蓿地分别选用人工肥料和猪粪进行施肥：这些猪采用大麦粉、麦麸以及少量的肉粉、黑麦、麦芽、骨粉以及鱼肝油混合物来喂养。在施人工肥料的地里，植株生长得更茂盛，苜蓿的产量也更高；但是在选用猪粪进行施肥的地里，苜蓿植株生长得更大。对大鼠进行预防性试验和治愈性生长试验的结果显示：维生

of the seeds from the manured patch was much less than that of those from the dunged patch. In further experiments vitamin B was extracted from pigs' dung by means of alcohol.

There is evidence that lower organisms can synthesise vitamin B or similar growth factors, and that this synthesis may occur also in the intestinal tract in higher animals. Thus, Reader has found that the meningococcus can synthesise a growth factor for a streptothrix, all the vitamin B_1 being previously removed from the medium; and G. L. Peskett[4] has shown that yeast can synthesise vitamin B_1. Intraintestinal synthesis may be the explanation of "refection" which has been described by L. S. Fridericia and H. Chick and M. H. Roscoe[5]. In this condition rats maintained on a vitamin B free diet containing uncooked rice starch passed bulky white faeces, and at the same time were cured of their symptoms and put on weight. The faeces contained abundant vitamin. The condition appeared to depend on the presence of uncooked starch in the diet and a virus in the intestine.

W. R. Aykroyd and M. H. Roscoe[6] have investigated the distribution of vitamin B_2. Wheat and maize were poor sources: the germ and bran of wheat contained more than the endosperm, but maize germ contained less than wheat germ: dried peas also contained little. Dried yeast and ox liver and fresh milk were excellent sources, and egg-yolk and dried meat good. It was possible to cure rats suffering from the dermatitis of vitamin B_2 deficiency, as well as to stimulate their growth.

The physiological functions of the vitamin B complex are incompletely understood: in its absence the metabolic processes of the tissues are imperfectly performed, and investigations have thrown some light upon the details of the defects. Thus the vitamin is related to both protein and carbohydrate metabolism. G. A. Hartwell has found that young rats die, with engorgement of the kidneys, when the synthetic diet contains 20 percent edestin and 5 percent yeast extract, although older animals thrived on the diet even with a lower allowance of yeast[7]. Increasing the amount of yeast extract permitted normal growth: the factor responsible was found to be thermostable. Caseinogen and egg-albumin required less yeast extract than edestin for normal metabolism.

H. W. Kinnersley and R. A. Peters have investigated the relation between the lactic acid content of the brain and the symptoms of head retraction in pigeons fed on a diet of polished rice[8]. Using a special technique, it could be demonstrated that birds showing opisthotonos had more lactic acid in their brains than normal birds, and that this increase was most marked in the parts below the mid-brain and occurred here first at a time when symptoms were threatening. The increase was not observed after cure by a dose of vitamin B_1 concentrate. The symptoms appear to be due to this accumulation of lactic acid, and the fact that it is localised indicates that vitamin B_1 is intimately concerned in the intermediary metabolism of carbohydrates, apparently with the oxidative removal of lactic acid. In this connexion it might be remarked that H. Yaoi found that muscle from polyneuritic pigeons reduced methylene blue more feebly than normal muscle, but

素 B 在人工肥料施肥地块的种子中的含量，远远少于在猪粪施肥地块的种子中的含量。在进一步的试验中，采用醇萃取法从猪粪中得到了维生素 B。

有证据证明低等生物可以合成维生素 B 或者类似的生长因子，这种合成过程或许也会发生在高等动物的肠道内。因此，里德发现脑膜炎球菌可以合成一种链丝菌生长所需的生长因子，而培养介质中的维生素 B_1 已经被提前移除了；佩斯凯特 [4] 的研究结果表明，酵母可以合成维生素 B_1。肠内的合成过程可以解释为弗里德里西和奇克以及罗斯科描述的"点心" [5]。在这种环境下，长期用不含维生素 B 的食物（其中包含有未煮过的大米淀粉）喂养的大鼠会排出大量的白色粪便，病症得到了治愈，体重增加，并且粪便中含有丰富的维生素。这种情况似乎与食物中未烹饪过的淀粉以及肠道中的细菌有关。

艾克罗伊德和罗斯科 [6] 研究了维生素 B_2 的分布情况。小麦和玉米中的含量都很低：麦芽和麦麸中维生素 B_2 的含量要高于小麦胚乳中的含量；但是玉米胚芽中的含量要低于麦芽中的含量；干豆中含量也很少。干酵母、牛肝以及鲜牛奶是维生素 B_2 的最佳来源，蛋黄和干肉也是不错的来源。也许能治愈由于维生素 B_2 缺乏而患皮炎的大鼠，还能促进它们的生长。

我们对 B 族维生素的生理学功能了解得还不够透彻：当它们缺失时，组织的代谢过程不能很好地完成；目前的研究已经揭示出了在维生素缺失时的一些具体表现。因此维生素与蛋白质和碳水化合物的代谢都有关系。哈特韦尔发现，当选用包含 20% 麻仁球蛋白和 5% 酵母提取物的人工合成食物进行喂养时，幼年大鼠就会因为肾充血而死亡；而成年鼠即使在酵母添加量更低的情况下也能健壮生长 [7]。当增加酵母提取物的含量时，幼年大鼠恢复正常生长；与之相关的因子被认为具有耐热性。对于正常的代谢过程，酪蛋白原和鸡蛋清蛋白代谢所需的酵母提取物要少于麻仁球蛋白。

金纳斯利和彼得斯研究了当采用精制大米作为食物喂养时，鸽子产生脑萎缩症状和脑中乳酸含量之间的关系 [8]。利用一种特殊的技术，可以证明出现角弓反张症状的鸟类的脑中所含的乳酸要高于正常鸟类，在中脑偏下部分的含量增加是最为明显的，当症状产生时，最先表现出来的也是这一位置。当用一定量的维生素 B_1 浓缩物进行治疗后，乳酸浓度增加的现象就消失了。这种症状的产生似乎与乳酸的积聚相关，而且它局部发作的事实也说明维生素 B_1 与碳水化合物的中间代谢过程是密切相关的，显然是因为维生素 B_1 的氧化消耗了乳酸。在谈到这方面的联系时，我们有必要谈及矢追的研究工作，他发现患有多神经炎的鸽子的肌肉代谢亚甲基蓝的能力小于正常

that there was no difference in the glutathione contents[9]. Peters in his Harben Lectures has adduced some evidence that vitamin B_3 may be concerned with the mobilisation of water, and that in its absence together with that of vitamin B_1 oedema accompanies the polyneuritis in its terminal stages.

C. W. Carter and A. N. Drury have examined the nature of the slowing of the heart beat in rice-fed pigeons[10]: it appears to be due to an overaction of the vagal centres producing a heart block. The condition is cured by whole wheat, so that the factor responsible may be that described by Williams and Waterman.

G. F. Marrian, L. C. Baker, J. C. Drummond, and H. Woollard[11] noticed changes in the adrenal glands of pigeons starved or fed on rice only, and Marrian has investigated these alterations in more detail[12]. Hypertrophy was found in inanition, even though vitamin B_1 was given, and in vitamin B deficiency, whether accompanied or not by inanition. Oedema accounted for half the hypertrophy in inanition. The adrenaline content was increased in the latter condition, but was relatively low in vitamin B deficiency. It appeared that the hypertrophy in inanition affected chiefly the medulla, and in vitamin B deficiency, the cortex of the gland.

It is now well known that vitamin B deficiency is associated with loss of appetite. B. Sure has made a detailed study of the anorexia in the rat and found that it is promptly cured by the administration of a vitamin B concentrate[13]. The loss of appetite may be associated with the failure of the gut to empty itself, and a decrease in the digestive secretions. J. L. Rosedale and C. J. Oliveiro[14] found that in pigeons suffering from beri-beri the pancreas failed to form the enzymes required to digest protein and fat.

It might be expected that animals suffering from vitamin B deficiency would show derangements of the sexual function. H. M. Evans, however, found that in male rats, provided vitamin E was supplied, fertility was unaffected and sex interest was decreased only a few days before death[15]. In the female rat the oestrous cycle stopped abruptly after about four weeks on the deficient diet; loss of weight followed immediately[16]. Injections of oestrin produced the signs of oestrus during the anoestrus, but without stimulating the ovaries, which had become much atrophied.

W. Nakahara and E. Sanekawa have found that chicken sarcoma and rat sarcoma and carcinoma do not apparently require vitamin B_1, and contain little of it[17]. In the first set of experiments, chickens were fed on polished rice and a salt mixture; the livers from healthy birds, and those carrying growths of the Rous sarcoma, were found to contain equal amounts of vitamin B by test on rats, indicating that the tumour did not deplete the birds' store of vitamin. In the second set, the rat tumours were fed to pigeons and rats maintained on vitamin B free diets: only minimal amounts of the vitamin were found to be present.

(**127**, 204-205; 1931)

个体；但是两者的谷胱甘肽含量并没有明显的差别 [9]。彼得斯在他的哈本演讲中已经举出一些证据证明维生素 B_3 可能和水的代谢有关；当它和维生素 B_1 同时缺失时，在多神经炎发生的末期，会伴有水肿症状的出现。

卡特和德鲁里研究了采用稻米喂养的鸽子心跳会放缓的本质 [10]。它似乎与迷走中枢神经的过度反应导致了心传导阻滞相关。喂食全麦可以治愈这种病，因此发挥作用的因子可能就是威廉斯和沃特曼曾经描述过的那种物质。

马里安、贝克、德拉蒙德以及伍拉德 [11] 注意到了饥饿的或只喂养稻谷的鸽子在肾上腺器官上的一些变化，马里安对这种改变在更深的层面上进行了研究 [12]。研究发现在饥饿的情况下鸽子的肾上腺变得肥大，即使补充了维生素 B_1 仍然如此；而当维生素 B 缺乏时，无论是否伴随有饥饿，都会出现肾上腺肥大的症状：在饥饿时，水肿是造成肾上腺肥大的部分原因。在饥饿的情况下，肾上腺素的含量会增加；但在维生素 B 缺乏时，其含量会相对较低。研究结果显示，似乎在饥饿的时候，肾上腺肥大主要影响的是肾的髓质；而当维生素 B 缺乏时，肾上腺肥大主要影响的是肾的皮质。

现在大家都知道，维生素 B 的缺乏会导致食欲的减退。休尔仔细研究了大鼠食欲减退的情况，他发现服用维生素 B 浓缩物能很快治愈食欲减退 [13]。食欲的减退也许与肠道无法有效排空和消化液分泌减少有关。罗斯黛尔和奥利维罗 [14] 发现，对于患有脚气病的鸽子，其胰腺无法产生消化蛋白和脂肪所需的酶。

我们可以认为，患有维生素 B 缺乏症的动物会出现性功能的紊乱。然而，埃文斯发现，对于雄性大鼠，在给予维生素 E 的前提下，其生育能力不会受到影响，其性欲只有在临死的前几天才会发生减退 [15]。对于雌性大鼠，用缺乏维生素 B 的食物喂养大概四周之后，其发情周期会突然停止；接着很快出现体重下降 [16]。注射雌激素可以诱发处于不动情期的雌性大鼠发情，但不会刺激已经发生了严重萎缩的卵巢。

中原麻衣和实川发现，小鸡肉瘤和大鼠肉瘤以及癌症的发生与维生素 B_1 的存在与否没有明显的关联，肿瘤中只含有极少量的维生素 B_1[17]。在第一组实验中，采用精制大米和盐的混合物喂养小鸡；在对大鼠的实验中发现：健康小鸡的肝脏和长有鲁斯肉瘤的肝脏含有相同量的维生素 B，从而说明肉瘤的存在不会耗尽小鸡体内贮存的维生素。在第二组实验中，用大鼠肉瘤来喂养鸽子，该大鼠持续摄取不含维生素 B 的食物；结果发现只有很少量的维生素出现在喂给鸽子的食物中。

（刘振明 翻译；刘京国 审稿）

References:

1. Plimmer, R. H. A., Raymond, W. H., Lowndes, J., and Rosedale, J. L., *Biochem. Jour.*, **21**, 1141 (1927): **23**, 545 (1929).

2. Bacharach, A. L., and Allchorne, E., *Biochem. Jour.*, **22**, 313 (1928).

3. Rowlands, M. J., and Wilkinson, B., *Biochem. Jour.*, **24**, 199 (1930).

4. Peskett, G. L., *Biochem. Jour.*, **21**, 1102 (1927).

5. Fridericia, L. S., Chick, H., and Roscoe, M. H., *Lancet*, **1**, 37 (1928).

6. Aykroyd, W. R., and Roscoe, M. H., *Biochem. Jour.*, **23**, 483 (1929).

7. Hartwell, G. A., *Biochem. Jour.*, **22**, 1212 (1928).

8. Kinnersley, H. W., and Peters, R. A., *Biochem. Jour.*, **23**, 1126 (1929): **24**, 711 (1930).

9. Yaoi, H., *Proc. Imp. Acad. Tokyo*, **4**, 233 (1928).

10. Carter, C. W., and Drury, A. N., *Jour. Physiol.*, **68**, *Proc.*, p. i. (1929).

11. Marrian, G. F., Baker, L. C., Drummond, J. C., and Woollard, H., *Biochem. Jour.*, **21**, 1336 (1927).

12. Marrian, G. F., *Biochem. Jour.*, **22**, 836 (1928).

13. Sure, B., *Jour. Nutrition*, **1**, 49 (1928).

14. Rosedale, J. L., and Oliveiro, C. J., *Biochem. Jour.*, **22**, 1362 (1928).

15. Evans, H. M., *Jour. Nutrition*, **1**, 1 (1928).

16. Parkes, A. S., *Quart. Jour. Exp. Physiol.*, **18**, 397 (1928).

17. Nakahara, W., and Sanekawa, E., *Proc. Imp. Acad. Tokyo*, **5**, 55 (1929): **6**, 116 (1930): *Scient. Pap. Instit. Physic. and Chem. Res.*, **10**, 211 (1929).

Protein Structure and Denaturation

C. Rimington

Editor's Note

It was known since the nineteenth century that heat, acidity or chemical reagents could make soluble proteins coagulate, a phenomenon called denaturation. But no one knew what caused this change. Here industrial chemist Claude Rimington ponders the question in the light of two recent findings: the changes in structure of keratin, the main component of wool, when stretched, and the observation that the denaturing of haemoglobin can be reversed. Previously, denaturation was thought to be one-way, as it is for boiled egg white (albumin). Rimington offers the first intimations that denaturation is a crucial aspect of the issue of protein folding, the process in which a peptide chain reversibly collapses into its enzymatically active form with a specific three-dimensional shape.

ASTBURY and Woods' fundamental work upon the micellar structure of the protein of wool fibres,[1] and the hypothesis they put forward as an explanation of the changes observed in the X-ray pattern when such fibres are stretched, would seem to be full of significance for protein chemistry in general.

Within the last ten years, different lines of evidence have been converging upon the view that some regularity, as regards pattern and molecular size, underlies the disordered confusion of data we possess relating to the proteins of the animal and vegetable kingdoms. The two most striking demonstrations in recent years of such uniformity are afforded by Svedberg's brilliant application of the ultracentrifuge to determine the particle mass of soluble proteins,[2] classes of "molecular weight" 1, 2, 3, and 6 times the common factor 34,500 being distinguished, and Gorter and Grendel's demonstration[3] that under appropriate conditions soluble proteins exhibit the phenomenon of surface spreading on liquids, and that all occupy the same surface area irrespective of particle mass (1, 2, 3, or 6 times 34,500). Using Svedberg's common factor 34,500 for the basis of their calculations, the Dutch workers obtain a value for the radius of the unit particle (22.5 A.) identical with that determined by Svedberg experimentally.

The most significant feature of Gorter and Grendel's work, however, is that their results imply a loosening, brought about by the surface forces, of the cohesive attraction holding the units of the aggregated proteins together. Astbury and Woods' investigations reveal a somewhat similar, although internal, deformation of the keratin structure of the wool fibre, brought about by purely physical means. Our conceptions of the chemical reactivity of protein structures clearly need revision in an attitude of greater attention to modern valence conceptions.

蛋白质的结构与变性

编者按

从19世纪开始人们就已经认识到热、酸或某些化学试剂可以导致可溶性蛋白质凝固，这种现象被称为蛋白质的变性。但没有人知道为什么会发生这种变化。工业化学家克劳德·里明顿根据最近的两个发现思索这个问题的答案：一个发现是羊毛的主要成分——角蛋白在被拉伸时会发生结构变化；另一个发现是血红蛋白的变性可以逆转。以前人们认为变性过程是单向的，就像煮熟的蛋白（白蛋白）无法再回到煮之前的状态。里明顿第一个指出变性是蛋白质折叠问题的重要方面，在蛋白质折叠过程中，肽链可逆地折叠成具有特定三维形状并可行使酶功能的活性形式。

在我看来，阿斯特伯里和伍兹在羊毛纤维蛋白胶束结构上的基础性工作[1]，以及他们为解释在羊毛纤维被拉伸时从X射线图谱中观察到的结构变化而提出的假说，对于蛋白质化学的总体发展具有非常重大的意义。

在过去的 10 年里，来自各方面的证据都表明这样一个观点：在我们研究动植物蛋白时得到的扑朔迷离的数据背后，显示出衍射图谱与蛋白质分子大小之间存在着某种规律性的联系。近年来，有两个最显著的成果证明了这种一致性。一个是斯韦德贝里用超速离心机成功地测定了可溶性蛋白粒子的质量[2]，在他的研究中"分子量"是按照公因子 34,500 的 1、2、3、6 倍进行分级的；另一个是戈特和格伦德尔发现[3]可溶性蛋白在适宜条件下可在液面铺展的现象，而且它们占据相同的表面积，不受粒子质量（34,500 的 1、2、3 或 6 倍）大小的影响。荷兰的研究人员以斯韦德贝里的公因子 34,500 为基础计算出了单位粒子的半径值（22.5 Å），这一结果与斯韦德贝里用实验测得的结果相同。

然而，在戈特和格伦德尔的研究成果中最重要的意义是：他们的结论暗示促使蛋白颗粒单元聚集在一起的内聚力会因表面力的作用而减弱。阿斯特伯里和伍兹在应用纯物理学方法研究羊毛纤维角蛋白结构时也发现了一些发生在蛋白质内部的类似结构变化。显然，随着人们对现代价态理论关注度的提高，我们需要对蛋白质结构的化学反应性的概念进行修正。

One more point cannot be too clearly emphasised which is common to the essential findings of Svedberg, Astbury and Woods, and Gorter and Grendel: the changes observed by these workers are strictly *reversible*.

In conclusion, I should like to touch upon the problem of protein denaturation, and to inquire whether it is not in the direction of such work as that of Astbury and Woods that we have to look for a solution of this problem? Denaturation of proteins, which can be brought about by mechanical as well as by chemical forces, is characterised by a loss of solubility at the isoelectric point. It was always thought to be an irreversible change, but Anson and Mirsky[4] have recently demonstrated its reversible nature in the case of globin. Some internal alteration takes place during denaturation, as evidenced by the change in reactivity of the sulphur groups,[5] but neither acid or base binding capacity[6] nor osmotic pressure[7] are affected — that is to say, there is no scission. Clearly, loss of isoelectric solubility must be due to change in some internal tautomeric configuration.

It is difficult to avoid the suggestion that a change, similar to that postulated by Astbury and Woods in explanation of the behaviour of the stretched and unstretched wool fibre, may in reality be the essential happening attending denaturation. The –CO–NH– group possesses strong polarity, but, by the rearrangement of peptide linkages into what are virtually closed ring systems, affinity for water would be enormously diminished. At present there exists no satisfactory hypothesis offering an explanation of denaturation. Such a scheme as the above may reasonably be entertained until further evidence can be brought forward of a chemical or physico-chemical nature which will throw more light upon the problem. Considering the remarkable and wholly unexpected results, mentioned above, of Gorter and Grendel, working upon protein surface films, it would seem that quantitative data bearing upon denaturation is likely to be obtained most readily by studies having a similar approach. The forces at play within the liquid and at the interface possess no mean magnitude. They are, however, susceptible of more precise control and exact manipulation than those involved in, let us say, heat coagulation or the application of vigorous chemical reagents. From a study of the surface phenomena exhibited by proteins under varying conditions, coupled possibly with an application of the X-ray method to films of such proteins as can be made to give readily detectable diffraction photographs,[8] a solution not only of the denaturation process but also of the structure of native proteins may, in the future, be obtained.

(**127**, 440-441; 1931)

C. Rimington: Biochemical Department, Wool Industries Research Association, Leeds, Feb. 13.

References:
1. Astbury and Woods, *Nature*, **126**, 913 (1930).
2. Svedberg, *Koll. Zeit.*, **51**, 10 (1930).

还有一个无论如何强调都不为过的观点，也是斯韦德贝里、阿斯特伯里和伍兹以及戈特和格伦德尔这三个研究小组各自重要发现的共同点：他们观察到的蛋白质结构的变化，严格说来都是**可逆**的。

最后，我想谈一谈蛋白质变性这个问题，探讨一下此问题是否与阿斯特伯里和伍兹的研究方向不一致，因而我们必须为解决这个问题另找一个答案？机械力或化学力可能会引起蛋白质变性，变性的标志是蛋白质在等电点处丧失水溶性。人们以前一直认为蛋白质变性是不可逆的，但安森和米尔斯基 [4] 最近在研究球蛋白时发现蛋白质的变性是可逆的。在变性过程中蛋白质分子内部会发生一些改变，例如含硫基团反应活性的变化 [5]，但其酸碱结合能力 [6] 和渗透压 [7] 都不会改变，也就是说，蛋白质分子中的肽键没有发生断裂。很明显，蛋白质在等电点处水溶性的丧失必定与其内部互变异构体构型的改变有关。

一个不容忽视的观点是，变化，类似于阿斯特伯里和伍兹在解释拉伸状态和非拉伸状态的羊毛纤维时所提到的变化，实际上可能是伴随蛋白质变性的基本事件。蛋白质的肽键，即 –CO–NH– 基团具有很强的极性，但是在经过肽链的重排，形成几乎封闭的环状结构后，其亲水性会大大降低。目前还没有一个能够圆满解释变性现象的假说。我们有理由暂且接受上述说法，直到能提供更多证据揭示蛋白质的化学或物理化学特性，才能更好地阐明这一问题。前面提到戈特和格伦德尔在研究蛋白表面膜时得出的一些值得关注并且完全出乎意料的结果，表明通过类似的研究方法将可能很容易地获得有关蛋白质变性的定量数据。尽管在液体内部和界面处的蛋白质间的相互作用力是不弱的。然而，这种相互作用力比热凝聚力或强化学试剂的作用力更易被严格控制和准确操纵。通过研究一些蛋白质在不同条件下的表面特性，结合应用 X 射线法获得这些蛋白膜的衍射图像 [8]，我们就可能在不久的将来揭示出蛋白质的变性过程，或许还能破解天然蛋白的结构。

（韩玲俐 翻译；周筠梅 审稿）

3. Gorter and Grendel, *Proc. Acad. Sci. Amsterdam*, **32**, 770 (1929).

4. Anson and Mirsky, *J. Gen., Physiol.*, **13**, 469 (1930).

5. Harris, *Proc. Roy. Soc.*, B, **94**, 426 (1923).

6. Booth, *Biochem. J.*, **24**, 158 (1930).

7. Huang and Wu, *Chinese J. Physiol.*, **4**, 221 (1930).

8. Ott, *Kolloidchem. Beih.*, **23** , 108 (1926).

The Molecular Weights of Proteins

W. T. Astbury and H. J. Woods

Editor's Note

Although the importance of proteins in living things was widely recognised in the 1930s, there were only rudimentary ideas of how these molecules were constructed. William Astbury at the University of Leeds had made a special study of natural proteins, such as the keratin of which hair is made, using X-ray diffraction. As he explains here, his attention had been captivated by the work of The Svedberg in Uppsala in Sweden, who had developed an ultra-centrifuge for measuring the molecular weights of complex molecules. Astbury and Henry John Woods, his colleague, used this as a starting point for an entirely speculative (and mistaken) account of how protein molecules in general might be constructed.

ONE of the most satisfactory features of recent advances in the X-ray analysis of compounds of high molecular weight has been the degree of co-ordination between the efforts of the structure analyst and those of the chemist. Especially is this true in the case of investigations of the structure of cellulose and its derivatives. The question of protein structure, however, appears to bring in its train problems of quite another order of complexity, and it does not seem to be at all clear what is connoted by the phrase "molecular weights of proteins". Such X-ray photographs of fibrous proteins as have been obtained point to the periodic repetition of comparatively simple units with imperfect or variable side-linkages. In the quest for chemical data to correlate with these results, the crystallographer is at once brought up against the remarkable observations of Svedberg, that there are groups of soluble proteins of "molecular weights" which are simple multiples of 34,500. The present situation is most simply described by quotations from two recent letters[1, 2] to *Nature*: —

1. "The two most striking demonstrations in recent years of such uniformity are afforded by Svedberg's brilliant application of the ultracentrifuge to determine the particle mass of soluble proteins, classes of "molecular weight" 1, 2, 3, and 6 times the common factor 34,500 being distinguished, and Gorter and Grendel's demonstration that under appropriate conditions soluble proteins exhibit the phenomenon of surface spreading on liquids, and that all occupy the same surface area irrespective of particle mass (1, 2, 3, or 6 times 34,500). Using Svedberg's common factor 34,500 for the basis of their calculations, the Dutch workers obtain a value for the radius of the unit particle (22.5 A.) identical with that determined by Svedberg experimentally."

2. "Three determinations of the sedimentation equilibrium of insulin at a pH of 6.7–6.8 gave as a mean value for the molecular weight 35,100, which within the limits of experimental error is the same as that for egg albumin, 34,500, and for Bence Jones

蛋白质的分子量

威廉·阿斯特伯里，亨利·约翰·伍兹

编者按

尽管在 20 世纪 30 年代蛋白质在生物体中的重要作用就得到了广泛的认可，但当时人们对于蛋白质分子是如何组成的还只有一些初步的概念。利兹大学的威廉·阿斯特伯里应用 X 射线衍射法对包括构成头发的角蛋白等多种天然蛋白进行了专门的研究。阿斯特伯里在这篇文章中指出，瑞典乌普萨拉的斯韦德贝里发明了一种能够测量复杂分子分子量的超速离心机，这一工作引起了他的注意。阿斯特伯里和他的同事亨利·约翰·伍兹正是以此为出发点推测了（尽管是错误地）总体上蛋白质分子的可能构建方式。

在利用 X 射线衍射方法分析高分子化合物方面的最新进展中，其中一个最令人满意的方面是结构分析学家和化学家之间的高度合作，尤其是对纤维素及其衍生物结构的分析。然而，在对蛋白质结构问题的研究中似乎还会遇到一系列另一种层面上的难题，并且我们也根本不清楚"蛋白质分子量"一词的确切内涵。已经得到的纤维蛋白的 X 射线照片显示，蛋白质是由一些相对比较简单的单元通过周期性重复构成的，这些单元带有一些可变的侧链基团。在寻找与这些结果相对应的化学数据时，晶体学家要想办法解释斯韦德贝里的怪异实验结果：许多可溶性蛋白质的"分子量"都是 34,500 的整数倍。对目前研究情况的最简单的描述可以从《自然》杂志的两篇快报文章 [1, 2] 中得知：

1."近年来关于这种一致性的两项最激动人心的成果分别来自斯韦德贝里与戈特和格伦德尔。斯韦德贝里巧妙地用超速离心机测定了可溶性蛋白质的分子量，并按"分子量"是公因子 34,500 的 1、2、3、6 倍而对它们进行了分级。戈特和格伦德尔的研究则表明，可溶性蛋白质在合适的条件下会铺展于液体表面，此时不管蛋白质的分子量有多大（公因子 34,500 的 1、2、3、6 倍），它们都占据相同的表面积。此外，荷兰的研究者以斯韦德贝里得到的公因子 34,500 为计算的基础，测算出了单元粒子的半径值（22.5 Å），这一数值与斯韦德贝里通过实验测定的结果完全一致。"

2."在 pH 值为 6.7 ~ 6.8 的条件下，采用沉降平衡法对胰岛素的分子量进行测定，三次测定结果的平均值表明该蛋白的分子量为 35,100，考虑到实验误差范围，可以认为这一数值和其他一些蛋白质的分子量是相同的，比如，卵清蛋白的分子量为 34,500，

protein, 35,000... The sedimentation equilibrium determinations show that crystalline insulin is homogeneous with regard to molecular weight, that is, the molecules in the sample studied were all of the same weight."

If now we consider this problem from the purely crystallographic point of view—and it has been demonstrated that proteins under certain conditions can give rise to X-ray crystal photographs—the numbers 1, 2, 3, and 6 immediately invite attention as being possible numbers of "molecules" which can go to form a unit of pattern. The suggestion thus arises that, provided we can explain the occurrence of the weight 34,500, the rest may be merely another aspect of that grouping of molecules which is called crystalline. But if this is so, we have to account for the non-occurrence of the number 4, and the explanation of this gap must be given in terms of some outstanding characteristic of proteins in general.

In order to explain the sequence of numbers observed, it does not seem necessary to invoke anything more unfamiliar than the ordinary peptide chain, $-CO-NH-CHR-$ $CO-NH-CHR-$, which is built up of a succession of triads of which the $-CO-$ and $-NH-$ groups are unsaturated; for if we postulate that the $-CO-$ and $-NH-$ groups of neighbouring chains can be linked together by secondary valences, the following simple crystallographic combinations[3] are at once available (Figs. 1 a, 1 b, 1 c, 1 d).

Fig. 1 a — Space Group C_1.
Fig. 1 b — C_2^1.
Fig. 1 c — C_3^2.
Fig. 1 d — D_3^3 & D_3^4.

Fig. 1.

In Fig. 1 b, corresponding to the crystallographic space-group C_2^1, the unit of pattern is a pair of chains pointing in opposite directions, while the basis of Fig. 1 c, space-group C_3^2, is a self-contained threefold screw of chains all pointing in the same direction. In Fig. 1 d one of the chains has been omitted to avoid confusion, but it will be seen that it is a grouping which is a combination of (b) and (c) based on the space-groups D_3^3 and D_3^4, and is also a self-contained threefold screw, but this time not of single chains, but of pairs of chains such as are shown in Fig. 1 b. All these molecular associations are well-defined crystallographic types—the arrangement shown in Fig. 1 d, for example, corresponds to the structure of such a common crystal as quartz—which might be expected to undergo reversible dissociation into their constituent units, or sub-groups. That such a process actually does take place is best illustrated by the words[4] of Svedberg himself: —"The protein molecules containing more than one group of weight 34,500 are, as a rule,

306

本周蛋白的分子量为 35,000 等。沉降平衡的测定结果表明从分子量的角度上说结晶胰岛素是均一的，也就是说，被测样品中的所有分子都具有相同的分子量。"

如果现在我们完全从晶体学角度来思考这个问题，或者说我们已经证实蛋白质在某些条件下可以产生 X 射线晶体衍射图，那么我们会立刻联想到上述数字 1、2、3、6 可能就是能够产生某种衍射图样单元的"分子"的个数。如果这种猜想成立，而且若是我们能够解释分子量为 34,500 的结构单元的存在，那么剩下的可能就只是晶体分子中单元结构如何分组的问题了。但是，即使这样，我们还是必须说明数字 4 为何没有出现，对这个缺口的解释必须考虑蛋白质总体上的一些突出特点。

为了解释观测到的数字序列，我们没有必要引入任何比普通肽链更不熟悉的东西。我们知道，肽链 –CO–NH–CHR–CO–NH–CHR– 是由一系列三联体组成的，其中的 –CO– 和 –NH– 是不饱和基团。如果我们假设相邻肽链中的 –CO– 和 –NH– 可以通过副价键连接起来，那么就可以立刻得到以下这些简单的晶体组合样式 [3]（图 1*a*、1*b*、1*c*、1*d*）。

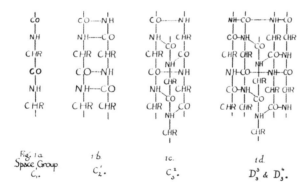

图 1

在图 1*b* 中，样式单元是方向相反的一对肽链，这对应于晶体学上的空间群 C_2^1，而图 1*c* 中的样式单元是由三条方向一致的肽链组合成的独立的三螺旋，这对应于晶体学上的空间群 C_3^2。在图 1*d* 中，为避免过于混乱我们省略了其中的一条肽链，但可以看出这是在空间群 D_3^3 和 D_3^4 的基础上形成的一种晶体组合样式，它结合了图 *b* 和图 *c* 的特点，其样式单元也是独立的三螺旋，不过每股螺旋并不是单条肽链，而是如图 1*b* 中所示的三对肽链。所有这些分子缔合都有明确的晶型。比如，图 1*d* 所示的结构排列就是一种非常常见的晶型结构——石英，也许可以认为这种晶型结构能够可逆地解离成其组成单元，或者叫亚基。斯韦德贝里本人的一段话 [4] 很好地描述了实际发生的这种过程："一般来说，当溶液 pH 升高到超过某一特定值时，由多个分子

dissociated into molecules of lower numbers of groups of 34,500 when the pH of the solution is raised over a certain value. Thus the proteins of weight (6×34,500) split up into molecules of 1/2, 1/3, and 1/6 of the original molecule, but never into molecules of 1/4 or 1/5 of the original. This is in line with the fact that proteins possessing these latter weights at or near their isoelectric point have not been met with. At sufficiently high alkalinity all proteins have the same molecular weight, viz., 34,500."

The problem embodied in the last sentence still remains for discussion, and we should like to suggest as the interpretation of this, the most fundamental difficulty of all, that the observed constancy of unit molecular weight is simply a case of the vibrational instability of peptide chains when their length exceeds a certain value. If we accept the X-ray indications that the fibrous proteins, such as hair[5] and silk,[6] are based on the periodic repetition of comparatively simple units, then the probability of disruptive resonance occuring among the constituents of the peptide chain will continually increase with the length, so that excessively long chains would be liable to spontaneous decomposition into shorter chains. We may imagine some such process taking place in the laboratory of the living cell as the amino-acids are laid down in long chains at a surface and consolidated by crystallographic groupings in the manner suggested above, or by intra-molecular folding such as has been demonstrated in the case of wool and hair.[5] From this point of view, it does not seem likely that the unit molecular weight of proteins is strictly constant—this, too, is in agreement with experiment—but there is a strong probability that, given the appropriate conditions, many proteins will be based on a roughly constant weight of peptide chain.[*]

A phenomenon which appears to involve analogous reasoning is the decay of tension at constant length which takes place in stretched hair containing moisture, and which has been investigated by Speakman.[7] A large part of this loss of tension is quite permanent,[8] in spite of the fact that the stretched hair still retains its power of recovering at least its original length in water. The rate of decay of tension varies with the type of wool or hair and with the nature of the wetting agent, and increases with rise of temperature. It is extremely rapid in steam, a short treatment with which permanently alters the load/ extension curve, and so loosens the internal structure of the fibre that it may be caused to contract to two-thirds of its original length.[5]

It is clear, of course, that the wetting agent plays an important part in this permanent destruction of internal tension, but it seems not at all unlikely that vibrational instability

---

[*] If we assume the essential correctness of the structure proposed,[5] we may make an estimate of the length of peptide chain in animal hairs. The average molecular weight of the chief amino-acids in wool (which are present in roughly equal molecular proportions) is about 121, and three amino-acids occupy a length of 5.15 A. along the fibre axis. The length, corresponding to 34,500, is thus about 500 A. It is a striking fact that this is approximately the length which is the minimum possible to give the observed X-ray diffraction effects[9], That it is also near the actual length is indicated by the fuzziness which appears in X-ray photographs of hairs which have developed pronounced permanent decay of tension.

308

量为 34,500 的基团组成的蛋白质分子会降解为由较少分子量为 34,500 的基团组成的小分子。这样，分子量为 6×34,500 的蛋白分子就会分解为分子量只有原分子的 1/2、1/3 或 1/6 的小分子，但绝不可能产生分子量为原分子的 1/4 或 1/5 的小分子。这和实验事实是相吻合的，即在相应的等电点处及其附近并没有发现具有后两种分子量的蛋白质。在碱性足够强的条件下，所有蛋白质都具有相同的分子量，即 34,500。"

前段最后一句话涉及了一个仍有待讨论的问题，这是个最基本也最难的问题，我们倾向于这样来解释，实验中发现的单元分子量的恒定性，仅仅是由于当肽链长度超过某一值时产生的振动不稳定性造成的。如果我们认可从 X 射线衍射结果得到的推论，即认为组成头发 [5] 和蚕丝 [6] 等的纤维蛋白都是由相对简单的单元结构周期性重复构成的，那么，随着肽链长度的增加，组成肽链的各部分之间的共振造成肽链断裂的可能性也会增加，这样，过长的肽链肯定会自发地分解为短一些的肽链。我们可以设想发生在活细胞"加工厂"中的某些类似的过程，在合成蛋白质的细胞器的膜表面，氨基酸被添加到长的肽链上，然后按照前面所述的晶体组合样式，或者通过像羊毛和头发中那样的分子内折叠 [5] 组合在一起。从这种观点来看，蛋白质结构单元的分子量应该不会是严格恒定的，这和实验结果也是吻合的。但很有可能的是，在给定的适宜条件下，许多蛋白质将是由分子量大致恒定的肽链组成的。*

另一个可能由类似原因导致的现象是关于毛发张力消退的。斯皮克曼发现，被拉直的湿发在保持长度不变的同时张力会逐渐消退 [7]。尽管这些被拉直的头发至少在水中仍能恢复到它原来的长度，但大部分张力的损失是永久性的 [8]。张力消退的速度随着羊毛或头发的类型及浸润溶剂的性质的变化而变化，随着温度的升高而加快。在蒸汽中张力消退的速度非常快。一个短暂的蒸汽处理就可以永久性地改变毛发的载荷 – 伸展曲线，使纤维的内部结构变松弛，以至于可能使毛发的长度收缩到原长的 2/3[5]。

当然，现在已经很清楚，在内部张力被永久性破坏的过程中浸润溶剂起着重要的作用，但是，肽链振动的不稳定性可能也是这一过程中的一个必需因素。对拉直

---

* 如果我们假设前面提出的结构基本上是正确的[5]，那么我们就可以估计动物毛发中肽链的长度了。羊毛中主要氨基酸（这些氨基酸在羊毛中的分子比例基本相等）的平均分子量大约为121，另外知道3个氨基酸沿纤维轴方向的长度为5.15 Å，那么分子量34,500对应的长度大约为500 Å。令人吃惊的是，这大致等于能够观测到X射线衍射效应的最小长度[9]，同时也与由发生张力永久性显著消退的毛发的X射线衍射照片中的模糊衍射斑推测得到的毛发中蛋白质的实际长度相接近。

also is an essential factor in the process. After treatment of stretched hair with steam, the longitudinal swelling of the fibre in water is considerably increased, a fact which, taken in conjunction with the observation that X-ray photographs of hair which has been held stretched in water for several weeks show a definite fuzziness of the reflections associated with the length of the peptide chains, suggests that the average length of the chains is decreased by sustained tension in the presence of water.

We have recently commenced an investigation of the influence of radiations, such as ultra-violet light and X-rays, on the elastic and other properties of animal hairs, so that in this connexion it is convenient to mention here some remarkable observations which we have made on *unstretched* wool exposed for some sixty hours to the full beam of a Shearer X-ray tube (copper anticathode). After this treatment the fibres show many of the properties which are characteristic of wool which has been exposed *in the stretched state* to the action of steam. For example, they have the property of contracting in steam by as much as 37 percent below their unstretched length, and their longitudinal swelling in water after steaming is found to be increased from the 1 percent of normal wool to as much as 10 percent. This seems to be a clear case of the disruptive action of high-energy quanta on the length and cohesion of peptide chains, and must be closely related to the influence of various radiations on biological activity.

These experiments are being continued and will be reported in detail in due course.

(**127**, 663-665; 1931)

W. T. Astbury, H. J. Woods: Textile Physics Laboratory, University, Leeds, Mar. 27.

References:

1. Rimington, C., *Nature*, **127**, 440 (1931).

2. Svedberg, T., *Nature*, **127**, 438 (1931).

3. Astbury, W. T., and Yardley, K., *Phil. Trans. Roy. Soc.*, A, **224**, 221 (1924). See Plate 5 (1), (7), Plate 16 (144), Plate 18 (157 and 158).

4. Svedberg, T., *Trans. Faraday Soc.*, *General Discussion on Colloid Science applied to Biology*, 741 (1930).

5. Astbury, W. T., *J. Soc. Chem. Ind.*, **49**, 441 (1930): *J. Textile Science*, **4**, 1 (1931). Astbury, W. T., and Woods, H. J., *Nature*, **126**, 913 (1930). Astbury, W. T., and Street, A., *Phil. Trans. Roy. Soc.*, A, **230**, 75 (1931).

6. Brill, R., *Ann. Chem.*, **434**, 204 (1923). Meyer, K. H., and Mark, H., *Ber.*, **61** (1932; 1928). Kratky, O., *Z. phys. Chem.*, **B5**, 297 (1929).

7. Speakman, J. B., *Proc. Roy. Soc.*, B, **103**, 377 (1928).

8. Speakman, J. B., *Trans. Farad. Soc.*, **25**, 169 (1929).

9. Hengstenberg, J., and Mark, H., *Zeit. f. Krist.*, **69**, 271 (1928).

的头发进行蒸汽处理后，该纤维在水中的纵向延伸显著增加。另外，在水中被拉伸了几个星期的头发的 X 射线衍射照片显示，存在一个与肽链长度有关的边界明确的模糊图案。以上这些事实提示我们在水中毛发肽链平均长度的减少可能是由于持续的张力引起的。

我们最近开始了一项关于紫外线、X 射线等辐射对动物毛发弹性和其他特性的影响的研究，这里我们要提及一些相关的值得注意的实验结果。我们对在希勒 X 射线管（铜对阴极）的最大强度束下照射约 60 个小时的**未被拉伸**的羊毛进行了观察。结果发现，经过这样处理后的羊毛纤维表现出了很多与**拉伸状态下**接受蒸汽处理的羊毛一样的特征。例如，它们在蒸汽中都具有长度可以收缩到比未被拉伸时短 37% 的特性，另外，蒸汽处理后它们在水中纵向的延伸都会增加，增加幅度从普通羊毛的 1% 到 10% 不等。这也许就是高能量子对肽链长度和内聚性具有破坏作用的实例，各种辐射对生命活动的影响一定与此密切相关。

这些实验正在进行之中，详细的实验结果将在适当的时候公布出来。

（李飞 翻译；周筠梅 审稿）

Cytological Theory in Relation to Heredity*

C. D. Darlington

Editor's Note

Cyril Dean Darlington was trained as a botanist but converted himself into a cytologist after it had been recognised that chromosomes, located in the cell nuclei of plants and animals, carry genes that to determine the organisation of living things. This article is an attempt to explain the distinction between the two forms of cell division that occur in living things: the division of body cells into two essentially identical cells, called mitosis, which is the basis for the growth of tissues in plants and animals, and the process of meiosis, which gives rise to cells containing half as many chromosomes as the original and which is the means by which germ cells are produced (in animals as well as plants). At this early stage in the development of genetics there were no known means of identifying of even locating the genes carried on the chromosomes.

THE chromosome theory of heredity, by relating chromosome behaviour with the phenomena of inheritance, has obviously made it possible to apply the cytological method to the study of inheritance. With this profitable field before them, geneticists and cytologists have not hesitated to draw conclusions in the one field from observations made in the other, but in order to do so they have had to apply certain rules of interpretation. Their method has naturally been to assume, so far as possible, a direct relationship between cytological and genetical observations. The geneticist has therefore not only assumed that the material of every part of the chromosome has a specific genetic effect, which is a widely verified assumption; but also that the capacity of the chromosome for variation is equally specific, so that it is possible to refer to hereditary differences and to particles of chromosome alike as "genes". This second assumption is also widely verified; but it is subject to serious exceptions in that two different kinds of change have been shown to befall the same particle, namely, internal change and external change such as loss or re-arrangement. This constitutes no primary objection to the theory of the gene but rather indicates a necessary enlargement of its scope.

Cytologists, on the other hand, in translating their observations into genetical terms, have sought to apply the chromosome theory to the interpretation of meiosis. With the help of the simple rule that the pairing of chromosomes is a criterion of their relationship, they have set to work to examine meiosis in hybrids and in ring-forming plants (such as various species of *Oenothera*). The results of these studies have been confusing because investigators have not first examined the principles they were applying to see if they were indeed principles or merely empirical rules of special derivation and therefore of limited application. We now have evidence by which to test them.

---

* Substance of three lectures given at the Royal Institution on Mar. 10, 17, and 24.

遗传的细胞学理论 *

西里尔·迪安·达林顿

编者按

西里尔·迪安·达林顿原本是要被培养成为一名植物学家的，然而当人们认识到位于植物和动物细胞核中的染色体是携带基因的载体，基因决定生物的组织结构之后，他便转向这一领域，成为了一名细胞学家。本文试图阐释生物体内两种细胞分裂形式之间的差别：即有丝分裂和减数分裂。前者是指体细胞分裂成两个含有相同遗传信息的细胞，是动植物组织生长的基础；而后者分裂得到的细胞所含的染色体数目只有原来的一半，是动植物产生生殖细胞的方式。那个时候还处在遗传学发展的早期阶段，尚未出现可以鉴别基因的方法，甚至连染色体上基因的定位都无从下手。

遗传的染色体理论将染色体行为与遗传现象联系起来，这使得将细胞学方法应用于遗传研究成为可能。在这一回报丰厚的领域里，遗传学家和细胞学家们可以根据一个领域内的观察结果毫不迟疑地推出另一个领域中的结论，但为了能够顺利地进行这一工作，就必须建立某些解释规则。他们的方法自然是尽可能地设想出细胞学和遗传学观察二者之间存在的直接关联。因此遗传学家不仅假设染色体的每部分都具有特定的遗传作用（该假设已被广泛证实），而且还假设染色体具有特定的变异能力，因此就可以将遗传差异和诸如染色体颗粒之类称为"基因"。第二个假设也已被广泛证实，但存在严重异常的情况，即缺失或重排等内部变化和外部变化发生在同一颗粒上时。但是这些例外不足以构成反对基因学说的条件，相反表明该学说的作用范围需要进一步拓展。

另一方面，当细胞学家用遗传学词汇描述他们的观察结果时，他们已经是在尝试用染色体学说来解释减数分裂了。在这一以染色体配对为关系准则的简单规则的指导下，细胞学家们开始研究杂交体及成环植物（例如月见草属的各种植物）的减数分裂情况。这些研究的结果是令人怀疑的，因为研究者们并没有首先验证他们所参照的原则，以证实这些原则确实是真正的法则，还是仅仅是根据特定情况推导出来的适用范围有限的经验法则。现在，我们有了可以检验它们的证据。

---

* 本文是作者分别于 3 月 10 日、17 日和 24 日在英国皇家科学院所发表的演讲的主要内容。

Meiosis consists in the occurrence of two successive divisions of a nucleus in the course of which the chromosomes divide once instead of twice as they would in two ordinary mitoses. Where the distribution of the chromosomes is regular, the four daughter nuclei therefore have half the number of chromosomes of the parent nucleus (Fig. 1).

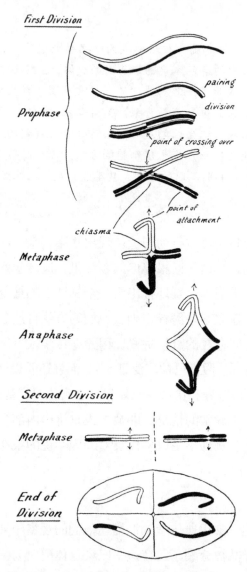

Fig. 1. Diagram to show the development of one pair of chromosomes at meiosis, and their relationship on the assumption that crossing over is the cause of chiasma formation. The four stages of prophase shown are: (1) leptotene, (2) pachytene before division, (3) pachytene after division, (4) diplotene to diakinesis.

At the first division, the chromosomes come together in pairs, and a whole chromosome of each pair passes to one pole to divide at the second division of the nucleus. To express this comparatively with regard to mitosis, we may say that while two half-chromosomes (or "chromatids") are associated in pairs at a mitosis, four are associated at the first metaphase

在减数分裂过程中，细胞核连续分裂两次而染色体只分裂一次，这与有丝分裂不同，在两次正常的有丝分裂过程中，细胞核分裂两次染色体也分裂两次。因而当染色体正常分配时，减数分裂产生的四个子细胞核具有母细胞核染色体数的一半（图1）。

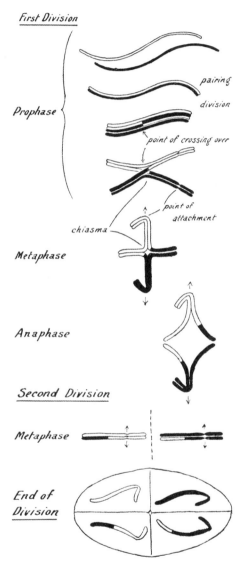

图 1. 一对染色体发生减数分裂的过程及这对染色体的关系示意图（假定形成交叉的原因是基因交换）。
前期包括的四个阶段是:(1) 细线期,(2) 分裂前的粗线期,(3) 分裂后的粗线期,(4) 双线期至终变期。

第一次减数分裂时，染色体成对地聚集在一起，每对中都有一整条染色体移向细胞的一极，在细胞核第二次分裂时，该条染色体将发生分离。如果要以与有丝分裂进行对比的方式来表达，我们可以说：在有丝分裂过程中是两条"半染色体"（或

of meiosis. A numerical reduction in the chromosomes must be attributed directly to the lack of any splitting of the chromosomes in the interval between the two divisions of the nucleus such as ordinarily occurs. But this is readily related to the fact that each chromosome is already split into the two chromatids which have passed together to one pole. This in turn is related to the pairing of the chromosomes.

It has therefore seemed natural (since 1890) to regard the essential difference between meiosis and mitosis as consisting in the pairing of the chromosomes. Since different pairs of chromosomes pass at random to the two poles (so that A_1–A_2 and B_1–B_2 may give daughter nuclei A_1B_1 and A_2B_2 or, equally, A_1B_2 and A_2B_1), and since the chromosomes are qualitatively differentiated, it follows that those which pair and pass to opposite poles must be similar if meiosis is to yield similar reduced nuclei (Boveri). Clearly, likeness is a condition of pairing. But since the chromosomes that pair can be seen to be morphologically alike and therefore to be corresponding structures derived (so far as observation then showed) from opposite parents (Montgomery), it seemed enough to say that this pairing was due to the likeness of the chromosomes. An "incipient" association is often to be seen at mitosis in the somatic cells. Perhaps, therefore, meiosis was the final step in the sexual process in which the maternal and paternal elements at last united.

Such is, in a general way, the "explanation" of meiosis that is current today. To be sure, we now know that the association cannot be attributed to an attraction between chromosomes derived from opposite gametes, since pairing has been found in meiosis in parthenogenetic organisms,[1,2] and very often between chromosomes derived from the same gamete in polyploid plants. It may also be objected that this is merely to explain *ignotum per ignotius*. But it is still taken to be a satisfactory basis for cytological, genetical, and evolutionary deduction. Incompatible observations are freely ascribed to "mechanical" or "physiological" conditions.

There are many recent observations of this kind. There are tetraploid plants (such as *Primula sinensis*[3]), the nuclei of which contain four identical chromosomes of each of the twelve types that are represented twice in the diploid. These chromosomes usually associate in fours at meiosis, as they would be expected to do if likeness were the sole condition of pairing. But nearly always one, two, or three of these groups fail to be formed and their chromosomes appear merely paired. This is not explicable on the affinity theory. The chromosomes should be either *all* in fours or *all* in pairs.

Other observations of the same type are : (1) The occurrence of unpaired chromosomes in triploids, instead of all three identical chromosomes of each type being associated (*Zea*[4], *Tulipa*[5], *Lilium*[6]). (2) The occurrence of unpaired fragment chromosomes, although these have identical mates with which they can pair (*Secale*,[7] *Matthiola*,[8] *Tradescantia*[9]).

称"染色单体")连接在一起成对存在；而在第一次减数分裂中期是四条染色单体聚集在一起。染色体数目减少的直接原因是在细胞核两次分裂的间期，染色体通常会少分裂一次。但这显然与每条染色体已经分裂成了两条染色单体，这两条染色单体共同移向细胞的一极有关，反过来后者又与染色体配对相关。

因此大家很自然地（从 1890 年开始）认为减数分裂和有丝分裂的本质区别就存在于染色体配对的过程中。因为不同对染色体是随机地移向两极的（因此 A_1-A_2 和 B_1-B_2 可以产生子细胞核 A_1B_1 和 A_2B_2，也同样可以产生 A_1B_2 和 A_2B_1），并且由于染色体是定性分化的，如果减数分裂产生的是相似的但染色体数目减少的细胞核，那么进行配对并移向相反两极的染色体也一定是相似的（博韦里）。这显然说明，相似性是染色体得以配对的一个条件。由于配对的染色体在形态学上是相似的，所以可以将它们看作是来自（就继后进行的观察而言）异性双亲的对应结构（蒙哥马利），这似乎足以说明配对的原因是染色体的相似性。在体细胞有丝分裂时就经常可以看到"萌芽状态"的联会现象，因此，减数分裂很可能是有性生殖过程的最后一步，母本元素和父本元素在有性生殖过程中得以最终结合。

这是目前普遍接受的对减数分裂的"解释"。诚然，我们现在知道联会的发生并不是由于来自异性配子的染色体间存在某种吸引力，因为已经发现孤雌生殖的生物也存在减数分裂配对现象 [1,2]，而且在多倍体植物中，配对总是发生在来自相同配子的染色体间。也可能有人提出反对意见，认为这种释义解释得比原来需要解释的事物更难懂，但人们仍然把它当作是细胞学、遗传学和进化演绎的合理依据。而把与之矛盾的现象随意归咎于"机械条件"或"生理状态"的影响。

最近有许多这样的研究，例如四倍体植物（如藏报春 [3]）的细胞核含有四套相同的染色体组，每组含有 12 条不同类型的染色体，即二倍体的两倍。这些染色体在减数分裂时通常是以四条为一组进行联会，如果相似性是配对的唯一条件，则这和预期的结果是一致的。但是这些染色体中几乎总会有一组、两组或三组不能联会成功,而只是以成对形式出现。这种情况用亲近度理论是解释不通的,因为根据该理论,这些染色体应该**全部**是四条一组，或者**全部**是两条一组。

同类的研究还有：（1）三倍体中存在着不发生配对的染色体，而不是三套染色体组中每种类型的染色体都进行联会（例如玉蜀黍属 [4]、郁金香属 [5]、百合属 [6]）。（2）存在未配对的零散染色体，尽管在这些染色体上存在着完全相同的可以配对的对象，（如黑麦属 [7]、紫罗兰属 [8]、紫露草属 [9]）。

The only difference between these fragments and the other chromosomes which pair regularly appears to be their smaller size. If the triploids are examined, it is similarly found that the chromosomes which fail to associate regularly in threes are the small ones (*Hyacinthus*[10]). Therefore, not only *likeness* but also *size* bears some relation to the pairing of chromosomes.

If now we turn to consider the structures of the paired chromosomes at meiosis we find a variety of form that shows, at first sight, neither a rule in itself nor any clear relationship with ordinary mitosis. The two processes must be studied in their development in order to be seen in relationship.

The prophase of mitosis is characterised by a linear contraction of two threads, associated side by side, to become the two cylindrical rods which constitute the metaphase chromosome. At meiosis we find at the earliest stage a difference. The threads observed are single. They soon come together in pairs side by side and reproduce the conditions observed at the prophase of mitosis very closely indeed. But on account of their pairing they are present at this pachytene stage in half the number found at the prophase of a mitosis in the same organism. Evidently, therefore, the single threads at the earlier stage were chromosomes still undivided although in the earliest visible stage in mitosis they have already divided.

After an interval, splits appear in the pachytene thread, separating it into two threads, each of which is now seen to be double. But instead of these splits passing right along the paired chromosomes and separating them entirely, it is found, when they meet, that the double threads that separate in one part are not the same pair of threads that separate in another. The separated pairs of threads therefore change partners, and the points at which they change partners (there are often several distributed along the paired chromosome) are called "chiasmata". This stage is diplotene (Fig. 1).

Between diplotene and metaphase there is further linear contraction, and the structure of the paired chromosomes may remain the same in regard to the relationships of the four threads of which they are composed: that is, the chiasmata may remain stationary. But they may undergo a change which consists in the opening out of the loop that includes the spindle-attachment, at the expense of the adjoining loops, as though the spindle-attachments of the chromosomes were repelling one another. In other words, the chiasmata appear to move along the chromosome towards the ends: finally, the chromatids are associated in pairs with changes of partners only at the ends. Such changes of partners are called "terminal chiasmata", and the frequency of the end-to-end unions at metaphase corresponds with the frequency of the chiasmata seen earlier, when they were still interstitial, in small chromosomes (fragments) which only have one chiasma at most.[11] Further, in organisms with large chromosomes it is still possible to see the change of partner: at the end the association is double; it is between the ends of two pairs of chromatids, not merely between the ends of one pair of chromosomes.

这些不配对的染色体片段和其他进行正常配对的染色体的唯一区别就是它们会比较短小。对三倍体的联会研究同样可以发现，那些未能正常形成三倍体的染色体往往长度较短（如风信子属[10]）。因此，除了**相似性**之外，染色体的**长短**也与其能否发生配对有一定的关联。

如果我们转而考虑减数分裂中配对染色体的结构，我们会发现染色体配对的类型存在着多种方式，乍看起来，这些方式本身并无规律可循，且与普通有丝分裂的关系也不十分明确。要想了解有丝分裂和减数分裂的关系，就必须深入研究它们的详细过程。

有丝分裂前期的特点是，两条紧密相连的染色质丝发生线性收缩，从而形成了分裂中期的圆棒状染色体。我们发现减数分裂的起始期与此完全不同，它的染色质丝是单条的。不久之后，它们会聚集到一起，两两成对紧密排列，随后的过程的确与有丝分裂前期特别相似。但是在同一生物体内，染色体配对使得在减数分裂粗线期所观察到的染色体数目仅为有丝分裂前期的一半。由此可见，减数分裂初期的单条线状结构其实是尚未分开的染色体。而在有丝分裂中，染色体在可见阶段的最初就已经分开了。

减数分裂进行一段时间之后，粗线期的染色质丝开始出现分离，每条染色质丝分成两条，从而可以观察到双线状的结构。研究发现这些分裂并没有将配对的染色体完全分离，两个双线状结构分开的部位发生在染色体的不同区域。因而分离产生的一对双线状结构是染色体片段变换过的重新组合，它们发生交换的位点（一般认为配对染色体上分布着数个这样的位点）叫做"交叉"。这一时期被称为双线期（图1）。

在双线期和中期之间，染色体会发生进一步的线性收缩，这时配对染色体四条线的结构关系还是相同的。也就是说，交叉可能尚处于静止状态。但似乎因为染色体上纺锤丝的附着部位相互排斥，导致相邻的环状结构被打开，并且分开的区域逐步扩大，预示着此时的染色体可能正经历着某种变化。换言之，交叉是沿着染色体向其末端移动的，最后染色单体成对连接在一起，仅在末端发生了交换，这种变化被称为"末端交叉"。研究表明，小染色体（片段）至多只发生一次交叉，它们在中期的末端结合频率与其交叉频率是一致的[11]。此外，在大染色体生物中，也可能会观察到交换发生在配对的染色体之间，这时染色体末端存在着两处连接，即不只是一对染色体的末端之间发生了交换，而是两对染色单体的末端都发生了交换。

These observations point to the chiasmata being the immediate cause of pairing between chromosomes. How can such a hypothesis be tested? It is found that given pairs of chromosomes have a constant range in the number of chiasmata formed. For example, in the M chromosome of *Vicia Faba*[12,11] from 3 to 13 chiasmata are found at the metaphase, with a mean of 8.1. The M chromosome, which is much shorter has a range of 1 to 6, with a mean of 3.0. If we suppose that small chromosomes arising by fragmentation have a chiasma frequency proportionate to their length as compared with their larger neighbours, then we can predict from observations of their size and of the observed frequency of chiasmata in the large chromosomes what their frequency of pairing will be, on this hypothesis. Thus, in the variety "Yellow" of *Fritillaria imperialis* it was found that the chiasma-frequency was 2.58 in the large chromosomes. The fragments were about one-ninth of the length of the large chromosomes. They should therefore have chiasmata in a frequency of 2.58/9 per pair, or 0.29. This means that they should pair in 0.29 cases (neglecting the frequency of one pair forming two chiasmata, which should be slight). They were observed to pair in 0.22 of cases.[11] Here is an example of the type of observation which is susceptible of statistical analysis and supports this hypothesis.

Now, if we admit chiasmata as the condition of chromosome pairing, a considerable simplification is possible in stating the relationship of mitosis and meiosis. Throughout the prophase of mitosis, the threads are held together by an attraction in pairs. The same rule applies to meiosis, for the evidences of failure of pairing of fragments, of odd chromosomes in triploids, and of the four chromosomes of a type in tetraploids all point to the chromosomes having no present attraction at metaphase. They are merely held together by the chiasmata—that is, by the attraction between the pairs of half chromosomes and the exchanges of partners amongst them; and this attraction exists equally at mitosis.

This being so, we must look to the earliest stage of prophase to find the essential difference between the two types of nuclear division. It evidently lies in the time at which the chromosomes split into their two halves. At mitosis, it is probable that this has already happened before the chromosomes appear at prophase. At meiosis, it does not happen until pachytene (possibly at the moment at which the diplotene loops appear). The prophase of meiosis therefore starts too soon, relative to the splitting of the chromosomes. If we consider that there is a universal attraction of threads in pairs at the prophase of any nuclear division, as we see it at mitosis, it follows that this condition is fulfilled by the pairing of chromosome threads when they are still single, and their separation at diplotene when they have at last come to divide. The decisive difference would therefore appear to be in the singleness of the early prophase threads in meiosis. This singleness may be attributed to one or both of two causes: (i) a delayed division of the chromosomes, (ii) a precocious onset of prophase. The second of these seems the more likely explanation, on account of the short duration of the pre-meiotic prophase in some animals. Either assumption would account for the most characteristic of all secondary features of meiosis, namely, the exaggerated linear contraction of the chromosomes, paired or unpaired, if the time relationship of metaphase to the division of the chromosomes remains the same.

上述观察表明交叉才是染色体配对的直接原因。那么怎样才能验证这一假说？研究发现给定染色体对中形成的交叉数都有一个恒定的范围。例如蚕豆的 M 染色体 [12,11] 在中期可以形成 3 到 13 个交叉，平均值为 8.1。M 染色体要短很多，可以形成 1 到 6 个，平均值为 3.0。如果我们假设，断裂产生的小染色体的交叉频率和它们的长度所成的比例与邻近的长染色体一致，那么依据这一假说，我们就可以根据由观测得到的染色体片段长度以及长染色体发生交叉的频率来预测它们的配对频率。因此，我们发现冠花贝母的"黄花"品种中长染色体的交叉频率为 2.58，染色体片段的长度约为长染色体的 1/9，所以每对染色体片段的交叉频率应为 2.58/9，即 0.29。这意味着这些染色体片段之间发生配对的比例为 0.29（每对染色体形成两个交叉的可能性很小，因而可以忽略不计）。实际观察到的配对比例为 0.22[11]。上述这类观察结果就是一个易受统计分析影响并能印证本假说的例子。

在接受交叉是染色体配对条件的前提下，再来描述有丝分裂和减数分裂的关系就相当简单了。在有丝分裂前期，线状结构通过某种引力而结合在一起。同样的规则也适用于减数分裂，对于不能配对的染色体片段、三倍体中不能配对的剩余染色体以及四倍体中不能配对的同一类型的四条染色体，它们的存在都说明在减数分裂中期，这些染色体间缺乏这样的吸引力。它们仅仅由交叉维系在一起，即它们是由半染色体对之间的吸引力和交换维系在一起的，这种吸引力同样存在于有丝分裂中。

在这种情况下，要想找到这两种核分裂类型的本质区别，就必须从前期的最初阶段入手。很明显，区别就在于染色体是何时解离成两条染色单体的。在有丝分裂中，染色体很可能在前期之前就已经完成了这一解离；而在减数分裂中，染色体直到粗线期才发生解离（也可能是在双线期的交叉环出现的时候）。因此，相对于染色体的分离而言，减数分裂前期开始得太早了。假如我们认为在任何核分裂的前期，成对的染色质丝之间都普遍存在吸引力，正如在有丝分裂中看到的那样，则当减数分裂单体状态的染色质丝配对的时候，以及它们在双线期最终分离的时候就满足这一条件。因此决定性差异应该是在减数分裂前期的早期染色体为单体结构。造成单体结构的原因可能是下述两个中的一个或者兼而有之：(1) 染色体的延迟分裂，(2) 前期的早发性。用第二个理由来解释似乎更易被接受，因为在有些动物中，减数分裂前期之前的阶段持续时间比较短。如果中期和染色体分裂的时间关系也是一样的，那么在减数分裂的所有次要特征中最为特殊的一个特征就可以用这两个假设之中的任意一个来说明，即配对或未配对染色体的超常线性收缩。早熟论 [13] 可以通过观察减

This hypothesis of precocity[13] may be tested by the observation of a correlation between irregularities in meiosis and (a) abnormality in the timing of meiosis, and (b) diminished contraction of the chromosomes at metaphase.

The first of these tests is applicable to many organisms with occasional suppression of reduction; the aberrant nuclei enter on the prophase of meiosis either earlier or later than the normal nuclei.[14,15,16,17] When they are too early, it may be supposed that a premature division of the chromosomes has precipitated the prophase; when too late, it may be supposed that the prophase has been delayed. In either case, the chromosomes would no longer be single at early prophase and the condition of their pairing would be lost.

Such a cause of failure of pachytene pairing may be expected to be distinguishable by its effect on the contraction (the second kind of test). For when failure of metaphase pairing is not due to an upset in the timing of prophase but merely to failure of chiasma formation, we might expect normal meiotic contraction; this is the case in maize.[18] Where the prophase has been delayed, we might expect an approach to mitotic conditions; this is the case in *Matthiola*.[19, 20] Other critical evidence in favour of the hypothesis has already been quoted in these columns.[21]

By trying to define in this way the relationship of meiosis to mitosis, we find out what is essential and therefore universal in meiosis, and what is unessential and secondary. Only when the direct interpretation of events in the nucleus is clear (as it now seems to be) can we attempt their genetical interpretation on a satisfactory basis.

Two examples of the genetical interpretation of chromosome behaviour at meiosis are of immediate importance. It has been shown in every organism that has been adequately tested that crossing-over can occur between corresponding parts of the paired chromosomes at meiosis, actually between the chromatids, so that crossing-over in the region between C and D in a pair of chromosomes $ABCDE$ and $abcde$ will give four kinds of chromatid: $ABCDE$, $ABCde$, $abcDE$, and $abcde$ (Fig. 1). We may suppose that this crossing-over has no relation with anything observable cytologically; that it takes place when the chromosomes are intimately associated at pachytene and has no connexion with later behaviour. This view can only be taken when other possibilities are eliminated. We may also assume that crossing-over has some relationship with chiasmata, either as a cause ("chiasmatypy")[22, 23] or as a consequence, through breakage and reunion of new threads.[10,9] The last possibility has been eliminated by the statistical demonstration that terminal unions correspond in frequency with interstitial chiasmata,[11] and that the number of terminal chiasmata increases *pari passu* with the reduction of interstitial chiasmata.[3,4,20] The first possibility, that the chromosomes fall apart as they come together, and that the exchanges of partners at chiasmata are therefore due to exchanges in linear continuity or crossing-over between the chromatids, has been demonstrated in two ways.

数分裂中染色体出现的异常行为与以下两者之一的相关性来检验：(a) 减数分裂在时间上的异常，(b) 减数分裂中期染色体凝聚强度减弱。

上述两种检验中，前者适于出现在选择压力降低状况下的许多生物，这时异常核会比正常核或早或晚进入减数分裂前期 [14,15,16,17]。当细胞核过早进入前期时，可以假设染色体的过早分裂会使前期加速；而过晚时，可以假设前期就会延迟。无论过早还是过晚，在分裂前期的早期染色体都不再是单体状态，因而失去了配对的条件。

导致粗线期配对失败的原因可以通过其对染色体凝聚状态的影响加以辨别（上述两种检验方法中的第二种）。因为当中期配对失败不是由于前期的时间出现紊乱，而仅仅是由于不能形成交叉所致时，我们可以推测减数分裂的凝聚状态仍是正常的，玉米就属于这种情况 [18]。当前期延迟时，我们猜测可能有一种途径可以令其转而进行有丝分裂，紫罗兰属便是如此 [19,20]。支持该假说的其他关键性证据已经在专栏文章中被引用过 [21]。

在试图通过这种方式确定减数分裂和有丝分裂的关系时，我们发现了何为减数分裂必要而普遍的特征，何为减数分裂非必要且次要的特征。只有当细胞核中发生的所有事件都获得明确解释时（起码现在需要如此），我们才能在一个良好的基础上去尝试对这些细胞学观察做出遗传学解释。

有两个对减数分裂中染色体行为进行遗传学解释的例子具有非常重要的现实指导意义。已有研究显示，在被认真检验过的每一种生物中，减数分裂过程中的交换都会发生在配对染色体的对应部位，实际上是发生在染色单体之间，因此 $ABCDE$ 和 $abcde$ 这对染色体如果在 C 和 D 之间的区域发生交换，那么就可以产生四种染色单体：$ABCDE$，$ABCde$，$abcDE$ 和 $abcde$（图 1）。我们可以假设这一交换与任何可观察到的细胞学现象都没有关系，它在粗线期染色体紧密联会时发生，并与染色体后来的行为毫无关联。这一观点只有在排除了其他可能性之后才能被接受。我们也可以假定交换通过断裂及重接成新的线状结构而与交叉存在某种关系，或者是其原因（染色体交叉）[22,23]，或者是其结果 [10,9]。统计论证表明末端结合与中间交叉的频率是一致的 [11]，而末端交叉增加的数目与中间交叉减少的数目是成相同比例的 [3,4,20]，这就排除了上面提到的后一种可能性。第一种可能性认为当染色体聚集到一起时，原来的染色体就不复存在了。所以染色体在交叉处发生交换或者是由于沿着染色体全长发生线性连续性的交换，或者是由于染色单体之间发生了交换。这一可能性已经从两方面得到了证实。

In tetraploid *Hyacinthus* and *Primula* associations occur with such a spatial relationship that they can only be interpreted as the result of crossing-over.[3,25] In ring-forming *Oenothera*[28], chiasmata occur interstitially between a pair of chromosomes associated terminally with two others to give a "figure-of-eight". Such an arrangement also can arise only on the assumption of crossing-over. These demonstrations confirm Belling's interpretation of the *Hyacinthus* trivalents, which was not in itself indisputable.[5] Whether the observations are of universal application (the simplest assumption) or not, can only be shown by cytological tests of organisms which have been studied genetically.

A second problem is that of ring formation. Since, on the present hypotheses, the pairing of chromosomes at metaphase is conditioned by the formation of chiasmata at prophase between *parts* of chromosomes of identical structure, it follows that ring formation (where one chromosome pairs in different parts with parts of two others) must always be due to different arrangement of parts, that is, different structure, in the chromosomes contributed by opposite parents.[26,27,9] Thus the relationship of the chromosomes of two organisms can always be specified from the observation of the pairing behaviour of the chromosomes at meiosis in the hybrid. It is therefore possible to study differences of such a magnitude as will sterilise a hybrid and are therefore not susceptible of genetical analysis. This method is now being widely applied.

(**127**, 709-712; 1931)

References:

1. Seiler, J., *Zeits. f. indukt. Abstamm. u. Vererb Lehre*, **31**, 1-99 (1923).

2. Belar, K., *Biol. Zentrabl.*, **43**, 513-518 (1923).

3. Darlington, C. D., *Jour. Genet.*, **24**, 65-96 (1931).

4. McClintock, B., *Genetics*, **14**, 180-222 (1929).

5. Newton, W. C. F., and Darlington, C.D., *Jour. Genet.*, **21**, 1-16 (1929).

6. Takenaka, Y., and Nagamatsu., S., *Bot. Mag. Tokyo*, **44**, 386-391 (1930).

7. Gotoh, J., *Bot. Mag. Tokyo*, **38**, 135-152 (1924).

8. Lesley, M. M., and Frost, H. B., *Amer. Nat.*, **62**, 21-33 (1928).

9. Darlington, C. D., *Jour. Genet.*, **21**, 207-286 (1929).

10. Darlington, C. D., *Jour. Genet.*, **21**, 17-56 (1929).

11. Darlington, C. D., *Cytologia*, **2**, 37-55 (1930).

12. Maeda, T., *Mem., Coll. Sci. Kyoto*, B, **5**, 125-137 (1930).

13. Darlington, C. D., *Biol. Rev.*, **6** (in the press) (1931).

14. Rosenberg, O., *Hereditas*, **8**, 305-338 (1927).

15. Rybin, V. A., *Bull., Appl. Bot.* (Leningrad), **17**, 191-240 (1927).

16. Darlington, C. D., *Jour. Genet.*, **22**, 65-93 (1930).

17. Chiarugi, A., and Francini, E., *Nuo. Gio. Bot. Ital.* n.s., **37**, 1-250 (1930).

18. Burnham, C. R., *Proc. Nat. Acad. Sci.*, **16**, 269-277 (1930).

19. Lesley, M. M., and Frost, H. B., *Genetics*, **12**, 449-460 (1927).

20. Philp, J., and Huskins, C. L., *Jour. Genet.* (in the press).

21. Darlington, C. D., *Nature*, **124**, 62-64, 98-100 (1929).

22. Janssens, F. A., *La Cellule*, **34**, 135-359 (1924).

在四倍体植物风信子属和报春花属中，联会是伴随着一定的空间关系发生的，因此只能将其理解为交换的结果 [3,25]。在成环植物月见草属中 [28]，联会的一对染色体中有两条发生了中间交叉，另两条则发生了末端交叉，从而呈现出了"数字 8 的形状"，只有在交换时才可能出现这种排列。这些事例证实了贝林对风信子三价染色体的解释是无可置疑的 [5]。这些观察是否具有普遍适用性（最简单的假设），只能通过对已用遗传学方法研究过的生物进行细胞学试验来检验。

第二个问题是关于成环作用的。因为，根据现在的假说，减数分裂中期染色体配对是以前期染色体在相同**部位**形成交叉为条件的，所以成环作用（一条染色体与另外两条染色体在不同部位配对）肯定是由于来自异性双亲染色体配对部位的排列不同，即结构不同造成的 [26,27,9]。因此，通常可以通过观察杂合体在减数分裂过程中染色体的配对行为来具体说明两个生物体的染色体关系。因此有可能在杂合体不育的层面上来研究差异，这样就不易受遗传分析的影响了，目前这一方法正在被广泛地应用。

（刘皓芳 翻译；程祝宽 审稿）

23. Belling, J., *Univ. Calif. Pub. Bot.*, **14** (18), 379-88 (1929).

24. Erlanson, E. W., *Cytologia* (in the press) (1931).

25. Darlington, C. D., *Proc. Roy. Soc.*, **107**, 50-59 (1930).

26. Belling, J., *Jour. Genet.*, **18**, 177-205 (1927).

27. Darlington, C. D., *Jour. Genet.*, **20**, 345-363 (1929).

28. Darlington, C. D., *Jour. Genet* (in the press).

The Peking Skull[*]

G. E. Smith

Editor's Note

After the discovery of the first skull of Sinanthropus in 1928, responsibility for extracting the fragments of bone from the travertine rock in which the bone was embedded fell to Davidson Black, who was trained as an anatomist in Toronto, Canada, and who found his first job as an anatomy teacher at the Peking Union Medical College. (Black had raised funds for the excavation at Chou Kou Tien from the Rockefeller Foundation in New York, which continued to support the work after a skull had been found.) Black was a firm sinophile; born in 1884, he died at the early age of 59 in 1943.

WITH characteristic promptitude, Prof. Davidson Black has now provided us with a full report upon the features of the Peking skull, giving a detailed description of its external form, illustrated by 16 photographic plates (each photograph provided with a transparent explanatory drawing) and 37 text-figures. The drawings represent exact orthogonal projections not only of the type skull but also of the second skull of *Sinanthropus*, the finding of which was discussed in *Nature* of Aug. 9, 1930 (p. 210), and of a series of other fossil human skulls. The purpose of this comparision is to define the distinctive characters of *Sinanthropus* and to emphasise the contrasts in size and proportions that differentiate it from *Pithecanthropus* and the series of Neanderthal skulls. An elaborate series of measurements is provided, together with a statistical analysis of the significance of the figures, in comparison with those of other fossil human types, as well as of representatives of modern races of men. Hence complete data are now available to enable the anthropologist to realise the distinctive features of the Peking skull and the reasons which induced Prof. Davidson Black to differentiate it from all other known human types and assign it a distinctive generic rank.

The history of the finding of the skull by Mr. W. C. Pei on Dec. 2, 1929, has already been told in *Nature* (Mar. 22, 1930, p. 448). It was not until four months later that Dr. Black completed the process of removing from the surface of the skull the hard mass of travertine in which it was embedded. He then began to make casts and photographs of the specimen and to prepare the preliminary reports. After this was accomplished, he set to work to expose the interior of the skull, and in this he was inspired by the motive of preserving if possible the natural endocranial cast. Fortunately, this was possible because the braincase was fractured, enabling the bones to be removed piecemeal. Moreover, the

[*] Davidson Black , "On an Adolescent Skull of *Sinanthropus Pekinensis* in Comparison with an Adult Skull of the same Species and with other Hominid Skulls, Recent and Fossil", *Palaeontologia Sinica*, Series D, vol. 7, Fascicle 2. (Peiping: Geological Survey of China, Peiping, April 28, 1931.)

328

北京人的头骨[*]

埃利奥特·史密斯

编者按

自从 1928 年发现了北京人的第一具头骨之后，从埋藏该头骨的石灰华中取出骨头破片的责任就落到了步达生身上。他曾在加拿大的多伦多学习解剖学，并在北京协和医学院找到了他的第一份工作——担任解剖学教师。（步达生已从纽约的洛克菲勒基金会筹集到了用于周口店挖掘工作的资金，在发现了一具头骨之后，该基金会便继续资助此项挖掘工作。）步达生是一名坚定的亲华人士；他出生于 1884 年，卒于 1943 年，终年仅 59 岁。

步达生教授特别及时地为我们提供了一份完整的关于北京人头骨特征的报告。在这份报告中，他用 16 张照相图版（每张照片都附有清晰的说明图）和 37 张插图详细描述了该头骨的外部形态。这些图片不仅精确展示了这具正型标本头骨的正交投影图，也呈现了第二具北京人头骨（该头骨的发现曾在《自然》杂志 1930 年 8 月 9 日版的第 210 页讨论过）和一系列其他人类头骨化石的正交投影图。这一比较旨在确定北京人的区别性特征，并强调北京人与爪哇猿人及尼安德特人头骨系列在大小和比例方面的差别。在报告中，步达生教授提供了北京人与其他类型的人类化石以及与现代人的各个种族代表相比所得到的一系列精确的测量值以及这些数值的统计学分析结果。现在，这些完整的数据有助于人类学家搞清北京人头骨的区别性特征，以及步达生教授为什么要将北京人与其他已知的人类类型区别开来，并将其定为一种特别的属。

裴文中先生于 1929 年 12 月 2 日发现了第一具北京人头骨，其发现的历史早已在《自然》杂志上（1930 年 3 月 22 日，第 448 页）介绍过。可是直到 4 个月后，步达生博士才将覆盖在该头骨表面的石灰华硬块清除干净。接着他便开始制作这个标本的模型，为它照相，并准备对其进行初步报道。完成这些工作之后，他便开始着手暴露头骨的内部。在这一过程中，他一直把尽可能保全这具头骨的天然颅内模作为目标激励自己。很幸运，由于这具头骨的脑腔是断裂的，因而骨头可以被一片

---

[*] 步达生，"比较青年北京人头骨和成年北京人头骨以及其他原始人类的头骨，根据全新世的化石"，《中国古生物志》，丁种，第 7 卷，第 2 分卷（北平：中国地质调查所，北平，1931 年 4 月 28 日）

skull is that of a young adolescent, whose age, in Dr. Black's opinion, corresponds to that of a modern child between the time of eruption of the second permanent molar teeth and the attainment of adolescence, say 15±2 years. Thus it was possible easily to disarticulate the constituent bones. This work lasted until well into the summer of 1930, when Dr. Black succeeded in removing the cranial bones from the surface of the endocranial cast and then reconstituted each individual bone, and eventually rearticulated the skull with more precision than it had at the time when it was found. Before doing so, however, he made photographs and models of each separate bone, and took X-ray photographs to display the sinuses and other details in the texture of the bones, such, for example, as the labyrinth in the temporal bone. Then the skull was rearticulated and an artificial cast made of the cranial cavity.

The present monograph describes the external surface of the skull and each individual bone. The description of the endocranial cavity and cast which Prof. Davidson Black has obtained of it will be discussed in a second monograph that is now in course of preparation.

In July 1930 a large part of a second braincase was obtained from certain blocks of limestone which had been brought into the laboratory in October 1930. In his monograph Dr. Black gives full details of the comparison of the two skulls and the evidence upon which he relied to interpret the sexual characters and ages of the two individuals. The skull obtained in December 1929 he now regards (for reasons set forth in full in this report) as that of a youth in a stage of development between puberty and adolescence, and the second skull that of a woman. Partial obliteration of the left side of the coronal suture suggests that the latter was an adult, possibly more than ten years older than her companion. In the accompanying diagram (Fig. 1) Prof. Davidson Black's drawings of median longitudinal projections of the young skull (shaded) and of the adult (female) skull have been superimposed. The female skull is slightly thinner than that of the youth and is also larger, being somewhat higher and longer than the male skull (Fig. 1) and presenting other differences which are probably expressions of the difference in sex.

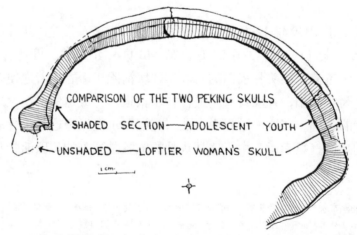

Fig. 1. Median longitudinal projections of the Peking skulls. × ½

一片地取下来，这使得保全这具头骨的天然颅内模成为可能。此外，该头骨是一位青少年的，步达生认为其年龄相当于一个现代孩子长出第二恒臼齿与青春期后期之间的时期，大约 15±2 岁。因此组成头骨的各块骨头很容易拆解开。这项工作一直持续到 1930 年的夏末，当时步达生博士已成功地将颅骨从颅内模的表面剥离，然后将每一块骨头重新拼接，最终将该头骨以比发现时更高的精度重新连接了起来。不过，在做这些工作以前，他对每一块分离下来的骨头都进行了拍照和模型制作，并且拍了 X 光片以看清骨窦及骨组织中的其他细节，例如，颞骨中的迷路。然后，他重新拼接了头骨，并且制作了颅腔的人工模型。

现在这篇专题报告描述了北京人头骨的外表面及每一块骨头。步达生教授在正在准备之中的第二本专著里将对他已得到的该头骨的颅内腔和颅内模进行描述。

1930 年 7 月，从石灰岩块中得到了第二具北京人头骨的大部分，并已于 1930 年 10 月被带回了实验室。在步达生博士的专著中，他对两具头骨进行了详细的比较，并给出了他用于判断两个个体的性征和年龄的证据。现在他认为：在 1929 年 12 月得到的头骨是处于青春期发育阶段的青年人（本报告完整地陈述了作出此判断的原因）的头骨，第二具头骨是一个女人的头骨。左侧冠状缝的部分消失意味着后者是一位成年人，可能比前者大十多岁。在附图（图 1）中，步达生教授将青年人的头骨（阴影部分）与成年女性的头骨的正中纵向投影图进行了重叠效果的展示。图 1 显示这位女性的头骨比青年人的稍微薄一些、大一些，比男性的头骨高一点、长一点，还有一些差异可能是性别差异的表现。

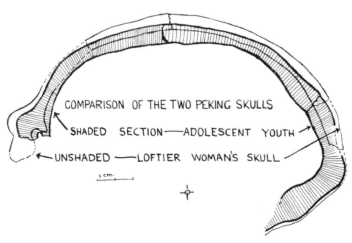

图 1. 北京头盖骨的正中纵向投影图。× ½

In view of the claims put forward by certain writers that the Peking skull should be included in the genus *Pithecanthropus* or, alternately, in the species *H. neanderthalensis*, Prof. Black has devoted a large amount of attention to the comparison of the projections of the skulls of *Pithecanthropus* and the various representatives of the Neanderthal species. By means of statistical comparisons he has made out a conclusive case in justification of the necessity of making a new genus and species for the reception of the Peking skulls.

While it is evident that the crania of *Sinanthropus* and *Pithecanthropus* resemble one another (Fig. 2) much more closely than they do any other human type, it is no less certain that they differ from one another in point of size, proportions, and detail to a degree amply sufficient to proclaim their generic distinction. It is a remarkable fact, Prof. Black adds, that in all its cranial parts *Pithecanthropus* shows "evidence of an archaic specialisation in marked contrast to the evidences of archaic generalisation so abundantly preserved in the crania and teeth of *Sinanthropus*". In other words, the apparent primitiveness of the Java fossil is in part probably due to degenerative changes responsible for the uncouth shape of the skull, which presents so striking a contrast to the elegant and undistorted braincase of the Peking man. Apart from its massive supraorbital torus and reduced third molar tooth, the Peking skull presents no highly specialised features. On the contrary, its general proportions, the morphology of the teeth, and the features of the tympanic and other individual elements of the skull, all provide evidence that *Sinanthropus* was a generalised and progressive type.

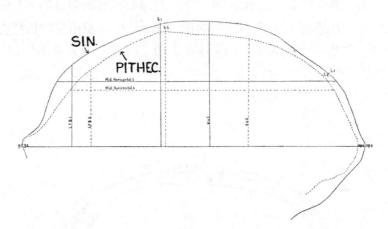

Fig. 2. Mid-sagittal skull contours of *Sinanthropus* and *Pithecanthropus*; left norma lateralis view. ×½

Prof. Black does not devote much attention to the comparison with the Piltdown skull. The purpose of the present work is to provide anthropologists with a detailed description of his specimen and an exact comparison with other specimens of unquestioned and generally recognised authenticity. For this reason, as well as to avoid partiality, he uses the data collected by Dr. H. Weinert in the case of *Pithecanthropus*, based upon the study of the actual fossil, with information provided by Prof. Dubois in amplification. Similarly, for the

　　鉴于某些作者主张应该将北京人头骨归到爪哇猿人属或者尼安德特人种中，步达生教授倾注了大量精力比较北京人头骨与爪哇猿人头骨和尼安德特种的不同代表的头骨投影图。通过具有统计意义的比较，他提出了确凿的证据，证实有必要建立一个新的属和种来接纳北京人头骨。

　　尽管很明显，北京人与爪哇猿人的头盖骨的接近程度（图2）远远高于它们与其他任何人类类型的头盖骨的接近程度，但是同样明显的是，它们彼此在大小、比例和细节上都有所不同，不同的程度足以表明它们在属一级上是有区别的。步达生教授补充道，一个值得注意的事实是爪哇猿人头盖骨的各部分都显示了"古老的特化迹象，这与北京人的头盖骨和牙齿中充分保留下来的古老的一般化证据形成了鲜明的对比"。换句话说，爪哇化石的明显原始性在某种程度上可能是由于发生过导致头骨形状怪异的退行性变化，其头骨形状与北京人头骨的精美无畸变形成了强烈对比。除了眶上圆枕较大以及第三臼齿有所缩小以外，北京人头骨没有表现出高度特化的特征。相反，该头骨的一般比例、牙齿的形态、鼓部和头骨其他个别成分的特征都证明北京人是一个一般化的、进步的类型。

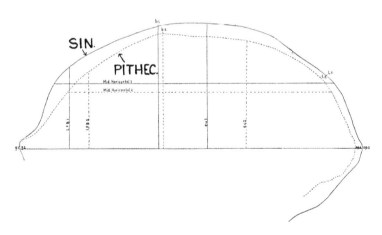

图2. 北京人和爪哇猿人头骨矢状面轮廓；左侧面观视图。× ½

　　步达生教授没有花费太多精力来比较北京人头骨和皮尔当头骨。目前工作的目的是为人类学家提供一份关于北京人头骨标本的详细描述，并将其与其他公认为真实可靠的头骨标本进行精确的比较。因为这个原因并为了避免偏见，步达生教授引用了韦纳特博士在研究爪哇猿人时根据对真实化石标本的研究搜集到的资料，以及杜波依斯教授在详细描述爪哇猿人时提供的信息。与此相似，步达生教授引用了莫

Neanderthal skulls Dr. Black relies on the data and figures provided by Dr. G. M. Morant. As, unfortunately, there is still considerable doubt in the minds of many anthropologists concerning the Piltdown skull and the mode of its reconstruction, Dr. Black does not make much use of it for comparison. He does, however, emphasise the fact that the peculiarly developed postero-inferior parietal boss in the Peking skull resembles in certain important features the similar, if less obtrusive, development of the Piltdown parietal. He also directs attention to the similar thickness of the skull in the genera *Sinanthropus* and *Eoanthropus*, but points out that the range of unevenness in thickness of the Peking skull presents a marked contrast to the more uniform Piltdown fragments.

Although Dr. Black himself has refrained from instituting detailed comparisons with the Piltdown skull, it is interesting to compare (Fig. 3) the transverse section he provides of the skull of *Sinanthropus* (the thick lines) with a section made in the corresponding plane (auditory meatus) of the reconstruction of the Piltdown skull (the larger shaded area) made by the late Prof. John Hunter. This section, like the view of the two skulls from the posterior aspect, brings out the essential identity of their architectural plan in a most striking way, and reveals a similarity of form and proportions which is unexpectedly close. Apart from the difference in thickness, the adult skull of *Sinanthropus* approaches even nearer to the Piltdown skull in some respects. Thus (see Fig. 1) it is loftier than the type skull and its height is identical with that of *Eoanthropus*, but the latter is considerably wider and correspondingly more capacious.

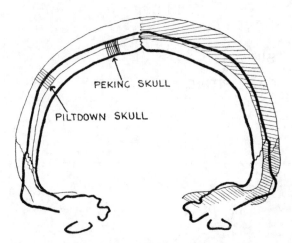

Fig. 3. Transverse sections of Peking and Piltdown skulls. ×½

The general form and proportions of the Peking skull, as well as the details of many of its constituent parts, are surprisingly modern in character. The man of China was clearly a very primitive and generalised member of the human family close to the main line of descent of *Homo sapiens*.

Prof. Black devotes particular attention to the unique character of the temporal bone, which presents a marked contrast to that of all other known men and apes. Of special

兰特博士提供的关于尼安德特头骨的资料和插图。遗憾的是，因为仍然有很多人类学家对皮尔当头骨及其复原头骨颇为疑惑，所以步达生博士没有充分利用它与北京人头骨进行比较。但是他强调北京人头骨中特殊发育的后下顶骨突起与皮尔当顶骨的发育在某些重要特征上有几分相像，只是前者不如后者凸出。他也注意到北京人和曙人头骨（即皮尔当头骨）厚度相似，但他指出北京人头骨的厚度不均匀程度与皮尔当头骨碎片厚度的相对均一形成了鲜明的对照。

　　尽管步达生博士自己已尽量避免对北京人头骨与皮尔当头骨进行详细的比较，但是将他提供的北京人头骨的横切面（图中用粗线表示）与已故的约翰·亨特教授复原的皮尔当头骨（图中较大的阴影区域）相应的（经过外面耳道的）截面进行比较（图3），还是很有趣的。这一截面就像这两具头骨的后面观一样，以一种非常引人注目的方式反映了二者结构比例的本质上的一致性，并且揭示了二者在形状和比例上的惊人相似性。除了厚度上的差异，成年北京人的头骨在某些方面与皮尔当头骨甚至更加接近。因此（见图1）成年北京人的头骨比正型标本的头骨更高，与曙人头骨的高度一致，但是曙人头骨要宽很多，相应地容积也更大。

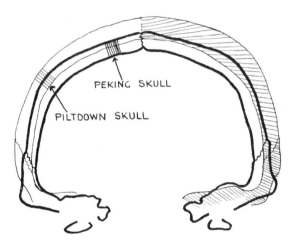

图3. 北京人和皮尔当头骨的横切面。×½

　　北京人头骨的一般形状和比例以及其众多组成部分的细节在特征上都惊人地与现代人相近。很明显中国人是人科中一个非常原始的、一般化的成员，与智人后代的主干很接近。

　　步达生教授对北京人头骨颞骨的独特特征给予了特别关注，因为该颞骨与所有其他已知人类和猿类的颞骨都形成了鲜明对照。他尤为感兴趣的是鼓部和乳突部的

interest are the distinctive features of the tympanic and mastoid portions, showing, not only in the case of the mastoid but also in the form of the auditory meatus and middle ear, characters which in modern man occur only in new-born infants and very young children—a widely open meatus terminating at an ear drum the inclination of which closely approaches the horizontal. This state of affairs is lost in *Homo sapiens* long before the age of puberty is attained. As Dr. Black remarks, the features of the tympanic region of *Sinanthropus* are admirably suited to serve as a starting-point for phylogenetic speculation. "With this generalised type before us it is not difficult to imagine developmental stages through which such an element in a stem-form may well have passed, leading to the modifications such as are characteristic of the Piltdown, Neanderthal, and modern men". On the other hand, since all the essentials of the tympanic morphology of the great anthropoid apes may be recognised in these elements of *Sinanthropus*, a comparison of the latter may serve to indicate in some measure the degree of their divergence from a common type. In spite of this provocative comment, Prof. Davidson Black refrains from discussing the intriguing problems he mentions. He does not depart from the admirable restraint that characterises all he has written upon this subject, which makes his monograph a reliable guide to those who want the data and prefer to form their own opinions as to their meaning.

The great importance of the discoveries in China lies in the fact, not only that the material is more abundant than the remains found at Trinil and Piltdown, and their geological age is unquestionable, but also that, while *Sinanthropus* is differentiated from the genera *Pithecanthropus* and *Eoanthropus*, it is much more generalised than either, yet definitely linked to both. While it is the most primitive type of human being so far discovered, its structural affinities with both the Javanese and the British genera link together all the known types of Pleistocene men and give cohesion to our knowledge.

It is fortunate that the information concerning this unique material has been so fully and so promptly supplied to anthropologists in a monograph which is distinguished by admirable clearness and impartiality. Once more Prof. Davidson Black deserves our congratulations on a great achievement.

(**127**, 819-821; 1931)

区别性特征，不仅在于乳突部，而且在外耳道形状和中耳方面，都显示出在现代人中只见于新生儿和很小的儿童中的特征——终止于鼓膜的外耳道开口广阔，鼓膜位置接近于水平。而这种情况早在智人的青春期到来之前就会消失。正如步达生博士所说，北京人鼓室区域的特征非常适合作为推测系统发育的出发点。"因为有了这一一般化的类型，我们就不难想象出鼓室可能经过的发育阶段，这些发育阶段导致了那些使皮尔当、尼安德特和现代人具有不同特征的改变"。另一方面，因为大型类人猿鼓室形态的所有特点都可以在北京人的相应部位找到，所以对鼓室特征进行的比较可能揭示他们从一个共同的祖先趋异进化的程度。尽管这一说法很具煽动性，步达生教授还是忍住不去讨论他所提到的这些引人入胜的问题。他在对这一主题的所有叙述中一直保持着令人钦佩的严谨，这使得他的专著给那些想要得到资料并且更喜欢对这些资料的意义形成自己的观点的人提供了可靠的指导。

在中国发现北京人头骨的巨大价值不仅在于这里的材料比特里尼尔和皮尔当的遗存骸更丰富，它们的地质年代也是确定无疑的，而且，还在于尽管北京人不同于爪哇猿人属和曙人属，它比后两者都更一般化得多，但肯定与后两者都存在联系。尽管北京人是迄今发现的最原始类型的人类，其与在爪哇和在英国发现的化石人属在结构上的亲缘关系将所有已知的更新世人类都联系在了一起，也使我们的知识融会在了一起。

很幸运，步达生教授将与北京人头骨这一独特材料有关的信息如此充分而且迅速地以一部条理清晰、论述公正的专著提供给了人类学家。步达生教授的伟大成就值得我们再一次祝贺。

（刘皓芳 翻译；林圣龙 审稿）

Recent Advances in the Chemistry of the Vitamins

Editor's Note

Despite the complaint of neglect that opens this piece by English biochemist Frederick Gowland Hopkins, president of the Royal Society, vitamin chemistry was being pursued with much energy in the 1930s. Hopkins had himself won a Nobel Prize two years earlier for his work in this field. Yet the molecular structures of none of the vitamins discussed here—A, B and D, the latter two being in fact several substances—was then known in detail, nor were their precise modes of biochemical action. Indeed, the rather technical and laborious chemical procedures detailed here are an indication of how painstakingly those answers had to be sought.

AT the meeting of the Royal Society held on June 18, the President, Sir Frederick Gowland Hopkins, opened a discussion on the chemistry of the vitamins. He said that although many discoveries had been communicated in the past to the Society, there had been very few papers dealing with the vitamins. He felt great satisfaction that this discussion should occur during the first year of his presidency, especially as the subject was still growing in interest. The Society were to be congratulated on the presence of a number of foreign workers, so that the discussion would have an international character. As the subject was so vast, he proposed to limit it to the chemistry of the vitamins, and suggested that vitamins D, A, and the B complex be taken in that order.

Vitamin D

Prof. A. Windaus said that up to the present, investigations had had to be carried out on impure substances and chiefly by means of physical experiments. No stable equilibrium was formed between ergosterol and the products of irradiation. By further irradiation two crystalline substances could be obtained, neither of which could be converted into the other; it appeared therefore that two series of products were formed on irradiation. Reerink and van Wijk had found that no matter whether 10 or 50 percent of the ergosterol was changed by long wave irradiation, the absorption spectrum of the product was always the same. He suggested that this was due to several substances being formed in constant proportion, and that the absorption spectrum was that of the mixture. None of the products of irradiation are precipitated by digitonin. Failure of precipitation, however, does not imply that a change has taken place in the hydroxyl group. On treatment with phenyl isocyanate the antirachitic activity of the irradiated ergosterol is destroyed. Treatment with warm caustic potash results in the reconversion of the phenyl urethane to vitamin D.

维生素在化学方面的新进展

编者按

尽管皇家学会主席、英国生物化学家弗雷德里克·高兰·霍普金斯在开篇时抱怨维生素化学领域没有得到足够的重视，但在 20 世纪 30 年代人们还是对此付出了很大的努力。两年前霍普金斯本人因在这个领域中的成就而赢得了诺贝尔奖。不过，这里讨论的几种维生素（维生素 A、B、D，后两者实际上是由多种物质组成的）的分子结构和准确的生化反应模式在当时并不清楚。的确，文中所描述的非常专业和繁复的化学过程足以说明探索答案的过程是多么艰辛。

在 6 月 18 日召开的皇家学会会议上，学会主席弗雷德里克·高兰·霍普金斯爵士在有关维生素化学的讨论中首先发言。他表示：尽管学会以往对许多新发现都进行过交流，但在维生素方面的文章却是少之又少。能在自己的第一年任期内举行这样的讨论，他感到非常满意，尤其是因为这一主题正在引起越来越广泛的关注。学会为有许多国外的研究者参会而感到庆幸，因此这将是一次国际性的盛会。鉴于这一主题非常宽泛，他提议将其限定在维生素的化学特性方面，并建议按照维生素 D、维生素 A 和维生素 B 复合物的顺序进行讨论。

维生素 D

温道斯教授说，迄今为止人们不得不对混合物进行研究，而且研究主要依靠的是物理实验。在麦角甾醇和其照射产物之间没有形成稳定的平衡，通过更长时间的照射可以得到两种结晶体，而这两种结晶体是不能相互转化的。因此可以得出结论：麦角甾醇在照射过程中生成了两个系列的产物。雷林克和范维克发现在经过长波照射后，无论发生变化的麦角甾醇是 10% 还是 50%，其产物的吸收光谱通常都是相同的。温道斯教授认为这是由于形成的几种物质具有恒定的比例，而得到的吸收光谱正是这个混合物的吸收光谱。没有一种照射产物能用毛地黄皂苷沉淀出来。然而不能沉淀出来并不意味着羟基发生了变化。在用异氰酸苯酯处理照射过的麦角甾醇后，其抗佝偻病的活性被破坏。而用温和的氢氧化钾进行处理，则会使苯氨基甲酸乙酯重新转化为维生素 D。

Prof. Windaus had found that the vitamin had the same molecular weight and formula as ergosterol and also contained three double bonds. The dihydro derivative obtained from irradiated ergosterol by treatment with sodium in alcoholic solution was inactive, but it is not certain that it is a derivative of vitamin D itself. The vitamin is more sensitive to a temperature of 180° than ergosterol; after heating, the absorption spectrum shows a band at 2,820–2,920 A.

Crude vitamin D is stable in oil, although the absorption spectrum and specific rotation rapidly change. The toxicity of the crude product varies with the potency, but it is still possible that the two properties may be due to different substances. The problem has not yet been solved. He had noticed that in vacuum tubes, in which crude vitamin D had been sealed, crystals appeared, but he had not been able to obtain them in a pure state by recrystallisation from cold acetone or by fractional precipitation. He had found, however, that when irradiated ergosterol was treated with maleic or citraconic anhydride in ethereal solution at room temperature for one to three days, a reaction occurred between certain of the substances present and the anhydride. When these inactive products were removed by solution in dilute caustic potash, crystals could be obtained from the ethereal layer on evaporation of the solvent. The yield was 50 percent of the crude product, or 60–70 percent of the material which failed to react with maleic anhydride.

Vitamin D crystallises in long needles of melting point 122°; $[a]_D^{18} = + 136°$ in acetone, and $[a]_{Hg}^{18} = + 168°$ in acetone.

The crystals show a band in the absorption spectrum at 2,650–2,700 A. Their potency was found to be 2–2½ times that of the M.R.C. standard.

He considered that the product obtained by Reerink and van Wijk was different from his, since it had a lower specific rotation and an absorption spectrum of a different shape, but that Bourdillon's crystals were probably the same. He agreed with the suggestion of the latter that there might be several compounds showing vitamin D activity, but considered that his crystals were responsible at any rate for the chief part of it. Vitamin D is an isomer of ergosterol in which there has been a structural rearrangement with an increase in the spatial size of the molecule.

Prof. B. C. P. Jansen said that Reerink and van Wijk had now obtained crystals of a melting point 140°. Their former product with a lower melting point had contained ether of crystallisation.

Dr. R. B. Bourdillon said that the discovery of the method of obtaining crystals of vitamin D by means of maleic anhydride was of outstanding importance and likely to lead to a solution of the problem of its constitution. He and his co-workers had obtained very similar crystals by distillation of irradiated ergosterol in a high vacuum. Their melting

340

温道斯教授发现维生素 D 具有与麦角甾醇相同的分子量和分子式，并同样具有三个双键。由照射过的麦角甾醇在乙醇溶液中用钠处理后得到的二氢衍生物是没有活性的，但不能确定该物质就是维生素 D 本身的一个衍生物。在 180℃下，维生素 D 比麦角甾醇更加敏感；加热后，在吸收光谱的 2,820Å ~ 2,920 Å 处会出现一条谱带。

天然的维生素 D 在油中是稳定存在的，尽管其吸收光谱和比旋光度会迅速地发生变化，此外其毒性也会随着效价而改变。但这两种特性仍有可能是基于不同的化学物质而产生的，这个问题一直没有得到解决。温道斯教授还发现将天然的维生素 D 封在真空管中会出现结晶，但是他尝试在冷的丙酮中进行重结晶或者采用分级沉淀的方法都未能获得纯的维生素 D 结晶。然而，他发现照射过的麦角甾醇在室温下用马来酸酐或者柠康酸酐的醚溶液处理 1 ~ 3 天后，其中的某些物质与酸酐发生了化学反应。当用稀释的氢氧化钾溶液除去这些无活性的产物后，蒸发溶剂，就可从醚层中得到结晶，其产率按照粗品计算为 50%，按照没有与马来酸酐发生反应的物质计算则为 60% ~ 70%。

维生素 D 结晶为长针状，熔点是 122℃；$[a]_D^{18} = +136°$（丙酮），$[a]_{Hg}^{18} = +168°$（丙酮）。

此结晶体的吸收光谱在 2,650 Å ~ 2,700 Å 处有一条谱带。其效价为英国医学研究理事会（M.R.C.）规定标准的 2 ~ 2.5 倍。

他认为雷林克和范维克得到的产物与他得到的不同，因为他们的产物比旋光度值较低，而且吸收光谱的形状也不一样，不过鲍迪伦获得的结晶体也许与他的是相同的。他同意鲍迪伦的看法，认为可能有好几种化合物都能表现出维生素 D 活性，但又把自己所获得的结晶体看作是起主要作用的部分。维生素 D 是麦角甾醇的一个异构体，是通过结构重排而得到的空间尺寸更大的分子。

詹森教授说，雷林克和范维克现在已经得到了一个熔点为 140℃ 的结晶体。他们先前制得的熔点较低的产物含有醚的结晶。

鲍迪伦博士认为发现用马来酸酐获得维生素 D 结晶的方法具有重大意义，很有可能帮助我们解决维生素 D 的构成问题。他和他的同事们通过在高真空下蒸馏照射过的麦角甾醇得到了极其类似的结晶体，其熔点为 123℃ ~ 125℃，在醇溶液中对汞

point was 123–125°. The specific rotation to the mercury line in alcoholic solution was +250 to +260°. The maximum absorption occurred at 2,700 A. and was greater than that of ergosterol. The potency was 18–22,000 M.R.C. U./mgm., that is approximately the same or slightly less than that shown by Prof. Windaus' crystals. They had been able to prepare an oxalate and acetate and reconvert these back to the original crystals. They had found the rotation and potency to remain constant for some weeks in dry air or in vacuo. Neither by distillation at a temperature of 160° nor by further irradiation with loss of two-thirds of the potency had they been able to separate their crystals into two separate compounds. Moreover, on irradiation the loss of antirachitic activity and the changes in the absorption coefficient and specific rotation were absolutely parallel. They therefore considered that their crystals were a definite chemical compound and had ventured to call it "Calciferol". However, the unity of the compound was not absolutely certain, owing to variations in the specific rotation of different preparations. They had recently obtained some crystals with a rotation of +290°. He suggested that there were at least two vitamins, isomorphic and with the same absorption spectra, the laevo form being very unstable, while the dextro form was stable. This suggestion would explain the discrepancies in his own work and that of other observers.

Dr. O. Rosenheim said he considered the work now reported was the most important since the original researches on vitamin D. He thought that Windaus' and Bourdillon's crystals were identical, although it was possible that there might be an impurity present as in Reerink and van Wijk's preparation. Spectroscopic methods had not led to much advance, but success had come from biological and organic chemical research, and especially from Bourdillon's method of fractional condensation. He pointed out that if ergosterol was irradiated until 20 percent had been changed and the 80 percent of inactive material removed, the activity of the residue was only the same as that of the original irradiated material. On distillation of the residue, 20 percent was obtained as crystals, but the activity was only twice that of the crude product.

Vitamin A

Prof. H. von Euler said that it was early noticed that fat soluble growth-promoting material was frequently associated with a red or yellow coloration but that the converse was not true. Working with Karrer, he had found that of all red or yellow substances examined, only carotene had vitamin A activity. Carotene was usually supposed to be optically inactive, but he, as well as Rosenheim and Kuhn working independently, had found that it could be fractionated into two forms, one melting at 170° and optically active, and the other melting at 183° and optically inactive. Both forms had growth-promoting power. The latter gave an earlier growth response, but after three weeks the differences between the two disappeared. A dose of 0.003 mg. daily would produce a daily increase in weight of one gram in the rat. Complete hydrogenation of carotene inactivated it, but reduction with aluminium amalgam at first increased the activity. Hydrocarotene containing eight double bonds is much more active than carotene, the daily dose required being only 0.0005 mg. It also gives a higher blue value with the Carr and Price colour test. The

342

线的比旋光度值为 +250°~+260°。它最大的光谱吸收出现在 2,700 Å，这比麦角甾醇的还要大。这个结晶体的效价为 18~22,000 M.R.C. U/mg，这与温道斯教授得到的结晶体的效价大致相同或者略微偏低。他们已经能够用这种结晶体制备草酸盐和醋酸盐，并能将它们重新转化为原来的结晶体。他们发现在干燥的空气中或者真空下结晶体的旋光度和效价会持续几个星期保持不变。既不用在 160℃ 下进行蒸馏，也不需要更长时间的光照射（这样会损失 2/3 的效价），他们就能将得到的结晶体分离为两个单独的化合物。而且在光的照射下，这两种化合物的抗佝偻病活性损失与吸收系数和比旋光度的变化完全一致。因此，他们认为自己得到的结晶体是一种明确的化合物，并大胆地把它称为"钙化醇"。然而，由于在不同的制备方法中这个化合物的比旋光度会发生变化，所以不能完全确定该化合物的性质是统一的。最近他们获得了一些旋光度为 +290° 的结晶体。鲍迪伦博士认为这些结晶体中至少存在着两种结构和吸收光谱相同的维生素，其中左旋形式的维生素非常不稳定，而右旋形式的维生素是稳定的。这样的观点也许能够解释他的研究与其他观测者的研究之间的差异。

罗森海姆博士认为现在报道的研究工作是自人们开展对维生素 D 的研究以来最为重要的工作。他认为温道斯和鲍迪伦的结晶体是相同的，尽管有可能会像雷林克和范维克的制备物一样存在一定的杂质。在光谱学方法的研究上并没有取得什么进展，但在生物学和有机化学方面却取得了成功，尤其是鲍迪伦的分级凝缩方法。他指出如果光照射麦角甾醇直到 20% 的物质发生了变化，然后将 80% 的无活性物质除去，那么残留物的活性才跟最初被照射的物质相同。将残留物进行蒸馏，可以得到占残留物 20% 的结晶，但其活性仅是天然产品的两倍。

维生素 A

冯欧勒教授说，人们很早就注意到脂溶性的促生长物质常常与红色或黄色相关联，但不能倒推。他与卡勒一起研究发现，在所有检测过的红色或黄色物质中，只有胡萝卜素具有维生素 A 活性。人们通常认为胡萝卜素是不具备光学活性的，但与罗森海姆和库恩的独立研究结果一样，冯欧勒教授也发现胡萝卜素可以分成两种：一种熔点为 170℃，具有光学活性；另一种熔点为 183℃，不具有光学活性。这两种胡萝卜素都有促进生长的作用。后者的促生长作用出现得比较早，但 3 周以后两者之间就没有什么差别了。每天以 0.003 mg 的剂量给大鼠喂食胡萝卜素，可以使它的体重每天增加 1 g。完全氢化胡萝卜素会消除它的促生长作用，但如果先用铝汞合金进行还原却会使它的活性增加。含有 8 个双键的氢化胡萝卜素比胡萝卜素的活性要高很多，所以要使大鼠体重增加 1 g，每天只需 0.0005 mg 的剂量。而且它在卡尔和普赖斯颜色试验（卡 – 普二氏试验）中表现出更高的蓝色值。这种氢化胡萝卜素的

absorption spectrum is very similar to that attributed to vitamin A in cod liver oil.

Moore considers that carotene is converted to vitamin A in the body of the rat. Prof. von Euler had tried to effect this conversion in vitro, but only with the serum of the hen had he found that carotene could be converted into a substance very similar to vitamin A although not spectroscopically identical. He considered it probable that the transformation occurs in the blood, and that vitamin A may act as a catalyser in oxidation. He had also examined the anti-infective action of carotene and had found that it did not affect haemolysis in vitro nor react with amboceptor, but that when it was given in excess to rabbits, the amboceptor in their blood was increased.

Dr. Rosenheim said that carotene is a mixture which has not yet been completely separated, and that it is dangerous to attribute activity to any particular isomer. The optical activities obtained by different observers vary considerably. It is possible that the transformation of carotene into vitamin A in the body may be only the accumulation of vitamin A present as a contaminant in the carotene. The band in the absorption spectrum of cod liver oil at 3,280 A. may not be due to vitamin A but to a substance accompanying it.

Dr. R. A. Morton said that if the band at 3,280 A. is to be attributed to vitamin A, then the latter cannot be present in carotene as an impurity. Dihydrocarotene is not vitamin A, since the band in its absorption spectrum is at 3,170 A., the blue colour given with antimony trichloride is not the same as that given by the vitamin, and the ratio of blue colour to intensity of absorption is different.

Vitamin B

Prof. B. C. P. Jansen said that when 100 kgm. of rice polishings were extracted with dilute acid alcohol, 30 kgm. went into solution. By treatment with acid clay it was possible to adsorb nearly all the vitamin B_1, but only 100 grams of contaminating solid material. The vitamin B_1 present accounted for only 1 percent of the adsorbed material. By fractional precipitation with silver nitrate and baryta it was possible to remove impurities at a strongly acid reaction and to precipitate two-thirds of the vitamin at pH 4–7: above this pH impurities were precipitated as well.

Further purification could be effected by precipitation with phosphotungstic acid, and decomposition of the precipitate with baryta, and precipitation with platinum chloride from alcoholic solution. When the platinum was removed with powdered silver, 1.4 gram was obtained, of which 0.4 gram was pure vitamin. By numerous fractionations with acetone from solution in absolute alcohol, 30 mgm. of pure vitamin were obtained. The formula by analysis was $C_6H_{10}ON_2HCl$. The process he had devised was therefore very wasteful. It could be improved by the use of Peters' method of fractional precipitation with phosphotungstic acid. Prof. Jansen had found that silicotungstic acid was as suitable. When the original 100 grams obtained by adsorption on acid clay were treated with this

吸收光谱非常类似于鱼肝油中维生素 A 的吸收光谱。

穆尔认为胡萝卜素是在大鼠体内转化成维生素 A 的。冯欧勒教授曾试图在体外完成这样的转化。但他发现只有用母鸡的血清才能使胡萝卜素转化为一种与维生素 A 非常类似的物质，尽管二者的光谱并不一致。于是，他认为这种转化很可能是在血液中发生的，而维生素 A 可能是氧化过程中的一个催化剂。他还对胡萝卜素的抗感染作用进行了研究，结果发现它既不影响体外的溶血，也不与溶血素发生反应，但当给兔子的剂量过多时，兔子血液中的溶血素会增多。

罗森海姆博士认为胡萝卜素是一种混合物，这种混合物到现在也没有被完全分离开，把它的活性归因于任何一种异构体都是有风险的。因为，首先，不同的研究者得到的光学活性存在着很大的差异。其次，胡萝卜素在体内转化成维生素 A 的过程可能仅仅是维生素 A 作为胡萝卜素中的一种杂质积聚在了一起。再次，在鱼肝油的吸收光谱中位于 3,280 Å 的谱带也许并不是由维生素 A 引起的，而是由于一种与之相伴的物质。

莫顿博士认为如果在 3,280 Å 的谱带是由维生素 A 引起的，那么维生素 A 就不可能是胡萝卜素中的杂质。二氢胡萝卜素不是维生素 A，因为它的光谱吸收带在 3,170 Å，并且由三氯化锑试验（译者注：就是上面提到的卡－普二氏试验）给出的蓝色值与维生素 A 的不同，此外蓝色值与吸收强度的比值也不一样。

维生素 B

詹森教授指出在用稀释的酸醇溶液萃取 100 kg 的米糠时，会有 30 kg 的米糠进入溶液中。再用酸性黏土进行处理，可能会吸附几乎所有的维生素 B_1，不过只得到 100 g 含杂质的固体物质。把全部的维生素 B_1 加起来也只占被吸附物质的 1%。通过硝酸银和氧化钡的分级沉淀，可能会在强烈的酸性反应中除去杂质，并在 pH 值 4～7 的条件下析出三分之二的维生素，pH 值高于 7 时，杂质也会一起析出。

进一步的纯化可以通过以下过程进行：用磷钨酸进行沉淀，然后用氧化钡分解此沉淀物，再用氯化铂从醇溶液进行沉淀。在用银粉除去铂之后，得到了 1.4 g 的产物，其中含有 0.4 g 纯的维生素。再将此产物溶于无水乙醇中，用丙酮进行多次分馏，最终可以得到 30 mg 纯的维生素。通过分析得知其分子式为 $C_6H_{10}ON_2HCl$。可以说，詹森教授所设计的这个制备方法是非常不经济的。也许可以用彼得斯的磷钨酸分级沉淀法对其加以改进。詹森教授已经发现硅钨酸也同样适用。在用硅钨酸处理从酸性黏土中吸收得到的 100 g 粗品时，有 2/3 的维生素会沉淀出来，不过析出的质量只

reagent, two-thirds of the vitamin were precipitated, but only 20 percent of the total solid. A further improvement had been effected by Seidell's benzoylation process, which removes impurities. By this method von Ween had obtained 140 mgm. of pure vitamin B_1 from 75 kgm. rice polishings.

Prof. A. Seidell said that Prof. Jansen had had success with a method that had failed in his hands. He had felt that precipitating agents ought to be avoided, and so was led to the method of benzoylation in chloroform solution, which removes a considerable amount of nitrogenous material. He had not succeeded in obtaining crystals, although his purest preparation was nearly as active on rats, but when tested on pigeons by Peters' method the activity was found to be only 1/5 to 1/10 that of the crystals.

Prof. R. A. Peters said that he had hoped to be able to use the animal as a test object only and not concern himself with the physiological side of the problem, but he had found that it was only possible to reach a final conclusion on the chemistry when the physiology was also taken into account. He had been able to confirm Prof. Jansen's process up to the platinum chloride stage. Working with yeast he had obtained purification by removal of inactive material with lead acetate and baryta followed by adsorption of the vitamin by charcoal. By fractional precipitation with phosphotungstic acid at pH 5–7, vitamin B_1 was obtained. Its activity was 0.012 mgm. per dose, while that of Prof. Jansen's crystals in his hands was 0.008 mgm. Both preparations gave the Pauly reaction with the same intensity. Miss Reader had shown that 25–50 percent of Prof Jansen's crystals had vitamin B_4 activity. She had now been able to isolate vitamin B_1 from B_4 by fractional adsorption on to charcoal in the earlier stages of the purification, but the separation of B_4 from B_1 was less complete.

Dr. B. C. Guha said that he had been able to separate B_1 from B_2 by electrodialysis. Vitamin B_1 appeared to be a strong base, while B_2 was a neutral substance.

The President, in summing up the discussion, said that the progress of the last few years should make them optimistic. He expected that we should soon know the constitution of vitamin D. The chemical problems involved were *a priori* difficult, but workers were finding that the compounds were amenable to the methods of organic chemistry. He thanked the visitors from overseas for taking part in the discussion.

(**128**, 39-40; 1931)

有总固体量的 20%。塞德尔用苯甲酰化方法去除杂质，使得纯化进行得更加彻底。通过这个方法，冯维恩从 75 kg 米糠中得到了 140 mg 纯的维生素 B_1。

塞德尔教授说，詹森教授用一个在他手上失败了的方法获得了成功。他曾认为应当避免使用沉淀剂，而正是如此才产生了在氯仿溶液中进行苯甲酰化的方法，这种方法能够除去大量的含氮物质。他还没有成功地获得结晶体，尽管他所制备的最纯物质在大鼠身上与维生素 B_1 结晶体有近似的活性，但在用彼得斯的方法对鸽子进行试验的时候，发现其活性只有维生素 B_1 结晶体的 $1/5 \sim 1/10$。

彼得斯教授说，他曾希望能够只把动物作为测试对象而不考虑其生理学方面的问题，但是他发现只有在同时考虑生理学问题的时候，才有可能在化学方面得出最终的结论。他能够确定詹森教授的制备过程直到用氯化铂溶液进行沉淀的阶段。他从对酵母的研究中得到了这样的纯化方法：用醋酸铅和氧化钡除去无活性物质，接着用活性炭吸附维生素。在 pH 值为 $5 \sim 7$ 下用磷钨酸进行分级沉淀就得到了维生素 B_1，其活性为每次剂量 0.012 mg，而詹森教授的结晶体活性为 0.008 mg。在这两种制备方法中都发生了剧烈程度相同的波利反应。里德小姐曾表示詹森教授的结晶体中有 $25\% \sim 50\%$ 具有维生素 B_4 的活性。现在她已经能够在纯化阶段前期用活性炭分级吸附的方法从维生素 B_4 中分离出维生素 B_1，但是从维生素 B_1 中却不能完全分离出维生素 B_4。

古哈博士说他已经能够用电渗析的方法从维生素 B_2 中分离出维生素 B_1。维生素 B_1 似乎是一种强碱，而维生素 B_2 是一种中性物质。

学会主席在总结这次讨论时说，最近几年的研究发展使他们感到很乐观。他预计我们应该很快就能确定维生素 D 的构成。其中所涉及的化学问题是推理上的难点，但是研究人员发现这些化合物可以用有机化学方法进行检验。他对参加这次讨论的海外人士表示感谢。

（刘振明 翻译；刘京国 审稿）

Oxidation by Living Cells[*]

J. B. S Haldane

Editor's Note

Since 1921, J. B. S. Haldane had been a reader in biochemistry at the University of Cambridge, where he concentrated on studies of enzyme kinetics. Here, after almost a decade of research into the action of enzymes and following a series of lectures delivered at the Royal Institution, Haldane summaries what is known about oxidation taking place in a cell. What is particularly surprising is just how much detail was still wanting, especially as it was only a matter of years before Hans Krebs worked out the biochemical steps behind the citric acid cycle. Krebs is known to have read and been influenced by Haldane's 1930 book *Enzymes*.

UNTIL recently our knowledge of the chemistry of respiration stopped abruptly at the boundary of the cell. We knew how the oxygen was carried to it in vertebrate blood, and the carbon dioxide carried away. We also knew that the rate of oxygen consumption by the body as a whole, and by certain organs, was a function of numerous variables, such as temperature, hydrogen ion concentration, nervous stimulation, and so on. A certain number of partially oxidised metabolites, such as β-hydroxybutyric acid, had been isolated. But such quantitative knowledge as existed with regard to the details of oxidation was mainly confined to reactions in which coloured molecules were involved: for example, the reduction of methylene blue to a colourless substance, or the oxidation of p-phenylene-diamine to a coloured one.

The modern period began with the work of Batelli and Stern, and of Bach and Chodat, in Geneva, and since the War the most important centres of research have been the laboratories of Thunberg in Sweden, of Warburg and Wieland in Germany, and of Hopkins in England. This work has led to the recognition of a number of distinct catalysts, each responsible for a different part in the process of respiration. Inorganic catalysts of oxidation may activate the oxidant, the reducer, or both. Thus, Langmuir concluded that when a hot platinum surface catalyses the union of hydrogen and oxygen, a layer of adsorbed O_2 molecules is so activated as to unite with H_2 striking them; but adsorbed H_2 does not unite with bombarding O_2. On the other hand, when the same reaction is catalysed by porcelain, both molecular species must be adsorbed side by side before they react.

We need not, therefore, be surprised to find in the cell catalysts which activate O_2, alongside of activators of reducers such as the lactate ion, and shall be prepared to steer a course between the unitary theories of Warburg[1] and Wieland[2], who respectively regard

---

[*] Substance of lectures delivered at the Royal Institution on Feb. 5, 12, and 19.

348

活细胞氧化*

約翰·波顿·桑德森·霍尔丹

編者按

从 1921 年起，霍尔丹一直在剑桥大学作生物化学专业的高级讲师，他在那里潜心研究酶动力学。在花了近十年的时间研究酶反应以及随后在英国皇家研究院作了一系列的报告之后，霍尔丹对细胞中发生的氧化反应进行了总结。令人格外惊讶的是：仅仅在几年之后，汉斯·克雷布斯就提出了柠檬酸循环背后的生物化学原理，而与这个发现有关的很多细节在这篇文章中都没有涉及。大家都知道克雷布斯曾经研读过霍尔丹 1930 年的著作《酶》并深受它的影响。

最近，人们对呼吸化学的认识在细胞周边区域突然止步不前。我们已经知道氧气是如何被运送至脊椎动物血液中的细胞里的，也知道二氧化碳是如何释放的。我们还知道机体整体和特定器官的氧气消耗率会受到很多因素的影响，如温度、氢离子浓度以及神经刺激等等。虽然人们已经分离得到了一些不完全氧化的代谢产物如 $\beta-$ 羟基丁酸，但是目前涉及氧化过程的定量研究还只局限于有色分子参与的反应，如亚甲基蓝从蓝色还原成无色产物的反应，或者对苯二胺被氧化成有色产物的反应。

新的发展阶段始于巴泰利和斯特恩以及巴赫和肖达在日内瓦的研究工作；而第一次世界大战以来，瑞典的通贝里实验室、德国的沃伯格和维兰德实验室以及英国的霍普金斯实验室成为世界上最重要的几个研究中心。他们的研究工作使人们可以识别很多种催化剂，每一种在呼吸过程中都起着不同的作用。无机的氧化反应催化剂可以活化氧化剂或者还原剂，或者既活化氧化剂又活化还原剂。据此，朗缪尔得出结论，当用高温的铂表面催化氢与氧的化合时，吸附态的氧分子层被高度活化，受到氢攻击后极易与其化合；但是反过来，吸附态的氢却很难与攻击态的氧发生化合反应。另一方面，当相同的化学反应被瓷器催化时，那么两种分子必须被吸附到彼此靠近的位置才能发生反应。

因此，我们无需惊讶于这样的发现，即细胞催化剂不仅能激活氧气，还可以作为乳酸离子等还原剂的激活剂；鉴于沃伯格 [1] 和维兰德 [2] 分别主张氧气和还原剂的激活

---

* 这是霍尔丹于 2 月 5 日、12 日和 19 日在英国皇家研究院发表的演讲。

349

the activation of oxygen and of reducers as the fundamental feature of respiration.

When an enzyme catalyses the reaction between two molecular types, one very restricted, the other very general, we describe it as specific for the former. Thus Dixon[3] and Coombs[4] found that xanthine dehydrogenase catalyses the reaction $AH_2+B=A+BH_2$, where A must be one of a small number of purine bases (it is possible that the same enzyme also activates aldehydes; if so, they are oxidised at about one percent of the rate of the purines). But B may be oxygen, iodine, nitrate, permanganate, or any of a large number of dyes, such as methylene blue. These latter are perhaps all held on the enzyme surface near the former, but it is difficult to imagine that there is a single molecular grouping responsible for activating all of them. For reaction it is not sufficient that a molecule should be united with the enzyme; it must be activated as well. Thus, uric acid unites with xanthine dehydrogenase at the same spot as xanthine, thus inhibiting its oxidation, but is not oxidised, though another enzyme can accomplish this process.

A large number of dehydrogenases are known which act in a similar manner, each causing the activation of one or more organic substrates. Thus, lactic dehydrogenase, which can be obtained in solution from a number of sources, activates several α-hydroxyacids; succinic dehydrogenase, another enzyme easily obtained in solution, activates succinic and methyl-succinic acids; and so on. The activity of these enzymes is generally measured by the rate at which they catalyse the reduction of methylene blue by their substrates. They usually have a wide range of optimal pH from about 7 to 10, instead of a small range like hydrolytic enzymes, and a fairly constant Q_{10} in the neighbourhood of 2, that is, a critical increment of about 12,000 calories. They are not inhibited by small concentrations of cyanide or sulphide, but are so by the usual enzyme poisons, such as heavy metals and nitrites, and oxidising agents. The formic dehydrogenase of *Bacillus coli* appears to be a copper compound, but there is no evidence that most dehydrogenases contain metals. Quastel[5] and his colleagues have made a very thorough study of the dehydrogenases on the surface of *Bacillus coli*. There are probably at least seven different ones, and possibly many more. In this case they can readily be shown to be concerned in oxygen uptakes. In certain conditions malonic acid inhibits methylene blue reduction by succinic acid, both uniting with the enzyme at the same point; and in high concentrations of both, the rate of reduction of methylene blue depends on the ratio of the two. Cook[6] found that oxygen is reduced at nearly the same rate as methylene blue, and malonic acid inhibits its reduction to about the same extent.

When a dehydrogenase and its specific substrate act directly on O_2 it is reduced to H_2O_2. Some anaerobic bacteria act in this way. Bertho and Glück[7] found that at least 90 percent of the O_2 consumed by *Bacillus acidophilus* is converted into H_2O_2, which damages the bacteria. But some dehydrogenases when separated will not reduce O_2, and a separate activator is required. We know rather less about the activation of O_2 than of H_2O_2. This latter can be activated by two different enzymes, catalase and peroxidase, and by heat-stable peroxidase-like substances such as cytochrome and haematins.

作用是呼吸的基本特征，我们将会在这两种理论之间找到一条前进的道路。

参与酶促反应的两类分子中，一种的范围非常有限，而另一种则具有很广泛的适用性，我们将前者称为专一性。狄克逊[3]和库姆斯[4]发现，在黄嘌呤脱氢酶催化的反应 $AH_2+B=A+BH_2$ 中，A 必须是少数嘌呤碱中的一种（这种酶也许还可以活化醛基，但是其氧化醛基的效率大约只有嘌呤的百分之一），而 B 可以是氧气、碘、硝酸盐、高锰酸盐，或者是为数众多的染料中的一种，如亚甲基蓝。虽然这些 B 类分子可能都会紧挨着 A 吸附在酶分子的表面，但很难想象只用一种分子就能把它们全部激活。分子与酶的结合不足以使反应发生，它们还必须要被激活。因此，尿酸能以与嘌呤相同的位点紧密结合在黄嘌呤脱氢酶上，以此抑制嘌呤氧化；但尿酸自身不被氧化，尽管另一种酶可以完成其氧化过程。

已知有大量的脱氢酶都以与上述相似的方式发生作用，它们都能激活一个或多个有机底物。例如，可以从多种原料的溶液中获取的乳酸脱氢酶能够活化好几种的 $\alpha-$ 羟基酸；另一种很容易从溶液中获取的琥珀酸脱氢酶能够活化琥珀酸和甲基琥珀酸等等。这些酶的活力通常以它们用催化底物还原亚甲基蓝的效率来衡量。脱氢酶常常拥有广泛的最适 pH 范围，可以从 7 一直到 10，这一点不同于最适 pH 范围很窄的水解酶；脱氢酶还具有相当稳定的 Q_{10}，大约为 2，即临界增量约为 12,000 卡路里。它们不会被低浓度的氰化物或硫化物所抑制，但是会被各种常规的酶失活剂，如重金属、亚硝酸盐，以及氧化剂抑制。虽然大肠杆菌的甲酸脱氢酶似乎是一种含铜的化合物，但是并没有证据说明大多数的脱氢酶都含有金属离子。夸斯特尔[5]及其同事对大肠杆菌表面的脱氢酶进行了彻底的研究。这些脱氢酶可能至少有七种或更多，在研究中很容易发现它们都参与了氧气摄入过程。在特定条件下，丙二酸能够抑制琥珀酸对亚甲基蓝的还原作用，因为二者都结合在酶的相同位点上；当两者浓度都很高时，亚甲基蓝被还原的速率取决于两者之比。库克[6]发现，氧气被还原的速率与亚甲基蓝几乎相同，而丙二酸对这个还原反应的抑制程度也大致等于对亚甲基蓝的抑制程度。

当脱氢酶及其特异性底物与 O_2 直接发生反应时，O_2 被还原成了 H_2O_2。一些厌氧细菌就是以这种方式发生反应的。贝尔托和格吕克[7]发现，嗜酸杆菌所消耗的 O_2 至少有 90% 转化成了对细菌有害的 H_2O_2。不过，有一些脱氢酶一旦被分离就失去了还原氧气的能力，必须有激活剂的协助才能完成还原反应。我们对 O_2 活化过程的了解比 H_2O_2 更少，后者不仅能够被两种不同的酶所活化，即过氧化氢酶和过氧化物酶，还可以被耐高温的过氧化物酶类似物所活化，如细胞色素和羟高铁血红素。

Catalase catalyses the reaction $2H_2O_2=2H_2O+O_2$. Zeile and Hellström[8] have shown that it is a derivative of haematin with a definite spectrum, and convertible into a haemochromogen, or into protophyrin. It unites with HCN to give an inactive compound. Under suitable conditions a catalase molecule can destroy more than 10^5 H_2O_2 molecules per second. Peroxidase catalyses the reactions $H_2O_2+X=H_2O+XO$, or $nH_2O_2+nX=nH_2O+nXO$, where X may be a large variety of molecules, generally aromatic, but including nitrite and HI. It can be very highly concentrated, and appears to be a coloured iron compound. Its extreme sensitivity to cyanide suggests that it is of a similar nature to catalase.

The oxygen activators (oxygenases or *Atmungsferment*) have been specially studied by Warburg.[1] They unite not only with O_2 but also with CO, for which they have a rather smaller affinity. Like CO-haemoglobins, the CO-oxygenases are generally sensitive to light. Thus the oxygen uptake of yeast in presence of glucose or alcohol is reduced to about 50 percent, in a mixture containing ten parts of CO to one of O_2, in the dark. In strong light it returns to almost normal values. By studying the relative efficiencies of different monochromatic lights, Warburg and Negelein[9] found that its spectrum is very similar to that of alkaline haematin, and still closer to that of iron-phaeophorbide-*b*.

Cook, Haldane, and Mapson[10] worked with *B. coli* in toluene-saturated buffer solutions. Under these conditions, succinic, lactic, and formic acids each lose two hydrogen atoms and no more. It is thus possible to study reactions much simpler than the complete oxidation of a substance such as glucose. They found that CO and HCN, which in moderate amounts do not prevent oxidations by methylene blue, inhibit oxygen reduction. In both cases the oxidation of lactate is more sensitive than that of formate, while that of succinate is intermediate. Hence there appear to be three oxygenases with specific relative affinities for CO and O_2, like those of the haemoglobins, and also with different affinities for HCN. Each dehydrogenase is associated with a particular set of oxygenase molecules, for oxygen uptakes in presence of formate and lactate are strictly additive, even when oxygenase activity has been reduced by HCN. If the various dehydrogenases could draw on the same common stock of oxygenase molecules for activated oxygen, this would not be so. This rather rigid organisation is probably exceptional, for oxidations by *B. coli* are largely carried out on its surface, instead of internally, and it does not contain all the three cytochromes. In the absence of toluene, similar but not quite so clear-cut results are obtained.

Cytochrome is the name for a group of metal-porphyrin compounds, the metal being probably iron, which are found in almost all cells, and have been studied by Keilin[11]. When the cell runs short of oxygen, through asphyxia, intense metabolism, or cyanide poisoning, a strong spectrum of cytochrome, resembling that of a mixture of haemochromogens, appears. If the supply of oxygen becomes adequate, the characteristic bands disappear, being replaced by a fainter spectrum of the alkaline haematin type. Cytochrome is not oxygenase, as it does not combine readily with CO or HCN. One of the three components of cytochrome, cytochrome *c*, has been obtained in fairly strong solution. It

过氧化氢酶催化的反应是 $2H_2O_2 = 2H_2O + O_2$。蔡勒和赫尔斯特伦 [8] 指出，在这个反应过程中羟高铁血红素产生了具有特定光谱的衍生物，并且转换为血色原或者原卟啉。与 HCN 的结合能够使该酶失去活性。在合适的条件下，一个过氧化氢酶分子每秒可以分解超过 10^5 个 H_2O_2 分子。过氧化物酶催化的反应是 $H_2O_2+X=H_2O+XO$，或 $nH_2O_2+nX=nH_2O+nXO$。在该反应中，X 可以是一大类分子，一般是芳香类分子，也包含亚硝酸盐和 HI。该酶可以高度浓缩，是一种有色的含铁化合物，对氰化物的高度敏感性表明它与过氧化氢酶具有相似的特性。.

沃伯格 [1] 专门研究了氧活化剂（加氧酶或者呼吸酶）。氧活化剂不仅与 O_2 结合，还可以与 CO 相结合，只是它们之间的亲和力稍弱。与 CO– 血红蛋白相似，CO– 加氧酶一般也会对光敏感。因此，将 CO 与 O_2 以 10∶1 的比例混合，如果有葡萄糖或乙醇存在，酵母在黑暗条件下对混合物中氧气的摄入量会减少到 50%，而在强光下会恢复到接近正常水平。通过研究不同单色光的相对效率，沃伯格和内格莱茵 [9] 发现此复合物的光谱与碱性羟高铁血红素很相似，而与铁–脱镁叶绿酸 –b 的光谱更为相近。

库克、霍尔丹和马普森 [10] 三人将大肠杆菌置于甲苯饱和的缓冲溶液中进行研究，在这种条件下琥珀酸、乳酸以及甲酸都只脱去了两个氢原子。这就使得对反应的研究比研究像葡萄糖那样的完全氧化要简单得多。他们发现适量的 CO 和 HCN 不会抑制亚甲基蓝的氧化活力，但是会抑制氧气的还原力。在这两种分子存在的条件下，乳酸比甲酸更容易被氧化，而琥珀酸氧化则位于两者之间，这表明可能有三种加氧酶存在，它们像血红蛋白一样对 CO 和 O_2 的相对亲和力各不相同，对 HCN 的亲和力也不相同。每一种脱氢酶都与一个特定系列的加氧酶相关联，因为在甲酸和乳酸存在的条件下，即使加氧酶的活力被 HCN 抑制，氧气的摄入量也是严格遵守相加性法则的。如果不同的脱氢酶能帮助相同的加氧酶携带等量的活化态氧，那么结果就不会如此。这种严格的组织很可能具有例外，因为大肠杆菌的氧化作用主要发生在其表面而不是内部，另外，大肠杆菌也不含有全部的三种细胞色素。在去除甲苯后，研究人员获得了虽然相似但不很确定的实验结果。

细胞色素指的是金属与卟啉之间形成的一类复合物，与卟啉结合的金属通常是铁，在几乎所有的细胞中都存在，而且基林 [11] 也对此进行了研究。当细胞缺氧并导致窒息、剧烈代谢或氰化物中毒时，会产生很强的细胞色素光谱，与血色原混合物的光谱类似。而当氧气的供应变得充足时，这个特征谱带就会消失，取而代之的是较弱的碱性羟高铁血红素类的光谱。细胞色素不是加氧酶，因为它不易于与 CO 或者 HCN 结合。目前研究人员已经得到了细胞色素的三大组分之一——细胞色素 c 的

is a red substance, only slowly oxidised by molecular oxygen, readily by mild oxidising agents. It can be reduced by reducing agents or living tissues, and is an iron-porphyrin compound. Keilin found that the oxygen uptake of a system composed of oxygenase from heart muscle, cytochrome, and cysteine behaves like that of a tissue to cyanide and CO. With this system he was able to show that oxygenase is heat-labile like an enzyme, which cytochrome is not. In plants, a particular type of oxygenase, which Keilin calls catechol oxidase, and the CO compound of which, where investigated, has been found to be insensitive to light, yields H_2O_2 when oxidising catechol and its derivatives, as shown by Onslow[12].

We can thus give a scheme (Fig. 1) which probably covers most of the oxidation process in the average cell. In anaerobes one or more of the catalysts is absent. Oxygen is activated by oxygenase, which is reduced by cytochrome, and the latter is reduced in turn by dehydrogenases of the common or anaerobic type, that is, those which cannot reduce O_2 directly. This process is occasionally simplified, as in *B. coli*, where cytochrome does not seem to intervene in certain oxidations. Oxygen can also be reduced to H_2O_2, either by catechol oxidase or by an aerobic dehydrogenase. This H_2O_2 is used for further oxidation with peroxidase, or is destroyed by catalase. The latter can act as a safety-valve owing to its low affinity. Whereas peroxidase acts most rapidly in a concentration of H_2O_2 which may be as low as 10^{-6} *M.*, catalase has an optimum H_2O_2 concentration of about 0.2 *M.*, and at 10^{-6} *M.* is working at only about 0.0004 of its maximum rate. Other substances may act as intermediates. Glutathione appears to remove hydrogen from certain groupings in proteins, the reduced glutathione being later re-oxidised. St. György's[13] hexuronic acid is apparently reduced by dehydrogenases and oxidised by peroxidase. Doubtless many more similar substances will be discovered in future.

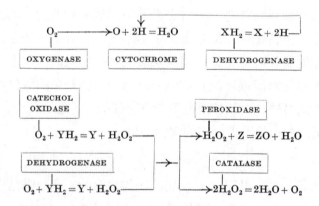

Fig. 1. The names of the catalysts are given in rectangles, the molecular species activated by each being indicated. The molecule YH_2 is catechol or a derivative in the case of catechol oxidase (Onslow's oxygenase) and is a purine base in the case of xanthine oxidase acting as a reducer of O_2. X and Z may be very varied.

高浓度溶液，它是一种红色物质，只能被分子态的氧缓慢氧化，易于被温和的氧化剂所氧化。它可以被还原剂或者活组织所还原，是一种铁–卟啉复合物。基林发现，由来自心肌的加氧酶、细胞色素以及半胱氨酸所组成的系统对氧气的摄入与一个组织对氰化物以及 CO 的摄入相似。通过这个系统，他可以说明加氧酶像多数酶一样具有热不稳定性，而细胞色素却不具有这样的特性。翁斯洛指出：在植物中有一种特殊的加氧酶，被基林称为儿茶酚氧化酶，它与 CO 的复合物对光不敏感，这种加氧酶在氧化儿茶酚及其衍生物时会生成 H_2O_2[12]。

因而我们可以画出一张图（见图 1），该图很可能涵盖了常规细胞内的大多数氧化过程。在厌氧微生物中会缺失其中的一个或多个催化剂。在这类体系中，氧气能被加氧酶活化，然后被细胞色素所还原，而细胞色素需要先被普通或厌氧型的脱氢酶所还原。也就是说，脱氢酶不能直接还原氧气。这个过程有时候也会被简化，比如在大肠杆菌中，细胞色素可能并不参与某些氧化过程。O_2 也可以被儿茶酚氧化酶或者需氧型脱氢酶还原为 H_2O_2，这些 H_2O_2 可以在过氧化物酶的作用下进一步氧化，或者被过氧化氢酶分解。后者因亲和力低而可以起到安全阀的作用。过氧化氢酶催化反应的最适 H_2O_2 浓度为 0.2 M，当 H_2O_2 浓度低至 10^{-6} M 的水平时，它的反应速率只有最大反应速率的万分之四；而即使在这样低的浓度下，过氧化物酶也能快速地与 H_2O_2 发生作用。其他物质可能是一些中间产物。谷胱甘肽似乎具有脱去蛋白质中某些特定基团上的氢原子的能力，被还原的谷胱甘肽随后还会重新被氧化。圣哲尔吉[13] 的己糖醛酸就显然能被脱氢酶所还原，也可以被过氧化物酶氧化。毫无疑问，人们将会发现越来越多的类似物质。

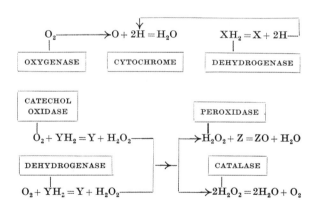

图 1. 长方格中给出的是各种催化剂的名称，图中标出了它们各自催化的分子种类。YH_2 在儿茶酚氧化酶（翁斯洛加氧酶）所催化的反应中代表儿茶酚或者儿茶酚的衍生物，在黄嘌呤氧化酶催化的反应中代表的是作为氧气还原剂的嘌呤碱。X 和 Z 所代表的分子范围则可能非常广泛。

We note the great importance of metal-porphyrin compounds (Table I.). It is fairly clear that their catalytic function is primitive. They have afterwards been modified to act as stores or carriers of oxygen in higher animals. Except in chlorophyll, a magnesium compound, the metal united with the porphyrin is usually, if not always, iron. As oxygenase, catalase, and cytochrome are almost universally distributed, we need not be surprised that haemoglobin and related pigments such as chlorocruorin have often been independently evolved, a suitable protein being combined with iron-porphyrin residue.

Table I. Metal-porphyrin in Compounds found in Cells.

1. Catalysts. Chlorophyll *a* and *b*. Oxygenases. Catalases. Probably peroxidases. Cytochrome *a*, *b*, and *c*.

2. Mainly concerned in oxygen storage. *Arenicola* haemoglobin.

3. Mainly concerned in oxygen carriage. Vertebrate haemoglobins.

4. Uncertain whether in group 2 or 3. Many invertebrate haemoglobins. Chlorocruorin. Helicorubin, etc.

We know little as to the immediate source of CO_2. Two obvious processes are available, the dehydrogenation of formic acid, and the decarboxylation of pyruvic acid and related compounds by the enzyme carboxylase according to the equation: $R-CO-COOH = R-COH+CO_2$. The enzymes concerned in both processes have a wide distribution.

Still less is known of how in detail the energy made available in oxidation is passed on, or of how the rate of oxidation is controlled, though both processes are evident enough in the whole organism. The control is largely exercised on the fuel supply. Thus a rise of blood sugar in man causes an increased oxygen consumption. Hormones such as thyroxin are also concerned, but their mode of action is not understood. The energy is not generally liberated directly, but is largely employed in building up compounds of high chemical potential. These may either form new tissue or be available for rapid energy production. Thus the oxidation of sugar or lactic acid in muscle provides energy for the resynthesis of glycogen from lactic acid, and of phosphagen from creatine and phosphate.

If these processes are to be efficient, two conditions must be fulfilled. The energy of oxidation must be made available in quanta somewhat, but not greatly, larger than those required for synthesis; and the molecules undergoing oxidation and synthesis must be united with the same catalyst, which must thus have a double specificity. The significance of the complicated oxidising systems here described will remain obscure until we know what syntheses are correlated with each of them. A beginning of such an analysis has been made by Wurmser[14]. But an indispensable preliminary is a study of the total and free energy changes in the various reactions which are linked. This is still in its very early stages.

But even when we know the stages in the oxidation of different substances, and the use to which the energy thus made available is put, we shall be faced with the problem of regulation. In the living cell the activity of peroxidase or lactic dehydrogenase is doubtless governed by laws as definite as those which govern that of the heart in the living organism, laws which can be stated both in terms of chemistry and of biological function. The question of whether these two types of explanation can be reconciled, or whether one

356

我们注意到金属－卟啉复合物具有相当的重要性（见表1）。相当明确的一点是它们的催化功能都很简单。后来又认为它们可以在较高等动物的体内储备和运输氧。除了叶绿素是金属镁的复合物外，在其他大多数情况下，与卟啉结合的金属都是铁。由于加氧酶、过氧化氢以及细胞色素几乎到处都有，所以我们不必对血红蛋白和相关色素如血绿蛋白往往会不受限制地转变成与铁－卟啉残基结合的蛋白而感到惊讶。

表 1. 细胞中发现的金属－卟啉复合物

1. 催化剂。叶绿素 *a* 和 *b*，加氧酶，过氧化氢酶，可能有过氧化物酶，细胞色素 *a*、*b*、*c*。
2. 主要与氧的储备有关。沙蚕的血红蛋白。
3. 主要与氧的运输有关。脊椎动物的血红蛋白。
4. 尚不确定在第 2 组或第 3 组中是否存在。许多无脊椎动物的血红蛋白，血绿蛋白，蠕虫血红蛋白等。

对于 CO_2 的直接来源我们知之甚少。已知的两个显而易见的相关过程是甲酸的脱氢反应和丙酮酸及其相关化合物在羧化酶作用下的脱羧，所依据的方程式是：$R-CO-COOH = R-COH + CO_2$。这两个过程所涉及的酶都有很多种。

对于在氧化过程中所获得的能量具体是如何传递的以及怎样控制氧化的速率我们还知之甚少，尽管这两个过程在整个机体中是显然存在的。控制主要是由燃料供应实现的。因此一个人血糖升高会导致氧气的消耗量增加。另外，荷尔蒙如甲状腺素也会受到影响，不过它们的作用方式还不为人所知。氧化过程中产生的能量通常不会直接释放出去，其中很大一部分要用于合成高能化合物。这些高能化合物可能用于构成新的组织，也可能用于快速地产生能量。因此肌肉中糖或者乳酸的氧化可以为乳酸再合成糖原以及肌酸与磷酸盐再合成磷酸原提供能量。

要使这些过程都能有效运行，需要满足两个条件：氧化产生的能量必需略高于合成所需的能量；参与氧化和合成的分子必须结合在相同的催化剂分子上，因此该酶分子必须具有双重特异性。关于这里所说的复杂氧化系统，在弄清它们分别与哪些合成有关之前，人们还不清楚它们的重要性。维尔姆塞首先开展了对这方面的分析研究 [14]。但是此工作的一个必要前提是需要研究各个相互连接的反应中总能量以及自由能的变化，而这些研究尚处于刚刚起步的阶段。

不过，即使我们知道不同物质在氧化反应中所扮演的角色，也知道由此产生的能量是如何使用的，我们还将面临有关调控的问题。毫无疑问，正如机体遵守一些法则来维持心脏的正常功能一样，活细胞中过氧化物酶或乳酸脱氢酶的活性也严格遵守着某些法则，这些法则可以同时满足化学和生物学的要求，而目前这两种解释是否能统一还是其中某一种解释是多余的问题仍需要通过讨论解决。当然，这一天

of them is superfluous, will then have to be fought out. However, that day is far distant; meanwhile the biochemist can continue to accumulate knowledge without committing himself on philosophical questions.

(**128**, 175-178; 1931)

References:

1. Warburg, *Über die katalytische Wirkungen der lebendigen Substanz* (1926).

2. Wieland, *Ergeb. Physiol.*. **20**, 477 (1922).

3. Dixon, *Biochem. Jour.*, **20**. 703 (1926).

4. Coombs, *Biochem. Jour.*, **21**, 1259 (1927).

5. Quastel and colleagues, *Biochem. Jour.*, 1924–1928 (Bibliography in *Jour. Hyg.*, **28**, 139; 1928).

6. Cook, *Biochem. Jour.*, 24, 1538 (1930).

7. Bertho and Glück, *Naturwiss.*, **19**, 88 (1931).

8. Bolle and Hellerüm, *Zeit. physiol. Chem.*, 192, 171 (1930)

9. Warburg and Negelein, *Biochem. Zeit.*, **202**, 202 (1928).

10. Cook, Haldane, and Mapson, *Biochem. Jour.*, **25**, 534 (1931).

11. Keilin, *Proc. Roy. Soc.*, B, **98**, 312 (1925): **104**, 236 (1929).

12. Onslow and Robinson, *Biochem. Jour.*, **20**, 1138 (1926).

13. St. György, *Biochem. Jour.*, **22**, 1387 (1928).

14. Wurmser, *Oxidations et réductions* (1930).

的到来会很遥远，在此期间生物化学家们可以不断积累相关知识，而不必考虑哲学上的问题。

（高如丽 翻译；刘京国 审稿）

Oestrus-producing Hormones

G. F. Marrian and A. Butenandt

Editor's Note

In the 1930s, scientists were occupied with the identification not only of vitamins but also of hormones. The latter can be synthesised by the animal bodies in which they are essential ingredients of life, whereas vitamins are obligatory food supplements. This paper described the isolation of a hormone that induces oestrus in animals. Adolf Butenandt was one of Germany's most respected chemists of the period and was awarded a Nobel Prize in 1939. Oestrus-producing hormones and their analogues are a common route to birth-control pills.

RECENTLY, Doisy and his co-workers (1931) have reported the isolation from the urine of pregnancy of a crystalline substance possessing oestrus-producing activity, which is distinct from the active substance theelin, previously described by them. The latter substance, to which they gave the formula $C_{18}H_{21}(OH)_2$, was shortly afterwards isolated by one of us (Butenandt, 1929)[1] and by Dingemanse and co-workers[2] (1930). It was shown afterwards (Butenandt, 1930) that this substance is represented by the formula $C_{18}H_{22}O_2$, and that it behaves either as a hydroxy ketone or as a dihydroxy alcohol.

There is no doubt that the second substance isolated by Doisy and his co-workers[3], to which they give the formula $C_{18}H_{21}(OH)_3$, is identical with that fully described earlier by one of us (Marrian, 1930)[4]. Although Prof. Doisy refers to the triol previously isolated, there is no suggestion in his papers that it had been characterised as a trihydroxy substance of the formula $C_{18}H_{21}(OH)_3$. His view that the substance described by one of us is a mixture of both active substances is apparently based solely on a difference between the *uncorrected* melting points. The evidence of the analytical data, which clearly shows this supposition to be untenable, is ignored.

A year ago when the presence in urine of two distinct oestrin-producing substances was clear to us, we were considerably puzzled over the relationship between them. The suggestion was tentatively advanced (Marrian, 1930) that the substance $C_{18}H_{22}O_2$ on treatment with hot alkali took up the elements of water to form $C_{18}H_{24}O_3$. This supposition was afterwards shown to be incorrect (Butenandt, 1930), since the former substance proved to be unchanged by such treatment. At the same time it was shown that both substances occur together in urine, and that by distillation in a high vacuum with potassium bisulphate, $C_{18}H_{24}O_3$ could be converted into $C_{18}H_{22}O_2$. Prof. Doisy has made no adequate reference to this work, and has advanced the earlier view, which has been shown to be untenable.

(**128**, 305; 1931)

催情激素

盖伊·弗雷德里克·马里安，阿道夫·布特南特

编者按

在 20 世纪 30 年代，科学家们不仅从事于维生素的鉴定，也进行激素的鉴定。它们是生命活动必需的基本化学物质，激素能由动物体自身合成，而维生素则必须通过食物供给。本文描述了一种诱导动物发情的激素的分离。作者阿道夫·布特南特是德国当时最受人尊敬的化学家之一，他曾于 1939 年获得了诺贝尔化学奖。催情激素及其类似物是计划生育药物的常规成分。

最近，多伊西及其同事（1931 年）报道说他们从孕妇尿液中分离出一种具有催情活性的结晶性物质，该物质不同于他们之前提到的活性物质雌酮。之前提到的物质的化学式被他们定为 $C_{18}H_{21}(OH)_2$，这种物质随后被我们中的一位作者（布特南特，1929 年）[1] 和帝格曼斯及其同事 [2]（1930 年）相继分离。后来发现（布特南特，1930 年）该物质的化学式应为 $C_{18}H_{22}O_2$，它的化学行为类似于羟基酮或双羟基醇。

毫无疑问，被多伊西及其同事 [3] 分离出来并将分子式定为 $C_{18}H_{21}(OH)_3$ 的第二种物质与之前我们中的一位作者曾完整描述过的物质完全相同（马里安，1930 年）[4]。尽管多伊西教授参考了早先分离出来的三醇，但是在其文章中没有提示它是一种化学式为 $C_{18}H_{21}(OH)_3$ 的三羟基物质。他认为我们中的一位作者所描述的物质是两种活性物质的混合物，这一论点显然仅仅是基于两者**未经过校正**的熔点的差异。而明显证明这种观点站不住脚的分析数据却被其忽略了。

一年之前我们就已经知道尿中存在两种不同的催情物质，但对于它们之间的关系则非常困惑。有人提出假设（马里安，1930 年）认为经过热碱处理的 $C_{18}H_{22}O_2$ 因夺取水分子而形成了 $C_{18}H_{24}O_3$。这个假设后来被证明是错误的（布特南特，1930 年），因为已证明前者在经过这种处理后不会发生改变。另一方面，研究显示这两种物质在尿中是同时存在的，而且当在高度真空的条件下与硫酸氢钾一起蒸馏时，$C_{18}H_{24}O_3$ 能转变成 $C_{18}H_{22}O_2$。多伊西教授并没有充分地参考这些工作，而是进一步推进其早期提出的、业已证明是站不住脚的观点。

（毛晨晖 翻译；田伟生 审稿）

G. F. Marrian and A. Butenandt: London and Göttingen, July 23.

References:

1. Butenandt, *Naturwiss.*, **17**, 879 (1929). *Deutsch. Med. Woch.*, **55**, 2171 (1929). *Zeit. für physiol. Chem.*, **191**, 140 (1930). *Abh. d. Ges. d. Wissensch. zu Gottingen* (1931). *Math. phys.* Kl. iii. Folge, Heft 2.

2. Dingemanse *et al.*, *Deutsch. Med. Woch.*, **56**, 301 (1930).

3. Doisy *et al.*, *Proc. Soc. Exp. Biol. Med.*, **28**, 88 (1930). *J. Biol. Chem.*, **91**, 641, 647, 653, 655 (1931).

4. Marrian, *Chem. and Ind.* (June 20, 1930). *Biochem. Jour.*, **24**, 1021 (1930).

The Biological Nature of the Viruses*

H. H. Dale

Editor's Note

As late as the 1930s, the properties of the viruses appeared to be a complete mystery. Although infectious diseases were an important public health problem at the time, it was clear that bacteria could explain some but not all of these infections. Common ailments such as smallpox and scarlet fever appeared to have no mechanism to account for their prevalence, but the concept of sub-microscopic entities responsible for disease was for many people a way of avoiding the problem. Sir Henry Dale was a pharmacologist and physiologist who began work as a scientist at the British drug manufacturer the Wellcome Foundation. In this talk to the British Association for the Advancement of Science, Dale firmly advocated the view that viruses were real entities and that understanding their nature should be one of the principal objectives of biological research.

THE viruses are a group of agents, the existence of which would certainly be unknown to us but for the changes produced by their presence in the bodies of higher animals and plants. They seem to have one property at least of living organisms, in being capable, under appropriate conditions, of indefinite reproduction. We know nothing of their intrinsic metabolism: it has even been asserted that they have none. Few of them have yet been rendered visible by the microscope; it is, indeed, a question for our discussion whether any of them have yet been seen or photographed. It is a question, again, whether any of them, or all of them, consist of organised living units, cells of a size near to or beyond the lowest limits of microscopic visibility; or whether, as some hold, they are unorganised toxic or infective principles, which we can regard as living in a sense analogous to that in which we speak of a living enzyme, with the important addition that they can multiply themselves indefinitely. Some, however, would attribute this, not to actual self-multiplication, but to a coercion of the infected cells to reproduce the very agent of their own infection.

The problems presented by the nature and behaviour of the viruses cannot fail to raise questions of the greatest interest to anyone concerned with general physiological conceptions. What is the minimum degree of organisation which we can reasonably attribute to a living organism? What is the smallest space within which we can properly suppose such a minimum of organisation to be contained? Are organisation, differentiation, separation from the surrounding medium by a boundary membrane of special properties, necessary for the endowment of matter with any form of life? Or is it possible to conceive of a material complex, retaining in endless propagation its

* From the presidential address introducing a discussion on the subject in Section I (Physiology) of the British Association in London on Sept. 28.

364

病毒的生物学本质[*]

亨利·哈利特·戴尔

编者按

在 20 世纪 30 年代以前，病毒的本质还完全是一个未解之谜。虽然传染病在当时是一个重要的公共卫生问题，但细菌显然只是引起一部分传染病而非全部传染病的原因。人们无法解释像天花和猩红热这样的常见疾病是如何流行起来的，但对许多人来说，把亚显微物质看作是致病因素的想法可以作为回避上述问题的一种方法。亨利·戴尔爵士是一名药理学家和生理学家，他以科学家的身份在英国韦尔科姆基金会制药公司工作。在这篇对英国科学促进会的报告中，戴尔坚定地主张病毒是真实的实体，而且了解其本质应该是生物学研究的主要目标之一。

病毒是这样一群物质，要不是能够发现它们在高等动植物体内生存所造成的改变，我们必然无法知道它们的存在。病毒似乎至少具有生命的一种特征，那就是在合适的条件下能够无限繁殖。我们对它们的内部代谢一无所知；甚至有人声称它们根本就没有代谢。病毒在显微镜下几乎看不到；实际上，它们是否已经被看到或者被拍摄到仍是一个需要讨论的问题。另外，它们中的一部分或者全部是否具备有组织的生命单位，即大小接近或者低于显微可见的最小的细胞仍是一个疑问；或者正如某些人所说，它们只是无组织的毒性或者感染性物质，我们可以认为它们以一种类似于我们所说的有活性的酶的方式生存，并且还具有一个更重要的特征是它们能够无限繁殖。但是，有些人认为这不是真实的自我繁殖，而是强迫被感染的细胞产生它们自己的具有感染性的物质。

病毒的本质和行为这类难题向关注普通生理学概念的人们提出了非常有意思的挑战。我们能够合情合理地归为生命的最低组织化程度是什么？能够包含生物体最小组织的最小空间有多大？有组织、能分化以及通过一层特殊性质的界膜与周围基质分离是否是具有任何形式的生命体所必备的特征？或者是否可以设想它是一种物质复合体，保持了无穷尽地复制的生理特征，正如被感染细胞做出的高度特异性的反应一样，尽管它不是组织有序的单位，而是均匀地分散在了含水的介质中？对那

---

[*] 本文来自作者作为主席在英国科学促进会 9 月 28 日于伦敦举行的 I 分会（生理学分会）讨论中引入这一主题时所作的报告。

365

physiological character, as revealed by the closely specific reaction to it of the cells which it infects, though it is not organised into units, but uniformly dispersed in a watery medium? Among those who study the viruses primarily as pathogenic agents, these questions provide matter for debate; I suggest that they are questions with which the physiologist may properly be concerned.

I cannot deal with the history of the subject; but it is of interest to note that Edward Jenner was dealing, in small-pox and vaccinia, with what we now recognise as characteristic virus infections, long before there was any hint of the connexion of visible bacteria with disease. Pasteur himself was dealing with another typical case of a virus infection in the case of rabies. The clear recognition, however, of the existence of agents of infection, imperceptible with the highest powers of ordinary microscopic vision, and passing through filters fine enough to retain all visible bacteria, begins with Ivanovski's work in 1892 on the mosaic disease of the tobacco plant, brought to general notice and greatly developed by Beijerinck's work on the same infection some seven years later; and with Loffler and Frosch's demonstration, in the same period, that the infection of foot-and-mouth disease is similarly due to something microscopically invisible, and passing easily through ordinary bacteria-proof filters. Since those pioneer observations the study of viruses has spread, until they are recognised as the causative agents of diseases in an imposing and still growing list containing many of the more serious infections of man, animals, and plants.

If we are to discuss the biological nature of the viruses, it is obvious that we should begin by attempting some kind of definition. What do we mean by a virus? And what are the tests by which we decide that a particular agent of infection shall be admitted to, or excluded from, the group? But a few years ago I think that we should have had no difficulty in accepting three cardinal properties as characterising a virus, namely, invisibility by ordinary microscopic methods, failure to be retained by a filter fine enough to prevent the passage of all visible bacteria, and failure to propagate itself except in the presence of, and perhaps in the interior of, the cells which it infects. It will be noted that all three are negative characters, and that two of them are probably quantitative rather than qualitative.

Such a definition is not likely to effect a sharp or a stable demarcation. We shall see that its failure to do so is progressive. Nevertheless it would still be difficult to refuse the name of virus to an agent which fulfils all three criteria; and we must therefore, in consistency, apply it, on one hand, to the filtrable agents transmitting certain tumours, and, on the other hand, to the agents of transmissible lysis affecting bacteria, and now widely known and studied as bacteriophages. But the strict application of such a definition, based on negative characteristics, must obviously narrow its scope with the advance of technique. We may look a little more closely at the meaning of these different characters.

366

些把病毒作为主要致病因子来研究的人来说，这些问题是争论的焦点；我认为这些是生理学家更应该关注的问题。

我无法阐述有关这个话题的历史；但值得注意的是，早在有任何提示显示可见的细菌和疾病有什么联系之前，爱德华·詹纳在天花以及牛痘治疗中就研究过这个问题，这种病我们现在认识到是典型的病毒感染病。巴斯德研究的是另一个典型的病毒感染性疾病，即狂犬病。但是，真正清楚地认识到存在一种具有感染性的物质，这种物质即使用最高能力的普通显微镜也看不到，而它能够通过足以阻拦任何可见细菌的滤器，是在 1892 年伊万诺夫斯基研究烟草的花叶病之后。7 年以后贝杰林克对同一种疾病展开了研究，使这种认识得到了广泛的关注和极大的发展。同一时期，洛夫勒和弗罗施证明感染口蹄疫的原因也是某些显微镜下不可见并能轻易通过普通滤菌器的物体。自从有了这些开创性的观察结果，对病毒的研究逐渐流行，直到它们被确认为是人类和动植物多种极其严重的传染病的病因，并且列入的重大疾病还在不断增加。

很显然，如果我们要讨论病毒的生物学本质，首先应该尝试着给出某种定义。我们对病毒的定义是什么？我们通过什么检查方法能够确定一种特定的传染病原是或者不是病毒？但是数年以前，我想我们都已经毫无疑问地接受了病毒的三个基本性质，即普通显微镜下不可见；能够通过足以阻挡所有可见细菌的滤菌器；只有在受感染细胞存在的条件下才能繁殖，而且繁殖很可能是在细胞体内进行的。我们注意到所有这三条都是否定性的特征，而且其中两条很可能是可量化指标，而不是定性指标。

这样一个定义不可能给病毒划出一个清晰或者稳定的界限。我们将看到没有给出明确的界限是一种进步。然而，当某种物体满足了这三种特性后，我们仍然很难说清它不是病毒。因此，我们必须保持一致地将这个定义一方面应用到能够传播某些肿瘤且可穿过滤菌器的物体上，另一方面应用到能够传染性地裂解受感染细菌的物体，后者现在已经众所周知并被作为噬菌体研究。但是由于定义采用的都是否定性特征，想严格应用就必须利用技术的进步大大缩小其范围。我们应当更加细致地研究这几个特性的意义。

Microscopic visibility is obviously a loose term. Rayleigh's familiar formula, in which the lower limit of resolution is equal to one-half the wavelength of the light employed, divided by the numerical aperture of the objective, only gives us the smallest dimensions of an object, of which, with the method of transmitted illumination habitually used in former years, a critical image can be formed. There can be no doubt that the separate particles of practically all the agents to which the term virus would be applied fall below this limit of size. To put it in plain figures, their diameter is less than 0.2 micron. On the other hand, progress has recently been, and continues to be, rapid in the direction of bringing into the visible range minute bodies associated with a growing number of viruses. This has been effected, on one hand, by improvement in staining technique, which probably owes its success largely to increase of the natural size of the particles by a deposit of dye on their surface; and, on the other hand, by forming visible diffraction images of the unstained particles with wide-aperture dark-ground condensers, and by photographing the images formed of them with shorter invisible rays. Mr. Barnard has obtained such sharp photographic images of the bodies associated with one virus, measurements of which give their natural size by simple calculation.

The reaction of a cautious criticism to such a demonstration seems to have taken two different directions. There has been a tendency, on one hand, to exclude an agent from the group of viruses as soon as the microscope could demonstrate it with some certainty. Many have for years thus excluded the agent transmitting the pleuro-pneumonia of cattle, though the status of this organism has been compromised even more by the success of its cultivation on artificial media. Visibility seems to have rendered doubtful the position of the Rickettsia group of infections, and, if the test is logically applied, the process of exclusion can scarcely stop before the agents transmitting psittacosis, fowl-pox, infectious ectromelia, and even vaccinia and variola, have been removed from the group of viruses into that of visible organisms.

In discussing the biological nature of viruses as a whole, however, we can scarcely begin by accepting an artificial and shifting limitation of that kind. The real task before us, rather, is to discuss to what extent the evidence of these recent developments, which appear to show that some of the agents, known hitherto as viruses, consist of very minute organisms, can safely be applied to other viruses which are still beyond the range of resolution. Do these also consist of organisms still more minute, or are any of them unorganised? Another line of criticism, sound in itself, while not excluding from the virus group these agents for which microscopic visibility has been claimed, demands more evidence that the minute bodies seen or photographed are really the infective agent, and not merely products of a perverted metabolism which its presence engenders.

It is obvious that complete evidence of identity cannot be obtained until a virus has been artificially cultivated in an optically homogeneous medium. Meanwhile it is a question of the strength of a presumption, on which opinions may legitimately differ. Let us recognise that the evidence is not perfect, but beware of a merely sterilising scepticism. I suspect

显微镜下的可见性明显是一个不精确的界限。在瑞利的著名公式中，分辨率的下限等于使用的光波长的一半除以物镜的数值孔径。该公式仅仅给出了在使用前几年常用的透射照明显微镜时，能够形成临界图象所需的物体的最小尺寸。毫无疑问，实际上那些可以被称为病毒的单个粒子都在这个下限以下。用简单的数字表示就是，直径小于 0.2 微米。另一方面，最近关于将微小颗粒与数量不断增长的病毒相附着使其进入可见范围的研究已经取得了快速的进展，并且这些研究还在继续。这项研究的进展，一方面，受到染色技术进步的影响，其成功的原因很可能主要在于染料在颗粒表面沉积导致其自然尺寸变大；而另一方面，影响因素也包括通过大孔径暗视野聚光镜形成未染色颗粒的可视性衍射图像，并用更短的不可见射线拍摄下这些颗粒所形成的图像。巴纳德先生已经获得了与某个病毒相黏附的颗粒物质的清晰图像，经过测量和简单的计算可以得出病毒的自然尺寸。

针对这一论证似乎有两类不同的慎重批评。一方面，一些人倾向于将显微镜可以在某种程度上辨认出的物质从病毒种类中排除。这样，传播牛胸膜肺炎的病原就被排除了，尽管在人工基质上获得成功培养后使得这种生物更加偏离了病毒的定义。可见性似乎使得立克次氏体的定位仍存有疑问，而且如果合乎逻辑地运用此方法进行检验，排除的过程会一直继续，直到传播鹦鹉热、禽痘、传染性脱脚病、甚至牛痘和天花的病原都从病毒转移到可见生物群体中。

但是，要从整体上讨论病毒的生物学本质，我们不能从一开始就接受一个人为的还在变动的限制。最新的研究进展提供证据显示一些迄今为止被认为是病毒的物体都是由非常微小的有机组织组成，而我们真正面临的任务是要讨论这些证据到底在多大程度上能够可靠地应用到那些仍然在分辨率所及范围以外的病毒中。它们也是由更小的有机组织组成的吗？或者其中的某些是无组织的？另一个本身非常合理的批评虽然没有将这些显微镜下可见的物体排除出病毒的范畴，但要求要有更多的证据来支持这些所见的和所摄到的微小物体就是真正的传染病原，而不仅仅是由于病原的存在而造成的异常代谢所产生的代谢产物。

显然，我们不可能获得病毒的完整鉴别性证据，除非这种病毒在光学均一性培养基中可以人工培养。同时，这是一个假设程度的问题，在假设上存在不同的观点是合情合理的。我们要认识到证据并不完美，但是也要谨防打倒一切的怀疑论。我

that the attitude of some critics is coloured by past history of the search for viruses and especially by that part of it concerned with the curious objects known as "inclusion bodies", which are readily demonstrated with relatively low powers of the microscope, in the cells of animals and plants infected with certain viruses. From the earlier and admittedly hasty tendency to identify them as infective protozoa, opinion seems to have swung too quickly to the opposite extreme, of dismissing them as mere products of the infected cell. It is so comparatively simple, in some cases, to separate these bodies, that it is surprising that so few efforts have been made to test their infectivity. However, the power of such a body to convey at least one virus infection has been demonstrated; and since they have further been shown, in several cases, to consist of a structureless matrix packed with bodies looking like minute organisms, the burden of proof in other cases seems to me, for the moment, to rest on those who suggest that they consist wholly of material precipitated by the altered metabolism due to the infection.

The physical evidence, obtained by filtration through porous fabrics and colloidal membranes, and by measuring rates of diffusion, is, of course, purely concerned with the size of the units of infective material, and must be taken in conjunction with the evidence provided by the microscope. The crude qualitative distinction between the filterable and non-filterable agents of infection has long since ceased to have any real meaning. There is no natural limit of filterability. A filter can be made to stop or to pass particles of any required size. It is now realised that the only proper use of a filter in this connexion is to give a quantitative measure of the maximum size of the particles which pass it. Evidence from failure to pass must always be subject to correction for the effects of electrostatic attraction and fixation by adsorption on the fabric of the filter. A large amount of filtration evidence has, further, been vitiated by reliance on determinations of the *average* pore size of the filter. In dealing with an infective agent, the test for the presence of which depends on its propagation under suitable conditions, it is obviously the maximal pore size which is chiefly significant.

For these reasons a good deal of the evidence showing that certain viruses can be detected in the filtrates, obtained with filters which will not allow haemoglobin to pass in perceptible quantities, must be regarded at least with suspicion. Dr. Elford has recently succeeded in preparing filter-membranes of much greater uniformity, with a small range of pore-diameters. His measurements, with these, of the sizes of the particles of different viruses, show a range approaching the dimensions of the smallest recognised bacteria, on one hand, and falling as low, in the case of the virus of foot-and-mouth disease, as about three or four times the size of the haemoglobin molecule; the latter being given not only by filtration-data, but also by other physico-chemical measurements, such as those obtained by Svedberg with the ultracentrifuge. It should be noted, as illustrating the difficulties of the problem and the uncertain meaning of some of the data, that Elford has regularly found a bacteriophage to be stopped by a membrane which allows the foot-and-mouth virus to pass; while, on the other hand, recent determinations of the rate of diffusion of bacteriophage, made by Bronfenbrenner, put the diameter of its particles at 0.6 of a millimicron, that is, only about one-fifth of the accepted dimensions of the haemoglobin

猜想一些评论家的态度受到过去探索病毒的历史，尤其是与被称为"包涵体"的神奇物体有关的那部分历史的影响。"包涵体"是在受某些特定病毒感染后的动植物细胞内发现的，在分辨率相对较低的显微镜下就可以轻易观察到。最开始大家草率地倾向于把它们看作是传染性的原虫，但很快，观点转换到了另一个极端，即认为它们不是单纯的受感染细胞的产物。在一些试验中，分离这些包涵体是如此的容易，以致于很奇怪几乎没有人研究它们的感染性。但是，已经发现这样一种包涵体具有至少可以传播一种病毒感染的能力；在其他一些试验中，进一步发现它们是由无结构的基质包裹着一群看起来像是微小生物的物体。在我看来，现在从其他例子获得证据的重担似乎要落在那些认为它们完全是由感染造成的代谢异常而积聚的代谢废物的人身上。

用多孔织物和胶质膜进行过滤以及测量扩散速率获得的这些具体证据当然是完全与感染性物体组成单位的大小相关的，并且这需要和显微镜提供的证据相结合。这种对可滤过和不可滤过传染性物体的粗略定性区分早已没有实际意义了，因为并不存在有关滤过性的自然分界线。我们能够制造出一种滤器用来阻拦或者放过任意大小的颗粒。人们现在认识到，在这种情况下滤器的最有用之处仅在于能定量测量通过它的颗粒的最大尺寸。当不能通过时，必须要修正静电吸引效应和吸附固着在滤器的织物上产生的效应。大量的滤过研究数据都因过于依赖计算滤器的**平均**孔径而失去意义。要检测某种在合适条件下能传播的传染性物体的存在，很明显具有重要意义的是最大孔径。

因为这些原因，大量的证据显示某些病毒能够在滤液中检测到，而所用的滤器在已知的范围内却不能滤过血红蛋白，这种试验结果至少应该值得怀疑。埃尔弗德博士最近成功地制备了均一性更好的滤膜，滤孔的直径变化范围很小。他用此滤膜测量了不同病毒颗粒的大小，结果显示：一方面最大的接近最小已知细菌的尺寸，而最小的病毒，如口蹄疫病毒，只有血红蛋白分子大小的 3~4 倍。后面的数据不仅是根据过滤试验得到的，也采用了其他的物理化学测量方法，比如斯韦德贝里提出的超速离心法。为了说明问题的难度以及一些数据的不确定意义，我们注意到埃尔弗德发现了一种能够被滤膜阻挡的噬菌体，而这种滤膜允许口蹄疫病毒通过；而在另一方面，布朗芬布伦纳最近确定了噬菌体的扩散速率，通过速率计算将噬菌体颗粒的直径锁定在 0.6 个毫微米，即大约一个血红蛋白分子大小的 1/5。如果接受了这种估计，得出的结论必须是，噬菌体不仅是无结构的，而且其分子比高分子蛋白质

molecule. If we accepted such an estimate, we should be obliged to conclude, I think, not merely that the bacteriophage is unorganised, but that its molecules are something much simpler than those of a high-molecular protein. It has even been suggested, though on very imperfect evidence, that it may be a moderately complex carbohydrate. Are we, then, to suppose that the foot-and-mouth virus is a similarly unorganised and relatively simple substance? It is difficult to do so, in view of the series of other agents, all conforming in many aspects of their behaviour to the classical type of the foot-and-mouth virus, and yet showing a range of dimensions up to that at which their units are apparently becoming clearly visible by modern microscopical methods.

It will be clear, indeed, that, if we accept the lowest estimates for the size of the units of some viruses, such as the bacteriophage and the agents transmitting some plant diseases, we cannot by analogy apply the conception of their nature, thus presented, to viruses consisting of organisms which are ceasing to be even ultramicroscopic; and we should be led to doubt the identity with the virus of the bodies which the microscope reveals. If, on the other hand, we regard the still invisible viruses, by analogy with those already seen, as consisting of even much smaller organisms, we can only do so by rejecting the conclusions drawn from some of the physical evidence. It is, of course, possible that some of the agents called viruses are organisms and others relatively simple pathogenic principles in solution; but to assume at this stage such a fundamental difference, among members of a group having so many properties in common, would be to shirk the difficulty.

The third negative characteristic of a virus, namely, its failure to propagate itself, except in the presence of living cells which it infects, may obviously again provide an unstable boundary, shifting with the advance of our knowledge and skill. We may regard it as not only possible, but even likely, that methods will be found for cultivating artificially, on lifeless media, some of those viruses at least which have the appearance of minute organisms. It would be playing with nomenclature to let inclusion in the virus group depend on continued failure in this direction. On the other hand, the dimensions assigned to the units of some viruses, representing them as equal in size to mere fractions of a protein molecule, might well make one hesitate to credit them with the power of active self-multiplication. Experience provides no analogy for the growth of such a substance by self-synthesis from the constituents of a lifeless medium; the energetics of such a process might present an awkward problem. To account for the multiplication of such a substance at all, even in cells infected by it, we should be driven, I think, to the hypothesis which has been freely used to account for the propagation of bacteriophage, on one hand, and of typical viruses like that of herpes, on the other; namely, that the presence of the virus in a cell constrains the metabolism of the cell to produce more.

Bordet has used the reproduction of thrombin by the clotting of the blood as an analogy for the suggested reproduction of bacteriophage in this manner. Another, and perhaps closer, analogy might be found in recent evidence that a culture of pneumococcus, deprived of its type-specific carbohydrate complex, can be made to take up the carbohydrate characteristic of another type, and then to reproduce itself indefinitely with this new, artificially imposed specificity. The response of the cells of the animal

还要简单。尽管证据不是很充分，有人甚至提出它可能是中度复杂的碳水化合物。那么，我们是否也可以假设口蹄疫病毒也是一种类似的无结构而且相对简单的物质？这样说恐怕很难，因为纵观这一类别的其他病原，它们行为的许多方面都和经典的口蹄疫病毒非常一致，但是其大小范围很广，其中大的病原利用现代显微技术可以清晰地观察到它们的组成单位。

很显然，如果我们接受了某些病毒单位的最小估计尺寸，比如噬菌体和传播某些植物疾病的病原，那就确实不能用类推的方法将其本质的概念应用到那些由不再是超显微结构单位组成的病毒中。我们应该怀疑那些在显微镜下看到的物体是否是病毒。另一方面，如果通过类推，我们认为这些在显微镜下不可见的病毒，和可见的病毒一样，是由更小的有机组织组成，那我们就只能推翻一些从具体证据中得来的结论。当然，有可能某些被称为病毒的物体实际上是溶液中的其他有机体或者相对简单的病原体。但是现阶段在这样一个具有如此多共同特性的类群中假设存在一个这样根本的差别是在躲避困难。

病毒的第三个否定性特征，即它不能自我繁殖，除非有受感染的活细胞的存在，这同样也是一种不确定的界限，随着知识和技术的进步会发生改变。我们会认识到，在无生命的培养基中人工培养病毒的方法不仅是也许，而且是非常可能成功的，尤其是培养那些有微小有机体外观的病毒。用这个特征来界定病毒常常会失败，也许只是玩玩命名的文字游戏。另一方面，某些病毒组成单位的大小仅相当于一部分蛋白质分子，这可能让人不敢相信它们有自我繁殖的能力。对于能否通过无生命培养基中的成分自我合成病毒类物质，我们没有任何经验；这个过程的能量学就是一个很棘手的问题。为了能够说明病毒这类物质的繁殖过程，即便是在受感染的细胞内部，我想我们也必须借助那个已有的假说，它一方面用来解释噬菌体，另一方面用来解释疱疹病毒的繁殖过程：即细胞内存在的病毒迫使细胞的代谢发生改变，使其产生更多的病毒。

博尔代使用血液的凝集产生凝血酶来类比噬菌体的繁殖方式。另一个类比，可能更贴切，是在最近的试验中发现的，即去除肺炎球菌培养液中的类型特异性碳水化合物后，该球菌能够获取另外一种类型的碳水化合物，然后以新的类型不断繁殖，这样就人为地产生了一种新的特异性类型。动物的免疫细胞在接触了外源蛋白后的

body to even a single contact with a foreign protein, by the altered metabolism producing immunity, and often persistent for the lifetime of the individual, may suggest another parallel; but here the protective type of the reaction is in direct contrast to the supposed regeneration by the cells of the poison which killed them.

Boycott, again, has emphasised the difficulty of drawing a sharp line of distinction between the action of normal cell-constituents, which promote cell-proliferation for normal repair of an injury, and the virus transmitting a malignant tumour, or that causing foot-and-mouth disease. I do not myself find it easy, on general biological grounds, to accept this idea of a cell having its metabolism thus immediately diverted to producing the agent of its own destruction, or abnormal stimulation. It is almost the direct opposite of the immunity reaction, which is not absent, but peculiarly effective in the response of the body to many viruses. It is difficult, again, to imagine that a virus like rabies could be permanently excluded from a country if it had such an autogenous origin. The phenomena of immunity to a virus, and of closely specific immunity to different strains of the same virus, are peculiarly difficult to interpret on these lines.

This conception, however, of the reproduction of a virus by the perverted metabolism of the infected cell has been strongly supported by Doerr, in explanation of the phenomena of herpes. There are individuals in whom the epidermal cells have acquired a tendency to become affected by an herpetic eruption, in response to various kinds of systemic or local injury. From the lesions so developed, an agent having the typical properties of a virus can be obtained, capable of reproducing the disease by inoculation into individuals, even of other species, such as the rabbit, and exciting, when appropriately injected, the production of an antiserum specifically antagonising the herpes infection. Such phenomena have a special interest for our discussion, in that they can be almost equally well explained by the two rival conceptions. One regards the herpes virus as a distinct ultramicroscopic organism, and the person liable to attack as a carrier, in whom the virus can be awakened to pathogenic activity and multiplication by injuries weakening the normal resistance of his cells to invasion. The other regards it as a pathogenic principle produced by cells in response to injury, and awakening other cells to further production when transmitted to them.

This forms a good example of the central difficulty in dealing with the group of agents at present classed together as viruses. They seem to form a series; but we do not know whether the series is real and continuous, or whether it is formed merely by the accidental association, through a certain similarity in effects, and through common characteristics of a largely negative kind, of agents of at least two fundamentally different kinds. If we approach the series from one end, and watch the successive conquests of microscopical technique, or if we consider the phenomena of immunity over the whole series, we are tempted to assume that all the viruses will ultimately be revealed as independent organisms. If we approach from the other end, or consider analogies from other examples of a transmissible alteration of metabolism, we may be tempted to doubt the significance of the evidence provided by the microscope, and to conclude that all viruses are

反应可能也是一种类似的机制，这种通过改变代谢产生免疫的反应经常会持续一生，但是这种保护性反应恰恰抑制了被病毒感染的细胞的复制。

　　博伊科特再次强调：要在能够促进细胞增殖以便修复损伤的正常细胞成分和传播恶性肿瘤或引发口蹄疫的病毒之间划一条清楚的界线是很困难的。我个人认为，在普通生物学背景下很难接受这样一个观点，即一个具有自身代谢能力的细胞会立即转向产生破坏自身的物质或者异常刺激。这几乎是免疫反应的直接反例，免疫反应在机体应对许多病毒的反应时不仅存在，而且相当有效。同样，如果狂犬病毒具有如此强大的自我繁殖能力的话，很难想象它能在一个国家被消灭。对病毒的免疫以及对同一病毒不同株系的高度特异现象，在目前尤难解释。

　　但是，多尔在解释疱疹病毒现象时强烈支持病毒的复制是被感染细胞异常的代谢造成的。在对各种系统的或者局部的损伤作出反应时，某些人的上皮细胞更容易受到疱疹的感染。从这些伤口中能够获得具有典型病毒特征的物体，如果接种到生物体内，即使是其他物种的生物，例如兔子体内，也可以致病，而令人兴奋的是：如果注射得当,能够产生针对疱疹感染的特异性抗血清。这种现象是我们讨论的焦点，因为以下两种相互对立的观点都能很好地解释这种现象。其中一种观点认为疱疹病毒是特殊的超显微生物，那些易于感染的人作为携带者，由于损伤减弱他们体内细胞的正常抵抗力而导致病毒被唤起进行病理性活动和复制。另一种观点认为病毒是细胞应对损伤产生的病理性物质，当传递给别的细胞时可以唤起这些细胞产生相同的物体。

　　这是目前在分析病毒类病原时遇到的主要难题之一。这些病原似乎形成了一个系列，但是我们不知道它们是否真实并且持续存在，或者它是否仅仅是由两种或两种以上本质不同的病原通过产生相似效应或者通过大量相同的否定性特征而偶然联系在一起形成的。如果我们从这个系列的一端入手，关注显微技术的发展，或者如果我们考虑的是整个系列的免疫现象，那么我们就会倾向于认为所有病毒最终都会被证明是独立的生物体。如果我们从另一端入手，或者考虑从传染性代谢改变的其他例子来进行类推，我们可能会怀疑显微镜所提供的证据的重要性，并得出结论所有的病毒是无结构的、自行产生的、有毒的物体。如果我们采取谨慎的态度假设两

unorganised, autogenous, toxic principles. If we take the cautious attitude of supposing that both are right, and that viruses belonging to both these radically different types exist, where are we going to draw the line? Is the test to be one of unit dimension? If so, what is the lower limit of the size of an organism? Are we to suppose that inclusion bodies can only be produced by viruses which are independent organisms? And if so, does this conclusion also apply to the "X" bodies associated with the infection of plant cells by certain viruses?

If we try to form an estimate of the lower limit of size compatible with organisation, I think we should remember that particles which we measure by filters of known porosity, or by photomicrographs, need not be assumed to represent the virus organisms in an actively vegetative condition. They may well be minute structures, adapted to preserve the virus during transmission to cells in which it can resume vegetative life. Attempts to demonstrate an oxidative metabolism in extracts containing such a virus, separated from the cells in which it can grow and multiply, and to base conclusions as to the non-living nature of the virus on failure to detect such activity, must surely be regarded as premature.

Our evidence of the vitality of its particles is, as yet, entirely due to their behaviour after transmission. They may accordingly contain protein, lipoid and other molecules in a state of such dense aggregation that comparisons of their size with that of the heavily hydrated molecules of a protein in colloidal solution may well give a misleading idea of their complexity.

Apart from their known function as the agents transmitting many of the best known among the acute infections, it is impossible, to anyone having even a slight knowledge of the recent developments which began with the work of Rous and Murphy, to doubt that in the advance of knowledge concerning the nature of the viruses in general lies the brightest hope of finding a clue to the dark secret of the malignant tumours. In unravelling what is still such a tangle of contradictions, the animal biologist needs all the help that can be given by concurrent study of the analogous phenomena in plants.

(**128**, 599-602; 1931)

者都是对的，病毒分别属于这两种完全不同的类型，那么我们如何划清这条界线？组成单位的尺寸是界定的标准吗？如果这样，一个生物的尺寸的下限是多少？我们应该假设包涵体只能由属于自主生物的病毒产生吗？如果这样，那么这个结论也能应用到与某些病毒感染植物细胞相关的"X"体上吗？

如果我们想估计出有机组织结构尺寸的下限，我认为我们应该想到，我们用已知孔隙度的滤器或者显微照相测量的颗粒不一定代表活着的病毒。它们很可能是微小的结构，用于在病毒感染细胞时保存病毒，进入细胞后，病毒可以继续植物性生活。病毒在细胞体内生长和繁殖，但试图在从细胞中分离出来的包含病毒的提取液中检测氧化代谢活性，并且基于没有检测到这种代谢活动就得出病毒不是一种生物的结论肯定是草率的。

迄今为止，我们关于病毒颗粒生命力的证据都来源于它们在传播以后的行为表现。因此，它们可能会含有蛋白质、类脂和其他分子，这些分子以一种非常紧密的方式聚集在一起，以至于如果将它们的大小和胶体溶液中高度水合的蛋白质分子大小进行对比会让我们对其复杂性产生误解。

除了知道病毒的功能是作为病原传播许多我们最熟悉的急性传染病外，任何一个对由劳斯和墨菲的工作开创的最新进展稍有一点了解的人，都不可能怀疑进一步了解病毒的本质最有可能为揭开恶性肿瘤背后深藏的秘密提供线索。为了解决这个矛盾重重的问题，动物生物学家们需要全面地借鉴在对植物学中类似现象的同步研究中得到的结果。

（毛晨晖 翻译；秦志海 审稿）

Progressive Biology[*]

F. G. Hopkins

Editor's Note

One of the most striking features of this article by Frederick Gowland Hopkins is the breadth of what it attempts: in two pages, nothing less than a survey of the status of all of biology. It is hard to imagine such a thing being meaningfully attempted now. Hopkins ranges from the emerging understanding of genetic inheritance, made quantitative by J. B. S. Haldane and Ronald Fisher, to the study of nerve action and the biochemistry of vitamins. Hopkins alludes to the prevailing suspicion that the fundamentals of biochemistry should shed light on life's origins. And he suggests that viruses offer a glimpse of the fuzzy boundary between the living and non-living worlds, a view that is very much upheld today.

IT is not going too far to claim that recent progress in experimental biology, though to a superficial view less impressive, has been not less significant, and indeed not less revolutionary than the progress of modern physics. I might support this claim in many ways. It is, I think, justified in that region of knowledge where cytology and genetical studies meet. The progress and the significance have become the greater since it was recognised that the material units of characters—the chromosomes and subdivisions of chromosomes—are "determinants" rather than "carriers" of genetic factors.

This domain of disinterested science is making many practical contacts. To mention but a single recent instance: Prof. R. C. Punnett, one of the original discoverers of the phenomenon of sex-linkage in inheritance, by applying his expert knowledge of that phenomenon, has produced what may be called a synthetic fowl, of which the qualities are such as to make it of extreme value to the now highly important industry of poultry breeding and egg production. The bearing of the same body of knowledge upon human affairs has been recently very ably discussed by Prof. L. T. Hogben.

The phenomena of heredity were long the stronghold of those who cling to the obscurantism of vitalistic doctrines. Infinitely complex as of course they are, we have now abundant proof that they are susceptible to analysis and that today they are yielding their secrets to well-controlled experimental studies. The results of these are becoming quantitative and are even yielding material for mathematical treatment, as the interesting writings of Dr. R. A. Fisher and Prof. J. B. S. Haldane have shown.

Another region in which accurate experimentation has removed, and is continuing to remove, inhibitions due to obscurantist assumptions is the physiology of the nervous

---

[*] Excerpts from the presidential address at the anniversary meeting of the Royal Society on Nov. 30.

生物学的进展*

弗雷德里克·高兰·霍普金斯

编者按

这篇由弗雷德里克·高兰·霍普金斯所写的文章的最突出的特点之一，是其论述内容的范围非常之广：他用 2 页纸的篇幅全面评述了生物学的所有领域。很难想象现在还有人有意去尝试这种事。霍普金斯从霍尔丹和罗纳德·费希尔最近对遗传学的定量分析开始讲起，一直讲到人们在神经活动和维生素生化性质方面的研究。霍普金斯提到了那个遭到很多人置疑的观点，即生物化学的基本原理应该能说明生命的起源问题。他还提出病毒的出现使生命体和非生命体之间的界限变得模糊不清。现在这个观点已被大家普遍接受。

毫不夸张地说，尽管从表面看并没有那么令人印象深刻，但实验生物学的最新进展在重要性甚至开创性上都毫不逊色于现代物理学的发展。我或许可以从多个方面提出证据支持这个观点。我想主要的证据来自人们对细胞学与遗传学交叉领域的认识。由于人们认识到人类性状的物质基础——染色体及其组成部分——是遗传因子的"决定因素"而不是"载体"，从而使这些进展及其意义变得更加显著。

这个被人们忽视的科学领域具有许多实用价值。举一个最近的例子：庞尼特教授是遗传中性连锁现象的最早发现者之一，他应用自己在性连锁现象方面的专业知识发明了一种可以被称为"合成家禽"的产品，其质量如此之好使得其在当前非常重要的家禽饲养和禽蛋生产业中极具价值。最近霍格本教授非常巧妙地阐述了这个领域的知识对于人类的意义。

遗传现象很早以前就是那些信奉生机论的蒙昧主义者的大本营。尽管这些现象非常复杂，但是现在已经有足够的证据证明：它们的秘密是可以在精心设计的实验研究中被逐渐揭开的。正如费希尔博士和霍尔丹教授在生动有趣的文章中所指出的，这些研究结果正在被量化，甚而可以通过数学处理得出结论。

另一个一直在并将继续通过精确试验解除蒙昧主义假说带来的禁忌的领域是脊椎动物神经系统生理学。也许现在提出巴甫洛夫在条件反射方面的研究已经太迟了，

---

* 本文摘自作者作为主席在 11 月 30 日英国皇家学会的周年纪念会议上所作的演讲。

379

system of vertebrates. It is, perhaps, too late in the day to refer to the work of Pavlov upon conditioned reflexes, though it is justifiable to emphasise its still growing influence upon thought. The work is a supreme proof of the success of the experimental method in analysing even such apparently transcendental phenomena as those which underlie the higher functions of the brain.

The nature of the transmission of events in the nervous system is receiving much illumination from the work of the Royal Society's Foulerton professors. Prof. E. D. Adrian, having developed a most admirable experimental technique for the purpose, is studying the nerve impulse and its origin with highly profitable results. He and his colleagues are now able, with great gain, to work with single nerve fibres and single isolated end organs.

A striking circumstance, brought to light by Adrian's work and that of his colleagues, is that the nervous structures so far examined exhibit such physical regularity in their behaviour that results can often be predicted within about one percent. His experiments on animals have shown how the phenomenon of a grading in the contraction of muscles is controlled by the frequency of impulses sent out from the central nervous system, and by the number of nerve and muscle fibres involved. Moreover, he has been able to observe the activity of a single nerve cell in his own spinal cord by needle electrodes placed in his muscles, and finds human voluntary contractions are regulated in exactly the same way.

Further, Adrian has found that slow potential changes occur in nerve cells and that these are connected with their discharge of impulses; so far this work only extends to isolated nerve ganglia from insects and to nerve cells in the brain stems of fish, but the phenomena he has found are extremely significant, and it seems possible that changes of potential may be of fundamental importance in the activity of nerve cells.

Adrian has found that damaged nerve fibres set up impulses at very high frequencies, and these perhaps play an important part in sensations of pain, though his more recent work has made it clear that impulses in the smaller slowly conducting nerve fibres must also be concerned in the physical mechanism responsible for pain. A most striking feature of all this work is the general similarity of behaviour of nervous structures from whatever animal they may be taken.

Prof. A. V. Hill is studying the nerve impulse from the point of view of the thermal phenomena which accompany it. So small, of course, is the heat production in the nerve that its measurement, like that of the potential changes, calls for great refinements. It is becoming clear, though we do not yet know the details of its nature, that the nerve impulse consists of a transmitted physicochemical event, probably involving changes of ionic concentrations at membranes with consequential changes of electric potential; the whole cycle of events, comprising activity and recovery in the nerve, being supported by the energy derived from metabolic oxidative processes which, very small in scale, are associated with the cycle.

尽管怎么强调该学说对于思维日益明显的影响都不过分。该学说是利用实验方法分析诸如大脑高级功能本质等超自然现象的一个最佳成功例证。

英国皇家学会福勒敦教授们的研究使神经系统传递事件的本质越来越清晰。阿德里安教授已经为此开发出了一种最令人钦佩的实验技术,他正在研究神经冲动及其起源并且得到了非常有用的结果。他和他的同事们目前能够研究单个神经纤维和分离的单个末梢器官,并有重大收获。

阿德里安及其同事的研究工作揭示了一个引人注目的事实,即目前研究的神经结构表现出一种行为上的物理规律性以至于预测结果的误差通常在一个百分点以内。他在动物身上进行的实验显示肌肉收缩的分级现象是如何受到中枢神经系统发出的脉冲频率以及与之相关的神经纤维和肌纤维数量调控的。此外,他通过在自己的肌肉里放置针电极来观察自己脊髓中单个神经细胞的活动,并且发现了人类肌肉的自主收缩也是以完全相同的方式进行调控的。

此外,阿德里安还发现了神经细胞中发生的慢电位改变,以及这种慢电位改变与脉冲放电的关联。到目前为止,这项工作还仅限于分离的昆虫神经节和鱼脑干中的神经细胞,但是他发现的这些现象非常有价值,而且电位改变在神经细胞的活动中很可能具有根本的重要性。

阿德里安发现损坏的神经纤维能够以非常高的频率发放脉冲,而且这可能在疼痛的感觉中起着重要的作用,尽管他的最新研究结果已经清楚地表明由更细的慢传导神经纤维发出的冲动也与疼痛的自然机制有关。所有这些工作的一个最显著的特征是:任何一种动物的神经结构都会表现出大体相似的行为。

希尔教授正在根据与神经冲动相伴而生的热现象来研究神经冲动。显然,神经中的产热量是如此微小以至于其测量需要非常高的精确度,就像测量电位的变化一样。尽管我们还不知道神经冲动的细节特性,但越来越确信神经冲动是由可以传导的物理化学事件组成,很可能包括膜上离子浓度的改变和由此引发的电位改变;事件的整个周期包括神经的活化和复原,其能量来源于与这个周期相关的小规模氧化代谢过程。

When we hope for an increase in our knowledge of the nervous system we are always accustomed to look to researches from the laboratory of Sir Charles Sherrington. An extended study of reflexes has shown that the centripetal impulses do not pass straight through the spinal cord, but at central stations in the cord they are transformed into an enduring excitatory state, which may in turn set up fresh nerve impulses yielding the reflex discharges. The nature of this central excitatory state is being studied and will link up, I think, with some of Adrian's observations upon cell potentials.

It has long been suspected that when a sympathetic nerve is stimulated, adrenalin is liberated at the nerve ending, and that the observed effects are immediately due to the action of that substance. Now Dr. H. H. Dale, in conjunction with a member of his staff, Dr. Gaddum, has investigated the case of the para-sympathetic nerves, the influence of which in general opposes that of the sympathetic group, and has obtained good evidence that when one of these is stimulated, the substance acetyl-choline, previously existing in some inactive form, is liberated at the nerve ending. The action of acetyl-choline when injected into the circulation resembles in general the effect of stimulating para-sympathetic nerves, and there is every reason to believe that the physiological activity of the substance, rather than the transmitted physical impulse itself, is immediately responsible for the observed effects of stimulation. I may say that Otto Loewi, of Graz, has shown that when the heart beat is inhibited by vagus stimulation, acetyl-choline is actually formed in the organ, and the same substance, when artificially injected, is known to produce effects like those of the vagus.

In rather unexpected circumstances we have thus brought before us an example of specific physiological effects due to the influence of the specific structure of an organic molecule. Such effects and such relations are being demonstrated in increasing diversity as fundamental factors of organisation in the animal body. This is illustrated most strikingly, of course, in the domain of the control of its activities by a group of hormones. We find in the cases of adrenalin and thyroxin, the constitution of each of which is accurately known, widely different influences depending on differences of molecular structure.

I may logically pass from hormones to devote a few words to vitamins. We now possess proof that vitamin A is closely related to the carotenes, and this knowledge may well lead, without long delay, to the artificial synthesis of the vitamin itself. With respect to vitamin D, it seems probable, if not yet quite certain, that its artificial production is already accomplished. Some four years ago the constituent of animal and vegetable substances, which is converted into the antirachitic vitamin D by ultra-violet radiation, was identified as ergosterol by Rosenheim and Webster at the National Institute for Medical Research, and concurrently by Windaus in Göttingen. A team of workers at the National Institute, led by Dr. R. B. Bourdillon, appear now to have arrived at the next stage, of isolating the vitamin itself, in crystalline form, from the mixed products of irradiation; and Prof. Windaus, following with his co-workers a different route, has again arrived simultaneously at the same goal.

当我们希望丰富一些关于神经系统的知识时，我们总是习惯于去关注查尔斯·谢灵顿爵士实验室的研究工作。一项关于反射的扩展研究显示向心性的冲动不是沿着脊髓径直传导的，而是在脊髓的中心站点转换成持久的兴奋状态，随后可能会发出新的神经冲动而产生反射放电。人们正在研究这个中心站点兴奋状态的本质，而且我认为这会与阿德里安对细胞电位的一些观察结果相互联系。

很久以前就有人猜想：当交感神经被刺激时，神经末梢就会释放肾上腺素，而我们观察到的效应直接与肾上腺素的作用相关。现在，戴尔博士与为他工作的加德姆博士一起研究了副交感神经，副交感神经的作用大体上和交感神经的作用相反。他们已经获得了有力的证据证明当副交感神经受到刺激时，之前以非活性形式存在的乙酰胆碱会被神经末梢释放。乙酰胆碱在被注射到血液循环中之后，产生的效应同刺激副交感神经产生的效应十分相似，而且所有的证据都表明与观察到的刺激效应直接相关的是该物质的生理活性，而不是传导的物理冲动本身。我要提到的是：来自格拉茨的奥托·勒维已经证实了当心跳被迷走神经刺激所抑制时，心脏内确实形成了乙酰胆碱，而且已经知道在人为地将乙酰胆碱注入体内后，也可以产生类似于迷走神经刺激的效果。

在一些意想不到的情况下，我们已经展示出了由一个有机分子的特定结构能产生特定生理功能的实例。这种效应和两者之间的关系正逐渐被越来越多的证据证实为动物体组织系统的基础。认为动物的行为由一组激素所控制的思想显然是对上述观点的最好阐释。我们发现在已经明确知道分子结构的肾上腺素和甲状腺素中所表现出来的生理效应的巨大差异是由它们各自的分子结构决定的。

我希望将话题从激素合乎逻辑地转移到对维生素的简要说明上。我们现在有证据说明维生素 A 与胡萝卜素密切相关，利用这个知识很可能会在很短的时间内实现维生素的人工合成。至于维生素 D，其人工合成可能已经完成，但目前还没有完全确定。大约在四年以前，动植物原料中的某些经紫外线照射可转变成抗佝偻病的维生素 D 的成分同时被英国国立医学研究所的罗森海姆和韦伯斯特以及格丁根的温道斯鉴定出来是麦角甾醇。由鲍迪伦博士领导的国立研究所的一组研究者现在已经开始了下一阶段的研究，即以结晶的形式从照射后的混合产物中分离出维生素本身；与此同时，温道斯教授和他的同事用另一种方法也达到了相同的目的。

There is no doubt that the substance which the British group now term "calciferol", and which they have isolated as a dinitrobenzoate from the mixed product, is identical with the "vitamin D_2" which Windaus and Linsert obtained by a different method; and there is little doubt that this substance, as obtained in either laboratory, is the essential vitamin D in a state of practical purity. One milligramme of calciferol has an antirachitic activity corresponding to 40,000 of the newly accepted international units.

Before I close I would like to refer to a certain aspect of the chemical dynamics of living cells, concerning which progressive studies have been made during the last few years. The cell is, of course, a seat of catalysed chemical reactions and would seem to possess a multitude of catalysts each highly specific in the influence it exerts. In the case of reductions and oxidations, however, there are other agencies of less specific activity which promote the final stages of oxidation. Of great interest among these are certain combinations of metallic iron with pyrrol groupings. One such compound is concerned, possibly with the intervention of yet another, in bringing molecular oxygen into the field of activity. What, however, is specially interesting about these associations of pyrrol groupings with a metal is the wide extent of their biological functions.

We have long known, of course, of the presence of one such association in the chlorophyll molecule, where it functions as part of the trap for solar energy, and we have been long familiar with the presence of another in haemoglobin, where its function is to hold oxygen during its transference from the lungs of vertebrates to their tissues. Further, we now know that, within the tissues, two others promote oxidation, and yet a third prevents by its presence any deleterious accumulations of hydrogen peroxide. For adjustment to each separate function there is some slight modification of a fundamental structure. Compounds of the type in question are found in many of the lower organisms. Just as Nature seems to have hit upon sound principles for nerve structure early in evolution, so she seems to have satisfactorily chosen, very early, the chemical materials for life. This same suggestion is carried by all the more important constituents of living stuff; fundamentally the same throughout, yet always with minor differences underlying specific morphological differences.

To return to cell dynamics. Knowledge of enzymic catalysts which, with highly specific relations, activate in a certain sense the molecules which are to suffer oxidation, is almost daily accumulating. It is because this specific activation must precede oxidation that the indiscriminate action of oxygen on the living cell is prevented. Although an understanding of the complex organisation of chemical events in the living unit is far beyond our present powers, we are, I think, beginning to see what kind of organisation it may be.

One last word, however. We have assumed that the living cells we have best known are the ultimate units in biology. But of late years the viruses have forced themselves into our thoughts. What are viruses? Do they merely simulate some of the properties of the living? Can we conceive of them as something between the non-living and the living?

　　毫无疑问，这种被英国的研究小组作为一种二硝基苯甲酸酯从混合产物中分离出来并命名为"钙化醇"的物质，与温道斯和林泽特以另一种方法获得的物质——"维生素 D_2"十分相似。基本可以肯定，在这两个实验室中得到的物质，就是达到可应用纯度的维生素 D。一毫克钙化醇所具有的抗佝偻病活性相当于新国际单位的 40,000 倍。

　　在结束之前，我想提到活细胞的化学动力学这一特定领域，在过去的几年内这个领域的研究取得了一定的进展。当然，细胞是催化化学反应的温床，细胞内含有大量的催化剂，每一种催化剂的作用范围都很专一。但是，在还原和氧化的过程中，有一些专一性不是很强的物质，它们能催化氧化反应的最终步骤。其中很值得注意的是某些金属铁与吡咯基团的化合物。其中一种这样的化合物有可能在另一种化合物的作用下会将分子氧带入到活性区域。但是，就吡咯基团与一种金属的缔合来说，格外引人注目的是它们有多种多样的生物学功能。

　　当然，我们在很早以前就已经知道有一种这样的缔合物存在于叶绿素分子中，其功能是捕捉光能，而且我们也早已熟知在血红蛋白中还存在着另一种这样的缔合物，其功能是把氧气从脊椎动物的肺转移到各组织中去。此外，我们现在知道在组织中另有三种这样的缔合物，其中两种可以促进氧化，而第三种物质能够通过自身的存在而阻止过氧化氢的有害蓄积。为了适应每一种物质的独特生理功能，它们在基本结构上都有一些微小的修饰。这种类型的化合物在许多低等生物中也能找到。正如自然界似乎在进化早期就为神经结构找到了合理的原则一样，她似乎也很早就选定了合适的化学物质来形成生命。生命中所有更重要的组成成分都支持了这个观点。所有物质在本质上都是相同的，只不过通常在特定的形态差异下存在着微小的区别。

　　再回到细胞动力学。关于酶催化剂的知识几乎每一天都在增长，这些酶在某种意义上以高度专一的相关性活化了将要进行氧化反应的分子。因为在氧化之前必须进行这种有专一性的活化，所以避免了氧气对活细胞的非选择性作用。尽管我们目前还没有能力去了解生物体内各种化学作用的复杂机制，但我认为我们已经开始认识到这种机制可能会属于哪一种类型。

　　最后还有一句要说的话。我们已经假设我们最了解的活细胞就是生命的基本单位。但是近年来病毒迫使我们改变了想法。病毒是什么？它们仅仅是在模仿生命的某些特性吗？我们可以假设它们是介于非生命和生命之间的某种东西吗？它们是活

Are they alive? We do not yet know. Research upon them is at any rate intensely active at the moment. Its results may make it necessary to modify some fundamental biological concepts, and indeed be as revolutionary in their effects as the breaking up of the atom.

(**128**, 923-924; 1931)

的吗？我们现在还无从知晓。无论如何，目前在这方面的研究都是非常活跃的。研究结果可能导致有必要修正一些基础的生物学概念，而且由此产生的效应确实可以和原子分裂一样具有革命性的意义。

（毛晨晖 翻译；秦志海 审稿）

Crystal Structures of Vitamin D and Related Compounds

J. D. Bernal

Editor's Note

The chemical compound whose crystal structure is discussed here by J. Desmond Bernal, ergosterol, is a biological precursor to vitamin D_2. It is a complicated natural product with a hydrocarbon backbone that has several "chiral" centres, where the three-dimensional arrangement of atoms has two mirror-image forms. Such structures were particularly hard to unravel with X-ray crystallographic methods. Bernal's suggested structure for ergosterol is along the right lines but defective in some important respects. Yet it shows the molecular complexity that crystallographers were starting to feel able to contemplate.

I have had the opportunity of examining by X-rays the crystals of ergosterol and certain of its irradiation products, recently described by a team working at the National Institute for Medical Research.[1] Though the results are only preliminary, they seem of sufficient interest to warrant publication at this stage. Five substances, all of composition $C_{27}H_{41}OH$ or $C_{27}H_{43}OH$, have been examined, with the results shown in the accompanying table.

Substance	a	b	c	β	$c \sin \beta = d_{001}$	Space Group	No. of Mol. per Cell	No. of Mol. in Asymmetric Unit	Orders of Basal Plane, Estimated Intensities									
									1	2	3	4	5	6	7	8	9	10
Ergosterol	9.75	7.4	39.1	65°	35.40	$C_2^2-P2_1$	4	2	vvs	vs	mw	a	vw	mw	s	vw	m	vw
α-Dihydroergosterol and ethyl alcohol	30.8	7.4	43.1	53	34.5	C_2^3-C2	12	3	..	vs	mw	a	vw	mw	s	a	m	..
Calciferol	20.8	7.15	38.5	68	35.65	$C_2^2-P2_1$	8	4	vvs	vs	s	a	vw	mw	ms	vw	ms	..
Pyrocalciferol calciferol	20.2	7.35	40.0	63	35.8	C_2^3-C2	8	1	..	w	a	a	a	ms	w	m	w	m
Lumisterol	20.3	7.25	20.4	60	17.8 $= \frac{1}{2} \times 35.6$	$C_2^2-P2_1$	4	2	..	..	..	w	a	s	a	m	a	w
Cholesterol	16.4	..	..	..	33.3	$C_1^1-P_1$	..	..	..	vs	a	a	a	w	ms	vw	ms	w

N.B.—Letters for intensities: vs, very strong; mw, medium weak; etc.

The most striking features of the crystals are their essential similarities of properties and the simple relation between their unit cells. All the substances except lumisterol occur in platy crystals of a long-chain paraffinoid type. All are monoclinic and show a distinct tendency to elongation along the b-axis. (In lumisterol the crystals are fine needles with

维生素 D 及其相关化合物的晶体结构

德斯蒙德·贝尔纳

编者按

在这篇论文中，德斯蒙德·贝尔纳讨论了维生素 D_2 的前体化合物——麦角甾醇的晶体结构。麦角甾醇是一种复杂的天然产物，在它的碳氢骨架上有多个"手性"中心，其原子的三维排列有两种镜像的形式。这样的结构很难用 X 射线晶体学方法拆分。贝尔纳认为麦角甾醇的原子基本上是沿直线排列的，但在一些关键位置存在缺陷。这说明晶体学家已经开始研究分子的复杂结构了。

我曾用 X 射线检测过麦角甾醇晶体和它的一些照射产物，最近英国国家医学研究所的一个小组描述了这项研究工作 [1]。虽然得到的只是一些初步的研究结果，但在此阶段它们似乎就已经具备充分的发表价值了。该小组对五种构成全部为 $C_{27}H_{41}OH$ 或 $C_{27}H_{43}OH$ 的物质进行了检测，得到的结果如下表所示。

物质名称	a	b	c	β	$c \sin \beta$ $= d_{001}$	空间群	晶胞中的分子数	不对称单元中的分子数	按照基面顺序排列，用光谱强度估计的强度									
									1	2	3	4	5	6	7	8	9	10
麦角甾醇	9.75	7.4	39.1	65°	35.40	$C_2^2–P2_1$	4	2	vvs	vs	mw	a	vw	mw	s	vw	m	vw
α–二氢麦角甾醇和乙醇	30.8	7.4	43.1	53	34.5	$C_2^3–C2$	12	3	..	vs	mw	a	vw	mw	s	a	m	..
钙化醇	20.8	7.15	38.5	68	35.65	$C_2^2–P2_1$	8	4	vvs	vs	s	a	vw	mw	ms	vw	ms	..
焦钙化醇钙化醇	20.2	7.35	40.0	63	35.8	$C_2^3–C2$	8	1	..	w	a	a	a	ms	w	m	w	m
光甾醇	20.3	7.25	20.4	60	17.8 $= ½×35.6$	$C_2^2–P2_1$	4	2	..	..	a	w	a	s	a	m	a	w
胆固醇	16.4	..	..	..	33.3	$C_1^1–P_1$	..	..	..	vs	a	a	a	w	ms	w	ms	w

注意描述强度的字母含义：vs，非常强；mw，中等偏弱；等等

这些晶体最显著的特征就是它们的性质基本相似，而且在它们的晶胞之间存在着简单的联系。除了光甾醇以外，其他物质的晶体都是与石蜡类似的长链片状晶体。所有这些晶体都属于单斜晶系，并且在沿着 b 轴方向表现出明显的延长趋势（光甾

b as needle axis.) All are optically positive with (010) as optic axial plane and the fast direction γ inclined at a moderate angle to the *c* face. All have the same *b* axis of 7.2 A. and their *a* and *c* axes are simple multiples of 10 A. and 20 A. respectively. The spacing of the *c* plane is remarkably constant at 35.5 A., agreeing with the value found by K. Wejdling and E. Bäcklin.[2] It differs significantly from that of cholesterol, which was examined for comparison. Lumisterol has a halved *c* spacing, and α-dihydroergosterol, owing to the presence of alcohol of crystallisation, deviates from the others.

From these observations certain conclusions can be drawn:

1. The unit cell of the molecular compound calciferol-pyrocalciferol contains four molecules of each kind, which is the number the symmetry demands for the space group C_2^3–$C2$. This proves either that calciferol is a simple substance or that it contains other substances indistinguishable by X-rays. The former conclusion is far the more probable. The association of molecules found in the other cases consequently does not show that any of them consists of more than one molecular species and is purely of geometrical intermolecular origin.

2. The molecule of ergosterol and its photo-derivatives has the approximate dimensions 5 A. × 7.2 A. × 17–20 A. These form a double layer similar to those of long-chain alcohols and acids. Such dimensions are difficult to reconcile with the usually accepted sterol formula

which would lead to a wider and shorter molecule, but agree much better with one where the carbon chain is attached to atom 17 in ring iv,

醇的晶体是细针状晶体，其 b 轴即为针轴）。而且，它们都具有光学活性，以（010）为光学轴面，其速射向 γ 与 c 面成适度的斜角。它们的 b 轴同样都是 7.2 Å，而 a 轴和 c 轴分别为 10 Å 和 20 Å 的简单倍数。它们的 c 面间距明显为常量 35.5 Å，这与魏德林和贝克林得到的数值非常一致 [2]。它与在检测中用于对比的胆固醇有很大的差别。同时，光甾醇的 c 面间距仅为该常量的一半，而 α-二氢麦角甾醇因其结晶中乙醇溶剂的存在而与其他物质不同。

从上述观察研究中可以得出这样一些结论：

1. 化合物钙化醇-焦钙化醇的单位晶胞中含有四个分子，这也是空间群 C_2^3-C2 的对称性所必需的分子数。这证明钙化醇要么是一种简单的物质，要么含有 X 射线所不能识别的其他物质，而前者的可能性要大得多。在研究其他几种化合物的分子排列规律时没有发现它们中的任何一个是由一种以上的分子构成的，也没有一个纯粹是源于分子间的几何结合。

2. 麦角甾醇及其光学衍生物的分子大小约为 5 Å × 7.2 Å × (17~20) Å。它们形成了一个类似于长链的醇和酸的双层结构。这样的尺寸很难用普遍认可的甾醇结构式

来解释，因为从公认的甾醇结构得到的应该是一个更宽更短的分子，但如果把甾醇结构上的碳链移至环 iv 的 C-17 原子上就可以很好地解释了。

3. The rings lie approximately in the bc plane, that of the larger refractive indices $\beta\ \gamma$, and their width 7.2 is approximately constant in all the compounds. This is borne out by the observations of sterol films by N. K. Adam,[3] who finds the molecule area of 36 sq. A. for ergosterol, as against 35 sq. A. in the solid crystal.

4. The differences between the compounds is due to a differences in the side groups or linkages in the rings, leading to a different form of association in the solid and to a small but distinct redistribution of scattering matter along the chain lengths, as shown by the intensities of the c plane spectra. The greatest similarity is shown between ergosterol and calciferol, the chief difference being a double association of molecules in the latter case. It may be significant that while the basal plane intensities of ergosterol and dihydroergosterol are practically indistinguishable in spite of the notable difference in spacing, those of ergosterol differ from them particularly in the third and ninth orders. This would seem to indicate that the change had affected the carbon skeleton, not merely the position of double bonds in the molecule. In lumisterol and pyrocalciferol the change in intensities is so much greater that the resemblance is almost obliterated. It is doubtful, however, in view of the extreme complexity of the molecules, whether any conclusive evidence of the actual intra-molecular change can be found by X-rays alone. The most hopeful method would seem to be the examination of the ultra-violet absorption and Raman spectra of single crystals with polarised light at liquid hydrogen temperatures.

(**129**, 277-278; 1932)

J. D. Bernal: Mineralogical Museums, Cambridge, Feb. 2.

References:

1. Askew, F. A., Bourdillon, R. B., Bruce, H. M., Callow, R. K., Philpot J. L., and Webster, T. A., *Proc. Roy. Soc.*, B, **109**, 488 (1932). See also *Nature*, **128**, 758 (Oct. 31, 1931). I am indebted to Dr. Callow for the actual specimens and private communication giving later values.

2. *Acta Radiologica*, **11**, 166 (1930).

3. *Proc. Roy. Soc.*, A, **126**, 25 (1930).

3. 分子中的环结构基本上都位于 bc 平面内，bc 平面具有较大的折射率 $\beta\gamma$，并且在所有这些化合物中它们的宽度几乎都是 7.2 Å。这是亚当在观测甾醇薄膜时发现的 [3]，他还发现液态麦角甾醇的分子面积为 36 Å^2，而其固态晶体的分子面积为 35 Å^2。

4. 这些化合物之间的不同之处在于环上的侧链基团或键有所不同，这使得固态晶体中的结合方式各不相同，而且在链长度方向上散布的基团会出现小而明显的重排，正如 c 平面的光谱强度显示。麦角甾醇和钙化醇最为相似，它们的主要差别是钙化醇分子中存在双重结合。尽管麦角甾醇和二氢麦角甾醇在间距上有显著的差别，但是它们的基面光谱强度实际上没有什么区别，而麦角甾醇的基面光谱强度在第三和第九位上明显不同于其他化合物，这一点也许非常重要。它似乎说明这种改变不仅仅影响了分子中双键的位置，也影响到了碳骨架。在光甾醇和焦钙化醇中，光谱强度的变化更大以至于几乎看不出它们之间的相似性。然而鉴于这些分子的高度复杂性，不能肯定是否仅仅用 X 射线就能得到关于分子内实际变化的确切证据。看起来最有希望的方法是紫外吸收光谱检测法和在液氢温度下使用偏振光检测单晶的拉曼光谱法。

（刘振明 翻译；吕扬 审稿）

Hexuronic Acid as the Antiscorbutic Factor

Editor's Note

In the 1930s, medical people first became aware that certain chemicals are essential for healthy life but cannot be made in the human body. These were called vitamins. The search for Vitamin C was led by Albert Szent-Györgyi, then at the University of Szeged in Hungary. He was awarded the Nobel Prize for medicine in 1937. This letter describes Szent-Györgyi's isolation of vitamin C, contained in citrus fruits and which famously helped prevent the development of scurvy among sailors if included in their restrictive ship-borne diet. The paper is accompanied by two others from the University of Birmingham dealing with the chemical structure of the vitamin.

EXPERIMENTS are being carried out in order to decide whether "hexuronic acid" is the antiscorbutic factor. So far as is known, the distribution of this acid in plants follows closely the distribution of vitamin C. In the animal body it can also be found in relatively high concentration in the suprarenal cortex. Its chemical properties closely agree with the known properties of the vitamin. It was discovered and isolated several years ago at the Biochemical Laboratory, Cambridge.[1]

The hexuronic acid used in the present series was prepared in crystalline form from beef suprarenal glands two years ago at the Chemical Department of the Mayo Clinic.[2] As is known, 1.5 c.c. of lemon juice is the minimum protective dose for guinea-pigs against scurvy. This quantity of lemon juice contains approximately 0.5 mgm. of hexuronic acid. 1 mgm. of the acid has been given to our test animals daily, since, owing to the long exposure to air, some of our hexuronic acid preparation may have been decomposed.

The general procedure used in studying the antiscorbutic activity of hexuronic acid was that recommended by Sherman and co-workers.[3]

The test period in the first experiment consisted of 56 days. At the end of that time the guinea-pigs which had been receiving hexuronic acid, as well as the positive controls which received 1 c.c. of lemon juice, were chloroformed. The positive controls showed mild scurvy on autopsy, while the animals receiving hexuronic acid showed no symptoms of scurvy at all. The negative controls, which received the basal diet only, had an average survival of 26 days and had typical symptoms of scurvy. In this experiment, however, only a small number of animals were used, and the animals receiving hexuronic acid, as well as the positive controls, were losing weight continually because the basal diet employed at that time contained no milk powder (it consisted of rolled oats, bran, butter fat, and salt). For this reason we decided to repeat the experiment.

作为抗坏血病因子的己糖醛酸

编者按

20 世纪 30 年代，医学界人士首次认识到某些化学物质对于生命健康是至关重要的，但却无法在人体内合成。这些物质被称为维生素。对维生素 C 的探索始于阿尔伯特·圣哲尔吉，当时他还在匈牙利塞格德大学。他获得了 1937 年的诺贝尔医学奖。这篇快报描述了圣哲尔吉分离维生素 C 的工作，维生素 C 存在于柑橘类水果中，它非常有助于防止坏血病在只食用有限的船载食品的海员中的蔓延。此外，文后还附列了两篇来自伯明翰大学的关于维生素化学结构的文章。

我们正在进行实验以确定"己糖醛酸"是否就是抗坏血病因子。就目前的结果来看，这种酸在植物中的分布总是与维生素 C 的分布密切相关。在动物体内的肾上腺皮质中存在着相当高浓度的该物质。它的化学性质与维生素的已知性质十分吻合。这种物质是几年前在剑桥大学的生物化学实验室中被发现并分离出来的 [1]。

我们在当前一系列实验中所使用的己糖醛酸，是两年前梅奥医疗中心化学部从牛的肾上腺中制备得到的晶体 [2]。大家都知道，1.5 毫升柠檬汁是防止豚鼠患坏血病的最小剂量。这一剂量的柠檬汁中大约含有 0.5 毫克己糖醛酸。由于长期暴露于空气之中，我们制备的己糖醛酸可能已经部分发生了分解，因此在实验中我们每天给测试动物喂食 1 毫克己糖醛酸。

我们在研究己糖醛酸的抗坏血病活性时，采用了由舍曼和他的同事们推荐的基本步骤 [3]。

第一次实验的检测周期为 56 天。在这期间，测试豚鼠一直摄取己糖醛酸，而对阳性对照组的豚鼠每天供给 1 毫升柠檬汁，喂食周期结束后，对所有豚鼠进行氯仿麻醉。解剖结果表明，阳性对照组的豚鼠出现了轻微的坏血病症状，而摄取己糖醛酸的豚鼠则完全没有表现出坏血病症状。只食用基础食物的阴性对照组豚鼠的平均存活期为 26 天，并出现了典型的坏血病症状。不过，这一实验只使用了为数不多的豚鼠，而且摄取己糖醛酸的豚鼠与阳性对照组的豚鼠都发生了体重持续减轻的情况，这是因为实验所用的基础食物中没有奶粉（基础食物只包括燕麦片、糠、乳脂和盐）。因此我们决定重新进行这一实验。

In the test which is in progress at the present time the defects mentioned above have been remedied. A large number of animals has been used, and skimmed milk powder has been added to the basal diet.

The test was composed of the following groups: (1) Negative controls receiving the basal diet only, 9 animals. (2) Positive controls, receiving 1 c.c. of lemon juice daily, 8 animals. (3) Test animals receiving the basal diet and 1 mgm. of hexuronic acid daily, 10 animals. (4) Controls receiving mixed diet, 10 animals.

The negative controls all died between the time limit of 20–34 days, with an average survival of 26 days, after a continuous and big drop of weight. They all had symptoms of severe scurvy.

At the end of 55 days all the animals receiving hexuronic acid, as well as the positive controls with lemon juice or mixed diet, were living apparently in good health and were gaining weight consistently. At this time three animals which received hexuronic acid and two animals which received lemon juice were chloroformed. Mild symptoms of scurvy were present in the positive controls with lemon juice, but no signs of scurvy in the animals receiving hexuronic acid.

The test will be continued until the ninety-day period is over, and full details will be published later.

This research was supported by the Ella Sachs Plotz Foundation.

J. L. Svirbely* and A. Szent-Györgyi

*   *   *

At the wish and by the courtesy of Prof. A. Szent-Györgyi, I arranged to examine in my laboratory the "hexuronic acid" which he isolated while working in the Biochemical Laboratory, Cambridge. At the end of 1929 he sent me 10 grams of the substance, which had been prepared in the chemical laboratory of the Mayo Clinic, Rochester, U.S.A. Owing to the value and scarcity of this material, it has been necessary to carry out each experiment with very small quantities, and to establish with much deliberation and care the experimental conditions and controls. This work is still in progress and is being directed to the elucidation of the constitution and the achievement of the synthesis of the substance; this has involved the study of its chemical properties, and the formation of a crystalline derivative. The preliminary results now communicated show that the hexuronic acid is most probably the 6-carboxylic acid of a keto-hexose, which does not appear to be

---

* Holder of an America-Hungarian Exchange Fellowship, 1931-32, from the Institute of International Education, New York.

在目前正在进行的实验中，上面提到的各种缺陷都已得到了补救。实验中，我们使用了大量的豚鼠，并在基础食物中加入了脱脂奶粉。

这次实验包括以下几组动物：(1) 只摄取基础食物的阴性对照组，有 9 只豚鼠；(2) 每天摄取 1 毫升柠檬汁的阳性对照组，有 8 只豚鼠；(3) 每天摄取基础食物和 1 毫克己糖醛酸的测试组，有 10 只豚鼠；(4) 摄取混合食物的对照组，有 10 只豚鼠。

阴性对照组豚鼠在 20~34 天内全部死亡，平均存活期为 26 天，死之前体重一直在急剧下降。这组豚鼠都出现了严重的坏血病症状。

在第 55 天结束时，所有摄取己糖醛酸的豚鼠和摄取柠檬汁或混合食物的阳性对照组豚鼠都明显处于良好的健康状态，并且体重稳定增加。这次我们对 3 只摄取己糖醛酸的豚鼠和 2 只摄取柠檬汁的豚鼠进行了氯仿麻醉。解剖结果表明，摄取柠檬汁的阳性对照组豚鼠中有轻微的坏血病症状，而摄取己糖醛酸的豚鼠中则完全没有坏血病的迹象。

这次实验还将一直持续到第 90 天结束，随后我们会发表完整的详细结果。

本研究受到埃拉·萨克斯·普罗茨基金会的资助。

史文贝力[*]，圣哲尔吉

*　*　*

承蒙圣哲尔吉教授的托付和惠赠，我在自己的实验室中检测了他在剑桥大学生物化学实验室工作时分离出来的"己糖醛酸"。圣哲尔吉教授于 1929 年底寄给了我 10 克由美国罗切斯特市梅奥医疗中心化学实验室制备的该物质。由于这一物质得来不易、非常稀少，因此在进行每一次实验时只能使用很少的量，而且还必须格外小心谨慎地设定实验条件和对照物。检测工作目前还在进行中，其目标是弄清该物质的组成并找到合成该物质的方法，这就要对该物质的化学性质进行研究并且要得到一种结晶态的衍生物。对目前初步结果的讨论表明，这种己糖醛酸很可能是一种己酮糖的 6-羧酸，看起来这种己酮糖与 d-果糖没有什么关系，也不是相当于 d-半乳

---

* 1931~1932 年美国 - 匈牙利交流奖金持有者，由位于纽约的美国国际教育协会提供。

related either to *d*-fructose or to the ketose corresponding to *d*-galactose. This work has been conducted by my colleague Dr. E. L. Hirst, assisted by Mr. R. J. W. Reynolds, whose report is given in the accompanying note.

W. N. Haworth

*   *   *

The "hexuronic acid" prepared from suprarenal glands by Prof. Szent-Györgyi was a cream-coloured micro-crystalline powder, m.p. 184°–187° (decomp.). On recrystallisation from methyl alcohol-ether the substance was obtained in irregular aggregates of rectangular crystals, which were almost colourless. No change in m.p., analysis, or other properties was observed even after several successive crystallisations. The crystals showed brilliant colours when observed between crossed nicols in a polarising microscope. Before and after recrystallisation the same analytical figures were obtained (Found: C, 41.0; H, 4.7, $C_6H_8O_6$ requires C, 40.9; H, 4.5 percent). Neither nitrogen nor methoxyl was present. In aqueous solution the rotation $[\alpha]_D^{20°}+23°$(c. 1.1) increased slowly to +31° (3 days) and then decreased to zero (11 days).

The hexuronic acid reduced Fehling's solution, neutral silver nitrate, and neutral potassium permanganate in the cold. It gave the Molisch test and the orcinol reaction, but failed to show the naphtoresorcin colour test characteristic of glycuronic acid.

The hexuronic acid was monobasic (40 mgm. of sodium hydroxide neutralised 172 mgm. of substance—calc. for $C_6H_8O_6$, 176).

Oxidation by atmospheric oxygen in slightly alkaline solution, with a trace of copper as catalyst, introduced one carboxyl group in place of a primary alcohol group and the product reduced Fehling's solution. Oxidation to the same stage occurred with remarkable rapidity when the hexuronic acid reacted with neutral, acid, or slightly alkaline potassium permanganate. The reaction, which required two atoms of oxygen per molecule of the substance, was thereafter much less rapid but proceeded regularly in the cold until one further atom of oxygen per molecule had been absorbed. The product was now non-reducing.

When heated with phenylhydrazine in dilute acetic acid, the hexuronic acid gave intractable, dark-coloured, amorphous products which could not be purified. The action of *p*-bromophenylhydrazine in dilute acetic acid on the barium salt of the hexuronic acid (compare Goldschmiedt and Zerner, *Monatsh.*, **33**, 1217, 1912) gave a dark red micro-crystalline powder which, after recrystallisation from alcohol, had m.p. 230°–

糖的酮糖。这一工作是由我的同事赫斯特博士在雷诺兹先生的协助下进行的，本文后附有他们的报告。

霍沃思

*　　*　　*

圣哲尔吉教授从肾上腺中制得的"己糖醛酸"是一种奶油色的微晶粉末，熔点为184℃～187℃（分解）。在甲基醇醚中进行重结晶时，可以得到该物质的矩形结晶体的不规则聚集物，这一聚集物几乎是无色的。即使是在多次连续结晶之后，也没有发现该物质的熔点、分析结果及其他性质发生变化。将晶体置于偏光显微镜的正交尼科尔棱镜之间观察时，可以看到明亮的色彩。重结晶前后得到的分析结果是一样的（分析结果是：C 占 41.0%，H 占 4.7%。而在 $C_6H_8O_6$ 中 C 占 40.9%，H 占 4.5%）。晶体中不存在氮或甲氧基。在水溶液中，比旋光度 $[\alpha]_D^{20°}$ 从 + 23°（c. 1.1）缓慢增大到 + 31°（3 天），随后又下降到 0（11 天）。

这种己糖醛酸可以还原冷的斐林试剂、中性硝酸银和中性高锰酸钾。它在莫利希试验中给出阳性结果，可以发生地衣酚反应，但是在葡萄糖醛酸特征性的间萘二酚显色反应中给出阴性结果。

这种己糖醛酸是一元酸（40 毫克氢氧化钠能中和 172 毫克该物质，按 $C_6H_8O_6$ 计算需要 176 毫克）。

当在弱碱性溶液中以痕量的铜作为催化剂时，空气中的氧可以氧化伯羟基而在相应位置引入一个羧基，得到的产物能够还原斐林试剂。当这种己糖醛酸与中性、酸性或弱碱性的高锰酸钾反应时，可以非常快速地发生同样程度的氧化反应。在该反应中，每一个己糖醛酸分子需要两个氧原子，此后反应明显减慢，但是在冷溶液中反应确实依然在进行，直到每个底物分子再接受一个氧原子。此时的产物失去了还原性。

当这种己糖醛酸在稀释的乙酸中与苯肼一起被加热时，能生成难以处理又无法纯化的黑色无定形产物。在稀释乙酸中的对溴苯肼与这种己糖醛酸的钡盐反应（可以对照戈尔德施密特和策纳的工作，《化学月刊》，第 33 卷，第 1217 页，1912年）会生成一种暗红色的微晶粉末，在用乙醇进行重结晶后，这种粉末的熔点为

235° (decomp.). Analysis showed it to be the *p*-bromophenylosazone of a hexose-uronic acid (Found: C, 40.8; H, 3.5; N, 11.1; Br, 30.3. $C_{18}H_{18}O_5N_4Br_2$ required C, 40.7; H, 3.4; N, 10.6; Br, 30.2 percent). Control experiments with glycuronic acid and galacturonic acid failed to give the above substance. With glycuronic acid Goldschmiedt and Zerner's barium salt of the *p*-bromophenylosazone of glycuronic acid was obtained, whilst the galacturonic acid gave a yellow powder which appeared to be mainly the barium salt of the corresponding galacturonic acid derivative.

The above reactions, together with Prof. Szent-Györgyi's observation that oxidation by iodine in the cold removes two atoms of hydrogen, which are easily replaced by mild reducing agents, can be understood most readily on the following basis:

$$
\begin{array}{c}
\overset{\displaystyle \underset{\rule{3cm}{0.4pt}}{O}}{\text{CO}-\text{CHOH}-\text{CHOH}-\text{CH}-\text{CO}-\text{CH}_2\text{OH}} \;\underset{\text{H}_2}{\overset{}{\rightleftharpoons}}\; \text{HOOC}-\text{CHOH}-\text{CHOH}-\text{CHOH}-\text{CO}-\text{CHO} \\[4pt]
\downarrow 2O \qquad\qquad\qquad\qquad\qquad\qquad\qquad\qquad O \\[4pt]
\text{HOOC}-\text{CHOH}-\text{CHOH}-\text{CHOH}-\text{CO}-\text{COOH} \nwarrow \\[4pt]
\downarrow O \\[4pt]
\text{HOOC}-\text{CHOH}-\text{CHOH}-\text{CHOH}-\text{COOH}
\end{array}
$$

Inasmuch as the configuration and ring-structure of the lactone of the hexuronic acid have not yet been established, the structural formulae are given above in the open chain form, although the sugar-ring is most probably that of a keto-furanose.

E. L. Hirst and R. J. W. Reynolds

(**129**, 576-577; 1932)

J. L. Svirbely and A. Szent-Györgyi: Institute of Medical Chemistry, University Szeged, Hungary.
W. N. Haworth: University of Birmingham, March 28.
E. L. Hirst and R. J. W. Reynolds: Chemistry Department, University of Birmingham, March 28.

References:
1. Szent-Györgyi, A., *Nature* (May 28, 1927): *Biochem. J.*, **22**, 1387 (1928).
2. Szent-Györgyi, A., *J. Biol. Chem.*, **90**, 385 (1931).
3. Sherman, H. C., La Mer, H. K., and Campbell, H. L., *J. Am. Chem. Soc.*, **44**, 165 (1922).

230℃~235℃（分解）。分析表明，它应该是某种己糖醛酸的对溴苯脎（分析结果是：C 占 40.8%，H 占 3.5%，N 占 11.1%，Br 占 30.3%。而在 $C_{18}H_{18}O_5N_4Br_2$ 中 C 占 40.7%，H 占 3.4%，N 占 10.6%，Br 占 30.2%）。用葡萄糖醛酸和半乳糖醛酸进行的对照实验不能生成上述物质。用葡萄糖醛酸进行实验得到的是葡萄糖醛酸的对溴苯脎的戈尔德施密特和策纳氏钡盐，而用半乳糖醛酸进行实验则得到了一种黄色粉末，它似乎主要是对应于半乳糖醛酸衍生物的钡盐。

另外，圣哲尔吉教授观察到在低温下这种己糖醛酸能被碘氧化而失去两个很容易被温和的还原剂取代的氢原子，这一结果和前面所述的所有反应，都可以在下面的反应关系图的基础上得到顺利的理解：

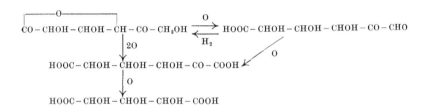

由于这种己糖醛酸中内酯的构型及其环状结构尚未确定，所以上图采用了开链形式的结构式，尽管其中的糖环很可能是一种酮基呋喃糖。

赫斯特，雷诺兹

（王耀杨 翻译；金城 审稿）

A Synthetic Oestrus-exciting Compound

J. W. Cook *et al.*

Editor's Note

Following the isolation of estrogenic hormones a few years earlier, James Wilfred Cook and his colleagues conjecture at the chemical structure for ketohydroxy-oestrin (estrone). Acknowledging the value of being able to synthesise such estrogenic compounds to order, they test whether a related chemical can stimulate estrus in ovariectomised rats. At lower doses, it has no effect on their experimental subjects but as they increase the dose the injected chemical induces full-blown estrus. Once such hormones could be manufactured, it was only a matter of time before researchers could produce contraceptive pills and hormone replacement therapies. The authors also note that certain chemical naturally present in the environment can have estrogen-like activity, something that caused increasing alarm in the coming decades.

IN conformity with the hypothesis, for which there is at present no experimental basis, that the ovarian hormones are formed by degradation of sterols, and in the light of recent developments in the chemistry of the sterols, ketohydroxy-oestrin is possibly represented by formula (i).

(i)   (ii)   (iii)

This accords with all the facts supplied by the work of Butenandt[1], Marrian[2], and others, and we decided that the arguments in favour of this formula were sufficient to justify attempts to synthesise compounds of this nature. By analogy with other physiologically active compounds, it seems likely that a whole group of substances of related chemical constitution will be found to have oestrus-exciting properties, and the synthetic production of such substances would probably be of considerable clinical value.

We have found that 1–keto–1:2:3:4–tetrahydrophenanthrene (ii), which we propose to utilise as a starting point in the synthesis of a substance of formula (i), has itself very definite oestrogenic action, although the does required is very large in comparison with oestrin. The oestrus-producing activity of the substance was examined by the Allen and Doisy procedure. The technique followed was that described by Allan, Dickens and Dodds[3]. The material was dissolved first in olive oil, and later in sesame oil. It was found that the substances were not readily soluble in olive oil, with the result that large volumes

402

一种合成的催情化合物

詹姆斯·威尔弗雷德·库克等

编者按

在人们成功分离出雌激素后不久，詹姆斯·威尔弗雷德·库克和他的同事们就开始推测酮羟基雌激素（雌酚酮）的化学结构了。为了确定合成这种雌激素化合物的价值，他们对相关化学品是否能刺激摘除卵巢的大鼠发情进行了检验。低剂量药剂对试验老鼠无任何作用，加大剂量以后，注射的化学药品诱使这些动物出现明显的发情症状。一旦这样的激素被制造出来，研究人员要制作避孕药片和实施激素补充疗法就只是时间上的问题了。作者还指出：环境中某些天然存在的化学品也可能有类似于雌激素的作用，这在随后的几十年里使人们越来越感到恐慌。

根据卵巢激素是由甾醇降解形成的假说（目前尚无实验基础）和甾醇化学的最新进展，酮羟基雌激素可能的结构式如（i）所示。

这与布特南特[1]、马里安[2]以及其他人研究得到的全部结果都很吻合。我们确信支持这个结构式的论据足以证明尝试合成具有这种特性的化合物是有必要的。类比其他生理活性物质，似乎有一大类具有相关化学组成的物质都具有催情特性，因而人工合成这类物质将会具有很大的临床价值。

我们发现我们计划用于合成结构式（i）的起始原料——1-酮-1，2，3，4-四氢菲（ii）本身就具有一定的催情功能，尽管所需的剂量比雌激素要大很多。以前曾用艾伦和多伊西的步骤检测过该物质的雌激素活性。以下是阿伦、迪肯斯和多兹所描述的检测方法[3]。起先用橄榄油溶解1-酮-1，2，3，4-四氢菲，然后用芝麻油。我们发现该物质不易溶于橄榄油中，因此在对摘除卵巢的动物用药时只能采取大剂量皮下

had to be administered subcutaneously to the ovariectomised animals. This proved to be unsatisfactory owing to leakage from the site of injection and intolerance to the oil, but with sesame oil the volume could be kept down to 2 c.c., and these adverse effects avoided. 25 mgm. of the substance in olive oil administered to ten ovariectomised rats produced no sign of oestrus, the animals remaining in a state of di-oestrus throughout the experiment. A batch of twenty animals injected with 50 mgm. dissolved in olive oil showed seven full oestrus responses, with three animals just short of the definition (a few leucocytes). In a series of twenty animals injected with 100 mgm. dissolved in sesame oil, a very much better response was obtained, all twenty animals going into oestrus. The oestrus in each case was complete.

In the case of the 50 mgm. dosage, oestrus appeared after 54 hours and terminated 150 hours after injection. In the case of the 100 mgm. in sesame oil, oestrus appeared after 52 hours. At the present moment, it is impossible to state the activity of the material in terms of oestrin since the relatively difficult solubility of the material together with the consequent difficulties of administration and absorption make a comparison impossible. Some form of "cross-over" method must therefore be evolved. There can be no doubt that a repetition of the standardisation experiments with 50 mgm. dissolved in a small volume of sesame oil would indicate much greater potency than a similar experiment conducted with olive oil as the vehicle.

The observations show that 1–keto–1:2:3:4–tetrahydrophenanthrene is capable when injected into castrated animals of inducing oestrus of an exactly similar type to that obtained by the injection of oestrin. This result is of importance, for 1–keto–1:2:3:4–tetrahydrophenanthrene is the first compound of known chemical constitution found to have definite oestrus-exciting activity and furthermore, its molecular structure has many points of resemblance to the structure suggested for ketohydroxy-oestrin. There is thus provided the first step in the task of defining the molecular conditions necessary for this type of physiological activity, and there are grounds for hoping that substances of a much higher order of activity will be found before very long.

The observation[4] that oestrogenic properties of a low order are possessed by suitable extracts of such a variety of materials as peat, brown coal, lignite, coal tar and petroleum is of interest, but in view of the fact that many such materials are known to contain carcinogenic constituents, the clinical use of such extracts without very stringent refinement is scarcely to be entertained.

We have also examined 4–keto–1:2:3:4–tetrahydrophenanthrene (iii) and 3–hydroxyphenanthrene; these gave no oestrus response when injected in doses of 50 mgm.

注射的方法。由于药物在注射位置处泄漏和动物对橄榄油的耐受性差,这种方法的效果并不理想,但如果使用芝麻油,则注射量可降至 2 c.c. 且上述副作用可被避免。对 10 只摘除卵巢的大鼠注射 25 mg 1-酮-1, 2, 3, 4-四氢菲橄榄油溶液,未发现有发情征兆产生,这些动物在整个实验过程中一直处于间情期。对 20 只摘除卵巢的大鼠注射 50 mg 1-酮-1, 2, 3, 4-四氢菲橄榄油溶液,结果有 7 只表现出了充分的发情反应,还有 3 只出现了不完全的发情反应(有少量的白细胞)。在给 20 只大鼠注射了 100 mg 溶于芝麻油的 1-酮-1, 2, 3, 4-四氢菲时,取得的效果非常明显,全部 20 只大鼠都产生了发情反应。而且每只大鼠的发情反应都是完全的。

当剂量为 50 mg 时,发情反应在 54 个小时后出现,并于 150 个小时后终止。在以 100 mg 该物质溶于芝麻油的实验中,发情反应出现于 52 个小时之后。目前,我们还不能把该物质的活性与雌激素的活性进行比较,原因是该物质溶解性较差以及由此引发的给药和吸收上的困难。因此必须引入某种"交叉"的方法。毫无疑问,重复进行将 50 mg 该物质溶于少量芝麻油的标准化实验会比以橄榄油为介质的类似实验效果更显著。

观察表明:当给阉割的动物注射催情药物时,使用 1-酮-1, 2, 3, 4-四氢菲产生的效果与注射雌激素非常相似。这一结果意义重大,因为 1-酮 1, 2, 3, 4-四氢菲是人们发现的第一个有明确催情活性的、已知化学结构的化合物,而且它的分子结构在很多方面都类似于酮羟基雌激素可能具有的结构。这就向解析具备此类生理活性的分子应该具有什么样的分子结构迈出了第一步,而且我们也有理由确信人们很快就会发现活性更高的催情物质。

有报告 [4] 显示:从诸如泥炭、土状褐煤、暗色褐煤、煤焦油和石油等多种原料中经过适当提取得到的物质具有低水平的雌激素特性,这是一个有趣的发现,但因为此类原料中大多含有致癌成分,所以在这种提取物没有经过非常严格的精炼的情况下是很少用于临床的。

我们还测试了 4-酮-1, 2, 3, 4-四氢菲(iii)和 3-羟基菲的活性;在以 50 mg 的剂量注射后,它们都没有产生发情反应。

We are indebted to Dr. H. Allan for kindly checking over the animal experiments.

(**131**, 56-57; 1933)

J. W. Cook, C. L. Hewett: Research Institute, The Cancer Hospital (Free), London, S.W.3.
E. C. Dodds: Courtauld Institute of Biochemistry, Middlesex Hospital, London, W.1.

References:
1. *Z. physiol. Chem.*, **208**, 129 (1932).
2. *J. Soc. Chem. Ind.*, **51**, 277 *T* (1932).
3. *J. Physiol*, **68**, 348 (1930).
4. Schering-Kahlbaum: Fr. Pat., 710, 857.

感谢阿伦博士友好地对我们的动物实验进行了核对。

（刘振明 翻译；田伟生 审稿）

Structure and Division of Somatic Chromosomes in Allium

T. K. Koshy

Editor's Note

When T. K. Koshy published this account of *Allium* chromosomes separating in 1933, there were three competing theories of chromosome structure—the nuclear components were either granular, spiral or shaped like honeycombs. In order for cell division and heredity to be fully understood, Koshy knew the argument had to be resolved. So he looked at dividing cells and saw that metaphase chromosome arms were composed of two spirals coiled in opposite directions. As the spirals uncoiled, the chromosome separated and the progeny were pulled to opposite sides of the nucleus. He observed this spiral structure in all stages of mitosis, and so helped settle the debate on chromosome structure.

THREE divergent views have been expressed regarding the structure of chromosome. Pfitzner in 1882 suggested that a chromosome is made up of a row of granules embedded in an achromatic or less chromatic matrix. About the same time, Baranetzky found a spiral structure in certain stages in the meiotic cycle of the chromosomes of *Tradescantia virginica*. Vejdovsky (1912) has termed this spiral the chromonema. The chromonema theory conceives of a continuous, filiform, spirally coiled chromatic element in an achromatic matrix. The alveolar theory foreshadowed by van Beneden and worked out by Grégoire and his pupils contemplates the homogeneous chromosomes of the metaphase becoming during the following stages a honey-combed structure by the appearance in it of numerous alveoles. The supporters of this view assume that a longitudinally aligned central series of alveoles would account for the origin of the split in each chromosome.

The exact nature of the behaviour of chromosomes both in mitosis and in meiosis, as well as their rôle in heredity, can be understood only if their real structure is known. A study of the structure of the chromosomes of *Allium* has been undertaken under the kind guidance of Prof. R. Ruggles Gates, with the view of solving some of these problems. A preliminary account of the results so far obtained by cytological observations and the use of wire models is given below.

(1) At the early metaphase each chromosome is seen to be composed of two spirally coiled chromonemata and the duality of this spiral has been observed in all stages of the mitotic cycle.

(2) The spiral is coiled in opposite directions in the two arms of the chromosome, the null

408

葱属植物体细胞染色体结构与分裂

科希

葱属植物体细胞染色体结构与分裂

科希

编者按

在科希1933年发表这篇葱属植物染色体分裂的研究报告时，存在着三种彼此矛盾的染色体结构理论——其争论焦点是细胞核组成元件究竟是颗粒状的，螺旋形的，还是蜂窝状的？科希知道：要全面了解细胞分裂和遗传实质就必须解决这一分歧。因此他观察了处于分裂的细胞，发现在分裂中期染色体臂是由两条反向卷曲的螺旋组成。当螺旋打开时，染色体发生分离，并被牵引到细胞核的两极。这种螺旋结构存在于有丝分裂的各个时期，他的这一发现对于解决染色体结构的分歧有着十分重要的意义。

染色体结构存在着三种不同的假说。早在1882年，普菲茨纳就提出染色体是由一列镶嵌在不被着色或不易染色基质上的颗粒组成的。就在此前后，巴拉涅茨基在紫露草减数分裂过程中的某些时期也发现了螺旋结构。1912年，韦多夫斯基将这种螺旋结构命名为染色丝。染色丝理论假定在非着色基质中存在着一种连续、纤丝状且螺旋卷曲着的染色质成分。染色体泡状学说最初由范贝内登提出，随后得到了格雷瓜尔及其学生的证实。他们认为细胞分裂中期的染色体在继后的各个时期中变成了蜂窝状结构，从外形上看似许多小泡状结构凝集在一起。泡状学说的支持者们认为纵向排列的中央泡状系列组分是每一染色体分裂的起始点。

只有弄清楚染色体的真实结构，才能解释细胞有丝分裂和减数分裂过程中染色体行为的真实本质，以及它们在遗传中的作用。为解决这些问题，我们在拉各尔斯·盖茨教授的悉心指导下对葱属植物细胞染色体结构进行了研究。到目前为止，我们在细胞学观察基础上利用金属丝模型得到了初步的结论，其主要观点如下：

(1) 在有丝分裂早中期，每一染色体均由两条螺旋卷曲着的染色丝组成，而且在细胞分裂的各个时期都观察到了这种螺旋与卷曲的二重性。

(2) 每一染色体的两条臂朝着相反的方向螺旋卷曲，反向螺旋的分界点是染色体

409

point of the spiral being the attachment constriction of the chromosome.

(3) This form of double spiral permits of the separation of the chromosome into two by a simple uncoiling. This commences at the ends, as observed in *Allium*, but conceivably in some forms it may be initiated by the pull of the spindle fibres at the point of constriction.

(4) This unwinding causes rotation in the two arms of the chromosome; it may be possible that, as a result of this rotary motion in each arm of the chromosome, longitudinal cleavage is initiated in each chromonema at the metaphase stage.

(5) The separated chromonemata (daughter chromosomes), which now present a double spiral structure in each, remain parallel until they are finally pulled apart to the opposite poles of the spindle.

The doubleness of the anaphase chromosomes is clearly seen and the twisted appearance of these is due to slight loosening of the double spiral, which has been observed by Hedayetullah[1] in *Narcissus* and by Miss Perry[2] in *Galanthus*. The threads of this spiral retain their spiral structure in the subsequent stages of the mitotic cycle. The different appearances presented by the chromosomes in these stages are mainly due to the compact or loose nature of the double spiral. The dual threads remain closely associated together until their final separation at the next metaphase.

Further work on this subject is in progress and it is hoped that a detailed account may be published at a later date.

(**131**, 362; 1933)

T. K. Koshy: King's College, London, Feb.7.

References:

1. Hedayetullah, *J. Roy. Mic.* Soc., 51, 347-386.

2. Perry, *J. Roy. Mic.* Soc., 52, 344-356.

的附着缢痕。

(3) 通过简单的解螺旋作用，双螺旋染色体可以分成两部分。在葱属植物中这一过程发生在细胞分裂周期的末尾，但不难想象，在某种形式上该过程可能起始于缢痕点纺锤丝的牵引。

(4) 上述解螺旋作用使染色体的两臂发生旋转，该旋转可能导致每条染色丝在细胞分裂中期产生纵向分裂。

(5) 现在每一条分开的染色体丝（子染色体）都成为一个独立的双螺旋结构，它们在最终被牵引到纺锤体相反的两极前，一直保持着相互平行的状态。

海德亚图拉[1]和佩里小姐[2]分别观察了水仙花与雪花莲的细胞分裂周期，在细胞分裂后期，成双的染色体清晰可见；若双螺旋稍微松弛，染色体外形则呈扭曲状。这种螺旋卷曲着的染色丝在有丝分裂中期继后各阶段均保持螺旋状态，而不同时期所观察到的染色体不同形态，主要与双螺旋结构的紧缩或松弛状态有关。直到下一细胞分裂中期，紧密相连的双线结构才会分开。

关于这一主题的深入研究还在继续，希望将来还会有更详尽的研究结果发表。

（韩玲俐 翻译；程祝宽 审稿）

Number of Mendelian Factors in Quantitative Inheritance

R. A. Fisher

Editor's Note

Ever since 1896, agricultural scientists at the University of Illinois had been selectively breeding maize for high and low protein content and for high and low oil content. In a recent publication, Illinois scientist Floyd L. Winter reported the results up to 1924. From a foundation stock of maize, he and his colleagues had created massive variation in these two traits in just under 30 years, clearly demonstrating a process of selection. Crucially, however, their different protein and oil lines were no less variable than the original stock. This, geneticist Ronald Fisher argues here, has "killed" the idea that selection of small differences will only result in small evolutionary changes and that the act of selecting reduces variation, bringing evolution to a grinding halt.

IN a note in the current *Eugenics Review* entitled "Evolution by Selection", "Student" has directed attention to some statistical consequences of the inheritance of quantitative characters, in relation to the theory that these are due to the cumulative effect of a number of ordinary Mendelian factors.

"Student" refers in particular to the remarkable selection experiment carried out by F. L. Winter[1], in which a commercial variety of maize was exposed to mass selection from year to year in two diverging lines for high and low protein, and in two more for high and low oil content. For protein the initial value was about 11 percent with a standard deviation of a little more than 1 percent, but in the average of the last three years the mean of the high selection line is 16.82 percent while that of the low selection line is less than half that value, namely, 7.53 percent. The aggregate change produced by selection in both directions is thus 9.39 percent, or more than nine times the original standard deviation.

With respect to variability, it may be noted that the high line now varies from 13.4 to 19.8 percent, while the extremes for the low line are 5.7 and 10.5 percent; so that the two lines are now separated by a considerable gap, and therefore cannot possibly have any single genotype in common. The variability of the low selection line has shown a slight tendency to diminish, and that of the high selection line a slightly greater tendency to increase, so that no general tendency to a decrease in variability ascribable to selection is to be observed; thus, there is no reason to think that the selective potentialities of the material have been appreciably exhausted in producing the great modification which has been brought about.

"Student" contrasts these well-substantiated facts with the belief, widely held among

412

数量遗传学中孟德尔因子的数量

罗纳德·费希尔

编者按

自从 1896 年以后，伊利诺伊大学的农学家通过对玉米的选择育种可以得到蛋白质含量高或低和含油量高或低的玉米。伊利诺伊州的科学家弗洛伊德·温特在最近发表的一篇论文中总结了 1924 年以前的研究成果。在不到 30 年的时间里，他和他的同事从玉米的原种中培育出了蛋白质含量和含油量迥然不同的品种，这说明了选择的过程。不过最重要的是他们培育出的这些新品种可以和原种一样发生变异。遗传学家罗纳德·费希尔在本文中指出，这一点成功地"驳斥"了某些人的观点，即认为在差异小的群体中进行选种只会产生较小的进化变异，而且选种将使变异减少，使进化过程突然中断。

在近期《优生学评论》中的一篇名为《通过选择而进化》的短文中，"学生"（编者注："学生"即为英国统计学家威廉·戈塞特）把注意力转向了数量性状遗传的一些统计结果，并认为这些结果归于多个普通孟德尔因子的累积效应。

"学生"特别提到了温特进行的受人瞩目的育种实验[1]。该实验对一个玉米的商业变种进行年复一年的混合选择，选择方向为高蛋白质和低蛋白质两个品系，以及高油量和低油量两种类型。蛋白质含量初始值约为 11%，标准偏差略大于 1%，但在过去三年中，高蛋白质品系的均值为 16.82%；而低蛋白质品系的蛋白质含量均值还不到该值的一半，为 7.53%。于是，在两个方向上通过选择形成的累计变化为 9.39%，或者说是最初标准偏差的 9 倍多。

关于变异性，可以注意到目前高蛋白质品系在 13.4% 到 19.8% 的范围内变化，而低蛋白质品系的极端值为 5.7% 和 10.5%。因此，现在这两个品系之间相差很大，因而不可能有任何共同的单一基因型。低选择品系的变异性显示出了轻微的减少趋势，而高选择品系的变异性则出现了稍大的增长趋势，因此未能观察到因为选择而使变异性下降的总趋势；因而，没有理由认为在产生这种巨大改变时，会耗尽该材料的选择潜力。

"学生"将这些被充分验证的事实与几年前遗传学家们普遍持有的观念——即对

413

geneticists not so many years ago, that the selection of small differences (fluctuations) can only lead to unimportant evolutionary effects. They may also be contrasted with the oft-repeated statement that selection can do no more than select the best of the existing variety of genotypes, and with the commonly taught belief that the diversity available for selection is easily exhausted, from which it is inferred that evolutionary progress must wait upon the occurrence of mutations. It was, indeed, often represented as consisting in these occurrences.

The results obtained with oil-content have been even more striking; for the high oil line now contains nearly six times as much as the line selected for low oil content, and differs from it by more than twenty times the original standard deviation. "Student" uses these data, together with reasonable estimates of the intensity of selection, to obtain an estimate of the least possible number of factors which must be postulated to obtain the results up to the date of the report; he concludes that at least 100-300 factors would be needed; and, taking into account the complete lack of evidence that selection is nearing its limit, considers that it is more probable that the actual number of factors is measured in thousands.

Estimates of the number of factors needed to explain quantitative inheritance are beset with considerable difficulty, and "Student" has admitted to me in correspondence that his calculation fails from over-simplification. Other well-established phenomena in maize, however, such as the flood of recessive defects revealed by every plant which has been used to found a selfed line, combined with the inevitable rarity of each of these defects, taken individually, in the population from which the foundation plant was selected, force one to the conclusion that all commercial varieties must be segregating in hundreds, and quite possibly in thousands of factors influencing the normal development of the plant. This emphatic experience, has, I believe, killed among maize breeders all those doctrines concerning the supposed inefficacy of the selection of minute differences, with which the teaching of modern genetics was at first encumbered.

It should be emphasised that the result of importance for evolutionary theory is not that the number of factors must be very large, thousands for example, rather than hundreds, but the direct demonstration that selection has the exact effects that selectionists have ascribed to it, without the limitations by which its action has been supposed to be restricted, on the strength of an early misapprehension as to the number and variety of the Mendelian factors exposed to its cumulative action.

(**131**, 400-401; 1933)

R. A. Fisher: Rothamsted Experimental Station, Harpenden, Feb. 15.

References:
1. J. *Agric. Res.*, **39**, 451–476 (1929).

细微差异（波动）的选择只会导致微不足道的进化效应——进行了对比。与这些事实形成对比的还有那个被再三重复的论断，即选择只能从基因型的现存变异中挑选出最好的，以及那个被普遍讲授的观点，即可供选择的多样性很容易被耗竭，因而进化过程必须等待突变的发生。实际上，在这些事件中通常是一致的。

从含油量中得到的结果更令人震惊，目前高油品系的含油量是低油品系的近 6 倍，在两个方向上通过选择形成的累计变化是最初标准偏差的 20 倍以上。"学生"利用这些数据，加之对选择强度的合理估计，得到了要想获得本报告中的结果所必需的最低可能因子数量的估计值。他认为至少需要 100~300 个因子，而考虑到没有任何证据表明进化已经接近了极限，他认为实际因子的数量很可能数以千计。

在估计解释数量遗传所需的因子数量时遇到了很大的困难。"学生"在写给我的信中承认，他的计算因过于简单而失败。然而，玉米中的其他一些已经得到确认的现象——如用来建立自交品系的所有植株都显示出大量的隐性缺陷，并且从选择建群植株的种群中单独筛选出的每一个缺陷都必定十分稀有等——迫使人们得出结论，即认为所有商业变种必然是数百个，并且很可能有数千个影响该植物正常发育的因子。我相信，这个特殊实例驳斥了在玉米育种者中流传的认为细微差异对选择无效的种种学说，而这些论点在一开始就阻碍了现代遗传学的教学。

应当强调的是，进化理论的重要结论并不在于因子数量一定要很大，比如是数千个而非数百个，而是在于它直接证明了选择具有自然选择论者所认为的确切效果，它的作用并没有受到人们一向认为会有的限制，而这些限制是由于早期人们对累积效应下孟德尔因子的数量和变化的误解所产生的。

（周志华 翻译；刘京国 审稿）

The Physical Nature of the Nerve Impulse*

A. V. Hill

Editor's Note

Following a lecture he had recently given at the Royal Institution in London, Nobel Prize-winning physiologist Archibald Hill here explains the nature of nerve impulses. This is one of the earliest examples of a biological problem the understanding of which demanded a strong degree of knowledge in physics. Hill ends by reflecting on an extremely hot topic of the period that promised to revolutionize the study of nerve cell function: Russian scientists had recently produced evidence that living organisms give off radiation, particularly when active, that will induce mitosis (cell division) in other cells. As Hill suspects, however, the phenomenon of "mitogenic radiation" was not borne out by further research.

ALL our sensations, all our movements, most of the activities of our nervous system, depend upon a certain transmitted disturbance which we call the nervous impulse: this, in the study of nerve activity, is what the atom, the electron and the quantum are to chemistry and physics. A rapid reaction to events occurring at a distance is necessary for efficient working. Special nerve cells, therefore, have been developed in all the larger animals: from these the axon or nerve fibre runs out, which is only 3 μ to 25 μ in diameter but may be many metres in length. Along these fibres wave-like messages are sent.

The velocity of a nerve impulse varies greatly according to the fibre in which it runs and to the conditions affecting the fibre. In the medullated nerve of a mammal the velocity is of the order of 100 metres per second. (Compare this with 330 metres per second, the velocity of sound.) In the medullated nerve of the frog at 20°C. it is about 30 metres per second. In the non-medullated nerve of a mammal it is said to be about one metre per second; in the non-medullated nerve of the pike and of *Anodon* respectively it is stated to be 0.2 metre and 0.05 metre per second. The velocities in this list are in the ratio of 2,000 to 1.

The nerve impulse is an event, a wave, a propagated disturbance, not a substance or a form of energy. It is transmitted along a thread of protoplasm which, in medullated nerve, is surrounded by a protecting or "insulating" sheath. Its passage can be detected in several ways: (*a*) by its physiological effect on the organ to which it runs, (*b*) by the electric change which accompanies its transmission, (*c*) by a production of heat, and (*d*) by a consumption of oxygen and a liberation of carbon dioxide.

---

* Friday evening discourse delivered at the Royal Institution on Feb. 10.

神经冲动的物理学本质[*]

阿奇博尔德·希尔

编者按

最近诺贝尔奖得主生理学家阿奇博尔德·希尔在伦敦的英国皇家研究院作了一场报告，随后他在《自然》杂志上发表了一篇解释神经冲动本质的文章。这是其中一个用大量物理学知识来解释生物学问题的早期例子。希尔以反思当时一个有希望使神经细胞功能的研究发生革命性变化的热门话题作为结尾——俄罗斯科学家最近发现生物体可以放出辐射，而且在辐射时会引发其他细胞的有丝分裂（细胞分裂）。然而，正如希尔所料，这种"导致有丝分裂的辐射"效应并没有被后续的实验所证实。

我们所有的感觉、行为以及神经系统的大部分活动都依赖于一种特定的可以传递的扰动，我们称之为神经冲动：神经冲动之于神经活动的研究就如同原子、电子和量子之于化学和物理学。对远程事件作出快速反应是高效工作所必需的。因此，所有大型动物都进化出了特殊的神经细胞，从其上延伸出直径只有 3~25 微米、而长度可达数米的轴突或神经纤维。信息像波一样沿着这些纤维进行传递。

神经冲动传递的速度在很大程度上受到传递它的纤维以及纤维所处环境的影响。在哺乳动物的有髓鞘神经纤维中，传递速度的数量级可达 100 米 / 秒（可以将这个速度与 330 米 / 秒的声速进行比较）。温度 20℃时在青蛙有髓鞘神经中的传递速度约为 30 米 / 秒。据说在哺乳动物无髓鞘神经中的传递速度为 1 米 / 秒；而在梭子鱼和无齿蚌的无髓鞘神经中，传递速度分别是 0.2 米 / 秒和 0.05 米 / 秒。在上面列出的数据中，最快速度与最慢速度的比值为 2,000:1。

神经冲动是一个事件、一束波、一种可传递的扰动，而非一种物质或能量形式。它沿着原生质丝进行传递，在有髓鞘神经中，原生质丝外面包围着保护性的或"绝缘"的鞘。神经冲动的传递可以用多种方式进行检测：(a) 途经器官的生理响应，(b) 与传播过程相伴的电位变化，(c) 产生的热效应，(d) 氧气的消耗和二氧化碳的释放。

---

[*] 本文取自希尔于 2 月 10 日在英国皇家研究院所作的周五晚间演讲。

417

No difference of electrical potential can be detected in an uninjured resting nerve. If, however, the nerve be injured, for example, by cutting, a potential difference is found of the order of a few hundredths of a volt between the injured and the uninjured parts, in the sense that positive current runs in an external circuit towards the former. The injury does not *produce* the potential difference; it merely allows its normal presence across the fibre boundary to be manifested.

If two electrodes be placed on a nerve and the nerve be stimulated, a momentary change of potential travels along it which can be recorded with an oscillograph. At any given instant a certain length of the nerve of the order of a few centimetres is found to be the site of a wave of negative potential. We are probably right in thinking that the impulse itself, whatever that be, occupies the same region and moves at the same speed as its electrical accompaniment.

Properties of the Nerve Impulse

Let us consider the properties of the impulse which is transmitted, or transmits itself, in nerve. The single impulse in the single fibre is the basis of nerve activity. Until recently, this individual impulse could not be separately examined and deductions had to be made from the results of stimulating many fibres in parallel. Of late, however, through improvements in electrical recording, it has become possible to register the form and movement of the electric change resulting from a single impulse in a single fibre.

(*A*) A single impulse has an "all-or-none" character. Its size cannot be varied by changing the strength of the stimulus which produces it: it makes no difference to the magnitude of the discharge how hard the trigger is pulled. (*B*) An absolute "refractory" period follows the passage of an impulse, a period during which no stimulus, however strong, can evoke a second response. In frog's nerve at 20°C. this absolute refractory period is about 0.001 sec. (*C*) As a consequence of (*B*), two nerve impulses going in opposite directions in the same fibre come each into the refractory region of the other and both are abolished. "One way traffic" alone is possible: separate sensory and motor systems are required. (*D*) After the absolutely refractory stage a relatively refractory stage persists, during which a stronger stimulus than usual is required to start an impulse, and the second impulse, measured by electric change or heat, is smaller than the first one. Discharge can take place before recharge is complete. (*E*) As a result of the refractory phase, the frequency of transmission is limited. Prof. H. S. Gasser of New York recently informed me that he had made mammalian nerve carry 1,500 impulses per second, and that at 1,000 per sec., impulses were not greatly subnormal. In the normal functioning of the nervous system nearly all messages consist of trains of impulses of varying frequency. (*F*) The impulse at any instant occupies a few centimetres of the fibre. This length is not much altered by a change of temperature and it is approximately proportional to the diameter of the fibre. There must be some simple physical reason for these facts, as also for the next one. (*G*) The velocity of the impulse is also approximately proportional to the diameter of the fibre. Consequently in a given nerve trunk which consists of fibres of various diameters, a

在未经损伤的处于静息状态的神经上是检测不到电位差别的。但是，如果神经受到损伤，比如被切割，在损伤部分和未损伤部分之间就会有几十毫伏数量级的电位差，从这个意义上来说，正向电流是沿着一个外部的环路流向损伤部分的。损伤并不是电位差**产生**的原因；它仅仅是让正常情况下就存在的跨纤维边界的电位差显现出来而已。

如果将两个电极接在一根神经上，神经就会受到刺激，可以用示波器记录沿神经传播的瞬间电位变化。研究发现：一段长度为几厘米的神经在任意一个瞬间都是负电位波的位点。我们认为神经冲动本身，无论其本质是什么，都和伴随它的电位处于同一个位置并以同样的速度进行传播，这个观点很可能是正确的。

神经冲动的特性

让我们来谈一谈在神经中传播的冲动或传播过程本身的特性。单个纤维中的单个冲动是神经活动的基础。直到最近，人们仍然不能单独检测这种单个的冲动，而只能根据刺激多个平行纤维得出的结果进行推算。但是现在随着电位记录技术的进步，人们已经有可能记录由单个纤维中单个冲动引发的电变化以及它的传播。

（A）单个冲动具有"全或无"的特点。冲动的大小不会随刺激强度的变化而变化：无论刺激多么强烈，也不会在放电量上产生差别。（B）在一个神经冲动通过后会出现一段绝对的"不应"期，在这个时期内无论多强的刺激都不能引发第二次响应。20℃时青蛙神经的绝对不应期约为 0.001 秒。（C）根据（B），两个在同一纤维中沿相反方向传播的神经冲动会使对方进入不应区而一起消失。只有"单向传播"是允许的：传播需要独立的感知和运动系统。（D）绝对不应期之后会有一个相对不应期，需要比平时更强的刺激才能在这个阶段产生冲动，而且第二个冲动的电位变化或热效应要小于第一个冲动。放电可以在充电完成之前发生。（E）由于存在不应期，所以冲动传播的重复频率是有限的。最近，纽约的盖瑟教授告诉我，他已经能让哺乳动物的神经在每秒内传递 1,500 个冲动，而且当达到每秒 1,000 个时，神经冲动并没有明显变弱。在功能正常的神经系统中，几乎所有的信息都是由一个个不同频率的神经冲动组成的。（F）神经冲动在任意一个瞬间仅占据神经纤维中数厘米的范围。这个长度基本上不随温度的改变而改变，并与神经纤维的直径大致成正比。用简单的物理学原理一定能解释这些现象，下一条也是如此。（G）冲动传播的速度也与神经纤维的直径大致成正比。因此，在给定的由不同直径神经纤维组成的神经干中，电波从一端向另一端传播时会表现为对应于频率曲线上若干极点的一系列波。

wave started electrically at one end gradually spreads out and appears as a series of waves corresponding to the several maxima in a frequency curve. (*H*) When the temperature is raised or lowered by 10°C. the velocity of transmission is increased or diminished in the ratio of 1.7 to 1. (*I*) The passage of an impulse is associated with a liberation of heat to which some special attention is necessary.

Heat Production of Nerve

No work is performed, no force is developed, no movement at all occurs in a nerve when it is active, and for long it was believed that in the transmission of nerve impulses no heat is evolved. Many attempts were made to measure the heat, all unsuccessful until 1925. Today, when the heat production of nerve can be measured almost as well as that of muscle could in 1920, it is hard to believe that for many years we argued as though there was no heat at all associated with nerve activity. It is true that the amount is small, that the *total* heat (which takes thirty to fifty minutes to appear) in a single impulse in a gram of nerve is only about one millionth of a calorie, and it is true still that we must have several impulses before we can measure the heat properly, but with that provision and with present-day arrangements, the measurement is comparatively simple and accurate. The sensitivity is such (i) that a galvanometer deflection of 1 mm. (readable to 0.1 mm.) corresponds to a rise of temperature of about three millionths of a degree, and (ii) that during a steady state of heat production one millimetre of steady deflection corresponds to a rate of heat production of about 2×10^{-8} calorie per second.

The heat is produced not only during the passage of the impulse *but for a long time afterwards*; some kind of breakdown occurs, presumably as the wave goes by. This breakdown has then to be reversed, the nerve allowed to recover, in a re-charging process of some kind which takes place afterwards. The fact that heat is produced, and the manner of its production, dispose of the possibility that the nerve impulse is, to use Bayliss's words, a "reversible physico-chemical process".

At 20°C. the initial heat in a single isolated impulse in a frog's medullated nerve lies between 10^{-7} calorie and 3×10^{-8} calorie per gram of nerve. The recovery heat, which continues for a long time after the impulse has passed, is about ten to thirty times as much. During continual stimulation, say of fifty shocks per second, a steady state is gradually reached in which recovery balances breakdown, the recovery heat in any given interval representing the process of restoration from all the impulses which passed in the preceding forty minutes. During the steady state, which is possible only in oxygen, the total heat production may occur at the rate of about 25 microcalories per gram per second. At a higher frequency the total heat rate may, for a time, be rather greater, but a genuine steady state is not possible; fatigue progressively sets in. It is striking that the *resting* rate of heat production of the same nerve is about 70×10^{-6} calorie per gram of nerve per second, so that moderate activity only increases the metabolism of a nerve by a comparatively small amount, even extreme activity does not double it. The mere maintenance of the machine in working order requires more energy than the excess required when it goes full speed during activity.

（*H*）如果温度升高或者降低 10℃，传播速率会以 1.7∶1 的比例增加或减小。（*I*）冲动的传播伴随着放热，这一点需要特别注意。

神经的放热

当神经具有活性时，既没有做功，也没有产生力，更没有移动位置，长期以来人们一直认为在神经冲动的传播过程中没有热量放出。在 1925 年之前，人们多次试图对神经冲动传播过程中可能产生的热量进行测量，但是都失败了。而现在，人们对神经放热的测量已经能够达到 1920 年对肌肉放热的测定水平，难以想象这么多年来我们一直相信神经活动中根本不会放热。当然这个热量非常小，1 克神经单次冲动产生的**总热量**（耗时 30~50 分钟）大约只有 1×10^{-6} 卡路里，而且必须在有若干个神经冲动之后我们才能够准确地测定这种热量，但在满足上述条件并运用现有技术手段的情况下，测量还是比较简单和准确的。其灵敏度是（i）检流计偏转 1 mm（最小刻度是 0.1 mm）对应于温度升高约百万分之三摄氏度；（ii）在稳定放热阶段，1 mm 的稳定偏移对应于约 2×10^{-8} 卡路里 / 秒的放热速率。

热量的产生不仅发生在冲动传过的瞬间，**在冲动传过之后很长一段时间内**也会发生；人们推测当电波通过时会造成某种程度的击穿，为了对击穿进行修复，神经会在之后的时间里进行某种形式的再充电过程以便恢复原状。放热现象本身以及其产生的方式说明：神经冲动很可能，用贝利斯的话讲，是一个"可逆的物理化学过程"。

20℃ 时，单个冲动在每克青蛙有髓鞘神经中产生的初始热量大约在 10^{-7} 到 3×10^{-8} 卡路里之间。恢复热在冲动传过之后很长一段时间内都会持续存在，其大小约为以上数值的 10~30 倍。在持续的刺激，比如 50 次 / 秒的电击下，就能逐渐达到恢复和击穿取得平衡的稳定状态，任何一段给定时间间隔内的恢复热代表了在之前40 分钟内传过的所有冲动的复原过程。在这种必须有氧参与才可能存在的稳定状态下，整个放热过程的放热速率大约为每克每秒 25 微卡路里。当频率较高时，总的放热速率可能会突然变大，但在这种情况下真正的稳定状态是不可能发生的；因为神经会渐渐疲劳。令人惊讶的是：在**静息**状态下同一条神经的放热速率约为每克神经每秒 70×10^{-6} 卡路里，所以中等程度的活动只会使神经代谢增加很少一部分，即便是强度最大的活动也不能使其翻倍。刚好能维持机器正常运转所需的能量要高于开足马力时所需的额外能量。

This initial heat is the result presumably of some chemical reaction involved in, or immediately after, the transmission of the impulse. If we suppose the reaction to occur throughout the substance of the axis cylinder, its heat is so small that it is difficult to picture any mechanism by which the change could be propagated. There are other grounds—for example, the relation between speed and diameter—for supposing that the reaction, whatever it be, is somehow connected with the surface of the fibres. It can be calculated that the area of the surface in a gram of nerve is of the order of 2,000 sq. cm., and that the energy in the transmission of a single effective impulse is from 5×10^{-3} to 2.5×10^{-4} erg per sq. cm. of fibre surface. This is still a very small quantity, the smallness of which can be realised by the statement that it is 1/4,000 to 1/80,000 of the surface energy of a water-olive oil interface.

One naturally asks, may not the initial heat be really due to the electrical disturbance transmitted in nerve? If we take the observed potential differences along the nerve and assume that these cause currents to flow through the conducting media inside and outside the sheath, then the Joule's heat of the currents can be calculated. The result is only a small fraction, less than one percent, of the observed initial heat. There is another possibility, however, namely, that the nerve is to be regarded as a charged electrical condenser which is discharged as the wave passes by. The energy of a condenser of capacity F microfarads, charged to V volts, is $5FV^2$ ergs. Taking V as 0.05 we should require a capacity of the order of half a microfarad per square centimetre of nerve fibre surface to give us the observed initial heat. There is some evidence that capacities of this size may possibly exist at the surface of living cells. If so, we should not need to look for a chemical reaction to explain the initial heat—the electrical discharge would be sufficient. Other evidence, however, makes it unlikely that this is really the source of the heat.

The recovery process by which the nerve is restored to its initial condition is of an oxidative nature, though deprival of oxygen does not immediately cause it to fail. Apparently the nerve possesses, maybe as a safeguard against asphyxiation, a store of oxygen in some form other than molecular. A nerve may go on functioning and carrying out its usual recovery for hours, or even days (depending on the temperature) before all its oxygen store is used up. Then, and only then, it fails. If a nerve asphyxiated by lack of oxygen is given oxygen once more, recovery rapidly occurs and excitability returns.

Other Effects of Oxygen

The difference of potential between an injured and an uninjured point of a crab's nerve is maintained for a long time in the presence of oxygen and the absence of stimulation. Stimulation rapidly reduces it, so to speak "depolarises" the nerve surface. The potential rises again to its full value if, and only if, oxygen be present. In the absence of oxygen it falls still further. The potential difference across the surface of a nerve depends for its maintenance on the continued presence of oxygen.

据推测，这些初始热量可能来源于传播神经冲动的过程或者紧随其后发生的化学反应。如果我们假设该反应是在整个轴突中发生的，那么这些热量就太小了以至于很难描述变化得以传播的机制。还有其他一些证据，比如传播速度和直径的关系，证明了化学反应不管怎么样都会在某种程度上与神经纤维的表面有关。计算表明：1克神经的表面积大约为 2,000 cm² 数量级，而单次有效冲动传播的能量为每平方厘米神经纤维表面 5×10^{-3}~2.5×10^{-4} 尔格。这个值仍然是非常小的，仅仅是水 – 橄榄油之间界面能的 1/4,000~1/80,000。

有人很自然地会问：这些初始热量难道不是由神经中传播的电扰动引起的？如果我们测量出沿神经的电位差，并假设电位差就是使电流在鞘内外传导介质中流动的原因，那么我们就能计算出电流的焦耳热。这个结果只占初始热测量值的很小一部分，还不到 1%。然而，还存在另一种可能性，即神经可以被看成是一个带电的电容器，当波通过时会发生放电。一个电容为 F 微法、充电到 V 伏特的电容所具有的能量为 $5FV^2$ 尔格。假设 V 等于 0.05，则电容的数量级要达到 0.5 微法每平方厘米神经纤维表面才能得到与观察到的初始热量相符的结果。有一些证据表明：在活细胞表面有可能存在这么大的电容。如果事实真的如此，那么我们就不再需要寻找化学反应来解释初始热量了，电容放电已经足以说明一切。不过其他证据否认了电容放电是热量来源的观点。

神经恢复到初始状态的复原过程是一种氧化过程，而去除氧气不会马上使其终止。显然，神经储存了大量非分子形式的氧，这也许是抵御窒息现象发生的一种手段。在它储备的氧耗尽之前，神经可以在数小时乃至数天内（取决于温度）继续发挥功能并且像往常一样进行复原。之后，也只有在氧气耗尽之后，神经才会失效。如果再次恢复对缺氧神经的氧气供应，它会很快复原并恢复活性。

氧气的其他作用

蟹神经损伤部位和未损伤部位之间的电位差在有氧但没有刺激的环境下能够保持很长一段时间。刺激可以使之很快降低，这就是所谓的神经表面"去极化"。当且仅当氧气存在时该电位才能再次达到最大值。没有氧气它就会降得更低。跨神经表面电位差的保持依赖于氧的持续存在。

Stimulation and lack of oxygen, however, are not the only means by which the potential difference can be reduced. There is normally a ratio of about 10 to 1 between the inside and the outside of a crab's nerve fibre in respect of potassium ion concentration. If we suppose that potassium is the only substance capable of penetrating the fibre surface of the crab's nerve, then this concentration ratio should lead to a potential difference of $\frac{RT}{F}\log_e 10$ which is about 58 mv. Values of 30 mv. in freshly dissected nerve are commonly observed, and sometimes more, and since a certain amount of short-circuiting must occur in the strongly conducting fluid between the fibres, the real value may well approximate to the 58 mv. required by the formula. If the hypothesis is right, the potential difference would be reduced by increasing the potassium ion concentration on the outside of the fibre. Cowan has found this to be the case. By soaking a nerve for a few seconds in sea-water to which potassium has been added, the potential difference may be varied from 30 mv. or 40 mv. down to nearly nothing as desired. The potential difference is clearly determined by the potassium ion concentration, and yet it is dependent also on the presence of oxygen. This paradoxical dependence on two such different factors has not yet been resolved: perhaps the oxygen is used in maintaining the normal properties of the surface membrane.

Another, and a most curious effect of oxygen, has recently been observed in the action of veratrine, a plant alkaloid, on frog's nerve. Applied to a muscle, this drug causes a prolonged response to a single impulse: applied to a nerve it has little effect, the action current is not greatly lengthened, the heat production is not largely increased, by soaking the frog's nerve for an hour in 1 in 50,000 veratrine solution. If, however, after the nerve has been soaked it be then asphyxiated for three or four hours until it is completely inexcitable, and if it be then revived by admitting oxygen, it shows in a striking manner a typical veratrine effect. The action current lasts hundreds of times as long, the heat production is thousands of times as great, as in the normal response of a nerve to a single shock.

Apparently the drug is usually unable to penetrate the nerve fibre sheath. When, however, the nerve has been asphyxiated, the condition of the surface has somehow been altered so that the substance which was previously held out is now able to penetrate. Oxygen is necessary for the maintenance of the normal impenetrable condition of the surface of the fibre. One naturally associated this effect with the presence of a thick sheath around the axis cylinder, and asked whether the veratrine effect could be manifested in other nerves in which no medullary sheath is present. The experiment was made by Cowan, and showed that in crab's nerve a typical veratrine effect is produced without any asphyxiation, and that the concentration of the drug required is only one thousandth of that for frog's nerve. Veratrine apparently can penetrate normally the very thin surface of a non-medullated nerve, but it is held out almost indefinitely, until the nerve is asphyxiated, by the sheath normally covering the axis cylinder of a medullated nerve.

This strange result suggested that the effect of curare should be similarly tested. Curare

424

然而，刺激和缺氧并不是使电位差下降的唯一方法。在正常情况下，蟹神经纤维内部与外部的钾离子浓度比为 10:1。如果我们假设钾是唯一能够透过蟹神经纤维表面的物质，那么这个浓度比将产生 $\frac{RT}{F}\log_e 10$ 的电位差，其大小约为 58 毫伏。人们在新鲜剥离的神经上测得的数值通常为 30 毫伏，有时会高一些，因为在纤维与纤维之间导电性很强的流体中必然会发生一定程度的短路，实际值可能刚好接近于由公式推算出来的 58 毫伏。如果这个假设是正确的，那么提高神经纤维外部的钾离子浓度就会降低电位差。考恩已经证实了这一点。他将一根神经浸泡在添加了钾离子的海水中，几秒钟后电位差从 30 或 40 毫伏降低到几乎为 0，这与预期是相符的。电位差明显取决于钾离子的浓度，并与氧气是否存在有关。人们至今也说不清为什么电位差会依赖于这两种完全不同的因素：也许氧气的作用是用来维持膜表面的正常特性。

最近人们在研究藜芦碱（一种植物碱）对青蛙神经的作用时发现了氧气的另一个非常奇怪的特性。把藜芦碱作用到肌肉上可以延长单个冲动的响应时间；但该药物对神经几乎没有什么影响，把青蛙神经放在 1:50,000 的藜芦碱溶液中浸泡 1 小时，动作电流并没有明显延长，放热也没有大幅增加。但如果使浸泡后的神经缺氧 3 小时或 4 小时直至完全失活，然后再供氧使其恢复活性，典型的藜芦碱效应就会以一种不同寻常的方式表现出来。此时的动作电流比一根神经对单个刺激的正常响应时间长几百倍，放热也增加了几千倍。

显然，药物在通常情况下是不能通过神经纤维的髓鞘的。但是当神经缺氧时，表面状态发生了某种变化，所以本来不能透过的物质现在可以透过了。氧气对于维持神经纤维表面在正常情况下的非通透性是非常必要的。人们会很自然地将这种效应与轴突外面包裹的厚鞘联系起来，并怀疑藜芦碱效应是否也能在没有髓鞘的神经中表现出来。考恩进行了这方面的实验，他发现即使不缺氧，蟹神经也能表现出典型的藜芦碱效应，而且所需药物的浓度仅为进行青蛙神经实验所需浓度的千分之一。在正常情况下，藜芦碱显然能够穿透很薄的无髓神经表面，但是几乎永远不可能穿透有髓鞘神经轴突外面的髓鞘，除非该神经经过了缺氧处理。

这个奇特的结果使人联想到应该用同样的方法研究一下箭毒的效应。箭毒能通

paralyses a muscle, in respect of impulses coming to it along its motor nerve. It was formerly thought to attack the neuro-muscular junction, though recently, on the strength of Lapicque's theories, it has been supposed to produce its effect by changing the "time scale of excitation" of the muscle fibre, putting it out of tune so to speak with its nerve. Recent work by Rushton has shown that this hypothesis of Lapicque's is untenable, so that the manner of action of the drug was still unsolved. It was possible that curare might be a potent nerve poison but normally unable to penetrate except at the neuro-muscular junction. At this point the medullary sheath is absent. Perhaps a nerve dosed with curare but apparently unaffected, might be found effectively poisoned after asphyxiation.

The experiment was performed by Fromherz: a muscle nerve preparation was paralysed by soaking in curare, the nerve was removed and still showed a normal action current: it was then asphyxiated in hydrogen and after asphyxiation allowed to recover (if it could) in oxygen. An unpoisoned nerve treated in this way immediately gives a large action current on readmission of oxygen. The curarised nerve gave only a small one and recovered very slowly. Curare, therefore, may render a nerve incapable of responding, once it is able—as during a state of asphyxia—to get in. This may be the solution of the problem: the effect of curare is not on the "chronaxie" of muscle, but on the part of the nerve which is exposed to the action of the drug, namely, its ending where the medullary sheath is absent.

Time Relations of Excitation

All living animals, organs, or organisms, have a certain characteristic scale of time. The fibres of the wing muscle of a fly can contract hundreds of times a second, those of the leg muscle of the tortoise, differing from the fly's in no other very obvious way, may take several seconds to give a single twitch. The mouse and the man differ greatly in many of their characters and the difference is largely that of the scale of time on which respectively they live. This difference of time scale is a necessary accompaniment of a difference of size: a very small motor may do 10,000 revolutions per minute, a large one must be content with a few hundreds. One of the most important problems of physiology, and it is a problem of general, almost of philosophical interest, is what determines the scale of time of an animal or cell.

In nerve, the time required at any point for the action current to go through its cycle, the time taken in the transmission of the impulse, and the time-scale (to which I will refer later) of electric excitation, all can be varied together as we alter the condition of the cell or pass from one cell to another. These are different aspects of the same phenomenon—the transmission of the impulse. At the end of the last century it was shown by Waller, employing stimulation by condenser discharge, that there is an "optimal" stimulus for any given tissue, in the sense that the energy in it is a minimum. He showed, and Keith Lucas following him showed, that the effectiveness of a condenser stimulus of given energy depends upon the rate of discharge of the condenser. If F be the capacity and R the resistance through which discharge takes place, the time of discharge is proportional to

过沿运动神经传入肌肉组织的冲动麻痹肌肉。之前人们认为它攻击神经与肌肉的结合部，但是按照拉皮克的最新理论，改变肌纤维"兴奋的时间尺度"使之不能与支配它的神经的时间尺度相匹配才是箭毒的作用机制。而拉什顿最近的研究成果表明拉皮克的这一假设是站不住脚的，所以该药的作用机制仍然没有解决。箭毒可能是一种潜在的神经毒剂，不过在一般情况下它不能穿透除神经肌肉接合部以外的部位。因为神经肌肉接合部不存在髓鞘。原本对箭毒没有明显反应的神经也许在缺氧后就会中毒。

弗朗姆荷茨进行了如下的实验：将一条肌肉神经样本浸泡于箭毒中使之麻痹，然后取出神经，仍然可以观测到正常的动作电流；再将其置于氢气环境中进行缺氧处理，随后在氧气中使之恢复（如果可以恢复的话）。一条没有经过箭毒处理的神经在重新供氧后会马上出现明显的动作电流。而用箭毒处理过的神经只会出现很小的动作电流并且恢复得很慢。因此一旦箭毒能够渗入，比如在缺氧条件下，它或许就能使神经丧失响应能力。这也许就是上述问题的答案：箭毒效应与肌肉的"时值"无关，而与暴露在药物中的那部分神经，即与没有髓鞘的神经末梢部位有关。

兴奋的时间关系

所有活的动物、器官或组织都具有特征性的时间尺度。苍蝇翅膀的肌肉纤维每秒能收缩数百次，除了需要几秒钟才能收缩一次以外，乌龟腿部的肌肉与苍蝇的肌肉没有太大的差别。小鼠和人类在许多特征上都有很大的不同，主要区别在于它们生存的时间尺度不同。时间尺度不同是体型大小不同的必然结果：一个非常小的马达也许每分钟能转 10,000 圈，而大马达只能转数百圈。生理学上的一个最重要的问题，同时也是一个具有普遍意义和哲学意味的问题，就是到底是什么决定了动物或细胞的时间尺度。

在神经中，如果我们改变细胞存在的环境或者从一个细胞传入另一个细胞，则动作电流完成整个回路所需的时间、传播冲动所需的时间以及电兴奋的时间尺度（我将在后面提到）都会发生改变。这些都是同一现象——冲动传播的不同表现形式。上个世纪末，沃勒在利用电容器放电的刺激实验中发现，对于任何特定的组织都有一个"最佳"的刺激条件，此时电容内的能量最小。他以及他之后的基思·卢卡斯都发现：特定能量电容器的放电刺激效率依赖于电容器的放电速度。如果 F 是电容器的容量，R 是放电时的电阻，则放电时间正比于 FR 的乘积，因此对于给定的刺激能量 $\frac{1}{2}FV^2$，其响应依赖于乘积 FR。

the product *FR*, and for given energy $\frac{1}{2}FV^2$ in the stimulus, the response depends on the product *FR*.

The use of condensers is very convenient and is generally adopted in studying the time relations of the excitatory process. The relation, however, is clearer if excitation be produced with the aid of a constant current of variable duration. If a constant current be led into an excitable tissue by two non-polarisable electrodes, in order that a stimulus of very short duration may be effective its intensity must be great, while if it lasts for a long time its intensity may be small. In the case of medullated nerve the times involved are very short, so that rather special methods of determining the duration of the constant current employed must be adopted. The relation between the duration of a constant current and the least strength required for excitation (the so-called strength–duration curve) is shown in Fig. 1.

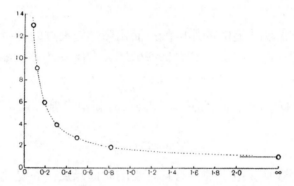

Fig. 1. "Strength duration curve" for nerve excitation. Horizontally, duration in σ (0.001 sec.): vertically minimum strength required.—From "Chemical Wave Transmission in Nerve", by A. V. Hill (Cambridge University Press, 1932).

In the case of nerve this curve is always approximately of the same form and may be defined by two parameters, the scale of time and the scale of current. The latter is of no particular interest, since most of the current is short-circuited in the fluids between the excitable elements and no strictly standard conditions for comparison can be defined. The scale of time, however, *is* important, and is very constant for a given nerve under given conditions, and the usual method of expressing it is by means of the minimum duration of a current of twice the threshold strength. This duration was called by Keith Lucas the "excitation time" and by Lapicque the "chronaxie", the latter term in its derivation meaning no more than the time scale of the process or tissue in question.

The excitation time of nerve depends upon many factors: (*a*) the nature of the nerve itself; it varies greatly from one fibre to another, in the same kind of way as the velocity of propagation of the impulse referred to earlier; (*b*) it is increased by a fall, decreased by a rise of temperature; (*c*) it is considerably affected by the nature of the ionic constituents in the solution around the nerve, particularly by the concentration of the calcium, and to a less extent of the potassium ion; (*d*) it depends upon the size of the fibre, being smaller in a fibre of greater diameter.

428

利用电容器研究兴奋过程的时间相关性是非常便利的，而且已经得到了广泛的认可。但如果使用持续时间不同的恒定电流来引发兴奋，其时间相关性会更加清晰。如果通过两个非极化电极将恒定电流引入可兴奋的组织中，为了使持续时间非常短的刺激能够产生效果，强度必须很大；而对于持续时间较长的刺激，强度可以小一些。在有髓鞘神经纤维中，这个时间非常短，以至于需要采用特殊的办法才能测定稳态电流的持续时间。稳态电流持续时间和兴奋所需最小强度之间的关系（被称为强度－时间曲线）如图 1 所示。

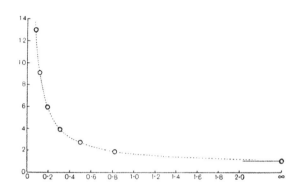

图 1. 神经兴奋的"强度－时间曲线"。横轴，持续时间 σ（0.001 秒）；纵轴，所需的最小强度。——摘自《化学波在神经中的传播》，希尔著（剑桥大学出版社，1932 年）。

就神经而言，这条曲线的形状基本上是大致相同的，也许用两个参数就可以对其进行定义，即时间尺度和电流尺度。后者的意义不是很大，因为大部分电流都会在可兴奋部件之间的基质液中发生短路，无法定义严格的比较标准。重要的是时间尺度，对于给定条件下的特定神经，时间尺度是非常恒定的，人们通常用强度是临界强度两倍的电流的最小持续时间来描述它。基思·卢卡斯把这个持续时间定义为"兴奋时间"，而拉皮克称之为"时值"，后面这个术语的引申意义指的是不超过所研究过程或者组织的时间尺度。

神经的兴奋时间取决于多个因素：(a) 神经本身的性质；一根纤维和另一根纤维之间差别很大，这同之前提到的冲动传播速率受神经纤维的影响类似；(b) 温度降低，兴奋时间增加；温度升高，兴奋时间减少；(c) 神经周围溶液的离子组成对兴奋时间影响很大，尤其是钙离子浓度，钾离子的影响程度略低于钙离子；(d) 取决于神经纤维的大小，兴奋时间随纤维直径的增大而减少。

The form of the strength–duration curve leads us to a discussion of the nature of electrical excitation. A constant current passed through an excitable tissue excites twice, once at the cathode at make, once at the anode at break. Unless the current be too strong, when secondary effects (for example, electrolysis and polarisation) occur, no propagated impulse is started off during the passage of a constant current. With an alternating current, however, of not too high a frequency the case is different. An impulse starts off from the cathode for each positive phase of the current, another from the anode for each negative phase. This is true up to, say, 300 a second for a frog's nerve, 700 a second for a mammalian nerve. When, however, we consider the case of a much higher frequency, say, of 100,000 to 1,000,000 per second, a different result is found. Such currents produce no response even though they be so strong that considerable warming occurs. This is commonly imagined to be due to the so-called "skin effect", but it can readily be shown that with the high specific resistance of living tissues the skin effect does not come in until a frequency is reached far higher than that we are considering.

Let us suppose that excitation occurs, that an impulse is started, when the current outwards through the sheath of a nerve fibre in the region of the cathode attains more than a certain density (Fig. 2). When the outward current is large enough something is rendered unstable, and the state of instability is propagated as a wave. We suppose that the sheath of a fibre possesses the properties of a dielectric of high but not infinite specific resistance. The first effect of a difference of potential applied along the nerve is to charge the capacities at the anode and the cathode. The nerve, in fact, acts like a cylindrical condenser, the sheath being the dielectric between the plates.

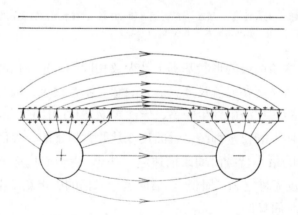

Fig. 2. Nerve fibre with two electrodes; current flow outside, inside and through sheath.

When an alternating current of high frequency is applied between electrodes resting on the nerve the effect of each cycle is to charge, alternately in opposite directions, the condensers lying near the electrodes. Unless the current be very strong, these condensers absorb it and prevent a potential difference from arising across the dielectric of sufficient intensity to drive any considerable current through the latter. With a lower frequency, however, or with a constant current, while the first effect is still to charge the condensers,

　　我们可以根据强度－时间曲线的形状来讨论电兴奋的本质。恒定电流通过可兴奋组织时会对它产生两次刺激，一次是当电路接通时在阴极发生，另一次是当电路断开时在阳极发生。除非电流太强，否则当有副反应（比如电解和极化）发生时，在恒定电流通过期间不会引发冲动的传播。但是如果换成频率不太高的交流电，情况就不同了。当电流处于正相位时会有一个冲动从阴极开始传播，而当电流处于负相位时会有一个冲动从阳极开始传播。对青蛙神经而言，频率要达到 300 次每秒才能出现这种情况；在哺乳动物的神经中，频率要达到 700 次每秒。然而，当我们考虑更高的频率时，比如 100,000~1,000,000 次每秒，就会发现不同的结果。尽管电流已经强到足以造成可观的加热效应，但不会产生任何响应。通常认为这一现象源于所谓的"趋肤效应"，但因为我们知道生物组织有很高的电阻，所以趋肤效应是不可能出现的，除非频率远远高于我们的预期。

　　我们可以假设当阴极区中通过神经髓鞘向外传播的电流超过一定量时，冲动会产生，兴奋也会出现（图 2）。当外向电流大到足以出现某种不稳定时，这种不稳定状态就会以波的形式传播开来。我们假设神经髓鞘具有很高的介电性，但电阻率没有达到无穷大。将电位差加到神经上后出现的第一个效应就是对阳极和阴极处的电容进行充电。实际上，神经就像一个圆柱形的电容器，髓鞘就是两个极板之间的电介质。

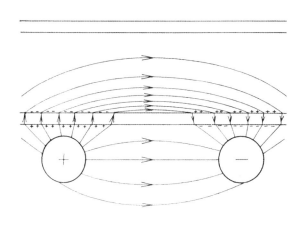

图 2. 有两个电极的神经纤维；电流从外部、内部和神经髓鞘中流过

　　当高频交流电加到神经上的两个电极之间时，其效果就是每一个周期都会交替地从相反的方向给电极附近的电容器充电。在电流不太强的情况下，电容器会吸收电流并阻止在足够强的电介质中产生电位差和可观的电流。但在频率较低或电流恒定时，尽管第一个效应还是使电容器充电，但在充电进行时会产生跨电介质的电位差，并最终导致电流通过电介质。当电流在阴极外达到足够高的强度时，就会产生不稳

as these are charged a difference of potential arises across the dielectric by which a current is caused to pass. When this current reaches a sufficiently high density outwards at the cathode, instability is produced and excitation occurs.

With constant currents, the shorter the duration the greater the current has to be to produce an excitatory effect. With a current of great duration the capacities in the surface of the fibre have ultimately no influence. With very short durations, however, the first effect is to charge the condensers and so to reduce the E.M.F.'s available across the dielectric. If then the applied current is cut off before the condensers are charged, the potential difference across it will be less than that which would ultimately be attained, and for excitation to occur the applied E.M.F. must be greater.

From this model an equation for the strength–duration curve can be deduced as follows:

$$C = \frac{R}{1 - e^{-t \big/ \frac{Fr(r_0 + r_i)}{2r + r_0 + r_i}}}$$

Here C is the current, t its duration, R a constant, F the capacity and r the resistance per sq. cm. of the sheath in the neighbourhood of the electrodes, r_i the resistance of the inside, r_0 that of the fluids on the outside, of the nerve between the electrodes.

For the case of electrodes very far apart where $(r_0 + r_i)$ is large compared with r, this equation approximates to

$$C = \frac{R}{1 - e^{-t/Fr}}$$

The important term in deciding the form of the relation is the product Fr, which is obtained by multiplying together the capacity and the resistance per square centimetre of the surface of the fibre. The relation experimentally observed in the strength-duration curve (Fig. 1) is fitted with sufficient accuracy by this equation.

If the excitation time be determined by the product of the capacity and the resistance per square centimetre of nerve fibre surface, then we should seek to explain the differences between different fibres, or between the same fibre under different conditions, by changes in the product Fr. The thinner the sheath the greater will F be, and the greater the excitation time: the lower the resistance, the shorter will be the excitation time. A rise of temperature presumably alters the excitation time by diminishing r. The absence of calcium causes a large increase in the excitation time, possibly through an increase in the resistance of the surface of the nerve. Such a change of resistance might be brought about by an alteration in the state of the emulsion of oil in water, or water in oil, of which it is possible that the sheath of the nerve is composed.

There are complications, however, in this story. It seems certain that the potassium ion

432

定性并出现兴奋。

就恒定电流而言，持续的时间越短，恒定电流就要越大才能产生兴奋效果。长时间的电流最终不能对神经纤维表面的电容产生影响。但在持续时间很短时，电容器会首先充电，从而降低了跨介电质的有效电动势。如果在电容充分充电之前将加在上面的电流切断，其产生的电位差就会小于最终能达到的电位差，此时需要使用更大的电动势才能产生兴奋。

根据这个模型，我们可以推导出如下的强度–时间曲线公式：

$$C = \frac{R}{1 - e^{-t\big/\frac{Fr(r_0+r_i)}{2r+r_0+r_i}}}$$

其中：C 是电流，t 是持续时间，R 是常数，F 是电容，r 是电极附近每平方厘米鞘的电阻，r_i 是两电极间神经的内部电阻，r_0 是两电极间神经外部液体的电阻。

如果两个电极相距很远，(r_0+r_i) 远大于 r，则上述公式可以近似表示为如下形式：

$$C = \frac{R}{1 - e^{-t/Fr}}$$

乘积 Fr 对于决定这个关系式的形式起着重要的作用，它是电容与在每平方厘米纤维表面积中的电阻的乘积。由实验得到的强度–时间曲线（图1）中的关系与这个公式的拟合结果非常一致。

如果兴奋时间由电容与在每平方厘米纤维表面积中的电阻的乘积决定，那么我们就应该试着用乘积 Fr 的变化去解释不同纤维之间的差别或者同一条纤维在不同情况下的差别。髓鞘越薄 F 越大，兴奋时间也就越长；电阻越低，兴奋时间越短。人们推测升温过程可以通过减少 r 来改变兴奋时间。缺钙会明显延长兴奋时间，这可能与增加了神经表面的电阻有关。这种电阻的变化也许是由于改变了油在水中或水在油中的乳化状态造成的，因为神经髓鞘有可能是由乳液组成的。

不过情况还很复杂。钾离子对神经纤维内外的电位差和电波传播时的动作电流

has some specific function in determining the potential difference which exists between the inside and outside of a fibre and the action current by which the wave is propagated. Possibly potassium has some specific solubility in the lipoidal substances of the nerve sheath, some specific power of penetrating which other ions have not. The current outward at the cathode by which we suppose excitation to occur is probably carried by potassium ions.

This may explain a phenomenon on which I have not yet touched, that of the gradual adaptation to a slowly increasing current. It has long been known that a current which would normally be strong enough to produce excitation may not do so if its full value be reached not suddenly but slowly. If potassium be the only means of carrying current through the sheath, its continued transfer outwards at the cathode would have the effect of depleting the inside and raising its concentration on the outside, so that a back E.M.F. would be generated (determined by $\frac{RT}{F} \log \frac{C_2}{C_1}$): this might effectively prevent the further transfer of current at a time when the externally applied E.M.F. at last reached the value at which, if suddenly applied, it would excite.

Similarly, we may explain the stimulus occurring at the anode at break of a long-continued constant current. During the prolonged passage of the current the potassium ions on the outside at the anode have been depleted by carriage through the sheath until either so few are available, or the back E.M.F. is so great, that no current can run. Breaking the circuit of the applied E.M.F., the constraint at the anode is released, the unusually high ratio (potassium inside): (potassium outside) immediately tends to right itself by the back transfer of potassium ions outwards through the membrane. This constitutes a current similar to that which occurs normally at excitation at the cathode; consequently, when the back rush of potassium ions is rapid enough, excitation occurs and an impulse starts off.

The normal function of motor nerves is to transmit impulses to the muscles, and for many years physiologists have discussed how the impulse in the nerve gets across to, and produces its effect in, the muscle fibre. Motor end-plates have been described by histologists, but their functions, and even their existence, are doubtful. It has been supposed that the electric change in the nerve which is an accompaniment of the impulse, starts the process of excitation in the muscle fibres, just as in the laboratory an impulse is started by an electric shock. This idea has led to the view that the muscle fibre and its motor nerve are normally "isochronous", that is to say, have the same "excitation time"; a muscle was supposed to respond to the impulse in a nerve if the latter was in tune with it, but not otherwise.

Some years before the War, Keith Lucas showed that in muscle there are two different excitable substances which he supposed were the muscle fibres themselves and the nerve twigs running to them. Lucas's experimental demonstration was denied by Lapicque who, employing different electrodes, found that the excitation time of both tissues was the same. Lucas's observations, however, have been reinstated by Rushton, and it seems

434

起着特殊的作用，这一点似乎是肯定的。也许钾离子在神经髓鞘的脂质成分中具有特殊的溶解能力以及其他离子所不具备的穿透能力。由阴极向外流动的电流很可能是由钾离子传输的，我们认为兴奋就是由这个电流引起的。

上述假设可以解释一个我尚未提及的现象，即对缓慢增加的电流的适应性现象。很久以前人们就知道：通常情况下强度足够产生兴奋的电流在缓慢而非很快达到最大值时是不能产生兴奋的。如果钾离子是使电流流向神经髓鞘的唯一方法，那么钾离子在阴极的持续外流就会耗尽鞘内的钾离子同时使外部的浓度增加，因而会产生一个反向电动势（大小由 $\frac{RT}{F} \log \frac{C_2}{C_1}$ 决定）；这个反向电动势也许会有效地阻止当从外部快速施加的电动势达到足以产生兴奋的值时进一步的电流外流。

我们也可以用类似的方法解释当持续的恒定电流中断时在阳极处产生的刺激。在电流持续通过的情况下，阳极处位于鞘外的钾离子会一直被向内输送，直到浓度远远不够或者反向电动势足够大以至于没有电流能够通过时为止。如果取消外加的电动势，阳极处的约束条件就解除了，这种不寻常的高浓度比（内部钾离子：外部钾离子）会立即引发钾离子跨膜向外输运从而使体系恢复到初始状态。这就形成了一种类似于在通常情况下处于兴奋状态时阴极处的电流；因此，当钾离子的回流速度足够快时，也会出现兴奋并开始冲动的传播。

运动神经的正常功能就是传递冲动到肌肉，多年来生理学家们一直在讨论神经中的冲动是如何通过肌肉纤维并在其中产生作用的。组织学家们提出了运动终板，但是运动终板的功能乃至它们的存在都还不能完全确定。人们假设神经中伴随着冲动的电变化引发了肌肉纤维中的兴奋过程，就像在实验室中冲动可以由电击产生一样。这个观点使我们联想到：肌肉纤维和它的运动神经在通常情况下是"同步"的，也就是说它们的"兴奋时间"相同；肌肉会对神经中与之同步的冲动作出响应，否则就不会有响应。

在第一次世界大战之前，基思·卢卡斯发现肌肉中有两种不同的可兴奋组织，他猜想这些组织可能是肌纤维本身以及延伸其中的神经分支。卢卡斯的实验结果被拉皮克否定，拉皮克在测试时使用了不同的电极，他发现两种组织的兴奋时间是相同的。但拉什顿重复出了卢卡斯的实验结果，这样拉皮克认为肌肉和神经在通常情况下同步的观点是否站得住脚就值得怀疑了。拉皮克理论的一个最精彩的应用是解释箭毒、

doubtful whether Lapicque's claim of normal isochronism between muscle and nerve can stand. The most beautiful application of Lapicque's theory, one which was perhaps just a little too convincing because of its beauty and because it appeared to explain so much, was that of the mechanism by which paralysis is caused by curare or other drugs or by such agencies as fatigue. Normally, the impulse from the nerve passes over into the muscle. A small dose, however, of curare, or the onset of fatigue, somehow breaks the connexion, and Lapicque maintained that this was due to the fact that the curare, or the other paralysing agency, had increased the excitation time of the muscle fibre until it was no longer isochronous with its motor nerve and therefore the impulse from the latter failed to affect it. The case was like that of two tuned electric circuits, sending and receiving—if the receiver were put out of tune with the transmitter, messages were not received.

Various experiments were adduced in support of this theory, and for a time it was accepted and was in danger of becoming a dogma. Unfortunately its experimental basis seems to be at fault. The experiments on which it was founded have been repeated by Rushton and their results denied. Other experiments have been made by which the "heterochronism" theory of curarisation has been made untenable. It is a strange thing in science to find a theory, so directly based upon apparent experimental facts, displaced by a direct denial of the facts: but so it seems to be.

The Propagated Disturbance

We have considered the manner in which excitation by an electric current occurs. With every adjustment made to get the most efficient stimulus, the energy in it is still very large when compared with that set free by the nerve itself as an impulse runs along it. Electrical stimulation is very wasteful compared with the natural stimulation from point to point by which an impulse is propagated. This is not difficult to understand. The chief part of the energy of an artificial stimulus is wasted in the fluid between the electrodes outside the active region of the nerve fibres. Only that fraction of the current which, according to our hypothesis, crosses the cathode region of the surface, is effective as a stimulus. In natural stimulation, that is from point to point in the propagation of the wave, there are no electrodes and there can be no short-circuiting in the different fluids: the stimulating current, therefore (if propagation be by means of the current), is far more efficiently used.

There is a tendency to assume, as I have assumed here, that propagation of the impulse from point to point occurs through the agency of the action current which can be detected at an active point. There is no doubt that the action current has the time relations of an efficient stimulus and, properly applied, it should have the magnitude requisite for excitation. We have no picture of the manner in which the excitatory disturbance is propagated, except that which supposes that the action current at any given point "stimulates" a neighbouring point, where in its turn a further action current is produced, which again stimulates a neighbouring point, and so on. The fact that the velocity of propagation runs parallel with the speed of development of the action current at a given point, and also inversely with the time scale of the process of excitation, strongly suggests that these three factors are linked together in some relation of cause and effect.

436

其他药物或者诸如疲劳等作用所导致的肌肉麻痹的产生机制，由于拉皮克的理论本身很完美，也似乎解释了很多现象，以至于这个理论有点太深入人心了。在通常情况下，冲动会从神经传到肌肉。但小剂量的箭毒或疲劳的出现也许会在某种程度上阻断它们的联系，拉皮克认为是箭毒或者其他可以导致麻痹的因素延长了肌肉纤维的兴奋时间，使之不再与它的运动神经同步，因此来自后者的冲动就不能在肌肉纤维中生效了。这就像两个调谐的电路，一个发送信号一个接收信号——如果接收方的频调与发送方不一致，就收不到信号了。

很多实验结果都证明了拉皮克的理论，在一段时间内它曾经被人们接受并差点成为公认的真理。不幸的是该理论的实验基础是错误的。拉什顿重复了证实该理论的实验，但得到的结果却是负面的。其他一些实验也说明了箭毒中毒的"异时性"理论是站不住脚的。这样一个直接建立在明确实验基础上的理论却因与事实完全不符而被取代，这在科学上是一桩怪事；但事实似乎就是如此。

传播出去的扰动

我们已经讨论了电刺激产生兴奋的方式。尽管为了达到最有效的刺激，人们对各个方面都进行了调整，但所需的能量仍比神经自身在冲动传播时释放的能量大很多。就传播冲动而言，电刺激的耗费要远远大于点对点的自然刺激。这一点并不难理解。人工刺激的大部分能量都浪费在了神经纤维活性区域之外两电极之间的液体中。根据我们的假设，只有跨过表面阴极区的那部分电流才会形成有效的刺激。自然刺激在以波的形式传播时是点对点的，没有电极，也不会在不同液体中出现短路，因此刺激电流（如果传播是通过电流进行的）的利用效率是非常高的。

正如我所设想的，人们倾向于假设冲动的点对点传播是在动作电流的作用下进行的，而这个动作电流可以在活动位点被检测到。毫无疑问，动作电流和有效刺激之间具有时间上的相关性，并且如果能够适当施加有效刺激，它可以达到产生兴奋所需的强度。我们不知道兴奋性扰动的传播方式，但可以假设任意给定点上的动作电流"刺激"了临近的点，随后产生了新的动作电流，后者又刺激了临近的点，如此反复进行。在任一给定位置，扰动的传播速度和该处动作电流产生的速度成正比，与兴奋过程的时间尺度成反比，这充分说明在这三个因素之间存在着某种因果关系。

It must not be imagined that this self-perpetuating electro-chemical wave is analogous to those waves in physics in which no new energy is required from point to point for the transfer of the wave. In sound, or in light, energy is forced into the medium at the source but no further energy is required for the propagation. Unquestionably in nerve, as the initial heat shows, energy is liberated at each point as the impulse passes by, and, moreover, in the next thirty or forty minutes several times as much energy is set free in restoring completely the *status quo*. With this qualification, however, we can think of the propagated disturbance as some type of self-transmitting electro-chemical wave.

The problem, therefore, of its nature ultimately resolves itself into two: one is that of the change which is produced at the cathode when a current of sufficient intensity causes conditions to become unstable and some mechanism to be fired off; the other is of the physico-chemical basis of the action current itself. If we could understand these two effects we could make a clearer picture of how the impulse is propagated. It seems likely that as the result of "excitation" an unstable state is reached, in which the potential difference normally held at the surface of the fibre is for a moment released. It discharges until some change sets in by which the instability is reversed and the initial condition realised once more. Looked at in this way, the action current is nothing more than a momentary discharge of the resting potential, which is normally to be seen between an injured and an uninjured point of a nerve. Our problem, therefore, comes to this, what is the nature of the instability which is produced by a sufficient outward current through the nerve surface, and how is this instability rapidly reversed and the nerve surface restored to its normal state?

The chemical reactions occurring in muscle are largely those of recovery. Lactic acid formation is involved in the restoration of creatine phosphoric acid which breaks down in activity; oxidation and the combustion of food-stuffs are involved in the restoration of the lactic acid. In nerve we know that in any case nine-tenths of the energy liberated is involved in recovery. It is not going very much further to suppose that the remaining tenth is involved in the immediate recovery process by which the instability produced by stimulation is reversed. I should picture the primary effect as a physico-chemical one transmitting itself along the surface. The surface is rendered somehow unstable by the passage of a current outwards across it, and the instability is propagated by means of the current it releases. The return of the surface to its normal state is the result of some chemical reaction involving the liberation of free energy. Without this return no further impulse could be propagated.

Mitogenetic Radiation in Nerve

The facts I have described so far are reasonably certain, though their explanation is not. I wish now shortly to refer to some others, of great importance if they are confirmed, but of which the evidence is as yet not quite convincing. During the last few years a number of papers have appeared from Moscow and Leningrad on the subject of so-called "mitogenetic radiation". The name implies that the radiation in question is able to cause mitosis in cells, and the approved method of detecting and measuring it is to determine the increase

438

千万不能认为这种持续不断的电化学波类似于物理学中的波，在物理学中，波的点对点传播不需要另外增加能量。虽然声波或者光波的产生需要能量，但在传播过程中并不需要额外补充能量。毫无疑问，在神经中，正如初始热所显示的，在冲动传过的每一点上都有能量释放出来，而且在之后的 30 或者 40 分钟内，数倍于此的能量将在体系复原时被释放。而按照这种分析方法，我们可以把传播的扰动看作是一种自动传播的电化学波。

因此，关于神经冲动本质的问题最终归结为了两个问题：其一是当电流强度大到足以引起状态不稳定时在阴极处发生的变化以及它的恢复机制；其二是动作电流本身的物理化学基本原理。如果我们能够弄明白这两个机制，我们就能更加清晰地认识到冲动是如何传播的。也许"兴奋"的结果就是达到一种不稳定态，这时通常存在于纤维表面的电位差会突然释放出来。放电过程会一直持续，直到这种不稳定态得到逆转并再次恢复到初始状态。从这个角度上看，动作电流不过是静息电位的暂时性放电而已，静息电位通常位于神经的损伤处和未损伤处之间。因此我们的问题就是：由足量的经过神经表面的外向电流导致的不稳定性的本质到底是什么，以及这种不稳定性是如何快速逆转并使神经表面恢复到正常状态的？

肌肉中发生的大部分化学反应都是可逆的。活动过程中分解的磷酸肌酸在重新生成时会导致乳酸的产生；而在乳酸的复原过程中会出现氧化和养分的燃烧。我们知道，在神经释放的总能量中，有 9/10 是在复原过程中释放的。由此很容易联想到剩下的 1/10 是由刺激产生的不稳定态在逆转后的快速复原过程中被消耗掉的。我认为初始作用应该来自沿神经表面传播的物理化学波。因为有向外流动的电流通过，神经表面变得有些不稳定，而这种不稳定会通过神经释放的电流传播开来。一些释放自由能的化学反应将使神经表面恢复到正常状态。在这种恢复完成之前，冲动无法继续传播。

神经中促进有丝分裂的辐射

至此为止我描述的事实都是确定的，尽管它们的解释并不确定。现在我想简单地讨论一下另外的几个问题，如果它们能够被证实，其意义是非常重大的，但目前证实这些问题的证据还不是很有说服力。近几年，在许多来自莫斯科和列宁格勒的文章中都提到了所谓的"促进有丝分裂的辐射"。这个名字意味着该辐射可以引起细胞的有丝分裂，而检测和测量它的有效方法是测定放置在辐射下的悬液中酵母细

in the number of yeast cells in a suspension subjected to the radiation. Living organisms themselves are said to give out this radiation, particularly when active, and the analysis of the radiation is held to indicate the type of chemical reaction involved in the activity.

The yeast cells prepared in a special manner are held in a suspension which is placed in two tubes, an experimental and a control. The experimental tube is exposed through its open end to the radiation in question, the control is kept without radiation. At the end of the exposure, samples are taken and incubated, and after three or four hours the cells are killed and counted; the excess of cells in the experimental suspension is expressed as a percentage of the control.

The radiation stated to be given out by living cells is in the ultra-violet region, chiefly between 1,900 A. and 2,500 A. Its amount is so relatively large that it can be split up by a quartz spectrograph into bands 10 A. wide and each band examined separately for its effect in producing the division of yeast cells.

This is not the occasion to deal with the general question of mitogenetic radiation, but a few months ago a series of papers appeared from the laboratory in Leningrad in which various results obtained from nerve are discussed. If it be true that excited nerve gives out a characteristic radiation which can be used to identify the chemical reactions involved in its activity, then indeed a new day has dawned in the very difficult problem of the physical nature of nerve activity.

In a figure in a recent paper by Kalendaroff (*Pflügers Arch.*, 231), successive spectra, analysed by a quartz spectrograph and the yeast cell indicator, refer to (1) a resting nerve, (2) ground up nerve, (3) mechanical stimulus, radiation from the point of stimulation, (4) electrical stimulus, between the electrodes, (5) injury, radiation 20 mm. from the place of injury, (6) electrical stimulus, radiation 20 mm. away, (7) mechanical stimulus, 20 mm. away. In the lower half of the figure there are spectra for (1) oxidation of pyrogallol in air, (2) glycolysis, (3) action of phosphatase, (4) splitting of creatine-phosphoric acid, (5) splitting off of ammonia from protein.

When we remember that maximal continuous stimulation of nerve does not double its resting metabolism, the variety and strength of the radiation emitted from active nerve, under the comparatively mild stimuli administered to it, are rather astonishing. A vague suspicion that the results are almost too good to be true is a little increased by a subsequent paper by Schamarina which shows an evident misunderstanding of the nature of nerve activity. It is known, and it is an obvious consequence of the existence of a refractory period following the passage of an impulse, that when two nerve impulses start at opposite ends of a nerve and meet in the middle they are unable to pass one another and both are wiped out. When a single impulse traverses the nerve the whole of the nerve goes through a phase of activity. When two impulses start at opposite ends of the nerve they meet in the middle and stop, but again the whole of the nerve has gone through the active phase.

胞数目的增加。据说活的生物体本身可以放出这种辐射，尤其是在它们活动的时候，而对辐射的分析则能够显示出该活动包含的化学反应类型。

将用特殊方法制备的酵母细胞置于悬液中，它们被分别装在两个试管内，一个是实验管，另一个是对照管。实验管的开口端暴露于辐射中，对照管则不受辐射。辐射结束后取出样品并进行孵育，3~4 个小时后杀死细胞并计数。实验悬液中增加的细胞数以相对于对照样品中细胞数的百分比来表示。

上述由活细胞发出的辐射位于紫外区，主要在 1,900 Å 到 2,500 Å 之间。该辐射强度很大，可以被石英摄谱仪分裂成一系列 10 Å 宽的谱峰，人们可以分别检测每一个谱峰对酵母细胞分裂的影响。

现在不是解决与促进有丝分裂的辐射的相关问题的时候，不过几个月前，来自列宁格勒实验室的一系列论文都讨论了从与神经相关的实验中得到的不同结果。如果兴奋的神经能够放出特征辐射，这种辐射可被用于鉴定在兴奋过程中发生的化学反应，那么人们就可以在新的起点上研究关于神经活性的物理本质这一非常困难的问题了。

在卡伦达洛夫最近的一篇论文（《欧洲生理学杂志》，第 231 卷）中有一幅图，图中有若干张用石英摄谱仪分析得到、以酵母细胞为指示剂的连续光谱，其中（1）静息状态的神经，（2）破碎的神经，（3）机械刺激状态下，刺激点发出的辐射，（4）电刺激状态下，电极间的辐射，（5）损伤状态下，距损伤点 20 mm 处的辐射，（6）电刺激状态下，距刺激点 20 mm 处的辐射，（7）机械刺激状态下，距刺激点 20 mm 处的辐射。在图的下半部分是下列反应的图谱：（1）焦酚在空气中氧化，（2）糖酵解作用，（3）磷酸酶的作用，（4）磷酸肌酸分解，（5）从蛋白中放出氨的反应。

我们知道即使是神经中最大的持续刺激也不能使它的静息代谢加倍，那么在比较温和的刺激下，从激活的神经中发出的辐射种类如此之多，强度如此之高是非常令人惊讶的。有人怀疑这些结果过于完美以至于很难让人相信是真的，沙玛丽娜后来发表的文章加深了人们对这些结果的怀疑，她所论述的内容其实是对神经活性本质的误解。我们知道：如果两个神经冲动分别从一条神经的两端开始相向传播并在中间汇合，它们将不能穿过对方而导致两者都消失掉，这显然是因为在冲动传过后存在一个不应期。当单个冲动传过神经时，整条神经都会经历一个兴奋期。当两个冲动从一条神经的两端开始相向传播并在中间汇合而终止时，整条神经同样也会经历一个兴奋期。如果发出的辐射是神经兴奋的结果，那么在上述两种情况下发出的

If radiation is given out as the result of nerve activity, its emission should occur equally in the two cases. Schamarina, however, expecting that because the two impulses destroy one another, therefore there should be no radiation from the point where they meet, has described experiments in support of the expectation. If the results are true, we need a new picture of the propagated disturbance in nerve.

The suspicion is strengthened by a further paper by Brainess describing the use of the same technique for the study of human fatigue. At the beginning of this remarkable paper it is stated that modern methods for investigating the phenomena of fatigue in man leave much to be desired, and that the author therefore took up the method of mitogenetic radiation in order to find a new and more accurate means of describing the state of fatigue in a factory worker. One hundred girls working in an electrical factory were examined, samples of their blood being taken at eight in the morning, at three in the afternoon and at five in the afternoon. The blood was dried on filter paper, then dissolved in distilled water and finally allowed to give out its radiation, which was measured as usual by the yeast cells. At eight in the morning the mean value of the radiation coming from their blood was 28, as measured by the percentage increase, over the control, of the number of cells in the yeast suspension. After seven hours work the girls were apparently completely exhausted, for their blood gave out no radiation at all, except in a few isolated cases. After two hours rest the radiation had risen to 28 again.

Of the social and industrial importance of these results, supposing them to be true, I need not speak, though I wonder if there are many British factory operatives who would be found completely exhausted, even of "radiation", after seven hours work. An even stranger result follows. Not only did haemolysed blood emit radiation which was abolished by seven hours work in a factory, but also the cornea and the conjunctiva did the same. The girls apparently had only to look at the yeast cells to set them dividing! At eight in the morning the radiation from the girls' eyes had a mean value of 24, after seven hours work it had a mean value of 4, after two hours recovery a mean value of 20. Finally—and most unromantically—it is stated that the spectrum analysis of the radiation given out by the girls' eyes showed that its only important component is that due to glycolysis!

It is not easy, in the case of the paper describing the results of the two nerve impulses meeting one another, to avoid the feeling that the expectation of a certain result has had something to do with its appearance, and it is difficult not to draw the same conclusion from the paper describing the new test for fatigue. The claims made are clearly most important if they can be verified, and one hopes that verification may soon be at hand. The difficulty in understanding nerve is largely that the changes in it are too small for ordinary chemical methods to detect. If the new methods elaborated by our Russian colleagues can throw real light on the subject, then we shall be deeply indeed in their debt. At present, however, one cannot stifle suspicion that the phenomena described may have more to do with the enthusiasm of those who describe them than with the physical nature of the nerve itself.

(131, 501-508; 1933)

辐射应该是相等的。但沙玛丽娜认为由于这两个冲动都破坏了对方，所以在它们汇合的地方应该没有辐射放出，她在文中描述了证实这一假设的实验。如果这个结论是正确的，那么我们就需要用新的方法来解释在神经中传播的扰动了。

布瑞内斯的一篇论文也加深了这个疑点，他把同样的方法应用于研究人类的疲劳。他在这篇著名文章的开头指出：目前用于研究人类疲劳现象的方法还有许多需要改进的地方，作者为了找到一种能够更加准确地描述工厂工人疲劳状态的新方法而采用了促进有丝分裂的辐射的方法。他对在电子工厂工作的 100 名女孩进行了测试，分别在早上 8 点、下午 3 点和下午 5 点采集她们的血液。血液在滤纸上吸干，然后溶解在蒸馏水中，使之能够放出辐射，并按常规用酵母细胞测定辐射强度。早上 8 点她们血液中辐射的平均强度值是 28，平均强度值用实验样本悬液中酵母细胞的数量相比于对照样本中增加的百分比表示。在工作了 7 个小时以后，女孩们显然已经疲惫不堪了，因为除了个别的几个人例外以外，大部分人的血液都不再发出任何辐射了。经过 2 个小时的休息后，辐射强度再次增加到 28。

尽管我怀疑是否有很多英国工厂的工人在工作了 7 个小时之后都会筋疲力尽，即使是仅仅表现在"辐射"上，但如果上述结果是正确的，那么这些结果对社会和工业的重要性我就不必赘述了。之后一个更加奇怪的结果出现了：不仅溶血的血液能够发出在工厂工作 7 小时后可能会消失的辐射，角膜和结膜也会产生同样的效果。女孩们只需看着这些酵母就可以使它们发生分裂！早上 8 点，从女孩们眼睛中发出的辐射平均值为 24，工作了 7 小时后平均值变成 4，休息 2 小时后平均值又达到 20。最后，也是最实际的结果是：对女孩眼睛中发出的辐射进行光谱分析后发现其唯一重要的成分来自糖酵解！

人们很自然地希望能从描述两个神经冲动相遇的文章中发现某种结论以解释一些现象，对于这篇描述疲劳检测新方法的文章也是一样。显然，如果前面提出的观点被证明是正确的，它们将变得非常重要，人们希望相关的验证工作能马上开始进行。在认识神经的过程中遇到的主要困难是其中的变化过于微小，用普通化学方法难以检测。如果俄罗斯同行们提出的新方法确实能够为这一领域的研究带来一线曙光，那我们真不知道要如何感谢他们才好。但是目前我们不能不怀疑他们所描述的现象在很大程度上是出于实验者的热情，而不是与神经本身的物理本质相关。

（毛晨晖 翻译；刘力 审稿）

Chemical Test for Vitamin C, and the Reducing Substances Present in Tumour and Other Tissues

L. J. Harris

Editor's Note

In the 1930s, chemists and physiologists were able to define and characterise chemicals known as vitamins, which are necessary for human well-being but which are not synthesised in the human body. Early in the decade, very little was known about these compounds, as papers from the time amply illustrate. This paper suggests how vitamin C can be recognised in foodstuffs by the change in the colour of an oxidation-reduction indicator.

IN previous communications a method has been described for estimating the hexuronic (ascorbic) acid content of foodstuffs, based on titration in acid solution with the oxidation-reduction indicator 2-6-dichlorophenolindophenol after preliminary extraction with trichloracetic acid[1,2,3]. Judging from the fact that this method when applied to some forty common sources—mostly fruit and vegetable materials—enabled the "minimal antiscorbutic doses" to be calculated to give results in excellent agreement with the values determined directly by biological tests, it is evident that the method has a considerable range of specificity. A number of necessary conditions and provisions were set out, which unfortunately there seems to have been some tendency to overlook, and it would appear advisable therefore to direct attention to certain considerations which must be borne in mind if the possibility of misleading conclusions is to be avoided.

As we have already pointed out[3], the reagent does not possess an *absolute* degree of specificity. Notably, free cystein (which may be present in stale or autolysed materials) was found to reduce it as readily as did the vitamin itself: this could easily be allowed for by a separate determination for cystein by the Sullivan method. Adrenalin also reduced the indicator, but much less intensely, so that in practice no ordinary natural source contains sufficient to interfere seriously. Products obtained by heating solutions of certain sugars, especially in alkaline media, tended to reduce the indicator; and we find that a number of proprietary baby foods and similar preparations give suspiciously high readings. Mr. A. L. Bacharach, of the Glaxo Research Laboratory, has titrated a series of specimens of malt-extracts by our method and found some of them to reduce the indicator strongly[4]. Among other materials of vegetable origin we found that the following also react appreciably with the indicator: yeast[3]; whole oats[3]; incubated pea mush[7]. Since these materials have not hitherto been regarded as sources of vitamin C, it would appear advisable to suspend judgment as to the precise nature of the reducing substance in such

444

肿瘤以及其他组织中所含的维生素C等还原性物质的化学检测

莱斯利·哈里斯

编者按

在 20 世纪 30 年代，化学家和生理学家已经能够定义和描述维生素这种化学物质了，维生素是人体维持健康所必需的物质，但人体自身不能合成。在那个年代，极少有文章会详细描述这类物质。而这篇文章就是其中之一，本文讲述了如何通过氧化还原指示剂的颜色变化来鉴别食物中是否存在维生素 C。

在以前的文章中，我们介绍了一种估算食物中己糖醛酸（抗坏血酸）含量的方法，即经三氯乙酸 [1, 2, 3] 初步萃取后用氧化还原指示剂 2,6– 二氯酚靛酚在酸溶液中进行滴定的方法。当这种方法被应用到约 40 种常见的含抗坏血酸的物质——主要是水果和蔬菜中时，计算出的"抗坏血病的最小剂量"与直接利用生物学实验得到的值十分吻合，很显然，该方法的特异性仍具有一定的局限性。而且，需要具备许多必要的前提条件。不幸的是，这些前提条件很容易被忽略。因此，我们应该注意一些需要考虑的事项，以免得出误导性的结论。

正如我们所指出的那样 [3]，这种指示剂并没有**绝对**的特异性。值得注意的是，自由的半胱氨酸（可能存在于不新鲜或自溶的物质中）可以像维生素一样轻易地还原指示剂，但用沙利文法可以很容易地单独测定半胱氨酸。肾上腺素也可以还原这种指示剂，但作用远没有那么强烈，实际上自然界中天然存在的物质，其还原性还不足以强烈干扰指示剂的作用。如果加热某种糖溶液，尤其是在碱性条件下加热得到的产物能还原指示剂。此外，我们还发现许多婴儿专用食品以及类似的制剂都给出了可疑的高还原性数值。葛兰素研究实验室的巴卡拉克先生用我们的方法测定了一系列麦芽提取物样品，发现其中一些可以很强烈地还原指示剂 [4]。在其他植物来源制品中我们发现以下几种物质也可以与指示剂发生明显的反应：酵母粉 [3]、全燕麦 [3]、孵育豌豆粥 [7]。由于至今还没有人认为这些物质含有维生素 C，所以我们最好等到目前进行的生物学试验有了结果，再来判断这些特殊条件下的还原性物质的确切本质。

445

special cases until the biological tests, now in progress, are concluded.

Turning to the animal kingdom, it might have been anticipated that the specificity of the test would be less certain. Nevertheless we found that the suprarenal gland (not hitherto recognised as an antiscorbutic) was very potent, the biological activity agreeing with the value determined chemically; and the same is true, approximately at least, for liver. A systematic survey of various animal tissues, initiated in this laboratory by Messrs. Birch and Dann[6], showed that many of them gave very substantial titres, often accounting for a large fraction of the total iodine-reducing value, hitherto held to be a measure solely of the glutathione content. In the case of one of these materials, the aqueous humour of the eye, the very surprising indication of the presence of large amounts of vitamin C has already been confirmed biologically[7].

Another material giving a high iodine value and of very obvious interest in this connexion is tumour tissue. Dr. E. Boyland of the Cancer Hospital Research Laboratory approached us for details of our method to apply to tumours. We are indebted to him for permission to refer here to his results, which show that tumour tissues of various kinds likewise reduce the indicator[8]. Our own independent observations confirmed this finding, although our experiences were limited only to the Jensen rat sarcoma. This we find to give a very constant titre, equivalent in terms of hexuronic acid to 0.4 mgm. per gm. of wet tissue. Biological tests have so far given somewhat inconclusive results as to whether the titre is due wholly to vitamin C. The freshly excised sarcoma (rendered available by the collaboration of Mrs. B. Holmes) was fed to a series of five guinea pigs in curative tests at the level of 3.5 gm. per day. If the indophenol titre were due *entirely* to vitamin C, 2.5 gm. per day would suffice as the minimal dose. However, the experimental animals receiving 3.5 gm. lost weight as rapidly and survived no longer than the negative controls, although at death the degree of scurvy appeared less severe. In such tests a complicating factor due to the possible toxic effect of relatively large amounts of animal tissue fed to a herbivorous species like the guinea pig has always to be borne in mind. Further assays by several alternative methods are in progress. In any case the presence in the tumour tissue of such high concentrations of an intensely reducing substance, hitherto unrecognised, seems of special significance, bearing in mind the distinctive character of the cell respiration of tumours. Furthermore, observations in another connexion with Dr. E. W. Fish seem to indicate that vitamin C is needed primarily for the maintenance of certain actively functioning cells, so that its apparent presence in tumour tissue seems additionally suggestive. It is proposed to investigate the effect of deprivation of vitamin C on tumour growth.

Returning to the question of the applicability of the chemical test, it may be concluded that, on all fours with the now well-known and extensively used antimony trichloride test for vitamin A, it furnishes a valuable if not absolutely infallible guide. Certainly for fruits and vegetables as ordinarily dealt with, the test seems to give perfectly reliable results

446

对于动物界，可以预测到上述检测方法的特异性就会更加不确定了。然而，我们发现肾上腺（至今还未被认为是一种抗坏血病的物质）有着较高的还原值，它的生物学活性与用化学方法测定的完全相符，其他组织也是一样，至少肝脏的情况大致如此。由我们实验室的伯奇先生和丹恩先生发起的一项关于各种动物组织还原性物质的系统调查[6]显示，许多动物组织都有很高滴度的还原性，并常常占全部碘还原值的大部分。而碘值法也是到目前为止测定谷胱甘肽含量的唯一方法。这些组织之一的眼球房水中含有大量的维生素C的奇怪现象，已经在生物学上得到了证实[7]。

在这项研究中，另一种具有很高还原值并具有重要意义的是肿瘤组织。肿瘤医院研究实验室的博伊兰博士在向我们咨询后，已尝试将这种方法应用到肿瘤研究上。非常感谢他允许我们在这里引用他的研究结果。其结果显示各种肿瘤组织同样可以还原指示剂[8]。尽管我们仅仅观察了詹森大鼠肉瘤，但我们的独立观察结果也可以证实上述结论。我们发现这种肿瘤有一个非常恒定的滴定度，相当于每克湿组织中有 0.4 毫克己糖醛酸。至于这一滴定度是否完全源自于维生素C，生物学试验至今未能给出明确的结论。我们把新切的肉瘤（由霍姆斯夫人协助提供）以每天 3.5 克的量喂给五只患坏血病的豚鼠以进行治疗试验。如果靛酚滴定度**完全**源自于维生素C，那么每天 2.5 克就足以作为试验抗坏血病的最小剂量。意外的是，尽管死亡时所患坏血病的程度看起来要轻一些，这些每天食用 3.5 克肉瘤的实验动物和阴性对照组一样很快地消瘦，而且并没有活的更长。在上述试验中，一定要始终牢记喂给像豚鼠这样的食草类动物相对大量的动物组织有可能会产生毒性反应这一复杂因素。因此，人们还在利用另外几种方法进行进一步的检测。无论如何，考虑到肿瘤细胞呼吸的显著特性，在肿瘤组织中存在如此高浓度的、至今尚未被探明的强还原性物质看起来具有特殊的意义。此外，菲什博士的另一项观察似乎表明，维生素C主要用于维持某些功能活跃的细胞，所以它在肿瘤组织中的存在似乎还有其他暗示。有人建议研究维生素C缺失对肿瘤生长的作用。

总之，考虑到化学试验的适用性问题，可以得出以下结论：与众所周知且广泛应用的利用三氯化锑测定维生素A的试验一样，这种方法即使不算绝对可靠，至少也有价值。当然，对于通常涉及的水果和蔬菜，这种试验似乎不需要进一步的检验

without further elaboration; when unusual types of material are under investigation, the test must be used with due understanding.

(**132**, 27-28; 1933)

Leslie J. Harris: Nutritional Laboratory, Cambridge, June 20.

References:

1. Birch, T. W., Harris, L. J., and Ray, S. N., *Nature*, **131**, 273 (Feb. 25, 1933).

2. Harris, L. J., and Ray, S. N., *Biochem*. J., **27**, 303 (1933).

3. Birch, T. W., Harris, L. J., and Ray, S. N., *Biochem. J.*, **27**, 590 (1933).

4. Bacharach, A. L., private communication.

5. Harris, L. J., and Ray, S.N., *Biochem. J.*, **26**, 2067 (1932);

6. Birch, T. W., and Dann, W. J., *Nature*, **131**, 469 (April 1, 1933).

7. Birch, T. W., and Dann, W. J., unpublished work.

8. Boyland, E., private communication; *Biochem. J.*, in the press.

就可以得到可靠性很高的结果；而当研究某些特殊物质时，这种试验只能在具备足够相关知识的前提下才可使用。

（李世媛 翻译；秦志海 审稿）

449

Mitosis and Meiosis

C. L. Huskins

Editor's Note

How chromosomes are partitioned when cells divide was still unclear in the 1930s. Here Canadian geneticist Charles Leonard Huskins from McGill University, Montreal, presents a unified theory of mitosis and meiosis (cell division that retains or halves, respectively, the number of chromosomes) that overcomes the pitfalls of its main predecessor, the so-called precocity theory of meiosis. English geneticist Cyril Dean Darlington had suggested that meiosis was precocious because the first stage (prophase) begins before chromosomes have divided into two identical parts (chromatids). Huskins neatly side-steps the problems associated with that idea by suggesting that the chromosome threads are attracted in pairs at all stages of meiosis and mitosis, but that pairs of pairs repel one another.

FROM observations made in this laboratory by S. G. Smith, E. Marie Hearne, Jane D. Spier, J. M. Armstrong, A. W. S. Hunter, and me on meiosis and both haploid and diploid mitosis in *Trillium*, *Matthiola*, a number of cereals and grasses, and in grasshoppers, it can be shown that, at all stages of mitosis and meiosis chromosome threads are attracted in pairs and that pairs of pairs are repulsed. A unified theory of chromosome behaviour thereby arises which seems adequately to explain the mechanism of both mitosis and meiosis, including the varied behaviour of univalents in the latter. Fig. 1 illustrates essential features, broken lines indicating stages during which splitting of the chromosomes can be seen to be occurring. (Throughout this note it is "effective" lateral splitting which is referred to; if Nebel's, unpublished, observations in *Tradescantia reflexa* are confirmed for other material the initiation of the split occurs one division cycle earlier.)

These observations were stimulated by Darlington's precocity theory of meiosis (which Dr. A. H. Sturtevant in his recent review in *Nature* of January 7, p. 5, states that he favours) and the new unified theory has, of course, features in common with it as well as with more generally accepted accounts. Briefly, according to our observations, chromosome behaviour in mitosis and meiosis is as follows: at the earliest prophase stage of mitosis the chromosomes are double; in meiosis they are single. The single threads pair in meiotic zygotene; during pachytene the "secondary" split occurs. Repulsion between pairs of chromatids begins with the "secondary split" but in *most* organisms the pairs are held together by changes of partner or chiasmata. So far we agree with Darlington, except that he has attributed universality to the chiasma mechanism. As I have pointed out

有丝分裂与减数分裂

查尔斯·伦纳德·赫斯金斯

编者按

20 世纪 30 年代，人们还不清楚细胞分裂时染色体是如何分开的。在这篇文章中，来自蒙特利尔麦吉尔大学的加拿大遗传学家查尔斯·伦纳德·赫斯金斯提出了一种关于有丝分裂和减数分裂（分别是指染色体数目保持不变或者减半的细胞分裂）的统一理论，这一理论克服了它的主要前任，即所谓的减数分裂早熟论的缺陷。此前，英国遗传学家西里尔·迪恩·达林顿提出，减数分裂是早熟的，因为在染色体分成完全相同的两部分（染色单体）之前，减数分裂的第一阶段（前期）就开始了。赫斯金斯提出，在减数分裂和有丝分裂的所有阶段，染色体丝都是配对相连的，不过这些配对的染色体丝再成对后，两组染色体丝配对之间则是互相排斥的，这样，赫斯金斯就巧妙地避开了与早熟理论相关的一些困难。

史密斯、玛丽·赫恩、简·施皮尔、阿姆斯特朗、亨特和我在实验室中对减数分裂以及延龄草、紫罗兰、大量谷物与草类和蝗虫中的单倍体与双倍体有丝分裂进行了研究，结果表明，在有丝分裂和减数分裂的所有阶段，染色体丝因相互吸引而配成对，但对与对之间是相互排斥的。因此就可以用一个统一的染色体行为理论来圆满解释有丝分裂和减数分裂的机制，包括在减数分裂过程中单价染色体表现出的各种行为。图 1 中的例子解释了有丝分裂和减数分裂的基本特征，间断的线说明该阶段是可以看到染色体发生分离的阶段。（图中所示的均为"有效"的横向分离；如果尼贝尔尚未发表的关于紫露草的观察结果对其他生命体也成立，那么分离过程应开始于一个分裂周期之前。）

因为我们是在达林顿的减数分裂早熟论的启发下开始这项研究的（斯特蒂文特博士在他最近发表于 1 月 7 日《自然》杂志第 5 页上的评论文章中表明了自己对该理论的支持），所以我们提出的新理论理所当然地与减数分裂早熟论以及更为广泛接受的理论有共同的特征。简言之，根据我们的观察，染色体在有丝分裂和减数分裂中的行为如下：在有丝分裂前期的最开始，染色体是双链的；而在减数分裂前期，染色体是单链的。两条单链染色体丝在减数分裂的偶线期配成对；到了减数分裂的粗线期，又会发生"第二级"分离。在"第二级"分离时，成对染色单体之间开始相互排斥，但在**大多数**生物体中，成对的染色单体通过互换链或交叉而聚集在一起。到此阶段为止，我们的观察结果与达林顿的理论是一致的，但我们不认同他认为交叉机制具有普适性的看法。我曾在另一篇文章中指出 [1]：威尔逊和施拉德斯的研究

451

elsewhere[1], the work of Wilson and the Schraders shows conclusively that it has not universal application, and almost equally certainly Darlington is wrong in applying it to *Drosophila*.

From diplotene on, our observations differ from previous accounts. As demonstrated at the Sixth International Congress of Genetics[2], a "tertiary" split develops before meiotic metaphase. (This was discovered independently and simultaneously by Nebel[3].) At a parallel stage of mitosis we find, in both root tips and the haploid pollen grains, a split occurring, as reported by Sharp[4], Hedayetuallah[5] and others in root tips. Darlington rejected Sharp's observations as "optical illusions". Some of ours being end-views of chromosomes cannot thus be disposed of. The "tertiary" split initiates a repulsion within the 4-partite daughter chromosomes, at anaphase causing their arms to separate widely and preparing them for the second meiotic division, in which we find the pre-metaphase split omitted. Omission of a split in one division is, of course, essential to reduction. It follows as a mechanical consequence of the preceding stages as here described. The splitting of univalents in both divisions, which occurs chiefly when there are many of them, is apparently due to their delaying the second division and thus permitting the ordinarily omitted premetaphase split to occur. It is inexplicable on Darlington's view that splitting occurs during the resting stage.

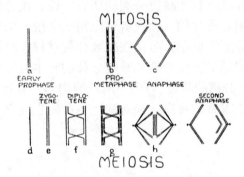

Fig. 1. Diagram of mitosis and meiosis. A univalent is included in h and i.

It is from this view that practically all the contradictions in Darlington's precocity theory arise. From it he is forced to assume that though at prophase single threads attract one another, they repulse one another at anaphase. To get around this duality of principle, he suggests[6] (p. 48) "It may be that the... spindle attachment... has not the property which the chromatids have, of associating in pairs". Again (p. 300) he assumes that though the chromosomes divide during the resting stage, their attachments divide at metaphase. Both these assumptions are theoretically inadequate and, according to us, observationally invalid. Finally, though the object of his precocity theory is to homologise mitosis and

已经证明交叉机制并不具有普适性，并同样明确地指出达林顿将交叉机制应用于果蝇是错误的。

从双线期开始，我们的观察结果就与达林顿的理论有了偏差。正如我在第六届国际遗传学大会上声明的[2]，"第三级"分离在减数分裂进入中期以前就已经形成了。（内贝尔在同一时间也独立地发现了这个现象[3]。）我们注意到，在有丝分裂中的对应阶段，根尖和单倍体花粉粒都发生了染色体的分离，这与夏普[4]和海德亚图拉[5]等人所报道的在根尖中发生的分离一样。达林顿不承认夏普的观察结果，认为它们是"视觉上的错觉"。但不能因此而否认我们观察到的一些染色体端视图。"第三级"分离开始于四分体中子染色体之间的相互排斥，在分裂后期排斥会引起染色体臂的大幅度分离，并为减数第二次分裂做准备，我们发现在减数第二次分裂时没有出现前中期的分离过程。当然，在一次减数分裂中省去一次分离过程是产生减数所必需的。正如本文所描述的，分离的缺省是前一步分裂过程自然发展的结果。在两次分裂中单价染色体的分离都主要发生在存在大量单价染色体的时候，这种分离的产生显然是由于减数第二次分裂的延迟，因而发生了在通常情况下不会出现的前中期分离。达林顿认为分离发生在分裂间期的观点是解释不通的。

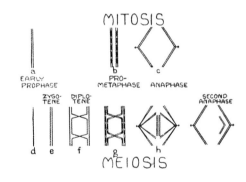

图 1. 有丝分裂和减数分裂示意图。在 h 和 i 中含有一个单价染色体。

这个观点可以让达林顿早熟论中几乎所有的矛盾都暴露出来。他不得不因此而假设尽管在分裂前期单链染色体丝之间是相互吸引的，但在后期它们又会相互排斥。为了避免在理论中出现二重性，达林顿指出[6]（第 48 页），"也许是因为……纺锤体着生区……不具备染色单体所具有的可以联接成对的属性。"他还假设（第 300 页）虽然染色体会在分裂间期分离，但它们的纺锤体着生区却在分裂中期才分开。上述两点假定都缺乏理论上的依据，也与我们的观察结果不符。最后的结论是，尽管达林顿提出早熟论的目的在于统一有丝分裂和减数分裂的机制，但他的基本原理在最

meiosis, Darlington is left at the end (p. 305) with a basic duality of principle: "Perhaps all nuclear division, *apart from the first division of meiosis* [my italics], is determined by division of the chromosomes."

The observations here summarised show this duality to be non-existent; the principle of attraction between single threads and repulsion between pairs of pairs, postulated by Darlington for the prophase only and denied by him at anaphase, really applies at all stages of mitosis and meiosis.

(**132**, 62-63; 1933)

C. Leonard Huskins: McGill University, Montreal.

References:

1. Huskins, C. L., *Trans. Roy. Soc. Canada*, **26** (Sect. V), 17-28 (1932).

2. Huskins, C. L., Smith, S. G., *et al.*, Proc. Sixth Internat. Cong. Genetics, **2**, 95-96, 392-393 and 396 (1932).

3. Nebel, B. R., *Z. Zell. Micro. Anat.*, **16**, 251-284 (1932).

4. Sharp, L. W., *Bot. Gaz.*, **88**, 349-382 (1929).

5. Hedayetullah, S., *J. Roy. Micro. Soc.*, **51**, 347-386 (1931).

6. Darlington, C. D., "Recent Advances in Cytology" (Churchill and Co., London) (1932).

后部分（第 305 页）出现了二重性："也许所有的细胞核分裂，**除了减数第一次分裂以外**[英文斜体是我加上的]，都是由染色体的分离决定的。"

根据本文中介绍的观察结果，这种二重性是不存在的；单链染色体丝之间相互吸引以及成对染色体之间相互排斥的原理完全可以应用于有丝分裂和减数分裂的所有阶段，而达林顿却错误地认为该原理只适用于分裂前期，不适用于分裂后期。

（管冰 翻译；刘京国 审稿）

The Genetics of Cancer*

J. B. S. Haldane

Editor's Note

Cancer is a genetic disease, resulting from mutations that cause a cell to become malignant. An individual's risk of developing cancer also has a strong genetic element, sometimes due to a pathogenic version of a single gene, or, more commonly, to the cumulative effect of many genes, which together influence the body's response to environmental carcinogens and ageing. In this paper the geneticist and evolutionary biologist J.B.S Haldane discusses all these factors. He concludes that "our knowledge is not sufficient to warrant interference with human breeding". This referred to the notion of using eugenics to control the incidence of "bad genes"; but the ability to genetically screen embryos is now forcing the question of other forms of intervention.

THE statement is occasionally made, and as frequently denied, that cancer is hereditary. It is, of course, clear that environmental influences can play a leading part in determining the production of cancer.

The question is whether nature, as well as nurture, is of importance. By far the most satisfactory evidence on this point comes from a study of genetically homogeneous populations of mice. By brother-sister mating for many generations (more than fifty in certain cases) pure lines can be built up in which the individuals are all homozygous for the same genes, apart from rare cases of mutation. By crossing two pure lines we obtain a population which is also genetically uniform, but not homozygous. Their progeny, however, is not genetically uniform.

In such a population, we can study three types of cancer:

(1) due to transplantation of a tumour from another animal.

(2) due to a carcinogenic agent. These include tar, the hydrocarbons shown to be carcinogenic by Kennaway and his colleagues, certain parasitic worms, and X-rays.

(3) spontaneous tumours; that is, tumours arising for no, at present, assignable cause.

Now the order of ease of study is the above. An inoculated tumour can be judged as a "take" or otherwise within a month. Tar painting may produce tumours within six months; but spontaneous tumours in many lines do not reach their maximum incidence until an age of eighteen or more months. Hence our knowledge of tumour etiology is in the above order,

---

* Substance of lectures delivered at the Royal Institution on February 2 and 9, 1933.

肿瘤遗传学[*]

约翰·伯登·桑德森·霍尔丹

编者按

癌症是一种遗传性疾病，是由细胞内基因的恶性突变造成的。遗传因素使一个人患癌症的危险大大增加；有时由于单个基因的突变，但更常见的是多个基因的累积突变影响了人体对环境中致癌物质和机体老化的反应。遗传学家和进化生物学家霍尔丹在这篇文章中讨论了所有的影响因素。他总结道："我们所掌握的知识尚不足以使我们敢于干预人类生育。"这一说法涉及利用优生学控制"不良基因"；但现有的大规模测试胚胎基因的能力正促使人们提出其他形式的干预手段相配合。

少数人认为癌症具有遗传性，而这一说法常常受到质疑。因为很明显环境因素在癌症发生中可以起到主导作用。

问题是，先天因素和后天因素是否同样重要。目前回答这个问题最令人满意的证据来源于对基因同源性小鼠群的研究。通过多代（在某些情况下要超过 50 代）的兄妹交配，能够建立起一个纯系种群，除了偶发的几个突变外，所有个体都是基因相同的纯合子。再将两个纯系种群进行杂交，我们可以得到一个基因完全一致的杂合子鼠群。而它们的后代，基因就不再完全一样了。

在这样一个种群中，我们能研究三种类型的肿瘤：

（1）从另外一个动物中移植过来的肿瘤。

（2）由致癌剂诱发的肿瘤。这些致癌剂包括焦油、由肯纳韦和同事发现的具有致癌性的烃类化合物、某些寄生虫和 X 射线。

（3）自发性肿瘤。即根据目前所知，诱因不明的肿瘤。

上面的顺序也反映了研究工作的难易程度。用一个月时间就可以判断接种过去的肿瘤能否"生长"。涂抹焦油在 6 个月内可以诱发肿瘤。而在许多品系的小鼠中，自发性肿瘤的发生率到第 18 个月或者更晚才会最高。所以我们对于肿瘤病因学的知识是通过以上顺序的研究获得的，但在临床实践中，就这些知识的重要程度来说，

---

[*] 本文取自 1933 年 2 月 2 日和 9 日作者在英国皇家研究院所作的报告。

457

though the order of practical importance for medicine is clearly the reverse.

It may be said at once that there are enormous differences between different lines as regards all three types of cancer. The genetics of reaction to transplantable tumours have been very fully worked out by Little and his colleagues. The laws disclosed are precisely similar to those which govern the transplantation of normal tissue or the transfusion of blood or of leukaemic corpuscles. A tumour arising spontaneously in one member of a pure line can be transplanted into all other members of it (actually more than 99 percent of successful "takes" can be achieved). Further, it can be transplanted into every F_1 mouse one of whose parents belonged to the pure line, but if these are mated together or outcrossed, only a minority of their offspring are susceptible. This at once suggests that susceptibility is due to the possession of certain dominant genes. This theory is fully confirmed by experiment. Supposing that a line X carries n pairs of genes $AABBCC$ ... which are needed for tumour growth, and are not found in a line Y, then the F_1 will be $AaBbCc$ ..., and 100 percent susceptible. Of the back-cross from mating of the F_1 with Y, only $\left(\frac{1}{2}\right)^n$ will carry all n genes, and of the F_2, $\left(\frac{3}{4}\right)^n$ will carry them, and thus be susceptible. In a number of cases, Cloudman[1] found that the two values of n so calculated were in agreement. The number of genes ranged from 2 to about 12.

Similarly, a tumour arising spontaneously in a F_1 individual, between two pure lines, in general requires in its host m genes contributed by one parent, and n contributed by the other. In such cases the tumour will grow in all the F_1, in $\left(\frac{1}{2}\right)^m$ of one back-cross, $\left(\frac{1}{2}\right)^n$ of the other, and $\left(\frac{3}{4}\right)^{m+n}$ of the F_2. Thus Bittner[2] found in one case $m=4$, $n=1$, while the observed value of $m+n$ was 5.

Occasionally a tumour in the course of transplantation changes its character, so that it becomes transplantable into a larger proportion of a mixed population (F_2 or back-cross). It is then found that one or more genes less are required for susceptibility in the host.

These facts can be explained if the host only reacts to a transplanted tumour so as to destroy it as the result of foreign antigens in that tumour, just as a recipient agglutinates the corpuscles of a donor if they carry foreign isoagglutinogens. On this hypothesis each gene is responsible for the manufacture of a particular antigen, as in the case of the red corpuscles. However, the genetical facts are quite independent of this hypothesis.

A thorough study of these phenomena is under way in Little's laboratory. The genes required for susceptibility to different tumours are being compared. Thus it was found that a number of tumours arising in the same line required the same basic gene for susceptibility, but each demanded a different assortment of extra genes. Some of these genes have been located, by means of linkage studies, in the same chromosomes as genes responsible for colour differences.

顺序显然是相反的。

需要立即指出的是在不同的品系中，这三种类型的肿瘤有很大的不同。利特尔和他的同事，已经把机体对移植肿瘤反应的遗传学机制进行了透彻的研究。他们发现调控移植肿瘤和调控那些正常组织移植、输血和输白血病细胞的机制非常相似。纯系种群中一个个体上自发发生的肿瘤能够被移植到该系所有的其他个体身上（实际上成功"接受率"能够超过99%）。此外，该肿瘤还能够被移植到任何一个 F_1 代小鼠身上，这些小鼠的亲本之一必须为上述纯系种群的成员。但是如果这些 F_1 代小鼠相互交配或者与品系之外的小鼠交配，它们的后代只有一小部分能够接受这些移植的肿瘤。这就暗示移植肿瘤的接受程度由是否拥有特定的显性基因所决定。这个理论完全被实验所证实了。假设一个品系 X 带有 n 对肿瘤生长所必需的基因 $AABBCC$ ……，而在品系 Y 中没有这些基因，那么 F_1 的基因型将是 $AaBbCc$ ……，它们将100%能接受移植肿瘤。如果将 F_1 和 Y 进行回交，那么只有 $(\frac{1}{2})^n$ 的个体会携带所有 n 个肿瘤生长所必需的基因，而对 F_2 来说，$(\frac{3}{4})^n$ 的个体会携带这些基因，因而能够接受移植的肿瘤。在多次试验中，克劳德曼[1]发现如此计算出来的两个 n 值是相同的，这些基因数在2到12之间波动。

与此类似，两个纯品系杂交后的 F_1 代个体要产生自发性肿瘤，总体来说，需要其亲本一方提供 m 个基因，另一方提供 n 个基因。这种情况下，这个肿瘤可以在所有 F_1 代个体、$(\frac{1}{2})^m$ 的 F_1 与带 m 个基因的亲本一方的回交后代、$(\frac{1}{2})^n$ 的 F_1 与带 n 个基因的亲本一方的回交后代和 $(\frac{3}{4})^{m+n}$ 个 F_2 代个体中生长。这样，比特纳[2]发现在这个试验中 $m=4$，$n=1$，而观察到的 $m+n$ 值正好是5。

偶尔的情况下，肿瘤在移植过程中发生了性质的改变，以至于该肿瘤能被移植到一个混合种群（F_2 或者回交种群）的更多个体中去。人们这时发现，肿瘤移植时宿主基因的要求会少一个或者多个。

这些现象可以这样解释，宿主只有在识别移植肿瘤中的外源性抗原后，才产生对移植肿瘤的排斥，正像如果供体的血细胞含有外源的同种凝集原，受体就会产生抗供体的凝集反应。根据这个假说可知，如同在红细胞中一样，每个基因负责编码一种特定的抗原。但是，遗传学的实际情况与这一假说并不相符。

在利特尔的实验室，一项针对这个现象的深入研究正在进行。实验室的工作人员正在对照不同肿瘤易感性所需的决定基因。他们发现对于在同一品系发生的多种类型的肿瘤，其易感性都需要相同的基本基因。但是每一种肿瘤的生长又需要不同种类的额外基因。通过连锁研究的分析方法，他们发现其中一些基因与决定肤色差异的基因定位于同一条染色体上。

The tendency to develop cancer as the result of tarring varies greatly in different lines. Thus Lynch[3] compared two lines A and B which differed in their spontaneous tumour rates, A having a higher incidence of spontaneous lung cancer than B. These tumours, however, never appeared before the age of 15 months. By tarring a large area of skin between the ages of 2 and 6 months, she induced lung tumours before the age of 13 months in 85 percent of line A, and 22 percent of line B. The difference was 6.3 times its standard error. On crossing A and B she found 79 percent susceptibility. The back-cross to A gave 81 percent susceptibility, while that to B gave 39 percent. These figures suggest that susceptibility is determined, among other things, by a number of dominant genes. Other workers have obtained essentially similar results. Their importance for non-genetical workers on cancer is considerable. It is clear that in comparing the efficacy of two different carcinogenic agents, far fewer mice need be used in a pure line than in a mixed population, and it is worth nothing that the variance in a mixed population is only about halved when litter mates instead of individuals taken at random are used as controls.

In the same way, Curtis, Dunning and Bullock[4] state that in the rat they have found that susceptibility to cancer, on infection with the cestode *Cysticercus*, is strongly inherited.

The problem of spontaneous cancer presents much greater difficulties. In no line can one obtain 100 percent of deaths from cancer, because over the long period necessary, deaths from other causes cannot be prevented. But some idea of the conditions in a highly cancerous line can be obtained from the work of Murray[5] . In a particular inbred line, 1,938 females lived to be more than 7 months of age. Of these, 65 percent died of mammary carcinoma, or were killed when severely affected with it. Above 80 percent of deaths in females more than twelve months old were due to this cause. None survived for 23 months and probably none would have reached two years without cancer had all deaths from other causes been prevented.

In such a line we can observe the effects on spontaneous cancer of prophylactic measures. Thus of 198 females ovariotomised at 7 months, only 40 percent died of mammary cancer, and one of these reached the age of 30 months. Still more striking is the fact that not a single operated female living beyond 22 months developed mammary cancer. It follows that in this stock the ovary plays an important part in the causation of mammary carcinoma. It is clear that a pure line (or the F_1 hybrids of two pure lines) furnish ideal material for the determination of factors in the environment favourable or otherwise to the development of cancer.

In contrast with such lines are others with an extremely low susceptibility to spontaneous tumours under ordinary laboratory conditions. A cross between such lines generally gives a hybrid generation with a cancer mortality nearly so high as that of the more susceptible line. Indications of linkage with colour genes have been obtained in one case.

It is important that the location of tumours is highly specific. One line has a high

焦油诱发性肿瘤发生的倾向，在不同品系中有很大的差异。因此林奇[3]对比了A和B这两个不同品系诱发性肿瘤发生的情况：它们自发性肿瘤的发生率不同，A自发性肺癌发生率要高于B。然而，这两个品系中的自发性肿瘤都是在小鼠15月龄后才出现。在2到6月龄小鼠的皮肤上大片涂抹焦油，到小鼠13月龄时A品系85%的个体中诱导出了肺癌，而B品系只有22%。差异是标准差的6.3倍。在A和B的杂交后代中，发现79%的个体有肿瘤易感性。子代与A回交的群体中，81%有易感性，而与B回交则只有39%。这些数据表明肿瘤易感性除了受其他因素影响之外，也受大量显性基因的影响。其他研究者也得到了基本类似的结果。对于那些不从事基因学的肿瘤工作者来说，他们的结论具有十分重大的意义。如果对比两种不同致癌物质的致癌效果，很明显，所需要的纯系小鼠比混合种群小鼠要少得多。当我们选择同窝出生的混合种群个体代替随机个体作为对照时，得到的数据应小心对待，变化仅为一半时，其意义可能并不大。

用同样的方法，柯蒂斯、邓宁和布洛克[4]宣称已经发现大鼠在感染囊尾蚴后可以诱发肿瘤，而肿瘤易感性在很大程度上是遗传因素决定的。

自发性肿瘤的研究面临更大的困难。没有一个品系会因癌症而100%死亡，因为在肿瘤发生的漫长过程中，其他原因导致的死亡是不可避免的。但是，默里[5]的工作提到了利用高致癌品系小鼠解决这个问题的观点。在一个特别近亲繁殖的品系中，1,938只雌鼠的生存时间超过了7个月。其中65%死于乳腺癌，或者由于乳腺癌严重影响生活而被杀掉。在生存超过12个月的雌鼠中，有80%以上死于乳腺癌。没有一只生存时间超过23个月。即使避免其他死因，这些小鼠也不可能无癌生存到2年。

在这个品系中，我们可以观察避孕措施对自发性肿瘤的作用。在长到7月龄时被切除卵巢的198只雌鼠中，只有40%死于乳腺癌，而且其中一只生存到了30月龄。更令人瞩目的是没有一只被切除卵巢的小鼠在生存超过22月龄时发生乳腺癌。这证明卵巢在该品系小鼠乳腺癌的发病中起重要作用。很明显，纯系动物（或者是两个纯系的F_1杂交后代）为确定环境中的各种因素是否有利于肿瘤的发生提供了理想的材料。

相对于这些易感品系来说，在正常实验室条件下，其他品系的动物对自发性肿瘤的易感性极低。但是这些品系之间的杂交后代常常具有与易感品系的后代同样高的肿瘤死亡率。在这项研究中，人们发现肿瘤基因可能与肤色基因有关。

肿瘤发生的部位具有高度的特异性，这一点非常重要。其中一个品系只是在雌

deathrate from mammary carcinoma in females only, and few tumours elsewhere. Another line has a heavy incidence of primary lung carcinomata in both sexes, and little mammary carcinoma. A third has few carcinomata of any kind, but sarcoma is not very rare. The genetics of spontaneous cancer will clearly be very complicated, and it is quite ludicrous to ascribe it to the activity of one gene, dominant or recessive.

Besides the work described above, a good deal has been done on stock which was not genetically homogeneous. From this work it is clear that, while a tendency to spontaneous cancer is hereditary, it is not due to a single gene, dominant or recessive, and also that a particular localisation of cancer may be hereditary. Thus Zavadskaia[6] found 13 out of 45 tumours in the occipital region in one particular line, and only 1 out of 212 in other lines. But all work with genetically heterogeneous material is unsatisfactory, because any individual may die before reaching the cancer age, and no other individual will be of just the same genetical make-up. Hence really exact work is impossible.

In the same way, human cancer tends to "run in families" to some extent, but precise analysis is only possible where, as with retinoblastoma and some sarcomata, its victims are attacked early in life. Here there is reason to believe that a single dominant gene is mainly responsible for the cancerous diathesis, though environmental and possibly other genetic factors may be concerned as well.

A particularly clear case of the interaction of nature and nurture in cancer production is found in the case of human xeroderma pigmentosum, almost certainly a recessive character. Here the skin becomes inflamed, and ultimately cancerous, under the influence of light. We could speak with equal logical propriety of the recessive gene or the light as the "cause" of the cancer, but as the former is rare, and the latter universal, it is more natural to regard the cancer as genetically determined.

Thus we have evidence of many different types of genetical predisposition to cancer, and although the data available on mice suggest that this predisposition is generally due to multiple dominant genes, it would certainly be incorrect to apply this theory to all human types of malignant disease.

The theory has been held by Boveri, Strong, and others, that the difference between a cancer and a normal cell is of the same character as that between the cells of two different varieties, that is to say, due to chromosomal aberration or gene mutation. This theory cannot of course be proved or disproved by genetical methods, as cancer cells do not reproduce sexually, and it is only by sexual reproduction that the geneticist can distinguish nuclear changes from plasmatic changes or virus infections.

The geneticist is concerned with the differences of "nature" (in Galton's sense of the word) which play their part, along with environmental differences, in determining whether a given animal will or will not develop cancer. He is not particularly qualified to determine whether the difference between a normal cell and a cancer cell is analogous to that

462

性动物中有很高的乳腺癌死亡率，而其他肿瘤则很少。另一个品系的动物无论雌雄均有很高发生率的原发性肺癌，而乳腺癌非常少见。第三个品系任何肿瘤的发生都很少，但是肉瘤却较常见。很明显，自发性肿瘤的遗传学机制是非常复杂的，绝对不可能完全归因于某一个显性或者隐性基因的作用。

除了上面所述的研究，还有大量的研究以基因不同源的动物为对象。这些研究结果很清楚地表明，尽管自发性肿瘤的倾向是有遗传性的，但不能归因于单个显性或者隐性基因，而特定的肿瘤发生部位可能也是遗传的。加瓦兹卡亚[6]在研究特定品系的动物中发现，45只动物中有13只患有枕部肿瘤，而在其他品系，212只动物中仅一只患有此病。但是所有以基因异源性动物为对象的研究都不能令人满意，因为任何个体都可能在达到癌症发病年龄之前死亡，而没有其他个体会有与其完全相同的基因组成。因此不可能获得真正准确的结果。

同样，人类肿瘤在一定程度上倾向于具有"家族聚集性"，尽管只有在例如视网膜母细胞瘤和某些肉瘤这些常常在生命早期发病的肿瘤中才有可能进行精确的分析。在此有理由相信，单个显性基因对肿瘤易感性起主要作用，尽管环境因素以及其他基因因素也有可能参与其中。

人类的着色性干皮病（该病几乎都是隐性遗传的）是一个能清晰说明在肿瘤发生过程中先天和后天因素如何相互作用的特殊例子。在光的影响下，皮肤会发炎并最终癌变。我们可以说这个隐性基因和光在逻辑上都能作为肿瘤的"起因"。但是由于这个隐性基因十分罕见，而光普遍存在，所以认为肿瘤是由先天基因决定的会显得更加自然。

这样，我们有大量关于各类肿瘤遗传易感性的证据，而且在小鼠身上得到的这些数据显示这种遗传易感性常常由多个显性基因决定，当然将这个理论运用到所有的人类恶性肿瘤中显然是不正确的。

博韦里、斯特朗和其他人提出的理论认为，癌细胞和正常细胞之间的差异与两个不同细胞变种之间的差异在本质上是相同的，也就是说，都是由染色体畸变或者基因突变引起的。这个理论当然不能被遗传学方法证实或者证伪，因为癌细胞不是通过性生殖的方式传代的，而遗传学家只能通过个体交配的方法将核改变与胞质改变或者病毒感染鉴别开。

遗传学家所关注的是，在决定一个特定动物是否会发生肿瘤的因素中，"先天"（以高尔顿的用词方式来说）因素和环境因素所起作用的不同。遗传学家们并没有特殊能力来确定正常细胞与癌细胞之间的差别，是否与减数分裂时产生的姐妹配子之

between sister gametes produced at meiosis, or to the difference which comes about at other cell divisions in the course of differentiation. The recognition of the importance of genetics for the study of cancer need not lead to any decision on this point.

To sum up, we can devise conditions under which either nature or nurture will play a predominant part in determining the incidence of cancer. Neither factor can possibly be neglected in a comprehensive survey. Except in a few cases, such as retinoblastoma, our knowledge is not sufficient to warrant interference with human breeding on eugenic grounds. Nevertheless, it is probable that in the ultimate solution of the problem of human cancer, eugenical measures will play their part.

(**132**, 265-267; 1933)

References:

1. Cloudman, A. M., *Amer. J. Cancer*, **16**, 568 (1932).

2. Bittner, J. J., *Amer. J. Cancer*, **15**, 2202 (1931).

3. Lynch, C. J., *J. Exp. Med.*, **46**, 917 (1927).

4. Curtis, Dunning and Bullock, *Amer. Nat.*, **67**, 73 (1933).

5. Murray, W. S., *Science*, **75**, 646 (1932).

6. Zavadskaia, *J. Genetics*, **27**, 181 (1933).

间的差别，或者与分化过程中细胞分裂的产物之间的差别一样。认识到遗传学在肿瘤研究中的重要性并不一定能对此有所帮助。

总体来说，我们能设计出各种不同的情况，在这些情况下，先天因素和后天因素中的一个会对肿瘤发生起主要决定作用。在一个综合性的调查中，两个因素都不能被忽略。除了视网膜母细胞瘤等少数例子以外，我们的知识不足以在保证优生的前提下干预人类生殖的过程。但要从根本上解决人类肿瘤的问题，很可能仍需要采取优生学的手段。

（毛晨晖 翻译；秦志海 审稿）

Vitamin A in the Retina

G. Wald

Editor's Note

The primary biological role of vitamin A is as the light-absorbing molecule retinol in the light-sensitive cells of the eye's retina. Here American biochemist George Wald reports a major step in the understanding of that function: the discovery of vitamin A in the retina of several animals. This association was not wholly unexpected, in view of the fact that a dietary lack of vitamin A was known to be linked to visual disorders. But Wald later went on to reveal the biochemistry involved: the presence of retinol in the light receptor protein rhodopsin and the different colour sensitivities of different forms of this protein in cone cells. For that work he was awarded a Nobel Prize in 1967.

I have found vitamin A in considerable concentrations in solutions of the visual purple, in intact retinas, and in the pigment-choroid layers of frogs, sheep, pigs and cattle. The non-saponifiable extracts of these eye tissues display in detail all of the characteristics of vitamin A-containing oils.

The blue antimony trichloride coloration given by retinal and choroid layer extracts, when observed spectroscopically, exhibits the sharp, strong band at 620 mμ specific for vitamin A. More concentrated preparations also display the characteristic fainter band at 580 mμ, recently shown to be due to a foreign material which in natural oils always accompanies vitamin A in varying concentrations[1]. Both bands fade rapidly after mixing with antimony trichloride, while a secondary absorption at about 500 mμ appears which is responsible for the red coloration in later stages of the reaction. This last phenomenon also is characteristic of impure vitamin A preparations.

Absorption spectra have been measured of the chloroform solutions of oils from the retinas and pigment layers of sheep and oxen, and from pig retinas. The extinction coefficient (log I_0/I) rises without inflection from 500 mμ to a single broad maximum between 320 mμ and 330 mμ. This is the characteristic vitamin A band. The smoothness of the absorption curves between 500 mμ and 400 mμ is an indication that no other carotenoids are present in these extracts, since all the other known carotenoids possess one or more absorption bands in this region[2]

Feeding experiments on rats suffering from avitaminosis have been performed at the Pharmacological Institute of Hoffmann – La Roche et Cie., Basle, using an extract of ox retinas, from which the sterins had been frozen. The results of these experiments with a first preparation have now been received. This oil, tested with antimony trichloride, had shown in the Lovibond tintometer a colour intensity of 20 c.l.o. units. The purest vitamin A preparations test at about 10,000 c.l.o. units. Therefore this oil contained

视网膜中的维生素A

乔治·沃尔德

编者按

维生素A在生物上的主要作用是作为视网膜上感光细胞中的光吸收分子。美国生物化学家乔治·沃尔德在本文中报导了理解这一功能的一个关键点：在几种动物的视网膜中发现了维生素A。这种关联的存在并不十分出人意料，因为人们通常把饮食中缺乏维生素A与视觉疾病联系起来。但沃尔德又揭示出了其中包含的生物化学机理：维生素A存在于光受体蛋白视紫红质上，这种蛋白在视锥细胞中有不同的种类，不同种类的光受体蛋白具有不同的光敏感性。由于这项工作他获得了1967年的诺贝尔奖。

我已经发现在青蛙、羊、猪和牛的视紫红质溶液、完整视网膜以及色素脉络膜层中都含有相当高浓度的维生素A。这些眼组织的非皂化提取物明确地显示出含维生素A油所具有的全部特性。

视网膜和脉络膜层的提取物能与三氯化锑发生显色反应呈现蓝色，如果用光谱法进行检测，会发现在620 nm（维生素A的特征吸收）处有尖锐、强烈的吸收带。浓度更高的提取物还会在580 nm处出现较弱的特征吸收带，最近的研究表明这是因为提取物中有一种外来物质，无论浓度如何变化它在天然油中总是与维生素A同时存在[1]。在与三氯化锑混合后，两条谱带迅速减弱，同时在500 nm附近出现了一个次级吸收，它是造成反应后期出现红色着色的原因。最后这个现象也说明维生素A的提取物中含有杂质。

我们已经测量了羊、牛视网膜和色素层提取物以及猪视网膜提取物在氯仿溶液中的吸收光谱。其消光系数（$\log I_0/I$）从500 nm开始直线上升，一直到320~330 nm之间的一个宽谱带最高点。这就是维生素A的特征谱带。吸收曲线在500 nm和400 nm之间没有什么起伏，这说明提取物中不存在其他的类胡萝卜素，因为所有已知的类胡萝卜素在这个区域都会有一条或多条吸收谱带[2]。

在巴塞尔的霍夫曼–罗氏公司的药理学研究所中，有人进行了一项用牛视网膜提取物（其中的硬脂酸甘油酯已经被冷冻）喂养患有维生素缺乏症的大鼠的实验。由首次提取物取得的实验结果已经得到了人们的认可。用三氯化锑检测此提取物，并用洛维邦德色辉计测量其颜色强度，结果为20个c.l.o.单位。而最纯的维生素A制备物的颜色强度约为10,000个c.l.o.单位。因此，该提取物约含有0.2%的维生素

about 0.2 percent vitamin A. Rats displaying the symptoms of avitaminosis were cured by administering a daily ration of 1 mgm. of the oil; 0.3 mgm. was found to be inadequate. By the antimony trichloride test, 1 mgm. of the oil contained 2γ vitamin A. The purest preparations of Karrer and his co-workers are capable of maintaining growth in rats when fed in a daily dosage of 0.5γ[3]. Since the vitamin requirement is appreciably greater for curing diseased rats than for maintaining growth in normal animals, the agreement is adequate.

Some time after these experiments had been begun, I learned of work from two other sources which, to a degree, anticipates the present results. Holm[4] has found that fresh calf retinas fed to rats suffering from avitaminosis are capable of curing xerophthalmia and restoring normal growth. Smith, Yudkin, Kriss and Zimmerman[5] have obtained similar results with dried pig retinas; choroid tissue proved ineffective. In neither contribution is the presence of vitamin A exclusively indicated. Of the other known carotenoids, carotene possesses all of the described characteristics. It is also not clear in the light of the present work why the latter authors found choroid tissue without effect; or why alcohol extracts of their preparations did not respond to the arsenic trichloride test, since alcohol is a good solvent for vitamin A.

The physiological significance of the presence of considerable quantities of vitamin A in the eye tissues will be discussed in detail in a more complete communication elsewhere. Most interesting is the relation of the presence of the vitamin in the eye to the optic disorders which are the specific symptoms of its absence from the diet: xerophthalmia, keratomalacia and—most pertinent to the present work—night-blindness.

(**132**, 316-317; 1933)

George Wald (National Research Fellow in Biology): Chemical Institute, University of Zurich, Aug. 6.

References:

1. Karrer, P., Walker, O., Schöpp, K., and Morf, R., *Nature*, **132**, 26 (July 1, 1933).

2. Von Euler, H., Karrer, P., Klussmann, E., and Morf, R., *Helv. Chim. Acta*, **15**, 502 (1932).

3. Karrer, P., Morf, R., and Schöpp, K., *Helv. Chim. Acta*, **14**, 1036 (1931).

4. Holm, E., *Acta Ophthal.*, **7**, 146 (1929).

5. Smith, Yudkin, Kriss and Zimmerman, *J. Biol. Chem.*, **92**, *Proc.*, xcii (1931).

A。患有维生素缺乏症的大鼠只要每天食入 1 mg 这样的提取物，就可以得到治愈；每天服用 0.3 mg 是不够的。根据三氯化锑试验，1 mg 的提取物中含有 2γ 维生素 A。如果用卡勒及其同事制备的最纯的维生素 A 以每天 0.5γ 的剂量喂食大鼠，就足以维持它们的生长 [3]。由于治愈患病大鼠所需的维生素剂量要略微大于维持健康动物生长所需的剂量，所以这与事实是相符的。

在这些实验开始后不久，我了解到其他两项研究工作在某种程度上预言了现在的结论。霍尔姆 [4] 发现：用新鲜的小牛视网膜喂养患有维生素缺乏症的大鼠，就可以治愈干眼症并能够恢复其正常的生长。史密斯、尤德金、克里斯和齐默尔曼 [5] 用干燥的猪视网膜也得到了类似的结果；但脉络膜组织被证明是无效的。上述两个文献并没有专门提到维生素 A 的存在。在其他已知的类胡萝卜素中，胡萝卜素具有上述的所有特性。以目前的研究成果而论，我们还不清楚为什么史密斯等人会认为脉络膜组织是无效的；抑或既然乙醇是维生素 A 的良溶剂，为什么他们制备的乙醇提取物对三氯化砷的检测却没有反应。

我将在其他论文中以更完整的形式详细地讨论大量维生素 A 存在于眼组织中的生理学意义。维生素在眼睛中的存在与因饮食中缺乏维生素而引起的视障碍疾病之间的关系最惹人注目：干眼症，角膜软化症以及与目前工作联系最紧密的夜盲症。

（刘振明 翻译；刘力 审稿）

Some Chemical Aspects of Life

F. G. Hopkins

Editor's Note

Scientific speculations about the origins of life go back at least as far as Charles Darwin. By the time of this address from Frederick Hopkins, a 1929 Nobel laureate for his work on vitamins, the question had received serious if tentative attention. Yet Hopkins' assertion that most biologists agreed on life's origin being both singularly important and singularly improbable very much remained true decades later. Hopkins' aim here is not to outline any explicit theories of how life began, but to provide a context for such discussions by describing life as a chemical process, for example in terms of enzyme catalysis and metabolism. After the later focus on genetics, biochemical energetics is today enjoying a revival as a fundamental aspect of living systems.

I

THE British Association returns to Leicester with assurance of a welcome as warm as that received twenty-six years ago, and of hospitality as generous. The renewed invitation and the ready acceptance speak of mutual appreciation born of the earlier experience. Hosts and guests have today reasons for mutual congratulations. The Association on its second visit finds Leicester altered in important ways. It comes now to a city duly chartered and the seat of a bishopric. It finds there a centre of learning, many fine buildings which did not exist on the occasion of the first visit, and many other evidences of civic enterprise. The citizens of Leicester on the other hand will know that since they last entertained it the Association has celebrated its centenary, has four times visited distant parts of the Empire, and has maintained unabated through the years its useful and important activities.

In 1907 the occupant of the presidential chair was Sir David Gill, the eminent astronomer who, unhappily, like many who listened to his address, is with us no more. Sir David dealt in that address with aspects of science characterised by the use of very exact measurement. The exactitude which he prized and praised has since been developed by modern physics and is now so great that its methods have real aesthetic beauty. In contrast, I have to deal with a branch of experimental science which, because it is concerned with living organisms, is in respect of measurement on a different plane. Of the very essence of biological systems is an ineludable complexity, and exact measurement calls for conditions here unattainable. Many may think, indeed, though I am not claiming it here, that in studying life we soon meet with aspects which are non-metrical. I would have you believe, however, that the data of modern biochemistry which will be the subject of my remarks were won by quantitative methods fully adequate to justify the claims based upon them.

生命的某些化学面貌

弗雷德里克·戈兰德·霍普金斯

编者按

人们对生命起源的探索可以一直追溯到查尔斯·达尔文时代。当因在维生素方面的研究而荣获 1929 年诺贝尔奖的弗雷德里克·霍普金斯发表这篇演说的时候，世人已经开始认真考虑生命起源的问题了，但也许只是尝试性的。然而霍普金斯指出：多数生物学家都认为生命起源问题非常重要也非常难以想象，在几十年后的今天仍然如此。霍普金斯这次演讲的目的不是为了总结任何一种能够清晰阐述生命如何起源的理论，而是为了引导大家去探讨生命过程的化学本质，例如，从酶催化和新陈代谢的角度上进行分析。生物学研究在后来的一大段时间内转向了遗传学，而如今生物化学能量学作为生命系统的基本特征又一次得到了复兴。

I

英国科学促进会这一次重返莱斯特必定会受到和 26 年前一样的隆重欢迎和盛情款待。在新一次的邀请与爽快的答复之中都谈及了双方在此前的交往经历中萌生出的相互欣赏。今天，主宾双方均有理由相互祝贺。在英国科学促进会的第二次来访中，我们发现莱斯特发生了一些重要的变化。如今，她已变成一座获得正式特许权的城市，成为主教辖区的中心。在这里，我们看到了一所研究中心、许多在初次来访时还不存在的精美建筑以及其他很多显示出城市进取心的标志。另一方面，莱斯特的市民也将发现，自从他们上一次招待科学促进会之后，该协会已经度过了自己的百年诞辰，还曾四次出访大英帝国的边远地区，并且在这些年中一直毫无懈怠地坚持开展那些有益而重要的活动。

在 1907 年，当时协会的主席是杰出的天文学家戴维·吉尔爵士，遗憾的是，他和很多曾聆听他讲演的人一样，如今已不在我们身边了。戴维爵士在那次演讲中谈到了用非常精确的测量方法所表征的科学现象。他所看重和称许的精确性此后被现代物理学发扬光大，现在更是达到了登峰造极的程度，其方法甚至已具备了真正的美学意义。与之相对，我必须要讨论的是实验科学的一个分支，由于它所涉及的是活的有机体，因此在实验观测方面还处于不同的层次。生物学体系的一个特定本质就是无法回避的复杂性，而精确观测所要求的条件是这里所不具备的。事实上，尽管我没有在这里宣称，但很多人也许都会认为在对生命的研究中我们很快就会遇到不可测量的现象。不过，我会让你们相信，在我将要谈论到的主题中，以定量方法测得的现代生化数据足以去证实基于它们的论断。

Though speculations concerning the origin of life have given intellectual pleasure to many, all that we yet know about it is that we know nothing. Sir James Jeans once suggested, though not with conviction, that it might be a disease of matter—a disease of its old age! Most biologists, I think, having agreed that life's advent was at once the most improbable and the most significant event in the history of the universe, are content for the present to leave the matter there.

We must recognise, however, that life has one attribute that is fundamental. Whenever and wherever it appears, the steady increase of entropy displayed by all the rest of the universe is then and there arrested. There is no good evidence that in any of its manifestations life ultimately evades the second law of thermodynamics, but in the downward course of the energy-flow it interposes a barrier and dams up a reservoir which provides potential for its own remarkable activities. The arrest of energy degradation in this sense is indeed a primary biological concept. Related to it and of equal importance is the concept of organisation.

It is almost impossible to avoid thinking and talking of life in this abstract way, but we perceive it, of course, only as manifested in organised material systems, and it is in them we must seek the mechanisms which arrest the fall of energy. Evolution has established division of labour here. From far back the wonderfully efficient functioning of structures containing chlorophyll has, as is well known, provided the trap which arrests and transforms radiant energy—fated otherwise to degrade—and so provides power for nearly the whole living world. It is impossible to believe, however, that such a complex mechanism was associated with life's earliest stages. Existing organisms illustrate what was perhaps an earlier method. The so-called autotrophic bacteria obtain energy for growth by the catalysed oxidation of materials belonging wholly to the inorganic world; such as sulphur, iron or ammonia, and even free hydrogen. These organisms dispense with solar energy, but they have lost in the evolutionary race because their method lacks economy. Other existing organisms, certain purple bacteria, seem to have taken a step towards greater economy, without reaching that of the green cell. They dispense with free oxygen and yet obtain energy from the inorganic world. They control a process in which carbon dioxide is reduced and hydrogen sulphide simultaneously oxidised. The molecules of the former are activated by solar energy which their pigmentary equipment enables these organisms to arrest.

Are we to believe that life still exists in association with systems that are much more simply organised than any bacterial cell? The very minute filter-passing viruses which, owing to their causal relations with disease, are now the subject of intense study, awaken deep curiosity with respect to this question. We cannot yet claim to know whether or not they are living organisms. In some sense they grow and multiply, but, so far as we yet know with certainty, only when inhabitants of living cells. If they are nevertheless living, this would suggest that they have no independent power of obtaining energy and so cannot represent for us the earliest forms in which life appeared. At present, however, judgment on their biological significance must be suspended. The fullest understanding of all the methods by

　　尽管关于生命起源的思考已经给很多人带来了智力上的愉悦，但我们关于这个问题所知道的一切就是我们一无所知。詹姆斯·金斯爵士曾经提出，虽然并非断言，这可能是一种病症——一种年深日久的病症！我觉得，大多数生物学家都同意，生命的出现随即成为宇宙历史中可能性最小和意义最深远的事件，而且他们都对目前暂且任这个问题自然发展的现状感到满意。

　　然而，我们必须承认，生命有一种根本性的属性。无论它出现于何时何地，宇宙中所有其余地方所呈现的熵的稳定增长都在此时此地被终止。没有可靠的证据能表明，生命的任何一种表现形式会最终超越热力学第二定律，但是它在能量流向下流动的过程中设置了一道势垒用于截流以储存能量，为自己的非凡活动提供势能。实际上以这种方式停止能量的消耗是一个基本的生物学概念。与之相关联并且具有同等重要性的是组织的概念。

　　要回避以这种抽象方式对生命进行思考和讨论几乎是不可能的，但是，显然，我们只有在有组织的物质系统中才能明确地感受到这一点，而且，我们必须要在这些系统中寻找遏止能量降低的机制。这里，进化导致了劳动的分化。我们都知道，早在很久以前，含叶绿素结构的极为有效的机能就已经为截留和转化辐射能提供了一个势阱——否则它便注定要消耗掉——从而为几乎整个生物世界供应了动力。但是，令人无法相信的是，这样一种复杂的机制竟与生命的最早期阶段相关联。现存有机体表现出一种可能是早期的方法。通常所称的自养型细菌通过将完全属于无机世界的原料催化氧化来获取生长所需的能量；例如硫、铁或者氨，甚至还有游离态氢。这些有机体不需要太阳能，但是它们在进化中被淘汰了，因为它们的方法缺乏效率。其他现存有机体，如某些紫色细菌，看来是向提高效率的方向前进了一步，但没有达到绿色细胞所拥有的程度。它们不需要游离态氧，但从无机世界获取能量。它们控制一个还原二氧化碳并同时将硫化氢氧化的过程。前者的分子靠太阳能来活化，它们的色素系统使这些有机体能够吸收太阳能。

　　我们是否应该相信，在那些由比任何细菌细胞简单得多的方式组织起来的系统中仍然存在生命呢？极为微小的滤过性病毒强烈地唤起了人们对于这一问题的好奇心，由于它们与疾病之间的因果关系，这些病毒如今已成为深入研究的主题。我们还不能确定地知道它们是否就是活的有机体。但是，我们现在可以断言：从某种意义上说，它们只有在寄生于活细胞中时才会进行生长和复制。如果这样它们也算是活的，那么这就意味着它们没有独立获取能量的能力，因而也就无法为我们展示出生命出现时的最初形式。但是，目前对它们生物学意义的判断必定还是悬而未决的。关于

which energy may be acquired for life's processes is much to be desired.

In any event, every living unit is a transformer of energy however acquired, and the science of biochemistry is deeply concerned with these transformations. It is with aspects of that science that I am to deal, and if to them I devote much of my address, my excuse is that since it became a major branch of inquiry, biochemistry has had no exponent in the presidential chair I am fortunate enough to occupy.

As a progressive scientific discipline, it belongs to the present century. From the experimental physiologists of the last century it obtained a charter, and, from a few pioneers of its own, a promise of success; but for the furtherance of its essential aim, that century left it but a small inheritance of facts and methods. By its essential or ultimate aim, I myself mean an adequate and acceptable description of molecular dynamics in living cells and tissues.

II

When the British Association began its history in 1831, the first artificial synthesis of a biological product was but three years old. Primitive faith in a boundary between the organic and the inorganic which could never be crossed, was only just then realising that its foundations were gone. Since then, during the century of its existence, the Association has seen the pendulum swing back and forth between frank physico-chemical conceptions of life and various modifications of vitalism. It is characteristic of the present position and spirit of science that sounds of the long conflict between mechanists and vitalists are just now seldom heard. It would almost seem, indeed, that tired of fighting in a misty atmosphere, each has retired to his tent to await with wisdom the light of further knowledge. Perhaps, however, they are returning to the fight disguised as determinist and indeterminist respectively. If so, the outcome will be of great interest.

In any event, I feel fortunate in a belief that what I have to say will not, if rightly appraised, raise the old issues. To claim, as I am to claim, that a description of its active chemical aspects must contribute to any adequate description of life is not to imply that a living organism is no more than a physico-chemical system. It implies that at a definite and recognisable level of its dynamic organisation, an organism can be logically described in physico-chemical terms alone. At such a level, indeed, we may hope ultimately to arrive at a description which is complete in itself, just as descriptions at the morphological level of organisation may be complete in themselves. There may be yet higher levels calling for discussion in quite different terms.

I wish, however, to remind you of a mode of thought concerning the material basis of life, which though it prevailed when physico-chemical interpretations were fashionable, was yet almost as inhibitory to productive chemical thought and study as any of the claims of vitalism. This was the conception of that material basis as a single entity, as a definite though highly complex chemical compound. Up to the end of the last century and even

对生命过程赖以获得能量的所有方法的最完整理解，是我们极为渴望获得的。

无论如何，每个活着的个体都是一个能量转换器——不管它是怎样做到的，而生物化学这门科学就是要深入考察这些转化过程。我将要讨论的就是科学的这些方面，而如果我的讲演过多地集中在这里的话，我的解释是，自从生物化学成为一个主要的研究分支以来，还没有一个像我这样有幸占据主席位置的人来阐释过。

作为一个不断发展的学科，它属于当前这个世纪。从上世纪的实验生理学家那里，生物化学获得了作为学科的认可，又从它自身的几位先驱者那里获得了成功的希望；但是就对其根本目标的促进而言，那个世纪留给它的遗产只是少许的事实和方法。而它的根本目标或者说最终目标，我本人认为，是对活细胞和组织中的分子动力学行为进行充分和可接受的描述。

II

当英国科学促进会于 1831 年成立时，第一个人工合成的生物制品只有三岁。人们就是在那个时候才认识到了，最初关于有机与无机之间存在着无法跨越的界限的信条，其根基已经不复存在。从那以后，在它存在的整个世纪中，协会一直在单纯利用物理化学概念解释生命与各种修正形式的生机论之间摇摆不定。当代科学之立场与精神的特征表现为：机械论者与生机论者之间长期以来的争论如今已经很少听到了。实际上，基本上可以看出来，每个人都已厌倦了在蒙昧气氛下的争执，各自返回营帐，明智地等待着更为先进的知识的指引。不过，也许他们将会分别伪装成决定论者和非决定论者重新加入战团。如果是这样的话，结果将会非常有趣。

无论如何，如果正确评价的话，我都会因如下信念而感到幸运——我将要讲述的内容不会是老调重弹。正如我即将谈到的，认为对生物化学面貌的描述必定会有助于充分地理解生命，这并不是要暗示一个活的有机体仅仅是物理化学系统。它意味着，在一个明确的和可认知的动力学组织水平上，可以只用物理化学术语对一个有机体进行合乎逻辑的描述。事实上，在这样一个水平上，我们可以希望最终形成一种完全自洽的描述，正如对组织的形态学水平上的描述可以是完全自洽的那样。也许在更高的水平上会需要用相当不同的术语进行讨论。

不过我希望能提醒你们，一种考虑生命之物质基础的思考模式，尽管它在物理化学解释流行时占据上风，却如同生机论的所有主张一样，对于诸多基于化学的思想与研究而言几乎是抑制性的。这种观念认为物质基础是一种单一实体，是一种尽管高度复杂但明确的化合物。直到上个世纪末甚至更晚些时候，"原生质"这个词对

later, the term "protoplasm" suggested such an entity to many minds. In his brilliant presidential address at the British Association's meeting at Dundee twenty-one years ago, Sir Edward Sharpey-Schafer, after remarking that the elements composing living substances are few in number, went on to say: "The combination of these elements into a colloid compound represents the physical basis of life, and when the chemist succeeds in building up this compound it will, without doubt, be found to exhibit the phenomena which we are in the habit of associating with the term 'life'." Such a compound would seem to correspond with the "protoplasm" of many biologists, though treated perhaps with too little respect. The presidential claim might have seemed to encourage the biochemist, but the goal suggested would have proved elusive, and the path of endeavour has followed other lines.

So long as the term "protoplasm" retains a morphological significance as in classical cytology, it may be even now convenient enough, though always denoting an abstraction. In so far, however, as the progress of metabolism with all the vital activities which it supports was ascribed, in concrete thought, to hypothetical qualities emergent from a protoplasmic complex in its integrity, or when substances were held to suffer change only because in each living cell they are fires built up, with loss of their own molecular structure and identity, into this complex, which is itself the inscrutable seat of cyclic change, then serious obscurantism was involved.

Had such assumptions been justified, the old taunt that when the chemist touches living matter it immediately becomes dead matter would also have been justified. A very distinguished organic chemist, long since dead, said to me in the late eighties: "The chemistry of the living? That is the chemistry of protoplasm; that is super-chemistry; seek, my young friend, for other ambitions."

Research, however, during the present century, much of which has been done since the British Association last met in Leicester, has yielded knowledge to justify the optimism of the few who started to work in those days. Were there time, I might illustrate this by abundant examples; but I think a single illustration will suffice to demonstrate how progress during recent years has changed the outlook for biochemistry. I will ask you to note the language used thirty years ago to describe the chemical events in active muscle and compare it with that used now. In 1895, Michael Foster, a physiologist of deep vision, dealing with the respiration of tissues, and in particular with the degree to which the activity of muscle depends on its contemporary oxygen supply, expounded the current view which may be thus briefly summarised. The oxygen which enters the muscle from the blood is not involved in immediate oxidations, but is built up into the substance of the muscle. It disappears into some protoplasmic complex on which its presence confers instability. This complex, like all living substance, is to be regarded as incessantly undergoing changes of a double kind, those of building up and those of breaking down. With activity the latter predominates, and in the case of muscle the complex in question explodes, as it were, to yield the energy for contraction. "We cannot yet trace," Foster comments, "the steps taken by the oxygen from the moment it slips from the blood into

476

于很多人来说还是指这样一种实体。在 21 年前英国科学促进会于敦提召开的会议上，时任主席的爱德华·沙比 - 谢弗爵士发表了精彩的演讲，他首先谈到构成有生命物质的元素数量很少，接着说道："由这些元素结合成的胶体化合物是生命的物理基础，一旦化学家成功地合成出这种化合物，毫无疑问，就可以发现它会表现出我们通常与'生命'这一术语联系在一起的现象。"这样一种化合物似乎对应于很多生物学家所说的"原生质"，尽管它可能只得到了很有限的关注。主席的这一断言似乎鼓励了生物化学家，但是他提出的目标却被证明是难以把握的，而努力的方向也已改弦易辙。

现在"原生质"这个名词被更加方便地保持着在古典细胞学中那样的字面意思，尽管它也会经常表示抽象意义。但是，如果新陈代谢和它所维持的所有生命活动的进行，在实际思想中被归因于一种原生质复合物在其完整状态下所呈现出来的假定属性，或者认为，物质经历转变只是因为它们在每个活细胞中都要以自身分子结构与特性的丧失为代价合成这种复合物，其本身是发生循环变化的神秘场所，那么就会深深地陷入到蒙昧主义之中。

如果上述假设被证明是合理的，那么以往认为化学家一旦接触生命物质就会把它变成死物的讥讽也就是合理的了。一位已故去多年的杰出有机化学家，曾在他快90 岁的时候对我说："研究生命的化学？那就是研究原生质的化学；那是超级化学；我年轻的朋友，去追寻其他的抱负吧。"

但是，在当前这个世纪所进行的研究中有很大一部分是在英国科学促进会最后一次访问莱斯特之后所完成的，由这些研究得到的知识印证了在那些日子里开始研究的少数人的乐观是有道理的。如果时间允许，我可以通过充足的实例来阐明这一点；但我相信只要一个例证就足以说明近年来的进展如何改变了生物化学的前景。我将请你们注意一下 30 年前在描述运动肌肉中的化学事件时所使用的语言，并且将之与现在所用的相比较。在 1895 年，一位富有洞察力的生理学家迈克尔·福斯特研究了组织的呼吸作用，尤其是肌肉运动对当时氧气供应情况的依赖程度。他详细阐述了当时流行的看法，这里可以简要概述一下。从血液进入肌肉的氧并不参与直接氧化，而是转变成肌肉物质。它消失在某些原生质复合物之中，以自己的出现为原生质带来了不稳定性。如同所有活体物质一样，这种复合物应该被看作是在持续地经历着一种具有合成和分解双重特性的转变。运动时分解是主要的，而在肌肉中，我们所讨论的复合物发生分解以产生收缩所需的能量。"我们还不能描绘——"福斯特谈道，"氧从血液进入肌肉物质直到与碳结合生成碳酸排出之间所经历的各个步骤。生命的全部奥秘就隐藏在这个过程之中，而目前我们必须满足于只知道开始

the muscle substance to the moment when it issues united with carbon as carbonic acid. The whole mystery of life lies hidden in that process, and for the present we must be content with simply knowing the beginning and the end."

What we feel entitled to say today concerning the respiration of muscle and of the events associated with its activity requires, as I have suggested, a different language, and for those not interested in technical chemical aspects the very change of language may yet be significant. The conception of continuous building up and continuous breakdown of the muscle substance as a whole, has but a small element of truth. The colloidal muscle structure is, so to speak, an apparatus, relatively stable even as a whole when metabolism is normal, and in essential parts very stable. The chemical reactions which occur in that apparatus have been followed with a completeness which is, I think, striking. It is carbohydrate stores, as distinct from the apparatus (and in certain circumstances also fat stores), which undergo steady oxidation and are the ultimate sources of energy for muscular work. Essential among successive stages in the chemical breakdown of carbohydrate which necessarily precede oxidation is the intermediate combination of a sugar (a hexose) with phosphoric acid to form an ester. This happening is indispensable for the progress of the next stage, namely, the production of lactic acid from the sugar, which is an anaerobic process.

The precise happenings to the hexose sugar while in combination with phosphoric acid are from a chemical point of view remarkable. Very briefly stated they are these. One half of the sugar molecule is converted into a molecule of glycerin and the other half into one of pyruvic acid. Now with loss of two hydrogen atoms glycerin yields lactic acid, and, with a gain of the same, pyruvic acid also yields lactic acid. The actual happening then is that hydrogen is transferred from the glycerin molecule while still combined with phosphoric acid to the pyruvic acid molecule, with the result that two molecules of lactic acid are formed[*]. The lactic acid is then, during a cycle of change which I must not stop to discuss, oxidised to yield the energy required by the muscle.

The energy from this oxidation, however, is by no means directly available for the mechanical act of contraction. The oxidation occurs indeed after and not before or during a contraction. The energy it liberates secures, however, the endothermic resynthesis of a substance, creatin phosphate, of which the breakdown at an earlier stage in the sequence of events is the more immediate source of energy for contraction. Even more complicated are these chemical relations, for it would seem that in the transference of energy from its source in the oxidation of carbohydrate to the system which synthesises creatin phosphate, yet another reaction intervenes, namely, the alternating breakdown and resynthesis of the substance adenyl pyrophosphate.

The sequence of these chemical reactions in muscle has been followed and their relation in time to the phases of contraction and relaxation is established. The means by which

* Otto Meyerhof, *Nature*, Sept. 2, p.337 and Sept. 9, p.373.

和结束的情况。"

如同我曾提议的那样，我们需要一种不同的语言来谈及目前已确知的有关肌肉和与其活动相关的事件的呼吸作用，而且对于那些对技术性化学内容不感兴趣的人来说，在语言上的这种转变也许还是很重要的。肌肉物质作为一个整体连续合成与分解的概念只含有很少量的真实成分。可以说，凝胶状肌肉结构，即使作为一个整体在新陈代谢正常时也是相对稳定的系统，而且其核心部分的稳定性极高。在我看来，该系统中发生的化学反应具有令人惊叹的完整性。与在系统中不同，碳水化合物储备（在某些情况下还有脂肪储备）进行的是稳定的氧化反应，并成为肌肉运转的最终能量源。碳水化合物在氧化之前必然要先发生化学分解，这个过程能够连续进行的关键在于由一分子糖（己糖）和磷酸形成一种酯类的中间化合作用。这件事对于下一个阶段，即从糖产生乳酸这一厌氧过程的进行来说是必不可少的。

从化学的角度上看，己糖在与磷酸结合时所发生的具体变化是很值得关注的。极为简略地讲，过程是这样的。糖分子的一半转变成一分子甘油，而另一半则转变成一分子丙酮酸。然后甘油失去两个氢原子，产生乳酸；而丙酮酸则得到两个氢原子，也生成乳酸。接下来的实际过程是，氢原子从仍与磷酸相连接的甘油分子转移到丙酮酸分子中，结果是形成了两分子乳酸 *。接着，在一个我不能停下来去讨论的变化过程中，乳酸发生氧化，产生了肌肉所需的能量。

但是，这一氧化过程所产生的能量还无法被机械性的收缩动作直接利用。实际上，氧化发生在收缩过程之后而不是之前或其间。不过，它释放出来的能量使重新合成磷酸肌酸的吸热反应得以进行，对于收缩过程来说，其能量更为直接的来源是磷酸肌酸在这一系列变化的较早阶段中的分解。这些化学反应甚至更为复杂，因为当能量从碳水化合物发生氧化的源头向合成磷酸肌酸的体系传输时，似乎还有另一个反应参与其间，即三磷酸腺苷交替进行的分解与重新合成。

我们已经知道了肌肉中上述化学反应的发生顺序，并且建立起它们与收缩阶段和伸展阶段的时间关系。目前尚不清楚能量从一个反应体系向另一个反应体系传输

---

* 奥托·迈耶霍夫，《自然》，9 月 2 日第 337 页和 9 月 9 日第 373 页。

energy is transferred from one reacting system to another has until lately been obscure, but current work is throwing light upon this interesting question, and it is just beginning (though only beginning) to show how at the final stage the energy of the reactions is converted into the mechanical response. In parenthesis, it may be noted as an illustration of the unity of life that the processes which occur in the living yeast cell in its dealings with sugars are closely similar to those which proceed in living muscle. In the earlier stages they are identical and we now know where they part company. You may be astonished at the complexity of the events which underlie the activity of a muscle, but you must remember that it is a highly specialised machine. A more direct burning of the fuel could not fit into its complex organisation. I am more particularly concerned to feel that my brief summary of the facts will make clear how much more definite, how much more truly chemical, is our present knowledge than that available when Michael Foster wrote.

Ability to recognise the progress of such definite ordered chemical reactions in relation to various aspects of living activity characterises the current position in biochemistry. I have chosen the case of muscle, and it must serve as my only example, but many such related and ordered reactions have been studied in other cells and tissues, from bacteria to the brain. Some prove general, some more special. Although we are far indeed from possessing a complete picture in any one case, we are beginning in thought to fit not a few pieces together. We are on a line safe for progress.

I must perforce limit the field of my discussion, and in what follows, my special theme will be the importance of molecular structure in determining the properties of living systems. I wish you to believe that molecules display in such systems the properties inherent in their structure even as they do in the laboratory of the organic chemist. The theme is no new one, but its development illustrates as well as any other, and to my own mind perhaps better than any other, the progress of biochemistry.

Not long ago a prominent biologist, believing in protoplasm as an entity, wrote: "But it seems certain that living protoplasm is not an ordinary chemical compound, and therefore can have no molecular structure in the chemical sense of the word." Such a belief was common. One may remark, moreover, that when the development of colloid chemistry first brought its indispensable aid towards an understanding of the biochemical field, there was a tendency to discuss its bearing in terms of the less specific properties of colloid systems, phase-surfaces, membranes, and the like, without sufficient reference to the specificity which the influence of molecular structure, wherever displayed, impresses on chemical relations and events. In emphasising its importance, I shall leave no time for dealing with the nature of the colloid structures of cells and tissues, all important as they are. I shall continue to deal, though not again in detail, with chemical reactions as they occur within those structures. Only this much must be said. If the colloid structures did not display highly specialised molecular structure at their surface, no reactions would occur; for here catalysis occurs. Were it not equipped with catalysts every living unit would be a static system.

时所采用的方式，但是最近的研究为这个有趣的问题提供了解释，而这只是表明在最后阶段反应能量是如何转变成力学响应的开始（虽然仅仅是开始）。插一句话，我们也许可以从以下过程中发现生命活动的统一性，即糖在活酵母细胞中发生的反应与在活体肌肉中发生的反应是高度相似的。在更早的阶段中，它们是完全相同的，而且我们现在知道它们是在哪里分道扬镳的。你们可能会惊讶于肌肉活动机理的复杂性，但是你们别忘了，这是一种高度专门化的机械装置。燃料更直接的燃烧不可能适合于它的复杂组织形式。我非常希望我对于这些事实的简要概述能使大家清楚，从明确性和纯化学性角度考虑，我们现在的知识到底在多大程度上超越了迈克尔·福斯特写作时所能达到的认识。

我们认识这些明确而有序的化学反应与活体行为诸多方面的关联性的能力标志着生物化学当今的发展状况。我曾选择肌肉作为案例，它也必定会成为我仅有的例子，但是很多这样有关联又有序的化学反应都在包括从细菌到脑在内的其他细胞和组织中得到了研究。有些结果被证实是普适的，另外一些则比较特殊。尽管我们实际上远远不能完整地解释任何一个案例，但还是在考虑将为数不少的碎片拼合在一起。我们正处在一条可以稳步发展的路线上。

我必须强行限定自己的讨论范围，下面我要讲的是一个特殊的主题，即分子结构对于决定生命体系之性质的重要性。我希望你们能相信，分子在这些体系中所表现出来的性质是由它们的结构决定的，甚至就像它们在有机化学家的实验室中所表现的那样。这并不是一个新主题，但是它的发展和其他方面的发展一样能说明生物化学的进步，我个人认为，它的发展比其他方面的发展更能说明这种进步。

不久之前，一位深信原生质是一种实体的著名生物学家这样写道："但似乎可以确定的是，活的原生质并不是一种普通的化合物，因此它不可能具有化学意义上的分子结构。"这种观点是大家所熟悉的。此外，有人会说，当胶体化学的发展第一次使其将对生物化学领域的理解作为自己的重要目标时，就有一种倾向，即根据胶体体系、相表面和膜等不那么专业的特性讨论结果，而没有充分考虑一旦发挥作用就会影响化学关系及化学反应的分子结构的特殊性。由于要强调它的重要性，我也就留不出时间来讨论细胞和组织中胶体结构的本质了，尽管它们都是重要的。接着我将继续讨论发生在这些结构中的化学反应但不会再那么详细了。这就是全部必须要说的。如果胶体结构并未在其表面呈现出高度优化的分子结构，就不会有化学反应发生；因为催化作用是在这里进行的。如果不具备催化剂的话，每个活体单元都将是一个静止的体系。

It is well known that a catalyst is an agent which plays only a temporary part in chemical events which it nevertheless determines and controls. It reappears unaltered when the events are completed. The phenomena of catalysis, though first recognised early in the last century, entered but little into chemical thought or enterprise, until only a few years ago they were shown to have great importance for industry. Yet catalysis is one of the most significant devices of Nature, since it has endowed living systems with their fundamental character as transformers of energy, and all evidence suggests that it must have played an indispensable part in the living universe from the earliest stages of evolution.

The catalysts of a living cell are the enzymic structures which display their influences at the surface of colloidal particles or at other surfaces within the cell. Current research continues to add to the great number of these enzymes which can be separated from, or recognised in, living cells and tissues, and to increase our knowledge of their individual functions.

A molecule within the system of the cell may remain in an inactive state and enter into no reactions until at one such surface it comes in contact with an enzymic structure which displays certain adjustments to its own structure. While in such association, the inactive molecule becomes (to use a current term) "activated", and then enters on some definite path of change. The one aspect of enzymic catalysis which for the sake of my theme I wish to emphasise is its high specificity. An enzyme is in general adjusted to come into effective relations with one kind of molecule only, or at most with molecules closely related in their structure. Evidence based on kinetics justifies the belief that some sort of chemical combination between enzyme and related molecule precedes the activation of the latter, and for such combinations there must be close correlation in structure. Many will remember that long ago Emil Fischer recognised that enzymic action distinguishes even between two optical isomers and spoke of the necessary relation being as close as that of key and lock.

There is an important consequence of this high specificity in biological catalysis to which I invite special attention. A living cell is the seat of a multitude of reactions, and in order that it should retain in a given environment its individual identity as an organism, these reactions must be highly organised. They must be of determined nature and proceed mutually adjusted with respect to velocity, sequence, and in all other relations. They must be in dynamic equilibrium as a whole and must return to it after disturbance. Now if of any group of catalysts, such as are found in the equipment of a cell, each one exerts limited and highly specific influence, this very specificity must be a potent factor in making for organisation.

Consider the case of any individual cell in due relations with its environment, whether an internal environment as in the case of the tissue cells of higher animals, or an external environment as in the case of unicellular organisms. Materials for maintenance of the cell enter it from the environment. Discrimination among such materials is primarily determined by permeability relations, but of deeper significance in that selection is the

482

众所周知，催化剂是一种只暂时性地参与化学反应却对化学反应起着决定和控制作用的试剂。反应完成后，它又恢复了原状。尽管催化现象早在上个世纪就已经被首先认识到，但很少有人对它进行化学上的研究和探索，直到几年前人们发现催化对于工业的重要意义之后。催化是自然界最重要的机制之一，因为它把能量转换者这一基本特征赋予了生命体系，而且所有的证据都表明，它从进化的最早阶段起到现在一直在生命体系中起着不可或缺的作用。

活细胞的催化剂是酶结构，它在胶体颗粒表面或在细胞中的其他表面上发挥着自己的作用。当前的研究使我们可以从活细胞和组织中分离出来的酶或者可以在其中识别出来的酶的庞大数目继续增加，同时还增加了我们对单个酶的功能的认识。

在细胞系统内的一个分子可以保持惰性状态而不参与任何反应，直到它在某个表面上与一种表现出根据其自身结构作出相应调整的酶结构相接触。而在这种结合中，惰性分子变成（用一个现在的词）"活化的"，接着便按照某种确定的路径发生变化。为我的主题考虑，我希望强调的有关酶催化的一个特性是它的高度专一性。一般地说，一种酶会调整到只对唯一一类分子或至多对与它结构密切相关的一类分子具有有效联系的状态。基于动力学的证据使我们有理由相信，酶与相应分子之间的某种化学结合会先于后者的活化过程，并且这种结合是与结构密切相关的。很多人会记得，很久以前埃米尔·费希尔就认识到酶在发生作用时甚至可以区分出两种光学异构体的不同，并指出它们之间必定有如锁和钥匙般密切的关系。

我要提请大家特别注意的是，在生物催化中的这种高度专一性有一个重要后果。一个活细胞是很多反应的发生地，而为了使它在给定的环境之中保持作为一个有机体的个体性特征，这些反应必须是高度有组织的。它们必须具有确定的性质而且要能够针对速率、反应次序以及所有其他关系进行相互调整。它们必须作为一个整体处于动力学平衡之中，并且必须在受到干扰后恢复平衡。那么，如果现在有任何一组催化剂，比如在一个细胞的器件中所发现的催化剂，其中的每一个都发挥着有限的并且是高度专一性的作用，这种特别的专一性必定是导致组织化的有利因素。

无论是在高等动物组织细胞中的内环境，还是在单细胞生物中的外环境，都应考虑与所处环境具有适当关系的任意个体的情况。维持细胞所需的原料从环境中进入细胞。对这些物质的分辨主要是靠渗透性的不同，但是在这个选择的过程中更为重要的是细胞催化剂的专一性。人们经常说，活细胞与所有无生命系统的不同在于

specificity of the cell catalysts. It has often been said that the living cell differs from all non-living systems in its power of selecting from a heterogeneous environment the right material for the maintenance of its structure and activities. It is, however, no vital act but the nature of its specific catalysts which determines what it effectively "selects". If a molecule gains entry into the cell and meets no catalytic influence capable of activating it, nothing further happens save for certain ionic and osmotic adjustments. Any molecule which does meet an adjusted enzyme cannot fail to suffer change and become directed into some one of the paths of metabolism.

It must here be remembered, moreover, that enzymes as specific catalysts not only promote reactions, but also determine their direction. The glucose molecule, for example, though its inherent chemical potentialities are, of course, always the same, is converted into lactic acid by an enzyme system in muscle but into alcohol and carbon dioxide by another in the yeast cell. It is important to realise that diverse enzymes may act in succession and that specific catalysis has directive as well as selective powers. If it be syntheses in the cell which are most difficult to picture on such lines, we may remember that biological syntheses can be, and are, promoted by enzymes, and there are sufficient facts to justify the belief that a chain of specific enzymes can direct a complex synthesis along lines predetermined by the nature of the enzymes themselves. I should like to develop this aspect of the subject even further, but to do so might tax your patience. I should add that enzyme control, though so important, is not the sole determinant of chemical organisation in a cell. Other aspects of its colloidal structure play their part.

III

It is surely at that level of organisation, which is based on the exact coordination of a multitude of chemical events within it, that a living cell displays its peculiar sensitiveness to the influence of molecules of special nature when these enter it from without. The nature of very many organic molecules is such that they may enter a cell and exert no effect. Those proper to metabolism follow, of course, the normal paths of change. Some few, on the other hand, influence the cell in very special ways. When such influence is highly specific in kind, it means that some element of structure in the entrant molecule is adjusted to meet an aspect of molecular structure somewhere in the cell itself. We can easily understand that in a system so minute the intrusion even of a few such molecules may so modify existing equilibria as to affect profoundly the observed behaviour of the cell.

Such relations, though by no means confined to them, reach their greatest significance in the higher organisms, in which individual tissues, chemically diverse, differentiated in function and separated in space, so react upon one another through chemical agencies transmitted through the circulation as to coordinate by chemical transport the activities of the body as a whole. Unification by chemical means must today be recognised as a fundamental aspect of all such organisms. In all of them it is true that the nervous system has pride of place as the highest seat of organising influence, but we know today that

它有从异质环境中选取正确的物质以维持自身结构与活性的能力。不过，不是生命行为而是它的专属催化剂的性质决定了它将有效"选取"的对象。如果一个分子获准进入细胞，并且没有受到能活化它的催化过程的影响，那么除了对某些离子和渗透进行调节之外，不会发生什么。当遇到一个适当的酶时，任何分子都不能免于变化，从而进入到某一条新陈代谢的路径之中。

这里必须还要记得，酶作为专属催化剂不仅促进了反应的发生，还决定了反应进行的方向。例如葡萄糖分子，尽管其固有的化学反应能力通常应该是一样的，但是它通过肌肉中的酶系统就会转化成乳酸，而通过酵母细胞中的另一个酶系统则会转化成酒精和二氧化碳。重要的是要认识到不同的酶可以依次起作用，并且专属催化性不仅具有选择性还具有导向性。即使是对于最难用这些路线来描述的细胞内的合成，我们也要记得，生物合成可以并且事实上也是由酶促进的，而且有充分的事实可以证明下述观点，即由专属酶构成的一条链能够沿着由酶自身性质所预先决定的路线引导一个复杂的合成过程。我还想更进一步谈谈本主题的这个方面，但是这样做可能会让你们的耐心经受很大的考验。我要补充的是，尽管酶控制是如此重要，但它并不是细胞内化学组织化的唯一决定因素。细胞胶体结构中的其他方面也各有其作用。

III

可以确定，在基于细胞内诸多化学反应精确协作的组织水平之上，一个活细胞对从外部进入自身的具有特定性质的分子表现出独特的敏感性。而很多有机分子所具有的属性使它们可以进入一个细胞却不产生任何影响。那些专属于新陈代谢过程的分子当然会遵循正常的变化路径。另一方面，还有少数分子则以极为特殊的方式影响细胞。如果这种影响是高度专属性的，那就意味着新进入的分子中的某些结构特征发生了调整并与细胞自身某处的分子结构的某一特点相适应。我们可以很容易理解，在这样小的一个体系中，即便是少量这种分子的侵入也能改变现有的平衡，从而在很大程度上影响了我们所观测到的细胞的行为。

这些关系，尽管绝不仅限于它们，在高等有机体中具有最重要的意义；在高等有机体中，单个组织虽然化学组成不同，功能有所区别，在空间中也是分离的，却能通过在循环中传输的化学物质而彼此影响，从而通过化学运输协调机体作为一个整体的活动。我们今天必须认识到，统一借助化学手段是所有这一类有机体的一个根本方面。在它们之中都有如下事实，神经系统作为组织化影响的最高场所具有首要地位，但是今天我们知道，即使是这种影响也经常（如果并非总是如此的话）通

even this influence is often, if not always, exerted through properties inherent in chemical molecules. It is indeed most significant for my general theme to realise that when a nerve impulse reaches a tissue the sudden production of a definite chemical substance at the nerve ending may be essential to the response of that tissue to the impulse.

It is a familiar circumstance that when an impulse passes to the heart by way of the vagus nerve fibres the beat is slowed, or, by a stronger impulse, arrested. That is, of course, part of the normal control of the heart's action. Now it has been shown that, whenever the heart receives vagus impulses, the substance acetyl choline is liberated within the organ. To this fact is added the further fact that, in the absence of the vagus influence, the artificial injection of minute graded doses of acetyl choline so acts upon the heart as to reproduce in every detail the effects of graded stimulation of the nerve.

Moreover, evidence is accumulating to show that in the case of other nerves belonging to the same morphological group as the vagus, but supplying other tissues, this same liberation of acetyl choline accompanies activity, and the chemical action of this substance upon such tissues again produces effects identical with those observed when the nerves are stimulated. More may be claimed. The functions of another group of nerves are opposed to those of the vagus group; impulses, for example, through certain fibres accelerate the heart beat. Again a chemical substance is liberated at the endings of such nerves, and this substance has itself the property of accelerating the heart. We find then that such organs and tissues respond only indirectly to whatever non-specific physical change may reach the nerve ending. Their direct response is to the influence of particular molecules with an essential structure, when these intrude into their chemical machinery.

It follows that the effect of a given nerve stimulus may not be confined to the tissue which it first reaches. There may be humeral transmissions of its effect, because the liberated substance enters the lymph and blood. This again may assist the coordination of events in the tissues.

From substances produced temporarily and locally and by virtue of their chemical properties translating for the tissues the messages of nerves, we may pass logically to consideration of those active substances which carry chemical messages from organ to organ. Such in the animal body are produced continuously in specialised organs, and each has its special seat or seats of action where it finds chemical structures adjusted in some sense or other to its own.

I shall here be on familiar ground, for that such agencies exist, and bear the name of hormones, is common knowledge. I propose only to indicate how many and diverse are their functions as revealed by recent research, emphasising the fact that each one is a definite and relatively simple substance with properties that are primarily chemical and in a derivate sense physiological. Our clear recognition of this, based at first on a couple of instances, began with this century, but our knowledge of their number and nature is still growing rapidly today.

486

过化学分子所固有的性质显现出来。对于我的一般主题来说，实际上最重要的是意识到，当一个神经脉冲到达组织时，位于神经末端处的一种确定化学物质的突然生成对于该组织对冲动的反应来说可能是至关重要的。

下面的情况是常见的：当一个脉冲通过迷走神经纤维到达心脏时，心跳减慢，或者因更强的脉冲而停止。当然，这是心脏动作正常控制的一部分。现在发现，只要心脏接收到迷走神经脉冲，乙酰胆碱这种物质就会在器官中被释放出来。进一步的事实对此给出了佐证，在没有迷走神经影响时，人工注射分级的小剂量乙酰胆碱也能对心脏起作用，以至于在每个细节上都能重现对神经进行分级刺激的效果。

此外，逐渐累积的证据表明，在与迷走神经属于相同形态族但供应其他组织的其他神经的情况下，活动时也同样会释放出乙酰胆碱，而且这种物质对于那些组织的化学作用会产生与神经在接受刺激时所观测到的效应相同的结果。还有一些可以说明的是，另外一组神经的功能与迷走神经族相对立；例如，通过某些纤维的脉冲加快了心脏跳动。也有一种化学物质在这些神经的末端被释放出来，而且该物质自身具有加速心跳的性质。接着我们发现，这些器官和组织对于凡是能到达神经末端的非专属性物理变化只有间接的反应。它们会在侵入者进入其化学系统时对具有基本结构的特定分子所造成的影响作出直接的反应。

由此可知，一种给定的神经刺激的作用可以不限于它最先到达的组织。它可能会传到肩部，因为释放出来的物质进入了淋巴和血液。这也会有利于组织内各种反应的协作。

根据临时在小范围内生成的物质和它们所具有的为组织转译神经信息的化学性质，我们可以合理地推测出那些携带着化学信息从一个器官到另一个器官的活性物质。在动物体内，它们在专门的器官中持续地产生，并且都有各自专属的一个或多个作用位点，在作用位点上它们能找到以这样或那样的方式适合于自身的化学结构。

下面我将来到熟悉的领域中，因为这种物质的存在以及被称为激素是大家都知道的常识。我只想说明最近的研究揭示出它们有多少种不同的新功能，并强调下面的事实，即每一种都是确定的并且相当简单的物质，具有主要是化学的而在衍生意义上还是生理学的性质。我们从本世纪就开始对这一点有了清晰的认识，最初的根据来源于一组实例，但我们关于它们的数量与性质的知识直到今天还在快速增长。

We have long known, of course, how essential and profound is the influence of the thyroid gland in maintaining harmonious growth in the body, and in controlling the rate of its metabolism. Three years ago a brilliant investigation revealed the exact molecular structure of the substance—thyroxin—which is directly responsible for these effects. It is a substance of no great complexity. The constitution of adrenalin has been longer known and likewise its remarkable influence in maintaining a number of important physiological adjustments. Yet it is again a relatively simple substance. I will merely remind you of secretin, the first of these substances to receive the name of hormone, and of insulin, now so familiar because of its importance in the metabolism of carbohydrates and its consequent value in the treatment of diabetes. The most recent growth of knowledge in this field has dealt with hormones which, in most remarkable relations, coordinate the phenomena of sex.

It is the circulation of definite chemical substances produced locally that determines, during the growth of the individual, the proper development of all the secondary sexual characters. The properties of other substances secure the due progress of individual development from the unfertilised ovum to the end of foetal life. When an ovum ripens and is discharged from the ovary a substance, now known as oestrin, is produced in the ovary itself, and so functions as to bring about all those changes in the female body which make secure the fertilisation of the ovum. On the discharge of the ovum new tissue, constituting the so-called *corpus luteum*, arises in its place. This then produces a special hormone which in its turn evokes all those changes in tissues and organs that secure a right destiny for the ovum after it has been fertilised. It is clear that these two hormones do not arise simultaneously, for they must act in alternation, and it becomes of great interest to know how such succession is secured. The facts here are among the most striking. Just as higher nerve centres in the brain control and coordinate the activities of lower centres, so it would seem do hormones, functioning at, so to speak, a higher level in organisation, coordinate the activities of other hormones. It is a substance produced in the anterior portion of the pituitary gland situated at the base of the brain which, by circulating to the ovary, controls the succession of its hormonal activities. The cases I have mentioned are far from exhausting the numerous hormonal influences now recognised.

For full appreciation of the extent to which chemical substances control and coordinate events in the animal body by virtue of their specific molecular structure, it is well not to separate too widely in thought the functions of hormones from those of vitamins. Together they form a large group of substances of which every one exerts upon physiological events its own indispensable chemical influence.

Hormones are produced in the body itself, while vitamins must be supplied in the diet. Such a distinction is, in general, justified. We meet occasionally, however, an animal species able to dispense with an external supply of this or that vitamin. Evidence shows, however, that individuals of that species, unlike most animals, can in the course of their metabolism synthesise for themselves the vitamin in question. The vitamin then becomes a hormone. In practice the distinction may be of great importance, but for an understanding of

　　当然，我们很久以前就知道，甲状腺对于保持机体内激素数量以及控制机体新陈代谢速率有着多么重要而深刻的影响。三年前，一项出色的研究揭示了甲状腺素的精确分子结构，该物质是出现这种效应的直接原因。这是一种并不怎么复杂的物质。我们在更早以前就已了解到肾上腺素的组成以及它对于维持多种重要生理调节过程的显著影响。但是它也是一种相当简单的物质。我只需再提醒你们注意的是分泌素，它是第一种获得激素之名的物质，还有胰岛素，我们现在对它如此熟悉是因为它在碳水化合物代谢过程中的重要性以及随之产生的在糖尿病治疗中的价值。关于这个领域的最新研究进展体现在对与协调性别现象密切相关的激素的研究上。

　　局部生成的确定的化学物质的循环在个体生长过程中决定了所有第二性征的适当发育。其他物质的性质保证了从未受精卵到胎儿期末期之间个体发育的应有进度。当一个卵子成熟并且从卵巢中释放出来时，一种现在我们称为雌激素的物质在卵巢内生成，并且发挥作用，导致能确保卵子受精的雌性机体出现所有相应的变化。随着卵子的释放，由通常所说的**黄体**组成的新组织在自己的位置上出现。接着，这里产生出一种特殊的激素，它的任务是唤起组织和器官中的所有相关变化，以保证了卵子在受精后的正确命运。很显然，这两种激素不会同时出现，因为它们必须轮流起作用，所以知道如何才能保证这种交替就成了一件非常值得关注的事情。此中真相是极为令人吃惊的。正如脑中的高级神经中枢控制并协调低级中枢的活动一样，可以这么说，在高级组织水平上起作用的激素也会这样去协调其他激素的活动。由位于脑基部的脑垂体的前部所产生的一种物质，通过到卵巢的循环，控制着卵巢激素活性的交替。我所提到的这些例子远没有穷尽现在已认识到的数量繁多的激素的作用。

　　要完整把握化学物质通过其特有的分子结构来控制与协调动物体内发生响应的程度，最好在考虑激素的功能时不要把它与维生素的功能过分地割裂开。它们一起构成了一大类物质，其中每一种都通过化学作用对生理反应施加着自己必不可少的影响。

　　激素是机体自身产生的，而维生素则必须要靠饮食来供应。这种区分方式在通常情况下是合理的。不过，偶尔我们也会遇到某种动物能够不需要外部供应这种或那种维生素。但是有证据表明，与大多数动物不同，这类物种的个体能够在新陈代谢过程中为自己合成该维生素。合成的维生素随即变成一种激素。在实践中这种区分可能是有重要意义的，但是从理解新陈代谢的角度来看，这些物质的功能比它们

metabolism the functions of these substances are of more significance than their origin.

The present activity of research in the field of vitamins is prodigious. The output of published papers dealing with original investigations in the field has reached nearly a thousand in a single year. Each of the vitamins at present known is receiving the attention of numerous observers in respect both of its chemical and biological properties, and though many publications deal, of course, with matters of detail, the accumulation of significant facts is growing fast.

It is clear that I can cover but little ground in any reference to this wide field of knowledge. Some aspects of its development have been interesting enough. The familiar circumstance that attention was directed to the existence of one vitamin (B_1 so called) because populations in the East took to eating milled rice instead of the whole grain; the gradual growth of evidence which links the physiological activities of another vitamin (D) with the influence of solar radiation on the body, and has shown that they are thus related, because rays of definite wave-length convert an inactive precursor into the active vitamin, alike when acting on foodstuffs or on the surface of the living body; the fact again that the recent isolation of vitamin C, and the accumulation of evidence for its nature started from the observation that the cortex of the adrenal gland displayed strongly reducing properties; or yet again the proof that a yellow pigment widely distributed among plants, while not the vitamin itself, can be converted within the body into vitamin A; these and other aspects of vitamin studies will stand out as interesting chapters in the story of scientific investigation.

In this very brief discussion of hormones and vitamins I have so far referred only to their functions as manifested in the animal body. Kindred substances, exerting analogous functions, are, however, of wide and perhaps of quite general biological importance. It is certain that many microorganisms require a supply of vitamin-like substances for the promotion of growth, and recent research of a very interesting kind has demonstrated in the higher plants the existence of specific substances produced in special cells which stimulate growth in other cells, and so in the plant as a whole. These so-called auxines are essentially hormones.

It is of particular importance to my present theme and a source of much satisfaction to know that our knowledge of the actual molecular structure of hormones and vitamins is growing fast. We have already exact knowledge of the kind in respect to not a few. We are indeed justified in believing that within a few years such knowledge will be extensive enough to allow a wide view of the correlation between molecular structure and physiological activity. Such correlation has long been sought in the case of drugs, and some generalisations have been demonstrated. It should be remembered, however, that until quite lately only the structure of the drug could be considered. With increasing knowledge of the tissue structures, pharmacological actions will become much clearer.

I cannot refrain from mentioning here a set of relations connected especially with the

的来源更要紧。

目前在维生素这一领域中的研究活动多得惊人。本领域中原创性研究论文发表量已达到每年约一千篇。目前已知的每一种维生素都得到了大量研究者的关注，关注点既包括其化学性质也包括生物学性质，而且，尽管许多出版物中所讨论的无疑都是一些细节问题，但重要事实的积聚量仍在迅猛增长。

很明显，关于这一宽广的知识领域，我只能够谈到其中很少的一部分。某些发展方向也一直是颇为有趣的。下列情况是我们所熟悉的：人们开始关注一种维生素（被称为 B_1）的存在，这是因为东方人习惯于吃去壳的而不是整粒的米；证明另一种维生素（D）的生理学活性与太阳辐射对机体的影响有关联的证据逐渐增多，这些证据说明它们之间确实是相关的，因为特定波长的射线可以将一种惰性前体转化成活性的维生素，作用于食物就如同作用于生物体表面一样；还有最近对维生素 C 的分离，人们从观察到肾上腺皮层呈现出强烈的还原性质开始不断积累有关维生素 C 性质方面的证据；或者还有下面的证据，有一种广泛分布于植物之中的黄色色素，虽然自身不是维生素，却能在体内转化成维生素 A；维生素研究中的这些和那些方面将会构成科学探索故事中趣味十足的篇章。

以上是我对激素和维生素的简短评述，到目前为止我只谈到了它们在动物体内所表现出来的功能。但是，能实施类似功能的同类物质是很多的，而且可能具有更为普遍的生物学价值。可以肯定，很多种微生物需要类似于维生素的物质以促进其生长发育，最近，有一类极为有趣的研究表明，高等植物中存在着某些由特殊细胞产生的能刺激其他细胞的生长，从而也能促进植物作为一个整体的生长的特殊物质。这种物质被称为生长素，本质上就是激素。

知道下面这件事对于我当前的主题来说格外重要，它也是我们得到极大满足的原因：我们对激素与维生素的实际分子结构的了解正在快速增多。我们已经对为数不少的物质有了确切的了解。实际上，我们有理由相信，在几年之内，这些知识就会拓展到足以使我们充分了解分子结构与生理学活性之间关系的程度。多年以来，人们一直在药物研究中寻找着这样的关联，而且某些推论也已得到阐明。但是应该记得，直到最近的时候，我们仍然只能考虑药物的结构。随着对组织结构的认识不断深入，药理作用将会逐渐变得更加清晰。

我在这里不可避免地要谈到一组与受到格外关注的组织发育现象有特殊联系

phenomena of tissue growth which are of particular interest. It will be convenient to introduce some technical chemical considerations in describing them, though I think the relations may be clear without emphasis being placed on such details. The vitamin, which in current usage is labelled "A" , is essential for the general growth of an animal. Recent research has provided much information as to its chemical nature. Its molecule is built up of units which possess what is known to chemists as the isoprene structure. These are condensed in a long carbon chain which is attached to a ring structure of a specific kind. Such a constitution relates it to other biological compounds, in particular to certain vegetable pigments, one of which, a carotene, so called, is the substance which I have mentioned as being convertible into the vitamin. For the display of an influence upon growth, however, the exact details of the vitamin's proper structure must be established.

Now turning to vitamin D, of which the activity is more specialised, controlling as it does the growth of bone in particular, we have learnt that the unit elements in its structure are again isoprene radicals; but instead of forming a long chain as in vitamin A they are united into a system of condensed rings. Similar rings form the basal components of the molecules of sterols, substances which are normal constituents of nearly every living cell. It is one of these, inactive itself, which ultra-violet radiation converts into vitamin D. We know that each of these vitamins stimulates growth in tissue cells.

Next consider another case of growth stimulation, different because pathological in nature. It is well known that long contact with tar induces a cancerous growth of the skin. Very important researches have recently shown that particular constituents in the tar are alone concerned in producing this effect. It is being further demonstrated that the power to produce cancer is associated with a special type of molecular structure in these constituents. This structure, like that of the sterols, is one of condensed rings, the essential difference being that (in chemical language) the sterol rings are hydrogenated, whereas those in the cancer-producing molecules are not. Hydrogenation indeed destroys the activity of the latter. Recall, however, the ovarian hormone oestrin. Now the molecular structure of oestrin has the essential ring structure of a sterol, but one of the constituent rings is not hydrogenated. In a sense, therefore, the chemical nature of oestrin links vitamin D with that of cancer-producing substances. Further, it is found that substances with pronounced cancer-producing powers may produce effects in the body like those of oestrin.

It is difficult when faced with such relations not to wonder whether the metabolism of sterols, which when normal can produce a substance stimulating physiological growth, may in very special circumstances be so perverted as to produce within living cells a substance stimulating pathological growth. Such a suggestion must, however, with present knowledge, be very cautiously received. It is wholly without experimental proof. My chief purpose in this reference to this very interesting set of relations is to emphasise once more the significance of chemical structure in the field of biological events.

Only the end results of the profound influence which minute amounts of substances with

的关系。用一些技术性化学的知识来描述它们会很便利，尽管我认为不强调这些细节该关系也可以是清晰的。当前被标记为"A"的维生素对于动物的一般发育是至关重要的。最近的研究提供了大量与这种维生素的化学本质相关的信息。它的分子由具有化学家们已知的异戊二烯结构的单元搭建而成。它们聚集在与一种特殊类型的环结构相连的长碳链上。这样的结构使它与其他一些生物化合物相关联，尤其是与某些植物色素相关联，其中一种被称为胡萝卜素，是我曾提到过的可以转化为维生素的物质。但是，要说明它对发育过程的影响，就必须了解维生素合理结构的精确细节。

现在转向维生素 D，它的作用更具专属性，尤其是在控制骨骼发育时更是如此，我们已经了解到它的结构单元也是异戊二烯基；但是不像在维生素 A 中那样形成一个长链，而是结合成一个稠环体系。类似的环构成了甾醇分子的基本结构，后者几乎是每个活细胞中的正常组分。甾醇是一种自身无活性但是能被紫外辐射转化为维生素 D 的物质。我们知道，这些维生素中的每一种都会刺激组织细胞的生长。

接下来考虑另一个刺激发育的例子，不同之处在于它的本质是病理学的。众所周知，长期与焦油接触会导致皮肤癌变的扩大。最近有一些极为重要的研究表明，焦油中的特定组分能独立地产生这种效果。进一步的结果显示，致癌能力与这些组分中的一种特殊类型的分子结构有关。这种结构与甾醇的结构相似，是一种稠环结构，其根本差别在于（用化学语言来说）甾醇环是氢化了的，而致癌分子中的环结构则不是。氢化作用实际上破坏了后者的活性。但是，回忆一下卵巢激素，雌激素。虽然雌激素的分子结构中含有甾醇的基本环结构，但是其结构中的一个环是没有氢化的。因此，在某种意义上，雌激素的化学本质将维生素 D 与致癌物质的化学本质联系起来。此外，目前已发现，具有显著致癌能力的物质在体内可以产生与雌激素相似的效果。

当面对这种关系时，很难不去猜测在正常情况下可以产生一种刺激生理发育的物质的甾醇代谢，是否会在极为特殊的环境中异常到足以使活细胞内产生一种刺激病变发育的物质。不过，用现在的知识来看，这样一种提法一定要十分谨慎地对待。它完全没有实验上的证据。我在此谈到这组极为有趣的关系的主要目的，是想再次强调化学结构在生物学活动领域中的重要性。

只有当具有适当结构的少量物质为活细胞和组织带来很大的影响时，我们才能

adjusted structure exert upon living cells or tissues can be observed in the intact bodies of man or animals. It is doubtless because of the elaborate and sensitive organisation of chemical events in every tissue cell that the effects are proportionally so great.

It is an immediate task of biochemistry to explore the mechanism of such activities. It must learn to describe in objective chemical terms precisely how and where such molecules as those of hormones and vitamins intrude into the chemical events of metabolism. It is indeed now beginning this task, which is by no means outside the scope of its methods. Efforts of this and of similar kind cannot fail to be associated with a steady increase in knowledge of the whole field of chemical organisation in living organisms, and to this increase we look forward with confidence. The promise is there. Present methods can still go far, but I am convinced that progress of the kind is about to gain great impetus from the application of those new methods of research which chemistry is inheriting from physics: X-ray analysis; the current studies of unimolecular surface films and of chemical reactions at surfaces; modern spectroscopy; the quantitative developments of photo-chemistry; no branch of inquiry stands to gain more from such advances in technique than does biochemistry at its present stage. Especially is this true in the case of the colloidal structure of living systems, of which in this address I have said so little.

IV

As an experimental science, biochemistry, like classical physiology, and much of experimental biology, has obtained, and must continue to obtain, many of its data from studying parts of the organism in isolation, but parts in which dynamic events continue. Though fortunately it has also methods of studying reactions as they occur in intact living cells, intact tissues, and, of course, in the intact animal, it is still entitled to claim that its studies of parts are consistently developing its grasp of the wholes it desires to describe, however remote that grasp may be from finality. Justification for any such claim has been challenged in advance from a certain philosophic point of view. Not from that of General Smuts, though in his powerful address which signalised our centenary meeting he, like many philosophers today, emphasised the importance of properties which emerge from systems in their integrity, bidding us remember that a part while in the whole is not the same as the part in isolation. He hastened to admit in a subsequent speech, however, that for experimental biology, as for any other branch of science, it is logical and necessary to approach the whole through its parts. Nor again is the claim challenged from the point of view of such a teacher as A. N. Whitehead, though in his philosophy of organic mechanism there is no real entity of any kind without internal and multiple relations, and each whole is more than the sum of its parts. I nevertheless find *ad hoc* statements in his writings which directly encourage the methods of biochemistry.

In the teachings of J. S. Haldane, however, the value of such methods have long been directly challenged. Some will perhaps remember that in an address to Section I, twenty-five years ago, he described a philosophic point of view which he has courageously maintained in many writings since. Dr. Haldane holds that to the enlightened biologist a

在完好无损的人或动物体内观测到最终的结果。毫无疑问，由于每个组织细胞内的化学反应具有精细且敏感的组织性，所以效果相应地也会非常显著。

生物化学的一项当务之急是探索这些活动的机制。必须要学会用客观的化学术语精确地描述维生素和激素等分子是如何以及从哪里闯入新陈代谢的化学过程的。事实上人们正在开展这方面的研究，但所使用的方法绝对没有超出这项研究本身的范畴。这方面的工作以及类似的其他工作不可避免地会和活有机体中整个化学组织领域知识的稳定增长相关，我们对这一增长的前景充满信心。这是我们的承诺。目前的方法还能更进一步，但是我确信，如果化学可以从物理学中借鉴一些方法，那么对这些新研究方法的运用将会极大地推动该项工作的进展，这些方法是：X 射线分析；当前对于单一分子表面薄膜和在表面发生的化学反应的研究；现代光谱学；光化学的定量研究。没有一个研究分支会比现阶段的生物化学从这些技术进步中获得的收益更多。尤其是对于生物系统的胶体结构更是如此，关于这一点我在本次演讲中很少涉及。

IV

作为一门实验科学，生物化学类似于古典生理学，并在很大程度上与实验生物学一样，其已经获得的大量数据均来自对有机体各分离部分的研究，而不是动力学反应继续进行的那些部分，并且必将继续通过这种方式获得大量属于自己的数据。尽管幸运的是，还有一些方法可以研究发生在完好无损的活细胞、完好无损的组织，以及完好无损的动物体内的反应，但是我们仍然有权宣称，生物化学对局部的研究仍在不断地扩充着它对想要去描述的整体的认识，无论这种掌握距离最终目的会多么遥远。类似于这样的说法曾经受到过某种哲学观点的抨击。这并非来自斯穆茨将军，不过在那次庆祝我们协会成立一百周年会议上所发表的强有力的演讲中，他像很多今天的哲学家一样，强调系统在处于完整状态时所具有的性质的重要性，他要我们记住，一个部分在处于整体中时并不等同于被分离开时的那个部分。不过，他很快便在后来的一次演讲中承认，对于实验生物学来说，如同科学的任何其他分支一样，通过部分到达整体是合乎逻辑的也是必需的。对这种说法的抨击也并非来自像怀特海那样的一位教师的观点，尽管在他的"有机哲学"中不存在任何一类没有内在和多重关系的真正的实体，而且每个整体都大于其各部分之和。但我还是在他的著作中找到了直接鼓励生物化学方法的专门陈述。

但是，在霍尔丹的教诲中，这种方法的价值却一直受到直接的质疑。可能有人会记得，在 25 年前的一次面向 I 分会的演讲中，他描述了一种其后他在多部著述中仍勇敢坚持的哲学观点。霍尔丹博士坚持认为，对于一位开明的生物学家，一个活

living organism does not present a problem for analysis; it is, *qua* organism, axiomatic. Its essential attributes are axiomatic; heredity, for example, is for biology not a problem but an axiom. "The problem of Physiology is not to obtain piecemeal physico-explanations of physiological processes" (I quote from the 1885 address), "but to discover by observation and experiment the relatedness to one another of all the details of structure and activity in each organism as expressions of its nature as one organism".

I cannot pretend adequately to discuss these views here. They have often been discussed by others, not always perhaps with understanding. What is true in them is subtle, and I doubt if their author has ever found the right words in which to bring to most others a conviction of such truth. It is involved in a world outlook. What I think is scientifically faulty in Haldane's teaching is the a priori element which leads to bias in the face of evidence. The task he sets for the physiologist seems vague to most people, and he forgets that with good judgment a study of parts may lead to an intellectual synthesis of value. In 1885 he wrote: "That a meeting-point between Biology and Physical Science may at some time be found there is no reason for doubting. But we may confidently predict that if that meeting-point is found, and one of the two sciences is swallowed up, that one will not be Biology." He now claims indeed that biology has accomplished the heavy meal, because physics has been compelled to deal no longer with Newtonian entities but, like the biologist, with organisms such as the atom proves to be. Is it not then enough for my present purpose to remark on the significance of the fact that not until certain atoms were found spontaneously splitting piecemeal into parts, and others were afterwards so split in the laboratory, did we really know anything about the atom as a whole.

At this point, however, I will ask you not to suspect me of claiming that all the attributes of living systems or even the more obvious among them are necessarily based upon chemical organisation alone. I have already expressed my own belief that this organisation will account for one striking characteristic of every living cell—its ability, namely, to maintain a dynamic individuality in diverse environments. Living cells display other attributes even more characteristic of themselves; they grow, multiply, inherit qualities and transmit them. Although to distinguish levels of organisation in such systems may be to abstract from reality, it is not illogical to believe that such attributes as these are based upon organisation at a level which is in some sense higher than the chemical level. The main necessity from the point of view of biochemistry is then to decide whether nevertheless at its own level, which is certainly definable, the results of experimental studies are self-contained and consistent. This is assuredly true of the data which biochemistry is now acquiring. Never during its progress has chemical consistency shown itself to be disturbed by influences of any ultra-chemical kind.

Moreover, before we assume that there is a level of organisation at which chemical controlling agencies must necessarily cease to function, we should respect the intellectual parsimony taught by Occam and be sure of their limitations before we seek for super-chemical entities as organisers. There is no orderly succession of events which would seem less likely to be controlled by the mere chemical properties of a substance than the

的有机体不会呈现出任何可供分析的问题；作为有机体，这是不言自明的。它的基本属性是自明的；例如，遗传对于生物学来说不是一个问题而是一个公理。"解决生理学问题不在于获得对生理过程的支离破碎的物理解释"（引自 1885 年的演讲），"而在于通过观察和实验来发现每个有机体中的结构与活动之全部细节彼此间的关联性，作为一个有机体来描述它的性质"。

我不能如此斗胆在这里讨论这些观点。它们经常为其他人所论及，也许不总是在足够理解的情况下。其中的真实部分是微妙的，而且我怀疑提出上述观点的人是否曾经找到过恰当的字眼来使大多数其他人确信它的真实性。它涉及世界观的问题。我认为在霍尔丹的教论中所存在的在科学意义上有缺陷的内容，正是面对证据时导致偏见的先验因素。他为生理学家安排的任务对于大多数人来说似乎是含糊的，而且他忘记了，基于准确的判断，对于局部的研究可以合乎理性地拼合成有价值的信息。在 1885 年他写道："可能在某些时候会发现生物学与物理科学之间的交汇点是无可置疑的。但是我们可以满怀信心地预言，如果找到了那个交汇点，并且两门科学中的一门要被吞并掉，那么被吞并的将不会是生物学。"现在他宣称，生物学实际上已经啃掉了硬骨头，因为物理学已经被迫不再处理牛顿式实体，而是像生物学家一样，处理像原子一样的有机体。我现在还不足以去评论这件事的重要意义，只有在我们发现某些原子能够自发破裂成几个部分，随后发现另外一些原子在实验室中也能这样分裂之后，我们才能真正了解到原子作为一个整体的所有情况。

然而就这一点而言，我希望你们不要猜测我会宣称：生物系统的所有属性乃至于其中较为明显的部分必定只能建立在化学组织的基础之上。我已经表达过我自己的观点，化学组织足以解释每个活细胞所具有的惊人特性——即在各种环境中保持其动态个性的能力。活细胞还呈现出其他更具自身特性的属性；它们生长、繁殖、继承品性并将其传承下去。尽管要在这样的系统中区分组织水平可能要对现实加以抽象，但我们仍然有理由相信，这些属性是基于在某种意义上高于化学水平的组织水平之上的。于是从生物化学的角度来看，主要任务就是要确定实验研究结果是否在其自身水平之上仍然是自成体系和前后一致的，这一点显然可以说清楚。生物化学现在所获得的数据肯定满足这一条件。化学一致性在其发展过程中从未表现出它会受到任何超化学因素的干扰。

此外，在我们假定存在某种高级别的组织使得化学控制试剂必定会停止作用之前，我们应该遵从奥卡姆关于理智节俭性的教诲，并且在寻找超化学实体作为组织者之前先要明确它的局限。在所有有序的连续变化中，介于受精卵与完整胚胎之间的细胞分裂与细胞分化最不可能只被一种物质的化学性质所控制。不过看起来，一

cell divisions and cell differentiation which intervene between the fertilised ovum and the finished embryo. Yet it would seem that a transmitted substance, a hormone in essence, may play an unmistakable part in that remarkable drama. It has for some years been known that, at an early stage of development, a group of cells forming the so-called "organiser" of Spemann induces the subsequent stages of differentiation in other cells. The latest researches seem to show that a cellfree extract of this "organiser" may function in its place. The substance concerned is, it would seem, not confined to the "organiser" itself, but is widely distributed outside, though not in, the embryo. It presents, nevertheless, a truly remarkable instance of chemical influence.

It would be out of place in such a discourse as this to attempt any discussion of the psychophysical problem. However much we may learn about the material systems which, in their integrity, are associated with consciousness, the nature of that association may yet remain a problem. The interest of that problem is insistent and it must be often in our thoughts. Its existence, however, justifies no pre-judgments as to the value of any knowledge of a consistent sort which the material systems may yield to experiment.

V

It has become clear, I think, that chemical modes of thought, whatever their limitation, are fated profoundly to affect biological thought. If, however, the biochemist should at any time be inclined to overrate the value of his contributions to biology, or to underrate the magnitude of problems outside his province, he will do well sometimes to leave the laboratory for the field, or to seek even in the museum a reminder of that infinity of adaptations of which life is capable. He will then not fail to work with a humble mind, however great his faith in the importance of the methods which are his own.

It is surely right, however, to claim that in passing from its earlier concern with dead biological products to its present concern with active processes within living organisms, biochemistry has become a true branch of progressive biology. It has opened up modes of thought about the physical basis of life which could scarcely be employed at all a generation ago. Such data and such modes of thought as it is now providing are pervasive, and must appear as aspects in all biological thought. Yet these aspects are, of course, only partial. Biology in all its aspects is showing rapid progress, and its bearing on human welfare is more and more evident.

Unfortunately, the nature of this new biological progress and its true significance is known to but a small section of the lay public. Few will doubt that popular interest in science is extending, but it is mainly confined to the more romantic aspects of modern astronomy and physics. That biological advances have made less impression is probably due to more than one circumstance, of which the chief, doubtless, is the neglect of biology in our educational system. The startling data of modern astronomy and physics, though of course only when presented in their most superficial aspects, find an easier approach to the uninformed mind than those of the new experimental biology can hope for. The primary

种本质上属于激素的传导物质可以在这出引人关注的大戏中扮演明白无误的角色。若干年前我们就知道，在早期的生长阶段，一组被施佩曼称为"组织者"的细胞能诱导其他细胞进入分化的后续阶段。最近的研究似乎表明，从无细胞的提取物中也可以得到能实现同样功能的"组织者"。看起来，我们所说的这种物质并不只限于"组织者"本身，而是广泛地分布在胚胎的外部而不是内部。不过，它确实仍是化学影响中的一个典型例子。

在这样的演讲中尝试对心理物理学问题进行讨论恐怕是不恰当的。无论我们对于这个在其完整状态时与意识相关的物质系统了解多少，这种关系的本质可能仍然是一个问题。对这一问题的关注是持久的，并且必定会经常出现在我们的思考中。但是，它的存在使我们不能预先判断该物质系统可以给实验提供的一致性知识的价值。

V

我认为，现在已经清楚，无论其局限在哪里，化学的思维模式都注定要深刻地影响生物学的思维。但是，如果生物化学家要在任何时候都倾向于高估他对生物学贡献的价值，或者低估自己所在领域之外的问题的重要性，那他不如为了这个领域暂时离开实验室，或者索性到博物馆去寻找关于生命所能具有的无限适应性的提示。这样他就会以谦卑的心态来工作，不管他认为自己的方法有多么重要。

但是，下面的宣称肯定是正确的：通过从早期关注死的生物产物过渡到当前关注活有机体内的活动过程，生物化学已经真正成为前进中的生物学的一个分支。它开启了在一代人以前还根本无法想象的关于生命具有物理基础的新型思维模式。这样的数据和它现在所提供的这种思维模式是渗透性的，而且必定会在生物学研究的所有方面都呈现。当然，目前这些方面中还只有部分受到了影响。生物学的所有方面都显示出快速的发展趋势，而且它对于人类福利的意义也越来越明显。

遗憾的是，这些生物学新进展的本质及其真正意义只为一小部分一般公众所认识。很少有人会怀疑，大众对科学的兴趣日益增长，但却主要局限于现代天文学与物理学中更为浪漫的方面。生物学进展的影响较小，这可能是由多种原因造成的，其中最主要的原因无疑是我们的教育系统对于生物学的漠视。现代天文学和物理学为未受教育的人了解它们的惊人成就找到了一条更为省力的方式，尽管肯定只能限于最表层的一些方面，但这也是新兴实验生物学所望尘莫及的。与之相互矛盾的是，现代天文学和物理学中涉及的主要概念却不那么令人熟悉。此外，作家们一直在以

concepts involved are paradoxically less familiar. Modern physical science, moreover, has been interpreted to the intelligent public by writers so brilliant that their books have had a great and stimulating influence.

Lord Russell once ventured on the statement that in passing from physics to biology one is conscious of a transition from the cosmic to the parochial, because from a cosmic point of view life is a very unimportant affair. Those who know that supposed parish well are convinced that it is rather a metropolis entitled to much more attention than it sometimes obtains from authors of guide-books to the universe. It may be small in extent, but is the seat of all the most significant events. In too many current publications, purporting to summarise scientific progress, biology is left out or receives but scant reference. Brilliant expositions of all that may be met in the region where modern science touches philosophy have directed thought straight from the implications of modern physics to the nature and structure of the human mind, and even to speculation concerning the mind of the Deity. Yet there are aspects of biological truth already known which are certainly germane to such discussions, and probably necessary for their adequacy.

VI

It is, however, because of its extreme importance to social progress that public ignorance of biology is especially to be regretted. Sir Henry Dale has remarked that "it is worth while to consider today whether the imposing achievements of physical science have not already, in the thought and interests of men at large, as well as in technical and industrial development, overshadowed in our educational and public policy those of biology to an extent which threatens a one-sided development of science itself and of the civilisation which we hope to see based on science". Sir Walter Fletcher, whose death during the past year has deprived the nation of an enlightened adviser, almost startled the public, I think, when he said in a national broadcast that "we can find safety and progress only in proportion as we bring into our methods of statecraft the guidance of biological truth". That statecraft, in its dignity, should be concerned with biological teaching, was a new idea to many listeners.

A few years ago the Cambridge philosopher, Dr. C. D. Broad, who is much better acquainted with scientific data than are many philosophers, remarked upon the misfortune involved in the unequal development of science; the high degree of our control over inorganic Nature combined with relative ignorance of biology and psychology. At the close of a discussion as to the possibility of continued mental progress in the world, he summed up by saying that the possibility depends on our getting an adequate knowledge and control of life and mind before the combination of ignorance on these subjects with knowledge of physics and chemistry wrecks the whole social system. He closed with the somewhat startling words: "Which of the runners in this very interesting race will win it is impossible to foretell. But physics and death have a long start over psychology and life!" No one surely will wish for, or expect, a slowing in the pace of the first, but the quickening up in the latter which the last few decades have seen is a matter for high

500

极为美妙的方式将现代物理学介绍给受过教育的公众，他们的书籍起到了很大的启蒙作用。

罗素勋爵曾大胆地表达了以下论点：从物理学到生物学，人们经历了从整个宇宙到局部地区的转换，因为从宇宙的角度看来，生命只是一件无关紧要的事。那些熟悉某个指定区域的人相信，比起偶然从宇宙指南手册的作者那里所获得的关注来说，大都市其实是一个更有资格被关注的领域。它的范围也许很小，但所有最重要的事件都在这里发生。在太多号称概述了科学进展的现行出版物中，生物学都被排除在外，或者只是被略有提及。如果要对现代科学与哲学相交的领域可能遇到的所有内容进行精辟的说明，就必须将思考从现代物理学的推论引向人类思维的本质与结构，甚至引向对于神性思维的思索。还有一些目前已知的生物学现象，它们与这次演讲无疑也是密切相关的，而且就恰当性而言可能是必要的。

VI

然而，由于生物学对于社会发展极其重要，所以公众对它的漠视就更加令人感到遗憾。亨利·戴尔爵士曾经谈到"今天值得考虑的是：就普通人的思维方式和关注点以及对技术和工业发展的意义而言，物理科学的卓越成就是否并未在我们的教育与公共政策中掩盖了生物学的成果，以至于达到了科学自身以及我们希望建筑在科学基础之上的文明有可能出现片面发展的程度"。沃尔特·弗莱彻爵士在去年的离世使这个国家失去了一位贤明的导师，他生前曾在一次面向全国的广播中讲了一句在我看来几乎使公众震惊的话，他说"只有我们越来越将生物学真理的指导引入治国方法时，才能找到安全与进步"。高高在上的治国之术会与生物学理论有关，这对于很多听众来说是一种新观念。

几年之前，剑桥哲学家布罗德博士——他对科学领域的熟悉程度远远超过许多其他的哲学家——曾评论过科学的不平衡发展会带来的灾难；我们对于无机自然界的控制程度之高与对于生物学和心理学的相对无知形成了鲜明的对比。在一次关于全球思想水平持续发展之可能性的讨论的结尾，他用下面的话作出总结：在对这些学科的无知与物理和化学知识的结合破坏整个社会系统之前，这种可能性取决于我们对充足知识的把握以及对生命和思维的控制。他用多少有点令人震惊的话作为结束："很难预言在这场极为有趣的竞争中哪位奔跑者能够获胜。但是物理学和死亡在起跑时就已大大领先于心理学和生命！"当然，没有人会希求或者期望在比赛一开始就落后一步，但是后者在最近几十年中的奋起直追令人非常满意。不过，要重复说明的是，有必要认识到生物学真理对于个人行为乃至于治国之术与社会政策都是

satisfaction. But, to repeat, the need for recognising biological truth as a necessary guide to individual conduct and no less to statecraft and social policy still needs emphasis today. With frank acceptance of the truth that his own nature is congruent with all those aspects of Nature at large which biology studies, combined with intelligent understanding of its teaching, man would escape from innumerable inhibitions due to past history and present ignorance, and equip himself for higher levels of endeavour and success.

Inadequate as at first sight it may seem when standing alone in support of so large a thesis, I must here be content to refer briefly to a single example of biological studies bearing upon human welfare. I will choose one which stands near to the general theme of my address. I mean the current studies of human and animal nutrition. During the last twenty years—that is, since it adopted the method of controlled experiment—the study of nutrition has shown that the needs of the body are much more complex than was earlier thought, and in particular that substances consumed in almost infinitesimal amounts may, each in its way, be as essential as those which form the bulk of any adequate dietary. This complexity in its demands will, after all, not surprise those who have in mind the complexity of events in the diverse living tissues of the body.

My earlier reference to vitamins, which had somewhat different bearings, was, I am sure, not necessary for a reminder of their nutritional importance. Owing to abundance of all kinds of advertisement, vitamins are discussed in the drawing-room as well as in the dining-room, and also, though not so much, in the nursery, while at present perhaps not enough in the kitchen. Unfortunately, among the uninformed their importance in nutrition is not always viewed with discrimination. Some seem to think nowadays that if the vitamin supply is secured the rest of the dietary may be left to chance, while others suppose that they are things so good that we cannot have too much of them. Needless to say, neither assumption is true. With regard to the second indeed it is desirable, now that vitamin concentrates are on the market and much advertised, to remember that excess of vitamin may be harmful. In the case of that labelled D at least we have definite evidence of this. Nevertheless, the claim that every known vitamin has highly important nutritional functions is supported by evidence which continues to grow. It is probable, but perhaps not yet certain, that the human body requires all that are known.

The importance of detail is no less in evidence when the demands of the body for a right mineral supply are considered. A proper balance among the salts which are consumed in quantity is here of prime importance, but that certain elements which ordinary foods contain in minute amounts are indispensable in such amounts is becoming sure. To take but a single example: the necessity of a trace of copper, which exercises somewhere in the body an indispensable catalytic influence on metabolism, is as essential in its way as much larger supplies of calcium, magnesium, potassium or iron. Those in close touch with experimental studies continually receive hints that factors still unknown contribute to normal nutrition, and those who deal with human dietaries from a scientific point of view know that an ideal diet cannot yet be defined.

必不可少的指导，这一点到今天仍应加以强调。如果坦诚地接受下面的事实，即人类自身的本性与生物学所研究的整个自然界的所有方面是大体一致的，再加上对生物学理论的深刻理解，那么人类将会从源于对以往历史和当今的无知的无数压抑中解脱出来，并且为了更高水平上的努力和成功而武装自己。

尽管乍看起来，这样做似乎不足以独立支撑如此宏大的一个主题，但在这里我必须满足于只简要地介绍一个关于生物学造福于人类的例子。我将选择一个与今天演讲的主要论题相接近的例子。我指的是当前关于人类与动物的营养的研究。在过去的20年中，也就是说，自从采用对照实验的方法以来，人们对营养的研究表明，机体的需求远比以前所认为的要复杂得多，尤其是那些消耗量极微的物质可能会，以各自的方式，和那些构成足量食谱中主要部分的成分一样重要。说到底，这种需求上的复杂性并不会使那些知道机体内不同组织中的反应非常复杂的人感到惊讶。

之前我对于维生素的讨论多少有点偏重于其他方面，我知道它们对于强调维生素的营养价值来说并不是必要的。由于有大量各种各样的广告，维生素会在客厅和饭厅之中被人们谈起，尽管不是那么经常但有时在托儿所中也会听到这种讨论，不过到目前为止，可能在厨房中还是讨论得不够。遗憾的是，在未受教育的人那里，它们在营养中的重要作用并不总是能分辨明晰。现在似乎还有一些人认为，如果保证了维生素的供应，那么食谱中的其他成分就可以听之任之了，而另外一些人则认为，它们是那种我们无论吃多少都不过分的好东西。毋庸赘言，这两种想法都不是正确的。就第二点来说，由于市场上有维生素提取物出售而且做了很多广告，所以确实需要提醒大家，维生素过量可能是有害的。至少对于标记为 D 的维生素来说，我们有关于这一点的明确证据。不过，越来越多的证据证明：每种已知的维生素都有极为重要的营养功能。有可能但也许还不能完全确定的是，人体需要所有已知的维生素。

在考虑到机体对一定矿物质的需求时，细节的重要性也是明显的。最重要的是，被大量消耗的各种盐分之间的适当平衡，但是对于在普通食物中只含微量的某些元素，它们的数量也是必不可少的，这一点逐渐得到了确认。只举一个例子——痕量铜元素的必要性，它在机体中的某处行使对于新陈代谢过程不可或缺的催化功能，与需要量更大的钙、镁、钾或铁的一样，它们都是必不可少的。那些与实验研究有密切接触的人不断地得到这样的暗示，即，仍有某些未知因素会影响正常的营养；而那些从科学角度考虑人类食谱的人则知道，目前还不能界定出一种理想的饮食。

This reference to nutritional studies is indeed mainly meant to affirm that the great attention they are receiving is fully justified. No one, I think, need be impressed with the argument that, because the human race has survived until now in complete ignorance of all such details, the knowledge being won must have academic interest alone. This line of argument is very old and never right.

One thing I am sure may be claimed for the growing enlightenment concerning human nutrition and the recent recognition of its study. It has already produced one line of evidence to show that nurture can assist nature to an extent not freely admitted a few years ago. That is a subject which I wish I could pursue. I cannot myself doubt that various lines of evidence, all of which should be profoundly welcome, are pointing in the same direction.

Allow me just one final reference to another field of nutritional studies. Their great economic importance in animal husbandry calls for full recognition. Just now agricultural authorities are becoming acutely aware of the call for a better control of the diseases of animals. Together these involve an immense economic loss to the farmers, and therefore to the country. Although, doubtless, its influence should not be exaggerated, faulty nutrition plays no small share in accounting for the incidence of some among these diseases, as researches carried out at the Rowett Institute in Aberdeen and elsewhere are demonstrating. There is much more of such work to be done with great profit.

VII

In every branch of science the activity of research has greatly increased during recent years. This all will have realised, but only those who are able to survey the situation closely can estimate the extent of that increase. It occurred to me at one time that an appraisement of research activities in Great Britain, and especially the organisation of State-aided research, might fittingly form a part of my address. The desire to illustrate the progress of my own subject led me away from that project. I gave some time to a survey, however, and came to the conclusion, among others, that from eight to ten individuals in the world are now engaged upon scientific investigations for every one so engaged twenty years ago. It must be remembered, of course, that not only has research endowment greatly increased in America and Europe, but also that Japan, China, and India have entered the field and are making contributions to science of real importance. It is sure that, whatever the consequences, the increase of scientific knowledge is at this time undergoing a positive acceleration.

Apropos, I find difficulty as today's occupant of this important scientific pulpit in avoiding some reference to impressive words spoken by my predecessor which are still echoed in thought, talk and print. In his wise and eloquent address at York, Sir Alfred Ewing reminded us with serious emphasis that the command of Nature has been put into man's hand before he knows how to command himself. Of the dangers involved in that indictment he warned us; and we should remember that General Smuts also sounded the

在这里提及营养学研究的主要目的实际上是为了证实该研究得到的深切关注是合理的。我认为，不需要向任何人强化这样的论点：由于人类已经在完全不了解所有这些细节的情况下存活到了今天，所以我们获取的知识必定只具有学术上的意义。这种论点是非常陈旧的，也是绝对错误的。

由于从人类食物以及最近对营养学研究的认识中得到了越来越多的启发，我能确定有一件事是可以说的。一系列证据表明：营养品对人类的辅助程度已经达到了几年前没有得到普遍认可的程度。那是一个我希望自己能够去追求的主题。我本人毫不怀疑来自多方的证据都会指向同样的方向，它们全部都应该是深受欢迎的。

请允许我最后一次谈及营养学研究的另一个领域。它们在畜牧业中的巨大经济价值还有待于全面的了解。农业专家在不久之前开始敏锐地认识到有必要对动物疾病进行更好的控制。动物染病会给农场主带来巨大的经济损失，也会给国家带来损失。无疑地，尽管不该夸大它的影响，根据阿伯丁郡洛维特研究所以及在其他地方所进行的研究，不合理的食物对于引发某些疾病起着不小的作用。还有更多能带来巨大效益的工作有待于进行。

VII

在最近这些年里，每个科学分支中的研究活动都明显增加了。所有的人都会认识到这一点，但只有那些能够仔细考察局势的人才能估计出增长量的大小。我曾想到有一份关于在英国进行的研究活动，尤其是有关国家资助的研究组织的评估，它也许很适合作为今天演讲的一部分。但想描述自己工作进展的欲望使我放弃了这个打算。不管怎样，我还是花了一些时间来考察并且得到了结论，其中包括，现在全世界参与科学研究的独立团体数量是 20 年前的 8 到 10 倍。当然我们必须要记得，不仅美国和欧洲大幅增加了研究基金，日本、中国和印度也已加入了这一领域，并且正在为科学作出真正有价值的贡献。可以确定，无论结果会是什么，目前科学知识的增长速度在明显地加快。

顺便说一句，今天，作为这个有重要影响的科学讲坛的主讲者，我很难避免提到一些我的前任曾说过的惊人词句，它们至今仍在思想、言谈和出版物中反复出现。那次在约克所作的智慧而生动的演讲中，艾尔弗雷德·尤因爵士非常严肃地提醒我们，人类在知道如何掌控自己之前，已经得到了对自然界的掌控权。他就这种指控中隐含的危险对我们提出了警示；而且我们应该记得，斯穆茨将军在伦敦也提出了

same note of warning in London.

Of science itself it is, of course, no indictment. It may be thought of rather as a warning signal to be placed on her road: "Dangerous Hill Ahead", perhaps, or "Turn Right"; not, however, "Go Slow", for that advice science cannot follow. The indictment is of mankind. Recognition of the truth it contains cannot be absent from the minds of those whose labours are daily increasing mankind's command of Nature; but it is due to them that the truth should be viewed in proper perspective. It is, after all, war, to which science has added terrors, and the fear of war, which alone give it real urgency; an urgency which must of course be felt in these days when some nations at least are showing the spirit of selfish and dangerous nationalism. I may be wrong but it seems to me that, war apart, the gifts of science and invention have done little to increase opportunities for the display of the more serious of man's irrational impulses. The worst they do perhaps is to give to clever and predatory souls that keep within the law, the whole world for their depredations, instead of a parish or a country as of yore.

But Sir Alfred Ewing told us of "the disillusion with which, now standing aside, he watches the sweeping pageant of discovery and invention in which he used to make unbounded delight". I wish that one to whom applied science and Great Britain owe so much might have been spared such disillusion, for I suspect it gives him pain. I wonder whether, if he could have added to "An Engineer's Outlook" the outlook of a biologist, the disillusion would still be there. As one just now advocating the claims of biology, I would much like to know. It is sure, however, that the gifts of the engineer to humanity at large are immense enough to outweigh the assistance he may have given to the forces of destruction.

It may be claimed for biological science, in spite of vague references to bacterial warfare and the like, that it is not of its nature to aid destruction. What it may do towards making man as a whole more worthy of his inheritance has yet to be fully recognised. On this point I have said much. Of its service to his physical betterment there can be no doubts. I have made but bare reference in this address to the support that biological research gives to the art of medicine. I had thought to say much more of this, but found that if I said enough I could say nothing else.

There are two other great questions so much to the front just now that they tempt a final reference. I mean, of course, the paradox of poverty amidst plenty and the replacement of human labour by machinery. Applied science should take no blame for the former, but indeed claim credit unfairly lost. It is not within my capacity to say anything of value about the paradox and its cure; but I confess that I see more present danger in the case of "Money versus Man" than danger, present or future, in that of the "Machine versus Man"!

With regard to the latter, it is surely right that those in touch with science should insist that the replacement of human labour will continue. Those who doubt this cannot realise the meaning of that positive acceleration in science, pure and applied, which

同样的警告。

当然，科学自身是无可指责的。它可以被更恰当地认为是放置于科学道路上的一个警示信号：也许是"前方有险坡"，或者是"右转"；但不会是"减速"，因为科学不会采纳这个建议。指责是针对人类的。那些每天都在以自己的劳动增加人类对自然的支配能力的人不可能不认可其中所含的事实；但正是因为有了他们，那些事实才会被正确地对待。毕竟，科学增加了战争的恐惧，而只有对战争的畏惧才能真正给科学带来紧迫感；总会有一些国家在正表现出自私倾向和危险的民族主义时要感受到这种紧迫感。可能我是错的，但在我看来，除了通过战争，科学和发明的馈赠几乎不会为人类表现其更为严重的非理性冲动增加多少机会。也许它们所做的最糟糕的事情也就是使一些狡猾而贪婪的人能够在法律允许的范围内对整个世界进行掠夺，而不是像以往那样仅限于一个教区或者一个国家。

但是艾尔弗雷德·攸英爵士告诉我们，"带着幻想的破灭，他站在一旁，看着过去常常带来无限喜悦的发现和发明的大规模庆典"。我希望那些应用科学并且为英国做出过很大贡献的人也许可以免受这种幻灭的伤害，因为我猜想这会给他带来痛苦。我很好奇，如果他能在生物学家的视角之上叠加"一个工程师的视角"，这样的幻灭是否还会出现。作为一个刚刚还在为生物学谋取权利的人，我很想知道。不过可以确定，工程师对于全人类之赠予要远远超过他可能给予破坏力量的帮助。

关于生物科学我们可以这样讲，尽管隐约听到有诸如细菌战这类的提法，但助纣为虐毕竟不是它的本质。我们现在还没有充分认识到生物学可以为使人作为一个整体配得上大自然的馈赠做些什么。关于这一点我已经说过很多。而在生物学对人类体质改善的贡献方面，是不存在什么疑问的。在这次演讲中，我曾经但只是大略提及生物学研究给予医学技术的支持。我原本想说的比这多得多，但我发现若是充分论述这个问题，就没有时间再说别的了。

还有两个最近非常引人注目的大问题使我在演讲的最后不得不提到。当然，我指的是富足中的贫穷悖论与机械对人力的代替。应用科学应该不会因前者而遭受指责，但实际上却不公平地失去了应得的荣誉。关于这一悖论及其解决方案，我没有能力说出什么有价值的东西；但我承认，我从"猴子 vs 人类"中看到的眼前的危机要比"机器 vs 人类"会在当下或未来造成的危机更多！

关于后者，必定正确的是那些与科学打交道的人会坚持认为对人力的替换还将继续。那些怀疑这一点的人无法认识到科学发展的积极作用，不管是纯粹科学还是

now continues. No one can say what kind of equilibrium the distribution of leisure is fated to reach. In any event an optimistic view as to the probable effects of its increase may be justified.

It need not involve a revolutionary change if there is real planning for the future. Lord Melchett was surely right when some time ago he urged on the Upper House that present thought should be given to that future; but I think few men of affairs seriously believe what is yet probable, that the replacement we are thinking of will impose a new structure upon society. This may well differ in some essentials from any of those alternative social forms of which the very names now raise antagonisms. I confess that if civilisation escapes its other perils I should fear little the final reign of the machine. We should not altogether forget the difference in use which can be made of real and ample leisure compared with that possible for very brief leisure associated with fatigue; or the difference between compulsory toil and spontaneous work.

We have to picture, moreover, in Great Britain the reactions of a community which, save for a minority, has shown itself during recent years to be educable. I do not think it fanciful to believe that our highly efficient national broadcasting service, with the increased opportunities which the coming of short wave-length transmission may provide, might well take charge of the systematic education of adolescents after the personal influence of the schoolmaster has prepared them to profit by it. It would not be a technical education but an education for leisure. Listening to organised courses of instruction might at first be for the few; but ultimately might become habitual in that part of the community which it would specially benefit.

In parenthesis allow me a brief further reference to "planning". The word is much to the front just now, chiefly in relation with current enterprises. But there may be planning for more fundamental developments; for future adjustment to social reconstructions. In such planning the trained scientific mind must play its part. Its vision of the future may be very limited, but in respect of material progress and its probable consequences, science (I include all branches of knowledge to which the name applies) has at least better data for prophecy than other forms of knowledge.

It was long ago written, "Wisdom and Knowledge shall be stability of Thy times". Though statesmen may have wisdom adequate for the immediate and urgent problems with which it is their fate to deal, there should yet be a reservoir of synthesised and clarified knowledge on which they can draw. The technique which brings governments in contact with scientific knowledge in particular, though greatly improved of late, is still imperfect. In any case the politician is perforce concerned with the present rather than the future. I have recently read Bacon's "New Atlantis" afresh and have been thinking about his Solomon's House. We know that the rules for the functioning of that House were mistaken because the philosopher drew them up when in the mood of a Lord Chancellor; but in so far as the philosopher visualised therein an organisation of the best intellects bent on gathering knowledge for future practical services, his idea was a great one. When

508

应用科学，这种发展趋势仍在继续。没人能说清楚闲暇时间的分布最终将达到什么样的平衡。无论如何，乐观地对待闲暇时间增加可能带来的影响也许是有道理的。

如果对于未来有一个真正的计划的话，其实并不需要涉及革命性的变化。梅尔切特勋爵前一阵子在上议院呼吁现在的思考应该为未来做打算当然是正确的；但是我认为，很少有务实的人会真的相信我们正在考虑的机器对人力的取代有可能会使社会产生新的结构。在某些关键环节上，这可能会大大不同于任何可供选择的社会形态，这些社会形态的名字增加了目前的对立。我承认，如果文明逃过了其他危机，那么我应该不必害怕机器的最终统治。我们不应该完全抹杀，真正充足的休息与可能是充满疲惫的、极为短暂的休息之间存在的差异；或者存在于强制性劳动与自觉工作之间的差异。

此外，我们不得不描绘一下英国社会的反应，在最近几年中，大多数英国人都表现出了一定的可塑性。我并不认为相信下面的事是不切实际的：我们高效的国家广播设施，以及短波长传输将为我们提供的越来越多的机会，在教师们通过个人影响帮助青少年学会利用它们之后，就可以很好地实现对青少年的系统教育。这并不是一种技术性的教育，而是追求闲暇的教育。最初可能只是少数人能聆听系统的教育课程；但最终将在这部分受益非常大的社会群体中成为习惯。

附带言之，请允许我再简短地谈一谈"计划"。眼下这个词非常热门，主要是与当前的企业界有关。但是还可以有针对更为根本性的发展的计划；针对今后社会改造的调整计划。在类似这样的计划中，训练有素的科学头脑必将发挥其作用。科学对于未来的洞见可能是极为有限的，但是就物质文明及其可能造成的影响而言，科学（我把与这个词有关的所有知识分支都包含在内了）至少比其他认识形式具备更好的可供预言使用的数据。

很久以前，有人写下了这样的话，"智慧与知识将成为你所处时代的稳定因素"。尽管政治家们可能会具有足够的智慧以解决他们注定要去处理的紧迫问题，但还应有一个可供他们利用的综合而明确的知识库。尤其是引领政府机构去与科学接触的方法，尽管近来已大幅改进，但还不够完善。无论如何，政治家都必然会去考虑当前而不是未来。最近我重读了培根的《新大西岛》，并且一直在思考他所说的所罗门宫。我们知道，所罗门宫的运作规程是错误的，因为哲学家是在陷入一位大法官的状态时完成它的；但是就一位哲学家能想象出一个由最出色的智者组成的致力于为未来发展收集知识的机构而言，他的想法是了不起的。当文明陷入危机而社会处于转型阶段时，是否可以有一座招收国家中最出色的智者而组成的宫殿，使其具有与

civilisation is in danger and society in transition, might there not be a House recruited from the best intellects in the country with functions similar (*mutatis mutandis*) to those of Bacon's fancy? A House devoid of politics, concerned rather with synthesising existing knowledge, with a sustained appraisement of the progress of knowledge, and continuously concerned with its bearing upon social readjustments. It is not to be pictured as composed of scientific authorities alone. It would be rather an intellectual exchange where thought would go ahead of immediate problems. I believe that the functions of such a House, in such days as ours, might well be real. Here I must leave them to your fancy, well aware that in the minds of many I may by this bare suggestion lose all reputation as a realist!

I will now hasten to my final words. Most of us have had a tendency in the past to fear the gift of leisure to the majority. To believe that it may be a great social benefit requires some mental readjustment, and a belief in the educability of the average man or woman. But if the political aspirations of the nations should grow sane, and the artificial economic problems of the world be solved, the combined and assured gifts of health, plenty, and leisure may prove to be the final justification of applied science. In a community advantaged by these, each individual will be free to develop his own innate powers, and, becoming more of an individual, will be less moved by those herd instincts which are always the major danger to the world.

It may be felt that, throughout this address, I have dwelt exclusively on the material benefits of science to the neglect of its cultural value. I would like to correct this in a single closing sentence. I believe that for those who cultivate it in a right and humble spirit, science is one of the humanities; no less.

(132, 381-394; 1933)

培根所设想的宫殿相似（在细节上需要作必要的改动）的功能呢？一座没有政治的宫殿，更愿意致力于对现有知识的综合，致力于对知识进步的持续性评估，以及始终坚持考虑它对于社会改良的意义。不该把它设想为只由科学权威所构成。它更可能是一个交流智慧的场所，在这里思想会超前于当前的问题。我相信，这样一座宫殿的功能在我们这样的时代中很可能会是真实的。现在我必须把它们留在你们的想象中，我很清楚，在很多人的头脑中，我可能由于这个直言不讳的提议而丧失了作为一个现实主义者的全部名誉！

现在我会尽快进入结尾部分。过去，我们中的很多人都有一种害怕把闲暇留给大多数人的倾向。要使人们相信闲暇可以是一种巨大的社会福利，就需要在思想上进行矫正，并且要抱有普通人可以被教化的信念。但是，如果国家的政治目标逐渐趋于理智，而且人为的世界经济问题得到了解决，那么确保同时得到健康、富足和休息就可以最终证明应用科学是正确的。在一个受益于上述因素的社会中，每个个体都将能够自由地表现自身特有的能力，而且由于变得更加个性化，所以更难于被群聚本能所驱动，后者常常是对世界的主要威胁。

可能有人认为，在整个这次演讲中，我只局限于介绍了科学的物质利益，而忽视了它的文化价值。我愿意用一句结束语来更正这一点。我相信，对于那些以健全和谦逊的精神耕耘的人来说，科学是一种人性；一如既往。

（王耀杨 翻译；刘京国 审稿）

The Activity of Nerve Cells*

E. D. Adrian

Editor's Note

In his presidential address to the British Association, Edgar Adrian here summarizes what is known about the activity and organisation of nerve cells. The specialized, energy-dependent cells convey electrical messages along thread-like extensions, like the "spread of a flame along a fuse". But the problem is how these cells interact to form the nervous system. Groups of neurons can, he says, orchestrate simple behaviours, but complex activities, such as learning and memory, defy reduction to single cells or distinct brain regions. And foreshadowing the beliefs of many modern neuroscientists, Adrian adds that the "ceaseless electrical pulsations" of nerve cells will help researchers unravel the workings of the nervous system.

SINCE the biologist seeks to understand life, he cannot be accused of lack of courage. But he can find out a great deal without approaching too near the central problem. He can find out how the living cell develops and how it behaves; he can follow many of the physical and chemical changes which take place in it, and could follow more if cells were not so inconveniently small. The immediate problems of the physiologist may be still further removed from the problem of life. They may deal, for example, with the mechanics of the vascular system or with the physical chemistry of blood pigments. But most physiologists aim at explaining the working of the body in terms of its constituent cells, and feel that this is a reasonable aim, even though we must take the cell for granted. Is it a reasonable aim when we are dealing with the working of the nervous system?

The nervous system is responsible for the behaviour of the organism as a whole: in fact, it makes the organism. A frog is killed when its brain and spinal cord are destroyed: its heart still beats and its muscles can still be made to contract, all the cells of its body but those of the brain and cord are as fully alive as they were before; but the frog is dead, and has become a bundle of living tissues with nothing to weld them into a living animal. This integrative action of the nervous system, to use Sherrington's classical phrase, we may be able to explain in terms of the reactions of the constituent nerve cells. We can at least discuss the point as physiologists. But the human organism includes a mind as well as a body. It may be best to follow Pawlow and to see how far we can go without bringing in the mind, but if the reactions of our nerve cells are to explain thought as well as action, we must face the prospect of becoming psychologists and metaphysicians as well. Fortunately, we need not yet go to such extremes.

---

* Presidential address to Section I (Physiology) of the British Association, delivered at Leicester on September 8.

神经细胞的活性*

埃德加·阿德里安

编者按

这篇文章来自时任英国科学促进会主席的埃德加·阿德里安对该学会所作的演讲，在这里他对神经细胞的组织结构及其活动进行了总结。这些分化的、依赖能量的细胞沿其丝状延伸结构传递电信号，就像"火苗沿着导火线的传播"一样。问题是，这些细胞是如何通过相互作用来组成神经系统的呢？阿德里安提出，神经元间的协同可以完成简单的行为，但是像学习和记忆这样的复杂活动是无法还原到单个细胞或者各个脑区的活动上的。另外，阿德里安还指出，神经细胞中"持续不断的电脉冲"将帮助研究者们破解神经系统的工作机制，这与现代神经学家们的观点不谋而合。

既然生物学家们在努力破解生命的秘密，我们就不能指责他们缺乏勇气。就算他们没能触及问题的最核心部分，也可以发现大量的事实。他们能发现活细胞是如何发育以及发挥功能的；他们也能追踪在细胞内发生的许多物理和化学变化；而且如果细胞不那么小的话，他们还可以研究得更深入。或许生理学家们目前研究的问题仍然与生命的本质问题相去甚远。例如，他们会去研究血管系统的机制或者血色素的物化性质。但是大部分生理学家的目标是用组成机体的细胞来解释机体的工作机制，并且认为这是一个合理的目标，尽管我们必须这么认为。而当我们研究神经系统的功能时，这个目标还合理吗？

神经系统从整体上控制着机体的行为：事实上是它构筑了有机体。当青蛙的脑和脊髓被破坏时，它就会死亡；虽然青蛙的心脏还在跳动，肌肉仍能收缩，除脑和脊髓之外的所有其他细胞都和以前一样充满活力；但青蛙死了，只剩下一大堆活组织而不能构成一个活的生物。按照谢林顿的经典说法，我们也许可以通过组成神经系统的神经细胞的反应活动来解释其整体功能。我们至少能从生理学家的角度讨论这个问题。但人类除了具有躯体之外还具有精神。或许最好的方法是沿着巴甫洛夫的脚步看看如果不考虑精神层面我们能够发现多少东西，如果用神经细胞的活动既解释思想又解释行为，那么我们就不得不面临同时成为心理学家和精神疗法家的可能。幸运的是，我们现在还不需要走这样的极端。

---

* 本文来自英国科学促进会主席于9月8日在莱斯特对I分会（生理学分会）发表的演讲。

The nervous system, the brain, spinal cord and peripheral nerves, is made up of a large number of living cells which grow, maintain themselves by the metabolism of food-stuffs, and carry out all the complex reactions of living protoplasm. In this there are enough problems for anyone; but we are concerned not with the general properties of living cells but with those special properties which enable the cells of the nervous system to perform their functions. Their function is to make the organism respond rapidly and effectively to changes in its environment, and to achieve this they have developed a specialised structure, and a complex arrangement in the body. They send out long threads of protoplasm which serve for the rapid transmission of signals, and they are linked to one another by elaborate branching connexions in the brain and the spinal cord.

Development of the Nervous System

The mapping of this network of paths was begun many years ago, and was the first step in the analysis. No progress could have been made without it, and its results are of vital importance to neurology. We are now witnessing a fresh period of interest in the geography of the central nervous system, but the problem is not how the nerve cells and their fibres are arranged, but why they are arranged as they are. R. G. Harrison, in his recent Croonian lecture before the Royal Society, recalled the time when he first cultivated living nerve cells outside the body. That experiment, made twenty-three years ago, marks the new epoch better than any other, for, besides introducing the method of tissue culture, it settled a long and bitter controversy as to the origin of nerve fibres. Nowadays the most elaborate transplantation experiments are carried out by the embryologists on amphibian larvae. Animals are produced with supernumerary limbs, eyes, noses, and even spinal cords. The growing nervous system is faced with these unusual bodily arrangements, and by studying the changes induced in it we can form some idea of the factors which determine its normal structure.

A review published this summer by Detweiler gives a vivid impression of the plasticity of the developing nervous system in the hands of the experimenter. As a rule it accepts the extra limb or sense organ, links it by nerve fibres to the rest of the organism and may develop more nerve cells to deal with it. The forces which mould the nervous system seem to come partly from within the central mass of nerve cells and partly from the body outside. These forces may be chemical or electrical gradients, and often the nerve fibres seem to grow in particular directions because they cling mechanically to structures already laid down, for example, to the main arteries of the limbs. It is unlikely that a simple formula will be found for such a complex arrangement, but the fact remains that the arrangement can be profoundly modified at the will of the experimenter. Its detail seems to depend not so much on the innate properties of particular cells as on the environment provided by the rest of the organism.

Reactions of the Neurons

This new embryological work supports the older in showing that the nervous system is

514

神经系统，即脑、脊髓和外周神经，是由大量活细胞组成的，这些活细胞利用食物代谢进行生长发育并维持活性，还要完成生物体的所有复杂行为。这里有太多的问题需要解决；然而我们关心的不是活细胞所具有的一般特性，而是那些使神经细胞发挥自身功能的特殊性质。神经细胞的功能是使生物体迅速而有效地对环境的改变作出反应，为了实现这个功能，它们形成了自己的特殊结构并在有机体内部形成了复杂的组织方式。它们发出长长的细胞质突起，用于快速传递信号，在脑和脊髓中它们通过精细的分支相互连接在一起。

神经系统的发育

许多年前人们就开始描绘这个复杂的网络，这是分析过程的第一步。如果没有这方面知识就无法开展下一步的工作，而且这项工作的研究成果对于神经病学也具有至关重要的意义。目前我们对中枢神经系统布局的研究进入了一个有趣的新阶段，不过问题不在于神经细胞及它们的纤维是如何分布的，而在于为什么它们会这样分布。最近，哈里森在英国皇家学会主办的克鲁尼安讲座上回顾了他第一次在体外培养活神经细胞的经历。这项在 23 年以前进行的实验在开创新纪元方面超越了其他实验，因为除了引进组织培养法外，它还平息了人们对神经纤维起源问题的长期激战。如今，胚胎学家们可以在两栖动物幼体中进行最精细的移植实验。人们可以制造出具有多余肢体、眼睛、鼻子甚至脊髓的动物。发育中的神经系统要面对这些不正常的身体构造，而通过研究其中发生的改变我们就能认识到一些决定其正常结构的因素。

在今年夏天发表的一篇综述中，德特韦勒生动地描述了发育中的神经系统在实验者手中的可塑性。在通常情况下，它会接受多余的肢体或感觉器官，通过神经纤维将其与身体其他部分联系起来，并发育出更多的神经细胞来支配它。塑造神经系统的力量似乎有一部分来自神经细胞的核心，而另一部分来自机体的外部。这些力可能来自化学或电位梯度，而且在通常情况下神经纤维似乎会向特定的方向生长，因为它们会机械地附着在已经长好的结构上，比如四肢的大动脉上。要在这么复杂的构建中找到简单的规律是不太可能的，但事实上这些构建是可以按照实验者的意愿发生很大变化的。有机体其余部分构筑的环境对构建变化的影响似乎要大于某些细胞的固有特性对它的影响。

神经元的反应

这项新的胚胎学研究结果证明了原有的理论，即认为神经系统是由"神经元"——

made up of "neurons", cells with thread-like extensions, and that they are the only active elements in it. These elements are all cast in the same mould, but are shaped differently by the forces of development. To this we can now add the fact that all neurons seem to do their work in much the same way. The activity which they show is in some respects remarkably simple. It is essentially rhythmic: a series of rapid alternations between the resting and the active state, due probably to rapid breakdown and repair of the surface. This at least is a fair description of the way in which the nerve fibres carry out their function of conducting messages, and we can detect the same kind of pulsating activity in the nerve cells of the brain.

The evidence comes from the analysis of minute electric changes, for cell activity sets up electrical eddies in the surrounding fluid, and these can be measured with a minimum of interference. The clearest results are given by the peripheral nerve fibres which connect the central nervous system to the sense organs and the muscles. The nerve fibres are conveniently arranged in bundles to form the nerve trunks: each fibre is an independent conducting path and there may be a thousand such paths in a fair-sized nerve, but it is not a difficult matter to study what takes place in the single fibre when it conducts a message.

We may begin with an external stimulus acting on a sense organ, a structure which includes the sensitive ending of a nerve fibre as an essential part. The ending is excited by the stimulus, the delicate equilibrium of its surface is upset and the disturbance tends to spread along the fibre. The spreading is an active process: it takes place because the fibre has a store of energy ready to be liberated at a moment's notice, and because the changes which attend its liberation at one point upset the balance at the next point and cause the same activity there. The spread of a flame along a fuse is a well-worn analogy. But the nerve fibre is so constituted that a disturbance at any point is almost immediately cut short. The change spreads along it as a momentary wave—a brief impulse followed inevitably by a brief interval of rest and recovery. If the sense organ remains excited, a second impulse passes up the fibre, and then another and another so long as the stimulus is effective.

The impulses in a given nerve fibre are all alike in magnitude, rate of travel, etc., but the frequency at which they recur depends on the intensity of the stimulus, rising sometimes so high as 300 a second in each fibre, or falling so low as 10. All the nervous messages take this form.

The conducting threads or nerve fibres are exceedingly insensitive to changes in their environment: their endings in the sense organs are exceedingly sensitive. The sole function of the ending is to act as the trigger mechanism for firing off the impulses, and the sole function of the nerve fibre is to carry the message without distortion. Both are specialised parts of the neurone with specialised reactions, but it is important to note that these reactions are not peculiar to the nervous system. Muscle fibres, developed from the mesoderm and specialised for contraction, conduct impulses which seem to differ merely in their time relations from those in nerve fibres, and they can also be made to behave like the sensory endings by treatment with various salt solutions. In sodium chloride, for

有丝状突起的细胞组成的，而且它们是系统内唯一的活性成分。所有神经元都是按同一种模式生成的，但在发育时长成了不同的形状。关于这一点我们现在又有了一个新的发现，即发现所有神经元都以大致相同的方式工作。在某种程度上可以说它们进行的活动非常简单。这个过程是重复出现的：处于静息状态和活动状态之间的一系列快速转换很可能是由膜表面的快速降解和修复导致的。这种解释至少能合理地说明神经纤维在实现信息传递功能时所采取的方式，而且我们在脑神经细胞中也检测到了同样的节律性活动。

证据来自对微小电位变化的分析，因为细胞活动会使周围液体形成电涡流，而这些涡流可以在最小的干扰下被测定。最清晰的结果是从把中枢神经系统连接到感觉器官和肌肉上的外周神经纤维中得到的。神经纤维集合成束形成神经干：每条纤维都是独立的传导通路，虽然在一条中等大小的神经干中可能有一千条这样的通路，但要研究单个纤维在传递信息时的情况并不是一件难事。

我们也许可以从作用于感觉器官的外部刺激入手，神经纤维的感觉末梢在感觉器官中是一个非常重要的部分。该末梢受到刺激后兴奋,其膜表面的微妙平衡被打破,而且这种扰动会沿着纤维传播下去。这种传播是一个主动的过程：它之所以能够发生是因为神经纤维提前储备好了能量以便在瞬间释放，而且因为在一个点上由能量释放引发的变化打破了下一个点的平衡并在那里产生了相同的变化。我们常用火苗沿导火线的传播来进行类比。然而神经纤维的结构如此完善以至于在任意点的扰动都可以马上被打断。变化以瞬时波的形式沿神经纤维传播——在一个短暂的冲动后必然紧接着短暂的休息恢复期。如果感觉器官仍处在兴奋状态，第二个冲动就又会沿着纤维传过来，只要存在有效的刺激，冲动就会不断地传过来。

特定神经纤维中的冲动在幅度和传播速度等方面都是相似的，但冲动产生的频率取决于刺激的强度，有时在每个纤维中可以达到 300/ 秒，有时低至 10/ 秒。所有由神经传播的信息都采用这种模式。

导电线或神经纤维对于周围环境的变化是非常不敏感的：但它们在感觉器官中的末梢却很敏感。末梢的唯一功能就是作为冲动的触发器，而神经纤维的唯一功能是准确无误地传递信息。两者都是神经元的特殊部分，具备特殊的反应，但值得一提的是这些反应并非神经系统所独有。从中胚层发育而来专门用来收缩的肌纤维也能传递冲动，与神经纤维传递的冲动仅存在时间关系上的差别，在各种盐溶液的作用下它们也能像感觉末梢那样工作。比如，在氯化钠溶液中，当肌肉受到牵拉时就会产生一系列的冲动，这些冲动就好像发生在一个唯一功能是充当"牵张感受器"

example, a series of impulses will be set up in a muscle fibre when it is stretched, as they would be in one of the sense organs the sole duty of which is to act as "stretch receptors". The muscle fibre makes a poor copy of the nervous mechanism, for it reacts jerkily and is often damaged in the process, but the ground-plan of the mechanism is the same.

Thus in the activities concerned in the rapid conduction and in the setting up of rhythmic trains of impulses, it does not appear that the cells of the nervous system have properties not shared in some degree by other tissues.

So far we have only considered what happens in nerve fibres. We can tap the messages which pass along the wires between the front line and headquarters, but this does not tell us how they are elaborated there. A great deal has been found out already by the analysis of reflexes—that is, by sending in a known combination of signals and finding what signals come out to the muscles; indeed, the great part of Sherrington's work on the spinal reflexes and Pawlow's on the brain has been carried out in this way. The results are so well known, however, that I shall deal here with a recent line of attack of an entirely different kind.

This method relies on the fact that nervous activity, in the central grey matter as in the peripheral nerves, is accompanied by electric changes. They seem to be a reliable index of the underlying activity, and by recording them we come a step nearer to the main problem. The chief difficulty is to interpret the records. In the cerebral cortex, for example, very large electric oscillations are constantly occurring, except in the deepest anaesthesia, but they vary from moment to moment and from place to place, and it is only in the visual cortex that they are under a fair degree of experimental control. Here they can be produced by shining a light in the eye (Fischer, Kornmüller and Tönnies) or stimulating the optic nerve (Bartley and Bishop), and the prospects of analysis are more hopeful.

At the moment, however, the most significant feature of these records from the brain lies in the appearance of the waves. Whenever a group of nerve cells is in action, in the cerebral cortex, the brain stem or the retina, and whether the nerve cells in question belong to a vertebrate, or an insect, the waves are alike in general form. Instead of the abrupt spikes which appear in a record from a nerve fibre when a train of impulses passes down it, we have more gradual potential changes which form a series of waves of smooth contour. In the simpler structures where most of the neurons are acting in unison, the waves may have a regular rhythm (5–90 or more a second), which rises and falls when the stimulus changes in intensity. It is often possible to make out both the abrupt nerve fibre impulses and the slower nerve cell waves, and to show that they occur together. In the cerebral cortex of an anaesthetised animal there is much more variety and less orderly repetition; the waves usually occur at irregular intervals; they vary in size and duration, and some of them may last for half a second or even longer.

的感觉器官之中。肌纤维并不是神经纤维的良好复制品，因为它只能间断性地作出反应而且冲动经常在传导过程中被损毁，但两者的基本作用机制是相同的。

因此在有关快速传递和冲动规律性形成的行为中，神经系统的细胞所具有的性质看上去与其他组织没有什么不同。

到目前为止我们只考虑了神经纤维中的情况。尽管我们能够从连接前线与指挥部之间的这段线路中截取传递的信息，但我们并不知道它们是怎样被制造出来的。人们在对反射的分析中已经取得了很大的进展——即通过发送一组已知信号并监测哪些信号会传递到肌肉中；实际上，谢林顿在脊髓反射方面的大部分工作和巴甫洛夫对脑的研究都是用这种方法进行的。不过他们取得的成果已经众所周知，我在这里要讲的是一种完全不同的新方法。

这种新方法基于一个事实，即无论是中枢灰质还是外周神经的活动都伴随着电位的改变。它们似乎是反映神经活动的可靠指标，通过记录它们，我们就又向解决主要问题迈进了一步。主要的困难在于对记录结果的解释。比如在大脑皮层，非常大的电振荡一直存在，除非处于最深度的麻醉之下，不过电振荡会随着时间和空间的不同不断发生变化，视觉皮层是唯一一个能让它们在一定程度下受到实验控制的地方。用光线照射眼睛（费希尔、科恩米勒和滕尼斯）或刺激视神经（巴特利和毕肖普）可以产生电振荡，这种分析方式是很有希望取得成功的。

但是现在，对于从大脑中获得的记录来说，波形是它们最重要的特征。只要一组神经细胞处于活动状态，我们就会得到大体类似的波，无论是在大脑皮层、脑干或者视网膜，也无论这些神经细胞是来自脊椎动物还是来自昆虫。与当一组冲动经过神经纤维时记录下来的尖峰不同，我们看到的多数是电位的逐渐变化，因而形成了一系列平滑的波。在大多数神经元同时起作用的简单结构中，波也许会有一个统一的节律（每秒 5~90 或者更多），周期性随刺激强度的变化有升有降。我们通常能辨认出突发的神经纤维冲动和平缓的神经细胞波，并且可以说明它们是同时发生的。在被麻醉动物的大脑皮层中，多为变化大且重复性不高的波；这些波出现的间隔通常是不规则的；它们的大小和持续时间不尽相同，有些可持续半秒或更长时间。

Nerve cell waves may be the wrong name, for they are probable due to the branching dendrites and not to the cell body of the neurone; but there can be no doubt that they represent a characteristic activity of the structures which make up the grey matter. They show that the same kind of rhythmic breakdown and repair of the surface takes place in this part of the neurone as in the nerve fibre, with the important difference that the changes develop and subside much less abruptly. The surface is not specialised for rapid conduction; the forces which restore the resting equilibrium are less powerful and there is more tendency to spontaneous breakdown and to long periods of uninterrupted activity. We know that the activity of the grey matter is far more readily influenced by chemical changes than is that of the nerve fibre with its elaborate fatty sheath and wrappings of connective tissue, and it seems probable that both chemical and electric changes may be concerned in the spread of activity from one neurone to another. How this spread takes place is still uncertain, and it is admittedly the most important problem we have to face.

In spite of this, we can claim to have some of the main outlines of neurone activity. Our nervous system is built up of cells with a specialised structure and reactions, but the reactions are of a type to be found in many other cells. The rhythmic beat of the heart is probably due to surface reactions not far different from those in the group of nerve cells which produce the rhythmic movements of breathing; and the factors, nervous and chemical, which regulate the heart beat are probably much the same as the factors which control the discharge of the neurone. We have a store of energy, replenished constantly by cell metabolism and liberated periodically by surface breakdown. The electrical gradients at the active point cause a spread of the breakdown to other regions, but sooner or later restoring forces come into play, the membranes are healed and the cycle is ready to be repeated. It is a long step from the mechanical precision of an impulse discharge in a nerve fibre to the irregularities of a record from the cerebral cortex, but there are many intermediate cases which will bridge the gap.

The Nervous System as a Whole

So far as the units are concerned, the prospect is encouraging. The difficulties begin when we come to the work of the nervous system as a whole. Many of its reactions are mechanical enough and can be explained in terms of the activity of groups of neurons, but there is much that resists this kind of treatment. It is perhaps encouraging that the difficulties are greatest when the reactions depend on the cerebral cortex, when they involve learning and memory, or, if you prefer it, habit formation and conditioning. They have been clearly stated by Lashley, and most of them can be reduced in the end to a simple formula, the failure of anatomical models of the nervous system. The revolt from the anatomical model has been growing for many years, though it may be doubted whether its sponsors ever believed in it as much as their critics suppose. It gave us diagrams of nerve centres and pathways which were valuable enough when they referred to known anatomical structure, but not when they referred, as they often did, to hypothetical centres and to pathways canalised by use. These too may exist, but they are not the whole explanation of cortical activity.

神经细胞波也许是一种错误的叫法，因为它们可能源自树突分枝而非神经元细胞体本身；但毫无疑问这些波代表了灰质结构的本质性活动。它们说明：神经元的这部分也和神经纤维一样发生着周期性相同的膜表面裂解和修复，但最大的不同在于变化的发生和消失更为平稳。这些膜表面并非仅能进行快速传导；膜表面保持静息平衡的力量不够强大，因而有自发裂解和进行长期不间断活动的趋势。我们知道化学变化对灰质活性的影响要远远大于有致密脂肪鞘和结缔组织包层的神经纤维，当兴奋从一个神经元传向另一个神经元的时候，化学性质和电位都有可能发生变化。我们现在仍不知道这种传播是如何进行的，大家都承认这是我们不得不去面对的最重要的问题。

尽管如此，我们可以认为自己已经掌握了神经元活性的主要概况。我们的神经系统由具有特殊结构和反应的细胞组成，不过在许多其他细胞中也能发现这样的反应。引起心脏律动的膜表面反应可能与产生规律呼吸运动的神经细胞群体的膜表面反应没有明显的不同；调节心脏律动的神经及化学因子很可能也与控制神经元放电的因子近似相同。我们所储存的能量，一方面由细胞代谢不断地补充，另一方面从膜表面的裂解中周期性地得到释放。激活部位的电梯度促使裂解传播到其他部位，但储存的能量迟早要发挥作用，在膜得到修复后就又可以进行下一轮的循环了。从神经纤维释放冲动的精确性到在大脑皮层中记录下来的不规则性之间还有很大一段距离，不过大量有关中间过程的研究将架起连接两者的桥梁。

神经系统整体观

就各个组成部分的情况而言，前景是乐观的。然而当我们把神经系统作为一个整体进行研究时就出现了困难。神经系统的许多反应高度机制化，可以用神经元群的活动来解释，但有很多因素与这种处理方法相抵触。也许值得庆幸的是，当反应依赖于大脑皮层，当涉及学习和记忆，或者，在你愿意的情况下形成习惯和条件反射时难度最大。拉什利已经对此进行了明确的阐述，大部分结果最终都归结成了一个简单的结论，神经系统的解剖模型是失败的。反对解剖模型的意见已经存在了许多年，尽管人们怀疑发起者们过去对它的支持是否和现在对它的批评一样坚定。模型为我们提供了神经中枢和通路的概况，但是仅当其所指的是已知的解剖结构时才有足够的价值，如果像过去那样指的是假设的中枢和改造的通路，那价值就不明显了。它们也可能是存在的，但不能完全解释皮质的活动。

Clinical neurology is partly to blame for the emphasis laid on exact localisation. The neurologist must locate brain tumours by analysing the disturbances they produce; consequently he welcomes the slightest evidence of localisation of function in the cortex, and finds the anatomical model valuable for correlating his observations. Undoubtedly there are well-defined nervous pathways, clear differences in cell structure and localised activity in different parts of the brain. As a modern addition to the evidence we have Foerster's recent work on the electrical stimulation of the human cortex, and his finding that stimulation of the temporal lobe may cause sounds and words to arise in consciousness, whilst stimulation of the occipital lobe gives lights or images. Bard has given another remarkable example of strict cortical localisation by his observations on certain postural reactions in the cat. These depend on a limited area in the frontal region, are not affected by damage to other parts of the brain, but are permanently lost if the frontal area is destroyed. The danger nowadays is that we may pay too little attention to such facts; but it is true, nevertheless, that the localisation is a matter of areas rather than of single neurons. This is shown by examination of habit formation, and by the remarkable way in which the nervous system adapts itself to injury.

It has often been pointed out that we learn to recognise shapes—the letters of the alphabet, for example—however they are presented to us. The pattern of black and white made on our retina by the letter A need not fall on a particular set of retinal endings connected with particular cortical neurons. We have learnt to recognise a relation of lines and angles, a pattern of activity in the cortex rather than an activity of specific points. This kind of reaction is not due to our superior intelligence. Lashley finds it in the rat, and psychologists of the *Gestalt* school have pointed out examples from all manner of animals. There is the same neglect of specific neurons in the formation of motor habits, for if we have once learnt to write the letter A with our right hand, we can make a fair attempt to write it with any group of muscles which can control a pencil.

The adaptations to injury present a different aspect of the same story. An insect which has lost a leg will at once change its style of walking to make up for the loss. This may involve a complete alteration of the normal method, limbs which were advanced alternately being now advanced simultaneously. The activities of the nervous system are directed to a definite end, the forward movement of the animal—it uses whatever means are at its disposal and is not limited to particular pathways.

When the central nervous system is injured, there is more evidence of localised function, but the localisation is no hard-and-fast affair. A rat uses its occipital cortex in the formation of certain visual habits. When this part of the cortex is destroyed the habit is lost, but it can be re-learnt just as rapidly as before with what remains of the brain. A monkey's arm is paralysed if the corresponding motor area of the cortex is destroyed, but the paralysis soon passes away although there is no regeneration of the motor cortex. What is more remarkable is that the recovered functions are not associated with the development of a new visual region or motor region in the brain. Though they were originally localised, there is no longer any one part of the cortex which is essential.

522

临床神经病学因为将重点放在准确的定位上而备受指责。神经病学家们只有通过分析症状才能定位脑肿瘤；因此他们乐意接受在皮层中进行最细小的功能定位，并且发现解剖模型与观察结果可以相互关联。毫无疑问，大脑中存在清晰的神经通路、明显不同的细胞结构以及不同部位的局部活动。弗尔斯特最近关于人类皮层电刺激的研究为我们增加了最新的证据，他发现刺激颞叶可以使意识中出现声音和语言，刺激枕叶可以出现光或图像。巴德在观察了猫的某些姿势反射之后，给出了另一个值得关注的皮层精确定位的例子。它们依赖于额叶区的有限范围，不受脑其他部位损伤的影响，如果额叶受到损伤，这一功能就会永久丧失。现在的危机在于我们很少关注这些事实；不过定位肯定是在一个区域内而不是在单个的神经元上。对习惯化的考查以及神经系统适应损伤的奇特方式都证实了这一点。

人们经常说我们要学习识别形状，如字母表里的字母，其实它们就在我们面前。字母 A 在我们视网膜上形成的黑白图案不需要落到与特定皮层神经元相联系的特定视网膜神经末梢上。我们已经学会识别线与角之间的关系，这是大脑皮层中的一种活动形式，而不是某些特殊点的活动。这类反应与人类超群的智力无关。拉什利发现大鼠也具有这一识别能力，**格式塔**学派的心理学家在各种各样的动物身上都找到了类似的例子。在运动习性的形成过程中也同样会出现忽略某些特殊神经元的情况，因为一旦我们学会了用右手书写字母 A，我们就可以通过一定的努力让其他能够控制铅笔的肌肉群也能书写字母 A。

对损伤的适应反映了这一问题的另一个方面。一只失去一条腿的昆虫很快就会通过改变走路的姿势弥补缺腿的不足。这样做可能会完全改变原来的方式，以前交替移动的腿现在要同时移动。要达到的特定目标支配着神经系统的活动使动物往前移动，采用什么方式可以自行决定而不受具体模式的限制。

在中枢神经系统受到损伤时，我们可以得到更多有关定位功能的证据，但定位并不是一成不变的。大鼠用枕叶皮层来形成某种视觉习惯。当这部分皮层受到损坏时，这种习惯也会丢失，但它能够利用剩余的大脑皮层和从前一样快地再次学会。如果运动皮层区受到损坏，猴子的上肢就会瘫痪，但瘫痪的上肢会很快恢复功能，尽管运动皮层不能再生。更神奇的是这些功能的恢复并不是因为在大脑中形成了新的视觉区或者运动区。虽然这些功能原先有定位，但任何一部分皮层都不是绝对必要的。

In reactions where there is no evidence of localisation (for example, the learning of maze habits in the rat), Lashley finds that the important factor is the total mass of the cortex and not the presence of particular regions. The effect of an injury depends on its extent and not on its situation. It depends, too, on the amount of grey matter (nerve cells and dendrites) destroyed, and not on the cutting of connexions between the different parts of the cortex. Thus the ability of the brain to form new associations, and generally to control the behaviour of the animal, depends primarily on the total area covered by the nerve cells of the cortex and their interlacing dendrites. For certain reactions it depends to some extent on the arrangement of pathways, but this arrangement is not essential. There is more localisation of function in the large brain of man than in the very small brain of the rat, for different cortical regions may be completely equivalent when they are separated by 5 mm., but not when they are separated by 100 mm. But apart from this difference in scale, it is likely that the human cortex has the same mass effect and plasticity of function.

How do the individual neurons combine to produce a system which can recognise a triangle or direct the movements of the organism with such disregard of detailed structure? If particular neurons or pathways are not tuned to triangularity, how can the whole mass be tuned to it, and why should the tuning be more certain when the mass is greater? Our data may be at fault and the mass effect an illusion, but there is certainly enough evidence for it to be taken seriously. Though there is no solution at the moment, I cannot believe that one will not be found—a solution which need not go outside the conceptions of physiology. It should be possible, for example, to find out how many neurons must be combined to give a system which reacts in this way and what kind of structure they must form. The nervous systems of insects may provide the clue, for these may contain a few thousand nerve cells in place of the ten thousand million in the human brain. It is possible also to study the reactions of isolated parts of the central nervous system, to see how far their behaviour can be explained in terms of the units which compose them. The retina is an interesting example of this kind, for it contains an elaborate structure of nerve cells and dendrite connexions, and has some of the reactions which we might expect from a mosaic of sensory endings, and some which depend on interaction between the different neurons.

Even now, however, we can form some idea of the way in which the grey matter can act as a whole. The electric oscillations in the cortex and in the grey matter generally are often due to a large number of units pulsating in unison. Sometimes there are several competing rhythms, and sometimes the collective action breaks down altogether, to reappear from time to time when some part of the system is stimulated to greater activity. When these collective rhythms appear, the neurons are already acting as though they formed one unit. There is no need to regard the dendrites as forming a continuous network—electric forces may well bridge the gaps between them—but they may form a system in which activity can be transmitted more or less freely in all directions. The patterns of activity in a system of this kind would be like the ripples on the surface of a pond, with the difference that some of the ripples may occur spontaneously, whilst others are due to incoming signals. Interference figures and nodes of vibration may then be all-important. They would at least

524

在没有定位证据的反应中（比如大鼠迷宫行为的学习），拉什利发现关键因素在于皮层的整体而不是特定区域的存在。一个损伤造成的影响取决于它的范围而不是位置。损伤造成的影响还取决于被损坏的灰质（神经细胞和树突）的数量，而与皮层不同区域之间的连接是否被破坏无关。因此大脑形成新联系以及整体控制动物行为的能力主要取决于皮层神经细胞和它们交错排列的树突所占的整体区域。对于某些反应来说，它在一定程度上取决于通路的排列，但这种排列并不是不可缺少的。人的大脑要远远大于大鼠的大脑，因而具有更多的功能定位，因为当分隔为 5 毫米时，不同的皮层区域可能是完全相同的，但是当分隔为 100 毫米时就不相同了。不过除了这种在脑大小上的差别，人类的皮层很可能也具有同样的整体效应和功能可塑性。

如果不考虑具体的结构，这些单独的神经元是如何组合在一起形成一个系统来识别三角形和指挥生物体的运动的呢？如果单个的神经元或者通路不能识别三角形，那么整体又是怎么识别的，而且为何当这个整体越大时其识别能力就越强呢？我们的数据可能是错误的，这种整体效应只是一个错觉，但确实有足够的证据使我们必须认真对待这个问题。尽管现在还没有解决的方法，但我不相信将来人们找不到一个答案，一个不会超出生理学范畴的答案。比如应该可以找到要想形成一个以这种方式发生反应的系统需要多少个神经元，以及它们需要形成什么样的结构。昆虫的神经系统或许能为我们提供线索，因为昆虫可能只含有几千个神经细胞，而人脑中有 100 亿个。也可以通过研究中枢神经系统各独立部分的功能来考察它们的行为在多大程度上可以用组成它们的单元来解释。视网膜是一个有趣的例子，因为它含有由神经细胞和树突联系组成的复杂结构，而且它的一部分行为可以用感觉末梢的嵌合体来解释，而另一部分行为则依赖于不同神经元之间的相互作用。

然而，即使是现在，我们也能了解到灰质作为一个整体的工作方式。皮层和灰质中发生的电振荡通常是由于大量组成单元一起振动的结果。有时候会出现几个竞争的振荡节律，有时候集体行动完全停止，当系统中某个部分被刺激到具有更大的兴奋性时，集体行动会再现。当集体节奏出现的时候，这些神经元表现得像是形成了一个整体。没有必要认为树突形成了一个连续的网络，因为电场力完全可以打通它们之间的间隙，但树突会形成一个能使兴奋在各个方向上自由传播的系统。这种系统中的兴奋形式就像池塘表面的涟漪，不同之处在于有些涟漪可能是自发产生的，而另一些是由于接受了传入的信息而产生的。因此，干涉图和振动节点都非常重要。至少它们提供了在不需要进行特定点激发的情况下识别成三角形或成正方形等的关

give a basis for the recognition of relations such as those of triangularity or squareness without the need for an excitation of specific points, and they might be formed with less distortion in a large pond than in a small one.

This does not take us vary far: in fact, the major problems of the central nervous system are left in greater obscurity than ever. But no one can observe these ceaseless electrical pulsations without realising that they provide a fresh set of data, and may give a fresh outlook on the working of the brain. The facts are still too uncertain to be worth treating in greater detail. But they accumulate rapidly, and several lines of evidence seem to lead in the same direction. For the present, it is enough to state our problem, that of the organisation of neurons into the nervous system. It is still a physiological problem, and I hope that a solution will be found on physiological lines. If it cannot be found, it will be extremely interesting to see where the breakdown occurs; and if it can, it will be even more interesting to see what light it throws on the relation of the nervous system to the mind.

(**132**, 465-468; 1933)

系的基础，而且在大池塘中可能比在小池塘中失真度更低。

我们并没有因此取得很大的进步：事实上，中枢神经系统的主要问题比以前更令人困惑。但是每一个看到这些持续不断的电振荡的人都会认为它们提供了新的数据，而且可能会为大脑的研究开辟新途径。这些事实仍然不是很可靠，不值得我们投入更多的精力。但是这方面的证据越来越多，而且好几条证据都指向了同一个目标。现在我们完全可以提出关于神经元在神经系统中的构成问题。这仍旧是一个生理学问题，我希望能在生理学范畴内找到答案。如果找不到，我们也非常想了解问题出在哪里；如果能找到，我们更愿意知道它是怎样解释神经系统和思维之间的联系的。

（毛晨晖 翻译；刘力 审稿）

X-ray Analysis of Fibres

W. T. Astbury

Editor's Note

The author here is William Astbury, whose work on the X-ray diffraction of fibrous biomolecules anticipated the structural insights of Linus Pauling, Maurice Wilkins, Francis Crick and James Watson. His work on keratin, the main protein component of hair, led him to postulate specific secondary structures in proteins: characteristic conformations adopted by the molecular chains of amino acids. In this summary of a meeting at the British Association, Astbury applauds the collaboration of botanists and physicists in addressing the structures of biological fibres—a collaboration that is still not unproblematic today. Astbury's description of proteins as "inflnltely variable and adjustable molecular patterns", with structural as well as enzymatic roles, foreshadows the puzzle of how these patterns are encoded in the molecules' chemical composition.

"PRESENT methods can still go far, but I am convinced that progress... is about to gain a great impetus from the application of those new methods of research which chemistry is inheriting from physics: X-ray analysis..."* So spake in general terms the president of the British Association in his address on the evening of September 6: on the following morning, Sections A (Mathematical and Physical Sciences) and K (Botany) foregathered to demonstrate the point in somewhat more detail. That physicists should hobnob with botanists—and not simply for the purpose of drinking tea—and on the following Monday† even be invited into the stronghold of vitalistically-minded zoologists is a very definite cause for congratulation, in spite of the dark mutterings of some that it is all very well to talk about the structure of molecules and adopt such an attitude of pitiable optimism in the face of "life" and all the tremendous tale of the activities of living organisms!

In opening the joint discussion of Sections A and K, Mr. W. T. Astbury outlined some developments in the X-ray interpretation of the properties of hair, feathers and other protein structures. Recent progress in our knowledge of the molecular structure of natural fibres arises largely out of the recognition by X-ray means that the solid state of fibres is a crystalline state, generally imperfect, it is true, yet nevertheless sufficiently organised to give valuable information about the form and properties of the giant molecules which orthodox chemistry suggests as their basis. The crystallites are sub-microscopic, but it can be seen at once from X-ray photographs that they are always *effectively* long and thin and lie with their long axes either roughly parallel to the fibre axis, as in silk and hair, or

---

* "Some Chemical Aspects of Life" (see *Nature*, Sept. 9, p. 389).

† Discussion on the structure of protoplasm, Section D, Sept. 11.

纤维的X射线分析

威廉·阿斯特伯里

编者按

本文作者威廉·阿斯特伯里在含纤维生物分子的 X 射线衍射方面的研究先于莱纳斯·鲍林、莫里斯·威尔金斯、弗朗西斯·克里克和詹姆斯·沃森对结构的分析。他通过对毛发中主要的蛋白质成分——角蛋白的研究推出蛋白质所特有的次级结构：氨基酸分子链的特殊构造。在这篇对英国科学促进会的一次会议的纪要中，阿斯特伯里高度评价了植物学家和物理学家在探索生物纤维结构时的良好合作，这样的合作在今天也未必那么容易实现。阿斯特伯里认为蛋白质"分子的结构变化多端并且可以改变"，可以构成一种组织，也可以是一种酶，这意味着人们在对这些结构按分子化学构成进行编码时会感到很困惑。

"目前的方法还可以更进一步，但是我确信这一进展……将会从新研究方法的应用中获得巨大的推动力，即在化学中借用物理学的方法：X 射线分析……"* 英国科学促进会主席于 9 月 6 日晚间所做的演讲中就是这样概括的；第二天早上，A 分会（数学和物理学）和 K 分会（植物学）在一起更详细地论证了这一观点。物理学家与植物学家在一起畅谈不仅仅是为了喝茶，在下一个周一时†他们甚至要被邀请到持活力论思想的动物学家的大本营中，这件事值得大加庆祝，尽管仍然有一些暗地里的非议，认为不该这样谈论分子的结构和在面对"生命"与所有关于活体之活动性的奇妙现象时采取如此卑微的乐观态度！

在 A 分会与 K 分会的联席讨论开始时，阿斯特伯里先生概述了 X 射线在解释毛发、羽毛和其他蛋白质结构的性质方面的一些进展。我们最近在认识天然纤维分子结构上取得的进展很大程度上来自借助 X 射线法得到的认识，即固态纤维处于结晶状态，一般是有瑕疵的，然而虽然如此，这对于给出有关巨型分子的形式和性质方面的有价值信息已经足够了，而巨型分子被认为是传统化学的基础。微晶是亚微观的，不过通过 X 射线照片就立刻可以看到，它们**实际上**是细长的，具有长轴，并且其长轴或者大致与纤维轴平行，就像在丝和毛发中那样，或者呈螺旋形环绕于纤

---

*《生命的某些化学面貌》（见《自然》，9 月 9 日，第 389 页）
† 关于原生质结构的讨论，D 节，9 月 11 日。

arranged spirally round it, as in ramie and cotton.

In the light of a mass of experimental evidence of one sort and another—and it must be emphasised that the study of the fine structure of biological subjects has advanced and will continue to advance only through a close alliance between all the various methods of attack—the conclusion seems irresistible that these crystallites, or organised aggregates, which make up the body of the fibre substance and which we must now identify with Nägeli's micelles, are simply bundles of long chain-molecules, bundles of varying size and degree of perfection of organisation, and probably without any particularly sharp demarcation one from another. In the case of fibres of cellulose and natural silk, when the X-ray data are submitted to detailed analysis and compared with the results of tensile experiments and the findings of organic chemistry, this concept leads further to the decision that the chain-molecules are stereo-chemically fully extended; but we immediately encounter difficulties when we try to apply these ideas to the study of protein structures other than silk, such as hair, collagen, muscle, etc. The main obstacle, however, is removed by the discovery and interpretation of the X-ray photograph of *stretched* hair, a photograph which shows that the molecule or complex of hair keratin, when pulled, undergoes a reversible intra-molecular transformation into an elongated stereo-isomer in which the polypeptide chains are analogous in form to those of silk fibroin, that is to say, are fully extended and correspond to the normal polypeptide chains of the chemist. It follows, therefore, that the chains in unstretched hair are in equilibrium in a folded state, so that the mechanism of its extraordinary long-range elasticity is inherent in the keratin molecule itself: by the application of tension in the presence of water the keratin molecule can be stretched to roughly twice its equilibrium length, to which it returns exactly when the tension is removed.

More recent work on this problem indicates now that the "unit" of the keratin complex is actually a polypeptide sheet or "grid" in which the main-chains are linked side-to-side by a long series of roughly co-planar cross-linkages formed by the side-chains of the various amino-acid residues incorporated in the structure. The folds in the main-chains of unstretched hair referred to above lie apparently in planes transverse to the side-chains, as one would perhaps rather expect: each grid simply flattens out when pulled in the direction of the main-chains, thereby giving rise to a complex system of stresses and strains which must be the basis of the observed long-range elasticity. From this point of view the elastic properties of keratin are in no way different in principle from those of the simpler molecules; the latter, too, are susceptible of distortion within the limits imposed by inter-bond angles, electrostatic attractions, rotation about bonds, and so on, but in keratin the possibilities are so enormously enhanced by the length and mobility of both main-chains and side-chains that at first sight we appear to be dealing with a new phenomenon.

The most beautiful example of this line of reasoning is afforded by feather keratin, which gives an X-ray photograph at the moment unique in crystal analysis. Besides revealing quantitatively and for the first time the truly heroic proportions of a protein molecule, this photograph shows also that the molecule or complex of feather keratin can be stretched

530

维轴周围，就像在苎麻和棉中那样。

根据大量这类或那类的实验证据——而且必须要强调的是，只有通过与各种攻坚方法紧密结合，对生物材料精细结构的研究才取得了今天的进步并且还将继续进步——结论看来是确凿无疑的，我们现在必须将这些构成纤维状物质实体的微晶或有组织的团聚体看作是内格里胶束，它们只是大小和组织化完美程度各不相同的长链分子束，而且彼此之间可能不具有任何鲜明的分界线。对于纤维素和天然丝中的纤维来说，在对 X 射线数据进行细致分析并与张力实验的结果和有机化学的结论进行比较时，由这一观念可以进一步得出链状分子在立体化学上是完全伸展的论断；但是当我们试图用这些思想研究除丝以外的蛋白质结构，如毛发、胶原蛋白和肌肉等时，便立即遭遇到了困难。不过主要的障碍已经排除了，因为发现和解释了**伸长**毛发的 X 射线照片，该照片显示，毛发角蛋白的分子或复合物在拉紧时发生了可逆的分子内转化，变成伸长的立体异构体，其中的多肽链具有与丝蛋白类似的形式，也就是说是完全展开的，相当于化学家所说的普通多肽链。由此可以知道，未拉伸毛发中的链处于折叠状态的平衡之中，因此具有非常高的弹性就成为角蛋白分子的固有特性：在有水存在时施加拉力，可以将角蛋白分子拉伸到约为其平衡长度的两倍，去掉拉力后分子又精确地回复到平衡长度。

目前对这一问题的最新研究结果表明，角蛋白复合物的"单元"实际上是多肽层或"栅格"，其中主链通过一长列基本上共平面的交联并排地连在一起，这些交联结构是由包含在结构中的各种氨基酸残基的侧链形成的。上面谈到的未伸展毛发中主链的折叠明显位于横截侧链的平面内，人们也许会更期待这样的结果：每一栅格在受到主链方向上的拉伸时只是变平，从而形成了一个复杂的应力应变系统，这必然就是已观测到的高弹性的基础。从这个角度来看，角蛋白的弹性性质与较简单的分子没有任何原则上的分别；后者也会在一定限度内易受由于键间夹角、静电吸引和相对于键的旋转等所造成的变形的影响，但在角蛋白中，由于主链和侧链的长度和运动性极大地增加了变形的可能性，以至于乍看起来我们就像是在面对一种新现象。

羽毛角蛋白是这一系列推理的最好实例，它的 X 射线照片在晶体学分析中是独一无二的。除去定量地以及首次真正全面地揭示蛋白质分子结构之外，这张照片还表明，羽毛角蛋白分子或复合物可以连续且可逆地伸长多达其平衡长度的 7%！看

continuously and reversibly up to as much as 7 percent of its equilibrium length! It seems clear, too, that we are again operating with a net- or grid-like system, a molecular device which we may feel sure is common in biological structures, and of which the elastic properties are of fundamental importance for our knowledge of the mechanism of both growth and movement.

The paper presented by Dr. J. B. Speakman on the co-ordination of chemistry and X-ray analysis in fibre research followed admirably on the above account, emphasising as it did once more the extreme fruitfulness of a union of branches which, alone in fields of such bewildering complexity, might well prove barren. There is a pronounced difference in the lateral swelling of wool or hair in weak and strong acids, the former being far more effective. Considerations based on the Donnan equilibrium indicate why this should be so, and the argument is given stereo-chemical form, so to speak, by the corresponding X-ray photographs, which show how, in hydrochloric acid, for example, the main outlines of the keratin complex are scarcely disturbed, though in quite a dilute solution of chloracetic acid the diffraction pattern is obliterated completely, only to return in all perfection when the acid is removed by washing and drying. It was pointed out by Speakman how this observation offers a possible means of estimating the size of the grid-like units of keratin suggested by X-ray analysis; for we should be able, from a study of heats of reaction and swelling, to measure the total inter-grid cohesion in the extended form (β-keratin) as compared with that in the single-chain protein of silk, to which X-rays have shown the main chains of β-keratin to be analogous (see above). Experiments to this end are in progress.

The study of the effects of de-aminating animal hairs provides a still more instructive example of the value of X-ray and chemical collaboration. Stretched hair, as is well known, can be "set" in the elongated form by exposure to steam, and X-rays show that this is due to a re-distribution of cross-linkages in the keratin grid, whereby a new equilibrium configuration is taken up with the main-chains in the extended state. The reversibility of the intra-molecular transformation is thus destroyed by prolonged steaming, and the photograph of β-keratin persists. The remarkable thing now is that it is found that de-aminated hair has lost this power of "permanent set", and to an extent depending on the degree of de-amination. The change can be followed throughout by means of X-rays, which show at once whether the β-photograph is "set" after a given amount of de-amination and steaming of the fibre in the stretched state.

The experiments are a most valuable contribution to our knowledge of the chemistry of keratin, and therefore of all proteins, for we may now feel confident not only that the process of "setting" wool and hair involves the $-NH_2$ groups of the basic side-chains, but also that the contractile power of keratin is by no means destroyed—rather is its range extended—on their removal. In view of the theory of K. H. Meyer that the contractile power of muscle arises from attractions between basic and acidic side-chains of one and the same main-chain, it is clear that this discovery may have far-reaching implications.

532

来同样明显的是，我们还是在处理一个网状或栅格状体系，我们可以确信这是一种在生物结构中常见的分子构成，而且其弹性性质对于我们认识生长和运动的机制具有非常重要的意义。

斯皮克曼博士提交的那篇关于在纤维研究中协同使用化学分析法和 X 射线分析法的论文很好地说明了上述观点，文中不止一次地强调，在这些令人困惑的复杂领域中独自做努力很可能一无所获，但把各个分支结合起来就会硕果累累。毛或发在弱酸和强酸中的侧向膨胀存在着明显差异，前者要更为显著。根据唐南平衡可以解释为什么会如此，其论述是以立体化学形式给出的，可以说，相应的 X 射线照片说明，例如在盐酸中，角蛋白复合物的主体轮廓几乎没有被打乱，而在相当稀释的氯乙酸中衍射图案则完全消失，直到通过清洗及干燥将酸去除后才得以完全复原。斯皮克曼指出，这一观察结果说明利用 X 射线分析法也许可以估计角蛋白栅格状单元的尺寸；因为我们应该可以根据对反应热和膨胀热的研究来测定伸展形式（β- 角蛋白）中栅格间的总内聚力，从而与丝中某种单链蛋白的内聚力进行比较，X 射线图显示 β- 角蛋白与该单链蛋白具有类似的主链结构（参见上文）。以此为目的的实验正在进行中。

对于动物毛发脱氨基效应的研究为我们提供了一个更有利于说明同时使用 X 射线法和化学分析法的价值的实例。大家都知道，拉伸的毛发会因暴露于水蒸气中而被"变"为伸长形式，X 射线显示这是角蛋白栅格中的交联结构进行重新分布的结果，此时处于伸展状态的主链转变为一种新的平衡构型。于是分子内转化的可逆性会因长时间的蒸汽处理而被破坏，而 β- 角蛋白的照片保持不变。现在值得关注的事情是已发现脱氨基的毛发失去了这种"永久应变"的能力，并且在一定程度上取决于脱氨基的程度。利用 X 射线法可以全程跟踪这种变化，它能马上显示出在对处于拉伸状态的纤维进行某种程度的脱氨基和蒸汽处理之后，β- 照片是否"变"了。

上述实验对于我们认识角蛋白的化学性质乃至所有蛋白质的化学性质都是很有价值的；因为我们现在不仅可以确信毛和发的"变化"过程涉及碱性侧链的 $-NH_2$ 基团，而且了解到角蛋白在移动时其收缩能力完全没有遭到破坏，其范围反而增大了。根据迈耶的理论，肌肉具有收缩能力是因为同一主链中酸性和碱性侧链之间的吸引，很明显这一发现可能具有深远的含义。

The botanists were offered an elegant piece of structure analysis by Dr. R. D. Preston who, continuing the work of Astbury, Marwick and Bernal which brought to light that the cell-wall of the alga, *Valonia ventricosa*, is constructed of two sets of cellulose chains crossing according to some regular plan at an angle near a right angle, described the present state of an X-ray exploration of the whole of the wall of a single complete cell. Since a normal photograph taken at any point of the wall gives the two cellulose directions at that point, the method adopted is to follow up one of the directions exactly as one follows lines of force with a small compass needle, the results being afterwards plotted both on the cell-wall itself and on a large-scale model made from a bladder.

The investigation is necessarily a prolonged one with the modest apparatus available, but already the findings are of a highly intriguing character. They show to date that the molecular structure of this cellulose *balloon* is built up in spiral fashion, exactly as are the cellulose *fibres* ramie, cotton, etc. The completion of the investigation will no doubt be eagerly awaited, for there is a widespread interest in the structure and metabolism of *Valonia*. The single cells of this alga are the first to be explored in detail by X-ray methods, and the discovery that it shares with the fibres a spiral architecture must be of deep significance for the problem of the mechanism of growth.

The mechanism of growth was also indirectly the subject of a fascinating contribution by Mr. J. Thewlis, who showed how X-rays have revealed the arrangement of the apatite crystals which constitute the enamel of teeth. Tooth enamel, like so many other biological structures, is of a fibrous nature, the hexagonal axis of the apatite crystals being the fibre axis. In human enamel there are two sets of fibres, one with the fibre axis inclined at about 20° to the normal to the tooth surface and on the same side as the tip, the other at about 10° and on the opposite side to the tip. In dog's enamel the fibre axis is at right angles to the surface of the tooth. Variations in the perfection of fibre orientation are observed, and three kinds of enamel can be distinguished. In human teeth it is found that one kind is associated with clinically immune teeth, and the other two with clinically susceptible teeth. Here again the verdict of X-ray analysis must ultimately prove of fundamental importance in the study of living things, and it is to be hoped that this most promising field of investigation will soon be extended so as to take in the effects of the action of vitamins.

The biological implications of recent advances in the X-ray analysis of protein fibres were again dealt with by Mr. W. T. Astbury at the discussion on the structure of protoplasm. No doubt some of the zoologists present were not a little shocked at such heresy, but nevertheless the message of X-rays seems clear enough. The proteins are infinitely variable and adjustable molecular patterns, exquisitely sensitive to changes in physical and chemical environment, and capable of functioning not only as enzymes but also as the material embodiment of the genes. Surely they are no other than the very patterns of life!

(**132**, 593-595; 1933)

普雷斯顿博士为植物学家们提供了一篇有关结构分析的精彩文章，他延续了阿斯特伯里、马威克和伯纳尔的工作——即发现单球法囊藻的细胞壁是由两组纤维素链按某种规则的方式以接近于直角的角度交叉构成的，描述了目前用 X 射线研究单个完整细胞的细胞壁整体的状况。由于从细胞壁任何一点所得到的普通照片都给出该点处的两个纤维素方向，可以采用一种方法准确跟踪其中的一个方向，就像用小磁针追寻磁力线的方向一样，之后将结果同时画在细胞壁上和用球囊制成的大尺度模型上。

这项研究必定是一项可以利用现成的简单设备来进行的长期工作，不过已有的发现已经引起了人们的广泛关注。目前的发现表明这个纤维素**气囊**的分子结构是依螺旋形构建的，恰如苎麻、棉等纤维素**纤维**一样。由于人们对法囊藻的代谢和结构都很感兴趣，所以大家在急切地等待着研究的完成。这种水藻的单个细胞是第一个用 X 射线方法进行精细研究的实例，研究结果表明它与纤维都具有螺旋形构造，这一发现对于生长机制问题必然会有深远的影响。

生长机制还是泽尔利斯先生在一篇众所关注的文章中间接提到的主题，他展示了如何用 X 射线法说明构成牙釉质的磷灰石晶体的排布方式。像许多其他生物结构一样，牙釉质具有纤维的性质，磷灰石晶体中六方晶系的轴就是纤维轴。人类的牙釉质中有两组纤维，其中之一的纤维轴与牙表面的法线方向大约呈 20° 角并与顶部位于同侧，另一组则大约呈 10° 角且与顶部位于异侧。在狗的牙釉质中纤维轴与牙表面成直角。通过观测纤维取向的精确度就可以区分出三类牙釉质。在人类牙齿中发现，其中一类与临床免疫牙有关，另外两类则与临床敏感牙有关。这个例子再次证明 X 射线分析最终一定会在生命体研究中发挥重要的作用，而且可以期待这一最具发展前景的研究领域将很快地拓展到对维生素作用效果的研究中去。

阿斯特伯里先生在讨论原生质结构时再次谈到了最近在用 X 射线分析蛋白质纤维上取得的进展对于生物学的意义。无疑，在场的一些动物学家会因这种异端言论而受到不小的触动，但 X 射线给出的信息毕竟已足够清晰。蛋白质分子的结构变化多端并且可以改变，对于物理和化学环境的变化极为敏感，不仅能够作为酶还能作为基因的物质化身发挥作用。确切地说，它们就是生命的具体形式！

（王耀杨 翻译；吕扬 审稿）

The General Nature of the Gene Concept*

R. R. Gate

Editor's Note

The discovery that DNA is the molecule of heredity was so profound that it is hard to imagine how people previously understood genes as concepts. When R. Ruggles Gates wrote this article, James Watson and Francis Crick's discovery was twenty years in the future, and the concept of the gene was nebulous in the extreme. Scientists were agreed that genes resided in chromosomes, but the nature of the genes themselves was completely unknown. Were they living things, or inorganic? Did they reproduce by duplication or fission? Were they indivisible, like elementary particles? How big were they? Were they all the same size, or all different? Could you hope to see genes arranged on a chromosome? In 1933, no answers were possible.

THE conception of the gene has resulted from two lines of biological evidence: (1) The amazing stability of the germ plasm, as expressed in the facts of heredity; (2) its occasional instability, as shown by the occurrence of mutations. That external forces, such as X-rays, impinging upon the germinal material should produce changes, is not surprising but inevitable. That the resulting effects are inherited, however, shows that the organism is incapable of regulating against changes in this particular part of its cell structure.

It appears that these phenomena of stability and inherited change can only be understood by recognising that some substances or structures in the chromosomes must maintain in general their spatial relationships and chemical nature, not only from one generation of organisms to another, but also with only minor changes through thousands, and in some cases even millions, of years. However protoplasm grows, these substances must be self-reproducing, with a permanence equal to that of the species itself, for when they change the species changes.

While emphasising these conclusions, which seem inevitable from the modern genetical work, I do not wish to minimise the importance of the cytoplasm. It has been shown, for example, by the investigations of embryologists (for example, Conklin, Lillie) that the visibly differentiated substances in various animal eggs can be displaced and rearranged by centrifuging, without affecting the development, yet if the fundamental hyaline ground substance of the egg-cell is disturbed, distortions of development will be produced. This and the facts of egg polarity argue strongly for a more or less determinate spatial arrangement of the cytoplasmic materials, at least in many animal egg-cells. It has also been shown by reciprocal crossing of plant species that some species are differentiated

---

* From a paper read on September 12 at a joint discussion of Sections D (Zoology), I (Physiology) and K (Botany) at the Leicester meeting of the British Association on "The Nature of the Gene".

536

有关基因概念的一般本质[*]

拉各尔斯·盖茨

编者按

能发现 DNA 是遗传分子真是一个了不起的成就，很难想象在此之前人们是怎样理解基因这个概念的。在拉各尔斯·盖茨写这篇文章的时候，詹姆斯·沃森和弗朗西斯·克里克的发现还是 20 年以后的事，所以基因的概念是非常模糊的。科学家们一致认为基因是存在于染色体中的，但对基因的本质却一无所知。它们是生命体，还是无机物？它们是通过复制或分裂方式增殖的吗？它们像基本粒子一样无法分割吗？它们有多大？它们具有同样的大小，还是各不相同？人们能看到基因在一条染色体中的排列吗？在 1933 年的时候，没有人能回答这些问题。

基因的概念基于以下两条生物学证据：(1) 种质具有惊人的稳定性，表现为遗传；(2) 它也有偶发的不稳定性，表现为突变。接触像 X 射线这样的外部因素会使遗传物质发生变化，这并不奇怪也无法避免。然而，这种变化导致的结果却是可遗传的，这表明，生物体没有能力对其细胞结构中这一特殊部分的变化进行调节。

遗传的稳定性和可遗传的变异这两种现象似乎只能这样来解释：染色体中的一些物质或者结构不仅会在生物体从一代传给下一代的时候大体保持原有的空间关系和化学性质，而且在跨越千年，甚至有时候可达几百万年的过程中，都只发生微小的变化。不管原生质如何生长，遗传物质都必须以保持自身物种永远不变的方式自我复制，因为当遗传物质发生变化时物种也就发生了变化。

上述结论似乎是现代遗传学研究必然得到的结果，然而在强调这些结论的同时，我并不想忽略细胞质的重要性。有例子显示，胚胎学家们（例如：康克林和利利）通过研究发现，在各种动物的卵中那些明显分化的物质可以用离心的方法移除和重排，并且不影响卵的发育。而如果将卵细胞中基本透明的基质搅乱，畸形的发育就会出现。这个现象和卵极性的事实有力地说明了细胞质中的物质具有或多或少的定向空间分布，至少在许多动物的卵细胞中是这样的。此外植物物种的正反交也表明，一些物种的区别仅仅在于它们的核内组分不同，而另一些物种还因细胞质的不同而

---

* 9 月 12 日英国科学促进会在莱斯特召开了关于"基因本质"的会议，本文是在 D 分会（动物学）、I 分会（生理学）和 K 分会（植物学）共同讨论时宣读的一篇论文。

only as regards their nuclear content, while in others the cytoplasm differs as well.

The spatial arrangement of the genic materials within the chromosomes is therefore not different in principle from that shown to exist in the cytoplasm of certain animal eggs. The main difference is that the chromosome is a thread-shaped structure and is believed to be differentiated only along its length, that is, its differentiation is regarded as one-dimensional rather than three-dimensional.

In what sense do genes exist? The gene is probably the last in the long series of representative particles beginning with Darwin's "gemmules" and the "pangens" of de Vries, which were formulated to account for the phenomena of heredity. With advancing knowledge, such conceptions have tended to lose their formal character as ultimate particles reproducing by fission, and to become more physiological and more closely related to the known structure and activities of the cell. They lost their morphological nature when the conception of the unit character was given up many years ago. Bridges's conception of genic balance is essentially physiological. As Sir Frederick Gowland Hopkins has said of all organic units, "The characteristic of a living unit... is that it is heterogeneous... The special attribute of such systems from a chemical point of view is that these reactions are organised". What is the nature of the organisation which leads us to the conception of the gene?

In 1915, I first pointed out that a gene represents a *difference*—a fact so obvious that its importance is in danger of being overlooked. Johannsen, who invented the term "gene", afterwards (1923) expressed the same point of view. Our actual knowledge of genes, apart from speculation, is derived entirely from their differential effects in development and from the phenomena of linkage and crossing-over. The visible difference in the developed organism is the product of an initial germinal difference which must have arisen at some time through a mutation. The great majority of biologists will agree in locating the genic materials in the chromosomes. In the endeavour to get a more intimate picture of the nature of the gene, we must therefore explore the structure of the chromosome. It is also necessary to remember that, like everything else in the organic world, the genes, as well as the chromosomes, must have had an evolutionary history.

There have been two main theories of chromosome structure. According to one theory, the core of the chromosome contains a continuous thread or chromonema, which takes on a spiral form in various stages of mitosis. Cytologists have brought strong evidence for the existence of chromonemata in plant cells. The investigations, particularly of Sharp and Kaufmann in the United States and of Hedayetullah (1931) and Perry (1932) in my laboratory, have given a clear and definite picture of the chromosome during the cycle of mitosis. These accounts agree in finding the chromosome to be a double structure throughout the mitotic cycle, containing two chromonemata which are spirally twisted about each other in anaphase, telophase and prophase, each chromonema splitting before the chromosome halves separate in metaphase. There is also much wider evidence for the existence of a chromonema as a continuous thread embedded in the matrix of the

538

不同。

因此，染色体中基因物质的空间分布与某些动物卵细胞质中基因物质的空间分布没有原则上的区别。主要的区别在于染色体是一种线状结构，而且只能沿着其长度方向进行分化，也就是说，染色体的分化是一维而非三维的。

基因的存在有何种意义呢？基因很可能是从达尔文的"芽球"与德弗里斯的"泛子"开始的一系列代表性微粒的延续，而这些微粒概念的提出都是用来解释遗传现象的。随着人们认知水平的不断提高，这些概念也开始逐渐失去它们作为分裂复制的终极微粒的正式特征，变得越来越生理学化，越来越接近细胞的已知结构和活性。当很多年前单位性状的概念被抛弃的时候，这些概念就已经失去了形态特征。而布里奇斯的基因平衡观点本质上是属于生理学范畴的。正如弗雷德里克·高兰·霍普金斯爵士对所有有机单元的描述那样："一个生命单位的特征……是其异质性……从化学角度来看，生命单位的这一特征表现为化学反应的有序性。"这种把我们引向基因概念的有序性到底具有什么样的本质呢？

在 1915 年，我首先提出一个基因代表一种**差异**——这一事实如此明显以至于它的重要性有被忽略的危险。不久以后（1923 年），"基因"这个名词的发明者约翰森也表达了同样的观点。我们对基因的真正认知，除去推测以外，完全来自它们在发育过程中的分化效应和连锁与交换的现象。在成熟生物体上看见的差异均来自原始胚种的差异，这种差异是在某个时期通过突变产生的。绝大多数生物学家都相信基因物质位于染色体中。因此为了更进一步地了解基因的本质，我们就必须探究染色体的结构。有必要提醒的是，像有机世界里的所有其他事物一样，基因以及染色体也一定会有进化的历史。

现在主要有两种有关染色体结构的理论。其中一种理论认为染色体的核心包含有连续的细丝或染色丝，在细胞有丝分裂的不同阶段呈现螺旋状。细胞学家们已经在植物细胞中发现了证明染色丝存在的有力证据。特别是美国的夏普和考夫曼，以及本实验室的海德亚图拉（1931 年）和佩里（1932 年）已经通过研究清楚明白地揭示出了染色体在细胞有丝分裂周期中的变化情况。这印证了在细胞有丝分裂周期中的染色体是一种含有两条染色体丝的双链结构：在有丝分裂后期、有丝分裂末期和有丝分裂前期两条染色丝螺旋卷曲在一起，在有丝分裂中期染色体平分之前，两条染色丝分离。还有更多的证据证明染色丝是存在的，它作为一种连续的细丝存在于

chromosome. The genes must then be contained in this thread, and they must undergo duplication into two series before these are separated by the longitudinal fission of the chromonemata. The duplication of the chromonemata must then be the fundamental process on which the phenomena of heredity depend.

Another theory of chromosome structure which has been much in vogue in recent years and has found perhaps its strongest support in animal cells is that of the chromomeres. According to this view, the chromosomes in prophase and telophase are made up of granules or chromomeres strung together on a fine connecting thread. Various attempts have naturally been made to identify these discrete chromomeres with the genes. They are perhaps most clearly demonstrated in such work as that of Wenrich on grasshoppers. The chromomeres in cytological preparations, however, differ greatly in size, and their number appears to be smaller than present estimates of the number of genes. Bridges has spoken of them as the houses in which the genes live. If this is the case, it would appear that whole families or even villages of genes must live in one house. Belling (1928) endeavoured to count the number of chromomeres in certain plant nuclei and has arrived at 1,400–2,500.

In a posthumous paper recently published, as well as in earlier papers, Dr. Belling strongly supports the chromomere theory, from observations of smear preparations of pollen mother cells in various lilies. Not only does he deny that the chromosomes are split in telophase, but he also holds the novel view that the prophase split in the chromomeres is not accompanied by division of the thread connecting them. Instead, he thinks connecting threads are formed *de novo* between the new daughter chromomeres, thus linking them up into a new chromosome. The chief merit of such a view appears to be that it would obviate many of the serious difficulties which still exist with regard to all current theories of chiasmatypy and crossing-over. The fact that such diverse views can be held by competent cytologists, shows the extreme difficulty of crucial observation in this field.

Recent observations now in progress in my laboratory indicate that chromomeres may not exist, at least in plant cells. We are finding that, in some cases at any rate, the appearance of a string of beads or a moniliform thread, when critically analysed, is due to the presence of two spirally intertwined chromonemata, the nodes and internodes of which give the superficial appearance of a single chain of chromomeres. It is therefore desirable that a re-investigation, particularly of animal chromosomes, be undertaken, to make certain whether chromomeres actually exist or whether they will bear the general interpretation here suggested. In the meantime, it appears that the core of many plant chromosomes is a continuous structure, not broken up into visibly discrete bodies. As the imagination of many genetical investigators has been caught by the idea of discreteness both in the gene and within the visible chromosome, it is well to emphasise this point.

The absolute discreteness of the genes within the chromonemata does not appear to be an essential part of the gene theory. It is well known that many of the Protozoa have

染色体基质中。那么基因一定包含在这条细丝中，并且一定是在染色丝纵向分裂将其分离之前复制为两份的。染色丝的复制一定就是遗传现象发生的基础。

另一个关于染色体结构的理论是染色粒理论，这个理论近年来非常流行，对其最强有力的支持多半来自动物细胞。按照这个理论，染色体在有丝分裂的前期和末期是由串在一条细丝上的颗粒或者染色粒组成的。自然会有各种各样的方法来证明这些不连续的染色粒带有基因。温里克在蝗虫上做的实验也许就是对上述理论的最清晰的证明。然而，用细胞学方法制备的染色粒在大小上差别很大，它们的数量也低于目前估计的基因数量。布里奇斯曾把染色粒比喻成房子，而基因就生活在里面。如果真是这样的话，那整个基因家族甚至整个基因村落就不得不生活在一间房子里了。贝林（1928 年）曾努力计算过某种植物细胞核中的染色粒数量，得到的结果是1,400~2,500 个。

在最近发表的一篇贝林博士生前写的文章以及一些早期的论文中，贝林博士强烈支持染色粒理论，他的根据来自对各种百合花粉母细胞涂片的观察结果。他不但否定了染色体在有丝分裂的末期分离，而且还提出了新的观点，即认为在有丝分裂的前期，连接染色粒的细丝并不随着染色粒的分离而分离。相反，他认为这些粘连的细丝在新生的子染色粒间重新形成，进而把它们组合成一条新的染色体。这种观点的主要优点在于它能消除存在于所有有关染色体交叉和交换的现存理论中的许多严重困扰。事实上，一些有水平的细胞学家们也可以持有不同的观点，这表明在该领域里得到关键的观察结果是极其困难的。

本试验室在近期的观察中得到的最新资料表明，染色粒也许并不存在，至少在植物细胞中是这样的。我们发现至少在某些时候，经过精密的分析，一串珠子或一条念珠形的线其实是由两条染色丝螺旋缠绕形成的，这些染色体丝上的节点和节间段在表现上类似于一条染色粒的单链。因此人们需要对此进行重新研究，尤其是对动物的染色体，以弄清楚染色粒是否确实存在，以及它们是否符合本文提到的普遍解释。同时，许多植物染色体的核心都具有连续的结构，并不会被破坏成可见的不连续部分。因为许多遗传学研究人员的想象力都已经被基因和可见染色体内部具有不连续性的观念所束缚，所以最好强调一下这一点。

染色丝中基因的绝对不连续性并不是基因理论的关键部分。众所周知，许多原生动物都拥有为数众多的染色体，这些染色体发生纵向分裂并展现出更高等生物中

numerous chromosomes which undergo longitudinal fission and exhibit the usual features of the chromosomes in higher organisms. Are we to suppose that these chromosomes are as highly differentiated along their lengths as the evidence of crossing-over leads us to believe they must be in higher plants and animals? I find it impossible to accept such a view, which would be virtually a denial of evolution except in the embryological sense. The alternative is to assume that, when the mitotic mechanism first evolved in the Protista, the chromosomes were perhaps differentiated from each other but each was uniform along its length. From this point of view the mitotic mechanism would be a striking example corresponding with Berg's idea of nomogenesis.

The development of the mitotic figure may be regarded as one of the main evolutionary achievements of unicellular organisms. We may reasonably suppose that it appeared there in its simplest form and that the chromosomes in these groups of organisms remained more or less longitudinally homogeneous. We may then think of the evolution of higher plants and animals as having taken place through internal differentiation of the chromosomes, combined with adhesion of the products of cell division into multicellular aggregates. Thus would gradually arise the condition which has been postulated for higher organisms as a result of experiments in crossing-over, that is, a set of chromosomes not homogeneous but longitudinally differentiated. According to this view, all the developments of evolution in multicellular groups were foreshadowed or at least made possible by the mitotic mechanism achieved by the Protista. Just as the simplest cell aggregates consist of undifferentiated cells, so their individual chromosomes are internally homogeneous, each containing a different type of genic material.

The current view of genes, as developed particularly in connexion with *Drosophila*, tacitly assumes that all genes are of the same kind. If the views here expressed have any validity, then it seems more reasonable to suppose that a portion of an original chromosome, not necessarily of minimum size, underwent a mutation. Later, a portion of this would undergo a different change, and so on until a series of genes or chemically different segments of various sizes would result. This would lead ultimately to some genes of minimum dimensions, although others might be larger, and segments of the original unchanged chromosome might remain. It would appear probable, however, that in this process the majority of genes would ere now have reached the minimum size. (I find that East in 1929 also emphasised the view that genes are probably of various sizes.)

Some workers have of course taken an entirely different view of the origin and history of genes, regarding them as the primordial bodies or organic units from and by which protoplasm has since been constructed. Numerous comparisons have been drawn between genes on one hand and bacteriophage and virus particles on the other, based on their supposed similarity in size and action. While such comparisons are suggestive, the view of the genes as differentiated at a later stage of evolution within the originally homogeneous chromosomes seems on the whole more probable, and on this view there is no need to regard them as indivisible, discrete bodies of uniform size and nature.

542

染色体所具有的一般特征。我们是否可以假设这些原生动物的染色体沿着其长度方向进行高度分化，而杂交实验的证据让我们有理由相信这么高的分化程度只能存在于更高等的动物和植物中？我发现接受这样的观点是不可能的，因为它除了在胚胎学上的意义以外，实际上是对进化论的一种否定。或者可以这样假设：当有丝分裂机制首次在原生动物中形成时，染色体也许就是可以互相区分的，但是每条染色体都沿其长度方向保持一致。从这样的观点来看，有丝分裂机制竟是印证贝尔格的循规进化学说的一个生动实例。

有丝分裂象的发展可以看作是单细胞生物进化的主要成就之一。我们可以合理地假设单细胞生物的有丝分裂具有最简单的形式，而且这些生物体的染色体或多或少地保持了纵向的相同性。然后我们可以认为更高级动物和植物的进化是通过染色体的内部分化实现的，并伴随着细胞分裂成多细胞簇的产物附着。由此就会逐渐出现通过杂交实验产生更高等生物的必要条件，即一组染色体不再相同而是在纵向上产生了分化。按照这种观点，多细胞簇中的所有进化发展都已被预示或者至少因原生动物所具有的有丝分裂机制而成为可能。同样地，最简单的细胞簇是由未分化的细胞构成，因此它们的每条染色体在本质上都是相同的，但包含着类别不同的遗传物质。

随着基因研究尤其是与果蝇有关的研究的发展，人们现在对基因的看法是默认所有的基因都属于同一类型。如果上述观点成立的话，那么假定在最初的染色体中有一部分（不必是最少的部分）发生了突变似乎显得更合理。接着，这部分染色体会经历不同的变化直至产生一系列基因，或具有各种不同大小、不同化学性质的片段。这也许最终会导致出现一些具有最小尺寸的基因，尽管其他一些基因的尺寸可能比较大，而最初未发生变化的染色体片断也许会保留下来。然而在这样的一个过程中大多数基因在此之前可能就已经达到最小尺寸了。（我发现在 1929 年伊斯特也强调过这样的观点：基因可能具有不同的大小。）

当然也有一些人对基因的起源和历史有着完全不同的观点，认为基因一直是构建原生质的原始模块或者有机单位。人们在基因与噬菌体之间以及基因与病毒颗粒之间进行了大量的对比分析，这些对比分析都基于认为它们在尺寸和行为上具有相似性。虽然上述对比对我们很有启发，但认为基因是在本来相同的染色体进化后期发生分化的观点似乎更有可能成立，而且根据这个观点也没有必要把它们看作是一种具有相同尺寸和本质、不可分割也不连续的模块。

Various estimates of gene size have been made in *Drosophila*. One of the latest, by Gowen and Gay (1933), arrives at a minimum size of 10^{-18} cm.^{3}, the number of loci in the nucleus being estimated at more than 14,000. This maximum size would only allow space for about fifteen protein molecules. There is at present a large margin of error in such estimates. From measurements of spermheads and chromosome lengths, these authors draw the interesting conclusion that the chromosomes are all arranged end-to-end in the *Drosophila* spermhead.

The view of gene origin within the chromosomes as sketched above appears to be supported by the fact that the genes are now known not to be uniformly distributed in the chromosomes. The Y-chromosome has long been recognised as nearly empty of genes, but later work of Dobzhansky (1933) and others shows that one third or more of the length of the X-chromosome at the right or proximal end near the spindle fibre attachment is also inert. In this region only one mutation, "bobbed bristles", is known to occur, and crossing-over apparently does not take place. Possibly these inert or "empty" segments may represent an earlier unmutated condition of the chromosomes. The bulk of the chromosome is probably composed of thymonucleic acid, but it does not necessarily follow that the genes embedded in the chromonema axis are derivatives of that substance.

Although Belling believed that each chromomere contains a visible gene, yet the bulk of evidence leads to the conclusion that the genes are ultramicroscopic, and Bridges (1932) has recently expressed the view that they are unimolecular. It has been more usual to picture them as definitely organised bodies containing a score or a few hundred molecules and reproducing either organically by fission or chemically by duplication. The idea that each gene is a single molecule, while avoiding the possibility of its divisibility, appears to add difficulties of another kind. It is difficult to see why a tenuous chain of single unlike molecules should persist in the core of the chromosome, as it would be necessary to assume. Chemical forces alone could scarcely be expected to hold such a chain together, even if we rely upon the properties of the carbon atom. On the other hand, whatever physical forces give the chromosome its unity as a structure, might also be concerned (1) in organising each group of like or unlike molecules into a gene, and (2) in maintaining their axial arrangement in the chromosome. Could one molecule exert its catalytic effect while maintaining its position undisturbed in the chromonema? And could it duplicate itself when the row of genes divided? The mere asking of such questions shows that we do not know whether genes should be regarded as organic or inorganic groupings, and it indicates also that the time-honoured phenomena of growth and reproduction formerly associated with such bodies are in some danger of being lost, although it must equally be said that they have not yet been eliminated.

The "scute" series of genes in *Drosophila* has become increasingly difficult to interpret on the prevalent conception of the gene as a body which can never be fractionated but can only undergo change (mutation) as a whole. On the other hand, the theory of step allelomorphism as developed by Dubinin (1932) and others is entirely in harmony with

　　各种估测基因大小的研究在果蝇上展开。其中最近的一个数据是由高恩和盖伊（1933 年）估测的，他们认为基因的最小尺寸可达 10^{-18} cm³，而细胞核中基因位点的数目估计在 14,000 个以上。基因的最大尺寸只能容纳大约 15 个蛋白质分子。目前这类估测都存在很大的误差。这些作者根据对精子头部和染色体长度的测量结束，得出了一个有趣的结论：果蝇精子头部的染色体全部都是端 – 端排布的。

　　以上简单描述了基因来源于染色体的观点。现在已经知道基因并不是均匀地分布在染色体中，这一现象似乎是对上述观点的有利佐证。Y 染色体一直被认为是几乎没有基因存在的。而多布赞斯基（1933 年）和其他人后来的研究工作表明 X 染色体的右侧或者靠近纺锤丝附着点的近端有三分之一或者更长的部分也是惰性的。在这个区域只有一种突变会发生，那就是"截刚毛突变"，而交叉互换不会发生。也许这种惰性或者"没有基因的"片段代表着一种染色体早期未发生突变的情况。大部分染色体很可能都是由胸腺核酸组成的，但镶嵌在染色丝轴上的基因并不一定是该物质的衍生物。

　　尽管贝林认为每个染色粒都含有一个可见的基因，但是大量的事实说明基因是普通显微镜无法看到的，最近布里奇斯（1932 年）提出了基因是单分子物质的观点。而人们通常把基因看作是有确定结构的实体，由二十个或几百个分子组成，通过有机性的分裂或者化学上的复制实现自身的繁殖。把基因看作是一个单分子虽然排除了它的可分割性，但似乎又增加了另一个麻烦。很难想象如果不是分子，一条脆弱的单链能稳定地保持在染色体的核心上，因而我们必须这样假设。仅靠化学力几乎不可能将这样的链结合在一起，即使我们信赖碳原子的特性。另一方面，如果使得染色体具有统一结构的是物理作用力，那么也许要考虑：（1）如何将每一个类似分子或不是分子的组分组织起来变成基因；（2）如何保持这些组分在染色体中沿轴向排列。当一个分子在染色丝中保持自己的位置不受干扰时它能发挥它的催化效应么？当一行基因被分开时它会进行自我复制么？提出这样的问题只能说明我们还不知道基因应该被归为有机类还是无机类。这些疑问还暗示，由来已久的生长和繁殖现象从前与这种物质相关，现在有可能会失去这种联系，尽管同样会有人认为它们目前还没有被消除掉。

　　当前流行的观念认为基因是一个不能被分开，而只能作为一个整体发生改变（突变）的实体，用这样的观念解释果蝇的"鳞甲"基因系列变得越来越困难。另一方面，由杜比宁（1932 年）等人发展的阶梯型对偶基因学说与我在上文中提到的基因进化观

the view of gene evolution which I have outlined above. On the assumption that genes are indivisible in all circumstances, it has been necessary to make them smaller and smaller, until the limit is now reached in the single molecule. But surely, if the atom itself can be disrupted by suitable forces, it is not unreasonable to suppose that something of a similar kind may happen to a group of molecules constituting a gene.

The genetic study of variegations in plants has also led to the view that the genes involved are compound structures, the somatic segregation of which results in the variegated condition. The studies of Emerson on the varieties of maize with variegated pericarp, of Baur on *Antirrhinum*, Eyster on maize and *Verbena*, Demerec on *Delphinium* and *Drosophila* are notable in this connexion. Eyster (1928) adopts the hypothesis that the genes causing variegation are compound structures composed of a constant number of "genomeres" which may or may not be of the same chemical or physical nature. Demerec (1931), however, explains the variegation in *Delphinium* as a result of highly mutable genes. It remains to be seen whether the divisibility of the gene in somatic tissues or the high mutability of such genes will supply the explanation.

We prefer to think of genes as differentiations of many kinds and sizes which have arisen in the core of the chromosome during its evolution, making it a nest of catalytic substances, most of them having specific effects mainly on the development of particular organs. By different processes of translocation in the nucleus, genes tend to become shifted from their original positions. The result is that genes affecting quite different organs come to occupy adjacent positions in the chromosomes. It seems quite likely that, from a historical point of view, mutation has been a much more orderly process than might be supposed from the present disorderly arrangement of the genes in *Drosophila*. The fact that the genes have been scrambled in this way seems to show that mere position within the chromosome is of little or no significance.

(**132**, 768-770; 1933)

点完全吻合。假定基因在任何环境下都是不可分割的，人们必然会不断地缩小它的尺度，到现在已经达到了单个分子的极限。但是当然了，如果原子本身可以被合适的力分解开来，那么假定类似的事情会在一群构成基因的分子中发生就没有什么不合理了。

对植物颜色变化的遗传学研究也证明了所涉及的基因具有复合结构，对其进行体细胞分离会导致杂色的情况。在此类研究中令人关注的有：用杂色果皮研究玉米多样性的埃默森、研究金鱼草的鲍亚、研究玉米和马鞭草的艾斯特以及研究飞燕草和果蝇的德梅雷茨。艾斯特（1928 年）采用的假说认为引起杂色的基因是一些由恒定数量"基因粒"构成的复合结构，这些基因粒的化学性质或物理性质可能相同，也可能不相同。而德梅雷茨（1931 年）认为飞燕草的杂色是由非常容易突变的基因引起的。体组织中基因的不可分割性或者这类基因的高度易突变性是否能解释这一现象还有待进一步的观察。

我们倾向于把基因看作是具有各种类型和尺寸的分化产物，这些分化产物出现在进化过程中的染色体的核心中，并使之成为催化物质的中心，这些催化物质中的大部分对于特定器官的发育具有特异性作用。通过在细胞核内发生的各种转运过程，基因易于从原来的位置移开。其结果是，控制完全不同的器官的基因在染色体上占据了相邻的位置。从历史的角度来看，突变过程很可能远比人们假定果蝇基因中目前的无序排列要有序得多。基因以这样的方式混杂排列在一起的事实似乎告诉我们，仅仅考虑基因在染色体上的位置是没有意义的。

（刘振明 翻译；刘京国 审稿）

Oxygen Affinity of Muscle Haemoglobin

R. Hill

Editor's Note

That the red colouring matter in muscle and blood is similar but not identical is borne out by this paper, with its evidence that the absorption of light differs between the two tissues. It had been suggested already that "muscle haemoglobin" has only half the molecular weight of blood haemoglobin; this paper shows also that the two have different affinities for binding oxygen (which is their biological function). In fact it turns out that "muscle haemoglobin", now called myoglobin, has only a quarter the mass of blood haemoglobin, something revealed in crystallographic work at Cambridge in 1958, which earned John Kendrew the 1962 Nobel Prize in chemistry.

THE fact that the haemoglobin within red muscle is a substance different from the haemoglobin in the circulation, has now been definitely established. The pigment has been isolated and prepared in a crystalline condition by Theorell[1], and measurements by Svedberg showed a molecular weight of half that of the blood pigment.

It is therefore interesting to consider properties indicating the function of this pigment in red muscle. It can be shown by a simple experiment that muscle haemoglobin has a higher affinity for oxygen than the haemoglobin of blood. Muscle haemoglobin shows its sharpest absorption band (α) at 5,800 A., that of ordinary blood being at about 5,770 A. A mixture of the two pigments in dilute solution is subjected to different tensions of oxygen under physiological conditions of temperature and pH. At higher oxygen pressures the α band occupies an intermediate position characteristic of the mixture; as the oxygen tension is lowered, the α band approaches 5,800 A. There must therefore be a considerable difference in affinity for oxygen between the two pigments under physiological conditions.

By means of a rapid spectroscopic method of measuring dissociation curves, the oxygen affinity of muscle haemoglobin has been compared with that of the haemoglobin of blood in a borate buffer at pH 9.3 and 17 °C. The method consists in introducing different amounts of a dilute (10^{-4} m.) haemoglobin solution into an evacuated vessel. The fluid is previously saturated with air, and the total amount of oxygen introduced into the vessel can be calculated. The percentage of oxyhaemoglobin is estimated using the principle of Krogh's method, by comparison with optical mixtures of oxyhaemoglobin and haemoglobin. The tension of oxygen in the vessel can then be calculated. For mammalian haemoglobin in borate buffer at pH 9 and 18 °C., the unoccupied volume of the vessel is of the same order as that of the fluid, and equilibrium can be quickly established. Such a rapid method is essential when dealing with haemoglobin under conditions in which it is likely to change to methaemoglobin, this being apparently the case with muscle haemoglobin.

肌肉中血红蛋白的氧亲和力

罗伯特·希尔

编者按

肌肉中的和血液中的红色物质虽然相似但不相同这一点主要被本文提供的数据——这两种组织在光吸收上存在差异——所证实。曾有人提出，"肌肉中血红蛋白"的分子量只是血液中血红蛋白分子量的一半，本文对它们的分析结果还表明，两者对氧有不同的亲和力（这一点与它们的生物学功能相关）。实际上后来证实：现在被称为肌红蛋白的"肌肉中血红蛋白"其实只有血红蛋白质量的四分之一。该结果由1958年剑桥大学的结晶学研究成果所证实，约翰·肯德鲁因此赢得了1962年的诺贝尔化学奖。

红色肌肉中的血红蛋白是一种与血液循环体系中的血红蛋白不同的物质，这一事实已经得到了确认。红色肌肉中的色素分子已经被特奥雷尔分离并制备成晶体形式[1]，斯韦德贝里的测定结果表明，红色肌肉中的色素分子的分子量是血液中色素分子的一半。

因此，认识那些反映红色肌肉中色素分子功能的性质自然是人们的兴趣所在。通过一个简单的实验就能揭示，肌肉中的血红蛋白比血液中的血红蛋白具有更高的氧气亲和力。肌肉中血红蛋白的最强光吸收带（α）出现在 5,800 Å，而通常血液的最强光吸收带则出现在 5,770 Å 附近。在与生理环境相同的温度和 pH 值条件下，两种色素分子的低浓度混合液被施以不同的氧分压。结果是，在氧分压较高的情况下，α 吸收带出现在一个比较靠中间的位置，这正是两种色素混合物的特征；随着氧分压的降低，α 吸收带的位置则逼近 5,800 Å。由此可以得出的结论是，在生理条件下这两种色素分子对氧的亲和力必然存在相当大的差异。

通过利用一种测量氧解离曲线的快速光谱学方法，我们在 pH 值为 9.3、温度为 17 ℃的硼酸盐缓冲液中比较了肌肉中血红蛋白与血液中血红蛋白的氧亲和力。具体方法是：向一个被抽成真空的容器中加入不同量的血红蛋白稀溶液（10^{-4} 摩尔）。这些溶液在加入之前是被空气饱和的，因此，加入到真空容器中的氧的总量是可以被计算出来的。根据克罗格方法的原理，通过与氧合血红蛋白和无氧血红蛋白的光学混合物进行比较，我们可以估算出氧合血红蛋白所占的百分比，继而我们可以计算出容器中的氧分压。对于溶解在 pH 值为 9.0、温度为 18 ℃的硼酸盐缓冲液中的哺乳动物血红蛋白而言，其溶液和容器中溶液上空的体积处于同一量级，因此氧气在二者之间的平衡能很快就被建立。当涉及血红蛋白有可能转变成高铁血红蛋白的条件时——这对肌肉中的血红蛋白而言是显然会发生的——这样一种快速检测方法就成为非常必要的了。

Fig. 1 shows two dissociation curves; one being that of the haemoglobin of ox blood, the other that of the muscle haemoglobin extracted from the perfused ox heart at pH 9.3 and 17 °C.

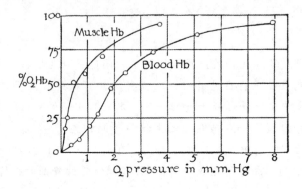

Fig. 1.

That the distinctive properties of muscle haemoglobin are due to its protein, globin, can be shown. If the globin is separated and combined with pure haemin from blood, the α band of the resulting oxyhaemoglobin, like the original muscle haemoglobin, still appears in the position 5,800 A. The combination of muscle globin with mesoporphyrin has a very sharp absorption spectrum identical in quality with that produced from globin of blood; the sharp band in the red region is, however, 15 A. displaced toward the red end of the spectrum. Thus the specificity of the globin is shown both in the case of the oxyhaemoglobin and after its reaction with porphyrin.

It seems clear that the presence of muscle haemoglobin within the muscle cells will be of definite advantage in oxygen transport. The actual amount of haemoglobin in the heart muscle is of the same order as the amount of haemoglobin in the capillaries. The dissociation curve is decidedly less inflected than that of the blood pigment. (This is in accord with the measurement of the molecular weight given in Theorell's paper.) In the middle range of the dissociation curve there is a large difference in the relative saturations at equilibrium, which will allow the muscle pigment to take up the oxygen from the blood. The respiration of the cells, containing in the case of red muscle a large amount of the oxidase-cytochrome system, can continue at very low pressures of oxygen. The muscle haemoglobin, with its relatively high affinity for oxygen, can be the intermediate carrier of molecular oxygen from the blood to the oxidase-cytochrome system in the cells.

(**132**, 897-898; 1933)

R. Hill: School of Biochemistry, Cambridge, Nov. 3.

References:
1. Theorell, A. H. T., *Biochem. Z.*, 1, 252 (1932).

图 1 展示了两条解离曲线，其中一条（右边的）代表的是牛血液中的血红蛋白，另一条（左边的）代表的是从在 pH 值为 9.3、温度为 17 ℃时灌注的牛心脏的肌肉中提取的血红蛋白。

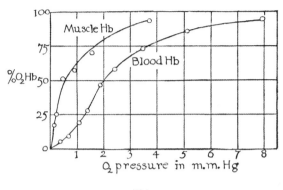

图 1.

肌肉中血红蛋白表现出来的不同特性是源自其组成中的蛋白质（球蛋白）部分，这一点是可以被证明的。如果把其中的球蛋白部分分离出来，并与来自血液中的氯高铁血红素结合的话，所产生的氧合血红蛋白的 α 吸收带仍会出现在 5,800 Å，这与原来的肌肉中血红蛋白是一样的。肌肉中血红蛋白的球蛋白部分与中卟啉分子的结合会产生一个非常尖的吸收谱，其情形等同于血液中的球蛋白。但是，在红光区的那个很尖的吸收带向光谱的红色末端偏移了 15 Å。以上结果表明，球蛋白的特异性无论是在其形成氧合血红蛋白时还是在其与卟啉辅因子结合后都会彰显出来。

似乎颇为清楚的是，肌细胞中血红蛋白的存在对于氧气的运输肯定是有利的。心肌中血红蛋白的实际含量与毛细血管中血红蛋白的含量处于相同的量级，其氧解离曲线的弯曲程度明显低于血液中所含色素的解离曲线。（这与特奥雷尔所发表论文中给出的分子量测定结果相吻合）。在这两种蛋白质的解离曲线的中间部分，处于平衡时的相对氧饱和度存在很大差异，这就使得肌肉中的血红蛋白能够从血液中摄取氧气。细胞的呼吸作用——就红色肌肉而言涉及大量的氧化酶-细胞色素蛋白质——能够在氧分压很低时仍然继续发生。具有较高氧亲和力的肌肉中血红蛋白可以作为中间载体把分子氧从血液转运到细胞中的氧化酶-细胞色素系统中去。

（姜薇 翻译；昌增益 审稿）

Possible Chemical Nature of Tobacco Mosaic Virus

E. Barton–Wright and A. M. McBain

Editor's Note

Are viruses alive or not? That question is still debated today, although it is perhaps best answered by recognising that viruses—genetic material in a protein coat, which hijacks the replication machinery of host cells—reveal these categories to be too vague at the molecular scale. Here plant virologists Eustace Barton-Wright and Alan McBain in Scotland help to initiate the debate with experiments that, in retrospect, offer a central part of the puzzle. They separate the tobacco mosaic virus into its protein and nucleic-acid (RNA) components, and find the latter is the key to infection. But lacking knowledge of RNA, and with biochemistry fixated on proteins, they cannot interpret the results other than to guess that viruses may be enzymatic.

IN a series of observations, Vinson and Petre[1] claim that the virus disease of tobacco mosaic behaves as a chemical compound. We have repeated this work in detail and confirmed it in every particular, and we are also of the opinion that the virus in this case behaves as a chemical compound and not as a living organism.

In our own investigations, we have employed Johnson's so-called No. 1 mosaic and we are indebted to Dr. Bewley, of the Cheshunt Research Station, for our source of the disease. The disease was transferred to *Nicotiana Macrophylla* by juice inoculation.

We have particularly examined the mixed phosphate eleuate described by Vinson and Petre. This was found to be highly infectious (plants inoculated, 10; plants diseased, 10). The eleuate was found to contain protein (xanthoproteic and biuret test). On gently warming, the protein was precipitated. It could also be precipitated by saturated, but not by half-saturated, ammonium sulphate solution. Contrary to the statement of Vinson and Petre, we found that when the eleuate was brought to pH 5 and acetone added (two volumes of acetone to one of eleuate) a heavy white precipitate fell. The acetone precipitate was found to contain protein and proved to be infectious (plants inoculated, 5; plants diseased, 5).

It was observed that the acetone precipitate could be separated into two fractions, a white crystalline solid and a gelatinous portion which proved to be protein. The question arose as to whether the virus was present in the protein fraction or in the white crystalline solid, or whether both were needed to bring about infection. Infection with protein alone induced disease (plants inoculated, 5; plants diseased, 5). We were quite unable to free the protein from virus.

烟草花叶病毒的可能化学本质

尤斯塔斯·巴顿－赖特，艾伦·麦克贝恩

编者按

病毒是不是活的呢？这是一个今天还在被争议着的问题，目前最好的回答可能是：我们需要认识到，以遗传物质被包裹在一层蛋白质外套中并劫持宿主细胞复制机器为特征的病毒说明，这样的生物分类法在分子尺度就显得过于模糊了。在本文中，苏格兰植物病毒学家尤斯塔斯·巴顿－赖特和艾伦·麦克贝恩通过他们的实验帮助启动了一场争辩。回过头去看，他们的实验结果为揭开此谜提供了关键内容。他们将烟草花叶病毒分离成了蛋白质和核酸（RNA）两种组分，并发现后者是导致感染发生的关键所在。但是，由于当时人们缺乏对 RNA（核糖核酸）的知识，生物化学的重点被聚焦在蛋白质身上，所以他们对上述结果的唯一解释只是猜测病毒可能属于一种酶类物质。

在一系列的观察中，文森和彼得[1]声称导致烟草花叶疾病的表现表明这种病毒为一种化合物。我们详细地重复了这项研究工作，完全证实了他们所有的观察结果，我们得出了同样的观点：在这里病毒表现为一种化合物，而非活的有机生命。

在我们自己的研究中，采用的是约翰逊的所谓的 1 号花叶。我们要特别感谢柴斯罕特研究站的比利博士为我们提供的病原。这种疾病被通过汁液接种的方式转移到烟草中。

我们特别检测了文森和彼得所描述的混合磷酸洗出液。这种洗出液被发现具有很强的感染性（接种植株：10；染病植株：10）。我们发现洗出液中含有蛋白质（黄色蛋白质反应和双缩脲检测）。当温和加热时，蛋白质发生了沉淀。饱和的（非半饱和的）硫酸铵溶液也可以使蛋白质发生沉淀。与文森和彼得所报道的观察不同的是：我们发现，如果将洗出液的 pH 值调节到 5 并加入丙酮时（每单位体积的洗出液中加入 2 单位体积的丙酮），就会出现大量的白色沉淀。我们发现这些丙酮沉淀物中含有蛋白质，而且已证明它们具有感染性（接种植株：5；染病植株：5）。

我们观察到，丙酮沉淀物可以被分离为两个组分，其中之一是白色结晶固体，另一部分的胶状物质被证明是蛋白质。接下来的问题是：病毒究竟是存在于蛋白质组分中，还是存在于白色结晶固体中，还是说导致感染时二种组分都是必需的。仅用蛋白质组分感染植物，也导致了染病（接种植株：5；染病植株：5）。当试图将蛋白质组分与病毒分开时，我们面临了相当的难度。

The white crystalline solid was purified by repeated precipitation with acetone, washed with ether and dried in a vacuum desiccator. It proved to be mainly composed of phosphate, but considerable organic matter was also present as it charred on heating. It was found to contain *no* nitrogen and proved infectious (first inoculation: plants inoculated, 5, plants diseased, 5; second inoculation: plants inoculated, 8, plants diseased, 8). The protein fraction apparently plays no part in infection. The following experiments confirm the fact that protein plays no part in bringing about the disease. The addition of 1 percent solution of safranin to the phosphate eleuate produced a slow precipitate. This was separated on the centrifuge, suspended in water and the safranin removed with normal amyl alcohol. The aqueous solution was found to be infectious and contained no protein, phosphate or nitrogen.

As a control, sap from healthy plants was treated in exactly the same way as that from diseased plants. The behaviour of the mixed phosphate eleuate with acetone was quite different. Instead of the heavy white precipitate described above, a faint opalescence appeared which did not settle for many hours.

That plant viruses are not living organisms has been previously suggested. It has been stated that they are possibly enzymic in nature. Vinson and Petre are of the opinion that tobacco mosaic virus is of the nature of a simple protein. The isolation by us of a white crystalline compound which contains no nitrogen and yet is highly infectious appears to us to preclude the possibility of tobacco mosaic virus being of the nature of a living organism. In its precipitation with safranin it shows affinities with the proteolytic enzymes, but until we have made further investigation of the substance we can make no definite statement as to whether or not it is enzymic in nature.

(**132**, 1003-1004; 1933)

E. Barton-Wright, Alan M. McBain: Scottish Society for Research in Plant Breeding, Corstorphine, Edinburgh, 12.

References:
1. *Contrib. Boyce Thompson Instit.*, 1, 479 (1929). 3, 131 (1931). See also Vinson, *Phytopath.*, 23, 35 (1933).

白色结晶固体被通过反复的丙酮沉淀而纯化，然后用乙醚清洗，并用真空干燥器进行了干燥。分析结果表明，这些白色结晶固体的主要成分为磷酸盐，但也含有相当量的有机物，因为加热时它会变焦。我们发现这些纯化的样品中**不含有**氮，但却被证实具有感染性（第一次接种：接种植株，5，染病植株，5；第二次接种：接种植株，8，染病植株，8）。其中的蛋白质组分显然在感染过程中不发挥什么作用。下面的实验结果也证实，蛋白质在导致疾病过程中并不发挥任何作用。在磷酸洗出液中加入 1% 的番红染料液会产生缓慢的沉淀。这样的沉淀被用离心机分离出来，重新悬浮于水，其中的番红染料再用标准的戊醇去除。所得到的水溶液被发现具有感染性，而且不含有蛋白质、磷酸盐或氮。

作为对照，来自健康植株的体液也用与以上处理染病植株体液精确一样的方法进行处理。用丙酮处理健康植株获得的混合磷酸洗出液表现出相当不同的行为。与前面描述的大量白色沉淀不同，这里出现的是清淡的乳白色沉淀，而且它们在过了很多个小时之后都不沉降下去。

植物病毒并非活的生物这一点先前已经被他人提出来过。有观点认为，它们可能本质上属于酶类物质。文森和彼得的看法是，烟草花叶病毒从本质上而言是一种简单的蛋白质。我们分离到了一种白色晶体化合物，它不含氮，但却具有很强的感染性。我们觉得，这些证据使我们能够排除烟草花叶病毒本质上属于一种活的生命体的可能性。在番红沉淀物中，表现出了与蛋白质水解酶的亲和性，但在对这些物质进行进一步研究之前，我们不能确切地陈述烟草花叶病毒从本质上是否真的属于酶类物质。

（韩玲俐 翻译；昌增益 审稿）

Recent Discoveries at Choukoutien[*]

D. Black

Editor's Note

Davidson Black and his team discovered the remains of a new kind of fossil man, *Sinanthropus* ("Peking Man", later assigned to *Homo erectus*) at the cave site of Choukoutien (now Zhoukoudian) in China. Here Black presents a brief report on excavations in younger strata at the same site, producing a rich fauna including *Homo sapiens*. This evidence, said Black, showed that Choukoutien had been occupied over a long period, successively by *Sinanthropus* and *Homo sapiens*.

Upper Palaeolithic Culture in "Upper Cave" Sediments

A detailed account of the results of the Choukoutien excavations up to May 1933 has already been presented in our memoir "Fossil Man in China" (*Mem. Geol. Surv. China*, Series A, No. 11). In that report it was noted that above the *Sinanthropus* deposits there occurred towards the top of the hill a pocket of grey sediments of apparently modern facies, the site being described as the "Upper Cave". During the past season, Mr. W. C. Pei has systematically investigated the deposits of the latter site, ably assisted by Mr. M. N. Pien. Their efforts have been rewarded by the discovery of much additional material of unexpected archaeological significance.

(1) *Sedimentary and lithological characters of Upper Cave deposits*. The "Upper Cave" was a true cave but became completely filled with a mixture of grey cave loam and angular flat limestone fragments, the latter being derived from the collapsed portion of its roof. The roof is preserved over a quite large recess of the cave which extends to a smaller lower chamber not yet completely excavated. Where exposed, the cave walls are covered with stalactites and stalagmites. The grey Upper Cave sediments are largely unconsolidated and are in contact only over a few square metres with the hard red beds and stalagmitic floors capping the *Sinanthropus* strata of Locality 1. Elsewhere the Upper Cave appears to be developed as an independent system.

(2) *Fauna of the Upper Cave*. Though not very abundant, the Upper Cave fauna is remarkably rich in types and includes a puzzlingly large number of almost complete skeletons, the bones of which lie in correct association and are but slightly fossilised. The most interesting forms are as follows:—*Hyaena* (an extinct species very different from that found in the *Sinanthropus* beds but similar to that of Sjara-osso-Gol); *Felis tigris* (entire

---

[*] Report of excavations during the field season 1933, presented at the annual meeting of the Geological Society of China on November 11.

周口店的最新发现[*]

步达生

编者按

戴维森·布莱克（步达生）和他的研究小组曾在中国周口店的岩洞里发现了一种新的化石人类，即"中国猿人"（"北京人"，后来被归为"直立人"）。在这里步达生简略地报告了同一地点更新层位的挖掘工作，这次发掘获得了丰富的动物群，其中包括智人。步达生认为这个证据表明很长一段时期内在周口店先后生活着中国猿人和智人。

"山顶洞"堆积中的旧石器晚期文化

在我们的研究报告《中国的化石人类》中已经详细记述了 1933 年 5 月前在周口店的挖掘工作结果（《中国地质专报》，A 辑，第 11 期）。这份报告指出了在中国猿人沉积物之上，靠近山顶的位置出现了一个明显具有现代特征的灰色堆积，这个地点被描述为"山顶洞"。在过去的一段时间内，裴文中先生在卞美年先生的得力帮助下，对后一遗址的堆积物进行了系统的研究。他们的努力换来了大量具有考古价值的意外发现。

（1）**山顶洞堆积物的沉积特点和岩石学特征**。山顶洞是一个真正的山洞，但是已完全被灰色坴姆土和扁状石灰岩碎片填满，后者来自洞顶塌落部分。洞内有一处洞壁向内形成一个较大的凹陷，其上方的洞顶保存了下来，而这个凹陷一直延伸至一个较小较低的尚未完全被挖掘的小室。露出的洞壁上覆盖着钟乳石和石笋。灰色的山顶洞堆积物大部分没有胶结，只有几平方米的面积与胶结坚硬的红色基底和石笋层接触，而这一层正是 1 号地点中国猿人地层的封顶。山顶洞的其他部分看起来是独立形成的一个系统。

（2）**山顶洞的动物群**。尽管山顶洞的动物群数量不是很大，但是种类丰富，令人费解的是，这其中包括大量近乎完整的骨架，这些骨架的骨骼正常关联，只是轻微石化。其中最令人感兴趣的种类如下：鬣狗（该物种现已灭绝，它与在中国猿人地层发现的物种存在很大差异，但与在萨拉乌苏发现的物种很相似）；虎（完整骨架）；

---

[*] 这是一篇 1933 年田野考古工作季关于发掘现场的报告，发表于 11 月 11 日中国地质学会的年会上。

557

skeleton); *Cynailurus*, which is now restricted to India (an entire skeleton); *Viverra* (no longer found in North China); the wild ass; *Equus hemionus;* and the deer, *Cervus elaphus* (an entire skeleton), having antlers curiously similar to the special form from Sjara-osso-Gol.

(3) *Human and cultural remains.* In association with this fauna there occur both human skeletal remains and traces of industry. The skeletal remains are of modern type (*Homo sapiens*) and so far comprise two almost complete but somewhat crushed skulls, other skull fragments and teeth, fragmentary lower jaws, bones of the upper extremity (including one clavicle displaying a healed fracture), vertebrae, leg and foot bones. Traces of fire (charcoal and ash) are abundant.

There are three stone implements in a beautiful black chert, a well-made scratcher in vein quartz and several flakes and nuclei in vein quartz, and also a needle (eye broken), a deer canon bone worked at both ends, some thirty or more fox canine teeth perforated for necklace, an ornamental cylindrical piece made from a long bone of a bird, and a considerable quantity of oolitic haematite probably imported from a considerable distance. So far, no trace of pottery, polished stone or microlithic industry has been encountered.

Conclusions. The material recovered will shortly be made the subject of a full report and the conclusions here offered are wholly tentative. (*a*) The Upper Cave deposits appear to be decidedly younger than the *Sinanthropus* layers of Locality 1, from which they are separated by stratigraphic and lithological disconformity and by a faunistic interval (absence of thick-jawed deer, occurrence of a special *Hyaena*, presence of *C. elaphus*, *E. hemionus*, etc.). (*b*) The Upper Cave deposit is, however, probably also Pleistocene in age (collapsed cave, loess-like sediments, presence of Hyaena; cf. *speloea*, *Cynailurus*, *Viverra*, *E. hemionus*, special deer, etc.). (*c*) In these circumstances, we are inclined provisionally to attribute the associated human remains to a Late Pleistocene, Palaeolithic culture. The latter would seem to correspond approximately to the same stage as the Upper Palaeolithic of Siberia and Europe. It appears, however, to be somewhat more advanced than the Ordos industries (Shui-tung-ko and Sjara-osso-Gol) in which no typically worked bones have thus far been found in certain association.

Cynocephalus Remains

In a cylindrical solution cavity about a metre in diameter in the limestone to the south of Locality 1, Mr. M. N. Pien discovered this season a considerable number of fossil bones imbedded in a peculiar red deposit containing a large proportion of small well-rounded pebbles. These bones are remarkably fossilised and heavy, many of them being water-worn and rounded. A few, however, are well preserved, among the latter being several teeth and limb bones of a large baboon, probably *Cynocephalus wimani*, Schlosser. Strikingly similar deposits containing the same type of heavy rolled bone fossils have already been encountered at the very base of the *Sinanthropus* deposits of Locality 1 (Lower Cave). At the present stage of excavation it remains an open question whether or not these beds represent a pre-Choukoutien stage or merely correspond to an early phase in the last filling of the clefts.

猎豹，现在只在印度有发现（一具完整骨架）；灵猫（在中国北方再也没发现过）；野驴；马鹿（一具完整骨架），令人奇怪的是，该鹿的鹿角与萨拉乌苏发现的那具鹿角形状非常相似。

（3）**人类和文化遗迹**。与动物群同时发现的既有人骨残骸，也有文化遗迹。人骨属于现代类型（智人），到目前为止发现的人骨残骸包括两个近乎完整但稍微有点被压破的头骨，此外还有一些头骨碎片和牙齿、下颌骨碎块、上肢骨（包括一块显示有一处骨折愈合迹象的锁骨）、椎骨、腿和趾骨。用火遗迹（木炭和灰）很丰富。

发掘物中包括三个用漂亮的黑燧石制成的石器，一个制作精良的脉石英刮削器，几个脉石英石片和脉石英石核，一枚骨针（针眼破损），一个两端都进行了加工的鹿胫骨，约三十个或者更多被钻孔用于制造项链的狐狸尖牙，一个由鸟的长管骨制成的装饰性圆柱状物件，还有很多可能是从相当远的地方带来的鲕状赤铁矿。到目前为止，尚未发现陶器、磨制石器或者细石器文化的任何痕迹。

结论。不久将会以修复的材料为主题完成一份详尽的报告，此处提供的结论都只是暂定的。（a）毫无疑问，山顶洞的堆积层看起来比1号地点的中国猿人地层更晚，地层学和岩石学的假整合性以及动物群的间隔（没有出现厚颌骨的鹿，出现了一种特别的鬣狗，还出现了马鹿、野驴等）使其与1号地点的中国猿人地层区别开来。（b）然而，山顶洞的堆积层可能在年代上也属于更新世（坍塌的洞穴，黄土样沉积物，存在鬣狗；同样也可以见到猎豹、灵猫、野驴、特殊的鹿等）。（c）在此情况下，我们暂时倾向于将相关的人类遗迹归属于一种更新世晚期旧石器文化。后者好像与西伯利亚和欧洲的旧石器时代后期的相同阶段基本对应。不过，它好像比鄂尔多斯文化（水洞沟和萨拉乌苏）更先进一些，鄂尔多斯文化中至今还没有发现类似的经过特别加工的骨头。

狒狒的遗迹

在1号地点以南有一个直径约1米的圆柱形石灰岩溶洞，卞美年先生于本季在该洞穴中发现了大量骨头化石，它们嵌于一块很特别的红色堆积物中，其中还含有大量磨圆的小鹅卵石。这些骨头石化很深且很重，其中许多受到了水磨作用变得光滑而没有棱角。但是，有几个保存得很好，其中有几颗大狒狒的牙齿和肢骨，可能就是施洛瑟命名的维氏狒狒。曾在1号地点（下洞）的中国猿人沉积层的基底处发现过与此惊人相似的、含有经过强烈滚动的同类骨头化石的沉积物。根据现阶段的挖掘进展，这些层位是代表一个前周口店时期还是只相当于裂隙最后填充的早期阶段仍是一个未解决的问题。

In any case it would seem that one must conclude from this latest discovery that the Choukoutien fissures have been successively inhabited by baboons, by *Sinanthropus* and by a modern type of *Homo*. However, such a coincidence appears less extraordinary when it is recalled that though Ordovician limestone is widely distributed along the Western Hills, at Choukoutien, on account of its low anticlinal structure at the borders of the plain, it is exceptionally well situated for dissection into fissures and caves.

(**133**, 89-90; 1934)

Davidson Black: F.R.S., Honorary Director, Genozoic Research Laboratory, Geological Survey of China.

不管怎样，根据这一最新发现似乎必然能够得出一条结论，即周口店裂隙先后被狒狒、中国猿人和一种现代类型的人属居住过。然而这种巧合似乎不足为奇，要知道，尽管奥陶纪石灰岩沿西山广泛分布，但周口店的平原边界处属于低背斜构造，因此正好有利于裂隙和洞穴的形成。

（刘皓芳 翻译；赵凌霞 审稿）

Ernst Haeckel

Editor's Notes

Professor Ernst Haeckel was an enthusiastic supporter of Darwin's theory of evolution and, from his base, at the University of Jena in Germany, did much to ensure the spread of this area of natural selection on the European mainland. This article was published on the centenary of Haeckel's birth. Before his attention was diverted by Darwin's thesis, Haeckel was best known as an embryologist: he is the originator of the notion, widely accepted in the nineteenth century, that the developing embryo recapitulates the evolutionary track of animals. W. H. Brindley added the following information about Haeckel's relationship with the University of Jena.

THE career of Prof. Ernst Heinrich Haeckel, the centenary of whose birth falls on February 16, belongs to the heroic stage of the history of the theory of evolution. In 1862, at the early age of twenty-eight, he was appointed to the chair of zoology in the University of Jena, a post which he held until his death in 1914.

Haeckel's life bears a strong resemblance to that of Huxley, for like Huxley his life's task was propaganda in favour of the theory of evolution against the then prevalent theory of the origin of species by a series of supernatural interpositions of the Divine Being. Like Huxley too, he was an ardent advocate of the animal origin of the human race. But there were marked differences between the two men; Haeckel was a harder hitter than Huxley, and withal a much more reckless one, since he was apt to make wild statements on the basis of insufficient data, as, for example, when he stated that if there were a line to be drawn between animals and men, the lower races must be included amongst the apes. The most recent anthropological studies seem to indicate that in the essential make-up of their minds the most primitive men are very like ourselves: the data and presuppositions from which they start are different and so are their customs and traditions, but granted these postulates the conclusions at which they arrive are natural enough. But on the whole, Haeckel was a sounder biologist than Huxley: whilst he embraced with enthusiasm Darwin's arguments about natural selection, he was never deceived into thinking that the mere survival of some and the death of others could account for progressive evolution: he saw quite clearly that the vital question was the origin of the "variations" which distinguished the survivor from his less fortunate brother, and in this matter he followed Lamarck. When he popularised his views in his famous "History of Creation" he dedicated the work to "Jean Lamarck and Charles Darwin".

Haeckel excelled Huxley also in the amount of actual zoological work which he accomplished. Thus he wrote a descriptive monograph of the Radiolaria collected by H. M. S. *Challenger*, giving the characters of no less than 3,600 new species. This work

562

恩斯特·海克尔

编者按

恩斯特·海克尔教授是达尔文进化理论的坚决支持者，他在其所任职的德国耶拿大学做了大量工作以确保自然选择理论在欧洲大陆的传播。本文发表于海克尔百年诞辰之际。在海克尔将兴趣转向达尔文学说之前，他是一位著名的胚胎学家，他最先提出了胚胎的发育过程会重现种系进化历程这一观点，这在 19 世纪被广泛接受。威廉·哈里森·布林德利在文后补充了一些关于海克尔与耶拿大学关系的信息。

恩斯特·海因里希·海克尔教授出生于一百年前的 2 月 16 日，他所生活的年代正好处于进化理论创建历史上的英雄时期。1862 年，海克尔刚满 28 岁就被任命为耶拿大学的动物学教授，并一直担任此职直到 1914 年去世（译者注：根据《不列颠百科全书》，海克尔逝于 1919 年）。

海克尔的一生与赫胥黎极为相似，像赫胥黎一样，他一生都以宣传支持进化论为己任，反对当时盛行的认为生物物种起源于某些神圣存在的超自然力量的学说，他也像赫胥黎一样热忱拥护人类起源于动物的学说。但是他们彼此之间也有明显的不同之处。同赫胥黎相比，海克尔更具攻击性，更加鲁莽，因为他动辄就在没有充分数据的情况下急于作出很不成熟的论断。例如，当他说如果在人和动物之间划一条界线的话，那么类人猿一定是与猿类划到一起的。最近的人类学研究似乎表明最原始的人类在思想的本质构成上与我们是非常相像的，尽管由研究结论得出的这些推测是非常自然的，但是最原始的人类的思想开始形成时所基于的信息和前提与我们不同。但是总的来说，与赫胥黎相比，海克尔可以称得上是更理性的生物学家。当他满腔热情地拥护达尔文的自然选择论断时，从来没有盲目地认为某些生物的生存和另一些生物的死亡可以解释为渐进性演化，他很清楚问题的关键在于"变异"的起源，正是这些变异将幸存者与被淘汰者区别开来，对待这一问题，他的观点与拉马克一致。当他在他著名的《自然创造史》一书中推广自己的观点时，他把这些工作献给了"让·拉马克和查尔斯·达尔文"。

在实际完成的动物学工作量方面，海克尔也胜过赫胥黎。他根据英国皇家海军舰艇"挑战者号"采集的资料撰写了一本描述放射虫的专著，介绍了不少于 3,600

occupied him for ten years. He also monographed the calcareous sponges, but the greatest task which he attempted was to sketch, assuming the truth of the evolution theory, the actual course which evolution had pursued in producing modern plants and animals. His conclusions were embodied in his "Allgemeine Morphologie", of which the "History of Creation" may be regarded as a popular edition. Of course, the state of zoological and botanical knowledge at the time that these books were written was far too incomplete to permit of any but the vaguest sketches of the course of evolution, but there can be nothing but admiration for Haeckel's bold adventure. In the circumstances, It was the right course to pursue: it summarised pre-existing knowledge and provided both a foundation and a framework for future work, and some of the most important and fundamental of Haeckel's ideas have stood the test of time. Thus he divided living beings into Animals, Plants and Protista; regarding the last group, which included the simple unicellular organisms, as the common seed-bed from which both animals and plants have sprung. The discovery of green ciliates like some species of *Stentor* and *Vorticella*, and of colourless carnivorous Dinoflagellates which devour young oysters, in addition to the ordinary brown species which live like brown seaweeds, has more than justified Haeckel's classification.

Haeckel's most far-reaching hypothesis was, however, his famous "biogenetic law". He invented the terms phylogeny and ontogeny—the first, according to him, designated the palaeontological history of the race, the second the history of the development of the individual from the egg to the adult condition. The law connecting these two was the "Biogenetic fundamental principle": stated in his own words, it ran thus: "Ontogeny is a short and quick repetition, or recapitulation of Phylogeny determined by the laws of inheritance and adaptation". Haeckel pointed out that if this principle be admitted, there is some hope of tracing, in outline at least, the actual course of evolution; whereas if we were to confine ourselves to palaeontological evidence, we should only see glimpses of evolution in special cases. The past history of the Vertebrata may be traced from fossils with considerable exactitude since vertebrates possess an internal skeleton which is often preserved and which gives in its scars and processes, evidence of the muscles which once accompanied it and consequently of the actions and habits of the animal which possessed the skeleton. The external skeleton of extinct Crustacea which clings tightly to every protuberance of the body, also reveals a good deal about the activities of its former possessor. But what scanty light do the shells of extinct Mollusca and the tests (testas) of ancient Echinoderms throw on the internal structure of their owners! Who would dream from their evidence that radiate Echinoderms were derived from bilateral ancestors?

In our judgment the formulation of this biogenetic law was the greatest service which Haeckel did to the science of zoology, and the more we reflect on it the greater the service will appear. Haeckel was, of course, aware that these reminiscences of ancestral life could be modified, blurred or occasionally completely obscured. He knew that for the elucidation of life-histories only the comparative method would avail; and just as in the comparison of two ancient documents the truth will shine through the errors peculiar to each one, so with life-histories.

个新种的特点。这一工作花费了他整整 10 年时间。他还写了一部关于钙质海绵类的专著，但是他试图完成的最伟大的工作是在假定进化论正确的前提下勾勒出进化在产生现代动植物的过程中所经过的真实路径。他的结论收录在他的《普通形态学》一书中，《自然创造史》可以被视为该书的科普版。当然，写这本书时动物学和植物学知识的发展程度还远不完善，这种状况决定了他能得到的只是进化过程的最模糊的轮廓，但是海克尔这种勇敢的冒险精神让我们非常钦佩。其实在这种情况下，概括已经存在的知识从而为今后的工作提供基础和框架，这是一条探寻真理的正确路线。海克尔的观点中最重要和最基本的部分已经经受住了时间的考验。例如他将生物划分成动物、植物和原生生物三大类，原生生物包括简单的单细胞生物，他认为动物与植物皆起源于这一类群。后来发现了绿色纤毛虫（如喇叭虫和钟形虫）、以幼牡蛎为食的无色肉食性沟鞭藻类、像褐海藻一样生存的普通棕色沟鞭藻类，这些发现已经证明了海克尔的分类方法是合理的。

然而，海克尔的假说中意义最深远的是著名的"生物发生律"。他发明了系统发育和个体发育这两个术语——根据他的说明，系统发育是指种系的古生物学历史，个体发育则指个体从卵子到成年的发育史。联系二者的法则就是"生物发生的基本原理"，用他自己的话说就是"系统发育由遗传定律和适应性法则共同决定，而个体发育是系统发育史的简短而迅速的重复或重演"。海克尔指出，如果这一原理成立，那么就有希望探索到真实的进化过程，至少也可以得到大致的轮廓。反之，如果我们把自己局限于古生物学证据的范围内，那么我们只能在特殊的例子中才能看到进化的发生。脊椎动物的内骨骼通常会以化石的形式保存下来，通过内骨骼上留下的疤痕和突起可以得到曾经与其结合在一起的肌肉的情况，进而获得具有这种内骨骼的动物的行为和习惯方面的信息，因此可以依据化石较为准确地追溯脊椎动物过去的历史。已灭绝的甲壳纲动物的外骨骼紧紧贴附在身体的各处结节上，这些外骨骼也揭示了关于其从前的主人活动的大量信息。但是，已灭绝的软体动物类的贝壳和远古棘皮动物的甲壳所揭示的关于其生物体内部结构的信息却少得可怜！谁能从这些外壳联想到辐射对称的棘皮动物是从两侧对称的祖先进化而来的呢？

依我们看来，生物发生律的提出是海克尔对动物科学最伟大的贡献，而且我们对其思考得越多，就越觉得其意义重大。当然，海克尔意识到了对于远古生命的这些追忆可能会被修改、可能会变得模糊，也可能会在不经意间就完全湮没。他知道只有比较法才有助于阐明生命史，正如比较两份古老的的档案，真相将通过它们各自特有的错误表现出来，生命史亦是如此。

The acceptance of this law as giving a picture of evolution drew with it certain conclusions as to the causes of evolution. Haeckel described variations as "adaptations". There were, he said, two classes of these, namely, (1) small ones which were the result of habits and which were *transmitted to posterity with greater certainty the longer they had lasted* (this is pure Lamarckian doctrine), and (2) great adaptations which appeared suddenly and the causes of which were unknown to us, though in some cases they appeared to have originated with intra-uterine influences. These latter are now, of course, called mutations, and it was the first category alone which Haeckel believed to be significant for evolution, for the growth of the individual suggests that evolutionary growth was slow, functional and continuous.

The biogenetic law proved a tremendous stimulus to zoological research. Of course, it encountered opposition; its enthusiastic votaries desired, like all enthusiasts, to reach the "promised land" at once: they failed to realise that ancestral history could only be elucidated by prolonged, careful and comparative research. They could not deny themselves the pleasure of making wild guesses as to ancestry based on the study of some one life-history and in time "Haeckelismus" became a term of reproach. But the principle was essentially sound; from all opposition it emerged triumphant: it has been transferred to ever wider fields and has been found to throw light even on the development of the mental life of man. A certain school of biologists at the present day affects to denigrate it and that for obvious reasons, for if it is sound then one thing is certain, mutations have played no part in evolution. But ancestral history stands out so clearly in some life-histories that none but the wilfully blind can deny its presence. Amongst the Ctenophora, for example, there are two aberrant forms, *Tjafiella* and *Coeloplana*. The first resembles a sponge, the second a flat-worm; yet both begin their free existence as typical little Ctenophores, globular in form with 8 meridional bands of cilia radiating from the upper pole. But if ancestral history is the foundation of some life-histories is it not reasonable to assume that it lies at the base of all?

The real originator of the theory that evolution proceeded by jumps and that "Discontinuity in variation was the cause of discontinuity in species" was the late Dr. Bateson. In his first and best work on the development of *Balanoglossus* he found himself driven to the conclusion that Echinoderms and Vertebrates had radiated from a common stock and his faith in "recapitulation" failed him, although it is interesting to record that this conclusion has been sustained by recent research and that from the most unlikely quarter, namely, biochemistry. He then made "*il gran rifiuto*" and fell back on sports and monstrosities as the material of evolution. At the meeting of the Zoological Congress in Cambridge in 1898, Bateson put forward his views. Haeckel was present at the meeting and some sentences of his still linger in our memory. He said that if views like these are to be accepted, "Kehren wir lieber zu Moses zurück".

E. W. MacBride

(**133**, 198-199; 1934)

认可生物发生律可以描绘进化的图景便可得出一些关于进化起因的结论。海克尔将变异描述为"适应性"。他说这种适应性包括两类：（1）小的适应性，这是习性的结果，**它们由父辈传给后代，这些习性持续的时间越久，传递给后代的可能性越大**（这属于纯粹的拉马克学说）；（2）大的适应性，这是突然出现的，尽管在有些例子中它们似乎是由母体内的某些影响引发的，但是其具体原因还不清楚。这些大的适应性现在被称为突变，而海克尔认为对于进化有意义的只有第一种适应性，因为个体生长表明进化的过程是缓慢的、功能性的和连续的。

生物发生律极大地刺激了动物学研究的发展。当然，也有人表示反对。像所有狂热者一样，该学说的忠实信徒们渴望能立刻"修成正果"，但却没有意识到祖先们的历史只能通过长久的、仔细的比较研究才能得以阐明。他们基于某一生命史的研究便对祖先展开漫无边际的遐想并沉浸在这样的乐趣中，最终，"海克尔主义"沦为耻辱的代名词。但是该学说的原理本质上是合理的，因而它以胜利者的姿态屹立于所有反对呼声之上，于是它被应用到了更加广阔的领域，人们甚至发现它对理解人类精神生活的发展也有帮助。现代生物学家中有一流派意在贬低海克尔的这一学说，他们指出如果这一学说是合理的，那么很显然可以确定突变对进化没有任何影响。但是祖先的历史在某些生命史中表现得如此明显清晰，除非故意对其视而不见否则就无法否认它的存在。例如，栉水母动物门存在 *Tjafiella* 和 *Coeloplana* 两种迥异的生命形式。第一种类似海绵，第二种则像扁形虫，而二者在生命开始时的状态与典型的小栉水母是一样的，即呈球形，具有自上端辐射发出的八行纤毛子午带。如果祖先历史确实是某些生命史的基础，那么假定在所有生命史中都存在祖先历史难道是不合理的吗？

进化是跳跃式进行的并且"变异的不连续性造成了物种的不连续性"，这一理论的真正创始人是已故的贝特森博士。其第一部也是最好的一部作品是关于柱头虫属发育史的研究，在这本著作中，他得到了如下结论：棘皮动物和脊椎动物从共同的祖先进化而来，他对"重演"学说的信仰导致了他的失败，然而有趣的是最近的研究以及从最不可能的生物化学角度得到的结果均支持了这一结论。后来他又来了个"大转折"，转而以突变体和畸变材料为研究进化的素材。在 1898 年召开的剑桥动物学大会上，贝特森提出了他的观点。海克尔也出席了那次会议，至今他的许多话我们仍记忆犹新。他说如果这样的观点都被认可的话，"那么我们宁愿相信摩西的存在"。

麦克布赖德

*　　*　　*

Many scientists will have read with keen interest Prof. MacBride's delightful sketch of Haeckel's work in *Nature* of February 10. As he points out, Haeckel's career belongs to the heroic stage of the history of the theory of evolution; certainly few men have been subjected to greater obloquy for promulgating that or any other doctrine. When his "General Morphology" appeared, it was met with "icy silence"—a reception which the impetuous and combative Haeckel could not tolerate. He would have preferred hostile criticism, rather than indifference; and to this indifference on the part of his fellow-scientists can be traced the commencement of that series of popular works on evolution which were met, not with "icy silence", but with fiery blasts from scientists and laymen alike.

At one period of the controversy, Haeckel felt that his presence at Jena was jeopardising the good name of his beloved university, so he offered to resign his chair; but the head of the governing body replied: "My dear Haeckel, you are still young, and you will yet come to have more mature views of life. After all, you will do less harm here than elsewhere, so you had better stop here." In point of fact, Jena never forsook Haeckel and Haeckel never forsook Jena, despite the flattering offers he received from the Universities of Vienna, Würzburg, Bonn and Strasbourg; and he died there, not in 1914 as mentioned by Prof. MacBride, but on August 8, 1919. An obituary notice appeared in *Nature* of August 21, 1919.

W. H. Brindley

(**133**, 331; 1934)

W. H. Brindley: 11, Millmoor Terrace, Glossop, Derbyshire, Feb. 9.

* * *

许多科学家都将怀着极大的兴趣拜读将发表于 2 月 10 日的《自然》上、由麦克布赖德教授执笔的对海克尔工作的精彩介绍。正如他所指出的，海克尔生活的年代正好处于进化理论创建历史上的英雄时期，的确很少有人因为发表了某一学说而招致比海克尔所受到的更多的责难。他的《普通形态学》问世时遭遇了"冰冷的沉默"——这是冲动好胜的海克尔最不能容忍的。他宁可受到怀有敌意的批评，而不愿别人对他毫不在意。其同行的科学家们对他的这种漠视一直持续到他出版了一系列关于进化方面的科普作品，与之前受到的"冰冷的沉默"相反，这些作品发表之初，就引发了来自科学家和外行们的猛烈抨击。

在争议四起的一段时间里，海克尔觉得自己的存在已经使他所深爱的耶拿大学的声誉受损，所以他提出辞去教授一职，但是管理机构的负责人回复道："亲爱的海克尔，你还年轻，你终会获得更加成熟的关于生命的观点。其实你不会对这里造成任何损害，所以希望你继续留在这里。"事实上，耶拿大学从未抛弃海克尔，而海克尔尽管收到过来自维也纳大学、维尔茨堡大学、波恩大学和斯特拉斯堡大学等众多大学的盛情邀请，但他也从未抛弃耶拿大学，直至 1919 年 8 月 8 日海克尔在此辞世（并非如麦克布赖德教授所言的 1914 年）。1919 年 8 月 21 日的《自然》刊登了一则他的讣告。

布林德利

（刘皓芳 翻译；陈平富 审稿）

Recent Developments of Sterol Chemistry in Relation to Biological Problems

J. Pryde

Editor's Note

Organic chemistry became an important economic activity in the nineteenth century with the development of the dyestuffs industry in Germany and later in Britain and the United States. In the twentieth century, organic chemists turned their attention to the chemicals produced by living things (some of which were also dyestuffs). This brief paper summarises what had been learned by the 1930s from the investigation of steroids and carcinogens (some of which are also steroids).

ONCE again there has been demonstrated in striking fashion the impetus which organic chemistry gains from biology, and how a field of organic research, formerly of purely academic interest, enters on a fresh phase of development in virtue of a new correlation with biological problems. The field in question is that of the sterols and the polycyclic aromatic hydrocarbons.

It is well known that the fundamental researches of Wieland, Windaus, Mauthner, Borsche, Diels and others on the sterols and bile acids received a new interest on the isolation of calciferol (vitamin D) from the products of irradiation of ergosterol, $C_{28}H_{44}O$, with which the vitamin is isomeric, and that our conceptions of the structure of these, and of other members of the cholane series to which they belong, have been re-oriented by the new formulae advanced by Rosenheim and King[1]. The structures below show the old (I) and the now accepted representation (II) of the cholane nucleus. The new, and at the time somewhat revolutionary, formulae conferred a great stimulus on the investigation of the whole series of compounds. They are based upon evidence which cannot be detailed here, but some of the more salient of the recent observations can be summarised.

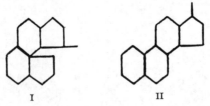

I   II

Thus, on drastic dehydrogenation with palladium-charcoal or zinc, cholesterol and cholic acid yield the fully aromatic hydrocarbon chrysene (III)[2], whilst less drastic dehydrogenation of these compounds and of ergosterol using selenium yields an interesting hydrocarbon of the composition $C_{18}H_{16}$, first obtained by Diels and his associates[3]. For this latter the constitution IV was suggested by Rosenheim and King[4].

与生物学问题相关的甾体化学的最新进展

普赖德

编者按

在 19 世纪，随着染料工业在德国以及而后在英国和美国的发展，有机化学研究成为一种重要的经济活动。到了 20 世纪，有机化学家将他们的注意力转向生命体产生的化学物质（其中一些也是染料）。这篇短文总结了 20 世纪 30 年代之前人们对甾体化合物和致癌物（其中一些也属于甾体化合物）研究所取得的成果。

有机化学可以从生物学获得发展的动力再次得到了证实：由于与生物学问题发生新的联系，之前仅具有纯粹学术意义的有机化学领域进入了崭新的发展阶段。本文要讨论的这个领域是有关甾醇和多环芳香烃的。

众所周知，由维兰德、温道斯、莫特纳、博尔舍、迪尔斯以及其他一些学者对甾醇和胆汁酸所作的基础研究，已引起了人们从麦角固醇 $C_{28}H_{44}O$（维生素 D 的同分异构体）的辐照产物中分离钙化醇（即维生素 D）的兴趣。此外，基于罗森海姆和金提出的分子式 [1]，人们对这些物质以及它们所属的胆烷系列的其他成员的结构也有了新的认识。下图所示为胆烷母核的旧结构式（I）和目前公认的新结构式（II）。这个新的、在当时带有几分革命性的结构式促进了人们对整个胆烷系列化合物的研究。在此我们要总结一下最近研究中的一些较为突出的成果，对于它们所依据的证据就不详细叙述了。

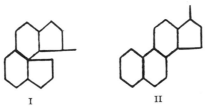

I　　　　　　II

首先，在用钯－炭或锌进行剧烈脱氢反应时，胆固醇和胆汁酸会产生"䓛"（III）这种完全芳香化的烃 [2]。当用硒进行不太强烈的脱氢反应时，上述化合物和麦角固醇都会产生一种令人感兴趣的、化学组成为 $C_{18}H_{16}$ 的烃，这种烃最早是由迪尔斯及其同事获得 [3]。罗森海姆和金提出这种烃具有图 IV 所示的结构 [4]。最近，科恩 [5] 通

Kon[5] has very recently proved the correctness of this suggestion by a synthesis yielding the desired 3-methyl*cyclo*pen-tenophenanthrene. It is therefore clear that the formation of chrysene in the more drastic process is due to ring enlargement associated with the migration of a methyl group, and the revised cholane formula of Rosenheim and King becomes firmly established upon fact.

III   IV

Secondly, the recent isolation and investigation of the female sex (oestrous-producing) hormone, mainly due to the efforts of Doisy in the United States, Marrian in Great Britain, and Butenandt in Germany, show that the hormone occurs in two forms—oestriol (V) and oestrone (VI), to adopt the nomenclature recently advanced in *Nature* by workers in this field[6]. Evidence is available which amply establishes the close relationship of the oestrane and cholane series, which may be inferred from the isolation of the same 1:2-dimethyl-phenanthrene from oestriol and from aetiobilianic acid of the cholane series[7]. Mention may also be made of the isolation from oestrone, after dehydrogenation in the presence of zinc, of a hydrocarbon of the same C_{18} series as that obtained from the cholane compounds. To this hydrocarbon Butenandt has ascribed the composition $C_{18}H_{14}$, but in all probability the compound is impure chrysene $C_{18}H_{12}$.

V   VI

Thirdly, it has been known for many years that the tars and pitches resulting from the pyrogenic decomposition of coal and other organic products frequently possess carcinogenic properties. Much patient work in Great Britain, with which the names of Kennaway and Cook and their collaborators are associated, has culminated in the isolation[8] from a soft coal-tar pitch of a pure actively carcinogenic hydrocarbon, namely, 1:2-benzpyrene (VII). This together with certain other but somewhat less active carcinogenic hydrocarbons [for example, 1:2:5:6-dibenzanthracene (VIII) and 5:6-*cyclo*penteno-1:2-benzanthracene (IX)][9] has been synthesised and the peculiar biological properties of these compounds have been amply proved.

过合成预期产物 3− 甲基环戊烯并菲证实了这种假设的正确性。由此可见，在较为强烈的反应中䓛的生成显然是由于发生了与甲基迁移有关的扩环反应，这为罗森海姆和金修正的胆烷分子结构提供了坚实的事实基础。

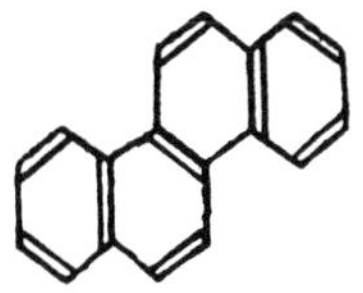

III 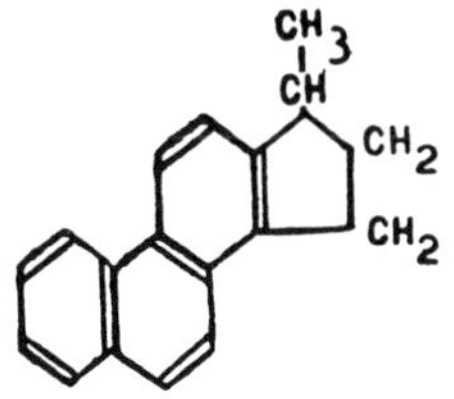   IV

其次，最近人们对雌激素（有引起发情的作用）进行分离和研究的结果表明，该激素以两种形式出现——用该领域研究人员最近在《自然》上提出的命名法 [6] 来表示，就是雌三醇（V）和雌酮（VI）。这主要归功于美国的多伊西、英国的马里安以及德国的布特南特等人的努力。有充分的证据证明雌烷和胆烷系列之间存在密切的联系，这一点可以从雌三醇和胆烷系列的原胆汁烷酸中都能分离出同样的 1,2− 二甲基菲 [7] 推断出来。此外，雌酮在锌的催化下发生脱氢反应后，可以分离出一种 C_{18} 系列的烃类物质，这与从胆烷系列的化合物中分离得到的 C_{18} 系列烃类完全相同。布特南特曾认为这种烃的组成是 $C_{18}H_{14}$，不过它很可能是不纯的"䓛" $C_{18}H_{12}$。

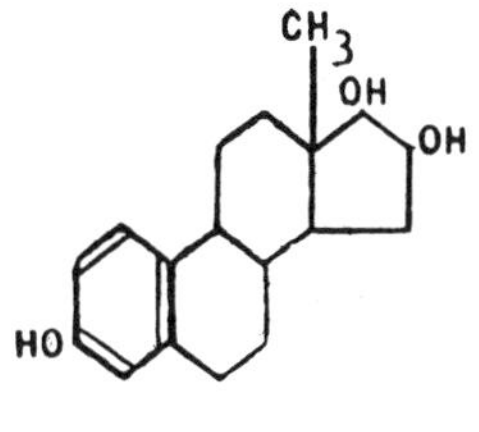

V 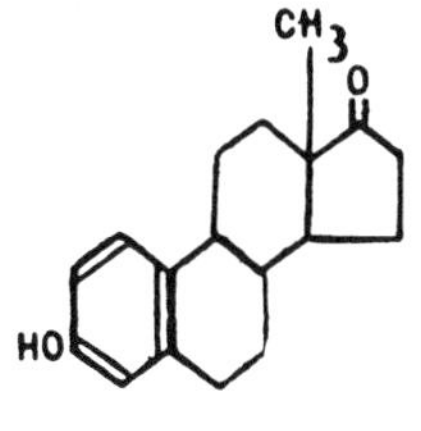   VI

第三，多年前人们就知道，由煤和其他有机物热分解产生的煤焦油和沥青往往具有致癌性。在英国，肯纳韦、库克以及他们的合作者们通过大量耐心细致的工作最终从软煤焦油沥青中分离出了一种纯的有致癌活性的烃类物质 [8]，即 1,2− 苯并芘（VII）。现在，人们已经合成了这种物质以及其他一些致癌活性较小的烃类物质，例如 1,2,5,6− 二苯并蒽（VIII）和 5,6− 环戊烯并 −1,2− 苯并蒽（IX）[9]。这些化合物特有的生物学性质也已经得到了充分的研究。

VII   VIII

It will therefore be realised that calciferol, oestrous-producing hormones, and carcinogenic hydro-carbons, all correlated with some phase of growth, all have the phenanthrene nucleus (X) in common. Lastly, the group of the cardiac-stimulating glucosides— strophanthin, digitoxin—yields aglucones in which the phenanthrene nucleus again occurs[10]. It may also be significant that some of the most powerful alkaloids, such as morphine, codeine, etc., of the opium group, the corydalis alkaloids and colchicine (meadow saffron) contain a phenanthrene nucleus. To this nucleus are added various cyclic and straight-chain substituents which confer on each group its characteristic biological activity.

IX   X

That these groups of compounds, of such apparently diversified physiological activities, should exhibit such fundamental constitutional similarities is sufficiently striking, but the story does not end here and indeed it would be bold to attempt to predict where it will end. Mention has already been made in these columns[11] of the oestrogenic action of certain synthetic hydrocarbons and their derivatives—either themselves carcinogenic or closely related to carcinogenic compounds—and of the similar activity of some of the sterols and calciferol. Amongst the former are 1-keto-1:2:3:4-tetrahydrophenanthrene (XI) and 1:2:5:6-dibenz-9:10-di-n-propylanthraquinol (XII). In reference to the activity of the latter compound, it is of interest to note that a series of diols derived from 1:2:5:6-dibenzanthracene was investigated[12]. Of these the dimethyl, di-n-amyl, and di-n-hexyl compounds are inactive, whilst the intermediate diethyl, di-n-propyl and di-n-butyl compounds are all highly active, the propyl derivative showing the maximum activity. The compounds mentioned above are the most active of those so far investigated; then follow in order of activity neoergosterol, 5:6-cyclopenteno-1:2-benzanthracene, 1:2-benzpyrene, calciferol and ergosterol. That behaviour characteristic of a specific hormone should be shared by other compounds of related structure, some possessed of physiological activities of their own, provides a remarkable extension of our conceptions of biological specificity. It suggests future developments of great interest in the chemistry and biology of the sterols and the polycyclic hydrocarbons.

574

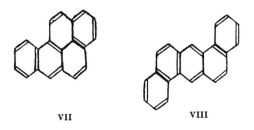

VII   VIII

由此我们会发现，钙化醇、能引起发情作用的激素以及具有致癌作用的烃类物质都与生长发育的某些阶段有关，并且它们都具有菲母核结构（X）。毒毛旋花子苷和洋地黄毒苷等具有强心作用的糖苷类物质的苷元也都含有菲母核结构[10]。某些非常强效的阿片生物碱如吗啡和可待因等、延胡索素类生物碱和秋水仙碱（藏红花）中都含有一个菲母核结构，这也是很值得注意的。与菲母核结构相连接的是各种不同的环状取代基和直链取代基，这些取代基赋予了每类化合物特有的生物学活性。

H_2C
H_2C CH_2

IX   X

这些生理活性明显不同的各类化合物却具有如此相似的基本结构，这是十分令人惊奇的。然而，故事远不止于此，现在就企图预言最终会如何恐怕还为时尚早。一些专栏文章[11]曾经提到过，某些合成的烃类物质及其衍生物具有雌激素的作用（这些烃类及其衍生物要么本身就是致癌物要么就是与致癌物有密切的关系），一些甾醇和钙化醇也具有类似的活性。属于第一类化合物的有 1-酮基-1,2,3,4-四氢化菲（XI）和 1,2,5,6-二苯-9,10-二正丙基蒽二酚（XII）。关于后一种化合物的活性，值得注意的是，人们已经对一系列由 1,2,5,6-二苯并蒽衍生出来的二醇类物质进行了研究[12]。其中，二甲基、二正戊基和二正己基化合物是无活性的，而中间的二乙基、二正丙基和二正丁基化合物则具有很强的活性，丙基衍生物的活性最强。以上提及的化合物是目前研究过的化合物中活性最强的。按照活性大小排序，依次是新麦角固醇、5,6-环戊烯并-1,2-苯并蒽、1,2-苯并芘、钙化醇和麦角固醇。一种特定激素的特征性功能也会被其他具有相关结构的化合物所共有，某些化合物还具有自身特有的生理活性。这极大地拓展了我们对生物学特异性的认识。这也预示着未来关于甾醇和多环烃类物质的化学和生物学将有非常引人注目的发展。

XI   XII

(**133**, 237-239; 1934)

References:

1. *J. Soc. Chem. Ind.*, **51**, 464, 954; 1932.

2. Diels and Gädke, *Ber.*, **60**, (B), 140; 1927.

3. *Annalen*, **459**, 1, 1927; **478**, 129; 1930.

4. *J. Soc. Chem. Ind.*, **52**, 299; 1933.

5. *ibid.*, 950.

6. *Nature*, **132**, 205, Aug. 5, 1933.

7. Butenandt, *J. Soc. Chem. Ind.*, **52**, 268, 287; 1933.

8. Cook, Hewett and Hieger, *J. Chem. Soc.*, 395; 1933.

9. *Proc. Roy. Soc.*, B, **111**, 455, 485; 1932.

10. Jacobs and Fleck, *J. Biol. Chem.*, **97**, 57; 1932.

11. *Nature*. 132. 1933.

12. Discussed at a meeting of the Royal Society on Nov. 16, 1933.

576

XI

XII

（王耀杨 翻译；田伟生 审稿）

Evolution of the Mind*

G. Elliot Smith

Editor's Note

In this transcript of a discourse at London's Royal Institution, Australian-born anatomist Grafton Elliot Smith reveals how neuroanatomy can help researchers understand the evolution of the mind. He highlights the importance of the neopallium (neocortex), which evolved in higher mammals to give them the ability to consciously direct movement. The structure interacts with other interconnected regions, including the cortex, thalamus and hypothalamus. In turn, the cortex influences muscular activity, shaping our experience of the world through motor control. "It is largely by doing things that experience is built up," says Elliot Smith, who sees other senses, such as vision and taste, as being equally important in the evolution of the mind.

IT may be asked by what right an anatomist, whose proper business is concerned with very concrete subjects, presumes to discuss so elusive and immaterial a subject as the evolution of the mind, even if it be admitted that the evolution of the chief organ of the mind comes within the proper scope of his field of work. I am encouraged, however, to embark on this hazardous attempt by the considered judgment of Prof. S. Alexander, who once expressed the opinion "that we are forced to go beyond the mere correlation of the mental with [the] neural processes and to identify them".

The great physiologist who is most competent to express an opinion on this issue has recently impressed upon us the need for caution in touching it. In the closing passage of his Rede Lecture on "The Brain and Its Mechanism", delivered in Cambridge on December 5, 1933, Sir Charles Sherrington used these words: "I reflect with apprehension that a great subject can revenge itself shrewdly for being too hastily touched. To the question of the relation between brain and mind the answer given by a physiologist sixty years ago was 'ignorabimus'. But today less than yesterday do we think the definite limits of exploration yet attained. The problem I have so grossly touched has one virtue at least, it will long offer to those who pursue it the comfort that to journey is better than to arrive, but that comfort assumes arrival. Some of us—perhaps because we are too old—or is it too young?—think there may be arrival at last." These opinions are even more appropriate to those who lack Sir Charles Sherrington's immense competence.

Hence I seize upon a confession made by Sir Charles elsewhere in his Rede Lecture:

"What right have we to conjoin mental experience with physiological? No scientific right;

---

* Friday evening discourse delivered at the Royal Institution on Jan. 19.

思维的进化*

埃利奥特·史密斯

编者按

在这份伦敦皇家研究院的演讲记录中，出生于澳大利亚的解剖学家格拉夫顿·埃利奥特·史密斯展示了神经解剖学如何帮助研究者理解思维的进化过程。他强调了新大脑皮质（新皮层）的重要性，高等哺乳动物新大脑皮层的进化使得它们能够有意识地指导运动。新大脑皮层与其他互相连接的区域（包括皮层和丘脑以及下丘脑）互相作用。然后，皮层影响肌肉的行为，通过对运动的控制来塑造我们对世界的体验。埃利奥特·史密斯说，"经验在很大程度上是通过做事建立起来的。"他认为，其他感觉，例如视觉和味觉，在思维的进化中具有同等的重要性。

即便承认思维主要器官的进化属于解剖学家正常的研究范围，我们仍可以追问，一个本职工作在于研究非常具体的事物的解剖学家有什么权力来探讨诸如思维的进化这样一个难以捉摸而又非物质性的课题呢。亚历山大教授曾表达过如下观点，"我们必须突破精神与神经过程之间简单的联系，必须要透彻地认识它们"。正是这一深思熟虑的论断使我受到鼓舞而从事这次有风险的尝试。

最近，在这个问题上最有发言权的一位卓越的生理学家提醒我们，在尝试研究思维的进化时需要保持谨慎。1933年12月5日，在剑桥里德讲坛，查尔斯·谢灵顿爵士在"脑及其机制"的演讲的结束语中说道："我所忧虑的是，如果过于草率地行事，那么这个重要的课题就会巧妙地报复我们。对于脑与思维的关联这个问题，60年前一位生理学家所给出的答案是'一无所知'。不过，我们认为现今探索这个问题遭遇的明确局限比过去少。这个我只是非常粗略地涉及过的问题至少具有这样一个优点，它将长期为那些致力于此的人们提供一种安慰，即，旅行的过程比到达终点更美好，但是这种安慰预设了到达的终点。我们中的一些人——或许是因为我们已经太老了——或者是这个问题太年轻？——认为终点最终是有可能到达的。"对于那些并不具备查尔斯·谢灵顿爵士那样卓越才能的人来说，上述观点就显得更恰如其分了。

因此，我要引用查尔斯爵士在里德演讲中所作的一则声明：

"我们有什么权利将精神体验与生理学实验结合起来呢？没有科学意义上的权

---

* 1月19日在英国皇家研究院周五晚上的演讲。

only the right of what Keats, with that superlative Shakespearian gift of his, dubbed 'busy common sense'. The right which practical life, naïve and shrewd, often exercises."

If scientific proof, however, is demanded, surely Sir Henry Head's investigation of sensation and the cerebral cortex supplies it by demonstrating in wounded soldiers the concern of the cortex with psychical functions—the dependence of mind on brain ("Studies in Neurology", 1920). Prof. Shaw Bolton, by comparative and clinico-pathological researches, has demonstrated the dependence of mind on the supragranular layer of the cerebral cortex.

With these assurances the mere biologist, while discussing strictly biological issues, can direct attention to certain psychological implications of anatomical facts and comment also on their neurological aspects for the interpretation of the mind and its working. In previous lectures at the Royal Institution I have discussed the significance of the heightened powers of vision in man's ancestors, which conferred upon them the ability to see the world in which they were living and appreciate something of what was happening in it, as well as to guide their hands to acquire skill, by the practice of which fresh knowledge and understanding were obtained.

Significance of Visual Guidance

We know enough of the comparative anatomy and palaeontology of the Primates to select a series of animals that can be taken to represent approximately the stages through which man's ancestors passed in their evolution towards man's estate, and by examining the connexions of the optic tracts in the brain, arrive at an understanding of what is involved in the acquisition of higher powers of visual discrimination (Fig. 1).

In this series of diagrams, it will be observed that at first the areas for touch, vision and hearing come into contact with one another but that eventually an area marked P (parietal association area) develops between them to provide a more efficient place of blending of the impulses from these three senses. At the same time there emerges from the front end of the brain a prefrontal area (F) which is essentially an outgrowth of the motor territory and an instrument whereby the activities of the whole cortex can in some way be concentrated on the process of learning to give motor expression to the total activities of the hemisphere. Certain poisons which exert a destructive influence on the supragranular layer of this part of the cortex lead to very significant mental results, such as are displayed in general paralysis of the insane, characterised at first by grandiose delusions and afterwards by a failure of the mental process altogether, profound dementia. The discussion of this evidence by Dr. J. Shaw Bolton ("The Brain in Health and Disease", 1914) affords another precise demonstration of the dependence of the mind upon particular parts of the brain.

580

利，而只是被具有超凡的莎士比亚式天赋的济慈称为'闲不下来的常识'的权利。这是一种现实生活（无论是幼稚还是精明）经常会使用的权利。"

然而，如果需要科学性的证据，那么毫无疑问，亨利·黑德爵士关于感觉与大脑皮层的研究已经提供了。亨利·黑德爵士的研究表明，受伤士兵的脑皮层状况与精神功能之间具有相关性——即思维对于脑的依赖性（《神经学研究》，1920年）。肖·博尔顿教授则通过比较研究和临床病理学研究，阐明了思维对大脑皮层的上颗粒层的依赖性。

因为有了这些论断，所以在严格讨论生物学问题时，纯粹的生物学家才能将注意力放在解剖学事实的明确心理学含义上，并且从神经学方面评论对于思维及其作用机制的解释。在英国皇家研究院此前的演讲中，我已经讨论了人类祖先视觉能力提升的意义，这一提升赋予了他们看清自身所处世界并评估发生于其中的某些事情的能力，同时也赋予了他们在实践过程中（新的知识和理解都借此而得）通过自己的双手掌握各种技能的能力。

视觉引导的意义

我们所掌握的关于灵长类的比较解剖学知识和古生物学知识，已经足以使我们选出一系列动物来近似代表人类祖先在其向人类阶段进化的过程中曾经历过的各个阶段，并且通过检查大脑中视觉神经束的连接，我们也可以理解与较高视觉辨别能力的获得有关的因素（图1）。

在这组图中，我们首先会观察到负责触觉、视觉和听觉的区域相互之间逐渐接触，但是最终一块标记为 P 的区域（顶部联系区）在它们之间发展起来，为来自上述三种感觉的冲动的混合提供了更为高效的场所。与此同时，在大脑的前端出现了一个前额区（F），它在本质上是运动区域的扩张，也是一种工具，整个皮层的活动能借此而以某种方式集中于学习过程，从而对大脑半球的全部活动给出运动表达。某些能对皮层此部分的上颗粒层施加破坏性作用的有毒物质能导致非常严重的精神后果，例如，精神全面麻痹所表现出来的症状，其特征最初为夸大妄想症，随后精神过程完全紊乱而发展为严重痴呆。肖·博尔顿博士对这一证据的讨论（《健康的大脑和患病的大脑》，1914年）给思维对于大脑特定部分的依赖性提供了另一番精确的论证。

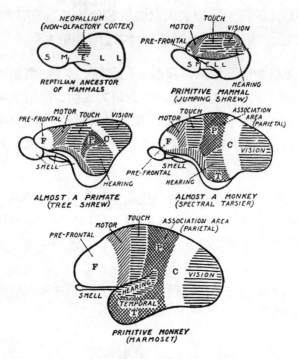

Fig. 1. A series of diagrams to suggest the origin of the neopallium in the ancestor of mammals; the rapid development of this cortical area in mammals, as touch, vision, hearing, as well as control of skilled movements, attain an increasing significance, the growing cultivation of vision which leads to the emergence of the Primates, the increased reliance on vision brings about an enhancement of skill in movement (and a marked expansion of the motor territory) and of tactile and auditory discrimination. (Based in part on the work of Profs. W. E. Le Gros Clark and H. H. Woollard.) From "Human History" (1930).

This is an example of the means whereby comparative anatomy can throw light upon the process of mental evolution, the structural changes in the eyes and brain which make possible not only the refinement of visual discrimination, but also the increasing participation of visual perception in the conscious life and in the guidance of the instruments (such as the hands) of muscular skill. The latter consideration is one of fundamental importance. For the study of the evolution of the nervous system impresses upon us the fact that one of its essential purposes is to make possible quicker, more complex and more purposive responses to changes in the animal's environment or the conditions in its own body.

It is a matter of real importance, therefore, that every advance in the powers of sensory perception and discrimination should be brought into relationship with this essential biological need of finding expression in action. Each of the major advances in vertebrate evolution is obviously correlated with differences in locomotion and muscular aptitude. When an amphibian emerged from a fish-like ancestor, the most obtrusive change was the substitution for swimming as a means of locomotion, the use of the newly-created "gadgets" which are represented by the limbs of a tetrapod land-living animal. The attainment of greater competence and agility in the control of the amphibian's four legs led to the

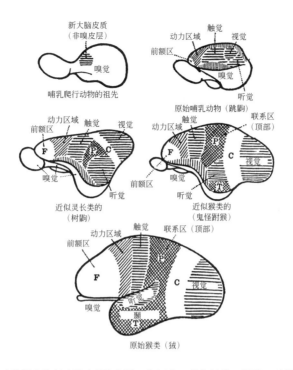

图 1. 暗示出哺乳动物祖先的新大脑皮质起源的一组图片；就像触觉、视觉、听觉以及对熟练动作的控制一样，哺乳动物这一皮层区域的快速发展获得了越来越显著的重要性。视觉能力的不断发展导致了灵长类动物的出现，对视觉不断增加的依赖性带来了运动能力的增强（以及运动区域的显著扩张）以及触觉与听觉分辨能力的增强。（部分依据勒格罗·克拉克和乌拉德两位教授的工作。）引自《人类的历史》(1930 年)。

这是比较解剖学能够以某种方式有助于阐明思维进化过程、眼睛和大脑的结构改变的一个实例，其中眼睛和大脑的结构改变不仅使视觉分辨能力的提高成为可能，还使视觉感知在意识支配的生活中以及在对肌肉技能部件（例如双手）的指挥中的参与程度逐渐增加。对于眼睛和大脑的结构改变的考虑具有根本的重要性。因为，对于神经系统的进化的研究使我们深切地感受到这样的事实：进化的本质目的之一就是对动物所处环境的变化或者其自身内部条件的变化的反应尽可能变得更快、更复杂、更果决。

因此，尤为重要的是，感官感知能力与分辨能力的每一次进步都应该与寻找有效表达这一生物学本质需求建立关联。脊椎动物进化过程中的每一个重要进展都与移动力和肌肉能力的差异有很明显的关系。当两栖动物从与鱼类相似的祖先中演变而来时，最突出的改变就是不再将游泳作为主要的移动方式，而是使用新生的"结构"——这可以被看作四足陆生动物肢体的原型。两栖动物在四肢的控制方面获得了更强的能力和灵敏性，这导致了爬行动物的出现，随着时间的推移，鸟类和哺乳

emergence of reptiles, from which in course of time birds and mammals were evolved; the former by high specialisation of the forelimbs by flight, and the latter by the acquisition of a cerebral instrument, the neopallium, which conferred the ability to attain unlimited powers of acquiring skill and to profit from experience. The highest powers of skill were made possible by the evolution of greater powers of visual guidance.

It is an obvious truism that man's mental superiority is largely the outcome of the perfection of the co-operation of hand and eye in the attainment of manipulative skill and dexterity. In the use of the hands for the expression of skill, the skin of the fingers acquires heightened powers of tactile discrimination, and thus becomes the special organ of the sense of touch and an instrument of perceptual knowledge second only to the eyes in significance.

The researches of Sir Henry Head and his collaborators have given us a new understanding of what is involved in tactile discrimination. The great sensory pathways in the spinal cord and brain-stem lead up to the thalamus in the forebrain, where they end in its ventral nucleus, the nerve cells of which transmit impulses in two directions—one to the cerebral cortex and the other to what Sir Henry Head calls the essential organ of the thalamus. The former is regarded by him as the mechanism for sensory discrimination, and the latter as the instrument for awareness to sensation and the appreciation of its affective qualities, its pleasantness or unpleasantness.

Hypothesis of a Thalamo–Cortical Circulation

In the *British Journal of Medical Psychology* in 1932, Mr. George G. Campion discussed the psychological implications of Head's clinical results. Emphasising the impossibility of separating from perception the affective factor, which is continually at work in our thought-processes, Mr. Campion gave expression to the view that the biological purpose of giving a meaning to experience is the essence of the comprehension of the nature of sensation. Mr. Campion has emphasised the further fact that the concept—the ultimate constituent element of what are called our cognitive dispositions—is not fixed and unchangeable, but is "a living plastic mental symbol subject to a process of organic growth, and that its growth is due to an affective factor which is constantly at work determining the selection of new sense data from the perpetual flux, interpenetrating the conceptual contents of our minds, and integrating all these various and varying constituents into the slowly maturing dispositions which constitute organised knowledge. The affective factor involved in this process has been variously called 'libido', 'love', 'interest', 'feeling', 'desire', 'liking', etc."

Mr. Campion further maintains that there is a continuous stream of neural impulses from the thalamus to the cortex and from the cortex to the thalamus, which keeps alive this living process of mental growth—the enrichment of the concept as the result of personal experience, the success or failure of the attempts to do things.

动物又从爬行动物进化而来；前者是通过飞行引起的前肢的高度特化，后者则是通过一种大脑结构——新大脑皮质获得的，新大脑皮质不但使生物能够通过学习技能而获得无限的能力，也赋予了他们从经验中受益的能力。更强的视觉指导能力的进化使生物具有最高能力的技能成为可能。

人类的精神优越性在很大程度上是手与眼在获得操作技能和灵活性的合作中完善的结果，这明显是真实的。在用手来表达技能的过程中，手指的皮肤获得了增强的触觉分辨能力，从而变成了触觉专用器官和在重要性上仅次于眼睛的知觉认识的结构。

亨利·黑德爵士及其同事的研究已经为我们提供了关于触觉分辨力的新理解。脊髓和脑干中的主要感觉通路导向前脑部分的丘脑并终结于其腹侧核，腹侧核的神经细胞沿两个方向传导脉冲——一个传向大脑皮层，另一个传向亨利·黑德爵士所称的丘脑核心结构。亨利·黑德爵士认为，前一个方向的传导是感觉辨别的机制，而后一个方向的传导则是作为认识感觉以及对其诸如喜悦或不快等情感属性进行评价的工具。

丘脑 – 皮质循环的假说

在 1932 年的《英国医学心理学期刊》中，乔治·坎皮恩先生讨论了黑德的临床研究结果的心理学含义。通过强调将情感因素——它一直作用于我们的思维过程——从感知中分离的不可能性，坎皮恩先生表达了以下观点：给经验赋予意义的生物学目的是对感觉本质的基本理解。坎皮恩先生还强调了更进一步的事实，即概念——我们通常所说的认知倾向的最终组成要素——并不是固定不变的，而是"一种受机体成长过程影响的、鲜活的、可塑的精神象征，它的发展可归因于情感因素，情感因素始终有效地决定着从连续的数据流中选取新的有意义的数据的过程，贯穿我们思维的概念性内容，并且将所有这些不同的变化的组分整合成缓慢成熟的倾向，它们构成了有组织的知识。这个过程中涉及的情感因素一直以来被冠以不同的名称：'感情冲动''爱''兴趣''情感''欲望''嗜好'等等。"

坎皮恩进一步认为：从丘脑到皮质以及从皮质到丘脑之间都存在持续的神经脉冲流，这可以保持思维成长过程的活性——由个人经历造成的概念的丰富，尝试去做某些事情的成功或失败的结果。

Developing this idea, Mr. Campion directs attention to the various parts of the cortex linked in an incredibly complicated way by association fibres and cortical association areas. The necessary implication of his hypothesis of the thalamo-cortical circulation of neural impulses (by means of the various thalamo-cortical and cortico-thalamic tracts of fibres), involves functional connexions of the various parts of the thalamus with one another by intercommunicating fibres. He predicts that as "the cortical association areas may be assumed to have a counterpart also in the thalami, it will be for neurologists to say whether these hypothetical association areas lie in and constitute a chief part of what Head has called the essential thalamic organs."

Since this prediction was made, Prof. Le Gros Clark, in the course of studies (*Brain*, vol. 55) in the comparative anatomy and physiology of the thalamus, has directed attention to the fact that such elements are actually found in the thalamus of the higher mammalia. There are cell masses (lateral nucleus (Fig. 3)) deriving their impulses from the main sensory part (ventral nucleus) of the thalamus, which merge sensory impulses of different kinds and establish direct connexions with those association areas of the cortex which link together the cortical sensory areas. This remarkable confirmation of Mr. Campion's hypothesis adds force to the argument that the mechanism of correlation in the thalamus is far more complicated than has hitherto been supposed, and represents what, following the lead of Sir Henry Head, one may suppose to be a mechanism for the integration of affective processes in the same way as the cortex effects the integration of the discriminative or cognitive aspects of experience.

In the process of acquiring knowledge and building up these vital mental elements, the concepts, to which reference has already been made, it is obvious that there must be a circulation of nervous impulses such as Mr. Campion assumes to maintain the cohesion and the integrity of the vital processes of thought. This circulation of impulses must be even more complicated than he has assumed, because the hypothalamus undoubtedly enters into the process and influences the activities both of the thalamus and the cortex, adding as its quota the visceral element which confers upon experience an emotional factor which is something more than the affective interest the thalamus is able to provide. Intimately intertwined with the whole of this complicated system—hypothalamus, thalamus and the sensory and association areas of the cortex—we have the complex mechanism for giving expression to their combined activities in actions which represent the biological purpose of the whole process. The powerful instrument of thought represented by speech affords an admirable illustration of the intimate correlation of muscular skill with cognitive aptitude to provide the essential currency of mind.

Almost every part of the cerebral cortex is intimately connected directly and indirectly with mechanisms in the central nervous system which are concerned with muscular activities, either those which directly effect movements, or on a vastly greater scale those which prepare and co-ordinate the state of the muscles of the whole body in readiness for prompt and efficient action. More than two-thirds of the fibres that leave the hemisphere have as their immediate purpose the establishment of connexions with the cerebellum,

通过发展这一观念，坎皮恩先生将注意力转向脑皮层的各个部分，它们是由联合纤维和皮层联系区以一种令人难以置信的复杂方式连接在一起的。他这种神经脉冲的丘脑－皮质循环(借助于各种丘脑－皮质和皮质－丘脑纤维束)假说的必然推论，就是丘脑各个部分彼此之间借助互相连接的纤维实现功能性联系。他预言，由于"可以假定丘脑中存在皮质联系区的一种匹配物，神经学家会疑问：这种假说性的联系区是否确实存在，这些联系区是否构成了黑德所称的丘脑核心结构的主要部分。"

上述预言出现之后，勒格罗·克拉克教授在对丘脑进行比较解剖学研究和生理学研究时（《脑》，第 55 卷）将注意力转向以下事实：在高等哺乳动物的丘脑中确实发现了上述要素。细胞团块（外侧核，图 3）从丘脑的主要感觉部位（腹侧核）获得脉冲，这些部位将不同类型的感觉脉冲综合起来并与连接皮层感觉区的各皮层联系区之间建立直接联系。这一引人关注的对坎皮恩先生假说的确证推动了以下观点的产生，即丘脑中相关性的机制远比此前所假定的要复杂得多，而且还指出，追随着亨利·黑德爵士的指引，也许可以假设这样一种机制：情感过程的整合与皮层达成经验的识别方面或认知方面的整合是以相同的方式进行的。

在获得知识和构建这些至关重要的思维基础即概念的过程中（我们已经谈论过它），很明显，必定存在一个像坎皮恩先生假定的那样的神经脉冲循环，以保持思维重要过程的内聚性和完整性。这种脉冲循环甚至必然要比他所假定的更为复杂，因为下丘脑无疑要参与到该过程之中，并且影响丘脑和皮质的活动，依其作用增加本能要素，这种要素给经验赋予了一个情感因素——后者超过了丘脑所能提供的情感兴趣。考虑到与整个复杂系统——下丘脑、丘脑以及皮质中的感觉区与联系区——的密切交缠，我们用一种复杂的机制来解释它们在体现着整个过程的生物学效果的活动中的协同行为。以语言为代表的强大的思想工具提供了一个极好的例证：肌肉技能与认知能力密切相关而保证思维的实质性流通。

几乎大脑皮层的每个部位都直接或间接地与中枢神经系统中控制肌肉活动的机构有密切的关联，这些肌肉活动要么直接影响运动，要么在更大的尺度上预备和协调全身肌肉的状态从而为发出迅速而有效的动作做好准备。在离开大脑半球的纤维中，有 2/3 以上是以与小脑建立联系为直接目的，而其功能则是为诸如躺倒等灵巧

and as their function, the rapid distribution of the muscular tone of the body in readiness for such skilled action as lies at the root of the brain's efficiency. The circulation of the thalamic and cortical currents maintains this constant state of readiness and is a vital and essential part of consciousness and mind.

The building up in the brain of concepts is dependent not merely on affective and cognitive experience based upon afferent impulses from the sense organs, but is also brought about as the result of muscular activity, the doing things with the hands, the gradual perfecting of the movements, the results of the success or failure of such efforts, and the afferent impulses which pour into the brain from the joints, the muscles and the skin areas to record the success or failure of particular muscular activities. It is largely by doing things that experience is built up. It is important therefore to recognise the very large part which such conative activities play in the building up of concepts. They are due not merely to the interaction of the affective and cognitive dispositions, but also to the dynamic factor which is conferred upon these processes by attempting to express in action the result of the discriminative activities of the cortex.

The Neopallium as the Essential Mental Instrument

More than thirty years ago, I directed attention to the fact (*J. Anat. and Physiol.*, p. 431; 1901) that with the evolution of mammals a new cortical instrument, which I called the neopallium, came into existence, and with its expansion provoked the vastest revolution that ever occurred in the cerebral structure. It came into being to form a receptive organ for fibres coming from the thalamus, whereby touch, vision, hearing and taste—in fact all the non-olfactory senses—secured representation in the cerebral cortex. To express this fact, Prof. Winkler, of Utrecht, calls the neopallium the thalamocortex.

In its earliest form the neopallium consists of a tiny area far forward in the hemisphere, where tactile impulses from the lips and tongue are brought into relationship with olfactory and gustatory impulses, and this area afterwards acquires the ability to control the movements of the lips and tongue. As the neopallium grows it establishes similar relations to the rest of the body and increases the range of its receptive powers not merely to the skin of the whole body, but also to the eyes and ears, and it establishes direct connexions with all the motor nuclei in the central nervous system. The neopallium not only gives the senses other than smell representation in the dominant part of the brain and a part in the control of behaviour, but it also provides a continuous territory in which co-operation between these various sensory influences can be established and their conjoint effects be brought to bear upon the mechanisms that control motor activities.

It is often supposed that there are in the cerebral cortex long association bundles to establish connexions between distant parts of the cerebral cortex. There has recently been published an important memoir by Dr. Stephan Poljak, a Jugoslav neurologist who began the research in question in my laboratory eight years ago, which disproves the existence of such long connexions. An impulse from one cortical area can only reach and influence

动作做准备时对机体中肌肉紧张的快速传送，这是脑效率的根本。丘脑与皮质的脉冲流的循环维持着这种准备式的恒定状态，而且这种循环是意识与思维的一个至关重要和本质的部分。

概念在脑中的建立不仅仅依靠于由来自感觉器官的传入脉冲产生的情感和认知体验，而且还取决于肌肉活动的结果，用双手所做之事，对运动的逐渐熟练，这些努力的成功或失败的结果，以及从记录特定肌肉活动的成功或失败的各关节、肌肉和皮肤区域注入脑的传入脉冲。经验在很大程度上是通过做事建立起来的。因此下面的认识是重要的，即这种意动活动在概念建立的过程中起了非常大的作用。它们不仅仅归因于情感和认知倾向的相互作用，还源于通过试图有效地表达皮质识别活动的结果而赋予这些过程的动力学因素。

作为基本精神结构的新皮层

三十多年以前，我曾关注过如下事实（《解剖学与生理学期刊》，第 431 页；1901 年）：随着哺乳动物的进化，一种新的皮质结构出现了，我称其为新皮层，它的扩张激发了在脑结构中曾经发生的最深远的一次变革。它逐渐形成了一个汇聚来自丘脑的纤维的接受器官，借助这些纤维，触觉、视觉、听觉和味觉——事实上是嗅觉以外的所有感觉——在大脑皮层中得到了固定表征。为了表现这一事实，乌得勒支的温克勒教授称新皮层为丘脑皮层。

在最早的形式中，新皮层由脑半球远前端的一小块区域构成，来自唇和舌的触觉脉冲与嗅觉和味觉脉冲在该区域产生了联系，后来这片区域获得了控制唇和舌运动的能力。随着新皮层的生长，它与机体其他部分建立了类似的联系，并且将接受能力的范围提升到不仅包括整个机体的皮肤还包括两眼和双耳，它还与中枢神经系统中的所有运动核建立了直接的联系。新皮层不仅在大脑主要区域和行为控制的一部分区域中对除嗅觉以外的其他感觉给予表征，而且还提供了一个连续区域，使上述各种感觉影响可以在这里建立协作，并且它们的协同效应可以影响控制肌肉运动行为的机制。

人们经常认为在大脑皮层中存在长的联系束，以便在大脑皮层中相距较远的部分之间建立关联。斯蒂芬·波利亚克博士最近发表了一份重要的论文集，这位南斯拉夫神经学家 8 年前就在我的实验室开始研究这个问题，他的研究证明这种长距离的联系并不存在。从一个皮层区域发出的脉冲只能通过在皮层自身中传输而到达

distant areas by travelling through the cortex itself. The act of correlation involves the whole cortex. Even in the simplest act of thought or skill, the whole neopallium participates. The manifold currents which circulate throughout the brain in the process of regulating muscular activities represent the means of integrating the cognitive, affective and curative activities in thought.

Not only the neopallium but also the brain as a whole adds its quota to the action—in particular the great mass of nervous matter at the threshold of the cerebral hemisphere known as the thalamus. It contributes the affective element, which is the interest, the stimulative of the whole complex process, to which it gives coherence. The cortex not only preserves the records of previous experience which provide the means for comparing present experiences with past happenings, but it also adds the spatial quality to sensation and the means of judging degrees of stimulation, and the afferent impulses which pour into the brain from the joints, the muscles and the skin areas, to record the success or failure of particular muscular activities. It is by doing things that experience is built up. It is important therefore to recognise the very large part which such conative activities play in the building up of concepts. They are due not merely to the interaction of the affective and cognitive dispositions, but also to the dynamic factor which is conferred upon these processes by attempting to express in action the result of the discriminative activities of the cortex.

For some years I have been attempting to demonstrate how vast a part the cultivation of visual discrimination has played, not simply in making it possible for human beings to see the world in which they live and appreciate some of the activities which are revealed to them by their eyes, but even more in contributing to conscious control of behaviour.

The earliest type of cerebral cortex necessarily has to perform both affective and cognitive functions. It enables its possessor to appreciate the attractiveness or unattractiveness of a particular scent, and to experience an interest in addition to the cognitive recognition of it.

The cortex, at first, however, exercises no immediate direction over the motor activities of the animal beyond provoking them and providing the initiative to action. This it accomplishes by transmitting to a mass of grey matter in its base (the corpus striatum) impulses which indirectly throw other parts of the brain and spinal cord into action to direct the movements that it starts. It is the impulses from the eyes, skin and "ears" (as yet organs not of hearing, but of recording movements in the water) which consciously direct the animal's movements, while its posture and equilibrium are being maintained by the automatic mechanism of the membranous labyrinth.

The tracts in the brain which convey the impulses from skin, eyes and ears are mainly concerned with transmitting to the various motor nuclei impulses that unconsciously influence and direct reflex movements, but they all send some of their impulses to a mass of grey matter in the forebrain, which lies immediately behind the striatum, to which it is intimately linked by many nerve fibres. This is the thalamus (Fig. 2). It confers upon all the

590

和影响远距离区域。关联过程涉及整个皮层。即使是在最简单的思考或技能过程中，也要全部新皮层参与。在调控肌肉活动的过程中，循环于大脑中的各种脉冲流表现了对思维中的认知、情感和治疗活动进行整合的方式。

不仅新皮层，脑作为一个整体也对该过程作出了自己的贡献——尤其是位于脑半球开端处的大量神经物质，通常称之为丘脑。它促成情感要素——这是关键所在，是整个复杂过程的刺激因素——它在这里提供了一致性。皮层不仅保留了先前体验的记录，从而为将当前体验与过去经历相比较提供了途径，它还给感觉增加了空间属性，增加了判断刺激强度的方式，增加了从关节、肌肉以及皮肤区域传入大脑的输入性脉冲，以记录特定肌肉活动的成功或失败。经验是通过做事建立起来的。因此，认识到这种意动活动在概念建立过程中所起的极大作用是很重要的。它们不仅仅归因于情感和认知倾向的相互作用，还源于通过试图有效地表达皮质识别活动的结果而赋予这些过程的动力学因素。

若干年来，我一直致力于阐释视觉辨别力的培养曾经起到的非常重要的作用，它绝不仅仅是使人类得以看到自身生活的世界以及认识一些通过自己的眼睛看到的某些活动，而更多的是有助于对行为进行有意识的控制。

大脑皮层的最初形态不可避免地要同时履行情感和认知的功能。它使得自己的拥有者能够评价一种特定气味的诱人之处或者不吸引人之处，并且除了对它的感知识别之外还体验到一种兴趣。

不过，皮层最初并不承担对于动物肌肉运动行为的直接指导，而只是激发它们和启发动作。这个功能的实现是通过将脉冲传输到位于基部的大量灰质（纹状体），间接地引发脑的其他部分和脊髓发挥作用以指导它所开始的运动。来自双眼、皮肤和"两耳"（那时还不是听觉器官，而是在水中记录运动的器官）的脉冲有意识地指导动物的运动，而它的姿势与平衡则通过膜迷路的自动机制得以保持。

脑中负责传输来自皮肤、双眼和两耳的脉冲的纤维束，主要负责将无意识地影响和指导反射运动的脉冲传输到各种运动核，但是它们都将自己的一些脉冲发送给前脑中的大量灰质——它紧挨于纹状体之后，二者通过很多神经纤维紧密相连。这就是丘脑（图2）。它给所有非嗅觉的感觉脉冲赋予了一种情感属性，这使它们获得

non-olefactory sensory impulses an affective quality which gives them a meaning and an influence in modifying behaviour. In other words, the effects of this sensory experience, when transmitted to the striatum, are to alter the animal's reactions to smell.

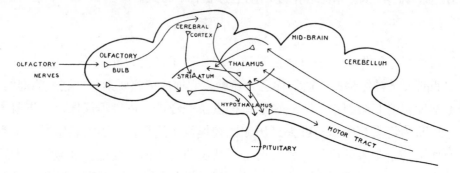

Fig. 2. Diagram of the primitive vertebrate brain to suggest the hypothalamic, thalamic, striatal and cortical connexions.

Emotional Factor in Mind

The activities of the striatum, when stimulated by the cerebral hemispheres and the thalamus, are expressed in impulses which proceed from it to the hypothalamus, a mass of grey matter lying beneath the thalamus. This surprising arrangement seems to confer upon the hypothalamus the decisive influence in translating into behaviour the initiative to action which lies in the cerebral cortex. The hypothalamus is the part of the brain which controls, by means of the sympathetic and parasympathetic systems, the most vital activities of the body itself, its visceral functions, its growth and metabolism, and even such appetites as those of sex. It is the essential instrument of emotional expression.

As the springs of action are profoundly influenced by hunger, thirst, sexual desire and other appetites and cravings, it is perhaps not surprising that in the most primitive vertebrates the instrument of the animal's vegetative needs should play a crucial part in shaping its conduct. To this part of the brain, impulses proceed from the olfactory tracts so as directly to control the activities of the alimentary and genital systems in anticipation of the realisation of the satisfaction of the respective appetites.

The study of the primitive brain impresses upon us the intimacy of the integration of the functions concerned with affective and discriminative knowledge and the translation of such information into appropriate action.

The higher type of brain distinctive of mammals, which opens up the possibility of the attainment of real conceptual knowledge and its biological application in increasingly complex acts of skill and thought, is distinguished by the growth of the thalamus and the transmission from it to the cerebral cortex of fibres in increasing numbers (Fig. 3).

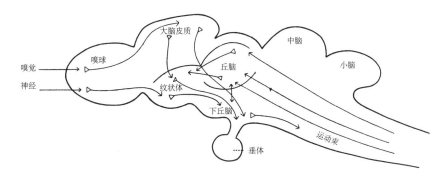

了一种意义和对于行为进行调整的影响力。换言之，这种感觉经验的效应一旦传输到纹状体，就会改变动物对于气味的反应。

图 2. 原始脊椎动物脑的示意图，揭示了下丘脑、丘脑、纹状体和皮层的联系。

思维中的情绪因素

在受到大脑半球和丘脑的刺激时，纹状体的活动就会表现为脉冲，并且从这里传输到下丘脑——位于丘脑之下的大量灰质。这种令人惊奇的排布方式似乎使下丘脑在将位于大脑皮层之中的动作发端转变为行动时具有决定性的影响。下丘脑是脑的一部分，它通过交感神经系统和副交感神经系统控制机体自身最重要的活动，包括它的内脏功能、生长与新陈代谢乃至像性欲这样的各种欲望。它是情感表达的基本结构。

由于行动的发生受到饥饿、口渴、性欲以及其他种种欲望与渴求的深刻影响，可能以下也不足以为奇了，在大多数原始脊椎动物体内，负责动物的植物性需求的结构应该在塑造其行为方面具有至关重要的作用。对脑的这部分来说，脉冲从嗅觉纤维束直行而来，以便直接地控制消化系统和生殖系统的活动，使种种欲望的满足得以实现。

通过对原始大脑的研究，我们对情感和分辨性知识相关功能的整合与将这些信息转化为适当的行动之间的密切关系留下了深刻的印象。

哺乳动物所特有的高级类型的脑，开启了获得真正概念性知识和这些知识在越来越复杂的技能与思考过程中的生物学应用的可能性，其不同之处在于丘脑的发育，以及有更多数量的纤维从这里传向大脑皮层（图3）。

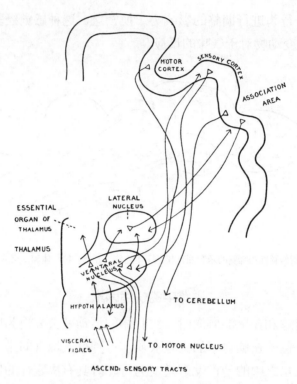

Fig. 3. Diagram of the thalamic, hypothalamic and cortical connexions of the human brain.

The recent progress in our knowledge of the structure and connexions of the thalamus and hypothalamus with the cerebral cortex, the hypothalamus and the sympathetic and visceral tracts of the organism had made it possible to carry Mr. Campion's suggestions a stage further than he himself has done. That this is possible is in large measure due to the illuminating researches of Prof. W. E. Le Gros Clark. The intensive studies which have recently been made by scores of investigators on the structure and connexions of the hypothalamus enable us to broaden the issues and consider the part played by these portions of the brain, which control the growth and metabolism of the body, and in particular visceral function, and how they are related to the thalamus and the cerebral cortex and provide the instrument for determining the emotional colour of experience and of regulating the manifestations of the appetites.

If Mr. Campion's views are correct, that the study of this neural machinery is essential for the understanding and interpretation of thought and behaviour, its structure and functions might be expected to be of great complexity. Hence it becomes essential to look at the whole issue from a much broader point of view than the mere connexions of thalamus and cerebral cortex.

Importance of Smell in the Primitive Vertebrate

In the brain of the most primitive vertebrate, the structural pattern is determined by

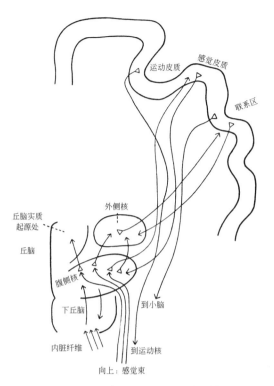

运动皮质　感觉皮质

联系区

外侧核

丘脑实质
起源处

丘脑

腹侧核

下丘脑

到小脑

内脏纤维　到运动核

向上：感觉束

图 3. 人脑中丘脑、下丘脑和皮层的联系。

　　关于丘脑、下丘脑与大脑皮层之间、下丘脑与交感神经和有机体内脏纤维束之间的结构和联系方面的认识的近期进展使我们有可能将坎皮恩先生的建议推进到比他本人所做的更进一步的阶段。这一可能性在很大程度上归功于勒格罗·克拉克教授很有启发性的研究。最近很多研究者对下丘脑的结构和联系进行的深入研究使我们得以拓宽这一论题，进而考虑脑中各部分——它们控制着机体发育和新陈代谢，特别是内脏功能——所具有的地位，以及它们如何与丘脑及大脑皮层相关联，如何促成经验的感情色彩的确定和如何促成调控欲望的表现形式。

　　如果坎皮恩先生的观点正确，即研究这一神经系统对于理解和解释思想与行为来说是不可缺少的，那么可以预期它的结构和功能具有很高的复杂性。因此，从一个更广阔的视角看待整个论题而不是仅仅考虑丘脑与大脑皮层的联系就显得十分必要了。

嗅觉对于原始脊椎动物的重要性

　　在大多数原始脊椎动物的脑中，其结构模式是由下列事实所决定的：嗅觉是主

the fact that smell is the dominant sense. The cerebral cortex is essentially a receptive instrument for impressions of smell, and the mechanism whereby consciousness of smell can influence the behaviour of the animal. When a primitive vertebrate such as a dogfish scents attractive food and pursues it, the culmination of the pursuit is represented by the seizure of the food and the appreciation of its taste. This is nearly akin to the initial olfactory experience which started the pursuit and dominated it, so that all the incidents of the pursuit become integrated into one experience, which is thus given coherence and meaning. Thus is initiated the ability to anticipate the result of a given course of action, and to recall in memory the connexion between the various incidents.

One must assume, therefore, that the primitive cortex is concerned not merely with the awareness of smell and the ability to discriminate between different kinds of smells, but also that it is concerned with the affective side of olfactory experience, with the attractiveness or repulsiveness of any scent and the influence of such affective experience in determining the nature of the response an individual odour can evoke. The cerebral cortex in such a primitive animal is incapable of directing movements, seeing that the sense of smell is utterly devoid of any spatial quality. When an animal scents an attractive food, it acquires from the sense of smell no idea as to the position in space of the object which provides the stimulus. It is merely stirred into action, and other neural mechanisms are responsible for controlling and directing the resulting activities. The cerebral cortex, so to speak, is the mere trigger which releases the activity of the brain and provokes and directs the movements.

The part of the cerebral hemisphere which translates these stimuli into action is the corpus striatum, and the striatum is connected with the thalamus, which receives from the body, that is through the skin, the eyes and the ears as well as the muscles and joints, impulses which modify and direct movements which result when the animal is thrown into action. The thalamus transmits the effects of these stimuli to the striatum and so modifies the motor activities. In the case of organs such as the eyes, the primary functions were concerned not merely with the awareness of illumination, but also of movements in the outside world, or rather movements of objects in the outside world in reference in its own body. The eyes have associated with them, in the brain, a complicated mechanism which enables them automatically to direct the movements of the body in relationship to events in the outside world. But quite apart from this, the eyes transmit to a part of the thalamus (the lateral geniculate body) impulses which are concerned with the awareness of the stimulus of light, and which influence these bodies and through them the thalamus as a whole, which in turn affects the functions of the striatum and the movements of the animal.

In the primitive vertebrate one must assume that the thalamus acts as an affective organ of all senses other than smell, and represents the instrument whereby the organism is pleasantly or unpleasantly affected by sensory experience, and that the cerebral cortex performs the analogous but more dominating aspect of the same function in relationship with smell. The dominant part of the cerebral cortex in the most primitive vertebrate is the hippocampal formation, and if one assumes the supreme function of the cortex

导性的感觉。大脑皮层在本质上是嗅觉印象的接受器件，也是使嗅觉意识影响动物行为得以实现的装置。当一只原始脊椎动物，比如狗鲨，闻到诱人食物的气味并去追逐时，追逐的结果是捕获食物并鉴定其气味。这与导致追逐开始并主导其过程的初始嗅觉体验十分类似，因而追逐过程中的所有事件逐渐整合成一次体验，进而给出了一致性和意义。从此开始有了预见给定行为过程之结果的能力，以及在记忆中回想各事件之间的联系的能力。

因此我们必须假定，原始皮层不仅仅与嗅觉意识和分辨不同类型气味的能力有关，而且还与嗅觉体验的情感方面有关，包括任何气味的令人喜欢或厌恶之处，以及这些情感体验对于决定某一具体气味所能引起的反应本质的影响。考虑到嗅觉完全不具备任何空间属性，因此原始动物的大脑皮层是不能指导运动的。当动物闻到诱人食物的气味时，它从嗅觉中得不到任何关于提供刺激的物体在空间中所处位置的信息。它仅仅是激发行动，由其他神经机制负责控制和指导因此产生的活动。可以说，大脑皮层仅仅是一个释放脑活动并且刺激和指导运动的扳机。

大脑半球中负责将上述刺激转化为行动的部分是纹状体，而纹状体是与丘脑相联系的，后者接受来自机体——即皮肤、双眼、两耳以及肌肉与关节——的脉冲，这些脉冲调节并指导动物展开行动时所产生的运动。丘脑将这些刺激的影响传输到纹状体，从而调整肌肉运动行为。对于诸如眼睛等器官来说，其首要功能并不仅仅是对光的意识，还包括对外部世界中运动的意识，或者更恰当地说，是外部世界中的物体相对于机体自身的运动。眼睛在脑中将它们整合，一种复杂的机制使它们得以自动地指导机体针对外部世界之事件的运动。但是又远不止于此，眼睛还向丘脑的一部分（外侧膝状体）传输脉冲，这些脉冲负责对光线刺激的意识，而且影响机体并使丘脑赖以成为一个整体，以及依次影响纹状体的功能和动物的运动。

我们必须假定，在原始脊椎动物中，丘脑的作用如同一个具有除嗅觉以外的所有感觉的情感器官，并且作为一种结构使生物体受到由感觉经验带来的愉快的或不愉快的影响，还要假定在与嗅觉相关的相同功能上大脑皮层起着类似的但是更具主导性的作用。在大多数原始脊椎动物中，大脑皮层的主导性部分是海马结构，并且，如果假设皮层的首要功能是决定动物的行为，那就有理由假设，原始海马体的目的

is to determine the behaviour of the animal, it is perhaps justifiable to assume that the purpose of the primitive hippocampus is to make possible the adequate association of the affective qualities of smell and to translate them into action by playing a dominant part in determining the animal's behaviour.

It is perhaps not without significance in this connexion that the efferent fibres from the hippocampal formation, after passing out of the cerebral hemisphere, terminate in the hypothalamus, that part of the brain which controls the visceral system (sympathetic and para-sympathetic) and thereby regulates the activity of the viscera. It is, in fact, that part of the brain which is intimately related to the functions of the appetites. Nor is it surprising that the particular part of the hypothalamus in which the hippocampal fibres terminate should be linked up with the thalamus, so as to provide a neural circuit in which the total affective qualities of all the senses are brought into relationship in such a way that they can influence through the striatum the motor responses of the body.

The researches of Prof. Le Gros Clark have established the fact that the thalamus contains three kinds of cell groups (Fig. 2). Those forming the ultimate termini of certain of the sensory pathways, which according to Sir Henry Head form the essential organ of the thalamus, are the instrument whereby we become aware of sensory experience and appreciate its affective qualities. Secondly, there is a group of cells (ventral nucleus) which receives the great sensory paths coming up from the other parts of the brain and the spinal cord, and transmits the impulses either to the corpus striatum or in mammals to the neopallium. In the third place, there is a group of nuclei in the thalamus which become well developed only in the higher mammals. They do not receive afferent impulses directly, but only from the intermediation of the ventral nucleus. The highest type of thalamic cells, known as the lateral nucleus (Fig. 3), establishes connexions with the parietal area of the neopallium, which intervenes between the sensory cortical areas for touch, vision and hearing (P, Fig. 1), and presumably confers upon this area the ability to provide sensory experience with spatial and discriminative qualities. All three categories of thalamic elements are intimately joined together by numerous fibre tracts so as to form a closely integrated functional whole the proper working of which is essential for cortical functions.

Integration of the Dispositions of the Mind

The common practice of psychologists of segregating the three dispositions of the mind, cognitive, affective and conative, and attempting to study them as isolated units, is devoid of justification. All three are indissolubly united in the working of the mind. To give them cohesion it is necessary to assume the existence of a circulation of nervous impulses from the thalamus to the cortex and to the widespread and complex mechanisms concerned with muscular activities.

In the growth of a concept conation plays a fundamental part. Man learns from experimentation. By the exercise of his manual dexterity he acquires knowledge of the properties of things, the nature of forces, and the means for interpreting (and in some measure understanding) the world in which he lives. The surprisingly large part of the

可能是使嗅觉的情感属性能够得以充分联系，并且通过在决定动物行为的过程中起主导性作用而将它们转化为行动。

下面这个联系可能不是毫无意义的，即由海马结构发出的传出纤维在经过大脑半球之后终止于下丘脑，下丘脑是大脑中控制内脏系统（交感神经和副交感神经）从而调控内脏活动的部分。事实上，正是脑中的这个部分与欲望的功能密切关联。同样不会令人惊奇的是，下丘脑中的特定部位，即海马纤维终止处，应该与丘脑相联系，这样才能提供一条神经循环，使得所有感觉的全部情感属性都能形成关联，以这样一种方式它们能够通过纹状体影响机体的运动反应。

勒格罗·克拉克教授的研究已确定了如下事实，丘脑中包含三种类型的细胞群（图2）。那些构成了某些感觉传导通路之最终目的地的细胞群——根据亨利·黑德爵士的看法，它们形成了丘脑的核心组件——是使我们得以意识到感觉经验并鉴别其情感属性的结构。第二，还有一群细胞（腹侧核）接收来自脑和脊髓其他部位的大量感觉传导通路，并且将脉冲传输到纹状体或者哺乳动物的新皮层。第三，丘脑中还存在一组核——只有在高等哺乳动物中丘脑才会变得如此发达。它们并不直接接收传入脉冲，而是只能借助腹侧核的媒介作用。最高级类型的丘脑细胞，即我们所说的外侧核（图3），建立了与新皮层顶部区域的联系，这个区域插入到触觉、视觉和听觉的感觉皮层区（P，图1）之间，据推测，这就使该区域具有了为感觉体验提供空间属性和分辨属性的能力。全部三类丘脑成分都通过大量纤维束紧密地联系在一起，以形成严密整合的功能性整体——它的正常运转对于皮层功能是至关重要的。

思维倾向的整合

心理学家的常见做法是分离认知、情感和意动这三种思维倾向并企图把它们作为独立单元加以研究，这些都是缺乏论证的。所有这三者在思维的运行过程中都是不能分解地结合在一起的。要使它们结合，就必须假定存在一个从丘脑到皮层的神经脉冲循环，以及与肌肉活动有关的一套普遍而复杂的机制。

在概念的产生过程中，意动起到了极为基础性的作用。人类从经验中学习。通过练习手的灵活性，人类获得了关于事物性质、力的本质和解释（在某种程度上是理解）其所处世界之方法的知识。大脑皮层中有大得令人惊讶的一部分来负责对肌

cerebral cortex that is concerned with the regulation of muscular functions and the multitude of its fibre-connexions with the cerebellum affords an impressive testimony of the vast significance of action in mind-making and emphasises what Prof. T. H. Pear has well called "the intellectual respectability of muscular skill". It is a truism that we learn by doing. In man, thought is a prerequisite for action, and action a corrective of thought. The biological justification for the evolution of the high degree of visual discrimination, whereby man knows the world and the society in which he lives, is the motor efficiency it makes possible.

The most significant factor in the evolution of the mind was effected when the direction of movements was transferred from the midbrain to the neopallium (see *Nature*, **125**, p. 820; 1930) and from being an unconscious automatism became a consciously directed process. For the neopallium not only established a direct control over the motor nuclei of the whole central nervous system, but it also became linked up with all the complicated machinery in other parts of the brain which are concerned with muscular activities.

This concentration of control in the neopallium implies a circulation of nervous impulses throughout the brain to effect cohesion between the living instruments of the conative dispositions with those of the affective (thalamus) and cognitive (neopallium) dispositions of the mind. A circulation such as Mr. Campion postulates is essential to the working of the mind.

This circulation in turn involves the hypothalamus, which presumably confers the emotional tone that plays a part in all mental and muscular activity in particular in artistic expression and the self-knowledge which is one of the most distinctive qualities of man and his thinking.

Anthropological investigations, the results of which I have summarised in chaps. v and vi of my "Human History" (1930), suggest that in primitive man there is an innate goodness and truthfulness, the awareness of which we call conscience. These qualities of the mind are responsible for character and personality. The terrible experiments which the incidence of diseases such as sleepy sickness (encephalitis lethargica) provides, has shown that these amiable qualities can be destroyed by minute injuries of certain parts of the brain in or in the neighbourhood of the hypothalamus. We must suppose that these parts of the brain are responsible for the maintenance of the innate goodness of human nature, the goodwill of normal man, seeing that their destruction causes so profound an alteration of character. Mr. Campion's hypothesis of a widespread circulation of nervous impulses provides an explanation of how these various dispositions of the mind and character may be integrated into the living human personality.

Before I close this discourse, I must express my gratitude to Mr. George Campion for his stimulating suggestions and to Prof. J. S. B. Stopford, of Manchester, for help in giving them neurological expression.

(**133**, 245-252; 1934)

肉功能的调节，而且它那些与小脑相连的大量纤维为思维发生行为的非凡重要性提供了令人瞩目的证据，并且强调了皮尔教授所称的"肌肉技能的智力品德"。我们通过做事而学习这是不言而喻的。对于人类而言，思考是行为的先决条件，而行为则可以修正思考。关于高程度的视觉辨别能力——人类藉此了解其生活的世界和社会——的进化的生物学依据，是由于进化而获得可能性的动作效率。

当对运动的指导从中脑转移到新皮层（参见《自然》，第 125 卷，第 820 页；1930 年）并且从一种无意识的自动作用转变为一种有意识的指导过程时，思维进化过程中最重要的因素就形成了。因为新皮层不仅建立了对于整个中枢神经系统的运动核的直接控制，而且还和脑中其他与肌肉活动有关的部位的全部复杂系统建立了连接。

新皮层中的这种集中控制意味着有一个遍布脑中的神经脉冲循环，它可以实现意动倾向的作用结构与思维中情感倾向（丘脑）和认知倾向（新皮层）的作用结构之间的联系。就像坎皮恩先生所提出的那样的循环对于思维的运行来说至关重要。

这个循环自然涉及下丘脑，据推测它负责感情属性，这种属性在所有的精神和肌肉活动中，尤其是在艺术表现和自我认识方面——这是人类及其思考所具有的最与众不同的属性之一——都有作用。

人类学研究——我在《人类的历史》（1930 年）的第五章和第六章中总结了其研究结果——指出，在原始人类中存在一种先天的善良与坦诚，这种意识就是所谓的良心。这种思维品质决定了品性与人格。由诸如昏睡症（昏睡性脑炎）之类的疾病的发生提供的很糟的实验表明，发生在脑的下丘脑之中或者其邻近区域中的某些部位的微小损伤就能摧毁这些美好的品质。考虑到它们的损坏会导致非常深远的品性变化，我们就必须假定，大脑的这些部位是与保持人性中先天善良品性（即普通人的善意）相关联的。坎皮恩先生关于神经脉冲普遍循环的假说对于上述各种思维与品性之倾向如何被整合成活生生的人性提供了一种解释。

在我结束这次演讲之前，我必须向乔治·坎皮恩先生表达感激之情，感谢他提供了启发性的建议，另外我还要感谢曼彻斯特的斯托普福德教授在向他们提供神经学表述方面给予的帮助。

（王耀杨 翻译；刘京国 审稿）

Hormones of the Anterior Lobe of the Pituitary Gland

Editor's Note

In the 1930s, biochemists spent much effort in identifying the vitamins necessary for healthy human life and which are not produced in the body, and also the hormones secreted by various glands in the body (such as insulin in the pancreas), which similarly sustain the whole body. This article, a digest of a book published in California, is a thumbnail account of the hormones of the anterior lobe of the pituitary gland.

IT is now generally admitted that the functions of the pituitary gland (or hypophysis) are mediated by the secretion of a number of hormones from its different parts; although no active principle has yet been isolated in the pure state, the fractionation of extracts has led to the preparation of solutions having only a part of the physiological activity of the original extract. Differences of opinion exist as to the number of hormones actually present, which can only be settled when they are finally isolated as chemical individuals. Our knowledge of the functions of the posterior lobe preceded that of the anterior, but within the last few years, with improvement in both chemical and surgical technique, and also following the discovery that hormones regulating certain of the sexual activities of the body are excreted in the urine, great advances have been made also in our knowledge of the functions of the anterior lobe.

It appears probable that a number of different hormones are secreted by this lobe, but attention has been directed especially to those stimulating growth and the sexual glands. One of the pioneers in this work has been H. M. Evans, of the University of California; the results of his researches, carried on over the last decade, are now available for study, in the form of a detailed monograph*. Although the association of overfunction of the pituitary with body overgrowth (gigantism or acromegaly) and of its underfunction with dwarfism has been frequently confirmed, it was not until 1921 that Evans and Long succeeded in preparing an extract of ox anterior lobes which stimulated growth in mammals. This was due to the facts that the growth hormone is a complex substance chemically resembling the proteins, is extraordinarily labile and can only be detected when administered frequently and parenterally to suitable animals. Adult female rats more than five months old (which have therefore ceased to increase in weight), are injected

---

* The Growth and Gonad-Stimulating Hormones of the Anterior Hypophysis. By H. M. Evans, K. Meyer and M. E. Simpson, in collaboration with A. J. Szarka, R. I. Pencharz, R. E. Cornish and F. L. Reichert. Memoirs of the University of California. Vol. 11. pp. 446. (University of California Press Berkeley, California, 1933.)

垂体前叶激素

编者按

20世纪30年代，生物化学家们花了大量的精力鉴定对于人类生命健康必需的维生素以及激素，前者在体内不能合成，后者由体内各种腺体分泌（比如胰腺分泌胰岛素）。这篇摘自于一部在加利福尼亚出版的图书的文章简略介绍了垂体前叶激素。

垂体的功能是通过其不同部位分泌的多种激素介导的，这一观点目前已经得到了普遍的认可；尽管目前还没有分离出纯的活性成分，但是对垂体提取物进行分级分离可以制备出具有原始提取物的部分生理活性的溶液。关于垂体实际分泌多少种激素存在不同的意见，只有将这些激素以化合物单体的形式分离出来之后，才能消除这些分歧。我们对于垂体后叶功能的认识领先于对垂体前叶的认识，但是，在过去的几年内，随着化学技术和外科手术技术的进步，以及调控机体某些性行为的激素会从尿液中排出这一现象被发现，我们对于垂体前叶功能的认识也已经取得了很大的进展。

看来垂体前叶很有可能分泌多种不同的激素，但是人们主要将注意力集中在那些促生长和促性腺的激素上。这个领域的先驱之一是加州大学的埃文斯，他在过去十年间所做研究的结果已经被编撰成一本详细的专著*供研究参考。尽管垂体功能亢进与身体过度生长（巨人症或者肢端肥大症）以及其功能减退与侏儒症之间的关系已经被多次证实，但是直到 1921 年，埃文斯和朗才成功地从牛垂体前叶中制备出了能够刺激哺乳动物生长的提取物。这是由于生长激素是一种化学结构类似于蛋白质的复杂物质，特别不稳定，只有当其被频繁地以肠道外给药的方式施加于适宜的动物时才能被检测到。向 5 月龄以上的雌性成年大鼠（体重已经不再增加）腹腔内注射生长激素，每日一次，持续 20 日；每组 4~6 只，每 5 天称一次体重。随给定的注

---

*《前垂体的生长和性腺刺激激素》，埃文斯，迈耶，辛普森，与绍尔卡，潘查兹，科尼什，赖歇特合著。《加州大学回忆文集》第 11 卷，第 446 页。（加州大学出版社：伯克利，加利福尼亚，1933 年。）

intraperitoneally daily for a period of 20 days; groups of four to six animals are used and they are weighed every five days. Gains in weight of 25–100 gm. can be obtained according to the dose given; the relationship between the logarithm of the dose and the gain in weight was found to be approximately linear. E. Bierring and E. Nielsen (*Biochem. J.*, **26**, 1015; 1932) have compared the composition of injected growing rats with that of normal growing rats and find that the former show a greater retention of water, but that the solid matter assimilated to the body tissues contains a much greater proportion of protein and less of ash and fat than that laid down by normal animals. About three quarters of the gain in weight of the injected animals is due to water retention, and three quarters of the dry matter deposited consists of protein.

The method of extraction recommended by Evans and his co-workers is briefly as follows: frozen ox anterior lobes are minced and extracted with water made alkaline with baryta; the mixture is centrifuged and the solution brought to pH8 with sulphuric acid and again centrifuged. The solution is then acidified and poured into excess of acetone; the precipitate is filtered off and dried. This powder is stable but still contains the gonad stimulating hormone. On extraction with 95–98 percent acetic acid, the latter is destroyed and the growth hormone can be precipitated from solution by acetone in the presence of quinine sulphate. Trichloracetic acid precipitates the growth hormone and part of the gonad stimulating hormone from aqueous solutions of the powder; in the supernatant fluid the latter can be obtained free from the former by precipitation with flavianic acid, which is then removed by 80 percent alcohol containing 1–2 percent ammonia. The purer growth hormone preparations are highly active in a daily dose of 5 mgm.

Hypophysectomised rats show a greater response to the growth hormone regardless of age or length of time after removal of the gland. Experiments with a hypophysectomised puppy are also described; the pituitary gland was removed when the animal was 8 weeks old; the operation was followed by complete cessation of growth. Daily intraperitoneal injections of the growth hormone resulted in a marked increase in weight and size, so that the animal finally became larger than its litter mate control. Signs of acromegaly, however, did not develop. The ovarian follicles showed considerable development and the thyroid was hyperplastic. Similar injections into a normal female resulted in the development of partial acromegaly, some gigantism and diabetes mellitus. A male, however, only developed adiposity. In dachshunds, the injections increased the size of the animals owing to increase in size of the skull and vertebrae, but the achondroplastic form of the short extremities was not altered; a male developed diabetes. The only outstanding acromegalic feature was a folding of the skin of the head and extremities. These results lend strong support to the generally accepted view that gigantism and acromegaly in human beings are due to over-secretion by the anterior lobe of the pituitary gland.

The gonad stimulating hormone (or hormones) is responsible for the normal development and maintenance in a state of functional activity of the sex glands, with the accessory organs and secondary sex characters. In the female the ovaries themselves respond readily

射剂量的不同，体重可以增加 25~100 g 不等；注射剂量的对数值与体重的增加几乎呈线性关系。比林和尼尔森（《生物化学杂志》，第 26 卷，第 1015 页，1932 年）对注射了生长激素的大鼠和正常生长的大鼠的组织进行了比较，结果发现：前者水潴留更加明显，但是被吸收到机体组织中的固形物中所含蛋白质的比例更高，所含矿物质和脂肪的比例更低。注射激素的大鼠增加的体重中 3/4 是由于水潴留，而且 3/4 的干物质沉积是由蛋白质组成的。

埃文斯及其同事推荐的提取生长激素的方法简述如下：将冰冻的牛垂体前叶切碎，用氢氧化钡碱溶液进行萃取；将混合物离心，上清液用硫酸中和到 pH=8，再次离心。将上清液酸化并倒入过量的丙酮中；过滤沉淀物并将其干燥。由此获得的粉末是稳定的，但是仍然含有促性腺激素。用 95% ~ 98% 的醋酸萃取后，促性腺激素会被破坏，在有硫酸奎宁存在的条件下，就可以用丙酮将生长激素从溶液中沉淀出来。三氯乙酸能够将生长激素和部分促性腺激素从粉末的水溶液中沉淀出来；从上清液中单独获取促性腺激素的方法是用黄胺酸进行沉淀，然后用含有 1% ~ 2% 氨的 80% 乙醇除去上清液中残留的酸。通过这种方法最终制备得到较纯的生长激素制剂，每天注射 5 mg，就能够发挥很高的活性。

垂体被切除后的大鼠对生长激素的反应更加明显，而且与年龄以及腺体移除后的时间长短无关。用垂体被切除的幼犬进行的实验如下所述：当幼犬 8 周龄时，摘除垂体；手术后，幼犬的生长完全停止。每日向其腹腔内注射生长激素会导致幼犬体重明显增加和体型明显增大，以至于最终比作为对照的同窝同伴长得更大。不过，并没有出现肢端肥大症的迹象。卵巢卵泡明显发育，并且甲状腺也出现增生。对一例正常的雌性个体进行类似的注射，该个体出现了一定程度的肢端肥大以及部分巨人症和糖尿病的症状。但是，经过同样处理的一例雄性个体则只出现了肥胖症。在对达克斯猎狗进行的实验中，激素的注射增加了颅骨和椎体的大小，从而增大了动物的体型，但是细小肢端软骨发育不全的状况没有改变；一例雄性个体患上了糖尿病。肢端肥大症唯一的显著特征是头部和肢端皮肤皱褶。这些结果强有力地支持了那个已被普遍接受的观点，即人类的巨人症和肢端肥大症是由垂体前叶的过度分泌造成的。

促性腺激素负责性腺的正常发育，并通过附属器官及第二性征使性腺功能活性维持在某个水平上。在雌性个体中，卵巢本身容易对该激素产生反应；在雄性个体

to the hormone; in the male the accessory organs show the most striking effects. Immature female rats were used by Evans and his colleagues for the assay of their preparations; injections were made on three days, the vaginal orifice examined on the fourth and fifth days, smears being taken as soon as it had opened, and the animals killed and examined about 96–100 hours after beginning treatment.

The hormone was prepared from ox anterior lobe (a poor source), from the serum of pregnant mares (a good source), and from the urine of pregnant women. The acetone powder from the alkaline extract of anterior lobes is dissolved in water and the reaction of the solution adjusted to give maximum precipitation: the greater part of the hormone remains in solution and is precipitated by flavianic acid, which can be removed afterwards by use of alcohol-ammonia mixture. Alternatively, the powder may be extracted with 50 percent pyridine, 50–60 percent alcohol or acetone containing 2–4 percent ammonia; the hormone is then precipitated by increasing the alcohol or acetone to 85 percent and adding a little acetic acid or salt. Pregnant mare serum was treated directly with acidified acetone and the powder purified by the methods used in the case of preparations from anterior lobes. From the urine of pregnant women the hormone (called prolan by Aschheim and Zondek its discoverers), was precipitated by excess of alcohol. The precipitate was extracted with dilute acid and the prolan reprecipitated with alcohol; the powder was purified by extraction with acetone-ammonia mixture. The minimum dose of the purest preparations was about 0.05 mgm.

Although preparations from these different sources all stimulate the gonads, yet they show differences in their chemical properties and biological effects. Prolan, for example, is more sensitive to both acid and alkali than preparations from pregnant mare serum; the latter, but not the former, give off hydrogen sulphide on treatment with alkali. However, the differences in chemical properties may be due to differences in the associated impurities. Differences in the biological effects produced are not so easily explained. Even though the minimal doses of different preparations may be the same, larger doses may have widely different effects on the ovary: thus increasing the dose of prolan increases the size of the ovaries at most four times, whilst with preparations from pregnant mare serum, there is a rough proportionality between dose and size up to about twenty-five times the minimal dose.

Evans and his co-workers have not been able to separate the gonad stimulating hormone into follicle stimulating and luteinising factors, corresponding to the prolans *A* and *B* of Aschheim and Zondek. A solution which is predominantly follicle stimulating at one dose level may produce corpora lutea at another level or when the injections are continued beyond the usual three-day period; the predominant effect may depend on the amount of purification to which the extract has been subjected. The type of response also depends in part on the time at which the examination is made after beginning the injections. There are indications that the presence of corpora lutea inhibits further development of the ovary: the occurrence of ovulation depends on the size of the ovary and the dose given. Hypophysectomised female rats were less sensitive than normal animals and

中，附属器官表现出最显著的反应。埃文斯及其同事用未成熟的雌性大鼠来检验他们的促性腺激素制剂；连续注射 3 天，在第四天和第五天时检查阴道口，阴道口一打开就立刻进行涂片，在开始处理之后大约 96～100 小时杀死这些大鼠并进行检查。

激素可由牛垂体前叶（比较差的来源）、怀孕母马的血清（比较好的来源）以及孕妇的尿液制备。将来自垂体前叶碱性萃取物的丙酮粉末溶解在水中，调节溶液中发生的反应使其能够产生最大量的沉淀物：大部分激素存留在溶液中，然后用黄胺酸使其沉淀，随后用乙醇－氨的混合溶液清除黄胺酸。另外也可以用 50% 的吡啶、50%～60% 的乙醇或者含有 2%～4% 氨的丙酮对此粉末进行萃取；然后提高乙醇或者丙酮的浓度到 85% 并加入少量醋酸或者食盐将激素沉淀出来。怀孕母马的血清则直接用酸化的丙酮处理，并用从垂体前叶制备激素的方法对粉末进行纯化。来自孕妇尿液的激素（其发现者阿什海姆和邹德克称之为绒毛膜促性腺激素）通过过量的乙醇进行沉淀。用稀释的酸萃取沉淀物，然后用乙醇将绒毛膜促性腺激素再沉淀下来；再用丙酮－氨的混合溶液进行萃取来纯化该粉末。所制得的最纯制剂的最小剂量大约是 0.05 mg。

尽管从这些不同的来源制备得到的样品都能刺激性腺，但是它们在化学特性和生物学效应上的表现并不相同。比如，绒毛膜促性腺激素比从怀孕母马血清中制备得到的激素对酸和碱都更加敏感；后者用碱处理时会释放出硫化氢，而前者不会。不过，化学特性方面的差异可能是由于含有的杂质不同。生物学效应方面的差异就不是这么容易解释了。尽管制备得到的各种样品的最小剂量几乎是相同的，但是，较大的剂量对于卵巢会有非常不同的作用：增加绒毛膜促性腺激素的量可以使卵巢的体积最多增大 4 倍，而对于从怀孕母马血清中制备得到的激素来说，剂量和卵巢的体积之间差不多成正比，卵巢的体积最大可达到最小剂量时的 25 倍左右。

埃文斯及其同事还不能将促性腺激素分离成卵泡刺激素和黄体生成素，分别对应于阿什海姆和邹德克所说的绒毛膜促性腺激素 A 和 B。激素溶液在一定剂量水平下可以显著地刺激卵泡的溶液，而在另一个剂量水平下或者连续注射超过通常的 3 天周期时就会产生黄体；其主导效应可能取决于提取物纯化的程度。产生效应的类型在某种程度上也依赖于开始注射后多长时间进行检查。一些现象提示黄体的存在抑制了卵巢进一步的发育；排卵的发生频率取决于卵巢的大小和注射剂量。垂体被切除的雌性大鼠的敏感性比正常大鼠低，而且对绒毛膜促性腺激素的反应要远远低

the response to prolan was much less than that to extracts of pregnant mare serum; simultaneous administration of the growth promoting hormone diminished the response. Substitution therapy failed to induce the rhythmic changes in the vagina characteristic of the oestrous cycle, a continuous oestrous reaction only being obtained. Pregnancy was not observed owing to failure of implantation, but it could be maintained in animals, hypophysectomised after implantation, by injection of mixtures of growth- and gonad-stimulating hormones.

In hypophysectomised female dogs, prolan had no effect on the genital system even in large doses, and when the system showed a marked degree of atrophy, an anterior lobe extract also had no effect. A mixture of the two preparations, however, stimulated the genitalia within ten days; the vulva increased to a size greater than that observed in normal oestrus in a litter mate control, the mammary gland and uterus showed marked development and the ovary was much enlarged and contained many corpora lutea. This result may be contrasted with some experiments on the hypophysectomised ferret recently published by M. K. McPhail (*Proc. Roy. Soc.*, B, **114**, 128; 1933). Anterior lobe extract alone produced extensive theca luteinisation of small follicles, but no development of large follicles: prolan alone caused many follicles to undergo partial growth, which, however, terminated in atresia: the vulva showed partial oestrous swelling. A mixture of the two preparations produced usually only theca luteinisation.

Without referring to other work in detail, it maybe stated that several workers in addition to Aschheim and Zondek have adduced evidence that the follicle stimulating and luteinising hormones from the anterior lobe are separate entities. The synergistic action with prolan may depend on the proportions of these factors present in different preparations. Apart from the chemical difficulties of preparing the hormones in a pure state, the facts that they act in succession, or if really a single entity initiate a series of reactions, introduces a complication into the evaluation of the biological tests, which only further work with a standardised technique can clarify.

In immature male rats, doses of gonad stimulating hormone sufficient to produce enlargement of the ovaries in immature females produced little or no increase in the weight of the testes, although the accessory organs grew markedly and attained the size characteristic of these organs in young adults. Larger doses of hormone, however, increased the weight of the testes. Senile males also responded by increase in weight of the accessory organs. In hypophysectomised males injections of the hormone caused regeneration of the atrophied testes, the seminal vesicles became enlarged and filled with fluid and spermatogenesis was resumed; the replacement therapy was complete since normal litters were sired, and the testes appeared normal on histological examination.

The atrophy of the thyroid and adrenal glands after removal of the pituitary was not repaired by injection of gonad stimulating hormone, but extracts containing the growth hormone maintained or restored the weight of these organs, although histologically the

608

于对怀孕母马血清提取物的反应；同时给予促生长激素可以减弱这种效应。替代治疗不能诱发发情周期中阴道特征的周期性变化，而只能获得持续的发情反应。因为胚胎不能着床，所以观察不到怀孕现象。但是如果在胚胎着床后切除垂体，则可以通过注射促生长激素和促性腺激素的混合物来维持动物的怀孕。

在垂体被切除的雌狗中，即使大剂量的绒毛膜促性腺激素也不能对生殖系统产生作用。当生殖系统表现出显著萎缩时，垂体前叶的提取物也不能发挥作用。但是，这两种物质的混合物能够在 10 天内刺激生殖器；外阴增大得比同窝作为对照的同伴正常发情期时还明显，乳腺和子宫明显发育，卵巢明显增大并含有大量黄体。这个结果和最近麦克费尔发表的在垂体被切除的雪貂上完成的一些实验的结果（《皇家学会学报》，B 辑，第 114 卷，第 128 页，1933 年）截然不同。单独用垂体前叶提取物能够使小卵泡的膜广泛黄体化，但是不能导致大卵泡的发育；单独用绒毛膜促性腺激素能够使许多卵泡部分发育，但是最终还是以闭锁告终；外阴表现出部分的发情期肿胀。两者的混合物常常只能引起膜黄体化。

不用再详细介绍其他研究工作，我们也可以说，除阿什海姆和邹德克以外其他一些研究人员也已经得到证据可以证明来自垂体前叶的卵泡刺激素和黄体生成素是两种不同的物质。它们与绒毛膜促性腺激素的协同作用可能与制备得到的不同样品中存在的这些物质的比例有关。除了制备纯净的激素存在化学方法上的困难之外，这些激素是顺次发生作用抑或是一种激素引发一系列的反应，这些事实都使得对生物学实验结果的评估变得更加复杂，只有通过标准化的实验技术进行更深入的研究才能澄清混乱。

将可使未成熟雌性大鼠卵巢增大的剂量的促性腺激素作用于未成熟的雄性大鼠后，其睾丸的重量增加很少或不增加，尽管附属器官的体积明显增大并且达到了年轻成年鼠的水平。不过，更大剂量的激素能够增加睾丸的重量。老年雄鼠也有反应，主要是附属器官重量增加。对于垂体被切除的雄鼠，注射激素能使萎缩睾丸再生，精囊增大并充满精液，精子发生过程重新开始；替代治疗获得了成功，因为已经能够生育出正常的幼崽，而且睾丸在组织学检查上也表现正常。

垂体被切除后，注射促性腺激素不能修复甲状腺和肾上腺的萎缩，但是含有生长激素的提取物则可以保持或者恢复这些器官的重量，尽管它们在组织学上不能完

normal structure was not completely regained. Evans's results do not show whether it is the growth hormone or some other active principles in the extracts which are responsible for these effects. The cachexia commonly observed in hypophysectomised rats was also relieved by injections of the growth hormone.

The data on which the workers in the University of California base the conclusions briefly reviewed above are available in detail in the monograph now under notice. The methods described should be of value to other investigators and their results should form the basis of further research in this important field.

(**133**, 401-403; 1934)

全恢复正常结构。埃文斯的结果没有揭示产生这些效应的到底是生长激素还是提取物中的其他活性物质。注射生长激素还能缓解垂体被切除后的大鼠中非常常见的恶病质。

加州大学的研究者们得出上述结论所依据的具体数据可以在那本受人瞩目的专著中查到。书中介绍的方法对于其他研究者来说应该是有价值的，而他们得到的结果则将成为在这个重要的领域中进行更深入研究的基础。

（毛晨晖 翻译；金侠 审稿）

A Rapid Test for the Diagnosis of Pregnancy

C. W. Bellerby

Editor's Note

In the 1930s, biologists were occupied with two important and novel searches, one for vitamins (essential constituents of the human diet usually manufactured by plants) and one for hormones (chemicals essential for the life of human beings but which have to be manufactured in the body from ingested food). This paper is one of the first to suggest that hormones may be effective in the diagnosis of pregnancy in women, based on the use of the South African clawed toad as test animal.

CURRENT biological tests for the diagnosis of pregnancy or detection of ovary-stimulating substances in gland extracts and body fluids have the main disadvantage that several days must elapse before a result can be obtained. Attempts have been made to remedy this by making use of the doe rabbit, because in this animal a response (ovulation) can be obtained in less than 14 hours[1]. The rabbit, however, requires a good deal of care in order to obtain consistent results. It is essential to know the previous history of does employed, and preferably only to use them a short time after parturition. Even so, variation in response to injection may be so great as to necessitate the use of more than one doe in order to be sure of the result.

The test described in the present note depends upon the observation by Hogben[2] that extraneous ovulation in the South African clawed toad (*Xenopus Loevis*) can be induced by injection of extracts of the anterior lobe of pituitary. *Xenopus* can be obtained easily and cheaply in large numbers. Several hundreds can be kept without difficulty at the sole cost of a few handfuls of raw meat once a week, provided that they are kept in a warm well-lit room and that their water is changed after feeding. Ovulation does not occur spontaneously in captivity. Ova shed as a result of injection are clearly visible and extruded in large numbers. No doubt exists, therefore, as to the validity of a response.

During the past two years, work has been carried out on the use of *Xenopus* for detecting and estimating ovary-stimulating substances in tissue extracts and body fluids such as pregnancy urine. The following main points have emerged[3].

(*a*) At a temperature of 20°C–25°C a single injection of an active preparation into the lymph sac is followed in the great majority of cases by complete ovulation within 9 hours. Very often a response is obtained in less than 6 hours.

(*b*) A given batch of toads can be used repeatedly, provided that a rest of at least one

一项快速诊断妊娠的实验方法

贝勒比

编者按

20 世纪 30 年代，生物学家致力于寻找两种重要而新颖的物质，一种是维生素（人类饮食必需的成分，通常由植物体合成），另一种是激素（人类生命必需的化合物，但是只能通过摄入食物在体内合成）。本文是最早提出激素可能可以用于诊断妇女妊娠的文章之一，其研究是以非洲爪蟾作为实验动物的。

当前用于诊断妊娠或者检测腺体提取物及体液中卵巢刺激物质的生物学方法，主要缺点在于检测结果需要等好几天才能获得。人们曾尝试用雌兔进行实验来弥补这个缺点，因为在雌兔体内不到 14 小时就可以检测到排卵反应 [1]。然而，要获得一致的结果，需要精心照料雌兔，必须清楚它们的成长过程，而且最好是只在分娩后很短一段时间内使用它们。即便如此，注射后排卵反应的差异也非常巨大，从而必须用多只兔子进行实验以确保结果的准确性。

本文所描述的实验基于霍格本 [2] 的发现，即注射垂体前叶的提取物可以诱发非洲爪蟾的体外排卵。非洲爪蟾很容易大批量低成本地得到。只要将它们养在温暖且光线充足的房间里，保证每次喂食后换水，只需每周一次放入少量生肉，数百只爪蟾就能很容易地生存下来。在这种养殖条件下，爪蟾不会自发排卵。对其注射刺激物质后，可以非常清楚地看到大量排卵的过程。因此，注射刺激物质可以诱发排卵反应这一点是毫无疑问的。

在过去的两年里，我们利用非洲爪蟾对组织提取物和体液（例如孕妇尿液）中的卵巢刺激物质进行了检测和估计。从这些实验中可以得出以下几点主要结论 [3]。

(*a*) 在 20℃~25℃下，往爪蟾淋巴囊内单次注射活性提取物后，绝大多数爪蟾在 9 小时内完全排卵。在 6 小时内排卵的情况也经常出现。

(*b*) 每一批爪蟾都可以重复使用，只要连续两次注射之间至少间隔一周。

week is allowed to elapse between successive injections.

(*c*) A definite quantitative relationship holds between dosage and response.

As a result of the first observation, a test for early pregnancy has been elaborated, the exact procedure of which depends upon the time which has elapsed from the last missed menstrual period :—

(1) If one month or more has elapsed, untreated urine from the suspected case is used. Ten toads are injected in the lymph sac with 1 ml. A positive diagnosis is made if ovulation occurs in at least 5 out of 10 animals within 9 hours. The correct temperature is obtained by keeping vessels containing the toads in a room heated to 20°C–25°C by means of an electric fire.

(2) If less than one month has elapsed, a sample of 100 ml. of urine is precipitated with acetone and centrifuged. The residue is suspended in 10 ml. of distilled water and 1 ml. of the suspension injected into each of 10 toads. A positive result is indicated as before. This procedure is necessary owing to the facts that in very early pregnancy there is an insufficient amount of ovary-stimulating substance in 1 ml. of urine to produce a response, and that a volume of fluid greater than 2 ml. cannot be injected into the lymph sac without risk of non-absorption.

A full account of this work will appear later. So far no incorrect diagnosis has been made. In view of the quantitative nature of the test, it is hoped to distinguish normal early pregnancy from ectopic pregnancy or conditions such as hydatidiform mole.

(**133**, 494-495; 1934)

C. W. Bellerby: Department of Social Biology, University of London, March. 19.

References:
1. Bellerby, C. W., *J. Physiol.*, **67**, Proc. xxxii; 1929.
2. Hogben, L. T., *Proc. Roy. Soc. S. Africa.*, March, 1930.
3. Bellerby, C. W., *Biochem. J.*, **27**, 615, 2025; 1933.

（c）注射剂量和妊娠反应之间存在明确的定量关系。

基于上述实验的结果，我们精心设计了一种检测早孕的方法。具体的操作方法取决于距离上次月经结束的时间有多长：

（1）如果时间已经超过 1 个月，那么可以直接用被检对象的尿液。在 10 只爪蟾的淋巴囊内分别注射 1 ml 尿液。如果在 9 小时内，10 只爪蟾中至少有 5 只出现了排卵，那么结果就是阳性的。将装有爪蟾的容器放置在 20℃~25℃的房间里，通过电热装置来保证温度的准确性。

（2）如果时间还未到 1 个月，那么就取 100 ml 尿样，用丙酮使其沉淀并离心。将沉淀物溶解于 10 ml 蒸馏水中。在 10 只爪蟾的淋巴囊内分别注射 1 ml 这种悬浮液。阳性结果的判断方法同前。这个操作过程是必需的，因为在怀孕后最初的时期，1 ml 尿液中所含有的卵巢刺激物质的量不足以引起排卵反应，而且为了保证吸收，不能向淋巴囊内注射 2 ml 以上的体液。

我们随后将详细叙述这项工作的过程。目前为止还没有出现错误的诊断。考虑到这种检测方法的定量性质，我们希望可以通过它将正常的早孕与异位妊娠或者诸如葡萄胎那样的其他情况区分开来。

（毛晨晖 翻译；金侠 审稿）

The Inheritance of Acquired Habits

E. W. MacBride

Editor's Note

By the mid-twentieth century, Lamarckism—the idea that organisms can pass on traits developed in response to their environments—was out of favour. Paul Kammerer, whose work William MacBride cites here, killed himself in 1926, shortly after his studies on Lamarckian inheritance in the midwife toad were shown to have been faked. MacBride remained a believer until his death in 1940, but he was not a crank. The inheritance of acquired characters is now called epigenetic inheritance. Its importance in evolution is controversial, but several laboratory studies have demonstrated that environmental effects can cause traits that organisms pass onto their offspring by chemically modifying DNA—changing gene activity, for example—without changing the DNA sequence itself.

FOR the last five years, experiments to test the heritability of acquired habit have been in progress in the Zoological Laboratory of the Imperial college of Science under my supervision; and an account of the work may be of interest to readers of *Nature*.

The first part of the results of these experiments has been published by the Royal Society: the second part is almost ready for publication. Miss Sladden, who carried out the work, began by rearing the young of *Salamandra maculosa* and the eggs of *Alytes obstertricans*, thus endeavouring to repeat Kammerer's work. It became evident, however, that we did not possess the equipment necessary to provide the conditions which would induce these animals to breed. We succeeded in confirming some of Kammerer's statements about the effect of the environment on the habits of one generation. Thus it is quite possible to induce *Alytes*. normally a land animal, to adopt an aquatic life; and in regard to *Salamandra* we were enabled to explain Herbst's failure to obtain Kammerer's results.

There are two distinct races of *Salamandra maculosa*, an eastern and a western. In the latter, which inhabits the Jura and the Vosges, the yellow pigment is arranged in two longitudinal bands on the back, over a general body colour of black. Miss Sladden has reared animals of this race from birth to an age of three years in boxes painted inside with bright yellow and also in boxes painted deep black inside. In neither case could we detect any alteration in the amount of yellow pigment as a result of the colour of the background. In the eastern race, however, which formed the subject of Kammerer's researches, the yellow pigment is arranged as a series of spots over a black background; and by experiments conducted by Mr. E. Boulenger, then curator of reptiles in the Zoological Gardens, and by myself, during the years 1919–1924, we were able to show that animals of this race exposed for long periods to a black environment do show definite reduction of the yellow

获得性习性的遗传

麦克布赖德

编者按

到 20 世纪中叶，拉马克学说——认为生物体能够传递为适应自身环境而发展来的性状——已经不再受到关注。威廉·麦克布赖德在此引用了保罗·卡默勒的工作。1926 年，保罗·卡默勒在他的关于产婆蟾的拉马克式遗传的研究被指证为造假后不久便自杀了。麦克布赖德直到 1940 年去世时仍然坚信卡默勒的观点，不过他对此并不狂热。获得性特征的遗传现在被称为表观遗传。人们对它在进化方面的重要性仍有争议，但是有些实验研究已经表明环境影响能够导致生物体产生某些性状，通过 DNA 的化学修饰——例如，改变基因的活性——而并不改变 DNA 本身的序列，生物体可以将这些性状传递给他们的后代。

过去 5 年中，在我的指导下，我们一直在帝国理工学院的动物学实验室进行实验来验证获得性习性的遗传。现对该工作加以介绍以飨对此感兴趣的《自然》杂志的读者。

这些实验结果的第一部分已由皇家学会发表；第二部分也即将发表。实施这项研究的斯莱登小姐从培养斑点蝾螈的幼仔和产婆蟾的卵子开始，致力于重复卡默勒的工作。然而，很显然我们并不具备必要的设备来创造诱导这些动物进行繁殖的条件。但我们成功证实了卡默勒关于环境影响一个世代的习惯的一些论断。因此，我们很可能可以通过诱导使通常情况下陆生生活的产婆蟾适应水生生活；对于蝾螈，我们已经能解释为什么赫布斯特没能得到卡默勒的实验结果。

斑点蝾螈有两个不同的种类，一种东方的和一种西方的。后者生活于侏罗山脉和孚日山脉，其通身为黑色，背上的黄色素呈两排纵向条带排列。斯莱登小姐将各个不同阶段（从出生直至 3 岁）的这种动物分别饲养在内壁涂成亮黄色和深黑色的盒子中。结果我们在两种颜色背景的盒子里都没有检测到黄色素的量有任何变化。然而，东方种类，即卡默勒的研究对象，其黄色素呈一系列点状排布在黑色的体色背景上；通过 1919~1924 年布朗热先生进行的实验和动物园爬行动物馆的馆长随后进行的实验以及我自己的实验，我们能够证明，如果将这种动物长期暴露于黑色环境中，那么他们确实可以表现出黄色素的明显减少。但是，即使斯莱登小姐已经可以成功地使她饲养的动物进行繁殖，所需的时间长度也不能允许她进行实验，因为

617

pigment. But even if Miss Sladden had been successful in getting her animals to breed, the length of time involved would have been prohibitive, since the adult condition is only attained after four years' growth. Therefore we sought for a convenient experimental animal in which the generations succeeded each other more rapidly.

Some years ago (1912–1915), in conjunction with another pupil (Miss Jackson, afterwards Mrs. Meinertzhagen), I conducted experiments on breeding the stick-insect, *Carausius morosus*, and I found that this insect, whose normal food in England is privet, could be forced by starvation to feed upon ivy. I therefore suggested to Miss Sladden that she should test the development of this ivy-feeding habit. This insect offers great advantages when used as an experimental animal. It is parthenogenetic: males only appear in small numbers every five or six generations and when they do appear they are at once recognisable by their smaller size and different coloration. The parthenogenetic insect produces about 150 eggs a year which take about three months to develop: there is no metamorphosis and as there are no wings the nymph is morphologically similar to the adult.

The plan adopted was to isolate the just hatched young, keeping each one in a separate box. These boxes were made of metal: they were circular and had glass covers. In each box was placed a small piece of ivy leaf. At the end of two days about ten percent of the insects had begun to eat ivy, the rest had not touched it. If we had reared from these insects alone we should have been accused of selection: but we adopted a different plan (suggested by my colleague, Mr. Hewer). The ninety percent which refused ivy were given a bit of privet leaf to eat and so rescued from starvation. Then after one day the privet was removed and the insect was again provided with ivy. This second provision of ivy was called the "second presentation". If after two days more the insect still refused ivy, it was again given privet for a day. The majority of the insects accepted ivy at the second presentation, but some held out until the third, fourth, or even fifth, presentation and one recalcitrant held out until the tenth presentation.

We started the experiments with 125 females. All the young which accepted ivy at the same presentation, to whatever mother they belonged, were classed together, and when they in turn became adult the eggs of each class were mixed together. From each mixture 100 eggs were selected in order to rear the next generation. In the second generation, in place of ten percent no less than eighty percent of the insects accepted ivy when first presented, that is, at the first presentation: in all, 800 insects were tested. In the third generation ninety-five percent accepted ivy at the first opportunity and 2,000 insects were tested.

Thus with these insects, we reached exactly the same conclusions as those arrived at by Prof. McDougall with regard to induced habits in rats, namely, that when members of one generation are compelled to adopt a new habit, a residual effect of this habit is carried over to the next generation, so that the young insects adopt the new habit more quickly than did their parents. We claim, however, that the stick-insect gives more conclusive

这种动物要经过 4 年才能达到成年状态。因此，我们要寻找一种方便的可以更加快速地完成传代过程的实验动物。

多年前（1912~1915 年），我与另外一个学生（杰克逊小姐，后来成为迈纳茨哈根太太）合作进行了饲养竹节虫（印度棒䗛）的实验。我发现，这种在英国通常以女贞为食的昆虫迫于饥饿会以常春藤为食，因此我建议斯莱登小姐研究这种常春藤食性的发展过程。作为一种实验动物，这种昆虫具有很大的优势。它们是孤雌生殖的：每五六代中仅出现少量的雄性昆虫，并且一旦出现雄性，我们立刻就可以通过它们较小的体型和不同的颜色将它们辨别出来。孤雌生殖的昆虫每年大概产 150 个卵，这些卵的发育需要大约 3 个月的时间：发育不经过变态过程，而且因为没有翅膀所以蛹虫与成虫在形态上很相似。

采用的实验设计是：隔离刚孵化的幼虫，将其分别饲养在不同的盒子里。这些盒子是用金属制成的圆形盒子，并且有一个玻璃盖子。在每个盒子中放置一小片常春藤叶子。在第二天快结束时，约有 10% 的昆虫已经开始吃常春藤了，其他的昆虫一直没有碰过常春藤叶子。如果我们当时再将这些已经食用常春藤的昆虫单独喂养，那么可能就会有人指责我们引入了选择：不过，我们采用了另一种不同的方案（根据我的同事休尔先生的建议）。我们给 90% 的那些拒绝食用常春藤的昆虫提供少许女贞叶子以使它们不至于饿死。一天后将女贞叶子取出，再给它们提供常春藤。常春藤的第二次供应被称为"第二次给食"。如果两天之后昆虫仍拒绝吃常春藤，就再喂它们一天女贞叶子。实验发现，大部分昆虫在第二次给食时就接受了常春藤，有些则一直坚持到第三次、第四次、甚至第五次给食时才接受，有一只顽抗的虫子直到第十次才屈服。

我们开始实验时使用了 125 只雌性昆虫。所有在同一次给食时接受常春藤的幼虫，不论它们是不是来自同一个母亲，都被归为一类。当它们变成成虫时，将每种类型的卵混合在一起。从每类混合中选出 100 个卵来繁殖下一代。第二代中，至少有 80% 而不是只有 10% 的昆虫在第一次喂食常春藤（即第一次给食）时就接受了常春藤：我们总共检测了 800 只第二代的昆虫。到第三代时，我们检测了 2,000 只昆虫，95% 的昆虫在第一次给食时就接受了常春藤。

因此我们通过这些昆虫得到的结论与麦克杜格尔教授从对大鼠的诱导习性的研究中得到的结论完全一致，即，当一代成员被迫接受一种新习性时，这种习性的残余效应会延续至下一代，因此幼虫会比它们的父母更快地接受这种新习性。然而我们要说明的是，对竹节虫进行实验得到的结果比大鼠更具说服力，因为尽管我们认

results than the rat, because although we think that Prof. McDougall has overcome all his difficulties, yet there were very serious objections to be faced with rats, such as possible mass-suggestion, parental training, etc., which are obviously inapplicable to insects.

What many people fail to realise, however, is that this transference of a residual effect of habit is the central principle of Lamarckism, clearly and unequivocally expressed by Lamarck himself. He said that "the environment produces no direct effect on the animal", but by making new needs (for example, the necessity of eating ivy or starving) it forces the animal to make new efforts to satisfy them, and "if these needs *continue for a long time* then the animal's efforts become habits" and habits by causing the use of some organs more and others less bring about the enlargement of the former and the diminution in size of the latter; and these changes are preserved by reproduction.

This article is written in the hope that other investigators will take up this question and repeat the experiments using other animals, especially other insects, as subjects; for only by such experiments can this fundamental principle be settled. Indeed, experiments with the larvae of moths were begun some years ago by Dr. Thorpe, of Cambridge. The attractive feature about such experiments is that the percentage of mortality is very low, so that the agency of "chance" or "natural selection" is excluded. Prof. Woltereck, whose great book "Grundzüge einer allgemeinen Biologie" was reviewed in *Nature* of December 17, 1932, removed Cladoceran Crustacea from northern lakes to Lake Nemi in Italy. When he examined the transported stock after twenty years he found them much altered in shape: when he again re-transferred some of this stock to the post-glacial lakes of their ancestry they reverted to their original shape—but only *gradually during the course of several generations*.

The Linnean Society recently had the privilege of hearing Prof. Woltereck deliver an address on the fauna of recent lakes in many lands. Summing up the evidence, Prof. Woltereck concludes that the time since the recession of the ice of the last phase of the glacial age, that is, about 10,000 years, has only sufficed for the production of new races: for the production of new species we must go back to pre-glacial times possibly 500,000 years ago. As I remarked in my comments on the lecture it would be hard lines on the experimenter if he had to live and experiment for 10,000 years, before he could hope to produce a new heritable *structure*, but heritable changes of *habit* in small rapidly breeding animals may be observed after experiments lasting from five to ten years.

Students of mutation, that is, "geneticists", will naturally inquire what is the relation between these changes of habit and mutations. That is a question for future study; here only certain tentative suggestions can be offered. From the study of the few cases in which mutations have been experimentally produced by such agencies as X-rays and heat, it may be concluded that they are due to some damage to the developmental machinery of the nucleus in the germ cells. They, and not the Lamarckian changes, are the results of the "direct action of the environment". So long as malign conditions surrounding early

620

为麦克杜格尔教授已经克服了他所遇到的所有困难，但大鼠实验仍面临着非常严重的缺陷，例如可能存在集体暗示及亲代训练等，而昆虫显然不存在这些情况。

然而许多人没有意识到的是，这种习性残余效应的传递是拉马克学说的中心原则，拉马克本人对此进行了清晰而又明确的表述。他说"环境并不会对动物产生直接的影响"，而是通过创造新的需求（例如，食用常春藤或者挨饿的必需性）来迫使它们作出新的努力以适应这些新的需求，"如果这些需求**持续很长一段时间**，那么动物的努力就会变成习性"，这些习性通过更多地使用某些器官和更少地使用另外一些器官而使得前一种器官体积变大而后一种器官体积缩小；这些变化通过繁殖得以保留下来。

写这篇文章是希望其他研究者能够去研究这一问题并使用其他动物（特别是用其他昆虫）作为研究对象来重复这一实验；因为只有通过这样的实验才能确定这一基本原则。实际上，多年前剑桥的索普博士就开始用蛾的幼虫进行实验。这种实验吸引人的特征是死亡率很低，因此"偶然"或"自然选择"的影响可以被排除。著有重要书籍《普通生物学概述》（1932 年 12 月 17 日《自然》上有对此书的评论）的沃尔特雷克教授，将大型蚤从意大利的北部湖泊迁移到内米湖。20 年之后当他检查这一迁移蚤群时，他发现它们的形状发生了很大改变：当他把其中一部分再次转运回原来它们祖先生活的冰后期湖泊时，它们又恢复了原始的形状——不过是**经历了几代的时间逐渐恢复的**。

林奈学会最近有幸聆听了沃尔特雷克教授所作的关于陆地上新形成湖泊中的动物群的演讲。通过总结证据，沃尔特雷克教授推断，自从冰河时代最后一个冰期的冰川消退以来的大约 10,000 年的时间只够产生新的种系：对于新物种的产生，我们必须回溯到大约 500,000 年前的前冰河时代。正如我在对此演讲的评论中所写的，如果实验者希望通过实验产生新的可遗传的**结构**，那么他必须生活且实验 10,000 年，这对实验者来说是极其困难的，但是如果用能够快速繁殖的小型动物作为实验对象，那么经过持续 5~10 年的实验之后这些小型动物的**习性**的可遗传改变也许就可以被观察到了。

研究突变的"遗传学家"会自然地探寻习性的这些变化和突变之间有什么联系。这是将来要研究的一个问题；这里只能提供一些尝试性的建议。从几个通过 X 射线和加热等实验手段产生突变的实例研究，可以推断出突变是由于生殖细胞的细胞核的发育结构受到了某种损伤而产生的。它们和非拉马克式的改变都是"环境的直接作用"的结果。只要早期发育阶段持续存在恶性条件，那么突变就会忠实地遗传下

development persist, the mutations are faithfully inherited, but if the organism can be replaced in its natural environment, then in a limited number of generations they pass off and the original constitution reasserts itself. In 1790 Capt. Cook introduced the English domestic pig into New Zealand in order to induce the Maoris to abstain from cannibalism. The animals escaped into the woods, and by 1840 had increased to herds of at least 40,000 in number and had assumed all the characters of the ancestral wild boar, including the fierce tusks—although in New Zealand there were no enemies which required such weapons to drive them off. Mutations seemingly are more surface phenomena than racial habits: they are indeed what Johanssen the inventor of the word "gene" called them, "superficial disturbances of the chromosomes", but racial habits belong to the inmost core of the heritable constitution.

(**133**, 598-599; 1934)

去；但是如果将生物重新放在其自然的生存环境中，那么突变就会在经历有限的几代后终止，原来的构成得以重现。1790 年，为了使毛利人不再自相残杀，库克船长将英国的家猪引入新西兰。结果这些家猪逃进了树林，到 1840 年时已经增加到至少 40,000 头，并且都呈现出了原始野猪的所有特征，包括凶猛的獠牙——尽管在新西兰并不存在需要它们使用这样的武器来驱赶的敌人。貌似比起种系习性来，突变是更加表面化的现象：事实上突变正是被"基因"一词的发明者约翰森称为"染色体的表面扰乱"，而种系习性则属于可遗传组成的最核心的部分。

（刘皓芳 翻译；刘京国 审稿）

X-ray Photographs of Crystalline Pepsin

Editor's Note

The division of proteins into two structural types—the compact or globular and the extended or fibrous—stems from these papers by J. Desmond Bernal and Dorothy Crowfoot at Cambridge and by William Astbury and R. Lomax at Leeds. Broadly speaking, globular proteins are enzymes, while fibrous proteins are structural, comprising biological tissues. Bernal, like Astbury a protégé of William Bragg, and Crowfoot (later a Nobel laureate for her work on biomolecular crystallography) express doubts here about whether the globular molecules of the enzyme pepsin consist of continuous polypeptide chains, but Astbury and Lomax suggest that the chain is "folded in some neat manner". This principle of protein folding is now central to the relation of structure to function in globular proteins.

FOUR weeks ago, Dr. G. Millikan brought us some crystals of pepsin prepared by Dr. Philpot in the laboratory of Prof. The Svedberg, Uppsala. They are in the form of perfect hexagonal bipyramids up to 2 mm. in length, of axial ratio $c/a=2.3\pm0.1$. When examined in their mother liquor, they appear moderately birefringent and positively uniaxial, showing a good interference figure. On exposure to air, however, the birefringence rapidly diminishes. X-ray photographs taken of the crystals in the usual way showed nothing but a vague blackening. This indicates complete alteration of the crystal and explains why previous workers have obtained negative results with proteins, so far as crystalline pattern is concerned[1]. W. T. Astbury has, however, shown that the altered pepsin is a protein of the chain type like myosin or keratin giving an amorphous or fibre pattern.

It was clearly necessary to avoid alteration of the crystals, and this was effected by drawing them with their mother liquor and without exposure to air into thin capillary tubes of Lindemann glass. The first photograph taken in this way showed that we were dealing with an unaltered crystal. From oscillation photographs with copper $K\alpha$-radiation, the dimensions of the unit cell were found to be $a=67$ A., $c=154$ A., correct to about 5 percent. This is a minimum value as the spots on the c row lines are too close for accurate measurement and the c axial length is derived from the axial ratio. The dimensions of the cell may still be multiples of this. Using the density measured on fresh material[2] as 1.32 (our measurements gave 1.28), the cell molecular weight is 478,000, which is twelve times 40,000, almost exactly Svedberg's value arrived at by sedimentation in the ultracentrifuge. This agreement may however be quite fortuitous as we have found that the crystals contain about 50 percent of water removable at room temperature. But this would still lead to a large molecular weight, with possibly fewer molecules in the unit cell.

624

胃蛋白酶晶体的X射线照片

编者按

蛋白质可以划分为两种结构类型——紧实型（或者说球形）和伸展型（或者说纤维状），这种划分来源于剑桥大学的德斯蒙德·贝尔纳和多萝西·克劳福特以及利兹大学的威廉·阿斯特伯里和洛马克斯的这些文章。一般来说，球形的蛋白质是酶，而纤维状的蛋白质是结构性的，用来构成生物组织。与阿斯特伯里（威廉·布拉格的学生）和克劳福特（由于在生物大分子晶体学方面的工作她后来获得了诺贝尔奖）一样，贝尔纳在这里对胃蛋白酶的球状分子是否包含连续多肽链表示了怀疑，阿斯特伯里和洛马克斯则认为链是"以某种巧妙的形式折叠的"。蛋白质折叠的这种原理是当前对球蛋白结构和功能之间关系研究的核心。

四个星期以前，米利肯博士带给我们一些由乌普萨拉的斯韦德贝里教授实验室的菲尔波特博士制备的胃蛋白酶晶体。它们具有完美的六方双锥形状，长度达到2毫米，轴率 $c/a=2.3\pm0.1$。当对处于母液中的晶体进行检测时，它们表现出适度的双折射和明显的单轴性，显示出良好的干涉图像。不过，当晶体被暴露于空气中时，其双折射迅速减弱。用一般方法得到的该晶体的 X 射线照片显示出一片模糊的黑影。这表明晶体彻底改变，从而解释了以前的研究者用蛋白质（仅就晶体形式而言）进行实验却得到阴性结果[1]的原因。不过，阿斯特伯里指出，改变后的胃蛋白酶是像肌球蛋白或角蛋白那样的链状蛋白，衍射结果表现出非结晶形的或纤维状的结构特征。

很明显，我们必须避免晶体的改变，而要做到这一点，则需要将其与母液一起提取出来，并在不与空气接触的条件下置于林德曼玻璃毛细管中。以此方法得到的第一张照片表明，我们所处理的是未改变的晶体。通过铜 $K\alpha$ 辐射得到的回摆照相结果表明，单位晶胞的大小为 $a=67\,\text{Å}$，$c=154\,\text{Å}$，准确至约 5%。这是最小值，因为 c 列上的点过于接近而无法精确测量，c 轴的长度是由轴率得到的。晶胞的大小可能仍然是它的倍数。对新制备的样品进行测量[2]得到其密度是 1.32（我们的测量结果是 1.28），利用此数值可以计算出晶胞的分子量为 478,000，是 40,000 的 12 倍，与斯韦德贝里用超速离心机通过沉降所得到的数值几乎完全一致。不过这可能是很偶然的，因为我们发现这些晶体中含有约 50% 可在室温下被除去的水。但这可能还会导致分子量更大，因为单位晶胞中可能只有更少的分子。

Not only do these measurements confirm such large molecular weights but they also give considerable information as to the nature of the protein molecules and will certainly give much more when the analysis is pushed further. From the intensity of the spots near the centre, we can infer that the protein molecules are relatively dense globular bodies, perhaps joined together by valency bridges, but in any event separated by relatively large spaces which contain water. From the intensity of the more distant spots, it can be inferred that the arrangement of atoms inside the protein molecule is also of a perfectly definite kind, although without the periodicities characterising the fibrous proteins. The observations are compatible with oblate spheroidal molecules of diameters about 25 A. and 35 A., arranged in hexagonal nets, which are related to each other by a hexagonal screw-axis. With this model we may imagine degeneration to take place by the linking up of amino acid residues in such molecules to form chains as in the ring-chain polymerisation of polyoxy methylenes. Peptide chains in the ordinary sense may exist only in the more highly condensed or fibrous proteins, while the molecules of the primary soluble proteins may have their constituent parts grouped more symmetrically around a prosthetic nucleus.

At this stage, such ideas are merely speculative, but now that a crystalline protein has been made to give X-ray photographs, it is clear that we have the means of checking them and, by examining the structure of all crystalline proteins, arriving at far more detailed conclusions about protein structure than previous physical or chemical methods have been able to give.

J. D. Bernal and D. Crowfoot

(**133**, 794-795; 1934)

*   *   *

It is now some time since we first took X-ray powder photographs of crystalline pepsin kindly sent by Prof. J. H. Northrop, but no really satisfactory interpretation of these photographs presented itself because they show features which we have learnt recently to associate with the fibrous proteins[3]: even single crystals, so far as we could judge with the minute crystals available, appeared to give results similar to those produced by many crystals in random orientation. The two chief rings have spacings of about 11.5 A. and 4.6 A. at ordinary humidity, corresponding to the "side-chain spacing" and the "backbone spacing", respectively, of an extended polypeptide[3].

It was difficult, of course, to reconcile such findings with external morphology and the Law of Rational Indices, but the photographs of Bernal and Miss Crowfoot, taken before the degeneration which we now see the crystals must have undergone on drying, clear up this long-standing problem at once. Furthermore, their photographs tend to confirm the suggestion[4] that the numbers 2, 3, 4, and 6 occurring in Svedberg's multiple particle weights are fundamentally of *crystallographic* significance, even though their conclusions to date appear to be against the chain mechanism proposed for the building-up of the various crystallographic groups[4].

这些测量结果不仅证实了蛋白质分子具有大分子量，而且给出了相当多的关于蛋白质分子性质的信息，在进一步的分析中无疑将会提供更多的信息。通过中心附近点的强度，我们可以推断蛋白质分子是比较密集的球体，也许通过共价桥结合在一起，但不管怎样，都是被相对较大的含水空间隔离开的。通过距离更远的点的强度，可以推断出蛋白质分子中内部原子的排布也是完全确定的（尽管没有纤维蛋白所特有的周期性）。观测结果与直径约为 25 Å 和 35 Å 的排列成六边形网格的扁圆球形分子是一致的，它们通过六方螺旋轴而彼此关联。利用这个模型我们可以想象，通过这种分子中氨基酸残基连接而形成链（就像聚甲醛的环链聚合作用那样），变性就可以发生。通常意义上的多肽链可能只存在于密度更高的蛋白或纤维蛋白中，而原生的可溶性蛋白质分子可能会将它们的重要组成部分更加对称地汇聚在非蛋白质基核周围。

现阶段，这些观点还仅仅是猜测，但既然已经制备出了一种结晶蛋白质并得到了它的 X 射线照片，那么很显然我们有办法来检测它们，并且通过考察所有结晶蛋白质的结构，就可以得到比以前用物理和化学方法所能给出的关于蛋白质结构的结果更为详细的结论。

贝尔纳，克劳福特

*　　*　　*

从我们第一次对诺斯罗普教授惠赠的结晶胃蛋白酶进行 X 射线粉末照相到现在已经有一段时间了，但是对于这些照片的真正令人满意的解释仍没有出现，因为我们最近才了解到照片呈现出的一些特征与纤维蛋白有关 [3]：即使是单晶——就我们根据现有的微小晶体所能做出的判断而言——看起来似乎也能给出类似于由多种随机取向的晶体得到的结果。在一般湿度条件下，两个主环具有约 11.5 Å 和 4.6 Å 的间距，分别对应于一条伸展的多肽链中的"侧链间距"和"骨架间距" [3]。

当然，要使这些发现与晶体的外观形态以及有理指数定律相一致是很困难的，但是，贝尔纳和克劳福特小姐在蛋白质变性——现在我们知道这是晶体在干燥过程中一定会发生的现象——之前照的照片立刻解决了这一长期存在的问题。此外，他们的照片还倾向于确认如下提议 [4]：出现在斯韦德贝里的多重粒子重量中的数字 2、3、4 和 6 具有重要的晶体学意义，即便目前为止他们的结论看起来似乎违背了针对各种结晶体群的构成而提出的链机制 [4]。

627

We are left now with the paradox that the pepsin molecule is both globular[5] and also a real, or potential, polypeptide chain system, and the immediate question is whether the chains are formed by metamorphosis and linking-up of the globular molecules, or whether the initial unit is the chain itself, which is afterwards folded in some neat manner which is merely an elaboration of the intra-molecular folding that has been observed in the keratin transformation[3]. What is either an exceedingly valuable clue or else only a fantastic coincidence is found in the fibre photograph of feather keratin[6], a study of which will be published shortly; for if, as Bernal thinks, the pepsin molecules are piled, perhaps in a screw, along the hexad axis, their length in this direction is 140/6, that is, about $23\frac{1}{2}$ A., which is almost exactly the strongest period along the fibre-axis of feather keratin, a period which is again repeated probably six (or a multiple of six) times before the fundamental period is completed! The innermost equatorial spot of the feather photograph also corresponds to a side-spacing of about 33 A. (though this is probably not the maximum side-spacing), which again is in simple relation to the side dimensions of the pepsin unit cell. As just said, these resemblances may be only accidental, but we cannot afford to overlook anything in such a difficult field, and it is not impossible that we have here an indication of how very long, *but periodic*, polypeptide chains can arise by the degeneration and linking-up of originally globular molecules.

W. T. Astbury and R. Lomax

(**133**, 795; 1934)

J. D. Bernal and D. Crowfoot: Department of Mineralogy and Petrology, Cambridge, May 17.

W. T. Astbury and R. Lomax: Textile Physics Laboratory, University of Leeds.

References:

1. G. L. Clark and K. E. Korrigan (*Phys. Rev.*, (ii), **40**, 639; 1932) describe long spacings found from crystalline insulin, but no details have been published.

2. J. H. Northrop, *J. Gen. Physiol.*, **13**, 739; 1930.

3. W. T. Astbury, *Trans. Faraday Soc.*, **29**, 193; 1933. W. T. Astbury and A. Street, *Phil. Trans. Roy. Soc.*, A, **230**, 75; 1931. W. T. Astbury and H. J. Woods, *Nature*, **126**, 913, Dec. 13, 1930. *Phil. Trans. Roy. Soc.*, A, **232**, 333; 1933. W. T. Astbury and W. R. Atkin, *Nature*, **132**, 348, Sept. 2, 1933.

4. W. T. Astbury and H. J. Woods, *Nature*, **127**, 663, May 2, 1931.

5. J. St. L. Philpot and Inga-Britta Eriksson-Quensel, *Nature*, **132**, 932, Dec. 16, 1933.

6. W. T. Astbury and T. C. Marwick, *Nature*, **130**, 309, Aug. 27, 1932.

现在留给我们的是一个悖论，即胃蛋白酶分子既是球形的 [5]，又是一个真正的或者潜在的多肽链体系，并且随之而来的问题是，链是否是由球状分子通过变性以及连接而形成的，或者是否最初单位就是链本身，随后链以某种巧妙的方式进行折叠，其方式不过是对我们在角蛋白的转变中所观测到的分子内折叠的精致化 [3]。羽毛角蛋白的纤维照片中的发现，可能是一条非常有价值的线索，也可能只是某种奇异的一致性 [6]，关于这一点的研究不久将会发表；按照贝尔纳的想法，如果胃蛋白酶分子沿着六次轴堆叠（也许是以螺旋形式），那么它们在这个方向上的长度是 140/6，也就是大约 $23\frac{1}{2}$ Å，这几乎恰好就是沿羽毛角蛋白纤维轴的最强周期，该周期可能会重复 6 次（或者 6 的倍数），直到基本周期完成为止！羽毛照片中最靠内的赤道点也对应于大约 33 Å 的旁侧间隔（尽管这可能不是最大的旁侧间隔），它与胃蛋白酶单位晶胞的侧向尺度也简单关联。如同刚才所说的，这些类同之处可能只是偶然的，但是在一个如此困难的领域中我们绝不能忽视任何事，而且对于如何通过最初的球状分子的变性以及连接形成非常长的、**但具周期性的**多肽链，我们也可能会从这里获得某种启发。

阿斯特伯里，洛马克斯

（王耀杨 翻译；周筠梅 审稿）

Crossing-over and Chromosome Disjunction

S. Gershenson

Editor's Note

Genetic recombination, which allows large segments of DNA to move from one chromosome to another, happens naturally during meiosis (cell divisions that halve the number of chromosomes) when chromosomes "cross over". But in 1935 the link between this cross-over and normal meiotic chromosome separation was far from clear. Sergey Gershenson from Moscow's Academy of Sciences here examines cross-over and separation in abnormal X-chromosomes from hybrid *Drosophila* females, and concludes that normal chromosome separation does not depend directly on cross-over. "A third, more general factor," he says, may be required. It wasn't clear what this could be.

BRIDGES, Anderson, Mather, Gershenson and many others have shown that there is a definite relation between crossing-over and the disjunction of chromosomes. Their work has made it evident that, during meiosis, crossover chromosomes are distributed between the daughter nuclei more regularly than non-crossover ones. However, the exact nature of this relation is still by no means clear. Darlington thinks that crossing-over is, in general, a necessary condition of regular chromosome disjunction. On the other hand, a number of general considerations as well as some recent facts obtained by Gershenson, Beadle and Sturtevant and Stone and Thomas, are opposed to such an interpretation in the case of *Drosophila*.

In order to decide the latter question, I undertook an experiment in collaboration with Miss Helene Pogossiants, in which, by using certain rearranged chromosomes, it seemed possible to obtain somewhat more conclusive results than in previous work. Crossing-over and non-disjunction were studied in females of *Drosophila melanogaster* carrying an X-chromosome with the *ClB* inversion and another X-chromosome (translocation X–IV or "Bar-Stone"), most of the genetically active part of which was translocated to the fourth chromosome. Data on single crossing-over in the non-inverted region and the number of recovered double crossovers in the inverted region enabled us to conclude, with a high degree of certainty, that single crossing-over in the latter region (undetectable by direct methods) is very low. Possible undetected crossing-over in the genetically inert parts of the X-chromosomes could influence only the disjunction of the *ClB* chromosome and the right (non-translocated) part of the other X-chromosome, as the left part, translocated to chromosome 4, does not carry any of the inert region of the X-chromosome. Non-disjunction of the left part of this X-chromosome and the *ClB* chromosome is much lower than would be expected if Darlington's hypothesis applies to *Drosophila*, and seems therefore to show that regular disjunction of chromosomes is not absolutely conditioned by crossing-over.

630

染色体分离与交换

格申森

编者按

遗传重组使大片段 DNA 从一条染色体移动到另一条染色体，自然情况下这一过程出现在染色体发生"交换"的减数分裂（染色体数目减半的细胞分裂）期间。但在 1935 年时人们对染色体的这种交换与正常减数分裂时染色体的分离之间的联系还很不清楚。莫斯科科学院的谢尔盖·格申森在这篇文章中分析了杂交的雌性黑腹果蝇中异常 X 染色体的交换和分离过程，并得出了正常染色体的分离并不直接依赖于染色体交换的结论。他认为可能需要"另外一个更普遍的因子"。但是这个因子是什么还不清楚。

布里奇斯、安德森、马瑟、格申森等人认为，染色体的分离与交换之间有明确的关系。他们的研究已经证实，在减数分裂期间，发生交换的染色体比不发生交换的染色体更经常性地分布于子细胞核之间。然而，出现这种关联的本质原因还不清楚。达林顿认为，一般而言，交换是染色体正常分离的必要条件。另一方面，大多数人的普遍观点以及最近格申森、比德尔、斯特蒂文特、斯通、托马斯关于果蝇相关方面的研究结果却与达林顿的观点相反。

为了解决上文中的最后一个疑问，我和海伦妮小姐共同合作利用某种重排的染色体进行实验，希望可以得到比先前的研究工作更有说服力的结果。我们利用雌性黑腹果蝇来研究染色体的交换和不分离现象，这种果蝇的一条 X 染色体上发生 ClB 倒位，而另一条 X 染色体上大部分的遗传活性部分被易位到第四条染色体上（X–IV 易位）。关于非倒位区的单交换数和倒位区恢复正常的双交换数的数据使我们可以确定无疑地得出一个结论，即倒位区发生单交换（不能通过直接方法检测到）的概率很低。X 染色体遗传惰性区域可能发生但无法被检测到的单交换只会影响 ClB 染色体和另一条 X 染色体的右臂（非易位区）的分离，因为被易位到第四条染色体上的左臂并不包含 X 染色体的任何遗传惰性区域。如果达林顿的假说也适用于果蝇的话，那么我们观察到的这条 X 染色体的左支与 ClB 染色体之间发生的不分离现象要少于它应有的比例，因此交换不是染色体正常分离的必要条件。

It seems clear that both crossing-over and chromosome disjunction are dependent on a third, more general factor, possibly on the intensity with which the conjugation of chromosomes takes place.

(**136**, 834-835; 1935)

S. Gershenson: Institute of Genetics, Academy of Sciences, Moscow.

很明显，染色体交换和分离都依赖于另一个更为普遍的因子，这个因子可能是染色体结合的强度。

（韩玲俐 翻译；王晓晨 审稿）

Crossing-over and Chromosome Disjunction

C. D. Darlington

Editor's Note

English geneticist Cyril Dean Darlington of London's John Innes Horticultural Institution replies to Gershenson's opinion (see the previous paper). The Russian geneticist had suggested that normal chromosome disjunction does not depend on cross-over and that the two processes depend on a third, as yet unidentified factor. The finding contradicted Darlington's view that crossing-over is part of normal chromosome disjunction, so he suggests that the third factor may be Gershenson's use of hybrids, which were "bound to have a complicated effect on disjunction." Darlington basically pins this paper's results down to experimental artefact. In later years he went on to outline the mechanics of chromosomal crossover and its role in inheritance.

DOBZHANSKY[1], Gershenson (above) and others have concluded that the disjunction of chromosomes does not depend directly on the occurrence of crossing-over between them, as I assume to be the case in all homozygous organisms[2]. They maintain that some "other factor" is concerned in their experiments. But these experiments have necessarily made use of hybridity, without which genetical tests are impossible; and the particular kind of hybridity (structural hybridity) they have used is bound to have a complicated effect on disjunction. It is itself the "other factor". Thus reciprocal crossing-over within two relatively inverted segments should give normal chromosomes with normal disjunction, other things being equal.

But when we consider the structure of the bivalent given by such crossing-over, we see that the disjunction expected in straightforward bivalents will not necessarily follow (Fig. 1). The pull which is to separate the paired chromatids is parallel to the plane of their association and not perpendicular to it. Resistance to separation is therefore not a certain constant minimum, but a function of the length between the two chiasmata. I have observed failure of disjunction in these circumstances in an inversion heterozygote in *Stenobothrus parallelus*. Other kinds of crossing-over in dislocated segments give rise to lagging, interlocking and irregular breakage. It is not therefore surprising that Grüneberg[3] finds the "non-disjunction" that has previously been attributed to non-pairing and random segregation is often due to loss of both partners.

染色体分离与交换

达林顿

编者按

伦敦约翰·英纳斯园艺中心的英国基因学家西里尔·迪安·达林顿对格申森的观点（见前一篇文章）给予了回应。俄罗斯基因学家格申森认为正常的染色体分离并不依赖于染色体交换，这两个过程都依赖于另一个尚不明确的因子。这一发现反驳了达林顿的染色体交换是正常的染色体分离的一部分的观点。达林顿认为，另一个因子可能源于格申森采用了"会对染色体分离产生复杂影响"的杂合体。达林顿从根本上认为那篇文章的结果是实验假象。之后的几年中，他继续总结了染色体交换的机制及其在遗传中的作用。

多布赞斯基[1]和格申森（上文的作者）等人认为染色体分离并不直接依赖于染色体交换的发生，而我认为在纯合生物中染色体分离是依赖于染色体交换的[2]。他们认为在他们的实验中需要考虑某种"其他的因素"。但他们的实验必须采用杂合体，否则就无法进行遗传检测。他们采用的特定的杂合形式（结构杂合）必定会对染色体分离产生复杂的影响。这本身就是所谓的"其他的因素"。因此在两个相对倒位的片段之间的相互交换将会产生能够正常分离的正常染色体，这种情况在其他的生物中也一样会发生。

但是当我们考虑到由这种交换形成的二价染色体的结构时，我们发现在简单的二价染色体中预期的染色体分离并不一定发生（图1）。配对染色单体分离时的拉力是与染色单体联会形成的平面相平行的，而不是相垂直的。因此，染色单体之间分离的阻力不是一个特定的恒定小量，而是两条交叉单体之间距离的函数。在这种情况下，我在草地蝗的杂合体倒位中观察到了染色体分离失败的例子。在染色体移位片段之间发生的其他种类的互换将使染色体片段发生滞后、互锁和不规则断裂等情况。因此，格吕内贝格[3]发现先前被认为是由染色单体之间不配对和随机分离导致的"非分离"片段通常是由这两个染色单体的丢失导致的，这就不足为奇了。

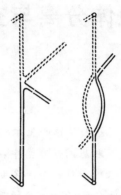

Fig. 1. The structure of bivalents with single (left) and double reciprocal crossing-over (right) between relatively inverted segments of homologous chromosomes.

This is merely one example of the special complications arising in structural hybrids. They have been described by Richardson for inversion hybrids and by myself for interchange hybrids in articles in the press[4]. They show the danger of arguing from the assumptions involved in an abstracted formal use of the terms "chromosome", "non-disjunction" and even "crossing-over" by the geneticist. They also show the difficulty the geneticist is faced with in dealing with the highly selected viable progeny of structural hybrids, a difficulty which can only be overcome by a close collaboration between those who are breeding the hybrids and those who are studying the structures found at meiosis in comparable material[2].

(**136**, 835; 1935)

C. D. Darlington: John Innes Horticultural Institution, London, S.W.19.

References:
1. *Z.I.A.V.*, **64**, 269–309.
2. *J. Genet.*, **31**, 185–212.
3. *J. Genet.*, **31**, 163–184.
4. *J. Genet.*, in the press.

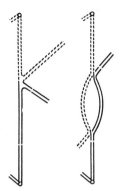

图 1.同源染色体相对倒位区域之间发生单互换（左）和双交叉互换（右）的二价染色体结构。

　　这仅仅是结构杂合体中存在特殊复杂性的一个例子。理查森曾论述过倒位杂合体中的这种特殊复杂性，我在即将发表的文章中 [4] 对互换杂合体的特殊复杂性也进行了描述。这些特殊复杂性表明，遗传学家们对"染色体""非分离"甚至"交换"这些术语抽象的正式用法所涉及的假设可能存在争议。同时，这些特殊复杂性也揭示了遗传学家们在处理结构杂合体的高选择性可存活后代时面对的困难。只有正在进行杂交育种的科学家们和正在研究那些相似物质的减数分裂时期出现的结构的科学家们 [2] 紧密合作，这些困难才有可能被克服。

（韩玲俐 翻译；王晓晨 审稿）

The Pattern of Proteins

D. M. Wrinch

Editor's Note

Although this paper presents an incorrect idea about protein molecular structure towards which biochemists now have a rather scathing attitude, it shows how unresolved that question was in the 1930s. The technique of X-ray crystallography was already being used by this time to look at proteins, notably by J. Desmond Bernal and his collaborators. Dorothy Wrinch was a mathematician with no formal training in chemistry, but her view, presented here, that protein backbones are built up of cyclic structures called "cyclols", gained support from some leading figures, especially the American chemist Irving Langmuir. This was partly due to Wrinch's energetic self-promotion, and *Nature* carried several subsequent elaborations of the idea, here presented only as a "working hypothesis".

ANY theory as to the structure of the molecule of simple native protein must take account of a number of facts, including the following:

(1) The molecules are largely, if not entirely, made up of amino acid residues. They contain $-NH-CO$ linkages, but in general few $-NH_2$ groups not belonging to side chains, and in some cases possibly none.

(2) There is a general uniformity among proteins of widely different chemical constitution which suggests a simple general plan in the arrangement of the amino acid residues, characteristic of proteins in general. Protein crystals possess high, general trigonal, symmetry.

(3) Many native proteins are "globular" in form.

(4) A number of proteins[1] of widely different chemical constitution, though isodisperse in solution for a certain range of values of pH, split up into molecules of submultiple molecular weights in a sufficiently alkaline medium.

The facts cited suggest that native protein may contain closed, as opposed to open, polypeptides, that the polypeptides, open or closed, are in a folded state, and that the type of folding must be such as to imply the possibility of regular and orderly arrangements of hundreds of residues.

An examination of the geometrical nature of polypeptide chains shows that, *since all amino acids known to occur in proteins are α-derivatives*, they may be folded in hexagonal arrays. Closed

蛋白质结构模型

林奇

编者按

虽然这篇文章在蛋白质分子结构方面提出的观点是错误的，但它体现了 20 世纪 30 年代时人们对这个问题的探索，而现在生物化学家对于蛋白质分子结构已经有了相当严谨和清晰的观点。当时，X 射线晶体衍射技术已经被用于观察蛋白质，尤为著名的是德斯蒙德·贝尔纳及其合作者所做的工作。作为一个在化学方面没有接受过正规训练的数学家，多萝西·林奇在本文中提出的蛋白质骨架由环状结构（环醇）形成的观点却受到一些权威人物的支持，特别是美国化学家欧文·朗缪尔。这在一定程度上是由于林奇积极的自我完善。林奇对于这个观点的多次加工与完善相继发表在《自然》上，这篇文章所呈现的还只是一个"正在完善中的假说"。

关于简单天然蛋白质分子结构的任何假说都必须考虑如下所述的许多事实：

（1）蛋白质分子主要——即使不是全部——由氨基酸残基构成。这些氨基酸残基通过肽键相连,但通常除了侧链之外只有很少的氨基基团,有时可能完全没有。

（2）化学组成大不相同的蛋白质具有普遍一致的结构，这表明了氨基酸残基的排列顺序遵循一个简单普遍的原则。蛋白质晶体具有高度对称性，通常是三角对称。

（3）许多天然蛋白质呈"球形"。

（4）许多化学组成大不相同的蛋白质,虽然在一定 pH 范围内的溶液中均匀分散，但在碱性足够强的溶液中会分解成分子量为蛋白分子量约数的分子 [1]。

上面列举的事实表明天然蛋白质可能包含闭合的而不是开环的多肽，并且无论是闭合的还是开环的，这些多肽都处于折叠状态，同时这种折叠类型必然意味着数以百计的氨基酸残基规则而有序的排列。

因为蛋白质中已知的氨基酸都是 α-衍生物，对多肽链的几何特性的研究表明多肽可能折叠成正六角形阵列。由 2、6、18、42、66、90、114、138、162……

polypeptide chains consisting of 2, 6, 18, 42, 66, 90, 114, 138, 162 ... $(18+24n)$ ... residues form a series with threefold central symmetry. A companion series consisting of 10, 26, 42, 58, 74, 90, 106, 122... $(10+16n)$... residues have twofold central symmetry. There is also a series with sixfold central symmetry: others with no central symmetry. Open polypeptides can also be hexagonally folded. The number of free $-NH_2$ groups, in so far as these indicate an open polypeptide, can be made as small as we please, even zero if we so desire. The hexagonal folding of polypeptide chains, open or closed, evidently allows the construction of molecules containing even hundreds of amino acid residues in orderly array, and provides a characteristic pattern, which in its simplicity and uniformity agrees with many facts of protein chemistry.

The stability of these folded polypeptide chains cannot be attributed to electrostatic attractions between the various CO, NH groups, for the appropriate distance between carbon and nitrogen atoms in these circumstances[2] lies between 2.8 A. and 4.2 A., whereas the distance in our case is at most 1.54 A. By using the transformation* suggested by Frank in 1933 at a lecture given by W. T. Astbury to the Oxford Junior Scientific Society,

$$\diagdown C = O \qquad H-N\diagup \quad \text{to} \quad \diagdown C(OH)-N\diagup$$

which has already proved useful in the structure of α-keratin[3], the situation is at once cleared up and we obtain (Fig. 1) the molecule "cyclol 6" (the closed polypeptide with six residues), "cyclol 18", "cyclol 42" (Fig. 2) and so on, and similarly open "cyclised" polypeptides (Fig. 3).

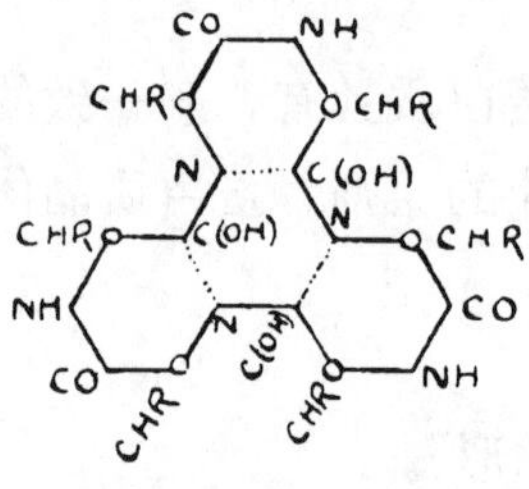

Fig. 1. The "cyclol 6" molecule.

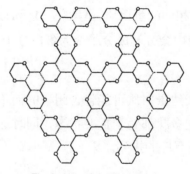

Fig. 2. A "cyclol 42" molecule.

---

* The application of this transformation to these molecules was suggested to me by J. D. Bernal.

（18+24*n*）……个氨基酸残基构成的一系列闭合多肽链具有三重中心对称性。由 10、
26、42、58、74、90、106、122……（10+16*n*）……个氨基酸残基构成的一系列多
肽链具有二重中心对称性。也有具有六重中心对称性的多肽链系列，其他多肽链没
有中心对称性。开环的多肽也可以折叠成正六角形阵列。就自由氨基的存在意味着
开环的多肽而言，我们可以按照自己的意愿使自由氨基的数目尽可能地减少，如果
确有必要甚至可以使其为零。开环和闭合的多肽链都可以折叠成正六角形阵列，这
就使得包含多达数百个氨基酸残基的分子的有序构建成为可能，同时也提供了一个
特征性的模式，这种模式的简单性和一致性符合许多蛋白质化学的研究结论。

　　折叠多肽链的稳定性不能归因于各种羧基和氨基基团之间的静电作用。处于静
电作用下的碳原子和氮原子之间的适宜距离[2]在 2.8~4.2 埃之间，而我们的研究中
这个距离最多只有 1.54 埃。在 1933 年参加阿斯特伯里为牛津初等科学学会所作的
一场报告上，弗兰克提出了转化的方法 *，

$$\diagdown C = O \qquad H - N \diagup \quad \text{缩合为} \quad \diagdown C(OH) - N \diagup$$

已经证明这种方法对于解析 α–角蛋白的结构是非常有用的 [3]。通过使用这种方法，形
势就立刻明朗起来了。我们得到了"六元环醇"分子（具有六个氨基酸残基的闭合
多肽）（图 1），"十八元环醇"分子和"四十二元环醇"分子（图 2）等等，同时也
得到了"开环"的多肽（图 3）。

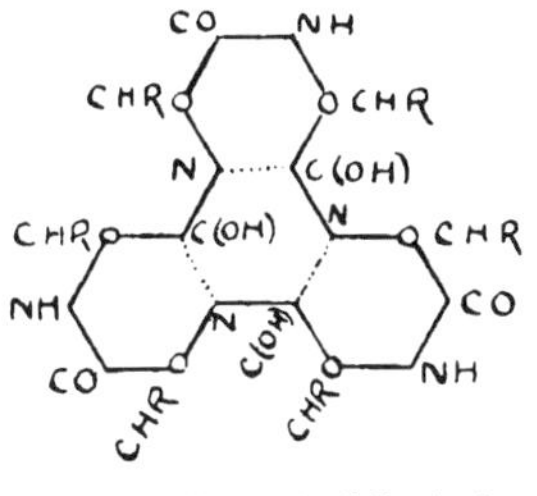

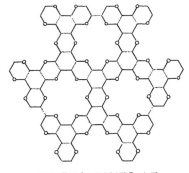

图 1. "六元环醇"分子

图 2."四十二元环醇"分子

---

* 贝尔纳建议我将转化的方法用在这些分子上。

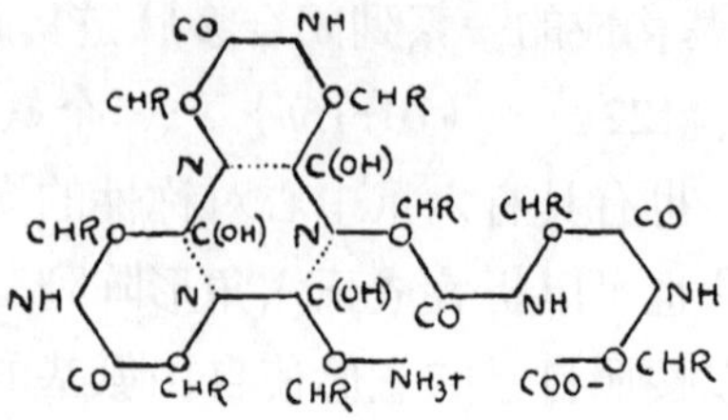

Fig. 3.

Hexagonal packing of polypeptides suggests a new *three dimensional* unit, $-\text{CHR}-\overset{|}{\text{C}}(\text{OH})-\text{N}{<}$, which may be used to build three-dimensional molecules of a variety of types. These are now being investigated in detail. At the moment we direct attention only to single cyclised polypeptides forming hexagons lying approximately in one plane. The cyclol layer molecule is a fabric the thickness of which is one amino acid residue. *Since all naturally occurring amino acids are of loevo type*[4] this fabric is dorsiventral, having a front surface from which the side chains emerge, and a back surface free from side chains. Both front and back carry trios of hydroxyls normal to the surface in alternating hexagonal arrays. Such a layer molecule and its polymers, formed also by the same transformation, can cover an area of any shape and extent. It offers suggestions as to the structure of the solid protein film when it is one amino acid residue thick. In its most compact form, the cyclol layer molecule gives an area per residue of about 9.9 A.² Less dense layers can be built, for example, from polymers of cyclol 18 and of cyclol 66 respectively, where the corresponding areas per residue are 13.2 A.² and 16.2 A.² respectively. The figures for unimolecular films of gliadin, glutenin, egg albumin, zein, serum albumin, serum globulin range from 1.724×10^{-7} gm./cm.² for serum globulin to 1.111×10^{-7} gm./cm.² for serum albumin[5,6,7]. With an average residue weight of 120, these densities give an area per residue ranging from 11.48 A.² to 18.82 A.² On the basis of the proposed hexagonal packing of polypeptides I therefore suggest that the upper limit of density of which a protein film is capable without buckling, provided that it is only one amino acid residue thick, is one residue per 9.9 A.²; further, that a higher density implies that the film, though it may still be unimolecular, is more than one amino acid residue thick.

Cyclol layers may also be used to build molecules and molecular aggregates with extension in three dimensions, since they may be linked front to front by means of the side chains, in particular, by cystine bridges, and back to back by means of hydroxyls[8]. The single-layer cyclol is a fabric capable of covering a two-dimensional area of any shape and extent; a three-dimensional array can then be built, layer upon layer, the linkage being alternately by means of side chains and hydroxyls. The idea that native proteins consist largely, if not entirely, of cyclised polypeptides therefore implies that some native proteins, including those of "globular" type, may have a layer structure.

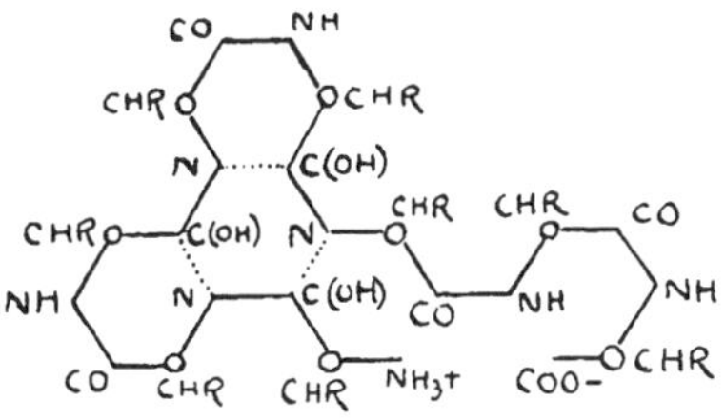

图 3

正六角形排列的多肽表明存在一个新的**三维结构**单元，$-CHR-\overset{|}{C}(OH)-N<$，可以用来构建多种类型的三维分子。现在正在对这些新的三维结构进行详细的研究。此刻我们关注的只是几乎在同一平面内形成六边形的单环多肽。这种层状环醇分子结构的厚度为一个氨基酸残基。**因为天然存在的氨基酸都是左旋型** [4]，因此这种结构呈背腹性，侧链出现在前表面而不是后表面。前后表面均有三个垂直于六角形表面的相间排列的羟基基团。这样的层状分子及其通过同样的转化方式形成的多聚体能够囊括蛋白质所有可能的形状。这为厚度只有一个氨基酸残基的固态蛋白质膜的结构提供了一些参考。在其最紧凑的结构中，层状环醇分子为每个氨基酸残基提供了约 9.9 平方埃的区域。在较为稀疏的结构中，例如，十八元环醇多聚体和六十六元环醇多聚体，每个氨基酸残基相应的区域大小分别为 13.2 平方埃和 16.2 平方埃。麦醇溶蛋白、麦谷蛋白、卵清蛋白、玉米醇溶蛋白、血清白蛋白和血清球蛋白形成的单分子膜的密度值的变化范围是从血清球蛋白的 1.724×10^{-7} 克 / 平方厘米到血清白蛋白的 1.111×10^{-7} 克 / 平方厘米 [5,6,7]。氨基酸残基的平均分子量为 120，因此从这些密度值推出每个氨基酸残基所占区域的变化范围是 11.48~18.82 平方埃。在前面已经提到的多肽的六角形排列的基础上，假如厚度只有一个氨基酸残基，那么我可以推算出在没发生折曲的蛋白膜中密度的上限为每 9.9 平方埃一个氨基酸残基；此外，较高的密度值意味着膜的厚度超过一个氨基酸残基，虽然它可能还是单分子的。

层状环醇结构也可以被用来构建在三维空间延伸的分子及分子聚集体。因为通过侧链（特别是胱氨酸桥）可以将环醇分子层状结构的前表面和前表面连接起来，也可以通过羟基把后表面与后表面相连接起来 [8]。单层的环醇结构可以囊括所有可能形状和程度的二维分子；通过环醇分子之间的侧链连接或者羟基连接，达到层与层的连接，我们就可以构建出三维阵列。天然蛋白质主要（即使不是全部）由环化多肽构成的观点意味着，包括那些"球形"蛋白质在内的许多天然蛋白质都可能具有层状结构。

Linkage by means of hydroxyls recalls the structures of graphitic oxide and montmorillonite, etc. Such a structure suggests a considerable capacity for hydration, an outstanding characteristic of many proteins. Further, since alternate layers are held together by means of hydroxyls, and contiguous molecules may also be held together in the same way, a protein molecular aggregate will, on this theory, necessarily be sensitive to changes in the acidity of the medium; in particular, a sufficiently high pH will cause such an aggregate to dissociate into single-layer units or into two-layer units joined by cystine bridges or side chains in covalent linkages. Svedberg's results, according to which a number of different native proteins break up into smaller molecules with sub-multiple molecular weights[1], here find a simple interpretation. The particular sub-multiples which occur may be regarded as affording evidence as to the type of symmetry possessed by the layers out of which the molecular aggregates are built.

The hypothesis that native proteins consist essentially of cyclised polypeptides thus takes account of the facts mentioned in (1), (2), (3), (4) above. Further, it derives support from the case of α-keratin, for with Astbury's "pseudo-diketopiperazine" structure[3] the polypeptides may be regarded as partially cyclised since they are cyclised at regular intervals, one out of every three (CO, NH) groups being involved. It is also suggestive in relation to a variety of other facts belonging to organic chemistry, X-ray analysis, enzyme chemistry and cytology. I cite the following:

(1) The rhythm of 18 in the distribution of amino acids in gelatin found by Bergmann[9], and the suggestion of Astbury[10] that in gelatin "the effective length of an amino acid residue is only about 2.8 A.".

(2) The low molecular weight not exceeding 1,000 found by Svedberg[11] for the bulk of the material from which lactalbumin is formed.

(3) Secretin[12], a protein with molecular weight of about 5,000, containing no open polypeptide chains.

(4) The nuclear membrane, which, consisting of proteins and lipoids, plays an important part in mitosis on account of its variable permeability.

(5) Bergmann's findings[13] with respect to dipeptidase; these suggest that the dipeptide substrate, upon which this enzyme acts, has a hexagonal configuration.

Finally, the deduction from the hypothesis of cyclised polypeptides, that native proteins may consist of dorsiventral layers, with the side-chains issuing from one side only, suggests that immunological reactions are concerned only with surfaces carrying side-chains. Hence, such reactions depend both on the particular nature and on the arrangement of the amino acids.

通过羟基连接形成分子的方式使我们想起了氧化石墨和蒙脱石等物质的结构。这种结构表现出了相当大的水合能力，而水合作用是许多蛋白质的典型特征。此外，因为相邻层是通过羟基相连的，所以相邻蛋白质分子之间也可以通过这种方式连接。根据这个理论，蛋白质分子聚集体必然对介质酸度的改变非常敏感；特别是足够高的 pH 值会导致聚集体分解成单层单元，或者由胱氨酸或侧链间的共价键连接在一起的双层单元。依据斯韦德贝里的研究，许多不同的天然蛋白质分解形成小分子，其分子量是原蛋白质分子量的约数 [1]，我们的理论对于这种现象提供了一种简单的解释。分子量为原蛋白质分子量约数的小分子的产生也可以为构成分子聚集体的单层单元所具有的对称性提供证据。

由此可见，天然蛋白质主要是由环化多肽构成的假说考虑到了前面（1）、（2）、（3）和（4）所述的事实。此外，基于阿斯特伯里的"伪二酮哌嗪"结构[3]，可以认为 α–角蛋白多肽是部分环化的，其环化发生的位置之间具有固定的间隔，每三个基团（羧基和氨基）中有一个参与这种环化，这对于我们的假说也提供了支持。对于有机化学、X 射线分析、酶化学和细胞学的一些研究来说，此假说也是有启发性的。现总结如下：

（1）伯格曼 [9] 发现凝胶中的氨基酸分布呈现十八节律，阿斯特伯里 [10] 提出在凝胶中"氨基酸的有效长度大约只有 2.8 埃"。

（2）斯韦德贝里 [11] 发现，构成乳清蛋白的大部分物质的分子量很低，不超过 1,000。

（3）分子量约为 5,000 的促胰液素 [12] 中没有开放的多肽链。

（4）由于具有可变的渗透性，由蛋白质和类脂构成的核膜在有丝分裂过程中发挥着重要作用。

（5）伯格曼发现了一些关于二肽酶的结果 [13]；这些结果表明这种酶作用的二肽底物具有六边形构型。

最后，环化多肽假说的一个推论是天然蛋白质可能由背腹层构成，侧链仅在其一侧出现。这一推论表明对于免疫反应来说只有含侧链的表面是重要的。因此免疫反应既依赖于氨基酸的特定性质也依赖于其排列方式。

Full details of the work, which is to be regarded as offering for consideration, a simple *working hypothesis*, for which no finality is claimed, will be published in due course.

(**137**, 411-412; 1936)

D. M. Wrinch: Mathematical Institute, Oxford.

References:

1. Svedberg, *Science*, 79, 327 (1934).

2. International Tables for the Determination of Crystal Structure.

3. Astbury and Woods, *Phil. Trans. Roy. Soc.*, **232**, 333 (1933).

4. Jordan Lloyd, *Biol. Rev.*, 7, 256 (1932).

5. Gorter, *J. Gen. Phys.*, **18**, 421 (1935); *Amer. J. Diseases of Children*, 47, 945 (1934).

6. Gorter and van Ormondt, *Biochem. J.*, **29**, 48 (1935).

7. Schulman and Rideal, *Biochem. J.*, **27**, 1581 (1933).

8. Bernal and Megaw, *Proc. Roy. Soc.*, A, **151**, 384 (1935).

9. Bergmann, *J. Biol. Chem.*, **110**, 471 (1935).

10. Astbury, Cold Spring Harbor Symposia on Quantitative Biology, **2**, 15 (1934).

11. Sjogren and Svedberg, *J. Amer. Chem. Soc.*, **52**, 3650 (1930).

12. Hammersten *et al.*, *Biochem. Z.*, **264**, 272 and 275 (1933).

13. Bergmann *et al.*, *J. Biol. Chem.*, **109**, 325 (1935).

我们的工作仅供参考，这只是一个**正在完善中的假说**，未成定论，此项工作的全部细节将会在适当的时间发表。

(赵凤轩 翻译；周筠梅 审稿)

A New Fossil Anthropoid Skull from South Africa

R. Broom

Editor's Note

For twelve years after Raymond Dart's discovery in 1924 of the infant *Australopithecus* at Taungs, the fossil record of human ancestry in Africa remained silent. Controversy reigned about whether the Taungs "baby" was a genuine intermediate in human evolution, or simply a juvenile ape, as many maintained. The impasse was broken by this announcement of the discovery of a fragmentary skull and brain-cast of an adult pre-human, by palaeontologist Robert Broom. It came from lime-workings near caves at Sterkfontein, soon to yield much more. Broom named the new form *Australopithecus transvaalensis*. The finding was the first step towards the acceptance of an ape-like intermediate in human evolution.

IT is nearly twelve years ago since Prof. R. A. Dart startled the world by the announcement of the discovery of a new type of fossil anthropoid found in a limestone cave at Taungs in Bechuanaland, South Africa. The specimen consists of most of the brain cast and the practically perfect face of a very young ape. The functional teeth are all of the milk set, though the first upper and lower molars have cut the gum but do not yet meet. Though the ape was only very young, Dart estimated the cranial capacity at more than 500 c.c., and considered that in an adult it might exceed 700 c.c. He believed that this little fossil ape is not very closely allied to either the chimpanzee or the gorilla, and that it is probably nearer to the ape from which man has been descended and thus to be practically the long sought for missing link.

Many European and American men of science considered that Dart had made a mistake, and that if he had had a series of young chimpanzee skulls for comparison he would have recognized that the Taungs ape is only a variety of chimpanzee. When after some years the lower jaw was detached from the upper and the crowns of the teeth could be examined fully, it was found that the milk teeth are not in the least like those of either the chimpanzee or gorilla, and that they agree entirely with those of man, though larger. In the gorilla and chimpanzee the first upper milk molars have each two cusps: in man and in *Australopithecus* there are three well-marked cusps in each. In the first lower milk molar of the gorilla there is only one large cusp; in the chimpanzee there is one large cusp and a second rudimentary cusp. In man and in *Australopithecus* there are four well-developed cusps.

I have constantly maintained since I first examined the skull in 1925 that Dart was essentially right in holding *Australopithecus* is not closely allied to either the gorilla or chimpanzee, and

648

发现于南非的一件新的类人猿头骨化石

布鲁姆

编者按

1924 年雷蒙德·达特于汤恩发现了幼年南方古猿，此后的 12 年间，再没有出现与非洲人类始祖的化石有关的报告。对于汤恩"幼儿"一直存在争议，即它究竟是人类进化过程中的过渡类型，还是仅仅只是一只幼年的猿类。古人类学家罗伯特·布鲁姆的发现打破了这个僵局，他发现了一件成年的史前人类的头骨残片以及脑模。第一件头骨残片是在斯泰克方丹岩洞附近的石灰岩矿场里发现的，随后又找到了更多的化石。布鲁姆将其命名为"南方古猿德兰士瓦属"。这一发现使人们开始接受人类进化过程中存在与猿相像的过渡种。

将近 12 年前，达特教授宣布，在南非贝专纳兰汤恩的一个石灰岩洞中发现了类人猿化石，当时这一消息震惊了世界。该标本来自一只非常年幼的猿，它保存有大部分脑模和几乎完美的面部部分。功能性牙齿都是乳齿，尽管第一上白齿和第一下白齿都已经破龈而出，但还不能咬合在一起。尽管这只猿非常年幼，但据达特估计，其颅容量超过了 500 毫升，并认为成年猿的颅容量可能超过 700 毫升。他相信这具小猿化石既不属于黑猩猩，也不属于大猩猩，可能更接近于一种进化成人类的猿类，实际上就是长久以来一直在寻找的、在整个进化过程中缺失的一环。

许多欧美学者认为达特犯了一个错误，如果他对一系列幼年黑猩猩的头骨进行比较的话，他就会意识到汤恩古猿不过是黑猩猩的一种而已。若干年后，当下颌骨与上颌骨被分开后，人们才得以充分地检查其牙冠，并发现其乳齿与黑猩猩和大猩猩的一点也不像，而与人类的相比，除大一点以外，其他方面完全相符。大猩猩和黑猩猩的每颗第一上乳白齿都有两个齿尖；人类和南方古猿的每颗上乳白齿则有三个明显的齿尖。大猩猩的第一下乳白齿只有一个大的齿尖；黑猩猩有一个大的齿尖和一个次尖。人类和南方古猿则有四个发育良好的齿尖。

达特认为：南方古猿与大猩猩和黑猩猩都没有亲缘关系，而是处于或接近于进化成人类的世系位置上。自 1925 年我初次研究汤恩古猿至今，我一直认为这一观点

is on or near the line by which man has arisen.

I do not know what is at present the opinion in Europe as to where *Australopithecus* ought to be placed. Gregory of New York regards it as fairly near to the origin of the human line; and Romer of Harvard says it is "clearly not a chimpanzee or a gorilla". But the most important thing to do seemed to be to get an adult skull. For the last three months, I have been busy working on the bone breccia of the limestone caves of the Transvaal largely in the hope of getting either a new "missing link" or a type of primitive man. I have so far found no trace of man, though I have discovered more than a dozen new species of fossil mammals, a number of which belong to new genera.

Two weeks ago [Dr. Broom's covering letter is dated August 8. —Ed.], when visiting the caves at Sterkfontein near Krugersdorp, Mr. G. W. Barlow, the very understanding manager of the lime works there and on whom I had impressed the importance of keeping his eyes open for a Taungs ape, handed me the brain cast of what appeared to be a large anthropoid (Fig. 1). It had been blasted out of the side of the cave a couple of days before. A search for some hours failed to find any other part of the skull, but we found the cast of the top of the head in the cave wall. A more extensive search on the following day with a large party of workers resulted in the discovery of most of the base of the skull, with the upper part of the face (Fig. 2). In the same matrix was found the detached right maxilla with three teeth, and the third upper molar was also found, though detached. The lower part of the face had been removed before fossilization; and so far no mandible or lower teeth have been found, though parts may yet be discovered in a mass of crushed and broken bones near the side of the head. As the bones are very friable, no attempt has as yet been made to remove them from the much harder matrix.

Fig. 1. Half side view of the brain cast resting on the imperfect base. The brow ridges are shown with parts of the frontal sinuses exposed. Part of the left cheek bone is also shown. About $\frac{1}{3}$ natural size. Photograph by Mr. Herbert Lang.

本质上是正确的。

我不知道当前欧洲对于南方古猿的进化归属持什么观点。纽约的格雷戈里认为南方古猿应该与人类世系起源的位置特别接近；哈佛的罗默说南方古猿"很明显既不是黑猩猩，也不是大猩猩"。然而最重要的事情可能就是找到一个成年头骨。过去的三个月，我一直忙于研究德兰士瓦的石灰岩洞中的骨角砾岩，主要是希望能得到一个新的"缺失的环节"或者得到一种原始人类类型。尽管我已经发现了十多个哺乳动物新物种化石，其中有许多属于新的属，但一直没有发现人类的踪迹。

巴洛先生是一位非常了解克鲁格斯多普当地情况的石灰工厂经理，我曾让他一定要留心寻找汤恩古猿。两个星期前（布鲁姆博士的附信上注明的日期是 8 月 8 日。——编者注），当我考察此地附近的斯泰克方丹山洞时，他把一件好像是大型类人猿的脑模（图 1）交给了我。几天前这个山洞已经从外面给炸掉了。我们寻找了几小时也没有发现该头骨的其他部分，但是我们在洞壁上发现了该头骨顶部的脑模。第二天我们发动更多工人进行更大范围的搜索，结果发现了与脸上半部分相连的大部分颅底（图 2）。在同一埋藏成分中，我们发现了带有三颗牙齿的右上颌骨和一个脱落下来的第三上臼齿。脸下半部分在石化之前就已经缺失；因此目前为止还没有发现下颌骨或下边的牙齿，但是在发现头部侧面的地方附近有大量破碎断裂的骨头，也许从中还会有一些发现。因为这些骨头都非常脆弱易碎，所以我们现在还没有尝试将它们从那些坚硬的母质中取出来。

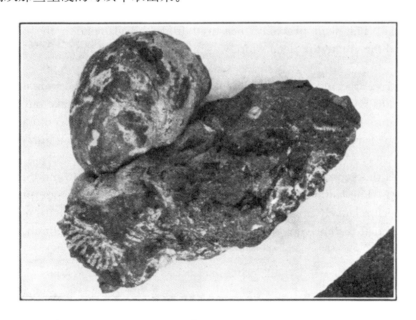

图 1. 置于不完整颅底上的脑模的半侧面图。图中展示了带有部分暴露的额窦的眉嵴。图中也可以看到部分左颧骨。该图大概是实际尺寸的 $\frac{1}{3}$。赫伯特·朗先生拍摄。

Fig. 2. Side view of right upper maxilla with the 2nd premolar and the 1st and 2nd molars. Parts of the roots of the canine and 1st premolar are shown. Slightly enlarged. Photograph by Mr. Herbert Lang.

Much of the cranial vault has been destroyed by the blast, but a large part of each parietal is preserved and a considerable part of the occiput. Unfortunately, the back of the brain cast is missing, and though the base of the skull is complete to the back of the foramen magnum, the contacts of the occipital fragment are lost.

The brain cast is perfect in its anterior two thirds. When complete it probably measured in length about 120 mm. and in breadth about 90 mm.; and the brain capacity was probably about 600 c.c. The skull probably measured from the glabella to the occiput about 145 mm., and the greatest parietal width was probably about 96 mm.

The brow ridges are moderately developed and there are fairly large frontal sinuses. The auditory meatus is 73 mm. behind the brow. It will be possible to make out much of the detailed structure of the base of the skull, but as yet no attempt has been made to clean it out as the bone is very friable and the investigation cannot be done in a hurry.

In the maxilla there are three well preserved teeth, the 2nd premolar and the 1st and 2nd molars (Fig. 3). The canine and 1st premolar are lost but the sockets are preserved. The canine has been relatively small. At its base it probably measured about 10 mm. by 8 mm. The 2nd premolar is somewhat worn. Its crown measured 11 mm. by 9 mm. Its pattern is well seen in Fig. 3.

图 2. 带有第二前白齿、第一和第二白齿的右上颌骨侧面图。可以从图中看到部分犬齿和第一前白齿的牙根。该图比实际尺寸略大。赫伯特·朗先生拍摄。

大部分颅顶都在爆炸中毁坏了，但是每块顶骨都有一大部分保留下来，此外还有一块相当大的枕骨残片。不幸的是，脑模的后面部分丢失了；尽管颅底直到枕骨大孔背部都是完整的，但是与枕骨片段相连接的部位丢失了。

脑模前部的 2/3 保存得很完好。如果头骨完整的话，长度可能达到约 120 毫米，宽度约 90 毫米；脑量可能约有 600 毫升。头骨从眉间到枕骨大约 145 毫米，顶骨的最大宽度可能有 96 毫米左右。

眉峰处于中度发育水平，并有非常大的额窦。耳道位于眉骨后 73 毫米处。将来还有可能弄清楚这件头骨颅底的大部分详细结构，但是因为骨头非常脆弱，所以目前还没有将骨头清理出来，因此暂时不能对其进行研究。

上颌骨中有三颗保存完好的牙齿，分别是第二前白齿、第一和第二白齿（图 3）。犬齿和第一前白齿缺失了，但是其齿槽保存了下来。犬齿相对较小，其基部可能约有 10 毫米 × 8 毫米大小。第二前白齿有些磨损。其牙冠大小为 11 毫米 × 9 毫米。从图 3 中可以清楚地看到它的形态。

Fig. 3. Crowns of right upper 2nd premolar and 1st and 2nd molars. × about $\frac{3}{4}$. Photograph by Mr. Herbert Lang.

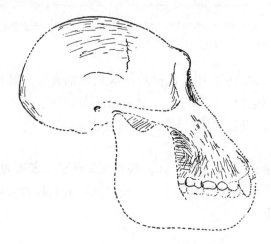

Fig. 4. Attempted restoration of skull of *Australopithecus transvaalensis* Broom. $\frac{1}{3}$ natural size. Sufficient of the cranium is preserved to show its shape with certainty. Most of the right maxilla is preserved, but it is not in contact with the upper part of the skull, and there is thus a little doubt as to the relations.

The 1st molar is moderately large. Anteroposteriorly it measures 12 mm. and transversely 13 mm. It is of the typical Dryopithecid pattern—four well-developed cusps with a little posterior ridge and a well-marked posterior fovea. The tooth agrees fairly closely with that of the first molar of *Dryopithecus rhenanus*. The 2nd molar is exceptionally large. It measures 14.5 mm. in antero-posterior length and is 16 mm. across. It has four large cusps with a well-marked posterior fovea. The 3rd molar has been detached from the bone but it is preserved in perfect condition and unworn. It has three well-developed cusps, but the hypocone is relatively small owing to the invasion of the large fovea. The tooth measures antero-posteriorly 13.7mm. and transversely 15.5 mm. The crown in this unworn condition is extremely wrinkled.

654

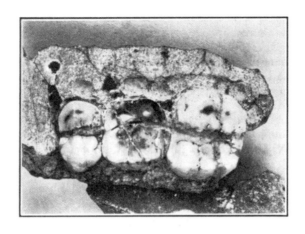

图 3. 右侧上部第二前臼齿和第一、第二臼齿的牙冠图。该图约为实际尺寸的 $\frac{3}{4}$。赫伯特·朗先生拍摄。

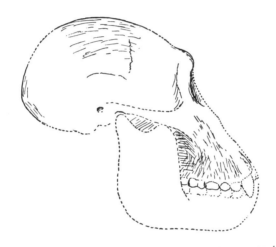

图 4. 布鲁姆对德兰士瓦南方古猿头骨的尝试性复原图。该图大小为实际尺寸的 $\frac{1}{3}$。完整保存的颅骨确切地展示出它的形状。右上颌骨大部分保存下来了，但是与头骨的上半部分没有衔接起来，因此对二者关系还有一点疑问。

    第一臼齿稍大，前后长 12 毫米，横向长 13 毫米。这属于典型的森林古猿型——有四个发育良好的齿尖并伴有一个小后嵴和一个明显的后凹。这颗牙齿与雷纳努斯森林古猿的第一臼齿非常相似。第二臼齿特别大，其前后长 14.5 毫米，横向长 16 毫米，有四个大的齿尖及一个明显的后凹。第三臼齿已经从上颌骨上脱落下来了，但是保存状况非常好，没有任何磨损。该牙齿有三个发育良好的齿尖，但由于有一个大的凹陷侵入，次尖相对较小。这颗牙前后长 13.7 毫米，横向为 15.5 毫米。未磨损的牙冠上布满了大量的褶皱。

The whole premolar and molar series measures 59 mm.

This newly-found primate probably agrees fairly closely with the Taungs ape, but the only parts that we can compare are the brain casts and the 1st upper molars. The brain cast of the new form is considerably wider, especially in the frontal region, and the molar teeth differ in a number of important details. Further, the associated animals found at Taungs are all different from those found at Sterkfontein. I think the Taungs deposit will probably prove to be Lower or Middle Pleistocene, while the Sterkfontein deposit is most probably Upper Pleistocene. I therefore think it advisable to place the new form in a distinct species, though provisionally it may be put in the same genus as the Taungs ape.

This discovery shows that we had in South Africa during Pleistocene times large non-forest living anthropoids—not very closely allied to either the chimpanzee or the gorilla but showing distinct relationships to the Miocene and especially to the Pliocene species of *Dryopithecus*. They also show a number of typical human characters not met with in any of the living anthropoids.

(**138**, 486-488; 1936)

R. Broom: F.R.S, Transvaal Museum, Pretoria.

整个前臼齿和臼齿齿列长 59 毫米。

这个新发现的灵长类动物可能与汤恩古猿的亲缘关系非常近，但是我们能进行比较的部位还只有脑模和第一上臼齿。这种新型的脑模相当宽，尤其在额部，臼齿也在许多重要细节上有所不同。此外，在汤恩发现的伴生动物与在斯泰克方丹发现的伴生动物完全不同。我认为汤恩沉积物可能是早更新世或者中更新世的，而斯泰克方丹沉积更可能是晚更新世的。因此我认为将这种新类型列为一种新的物种是比较合理的，尽管它可能暂时被归在与汤恩古猿相同的属中。

这一发现表明，更新世时期在南非已经存在不生活于森林中的大型类人猿——它们不是黑猩猩或大猩猩的亲缘种，但是与中新世特别是上新世的森林古猿具有明显的关联。它们还具有许多典型的人类特征，这是任何现存类人猿都不具备的。

(刘皓芳 田晓阳 翻译；冯兴无 审稿)

Discovery of an Additional *Pithecanthropus* Skull

G. H. R. von Koenigswald and F. Weidenreich

Editor's Note

Remains of a pre-human known as *Pithecanthropus* had been known from Java since Eugene Dubois's first discovery in 1891. In the early 1930s, fossils of a similar creature, *Sinanthropus*, were discovered in China. The relationship between the two forms was unclear until this brief announcement from Gustav von Koenigswald and Franz Weidenreich of discoveries from Java that looked much more like *Sinanthropus* than pithecanthropines typical of Java. This was an important step towards the recognition that all these east Asian forms belonged to *Homo erectus*, the earliest human definitely known to have ventured out of Africa.

DURING the systematic search for fossil man in Java, one of us (G. H. R. von K.) discovered, in 1937, in the Trinil formation of Sangiran (Central Java), an almost complete brain case of *Pithecanthropus*[1].

Amongst the material recently collected (July 1938) from the same area, a large fragment of an additional *Pithecanthropus* skull came to light. The fragment consists of the complete right parietal bone with the adjoining part of the left parietal bone and a small piece of the occipital bone. The three bones embrace in their original and entirely undisturbed arrangement a stone core composed of sandy tuft mixed with lapilli. The sagittal suture reaching from bregma to lambda is completely preserved. The right parietal bone also exhibits all the other sutures, only the sphenoidal angle being broken off. The coronal contour of the parietal bones is characterized by a very pronounced sagittal crest. Laterally there is a distinct depression reaching to the temporal line, from which the contour runs steadily outwards down to the squamous suture. The temporal line runs strikingly close to the sagittal suture.

These conditions entirely correspond to those which are characteristic of the Sinanthropus skulls. The pronounced flattening of the cap, so specific for the two *Pithecanthropus* skulls known hitherto, is completely missing in the case of this new *Pithecanthropus* skull. On the other hand, this skull has the following peculiarities in common with both *Sinanthropus* and *Pithecanthropus* skulls: the lowness of the entire cap and the position of the greatest breadth, the latter having undoubtedly been situated above the origin of the zygomatic arch, as is the case with all *Sinanthropus* and *Pithecanthropus* skulls.

According to the state of the sutures, the new skull belongs to a juvenile individual, in spite of the fact that the parietal bones show a thickness of more than 10 mm. near the bregma,

又一件爪哇猿人头骨的发现

孔尼华，魏登瑞

编者按

1891 年尤金·杜布瓦首次在爪哇发现了直立猿人化石，此后史前人类"爪哇猿人"化石便广为人知。20 世纪 30 年代初，在中国发现了一种与之类似的生物——"中国猿人"的化石。直到古斯塔夫·冯·柯尼希斯瓦尔德（孔尼华）和弗朗茨·魏登赖希（魏登瑞）发表了这篇简短的报告之后，这两种生物之间的关系才变得明晰，这篇报告中称，在爪哇发现了比典型的爪哇直立人看起来更接近中国猿人的化石。这对于认定所有的东亚猿人都属于"直立人"是非常重要的一步，而"直立人"是目前可以确定的从非洲迁徙出来的最早的人类。

我们对爪哇地区的化石人类进行了系统的搜寻，作者之一（孔尼华）于 1937 年在爪哇中部的桑吉兰特里尼尔地层组发现了一件几乎完整的爪哇猿人头盖骨 [1]。

最近（1938 年 7 月）在同一地区采集的材料中，发现了另一件爪哇猿人头骨的一大块碎块。这一碎块由完整的右顶骨、相邻的部分左侧顶骨以及一小块枕骨构成。这三块骨头胶结在混杂有火山岩的沙质石核中，没有错位。这件头骨完整保留有从前囟点到人字点的矢状缝部分。除蝶角破损之外，右顶骨上其他所有的骨缝都可以看到。顶骨的冠状轮廓有一个非常明显的矢状脊,侧面有一个延伸至颞线的明显凹陷。从颞线处，头骨轮廓向外平稳扩展向下至鳞缝区域。颞线位置非常接近矢状缝。

这些特征与中国猿人头骨的特征完全相符。 目前已知的两个爪哇猿人的头骨都明显地呈现出扁平的特征，但这一特征在本次新发现的爪哇猿人头骨上却没有出现。另一方面，这件头骨与爪哇猿人头骨以及中国猿人头骨在如下特征上是一样的：即整个头盖骨的低平程度以及最大宽度所在的位置。后者毫无疑问位于颧弓的起点之上，这与所有中国猿人和爪哇猿人头骨的情况都一样。

尽管新发现的这件头骨的顶骨在前囟处的厚度达十多毫米，并且其颞线很发达，但是根据骨缝的状态，它应该属于一个少年个体。枕骨圆枕区域仅有一小部分被保

and that the temporal line is well developed. The region of the occipital torus is preserved only to a very small extent, revealing only a faint swelling, apparently in correspondence with the age of the individual.

All the new *Pithecanthropus* finds demonstrate how important and promising it is to search for fossil man in Java, and to continue the work which has been made possible thanks to the generous support of the Carnegie Institution of Washington, D.C.

(**142**, 715; 1938)

G. H. R. von Koenigswald and Franz Weidenreich: Bandoeng, Java.

Reference:
1. *Proc. Kon. Akad. van Wetenschappen, Amsterdam*, 1938.

存下来，并显示出很微弱的隆起，这显然与该个体的年龄是相符的。

所有新爪哇猿人的发现证实了在爪哇寻找化石人类是非常重要和有前景的，能够继续该工作要感谢华盛顿卡内基研究院的慷慨支持。

（刘皓芳 翻译；刘武 审稿）

Structure of Proteins

Editor's Note

This anonymous report describes a lecture on protein structure delivered by American chemist Irving Langmuir, in which Langmuir voiced support for the "cyclol" theory of proteins advocated by Dorothy Wrinch. This theory posited a cage-like structure for proteins, composed of polyhedra made from rings of amino-acid units. Langmuir's recognition of the role of "hydrophobic interactions" in protein structure is now seen as pivotal, which makes his support of Wrinch's fanciful theory seem all the more perverse. Notable here is his admission that the cyclol theory did not follow from the crucial X-ray crystallographic data on proteins collected by Dorothy Crowfoot and J. Desmond Bernal; and indeed, Bernal is recorded as a sceptical voice at the end of Langmuir's talk.

DESPITE the cold weather and the proximity of Christmas some hundred and thirty people attended a meeting of the Physical Society on December 20 at which Dr. Irving Langmuir delivered an address on the structure of proteins. He began by enumerating the facts which pure chemistry had succeeded in accumulating about these substances, which play such an important part in living Nature. By the union of two amino acids, with the elimination of a molecule of water, the polypeptide unit is built up. Chemically, this has two valence bonds free, one at each end, and thus seems adapted solely for the formation of chain compounds. If proteins are indeed chain compounds, it would be difficult to understand why they have apparently definite molecular weights (often multiples of 18,000), or why the molecule behaves approximately as a spherical one.

A few years ago, Dr. Dorothy Wrinch suggested that, instead of the polypeptide unit (I),

where R is one of a number of radicals, the unit might really be (II), which differs from it only in the position of a proton within the molecule. It has, however, four free valency bonds, which (being attached to nitrogen or carbon atoms) make angles of $109°$ with each other. Investigation of the spatial arrangements possible with such a unit shows that they can form closed "cages", the plane sides of which are constructed of a lacework of hexagons. The discrete molecular weights (and indeed the ratios of these) thus receive a natural explanation, as does the approximately spherical shape of the molecule.

蛋白质结构

编者按

这份没有署名的报告描述了美国化学家欧文·朗缪尔所做的一次关于蛋白质结构的演讲。在这次演讲中朗缪尔声称支持由多萝西·林奇提出的蛋白质的"环醇"理论。这个理论假定蛋白质为笼状结构，是由氨基酸单元形成的环构成的多面体。朗缪尔对蛋白质结构中"疏水作用"的认可现在看来很关键，这使得他对林奇的奇怪理论的支持显得更不合常理。值得注意的是，他承认环醇理论与多萝西·克劳福特和德斯蒙德·贝尔纳收集到的有关蛋白质结构的重要的X射线晶体衍射数据并不相符；另外，在朗缪尔演讲的最后部分，他也谈到了贝尔纳对此理论的质疑。

尽管天气寒冷且临近圣诞节，但是仍有130多位科学家出席了12月20日举行的物理学会会议，会上欧文·朗缪尔博士做了一个关于蛋白质结构的演讲。他首先陈述了在纯化学领域积累得到的关于这些物质的研究成果，这些物质在生物界起着非常重要的作用。两个氨基酸连接在一起，同时脱去一分子水，就构建出了多肽单元。从化学角度上讲，该物质具有两个自由的价键，分子两端各有一个，从而使得分子似乎完全适合于形成链状复合物。如果蛋白质真的是链状复合物，就很难理解为什么它们具有如此明确的分子量（通常是18,000的倍数），以及为什么蛋白分子表现得更像是一个球状物。

几年前，多萝西·林奇博士就曾指出，除了多肽单元（I）之外，

这里的 R 代表多种氨基酸侧链基团中的一个，结构单元还可能是如（II）所示的那样，与前者相比，两者之间的差别只在于分子内质子的位置不同。然而，它却具有四个自由的化学键，彼此（连在氮原子或碳原子上）之间的夹角为109°。对这种结构单元可能形成的空间排布的研究显示，它们可以形成封闭的"笼状"结构，每一面由六边形构成。这样其分子量（事实上是它们的比率）离散分布以及其分子近似球形就都有了很自然的解释。

One cage in particular contains 288 amino-acid residues, which is equal to the number determined by physico-chemical methods in the case of egg albumin, within the limits of accuracy of these methods. In this particular case, the cage is a truncated tetrahedron which does not differ greatly from an octahedron.

Now proteins can be "denatured" by various mild treatments (shaking for three or four minutes, or heating to 65°C. for one or two minutes are sufficient in some cases) and then show markedly different properties from native proteins, but no change in molecular weight. It is suggested that these denatured proteins are in fact composed of the polypeptides which the chemist had discovered, since the treatment which he necessarily applies in order to make investigations would be sufficient to cause the change. The ease of denaturation, and the great alteration in properties, both follow naturally from this hypothesis, taken in conjunction with the "cyclol" hypothesis adopted for native proteins.

A particularly interesting fact which receives explanation by these theories is the insolubility in water of mono-molecular layers of proteins formed from soluble native proteins. The explanation lies in the positions of the side chains R, which in the cage structure are arranged around the edges of the hexagonal lacunae composing the plane sides of the cage, whereas in a mono-molecular layer, the cage is developed into a plane. When R is hydrophobe, the mono-molecular layer is formed with these radicals projecting; in the "cage" they project into the interior and are thus marked.

The last part of the lecture was devoted to the X-ray evidence, which is most complete in the case of insulin, since as shown by Crowfoot, this particular protein has but one molecule per unit cell, and the results are therefore more easily interpreted than would be the case with most proteins.

Dr. Langmuir stressed the point that the X-ray data, like many of the other facts explained by the theory, such as the appearance of α- but no β-amino acids, were not used as data in constructing it originally; nevertheless, the results of the X-ray investigation, in Dr. Langmuir's view, confirm many features of the structure to which the theory had led.

In the discussion which followed the address, Dr. Wrinch explained that the method used by Langmuir and Wrinch in interpreting the vector maps was a geometrical formulation of the classical picture of a crystal as a set of discrete units. Prof. J. D. Bernal uttered a word of caution against supposing that the theory was proved. He thought that certain features must be true, and others were possibly true, but in his view the X-ray data fail to confirm the proposed structure. He had had a vector map calculated for the latter, and found that it differed very markedly from that obtained experimentally. Prof. E. H. Neville showed the care that is needed in interpreting a vector map, by a simple example in which a map constructed from a given set of points appeared at first sight to correspond to quite a different type of distribution.

(**143**, 34-35; 1939)

特别是有一种笼状结构包含 288 个氨基酸残基，这与通过物理化学方法测定的卵清蛋白中氨基酸残基的数目是一致的，这些结果是在这些方法的精度范围内测定的。在此情况下，笼状结构是一个截顶四面体，和八面体没有太大的差别。

现在，可以通过各种温和的处理方法（振荡 3~4 分钟，在某些情况下 65℃ 加热 1~2 分钟足矣）使蛋白质变性，进而使其表现出与天然蛋白质截然不同的性质，但其分子量不发生改变。这表明这些变性的蛋白质实际上是由化学家已经发现的那些多肽链组成的，因为研究中必须用到的处理足以引起这些变化。综合考虑适用于天然蛋白质的"环醇"假说，蛋白质变性的易发生性及其性质上的巨大改变均可很自然地由此假说推导出来。

通过这些理论可以解释一个特别有趣的现象，即可溶性天然蛋白质在水中会形成不溶性的单分子层。其关键在于侧链 R 的位置。在笼状结构中这些侧链排布在六边形腔隙的边缘，构成笼状结构的侧平面，而在单分子层中，笼状结构转化为平面结构。当 R 基团是疏水基团时，单分子层由这些伸出的基团形成；在"笼状"结构中它们伸向内部，这也正是"笼状"结构的特征。

演讲的最后部分集中于 X 射线研究的证据，在这方面胰岛素的数据最为完整。就像克劳福特指出的那样，这种特殊的蛋白质在每一个单元晶胞中只含一个分子，因此相对于大多数蛋白质而言，胰岛素的结果更容易得到解释。

朗缪尔博士强调，就像许多其他可以被此理论解释的事实一样，例如 α-氨基酸而非 β-氨基酸的出现，在最初构建这种蛋白质结构理论时并没有用到 X 射线的研究数据。然而，在他看来，X 射线研究的结果验证了许多由此理论推导得出的结构特征。

在演讲之后的讨论中，林奇博士解释说，朗缪尔和林奇解析向量图时使用的方法是一种将晶体的典型图像看作一系列离散单元的几何学描述。贝尔纳教授说这个理论尚未被证实。他认为有些特征一定是真的，其他的则有可能是真的，但是依据他的观点，X 射线数据无法验证被提及的蛋白质的结构。他对后者的向量图进行了计算，发现与获得的实验结果有很大差别。内维尔教授通过如下一个简单的例子指出了在解析矢量图时要小心：从特定的一系列点出发构建得到的矢量图初看起来似乎对应着另一种非常不同的分布类型。

（刘振明 翻译；金侠 审稿）

Nature of the Cyclol Bond

I. Langmuir and D. Wrinch

Editor's Note

In the 1930s, the structure of protein molecules appeared to be a great mystery. That proteins are made from sub-units called amino acids (of which some 21 occur in natural proteins) was generally accepted, but the natural conformation of a protein molecule would have been a linear chain of amino acid "residues", which made it difficult for people to imagine how some protein molecules such as insulin appear to be roughly spherical in shape. One of the (incorrect) attempts to solve this problem was the so-called cyclol hypothesis, here advocated by Dorothy Wrinch, a British crystallographer, and Irving Langmuir, a distinguished American physical chemist.

I

THE confirmation by X-ray data[1,2] of C_2, the 288-residue cage structure proposed for the insulin molecule[3], makes it of interest to consider the nature of the cyclol bond, upon the postulation of which this structure and the cyclol theory of protein structure in general[4] depend. The making and breaking of a cyclol bond between an NH group of a polypeptide chain and a CO group of the same or of another polypeptide chain requires only the migration of an H atom thus:

$$\text{>NH} + \text{OC<} \rightleftharpoons \text{>N—(HO)C<}.$$

The making and breaking of a cyclol bond is therefore a prototropic tautomerism, inter- or intra-molecular, which indicates a special type of binding of the H atom. (A recent investigation has shown that a similar situation actually exists in crystalline diketopiperazine[5]). As a prototropic tautomerism, it is to be sharply contrasted with the making and breaking of a peptide link which, requiring the intervention of another molecule, say water, may be written:

$$\text{—NH—OC—} + \text{H}_2\text{O} \rightleftharpoons \text{—NH}_2 + \text{HOOC—}.$$

In order to direct attention to the fundamental difference between a prototropic tautomerism on one hand and a hydrolysis (or alcoholysis, etc.) on the other, it is convenient to talk in future of cyclol *bonds* and peptide *links*.

In current usage, the term protein is applied to anything having the chemical composition of amino acid condensation products. One class stands out as of paramount importance for the understanding of living matter, namely the crystalline globular proteins such as trypsin, pepsin and so on. These substances are already well defined in a number of ways, and it is with proteins belonging to this category that the cyclol theory is concerned.

环醇键的本质

朗缪尔，林奇

编者按

在 20 世纪 30 年代，蛋白质分子的结构仍是一个巨大的谜。当时人们普遍认为蛋白质是由被称为氨基酸（天然蛋白质中大约有 21 种氨基酸）的亚单元组成的。但这样一来，蛋白质分子的天然构象就应该是一条由氨基酸"残基"构成的线性直链，这样人们就很难想象为什么某些蛋白质分子（如胰岛素）是近似球形的。为了解决这一难题，人们进行了很多努力和尝试，其中之一就是由英国晶体学家多萝西·林奇和杰出的美国物理化学家欧文·朗缪尔提出的所谓环醇假说，但这种假说是错误的。

I

曾有人提出胰岛素分子具有由 288 个残基形成的 C_2 笼状结构 [3]，X 射线数据 [1,2] 对这一结构的确认使得人们对环醇键本质的关注变得很有意义，而胰岛素的分子结构以及蛋白质分子一般结构的假说 [4] 正是依赖于环醇键假说。一条多肽链中的 NH 基团与同一条或者另一条多肽链中的 CO 基团之间形成和断裂一个环醇键，只需要迁移一个 H 原子：

$$\rangle NH + OC \langle \rightleftharpoons \rangle N—(HO)C \langle$$

因此，环醇键的形成和断裂就是分子间或分子内质子转移的互变异构过程，这暗示了 H 原子的一种特殊键合方式。最近的一项研究显示，在晶态的二酮哌嗪（环缩二氨酸）中确实存在类似的情况 [5]。作为一种质子转移的互变异构过程，它与肽连接的形成和断裂形成了鲜明的对比。因为肽连接的形成和断裂需要另外一个分子（如水分子）的参与，可以写成下面的形式：

$$—NH—OC— + H_2O \rightleftharpoons —NH_2 + HOOC—$$

为了将注意力指向质子转移互变异构过程与水解过程（或者醇解等）二者之间的根本差异，简便起见，我们在以后的讨论中将它们称为**环醇键**和**肽连接**。

在目前流行的用法中，蛋白质一词可用于化学组成为氨基酸缩聚产物的任何物质。其中有一类对于理解生命物质最为重要，那就是具有结晶态的球蛋白，例如胰蛋白酶、胃蛋白酶等。这些物质在许多方面已有了详尽的定义，而环醇理论所关注的正是这一类蛋白质。

667

Among the properties by which the substances are already characterized are the following:

(1) They have definite molecular weights which are discretely arranged.

(2) They contain certain numbers of various particular amino acid residues and these numbers are frequently powers of 2 and 3.

(3) They are soluble in water or salt solutions, but their solubility is affected by changes in pH.

(4) They denature under very slight stimulus.

(5) They spontaneously form monolayers of extreme insolubility.

(6) They exhibit a high degree of specificity in biological reactions.

It has already been shown that the cyclol hypothesis explains (1), (2), (4). With regard to (3) and (5), we may point out that the extreme insolubility of the monolayer formed from a soluble protein[6,7,8] indicates that the globular protein has a structure in which hydrophobic groups can be completely masked, so leading directly to the idea of a cage structure which presented itself in the cyclol theory as a deduction from the geometry of polypeptide chains. The lacunae in the fabric allow ionized (and other) R groups to modify their positions profoundly in response to changes in pH, and indeed to lie inside or outside the cage. The spontaneous formation of protein monolayers from globular proteins in solution indicates that weak bonds only are broken. This we interpret to mean that, in the formation of monolayers, some or all of the cyclol bonds are opened, few or none of the peptide links being broken, so that protein monolayers consist of polypeptide chains partially or wholly decyclized, for the most part without open ends, the hydrophobic groups forming a separate phase in the surface. This type of structure explains many of their striking characteristics: for example viscosity, elasticity, etc.[7,8].

It may be emphasized that the breaking of cyclol bonds yields different degradation products according to the path of fragmentation. Thus cyclol-6 yields three diketopiperazine molecules or a single 6-residue chain (Fig. 1). In standard circumstances, a preferential path of fragmentation, probably largely determined by the R-groups, is to be expected[9], and it may be presumed that one and the same structure will under standard conditions break down in a unique way. Further, if preferential paths of breaking cyclol bonds be postulated, preferential paths of making cyclol bonds must also be postulated. Experiments have recently been reported purporting to disprove the cyclol hypothesis, in which glycyl leucine and leucyl glycine dipeptides were mixed and glycyl glycine and leucyl

下面列出的是这些物质已被确认的某些性质：

（1）它们具有呈离散分布的确定的分子量。

（2）它们包含一定数量的各种特定氨基酸残基，其数量往往成百上千。

（3）它们可溶于水或者盐溶液中，但 pH 的变化会影响其溶解性。

（4）它们会在很微弱的刺激下发生变性。

（5）它们可以自发地形成极难溶的单分子层。

（6）它们在生物反应中表现出高度的特异性。

以前已经证明，环醇假说可以解释（1）、（2）和（4）。至于（3）和（5），我们可以指出，可溶性蛋白质形成的单分子层的极度难溶性 [6,7,8] 意味着球蛋白具有一种能将疏水基团完全掩盖起来的结构，从而直接导致了在环醇理论中提出笼状结构的想法，这是从多肽链的几何特征中推导出来的。结构之间的缝隙允许离子化的（及其他的）R 基团随着 pH 的改变而大幅度地改变其位置，R 基团甚至可以位于笼的内部或外部。球蛋白在溶液中自发形成蛋白质单分子层，这表明只有弱键被打开了。我们对此的解释是，在形成单分子层的过程中，部分或全部环醇键被打开，但只有极少或没有肽连接断开，因而蛋白质单分子层是由部分或完全开环的多肽链组成，其中大部分没有开放的自由末端，疏水基团在表面形成了一个单独的相。这种结构解释了它们所具有的很多惊人特性：比如黏度、弹性等 [7,8]。

可能需要强调的是，根据断裂途径不同，环醇键打开后会产生不同的降解产物。因此六元环醇能够产生三个二酮哌嗪分子或者一条含有六个残基的链（图 1）。在标准条件下，可以预期一条优先的断裂途径（可能在很大程度上取决于 R 基团）[9]，并且还可以假定同一结构在标准条件下会以唯一的方式分解。另外，如果可以假定环醇键的优先断裂途径，那么相应地生成环醇键的优先途径也一定会被确定。最近报道的实验宣称推翻了环醇假说，实验中将甘氨酰亮氨酸与亮氨酰甘氨酸两种二肽混合，结果并没有形成甘氨酰甘氨酸和亮氨酰亮氨酸这两种二肽 [10]。即便有三嗪环形成（何况

leucine dipeptides were not formed[10]. Even if triazine rings were formed (and of this no evidence was offered), there seems no reason to suppose that the structure will break down to yield anything except the original molecules. We are therefore unable to accept these experiments as evidence against the cyclol hypothesis.

Fig. 1. Cyclol-6 (in centre) with its triazine ring dotted. The opening of this ring yields either three diketopiperazine molecules (shown on left) or a closed polypeptide chain of six residues (shown on right).

These suggestions of preferred paths of fragmentation apply equally to the globular proteins characterized, on the present view, by a specific set of amino acid residues, in a definite spatial interrelationship. They imply that for each protein there are one or more points at which the breaking of cyclol bonds normally starts. Thus an insulin molecule forming a monolayer would form a specific set of wholly (or partially) decyclized polypeptide chains. In this way, we can explain the highly characteristic films formed by different proteins, which permit an unknown protein to be recognized instantly as insulin or pepsin or papain and so on[6,7,8].

II

From the point of view of biology, cytology and immuno-chemistry, however, the most striking characteristic of the globular protein is its high degree of specificity. This, we suggest, indicates an organized structure with its own characteristic modes of vibration, which depend on the nature and arrangement of the various constituent amino acids. Since the breaking and making of cyclol bonds is a prototropic tautomerism, we anticipate that in appropriate circumstances there may be a kind of resonance in the molecule which accounts for the stability of the cyclol fabrics and the cyclol polyhedra. The cyclol fabric, a fragment of which is shown in Fig. 2, contains no double bonds and thus Frank[11] believed that "the cyclol molecule in itself offers no chance of constructing a resonant system". But each triazine hexagon has three parallel hydroxyls on the three carbon atoms, the next triazine hexagons in the fabric having their three hydroxyls on the other side of the fabric. With a symmetrical structure of this kind, it seems highly probable that the protons, instead of being attached to individual oxygen atoms, lie between the oxygen atoms serving as hydrogen bonds between them. There may thus be a marked resonance involving the three oxygens and the three hydrogens. Furthermore, resonance on a larger scale between neighbouring triazine groups may also be expected.

并没有证据支持这一点），似乎也没有理由推测该结构分解后会生成最初分子以外的任何其他物质。因此，我们无法接受把这些实验作为反对环醇假说的证据。

图 1. 六元环醇（位于中央）和它的三嗪环（由虚线表示）。如果把这个环打开，要么产生 3 个二酮哌嗪分子（示于左边），要么产生由 6 个残基组成的闭合多肽链（示于右边）。

　　根据当前的看法，关于优先断裂途径的说法同样适用于由具有明确空间互联关系的一组特定氨基酸残基构成的球蛋白。这些观点意味着，对于每种蛋白质来说，环醇键的断裂都有一个或多个正常的起点。因此，一个可以形成单分子层的胰岛素分子，会形成一组特定的完全（或部分）开环的多肽链。这样，我们就可以解释不同蛋白质所形成的高度特征化的薄膜，从而得以立即识别出一种未知蛋白质究竟是胰岛素、胃蛋白酶、木瓜蛋白酶，还是别的什么 [6,7,8]。

II

　　不过，从生物学、细胞学和免疫化学的角度来看，球蛋白最惊人的特征是它的高度特异性。我们认为，这意味着球蛋白是有序结构，并具有自身特定的振动模式，这种振动模式依赖于各种组分氨基酸的特性和排列方式。由于环醇键的断裂和形成都是质子转移的互变异构过程，我们预期，在某些适当的条件下分子中可能存在一种共振，这可以解释环醇织物与环醇多面体的稳定性。图 2 显示的是环醇结构的一个片段；环醇结构中不含双键，因此弗兰克 [11] 相信，"环醇分子自身没有构成共振体系的可能性"。但是，在每个三嗪六边形中，三个碳原子上都有三个平行羟基，而结构中下一个三嗪六边形也有三个羟基在结构的另外一侧。在这样一种对称结构中，看来极有可能的情况是质子并不是被束缚在单个氧原子上，而是位于两个氧原子之间并形成氢键将它们连接起来。因此可能存在一个涉及三个氧原子和三个氢原子的明显共振。此外，还可以预期，相邻三嗪基团之间可能存在更大规模的共振。

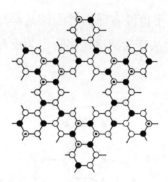

Fig. 2. A fragment of the cyclol fabric. The median plane of the lamina is the plane of the paper. The lamina has its "front" surface above and its "back" surface below the paper.

● = N.

◯ = C(OH), hydroxyl upwards.

⊙ = C(OH), hydroxyl downwards.

◯— = CHR, direction of side chain initially outwards

◯‒ = CHR, direction of side chain initially upwards.

This suggestion of intramolecular hydrogen bonds in the cyclol fabric is in line with the intramolecular bonds already postulated in a wide variety of compounds, including ice, alum, natrolite and other zeolites, oxalic acid dihydrate and formic acid, in which oxygen atoms are shown, by crystal structure data, to lie abnormally close together[12]. Such a close distance of approach is evidence of considerable mutual energy of oxygen atoms, which can only be due to the presence of hydrogen between them[12]. For certain compounds, for example, salicylaldehyde and o-nitrophenol, the presence of intramolecular hydrogen bonds has been confirmed spectroscopically by the non-appearance of the hydroxyl band[13]. In these cases the oxygen-oxygen distance is about 2.5 A., considerably shorter than the distance of 2.7 A., that represents double the radius of the ion O^{2-} and very much shorter than that for neutral oxygen. In the case of the cyclol fabric, hydroxyls are carried by three carbons in a triazine ring, which lie at the distance apart of $\frac{2}{3}a\sqrt{6}$, which is equal to 2.45 A. if a, a length intermediate between the C–C and C–N distances is taken(as has so far been done)as 1.5 A.

One of the most extraordinary properties of proteins is their specificity in biological reactions. The haemoglobin in each different type of animal seems to differ in spectrum and in details of its behaviour with oxygen. We are thus forced to conclude that, in proteins, certain features of the molecule can transmit some effect to other parts of the molecule, particularly to prosthetic groups. Such transmission of chemical influence, although observable to some extent in such compounds as sterols, seems to exist in a unique degree in the proteins: in long-chain compounds, on the other hand, there is practically no evidence of such transmissions to distances of more than a few atoms. It is well known that in aromatic chemistry the resonance, for example in the benzene molecule, causes substituents in different parts of the molecule to have an effect on one another very different from what would be expected if no resonance occurred. The high specificity of the proteins therefore in itself seems to demand resonance to a degree

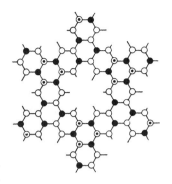

图 2. 环醇结构的片段。片层的中间平面为纸平面。片层的"前"表面在纸的上方,"后"表面在纸的下方。

● = N。
○ = C(OH),羟基向上。
⊙ = C(OH),羟基向下。
○— = CHR,侧链方向原本向外。
○- = CHR,侧链方向原本向上。

环醇结构中存在分子内氢键的说法,是与大量不同化合物中存在分子内氢键的假定相符合的,这些化合物包括冰、明矾、钠沸石及其他沸石、二水合草酸和蚁酸等。晶体结构数据显示,这些化合物中的氧原子彼此间非常靠近 [12]。如此近的距离是表明氧原子间具有相当强的相关能的证据,这只可能是由于两个氧原子之间存在氢原子的缘故 [12]。对于某些化合物如水杨醛或者对邻硝基苯酚来说,分子内氢键的存在已经通过光谱学方法得到了证实,因为光谱图中没有出现羟基带 [13]。在上述实例中,氧 – 氧原子间距约为 2.5Å,明显短于氧离子 O^{2-} 半径的两倍即 2.7Å,比起中性氧原子的半径就更是短很多。在环醇结构中,每一个三嗪环中的三个碳原子上都有一个羟基,间距为 $\frac{2}{3}a\sqrt{6}$,如果长度 a 的数值介于 C–C 与 C–N 间距之间(到目前为止人们都是这么处理的)即 1.5Å 的话,那么该间距等于 2.45Å。

蛋白质最不同寻常的性质之一就是其在生物反应中的特异性。每一种不同类型的动物体内,其血红蛋白似乎都具有不同的光谱,并且在与氧发生作用的细节上有所不同。由此必然会推出,蛋白质分子的某些特征可以将某些效应传递到分子中的其他部分,特别是传递给辅基。这种化学效应的传递现象,虽然在某种程度上也可以在诸如固醇类化合物中被观测到,但它在蛋白质中出现的程度似乎是独一无二的;另一方面,在长链化合物中,完全没有证据表明这种传递可以超出几个原子的距离。众所周知,在芳香族化合物如苯分子中,共振会使分子中不同部位的取代基彼此间产生相互影响,而如果没有共振的话,可以预测这种影响会是大不相同的。由此看来,蛋白质自身的高度特异性似乎比其他已知化合物需要有更大程度的共振。看来,

greater than that in other known chemical compounds. It would seem that the cyclol polyhedral structure with its sets of rings within rings and its multiple paths of linkage between atoms (cf. Fig. 2) should be capable of just such a type of resonance.

May we not perhaps regard each diazine ring in the cyclol fabric as the analogue of an electrical resonant circuit, the natural frequency of which is determined by the character of the side chains in this ring? Then n resonant circuits coupled together would be characterized by n^2 frequencies, no one of which being exactly what it would be if the resonant units were separated from one another. An electrical network, the resonant circuits of which have the geometry of the cyclol polyhedra, would give very close coupling between the adjacent hexagonal units. Thus the structure as a whole should possess a wide range of frequencies. The situation is somewhat analogous to that in the Debye theory where a crystal has a large number of modes of vibration. So in the protein the cyclol structure may be characterized by a whole spectrum of frequencies definitely correlated with the symmetry of the structure as a whole. Thus if a single factor disturbs the symmetry, it might have a profound effect upon some of the characteristic frequencies. If we imagine that these frequencies are important in the interactions between proteins and in funnelling the energy to certain parts of the molecule, we have a possible reason for the important effects produced by apparently minor changes in the structure of the protein molecule.

III

It appears that a strong case can be made out for the cyclol theory, since its implications, derived by simple geometrical arguments, fits the facts summarized above. The theory depends upon one postulate which thus becomes of special interest, particularly as its formulation has occasioned some uneasiness in chemical circles. The question then arises as to the possibility of proving directly the existence or non-existence of cyclol bonds.

The work on protein monolayers indicates that only weak forces are required to open cyclol bonds. We are of the opinion that many chemical techniques are sufficient to rip open the cyclol fabric into polypeptide chains, so that the chemist by his very operations may destroy the structure he seeks to study. On this view, any apparent contradiction between the cyclol hypothesis (which regards the globular proteins as polypeptide fabrics) and the chemical data relating to proteins (which appear to show them to be polypeptide chains) disappears. The situation is thus reminiscent of that in quantum physics, where the conditions required for the observation of the position and velocity of an electron themselves modify the phenomena under study. Similarly, it has recently been suggested[14] that enzyme studies do not permit the deduction that cyclol bonds do or do not exist in globular proteins. Direct knowledge of protein structure must therefore depend upon physical methods of investigation such as X-rays and spectroscopy, etc.

The picture of a globular protein with hydrophilic groups on its outer surface and hydrocarbon groups in contact within the cage is strikingly akin to the picture of the

环醇多面体结构及其一系列的环内环和原子间的多重连接方式（参见图2）应该是恰好可以产生这种共振的。

我们是否可以把环醇结构中的每个二嗪环视为一个共振电路的类似物，其天然频率由环内支链的特性决定？如果真是如此，那么偶连在一起的 n 个共振回路将由 n^2 种频率决定，并且其中没有一个会与这些共振单元彼此分离时呈现出来的频率严格一致。其共振回路具有环醇多面体几何特征的电网络会造成相邻六边形单元之间紧密偶连。因而其整体结构应具有大范围的频率。这有些类似于德拜理论中所提到的情况：一种晶体具有多种振动模式。因此，蛋白质分子中的环醇结构的特征就可以通过与整体结构对称性明确相关的一整套频率谱来明确描述。于是，如果有一个单一因素扰乱对称性时，那么它就可能对某些特征频率产生深刻影响。设想一下，如果这些频率对于蛋白质分子间的相互作用以及将能量集中于分子内特定部位的作用而言是重要的，那么我们就可能可以解释为什么蛋白质分子结构中看似很微小的变化也会导致重大的影响。

III

看起来似乎可以相信环醇理论了，因为从纯粹的几何学论据得出的该假说的推论与前面概述的各项事实都能吻合。环醇理论依赖于一个前提假设，这个假设因而吸引了人们的特别兴趣，一定程度上是因为它的表述方式引起了化学界的一些担忧，于是关于能否直接证明环醇键是否存在的问题就出现了。

对蛋白质单分子层的研究表明，只需要比较弱的作用力就可以打开环醇键。我们的看法是，有很多化学技术都足以使环醇结构断裂成多肽链，因此化学家所使用的特定操作很可能会破坏他试图研究的结构。这样一来，环醇假说（将球蛋白视为多肽结构）与蛋白质的化学数据（它们似乎显示球蛋白是多肽链）之间存在的任何矛盾都消失了。这种局面让人回想起量子物理学中的类似情况，即观测一个电子的位置与速度所需的实验条件本身会改变正在被研究的现象。与此类似，最近有人提出 [14]，对酶的研究不能推测出环醇键是否存在于球蛋白中。因此，关于蛋白质结构的直接知识必须依赖于诸如 X 射线及光谱学等物理学的研究方法。

球蛋白的亲水基团位于外表面，而烃基则位于笼中相互靠近，这与最近研究得

structure of certain micelles to which recent studies have led[15]. Ions, such as cetyl trimethylammonium sulphate, which contain long hydrocarbon chains form micelles in very low concentration. It is found that the size of such particles corresponds to a sphere of radius about equal to the length of the individual molecules and it is shown that the micelles are spherical particles, in which the tails are crowded together in the interior and the hydrophilic heads form the outer surface. The micelles in soap solutions have a similar structure.

Little or nothing is known at present as to the path or the nature of protein synthesis. We see no reason to suppose that a cyclol fabric forms spontaneously from polypeptide chains. If on occasion such cross linking occurs, this is probably a relatively unstable state. The non-formation of cyclol bonds between simple molecules[10,16,17] is thus irrelevant to the cyclol hypothesis and to the problem of the formation of complete globular protein molecules. Using the information obtained in recent studies of protein monolayers formed spontaneously, we deduce that, in the globular state, the CH_2 groups are completely masked from the aqueous medium. The globular protein in water is thus pictured as having its outer surface predominantly hydrophilic, the hydrophobic groups lying close together in the interior of the cage. A protein molecule of the size of insulin must contain several hundred CH_2 groups to account for the insolubility of the monolayer which it forms. When two CH_2 groups come into contact, energy of the order of 2,000 calories per gram molecule is involved[18], totalling for a protein molecule of the size of insulin, containing at least 300 CH_2 groups, an energy of upwards of 600,000 calories. This suggests one possible factor in the formation of a protein cage molecule.

We would also direct attention to the difference between the stability to be expected when a single cyclol bond forms within or between polypeptide chains and when three cyclol bonds form a triazine ring, and to the still more striking difference to be expected between the stability of a piece of cyclol fabric and of a complete cyclol polyhedron. Undoubtedly there are a number of factors favouring the complete polyhedral structure over and above any uncompleted cyclol structure. These may account for the existence of cyclol cages even if there should be an intrinsic instability in isolated cyclol bonds.

(**143**, 49-52; 1939)

References:

1. Wrinch, *J. Amer. Chem. Soc.*, **60**, 2005 (1938).

2. Wrinch and Langmuir, *J. Amer. Chem. Soc.*, **60**, 2247 (1938).

3. Wrinch, *Science*, **85**, 566 (1937); *Trans. Far. Soc.*, **33**, 1368 (1937).

4. Wrinch, *Nature*, **137**, 411 (1936) *et seq.*

5. Corey, *J. Amer. Chem. Soc.*, **60**, 1598 (1938).

6. Langmuir and Schaefer, *J. Amer. Chem. Soc.*, **60**, 1351 (1938).

7. Langmuir, Cold Spring Harbor Symposium on Proteins, **8**, 1938.

8. Langmuir, Pilgrim Lecture, Royal Society, 1938.

9. Wrinch, *Phil. Mag.*, **25**, 705 (1938); Cold Spring Harbor Symposium on Proteins, **8**, 1938.

10. Meyer and Hoheneiser, *Nature*, **141**, 1138 (1938).

到的某些胶束的结构 [15] 惊人地相似。某些具有长烃链的离子，例如十六烷基三甲铵硫酸盐，能在极低浓度时形成胶束。已发现这种胶束微粒的大小相当于一个半径约为单个分子长度的球体，并且结果表明胶束是球状微粒，其中各个分子的尾部挤在球的内部，而亲水的头部则形成了外表面。肥皂溶液中的胶束具有类似的结构。

目前，我们对于蛋白质合成的途径或其性质所知甚少，或者一无所知。我们没有理由假定环醇结构是由多肽链自发形成的。即使偶然发生交联反应，这也很可能只是一种相对不稳定的状态。简单分子之间不形成环醇键这种说法 [10,16,17]，与环醇假说以及完整球蛋白分子的形成问题并不相干。利用最近获得的关于蛋白质单分子层自发形成的研究结果，我们推测，在球形状态时，CH_2 基团是完全与水介质隔离的。由此可以描绘球蛋白在水中的形象，其外表面几乎完全是亲水的，而疏水基团则彼此紧靠着处于笼的内部。一个与胰岛素大小差不多的蛋白质分子必定含有几百个 CH_2 基团，这样才可以解释它所形成的单分子层的不溶性。当两个 CH_2 基团彼此接触时，涉及的能量的数量级为每克分子 2,000 卡 [18]，总计起来，一个与胰岛素大小差不多的蛋白质分子包含至少 300 个 CH_2 基团，总能量将达到 600,000 卡以上。这可能是形成蛋白质笼状分子的一个因素。

我们还应当把注意力转向可以预期的稳定性差异：如一个环醇键在多肽链内或多肽链间形成时的稳定性差异，三个环醇键形成一个三嗪环时的稳定性差异，以及可以料想到的存在于一个环醇结构与一个完整环醇多面体之间的更为惊人的差异。毫无疑问，与不完全的环醇结构相比，许多因素对完整的多面体结构的形成更为有利。这些因素也许可以解释环醇笼的存在，即便分离的环醇键也可能具有内在的不稳定性。

（王耀杨 翻译；顾孝诚 审稿）

11. Frank, *Nature*, **138**, 242 (1936).

12. Bernal and Megaw, *Proc. Roy. Soc.*, A, **151**, 384 (1935).

13. Hilbert, Wulf, Hendricks and Liddel, *Nature*, **135**, 147 (1935). Errera and Mollet *C.R. Acad. Sci.*, Paris, **200**, 814 (1935).

14. Linderstrøm-Lang, Hotchkiss and Johansen, *Nature*, **142**, 99 (1938).

15. Hartley and Runnicles, *Proc, Roy. Soc.*, A, 168, 401 (1938): Hartley *J. Chem. Soc.*, 1968 (1938).

16. Jenkins and Taylor, *J. Chem. Soc.*, 495 (1937).

17. Neuberger, Royal Society discussion on proteins, November 17 1938 (*Nature*, **142**, 1024; 1938).

18. Langmuir, *J. Amer. Chem. Soc.*, **39**, 1848(1917); Colloid Symposium Monograph 3, 48 (1925); "Colloid Chemistry", vol.1, 525 (1926).(Edited by J. Alexander, Chemical Catalog Co., N.Y.)

A Living Fish of Mesozoic Type

J. L. B. Smith

Editor's Note

Imagine discovering a dinosaur carcass on a lonely country road. That is not as fanciful as it seems, given the discovery over Christmas 1938 of a coelacanth off the coast of East London, South Africa—a fish believed to have been extinct since the age of the dinosaurs, some 80 million years before. By the time that ichthyologist James Leonard Brierley Smith could get to the scene to make the observations reported here, the single specimen had badly decomposed, but enough remained for him to be sure that the strange fish he named _Latimeria chalumnae_ was indeed a coelacanth similar to the Cretaceous form _Macropoma_. Smith's hope that another specimen might be captured was fulfilled—but not for another dozen years.

EX Africa semper aliquid novi. It is my privilege to announce the discovery of a Crossopterygian fish of a type believed to have become extinct by the close of the Mesozoic period. This fish was taken by trawl-net at a depth of about 40 fathoms some miles west of East London on December 22, 1938. It was alive when caught, and shortly after it died it was handed over to Miss Courtenay–Latimer, curator of the East London Museum. Miss Latimer wrote to me, enclosing a sketch and brief particulars of the specimen. Owing to the seasonal disorganization of the postal services, the letter did not reach me at Knysna, some four hundred miles away, until ten days later. It was obvious from the sketch and notes that the fish was of a type believed long extinct. Immediate telephonic communication with the East London Museum revealed that, owing to lack of preserving equipment at that Institution, the putrefied body had been disposed of beyond any hope of redemption, and the fish had been mounted by the local taxidermist.

Since the fish was unquestionably alive when caught, there is at least a possibility that this zoological tragedy may be ameliorated by the capture of another specimen. This is not so remote as might appear. After careful inspection of the mounted specimen, a responsible citizen-angler of East London stated that about five years ago he had found precisely such a fish, only considerably larger (_sic_), partially decomposed, cast up by the waves on a lonely part of the shore east of East London. When he returned with assistance, the monster had vanished with a risen tide. With regard to the present specimen, fortunately both Miss Latimer and the taxidermist were drawn to observe details of the carcass very closely, so that exhaustive independent questioning has left me with at least some definite information about the missing parts. Fortunately also, the terminal caudal portion of the vertebral column and part of the pectoral girdle remain. The skull is of course intact.

680

一种存活至今的中生代鱼类

史密斯

编者按

你是否设想过在一条荒凉的乡间小路上发现一具恐龙尸体。鉴于 1938 年圣诞节期间在南非的东伦敦海岸发现了腔棘鱼（人们原以为这种鱼在 8,000 万年前的恐龙时代就已经灭绝了），上述想法也就不像看上去那么荒诞了。当鱼类学者詹姆斯·里奥纳多·布莱尔利·史密斯到达现场进行本次观察时，这一标本已经腐烂得很厉害了，但是残余的躯体足以使他确信，这个被他命名为拉蒂迈鱼的奇怪鱼类确实与白垩纪大盖鱼属的腔棘鱼非常相似。史密斯希望能够捕获另一个标本的愿望最终得到了满足，但那已是十几年后了。

"在非洲总是可以发现新东西"。我很荣幸地宣布，我们发现了一种被认为在中生代末期就已经灭绝的总鳍鱼。它是在 1938 年 12 月 22 日于东伦敦以西几英里的深约 40 英尺的水中用拖网捕到的。刚被捕获时它还是活的，死后不久它就被移交到东伦敦博物馆馆员考特尼-拉蒂迈小姐的手中。拉蒂迈小姐给我写了一封信，随信附有该鱼标本的草图并简述了它的特征。由于邮政服务季节性的混乱，信件直到 10 天后才到达 400 英里外我所在的克尼斯纳。从信中的标本草图和记述可以明显看出，这条鱼属于一种之前被认为早已经灭绝的种类。我立即电话联系东伦敦博物馆，才发现由于博物馆没有相应的保存设备，腐烂的鱼体已经被处理过，并且没有修复的可能了，而鱼已经被当地的剥制师制作成标本了。

既然鱼在捕获时确实是活着的，那么至少存在这样一种可能性，那就是我们可以通过捕获另一条这种类型的鱼来弥补这一动物学遗憾。这件事情并不像表面看起来的那样难以实现。在对已经被制作成标本的鱼体进行仔细的观察之后，一位来自东伦敦地区的可靠的当地垂钓者声称，大约 5 年前他发现过一条一模一样的鱼，只不过那条鱼相对大一些，鱼体已经部分腐烂了，被海浪冲到东伦敦东部一个偏僻的海岸。当他带着帮手赶回那个地方时，那条鱼已经被上涨的潮水带走了。关于现在的标本，幸亏拉蒂迈小姐和动物标本剥制师都曾非常细致地观察过鱼体的细节，因此通过详尽的单独访问，我获得了一些有关标本缺失部分的明确信息。同样值得庆幸的是，脊柱尾部末端和部分胸带保留了下来。鱼的头骨也是完整无缺的。

The specimen is 1,500 mm. in total length, and weighed 127 lb. when caught. The colour was a bright metallic blue, which has faded to brown with preservation.

In major characters this remarkable specimen shows close relationship with the Mesozoic genus *Macropoma* Agassiz, of the family Coelacanthidae, order Actinistia. The gephyrocercal tail with protruding axial supplement, the normal first dorsal, the obtuse lobation of the remaining fins, the ganoin tubercle ornamentation on the scales and on some of the dermal bones of the head, the nature and arrangement of the dentigerous bones of the mouth, and the form of the dermal armour of the head, are all typically coelacanthid.

The skeleton was cartilaginous, the vertebral column apparently tubular, and the whole fish extraordinarily oily. The fish has small spiracles situated as shown in the accompanying illustration, and a definite though not very obvious lateral line, which continues uninterrupted to the end of the supplementary caudal. Other differences from the known coelacanthid fishes are the pronounced pedunculation of the lobate pectorals, the reduction of the dermal armour of the head, and the presence of two small heavily ornamented bones at the anterior lower corner of the opercular plate, which probably correspond with the more fully developed inter- and sub-opercula of teleosts, also a similar posterior post-spiracular ossicle. Dermal parafrontals are not visible. There is a free tongue composed of four fused segments covered with presumably ossified tubercles.

Coelacanthid Fish from East London, South Africa. The small arrow shows the position of the spiracle, and the dotted line indicates the position of a membrane behind the first dorsal fin.

It is probable that systematists will wish to propose a new family (some even a new order) for this fish, but I am at present satisfied that it is close enough to the Mesozoic Coelacanthidae to justify its inclusion in that family. It has been noted that certain coelacanthid fishes underwent little apparent change from the Devonian to the Cretaceous. It is therefore not surprising that this species, which presumably has survived from the Mesozoic, should still retain most of the features which characterize that family.

682

鱼被捕获时，全长 1,500 毫米，重 127 磅。全身原呈明亮的金属蓝色，但在保存过程中发生褪色而变为褐色。

这件珍贵标本的主要特征表明，它与中生代的腔棘目腔棘鱼科的大盖鱼属有着密切的关系。矛型尾具凸起的轴对称的中央尾叶，正常的第一背鳍，其他鳍呈钝圆形的缺刻结构，鱼鳞和鱼头部的一些真皮骨上装饰有硬鳞质结节，鱼口中带齿骨骼的特性和排列，及头部真皮骨的形态，这些都是腔棘鱼所具有的典型特征。

骨架为软骨，脊柱呈明显的管状结构，而整个鱼体格外油滑。鱼有小型的喷水孔，具体位置见本文的插图，一条确定的但并不明显的侧线一直延伸到中央尾叶的末端。与已知腔棘鱼相比，此鱼还具有如下不同：显著的胸鳍肉柄，头部真皮骨板减少，在鳃盖板的前下角有两块被严重修饰过的小骨，这两块小骨可能与硬骨鱼发育更全面的间鳃盖骨和下鳃盖骨相对应，此外还有一块类似于尾部后通气孔的小骨。真皮额顶骨不可见。游离的鱼舌由四个愈合部分组成，并覆有可能骨化的结节。

南非东伦敦发现的腔棘鱼。小箭头显示喷水孔，虚线指示位于第一背鳍后软膜的位置。

也许分类学家们希望为这种鱼建立一个新科（甚至新目），但是目前我确信这种鱼与中生代腔棘鱼科非常相似，甚至足以证明它们属于同一科。这个可能从中生代一直存活至今的鱼种，竟然仍保留有它所在科的大部分特征，但是鉴于一些腔棘鱼从泥盆纪到白垩纪几乎没有发生明显的变化，因此这就不足为奇了。

For the fish described and figured above I propose the name *Latimeria Chalumnoe* gen. et sp. nov.; the full account of the species and of its taxonomic relationships will be published in the *Transactions of the Royal Society of South Africa*.

(**143**, 455-456; 1939)

J. L. B. Smith: Rhodes University College, Grahamstown.

根据以上对这个鱼的描述和绘图，我建议将其命名为拉蒂迈鱼（新属新种）；关于这个物种的更详细全面的记述及其在分类学上的关系将刊登在《南非皇家学会学报》上。

（张玉光 翻译；陈平富 审稿）

The Structure of the Globular Proteins

D. Wrinch

Editor's Note

Here the pugnacious Dorothy Wrinch defends her (incorrect) theory of the molecular structure of proteins against its critics. Wrinch maintained that proteins weren't simple chains of amino acids linked by peptide bonds, but were joined into hexagonal ring structures that are linked into cage-like forms. The serious problems with this theory were already evident—in particular, the "side chains" on some amino acids simply wouldn't fit, a problem not addressed here. Wrinch's arguments here are all circumstantial, but they are a reminder of how difficult it was to say anything definite about protein structure until X-ray crystallography began, over a decade later, to reveal it in earnest.

O N recent occasions a number of objections to the cyclol theory have been raised, to many of which replies have already been given, explicitly or implicity, in a number of publications, notably in Langmuir's lecture to the Physical Society on December 20, 1938[1]. In view of the prominence which has been given to these criticisms, it seems desirable to summarize the replies.

It has been suggested that the number of theories such as the cyclol theory is so large that the *a priori* probability of any one is not sufficiently great for it to merit consideration[2,3]. It is difficult to give much weight to objections of this type until it proves possible to formulate, or failing this to establish the existence of, a sufficiently large number of independent theories, each of which leads to structures accounting for the definite and discretely arranged molecular weights of the native proteins, their capacity to denature to form spontaneously from solution highly insoluble monolayers and so on. So far as we are aware, no alternative structures have yet been proposed. Meanwhile, "the very large number of independent properties which are correlated by the cyclol theory and the extreme simplicity and the small number of postulates from which it is derived by purely logical processes"[1], seem to give it a considerable *a posteriori* probability.

It is suggested that there are grave objections on chemical grounds to the postulate of the four-armed unit >N–CHR–COH<[3] and indeed to the cyclol structure itself owing to the lack of precedent in organic chemistry for this diazine–triazine type of structure[4]. Proteins are so different from other substances that it is surprising that there is a reluctance to accept for them a structure for which no analogue in organic chemistry can be found. Globular proteins have not yet been synthesized in the laboratory. The fact that it has not proved possible to form ring structures from peptides[5] is therefore scarcely to be regarded as evidence against the cyclol hypothesis. Furthermore, we would not regard the lactam–lactim tautomerism of peptides as the chemical basis of the cyclol theory[4]. At present,

686

球蛋白的结构

林奇

编者按

针对批评，好斗的多萝西·林奇在本文中捍卫了她关于蛋白质分子结构的（不正确的）理论。林奇坚持认为蛋白质不是由氨基酸通过肽键连接形成的简单的链，而是由六边形的环状结构连接成为笼状的形式。这个理论面临的严重问题非常明显——特别是一些氨基酸上的"侧链"显然不能与这一理论吻合，但这一问题在本文中并未被提及。林奇在这里的辩解都是根据情况推测的，但是它提醒我们，要对蛋白质结构作出任何确定的描述有多么困难。直到十年之后 X 射线晶体学兴起，这一情况才得以改观。

最近在一些场合中，出现了一些反对环醇理论的意见；对于其中很多反对意见，已经有人在许多出版刊物上发表文章作出或明白或含蓄的答复，特别值得一提的是，1938 年 12 月 20 日朗缪尔在对物理学会的演讲中所作的回答 [1]。考虑到这些批评意见受到的重视，我觉得有必要在这里总结一下相关的答复。

有人提出，理论（例如环醇理论）的数目如此之多，以至于从先验的或然概率来说，任何一个理论都不足以引起人们的注意 [2,3]。要重视这类反对意见是很困难的，除非人们可以用公式表述足够大量的独立理论，或者如果做不到这一点的话，人们也要可以确证有足够大量独立理论的存在，其中每个理论都可以导出天然蛋白质的结构，这些结构都能解释天然蛋白质独立分布的分子量、变性能力（在溶液中自发形成高度不溶性的单分子层）等。就我们所知，目前还没有提出新的替代结构。同时，"与环醇理论相关的大量蛋白质的极其简单的独立性质，以及少数相关假设（环醇理论就是根据这些假设通过纯逻辑推理得到的）" [1] 似乎赋予它相当强的后验或然概率。

有人极力反对，认为对四臂单元 >N–CHR–COH< [3] 的假设、甚至对环醇结构本身的假设都缺乏化学基础，因为在有机化学里还没有出现过这种二嗪 – 三嗪类结构 [4] 的先例。蛋白质和其他物质之间的差异如此巨大，令人惊讶的是竟然有人不愿意接受它们作为可能存在的结构类型，仅仅是因为在有机化学中找不到它们的相似物。目前，实验室内还不曾合成球蛋白。因此，多肽 [5] 形成环状结构的可能性尚未被证实这个事实也不足以被视为反对环醇理论的证据。此外，我们也不会将多肽内酰胺 – 内酰亚胺的互变异构现象作为环醇理论的化学基础 [4]。目前我们还不知道在

we do not know whether polypeptide chains constitute an intermediate stage in protein synthesis, nor is it clear that the cyclization of chains, if it occurs, entails the postulate of lactam–lactim tautomerism of peptides.

Further criticisms of the cyclol hypothesis are as follows: (a) the theory is incompatible with the findings of enzyme chemistry[4,6]; (b) it cannot explain certain facts of denaturation[4]; and (c) it has not proved possible to synthesize in the laboratory certain "simple models of cyclol type"[4], or to induce the formation of cyclol links between simple molecules[6].

(a) In accordance with the current ideas in the kinetics of reactions, it was suggested by Langmuir[7] that there is a tautomeric equilibrium in the globular proteins between some of the four-armed units >COH–CHR–N< and the two-armed units –CO–CHR–NH–, by means of which cyclol bonds are continually opening and reforming. Thus at any one time there are some points at which a peptide link, CO–NH, can be attacked directly. The cyclol theory therefore does not, as has been suggested[4], need to make the assumption that proteolytic enzymes open cyclol bonds. We know of no evidence which conflicts with the view that the enzyme can take advantage of the situation arising from the opening of these bonds, so that, if it is a fact that proteolytic enzymes act only on CO–NH bonds, there is no incompatibility with the cyclol structure proposed for the globular proteins.

The objections (b) and (c) again raise the question of the nature of the cyclol bond. This bond in isolation may well be unstable; indeed the marked instability of the globular proteins in general and other considerations[8] require it to be so. Failure to synthesize in the laboratory so-called "simple models of the cyclol type"[4] (a phrase which may ultimately be found to be a contradiction in terms), or to induce the formation of isolated intermolecular cyclol links, is then no evidence against the cyclol hypothesis.

The stability of the globular proteins, under special conditions, in solution and in the crystal, we attribute to definite stabilizing factors[7,9]; namely, (1) hydrogen bonds between the oxygens of certain of the triazine rings, (2) the multiple paths of linkage between atoms in the fabric, (3) the closing of the fabric into a polyhedral surface which eliminates boundaries of the fabric and greatly increases the symmetry, and (4) the coalescence of the hydrophobic groups in the interior of the cage. If these stabilizing factors are withdrawn the cage structure collapses. Denaturation on this theory (apart from the special type due to the dissociation of structures held together by hydroxyl or hydrogen bonds or salt or covalent linkages between R-groups[10,11]), corresponds to any breakdown of the structure due to the opening of cyclol bonds[1,12]. When few bonds are open they may re-form because of the multiple paths of binding holding the parts in position. Reversible denaturation (of either type) is thus not excluded by the cyclol theory. In the case of the fibrous proteins in which some cyclol bonds may be present,[13] the first and second stabilizing factors may operate. The denaturation of myosin (which it is suggested is a difficulty for the cyclol theory[4]) thus seems to be easily interpretable.

688

蛋白质合成过程中多肽链是否会构成一种中间态，也还没有搞清楚多肽链的环化，如果环化发生的话，就意味着多肽的内酰胺 – 内酰亚胺结构互变异构的假设是必然的。

对于环醇假说进一步的批评如下：(*a*) 这个理论与酶化学发现的结果不相容 [4,6]；(*b*) 它无法解释蛋白质变性中的一些事实 [4]；(*c*) 尚未证明可以在实验室中合成某些 "环醇类的简单化合物" [4]，或者诱导简单分子形成环醇连接 [6]。

(*a*) 与反应动力学目前的观点一致，朗缪尔 [7] 提出：在球蛋白中，有些四臂单元 >COH–CHR–N< 和两臂单元 –CO–CHR–NH– 之间存在着互变异构平衡，因为环醇键不断地打开和再形成。所以在任何时刻，都可以从某些点上直接攻击多肽连接 CO–NH。因此，环醇理论没有必要像人们提出的那样 [4] 去假设蛋白水解酶可以打开环醇键。就我们所知，并没有证据与下述观点相悖，即酶可以在这些键打开的情况下发挥作用，因此，如果蛋白水解酶只作用于 CO–NH 键是事实的话，我们为球蛋白提出的环醇结构就不存在矛盾了。

反对的理由 (*b*) 和 (*c*) 再次提出了环醇键本质的问题。环醇键在孤立的情况下很可能是不稳定的；事实上，在通常以及其他一些情况下球蛋白所具有的显著不稳定性正需要它如此 [8]。那么无法在实验室中合成所谓 "环醇类的简单化合物" [4]（可能最终会发现这一说法是自相矛盾的）或者不能诱导孤立的分子间形成环醇连接的事实，就都不能作为证据来反对环醇假说。

我们将球蛋白在某些特定状态下（在溶液中或晶体中）的稳定性归因于特定的稳定因素 [7,9]；即 (1) 某些三嗪环上氧原子之间的氢键，(2) 纤维结构中原子之间的多种连接途径，(3) 纤维结构闭合成多面体的表面可以消除纤维结构的边界，并且极大地提高对称性，(4) 疏水性基团在笼状结构内部汇集。如果这些稳定性因素不存在了，笼状结构就会崩塌。在这个理论基础上的变性(不考虑由 *R* 基团间的羟键、氢键、盐键或共价键紧密维系的结构发生分解 [10,11] 这种特殊情况) 对应于环醇键打开造成的任何结构崩溃 [1,12]。当只有少数键打开时，它们可以重新再次生成，因为有多种结合途径使各部分维持原位。因此可逆变性（其中的任何一种类型）并没有被环醇理论排除在外。对于可能含有一些环醇键的纤维蛋白来说 [13]，第一个和第二个稳定性因素可以起到作用。肌球蛋白的变性（这被认为是环醇理论不能解决的一个难题 [4]）由此变得很容易解释。

The first stability factor also, it appears, may explain the fact that it has not proved possible to methylate or acetylate the peptide hydroxyls, which on the present hypothesis should be present in egg albumin and haemoglobin[14].

It may be emphasized that whereas the first three stability factors are uniform for all the globular proteins, the strength of the fourth factor varies with the specific amino-acid composition of the individual protein. This point is demonstrated by the case of insulin, in which the exceptional behaviour (its resistance towards heat and acid, the stability of its crystals in the dry state, etc.) can perhaps be correlated with its abnormally high content of hydrophobic groups. The fact that insulin "cannot be denatured in the usual sense of the word"[4] does not provide evidence against the cyclol theory of denaturation: it illustrates the importance of the fourth stability factor. Incidentally, insulin, like many other globular proteins, spontaneously forms very insoluble monolayers from solution[15], a property which in itself indicates the likelihood of a cage structure[1,10,11]. The spontaneous formation of monolayers from solution represents a very mild type of degradation, presumably involving no hydrolysis. It seems likely that much could be learnt about protein structure by studying the structure of the constituents of these protein monolayers. On the cyclol theory, fragments of fabric and closed polypeptide chains are to be expected when gentle methods of degradation are used which involve no hydrolysis[12]. The theory would not (as has been suggested[4]) make any predictions as to the nature of the products obtained when the usual gross methods of degrading proteins are used.

Other criticisms of the cyclol theory[3] (dealt with in detail in other communications[16,17]) relate to one particular point, namely, the suggestion that the insulin molecule has the cage structure C_2. It was shown at the outset[18] that this structure satisfies the chemical and physico-chemical facts and also those relating to cell molecular weight, space-group and cell dimensions obtained in an X-ray analysis of a dry zinc-insulin crystal[19]. Controversy has, however, arisen as to whether it is consistent with certain vector diagrams of a dry zinc-insulin crystal published later[20]. In view of authoritative opinions as to the inadequacy of the X-ray date[21] (which make it highly uncertain how far the vector diagrams give even an approximately correct picture of the situation in the crystal) and as to the feasibility of making any deductions from such data in the absence of a chemical analysis of the crystal[21,22], it seems necessary at present to abandon the hope of reaching any final conclusion on this special point. Arguments can only revolve round the plausibility of subsidiary hypotheses.

The Geometrical Attack on Protein Structure

It has been pointed out by Neuberger[4] that many of the arguments in favour of the cyclol hypothesis could be used equally well in favour of any "regular geometrical structure which joins peptide chains to form a globular molecule", a theme which has already been developed in considerable detail[23,24]. This remark, in focusing attention on the fundamental problem of protein structure, namely, the existence of megamolecules, puts into correct perspective the general investigations on protein structure of which those

第一个稳定性因素似乎也可以解释至今不能使多肽的羟基实现甲基化或者乙酰化这一事实。在目前的假说下，这种情况应当在卵清白蛋白和血红蛋白中存在[14]。

需要强调的是，尽管前三种稳定性因素对于所有的球蛋白是等同的，但第四种稳定性因素的强度会随着每种蛋白的特定氨基酸组成的变化而变化。这一点可以通过胰岛素的例子得到证明，它的异常行为（耐受热和酸的能力，晶体在干燥状态下的稳定性等）可能与其含有的疏水基团异常多有关。胰岛素"不会发生通常意义上的变性"[4]这一事实并不能作为反驳关于变性的环醇理论的证据：它只是进一步说明第四种稳定性因素的重要性。附带提一下，胰岛素像许多其他的球蛋白一样，可以在溶液中自发形成非常难溶的单分子层[15]，这个特性本身就暗示了它具有笼状结构的可能性[1,10,11]。在溶液中自发形成单分子层代表了一种非常温和的降解类型，据推测，这种类型不涉及水解作用。看来，可以通过研究这些蛋白单分子层组分的结构来获得许多关于蛋白质结构的知识。基于环醇理论，当采用不涉及水解在内的温和方法实现变性时，可以预期将会有笼状纤维片段和闭合的多肽链片段出现[12]。但是当采取降解蛋白惯用的粗糙方法时，这个理论不会（就像人们曾经建议的那样[4]）对获得产物的性质作出任何预测。

针对环醇理论的另外一些批评[3]（在其他通讯中已有详细的阐述[16,17]）是和一个特定的问题相关的，即认为胰岛素分子具有 C_2 笼状结构。我们在一开始就提到[18]，这个结构可以满足已知的化学和物理化学事实，以及通过对干燥的锌-胰岛素晶体[19]进行 X 射线分析所得到的细胞分子量、空间群和晶胞大小相关的诸多性质。然而，争论又出现了，即它是否与随后发表的干燥的锌-胰岛素晶体结构[20]的矢量图相一致。鉴于权威的观点认为 X 射线数据不充分[21]（这使得矢量图究竟能在多大程度上对晶体内真实情况给出即便只是近似正确的图像这一问题带有高度不确定性）以及在缺少对晶体进行化学分析的条件下[21,22]根据这样的数据进行任何推导并不可行，目前似乎必须放弃在这个特定问题上达到任何最终结论的希望。争议只能围绕着辅助性假说的合理性来进行。

对蛋白质结构的几何学上的攻击

纽伯格[4]曾经指出，支持环醇理论的许多论点同样适用于支持任何"能将多肽链连接成球状分子的规则几何结构"，这个主题已在很多细节方面都有了发展[23,24]。这个说法将注意力集中在蛋白质结构的基本问题上，即巨大分子的存在问题，把对蛋白质结构的一般研究纳入了正确的视野，对环醇理论的处理正是这些研究中的一部分。这些研究的范围广阔。如果蛋白质分子是氨基酸的多缩聚产物，由一个氨基

dealing with the cyclol theory form a part. These investigations have a wide scope. Given that protein molecules are polycondensation products of amino acids, interlinked by bonds between the nitrogen of one and the terminal carbon of another, there are only a certain mathematically determinate set of possible structures. The full and exhaustive formulation of these is the objective of this work[11,13,24].

Now the cyclol theory gives precise expression to this general point of view by postulating that these polycondensation products are built on the particular two-dimensional scheme exemplified by the cyclol network. No alternative comprehensive scheme has so far been forthcoming; but on the other hand, it has never been claimed that the cyclol pattern is the only design consistent with the chemical and stereochemical specifications[24]. The characteristic pattern must, of course, satisfy certain conditions; thus it must be such as to allow "the possibility of the regular and orderly arrangement of hundreds of residues"[25] (and herein lies the difficulty of attributing the fine structure of proteins to S–S, CO–S, NH–S or CO–NH–CO links as suggested by Neuberger[4]), and it must be capable of fragmentation into polypeptide chains. The cyclol postulate has been developed in considerable detail, since it was scarcely feasible to show the potentialities of the geometrical method by general statements alone.

It should, however, be recognized that there are broad deductions to be drawn from this point of view, irrespective of the particular form of the atomic pattern. Chief among these is the suggestion that the molecules of the globular proteins are cage-like structures, consisting of a characteristic fabric folded round to form a closed polyhedral surface[10,13,24]. This simple corollary of the idea of a protein fabric, though in point of fact suggested by the cyclol hypothesis, is independent of this or any other postulate. Whether or not the cyclol hypothesis proves to be correct, it certainly establishes the existence theorem in the mathematical sense, proving that there are no intrinsic impossibilities in the cage type of structure, showing how the atomic pattern can determine size and shape, and proving also that the cages may be rendered highly specific by the conditions of linkage imposed on the individual atoms.

The present controversy—wholly concerned with details of the cyclol pattern—makes it clear that the determination of the atomic pattern of proteins is beyond the range of present chemical and physical techniques. It then seems all the more important to direct attention to this general issue as to the general nature of structures of the globular proteins. Whatever the atomic pattern may be, the fact remains that the defining characteristics of the native proteins not only fit in with the idea of a cage, but even seem to demand a structsure of this kind for their molecules. On this point there seems indeed to be a growing consensus of opinion[2,4]. In any case, the central problem is to explain the existence of very large but chemically and physically well-defined molecules. The idea of a cage structure has been offered for consideration, as an explanation of this striking fact. Is it not possible, by means of experimental techniques already available, to recognize a cage when we see one?

(**143**, 482-483; 1939)

酸的氮和另一个氨基酸的末端碳之间形成的化学键相互连接，那么就只有某些特定的通过数学确定的可能结构存在。本项研究的目标就是对这些结构作出全面而详尽的表述 [11,13,24]。

现在环醇理论对上述这个一般性观点给出了准确的表达：它假定这些多缩聚产物是建立在以环醇网络为代表的特定二维系统之上。迄今为止还没有人提出其他可替代的全面方案；但另一方面，也从来没有人宣称过环醇模式是符合化学和立体化学特征的唯一模式 [24]。当然，这种特征模式必须满足某些条件；它必须要满足"数百个氨基酸规则有序地排列的可能" [25]（在这里存在的困难是如何像纽伯格 [4] 曾经建议的那样把蛋白质精细结构归因于 S–S、CO–S、NH–S 或者 CO–NH–CO 之间的连接），而且它还要可以断裂为多肽链。环醇假说在很多细节方面已经有了进一步的发展，因为仅依靠一般性陈述来表明几何学方法的潜能，几乎是不可行的。

然而应该认识到，不管原子排列的特定形式如何，从这个观点出发，都可以得到大量的推论。其中首要的建议是球蛋白分子为笼状结构，由特定的纤维结构折叠环绕构成一个封闭的多面体表面 [10,13,24]。关于蛋白质是纤维结构的观点，其简单的必然结果与环醇假说或任何其他假说无关，尽管事实上正是环醇假说提出了蛋白质的纤维结构。无论环醇假说的正确性是否被证实，它无疑在数学意义上建立了存在性命题法则，证实笼状结构没有内在的本质矛盾，显示了原子排列模式如何可以确定分子的大小和形状，另外也证实了笼状结构可以通过施加在各个原子上的连接条件而被赋予高度的特异性。

目前的争论——完全是关注环醇模式的细节——使人清楚，测定蛋白质内原子排列方式超出了目前化学和物理技术的能力范围。看来现在更为重要的事情是将注意力集中在球蛋白结构的一般性质这个更普遍的议题上。无论原子的排列方式如何，事实仍然是天然蛋白质的定义性特征不仅符合分子笼状结构的理念，甚至似乎还要求蛋白质的分子具有这样的结构。在这一点上，似乎达成共识的意见真的越来越多 [2,4]。无论如何，核心问题就是去解释那些非常大但是化学和物理上定义明确的分子的存在。作为这个引人关注事实的解释，笼状结构的理念已经被提出来供人们考虑。有没有可能在我们看到一个笼状结构的时候通过已经可用的实验技术手段把它识别出来呢？

（刘振明 翻译；顾孝诚 审稿）

References:

1. This lecture was summarized in *Nature*, **143**, 34 (1939): it will be published in full in the *Proceedings of the Physical Society*.

2. Bernal, Royal Institution, Jan. **27**, 1939.

3. Bernal, *Nature*, **143**, 74 (1939).

4. Neuberger, *Nature* [this issue P. 473].

5. Neuberger, *Nature*, **142**, 1024 (1938).

6. Bergmann and Niemann, "Annual Reviews of Biochemistry", 7, 11 (1938).

7. Cold Spring Harbor Symposium on Proteins, **6** (1938).

8. Wrinch, *Nature*, **138**, 241 (1936).

9. Langmuir and Wrinch, *Nature*, **143**, 49 (1939).

10. Wrinch, *Proc. Roy. Soc.*, A, **161**, 505 (1937)

11. Wrinch, *Phil. Mag.*, **23**, 313 (1938).

12. Wrinch, *Phil. Mag.*, **25**, 705 (1938).

13. Wrinch, *Proc. Roy. Soc.*, A, **160**, 81, 1937.

14. Haurowitz, *Z. physiol. Chem.*, **256**, 28 (1938).

15. Langmuir, Pilgrim Lecture, Royal Society, November 8, 1938.

16. *Nature* [forthcoming].

17. Languir and Wrinch, *Proc. Phys. Soc.* [in course of publication].

18. Wrinch, *Science*, **85**, 568 (1937); *Trans. Faraday Soc.*, **33**, 1368 (1937).

19. Crowfoot, *Nature*, **135**, 591 (1935).

20. Crowfoot, *Proc. Roy. Soc.*, A, **164**, 580 (1938).

21. Robertson, *Nature*, **143**, 75 (1939).

22. Bragg, *Nature*, **143**, 73 (1939).

23. Wrinch, International Congress of Physics, Chemistry and Biology, Paris, 1937.

24. Wrinch, Cold Spring Harbor Symposium on Proteins, **6** (1938).

25. Wrinch, *Nature*, **137**, 411 (1936).

Structure of Proteins*

J. D. Bernal

Editor's Note

Like several of the authoritative overviews in *Nature* at this time, this one is based on a talk at the Royal Institution in London. J. Desmond Bernal had worked there as a student of William Bragg, and by now he was the world's expert on the study of protein structure through X-ray crystallography. Here he presents the current state of play, in which impressive progress had been made despite the fact that proteins contain very complex structures built of hundreds or thousands of atoms. Bernal was still baffled by the apparent complexity of the process by which the chain-like molecules fold into a compact form (this is still not fully understood), but the important role of "hydrophobic" interactions between insoluble regions was already emerging.

THE structure of proteins is the major unsolved problem on the boundary of chemistry and biology today. We have not yet found the key to the problem, but in recent years a mass of new evidence and new lines of attack have enabled us to see it in a far more concrete and precise form, and to have some hope that we are near to solving it.

The problem of protein structure is twofold: the first is that of the form and properties of the protein molecule, and the second that of its internal structure. Owing to the extreme instability of the protein molecule, only the gentlest physical methods can be used; nevertheless three lines of attack—centrifugal, electrical, and X-ray—have already led to great success. The most fundamental has been the ultra-centrifuge, particularly the work of Svedberg and his school[1]. Protein molecules will sediment in sufficiently high centrifugal fields, and from their sedimentation constant it is possible to arrive at a fairly accurate value for the weight of the molecule, though there is some doubt as to whether what is measured is that of the complete protein molecule, because of its inevitable association with the solution in which it is suspended.

The most striking discovery of Svedberg was that the weights thus obtained seemed to fall into definite classes which were multiples of each other. This suggests very strongly that all proteins are built from some common unit. What that unit is is more difficult to determine. Originally taken as 35,000, proteins are now found of 17,000 and even of 10,000 molecular weight. Critical examination of the data, moreover, shows that there is no exact correspondence to certain weights, but rather a scattering of weights concentrated in certain regions. There is no doubt that in closely related proteins there are simple

---

* From a Friday evening discourse delivered at the Royal Institution on January 27.

蛋白质结构*

贝尔纳

编者按

就像同时期发表在《自然》上的其他几篇权威综述一样，这篇综述基于作者在伦敦皇家研究院的一次演讲。德斯蒙德·贝尔纳曾作为威廉·布拉格的学生在此工作，如今他已成为通过 X 射线晶体学研究蛋白质结构方面的世界知名专家。在此他介绍了这方面研究的现状，尽管蛋白质具有由成百上千个原子构成的非常复杂的结构，但是对其结构的研究仍取得了巨大的进展。链状分子折叠成一种紧密结构这一过程显而易见的复杂性难住了贝尔纳（这个问题到现在依然没有被彻底解决），不过他已经意识到了不溶性区域之间的"疏水性"相互作用在其中所起的重要作用。

时至今日，蛋白质的结构仍然是化学和生物学交叉研究领域尚未解决的一个主要问题。我们还未找到解决这个问题的关键点，但是最近几年大量新的证据和一系列新的研究使我们能够以更为具体和精确的形式去研究和看待它，同时也使我们有望解决这个问题了。

蛋白质结构的问题主要包括两个方面：首先是蛋白质分子的形式和性质，其次是它的内在结构。由于蛋白质分子极不稳定，因此只能采用最温和的物理方法进行研究；目前，三种研究方法——离心法、电化学法和 X 射线法，在对蛋白质的研究上已经取得了极大的成功。其中最为基础的研究工作是应用超速离心机获得的，特别是斯韦德贝里及其学院同事的研究工作 [1]。在足够高的离心场中蛋白质分子会发生沉降，依据它们的沉降系数就有可能推算出分子量相当精确的数值。由于蛋白质悬浮在溶液中时不可避免地与溶液结合，所以目前对于测定的对象是否是真正完整的蛋白质分子还存在一定的疑问。

斯韦德贝里最为著名的发现在于，这样得到的分子量似乎可以分成一些确定的等级，这些等级之间互为倍数。这一点非常强有力地说明，所有的蛋白质都是由一些共同的单元构建而成的。然而要确定这些单元是什么就更困难了。最初认为蛋白质单元的分子量是 35,000，不过现在又发现了分子量是 17,000 甚至 10,000 的蛋白质。此外，对数据的严格分析表明蛋白质的分子量并不是严格对应于某些确定的分子量，而是集中于某一区域内呈散点分布。毫无疑问的是，关系密切的蛋白质的分子量之

---

* 来自 1 月 27 日在英国皇家研究院星期五晚间演讲的内容。

relationships. Both haemoglobins and haemocyanins can, for example, be split reversibly into 2, 4, or 8 parts. But whether these relationships hold for all proteins is more doubtful; particularly as studies of viscosity have shown that proteins belonging to the same weight class such as insulin and gliadin may differ enormously in shape, the first being a flattened sphere and the other an elongated rod, and it is difficult to see what common physical structure they can have.

It has been claimed that this variation of weight of proteins of the same Svedberg class is to be expected on account of the known difference in amino-acid content. On the basis of existing analytical figures, it is as yet impossible to say whether this is or is not sufficient to account for the observed discrepancy. The figure of 288 amino-acid residues in a 35,000 molecular weight class protein cannot be taken at present as more than an inspired guess.

The second approach to protein structure comes from the electric properties of their molecules. The work of Cohn, Tiselius, and others gives us a picture of a protein molecule in solution as a particle covered with positive and negative charges due to the acid and basic nature of the amino-acid side-chains. The number of these groups depends on the condition of the medium. The protein molecule is, therefore, in its external relationship an essentially ionic structure, and probably carries with it an ionic atmosphere stretching out into the water in which it is dissolved.

The third line of attack is that of X-ray study. The beautiful crystalline forms exhibited by proteins, which have been known for at least a hundred years, were always a powerful attraction to X-ray crystallographers, but until the last five years the instability of protein crystals and their small size had defeated their efforts. Since 1934, however, it has been possible to examine a number of typical protein crystals[2]. This has been done by mounting them in small tubes in their mother liquor, for most, if not all, proteins suffer considerable breakdown if examined dry. From the beginning, the results of this examination revealed important new facts. In the first place, the pictures yielded by protein crystals were of exceptional perfection. They showed large unit cells with a great wealth of reflections (see accompanying illustration), and these reflections were found even at comparatively high angles corresponding to such low spacings as 2 A. This indicated that not only were the molecules of the proteins substantially identical in shape and size, but also that they had identical and regular internal structures right down to atomic dimensions.

From the size of the unit cells and their densities it was possible to compute the weight of matter in each repeat unit. This, however, does not immediately give the molecular weight, for two reasons. In the first place, the number of molecules in the cell is not known, and in the second it is difficult to determine how much of the cell weight is due to the protein and how much to the water in the cell. By measuring the loss of water on drying it is, however, possible to arrive at a figure for the—possibly fictitious—dry weight, which can be compared with those obtained by the centrifuge or by chemical means. Here it is found that there is excellent agreement on the assumption that there are only a few, 2, 4, or 8, such molecules per cell[3]. Thus the X-ray method furnishes an extremely accurate measure

间具有简单的关系。举例来讲，血红蛋白和血蓝蛋白都可以被可逆地分解为 2、4 或者 8 个部分。但是是否所有的蛋白质都有这种关系还存在着更多的质疑；特别是通过测定物质黏度的研究发现，属于相同分子量量级的蛋白质，例如胰岛素和麦醇溶蛋白，可能在形状上差别非常大；前者是一个扁平球体，而后者是伸长的棒状结构，很难看出它们具有什么共同的物理结构。

曾经有人认为属于同一个斯韦德贝里等级的蛋白质分子量的变化，可以被归结为已知的氨基酸成分之间的差异。然而，在现有分析数据的基础之上，至今都不可能说这是否就能充分地解释观察到的差异。因此，认为分子量在 35,000 这个等级的蛋白质是由 288 个氨基酸残基组成，这在目前只不过是个猜测而已。

研究蛋白质结构的第二类方法是基于这些分子的带电性质。科恩、蒂塞利乌斯以及其他一些研究者的工作给我们展示了由于氨基酸侧链具有酸性和碱性的特性，蛋白质分子作为携带正电荷和负电荷的粒子存在于溶液中的情形。这些带电基团的数目取决于溶液的条件。因而，从外表面观察，蛋白质分子是一个离子结构，进而溶解到水中时可能以一种离子态的形式分散到水中。

第三类方法是应用 X 射线衍射研究蛋白质的结构。至少在一百年前，人们就已经知道蛋白质能形成漂亮的晶体形式，这些晶体一直以来都深深地吸引着 X 射线晶体学家，但是直到最近五年，蛋白质晶体的不稳定性及其小的尺寸仍然一次次地挫伤研究者的信心。然而，自 1934 年起，研究一些典型的蛋白质晶体已经成为可能 [2]。这些研究取得成功的关键是将蛋白质置于装有它自身母液的小试管中，如在其干燥时进行检验，即使不是全部也是大多数的蛋白质晶体会出现相当程度的断裂。从一开始，这些研究结果就揭示了重要的新的事实。首先，蛋白质晶体产生的图像异常完美。通过大量的衍射，它们显示出了大的晶胞结构（参见随后的插图），即使在与 2 Å 这样的小间隔相对应的较大角度上也可以看到这些衍射。这些结果显示，蛋白质分子不仅在形状和大小上具有大体的一致性，而且在原子水平上也具有相同的、规则的内部结构。

从晶胞的尺寸和密度出发，我们有可能可以计算每一个重复单元中物质的质量。但是，这并不能直接给出分子的分子量，主要有两个原因。首先，晶胞中所包含的分子的数目是未知的；其次，很难测算出晶胞质量有多少是源于蛋白质，又有多少是源于晶胞中的水分子。然而，通过测量干燥过程中的失水量，就有可能可以得到（或者假定得到）其干重，这个数据可以与采用离心机或者化学方法得到的结果进行比较。研究结果与先前的假设有着近乎完美的一致，即每个晶胞中只包含少数这类分子，诸如 2、4、8 个 [3]。从而，X 射线方法提供了一种非常精确地测定蛋白质基本分子

for the basic molecular weight of proteins though it cannot determine its multiplicity. It may well be that many proteins are built of sub-units, which though of approximately equal weight are not chemically identical, and are more properly to be called molecular compounds than molecules.

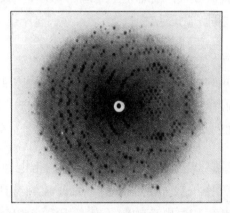

Horse methhaemoglobin, *b* axis, oscillation photograph, the ring coming through central spot showing (*hko*) zone.

The most striking feature of the cell measurements of crystalline proteins is the considerable change which occurs on drying, where in many cases shrinkage of nearly 50 percent occurs. But what is more remarkable is that the shrinkage is often confined to one or two directions in the cell. There are two possible explanations of this fact. First, that the molecules of the protein are linked together in extremely loose aggregates which collapse on the removal of water, leaving the skeleton of molecules in a now more closely packed array. The other explanation is that the molecules are held apart by their ionic atmospheres due to the charges on them, and contain, therefore, sheets of free water. It may be that both explanations hold, one more markedly for some crystals and the other for others. Thus for haemoglobin, where there is a marked shrinkage from 55 A. to 38 A. in one direction, it is difficult to imagine an ionic atmosphere, as the crystals are practically saltfree and at an isoelectric point of *p*H 6.8. On the other hand, it is difficult to explain the remarkable properties of tobacco mosaic virus on any other hypothesis. This virus has long thin particles which have a tendency to set equidistant and parallel, even down to concentrations of 13 percent and probably down to 1.5 percent. It is difficult to imagine what force other than that of ionic atmospheres can preserve this regularity. Quite recently we have shown that the equilibrium distance between the particles depends on *p*H and salt concentrations, varying, for example, between 320 A. at *p*H 7, and 206 at *p*H 3.4. It should be possible to find a quantitative theory to explain these changes and, indeed, the beginning of such a theory has been made by Langmuir[4] and Levine[5].

The general picture of the external character of protein molecules is beginning to be definite. The molecules are spheroidal bodies of dimensions varying from 30 A. to 100 A., are covered with hydrophil groups bearing charges of both signs, and probably carry with them in solution an atmosphere of ions.

量的方法，尽管该方法还不能检测它们的多样性。许多蛋白质都是由分子量近似相等但化学结构并不相同的亚基构成的，因此，更准确的是称其为分子复合物而不是分子。

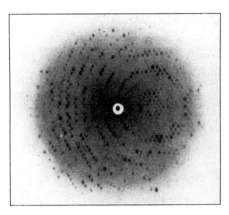

马的甲基血红蛋白，*b*轴，振动晶体图相，穿过中心点的圆环显示（*hko*）区。

结晶蛋白最为明显的一个特点是在逐渐干燥的过程中晶胞的大小会发生非常显著的变化。很多情况下，尺寸收缩可以达到 50% 以上。但是，更值得关注的是，这种收缩通常是被限定在晶胞的某一个或者某两个方向上。对于这种现象有两种可能的解释。其一，组成蛋白质的各个分子连接在一起形成了非常松散的聚集体，当除去水分子时此聚集体便会瓦解，只留下分子骨架，从而形成我们看到的这种更为紧密的堆积状态。另一种解释是将其原因归结为蛋白质分子带电而形成的离子氛对分子的隔离作用，以及由此而包含的大片的自由水。可能这两种解释都成立，其中一种解释非常适用于某些晶体，另一种解释可能适用于另外一些晶体。例如血红蛋白可以在一个方向上从 55 Å 显著地收缩到 38 Å，当晶体处在等电点的 pH 值 6.8 时，晶体几乎不带电，所以很难想象这种收缩是因为离子氛的存在而导致的。另一方面，任何其他假设都很难解释烟草花叶病毒的显著特征。这种病毒具有细长的颗粒，倾向于等距、平行地排列，即使是在浓度降低到 13%，甚至 1.5% 的情况下。我们很难想象除了离子氛之外什么力量能够保持这种规律性。就在最近，我们发现粒子间的平衡距离取决于 pH 值和盐浓度。例如，它可以在 320 Å (pH 为 7 时) 和 206 Å (pH 为 3.4 时)之间变动。应当有可能可以找到一种定量的理论来解释这些变化。事实上，朗缪尔 [4] 和莱文 [5] 已经给这样一种理论奠定了基础。

现在人们已经开始确定蛋白质分子外部特征的大致图像。这类分子的形状类似于球体，尺寸从 30 Å~100 Å 不等。分子的表面被亲水性基团覆盖，带有正负两种类型的电荷，并依靠它们溶解在溶液中形成离子氛。

The problem of the internal structure of protein molecules is one of enormously greater difficulty. Any effective picture of protein structure must provide at the same time for the common character of all proteins as exemplified by their many chemical and physical similarities, and for the highly specific nature of each protein type. It is reasonable to believe, though impossible to prove, that the first of these depends on some common arrangement of the amino-acids. The first hint of this comes from a study of the fibrous proteins. Here, owing to their technical importance and their much greater ease of handling, a very considerable advance has been made, due largely to the well-planned and persistent researches of Astbury[6].

The fully extended β-keratin fibre has a strong repeat at every 3.5 A., corresponding to the length of one amino-acid residue, and two other repeats at 10 A. and 4.5 A. respectively, corresponding to periodicities at right angles to the chain length. One of the most significant of Astbury's discoveries was that these two periodicities are also at right angles to each other. In other words, that the repeat unit of the fully extended protein has distinct characters in three dimensions. The significance of these repeats seems also beyond doubt, the 10 A. spacing corresponding to the length of the side-chains of the amino-acid residues, and the 4.5 A. to the so-called backbone spacing between the main chains themselves, which are held together by the CO and NH groups. In the shorter α form, the 10 A. spacing remains, while the 4.5 A. disappears, and this indicates that the chains are folded, but folded definitely in a plane and not as in rubber in two dimensions. What the precise mechanism of this folding is remains obscure, but it is of fundamental importance because it is probably very closely associated with the kind of folding that occurs in the corpuscular proteins.

The significance of these findings for the general protein problem is, first, that all soluble proteins hitherto examined give on denaturation a fibrous material which can be oriented and seems identical in basic structure with β-keratin, and secondly, that the actual change which a crystalline protein undergoes on denaturation must be very slight. This is shown by the close resemblance of powder photographs of crystalline proteins before and after denaturation, and also by the observations of Astbury, Dickinson, and Bailey[7] that a single crystal of excelsin gives on partial denaturation fibres orientated in the direction of the crystal axes. Further, recent work by Perutz[8] has shown by an entirely different method, namely, the study of the absorption spectra in different directions in crystals, that the position of the prosthetic group in haemoglobin is not altered by denaturation.

The evidence that the X-ray study of the crystalline proteins themselves provides for the elucidation of their structure is abundant, but it is extremely difficult to interpret. Photographs of crystalline proteins (see illustration) show hundreds of spots and marked differences of intensities stretching right out to reflections corresponding to interatomic distances. Unfortunately, however, direct analysis of these photographs is rendered impossible by the fact that we can never know the phases of the reflections corresponding to the different spots. The ambiguity introduced in this way can only be removed by some physical artifice, such as the introduction of a heavy atom, or the observation of intensity

解析蛋白质分子内部结构面临更大的困难。任何蛋白质结构的有效图像都必须同时提供对于所有蛋白质而言的共同特性，因为已确证它们不仅具有诸化学的和物理学的相似性，每一类蛋白质还具有高度的特异性。尽管无法证实，我们仍有理由相信这其中首要一点依赖于氨基酸的某种基本排布。关于这一点的最初线索来自对纤维蛋白的研究。由于这类蛋白在技术上的重要性以及它们具有的便于操作和处理的特性，此方面的研究已经取得了相当大的进展，这主要归功于阿斯特伯里等人的精心设计以及坚持不懈的研究 [6]。

全伸展的 β–角蛋白纤维每隔 3.5 Å 就会有一次重复，这个长度与一个氨基酸残基的长度相一致；另外两种重复的间隔分别是 10 Å 和 4.5 Å，与链长度的直角周期相一致。阿斯特伯里最为重大的发现之一是这两种周期性重复结构之间是彼此垂直的。换句话说，整个完全伸展蛋白的重复单元在三维空间上具有明显的特征。这些重复单元的重要性是毋庸置疑的，其中 10 Å 的间距对应于氨基酸残基侧链的长度；4.5 Å 的间距对应于主链之间我们称之为骨架的空间距离，它们彼此之间通过羧基基团和氨基基团连在一起。在较短的 α 类型结构中，10 Å 的间距仍然存在，4.5 Å 的间距却消失了，这充分说明这种链是折叠的，但折叠肯定发生在同一个平面上，而不像在橡胶中观察到的折叠发生在两个维度上。目前尚不清楚这种折叠的确切机制，但它具有根本的重要性，因为它可能与出现在血球蛋白中的折叠类型是紧密相关的。

这些发现对于蛋白质的基本问题的意义在于，第一，迄今为止观测过的所有可溶性蛋白质在变性后都会得到一种纤维状物质。这些物质能够被定向且似乎与 β–角蛋白的基本结构是一样的。第二，结晶蛋白在变性过程中经历的实际变化一定是非常微小的。结晶蛋白变性前后的粉末照片极度类同证实了这一点，阿斯特伯里、迪金森和贝利等人 [7] 对巴西果蛋白单晶的观测结果也证实了这一点，他们发现部分变性的巴西果蛋白单晶形成的纤维的取向沿着晶体轴的方向。除此之外，佩鲁茨 [8] 近来的研究工作通过一种完全不同的方法，即研究晶体在不同方向上的吸收光谱，证实了血红蛋白中辅基的位置不会因蛋白质的变性而改变。

对结晶蛋白本身进行的 X 射线研究为阐明它们的结构所提供的证据是很丰富的，但数据很难解释。结晶蛋白的图像（参见插图）展示了数百个点并且强度明显不同，这种不同直接反映蛋白质内部原子间的距离。然而，不幸的是，由于事实上我们无法确定不同点所对应的衍射相位，因此对这些图像的直接分析是不可能的。因此而引入的不确定性只能通过一些物理技巧来消除，例如引入重原子，或者通过观测脱水过程中强度的变化，迄今为止后者尚未在实践中得到实施。对于所有被检测过的

changes on dehydration, which have not hitherto been carried out in practice. For all but one of the proteins examined, the task is made still more difficult by the presence of more than one Svedberg unit in the cell.

The resulting X-ray pattern is such cases depends both on the position of the molecules in the cell and on their internal structure. Fortunately the one exception, insulin, has only one molecule per cell. An X-ray study of dry insulin crystals has been made by D. Crowfoot[9] and the intensities of the reflecting planes determined. It is to be regretted that it was not possible to examine the crystals wet because in the dry crystals no planes of spacing less than 8 A. are observed, and consequently no evidence exists as to the fine structure of the molecule. The evidence on the coarse structure is, however, striking enough; it can most easily be seen from a study of the Patterson projections and Patterson-Harker sections which Miss Crowfoot has computed for insulin.

It should be explained that these projections do not represent the distribution of the density of scattering matter in the cell, but merely the product of the densities at points separated by a constant vector. It is not surprising, therefore, that the interpretation of these diagrams has already given rise to acute controversy. It seems natural to assume, though it is in fact an arbitrary simplification, that the peaks in the Patterson projection correspond to distances between a small number of high concentrations of scattering matter. Attempts have been made to reduce the analysis of the pattern to that of finding a pattern of points in space that will give maxima at the places observed. All these attempts have in fact failed. It is easy to find a large number of point patterns which will give the correct basal plane projection, but none of these will at the same time fit the sections. The claims to have done so[10] are based on the arbitrary selection of a certain number of vectors, and the agreement disappears if all the vectors are taken[11]. The failure to find a point solution suggests that we are dealing with groups of size of the same order as their distance apart, and that the appearance of the enhancement of reflections at spacings 10 A. and 4.5 A. corresponds to an arrangement of side-chains or backbone spacings similar to those occurring in the fibrous proteins.

On the basis of present X-ray knowledge, it is clearly impossible to arrive at any detailed picture of protein structure. Indeed, up to now the chief value of the X-ray method has been to disprove such hypothetical structures as have been put forward. It is, however, possible to formulate in a broad way some of the possible modes of arrangement and to suggest working hypotheses as a guide to future work.

The formal problems of structure of the protein molecule are:

(a) What is the nature of the link between amino-acid residues?

(b) Does the linkage run through the whole of the molecule or only through parts of it? In other words, is the protein one unit held together by primary valence forces, or is it made of sub-units held together in some different way?

704

蛋白质（除了一个例外）而言，晶胞中存在不止一个斯韦德贝里单元，这使得研究工作依然是非常困难的。

在这些例子中，最终的 X 射线图案取决于晶胞中分子的位置以及它们的内在结构。幸好作为一个例外，胰岛素在每个晶胞中只有一个分子。克劳福特[9] 等人对干燥的胰岛素晶体进行了 X 射线研究，测定了反射平面的强度分布情况。遗憾的是，无法对带水晶体进行观测研究，因为在干燥的晶体中没有观测到任何间距小于 8 Å 的平面，因此无法获得关于分子的精细结构的证据。不过，关于粗略结构的证据就足以令人震惊了；这一点可以非常容易地从克劳福特女士对胰岛素计算得到的帕特森投影和帕特森 – 哈克截面中看到。

这里需要解释的是，这些投影并不代表晶胞中散射物质的密度分布情况，而仅仅是由一些被某个常向量分隔的点上的密度形成的。因此，丝毫不用感到奇怪的是，对于这些图像的解释早已经引发了激烈的争论。虽然事实上这是一种主观的简化，但我们会很自然地假设，帕特森投影上的峰对应于少数高浓度散射点之间的距离。目前已经有一些研究工作在试图简化对图案的分析，以发现空间中那些能够给出观测位置上最大值点的特征。事实上所有这些努力最终都宣告失败。虽然可以很容易地找到大量的点特征，它们可以得到正确的基本平面投影，但是它们其中的任何一个都不能同时满足区域的分布。导致这种情况的原因[10] 在于对一定数量的向量的主观选定，如果选择所有的向量，则一致性会消失[11]。无法找到解决问题的关键点的事实说明，我们处理的分子基团的大小与它们之间的间距处在同一个数量级上，因此在间距为 10 Å 和 4.5 Å 时反射强度的增加对应的是侧链或者主链间距的排布，就像在纤维蛋白中存在的情况那样。

基于当前的 X 射线知识，非常清楚的是，我们不可能获得关于蛋白质结构的任何细节图像信息。事实上，到目前为止，X 射线方法的主要价值是用于反驳那些已被提出的假设结构。然而，这种方法有可能在更广的范围内阐述可能的排列方式，以及提出有效的假设来指引未来的研究工作。

关于蛋白质分子的结构，真正重要的问题主要有：

(*a*) 氨基酸残基之间的连接的本质是什么？

(*b*) 这种连接贯穿于整个分子中还是仅仅存在于分子的某些部分？换句话说，是蛋白质作为一个整体依靠基本的价键力量结合在一起，抑或是它由亚单位构成，以另外的某种方式结合在一起？

(c) If such sub-units exist, what is the nature of the link between them?

As to the first question, the difficulty of accounting for the structure of spherical molecules out of a linear peptide chain has led to the idea of an alternate mode of linkage in which each amino-acid residue can be linked to four others and not merely to two. While this hypothesis at first sight has a theoretical attractiveness, it still lacks any chemical support, and has been subjected to serious criticism on chemical grounds[12]. There are enough unknown factors in protein structure already without employing doubtful chemical assumptions.

The question of the unitary nature of the protein molecule arises whether the inter-residue linkage is peptide or not. With a multiple link it is easy to construct models either of the cage or solid type, but as has already been pointed out, it is difficult to reconcile such continuous structures with the definite 10 A. discontinuities revealed by the X-ray analysis. With a peptide chain, however, the unitary solution becomes more difficult. It is difficult to imagine any kind of fold or coil by which a single chain can occupy the observable space and at the same time not be so intricate that its formation by any natural process would be enormously improbable. There is, however, much evidence that at least the larger protein molecules are not unitary in structure. In the first place, some of them can be split in solution down to particles of molecular weight of the order of 10,000, and this is probably not the lower limit, as smaller particles are difficult to isolate or measure.

Two lines of X-ray evidence also point to sub-units, the high symmetry of protein crystals and the 10 A. repetitions. The symmetry of protein crystals is much higher than would be expected statistically from compounds of such great complexity. This would seem to indicate that each molecule is built of sub-units themselves unsymmetrical but arranged in a symmetrical way. The size of the sub-units must lie between that of the smallest protein observed, that is, one with a molecular weight 9,000, and that of a single amino-acid of molecular weight averaging 120, that is, it must contain some sub-multiple of probably 72 amino-acid residues. The uncertainty arises from the fact that it is not necessary for all the sub-units to be the same, though some must be to account for the symmetry. The presence of trigonal symmetry suggests that the asymmetrical unit must be a third or less of this number, that is, it must contain 24, 12, or 8 amino-acid residues.

The question of the structure of the sub-unit would seem to raise at first sight the same difficulties as the structure of the molecule itself. Actually, however, with the smaller number of residues the difficulties and the improbabilities of the coiling become much less, particularly if we postulate—which is not unreasonable—that the sub-units are closed peptide rings. Such rings would necessarily curl up owing to the mutual attraction of the positive and negative charged amino and keto groups in the chain, and models of such coiled chains can be constructed which preserve the distances between such groups that have been found in other compounds. The difficulty is, however, that there is a very large number of such possible models, and as yet nothing to choose between them. The method of folding in the chains in the sub-units may well be similar to that of the contracted form (*c*) 如果这些亚单位存在的话，那么存在于它们之间的连接的本质又是什么？

对于第一个问题，解释由线性多肽链形成球形分子结构所面临的困难引出了交替式连接的概念，其中每一个氨基酸残基都可与其他四个或者不少于两个氨基酸残基连接起来。尽管这种假设初看上去具有理论上的吸引力，但是它仍然缺乏任何化学方面证据的支持，在化学领域 [12] 备受争议。即便不引入不确定的化学假设，在蛋白质结构研究方面也已经存在大量的未知因素了。

关于蛋白质分子整体本质的问题在于其残基之间的连接是否是肽键。通过多重连接的方式，很容易构建出笼状或者立体的模型，但是就像已指出的那样，使这些连续的结构符合于 X 射线分析揭示的确定的 10 Å 长度的不连续性是很困难的。就肽链而言，整体的解决方案变得更为困难。很难想象一条单链能够通过任何类型的折叠或者卷曲而占据整个可观测空间，另外不难理解的是，它也不可能通过自然过程而自发形成。然而，有相当多的证据表明，至少大的蛋白质分子在结构上不是单一的。首先，有一些大的蛋白质在溶液中分解成分子量在 10,000 这个量级上的粒子，而且，这可能还不是下限，只不过是由于更小的粒子难于被分离和测量。

来自 X 射线研究的两条证据同样证明了亚单位结构的存在，即蛋白质晶体的高度对称性以及 10 Å 的重复性。蛋白质晶体的对称性比我们根据同等复杂的化合物而统计推断出的要高得多。这似乎预示着每个分子都是由本身并不对称但却以一种对称的方式进行排布的亚单位构成的。亚单位的大小必然介于所观测到的分子量为 9,000 的最小蛋白质和平均分子量为 120 的单个氨基酸之间，也就是说，它必须包含大约 72 个氨基酸残基。其中的不确定性来自这样一个事实，即所有的亚单位不必完全一样，尽管其中的一些必须保证分子的对称性。三角对称的存在说明非对称单元中的氨基酸残基数必须是这个数的 1/3 或者更小，换句话说，它必须包含 24、12 或者 8 个氨基酸残基。

初看上去亚单位结构的问题与对分子本身结构的研究似乎有相同的困难。事实上，伴随着氨基酸残基数目的减少，其困难和卷曲的不可能性也越来越小，特别是如果我们假定——这不是不切实际的——亚单元是封闭的多肽环。由于链中氨基团和酮基基团之间正负电性的相互吸引，这些环必然会卷曲，并且这类卷曲链模型的构建可以基于从其他化合物中已获得的相似化学基团之间的距离。然而，困难在于，这类可能的模型非常多，无法在它们中间做出选择。亚单位中链的折叠方式可能会非常类似于上面提到的纤维蛋白的收缩模式，事实上，对于这些蛋白质的更细致的

of the fibrous proteins mentioned above, and indeed a more detailed study of these may give us a clue to the whole arrangement.

The postulation of sub-units, however, raises further questions in the molecular structure, for they must be bound together sufficiently tightly to hold the molecule together in aqueous and ionic solutions. For this purpose a limited number of bonds are available. Ionic bonds are plainly out of the question, as they would certainly hydrate. There remains the possibility of amino-carboxy links between the ends of side-chains, but this is somewhat liable to the same criticism. It seems more probable that the links are either or both of two kinds, S–S linkages and association of hydrophobe groups. With the small number of sub-units, the number of S–S linkages needed to hold the molecule together is for all proteins hitherto examined sufficiently provided for by the amount of sulphur present, and the extreme changes in activity that proteins undergo when S–S bonds are broken indicate that they may have this fundamental part to play. Whether this is so or not, however, the behaviour of the hydrophobe groups of the protein must be such as to hold it together. As Danielli[13] and Langmuir[14] have pointed out, on the basis of surface-film work, the protein molecule in solution must have its hydrophobe groups out of contact with water, that is, in contact with each other, whereas on the surface the molecule is broken up into a film of 10 A. thickness in which the hydrophobe groups are driven out of contact with the water. In this way a force of association is provided which is not so much that of attraction between hydrophobe groups, which is always weak, but that of repulsion of the groups out of the water medium.

Langmuir has used this picture as a justification of the cyclol cage hypothesis, but it is strictly quite independent of it, and the model outlined above has the advantage of accounting very satisfactorily for the phenomenon of denaturation, particularly on a surface. Once the sub-units come to the surface, their rings are brought into a plane and different rings can interact according to the familiar ring-chain polymerization process, which will result in the formation of the fibres which Astbury has shown to exist in such films. This polymerization process takes measurable time as Danielli's work has shown.

The picture thus presented is far from being a finished or even a satisfactory one. The crucial fact that requires elucidation is the precise mode of folding or coiling of the peptide chains, and for this we may have to wait for some considerable time, until the technique of X-ray and other methods have been advanced much further than at present. The problem of the protein structure is now a definite and not unattainable goal, but for success it requires a degree of collaboration between research workers which has not yet been reached. Most of the work on proteins at present is uncoordinated; different workers examine different proteins by different techniques, whereas a concentrated and planned attack would probably save much effort which is now wasted, and lead to an immediate clarifying of the problem.

(**143**, 663-667; 1939)

研究可以在整体排列上给我们一些提示。

　　然而，亚单位假设在分子结构方面提出了另外一些问题，因为如果存在亚单位，它们必然要非常牢固地结合在一起，从而保证分子在水中和离子溶液中聚集在一起。能用于达到这个目的的键是非常有限的。离子键显然要排除在外，因为它们必然与水结合。剩下的可能性是侧链末端的氨基和羧基之间的作用，但是它们在一定程度上也有与水结合的倾向。看上去更有可能的是 S–S 连接和疏水性基团之间的相互作用中的一种或两种。对迄今为止观测过的仅由少数几个亚单位组成的所有蛋白质而言，分子中存在的 S 的数量足以提供使整个分子结合在一起所需的 S–S 键，而且，S–S 键断裂时蛋白质活性所发生的巨大变化显示出这些二硫键有一些基本的功能。然而，无论它是否如此，蛋白质疏水性基团的特性决定了它们必然要结合在一起。就像丹尼利[13]和朗缪尔[14]基于表面膜工作指出的那样，溶液中的蛋白质分子必然会使其疏水性基团避免与水分子接触，而处于表面的分子则断裂形成厚度为 10 Å 的薄膜，其中的疏水性基团并不与水接触。通过这样一种方式，就提供了一种结合力，与其说这是疏水基团之间的弱的吸引力，不如说是使基团与水介质分离开来的一种排斥力。

　　朗缪尔曾经用这个图像来证明环醇笼状假设，但是从严格意义上来讲它是独立于这个理论的，上面所提到的模型的优势在于它可以非常好地解释变性现象，特别是在表面上发生的蛋白质变性。一旦亚单位到达分子的表面，它们的环就会破裂成平面，不同的环之间依据普通的环–链聚合过程发生相互作用，这会导致纤维的形成，就像阿斯特伯里发现纤维存在于这类薄膜中那样。这个聚合过程持续的时间是可测的，正如丹尼利的研究工作所展示的那样。

　　目前所得到的图像远不能说是最终的或者是令人满意的。需要阐明的至关重要的问题是肽链折叠或者卷曲的精细模型，为此我们可能不得不等待相当长的时间，直到 X 射线技术以及其他方法获得比现在更进一步的发展。蛋白质结构的问题目前是一个明确而且并非不可实现的目标，但是为了获得成功，它需要研究者之间某种程度的协作，然而现在的协作远未达到所需要的程度。目前绝大多数的蛋白质研究工作之间都是不协调的，不同的研究者采用不同的技术观测不同的蛋白质，也许集中而又有计划的研究可能会节省许多现在这种状态下浪费掉的时间和精力，从而有助于对问题的直接阐明。

（刘振明 翻译；周筠梅 审稿）

References:

1. Svedberg, T., *Proc. Roy. Soc.*, A, **170**, 40 (1939).

2. Bernal, J. D., and Crowfoot, D., *Nature*, **133**, 794 (1934).

3. Bernal, J. D., *et al.*, *Nature*, **141**, 521 (1938).

4. Langmuir, I., *J. Chem. Phys.*, **6**, 873 (1938).

5. Levine, S., *Proc. Roy. Soc.*, A, **170**, 145 (1939).

6. Astbury, W. T., "Fundamentals of Fibre Structure", *Phil. Trans.*, **232**, 333 (1933).

7. *Biochem. J.*, **29**, 2351(1935).

8. Personal communication.

9. *Proc. Roy. Soc.*, A, **164**, 580 (1938).

10. *J. Amer. Chem. Soc.*, **60**, 2247 (1938).

11. Bernal, J. D., *Nature*, **143**, 74 (1939).

12. Neuberger, A., *Proc. Roy. Soc.*, A, **170**, 64 (1939).

13. Danielli, J. F., *Proc. Roy. Soc.*, A, **170**, 73 (1939).

14. Langmuir, I., and Wrinch, D., *Nature*, **143**, 49 (1939).

Use of Isotopes in Biology

Editor's Note

Here *Nature* reports on developments in the application of radioactive elements and heavy non-radioactive isotopes as tracers in biology and medicine. The possibility of identifying single-atom decays made such methods very sensitive. At a recent meeting of the Chemical Society and the Physiology Society, researchers described the use of radioactive phosphorus in tracing the uptake of phosphorus by bones in the human body. Similar studies were probing the chemical activity of organs, and providing estimates of the timescales for protein formation. Other experiments were probing both human and plant physiology with tracers such as heavy oxygen or radioactive sodium. These experiments were generally safe, as the radioactivity involved was small compared to that naturally produced within the body.

THE discovery both of artificial radioactive elements and heavy non-radioactive isotopes, together with methods of concentration of the latter, has opened up new methods of examining the reactions and movements of substances in the body.

The isotopic indicator most frequently used in biological research is radioactive phosphorus (^{32}P) which may be prepared by the bombardment of carbon disulphide by neutrons from a radium–beryllium source. The sulphur atom takes up a neutron and gives out a proton, forming heavy, unstable, radioactive phosphorus (half-life period 14 days):

$$^{32}_{16}S + ^{1}_{0}n \rightarrow ^{32}_{15}P + ^{1}_{1}H.$$

The phosphorus is then oxidized to phosphate. Stronger preparations may be obtained by bombarding red phosphorus with deuterium ions using a cyclotron:

$$^{31}_{15}S + ^{2}_{1}D \rightarrow ^{32}_{15}P + ^{1}_{1}H.$$

For use, a small amount of the labelled sodium phosphate is added to ordinary sodium phosphate solution, and by this means the path of the phosphorus in the body can be traced. Estimations are carried out by observing the decay, at a given time, using a Geiger counter, and comparing directly with the decay of a similar standard at the same time, thus avoiding corrections for the rate of decay due to lapse of time. Thus if it is desired to determine the radioactive phosphorus content of the bone of an animal to which a labelled phosphate solution has been administered (and hence determine any exchange in the phosphate of the bone), a known weight of bone ash from the animal is placed under the Geiger counter and the strength observed. This is compared directly with the same weight of calcium phosphate, precipitated together with the labelled phosphate from a known weight of solution. This method has placed an extremely delicate method of estimation in the hands of the experimenter and has the great advantage that it is not necessary to purify the substance carefully from non-radioactive elements.

同位素在生物学中的应用

编者按

《自然》的这篇文章报道了在生物学和医学领域应用放射性元素和非放射性元素的重同位素作为示踪剂的进展。对单原子衰变的识别能力使该类方法具有很高的灵敏性。在最近召开的化学学会与生理学学会上,研究者们描述了在追踪人体内骨骼对磷的吸收时放射性磷的应用。类似的研究还有探测器官的化学放射性,评估蛋白质形成的时间。还有一些实验是利用诸如重氧或放射性钠等作为示踪剂研究人体生理及植物生理。与机体内自然产生的放射性相比,实验引入的放射性是较小的,因此这些实验基本上是安全的。

人工放射性同位素和非放射性元素的重同位素的发现,以及后者的浓缩方法的出现,为检测体内物质的反应和运动开辟了新途径。

在生物学研究中最常使用的同位素示踪剂是放射性磷(^{32}P),用来自镭–铍放射源的中子轰击二硫化碳可以得到。硫原子吸收一个中子并释放出一个质子,形成重的、不稳定的、放射性的磷(半衰期为 14 天):

$$_{16}^{32}S + _{0}^{1}n \rightarrow _{15}^{32}P + _{1}^{1}H$$

接着,磷被氧化成磷酸盐。借助于回旋加速器,用氘离子轰击红磷可以得到更强的制剂:

$$_{15}^{31}S + _{1}^{2}D \rightarrow _{15}^{32}P + _{1}^{1}H$$

使用时,将少量同位素标记的磷酸钠加入普通的磷酸钠溶液中,通过这种方法就能够追踪磷元素在体内的路径。在给定的时刻,使用盖革计数器,通过观测衰变进行估计,并且同时与类似标准物的衰变直接进行比较,这样可以不用再修正因时间流逝而产生的衰变速率。因此,如果想检测已被施用同位素标记的磷酸盐溶液的动物的骨骼中放射性磷的含量(并由此确定骨骼中任何磷酸盐交换),就要将已知重量的该动物的骨灰置于盖革计数器之下并观测其强度。将它立即与同等质量的磷酸钙进行比较,后者是从已知重量的溶液中与同位素标记的磷酸盐共沉淀得到的。这给实验者提供了一种极为巧妙的估计方法,其最大优势在于不必从非放射性元素中提纯物质。

Of the non-radioactive elements, deuterium (^{2}H) has been isolated in a state of purity; while heavy oxygen (^{18}O) and heavy nitrogen (^{15}N) have been concentrated sufficiently to make their use as indicators possible.

Heavy hydrogen and oxygen may be accurately estimated by conversion to water, and, after careful purification, determination of the density of the latter. Heavy nitrogen is determined by the mass-spectrograph. This method may also be used for the determination of heavy oxygen: it has the advantage that very careful purification is unnecessary.

It may be objected that the use of isotopes as indicators in the living body may disturb the normal conditions. Actually, the proportion of radioactive phosphorus which it is necessary to use is extremely small, and the radiation from it may be comparable with that from the potassium which is already present in the body, and on decaying it is converted to a sulphur atom. Heavy non-radioactive isotopes are also already present in small proportions in the body, and the use of a slightly greater concentration as an indicator should not be objectionable. In the case of deuterium, it may be best to avoid the use of concentrated preparations.

A joint meeting of the Chemical Society and the Physiological Society on February 9 last took the form of a discussion on the use of isotopes in biology, which was opened by Prof. G. von Hevesy, of the Institute of Theoretical Physics, Copenhagen.

Among examples of the use of radioactive phosphorus, Prof. Hevesy described the results of experiments in which labelled sodium phosphate was injected intravenously; blood samples of known volume taken at intervals showed a very rapid decrease at first, followed by a slow decrease, in the labelled phosphate. This was found to be taken up by the bones and organs of the body, but chiefly by the former. It can be shown, by shaking up solid calcium phosphate with labelled phosphate solution, that there is a rapid exchange between the phosphate ions and the phosphate in the solid, the rate depending on the concentrations of the ions and the surface area of the solid. In the case of the body, the weight of the solid phosphate in the bones is very great compared with the inorganic phosphate in the blood, and most of the labelled phosphate will exchange with ordinary phosphate on the surface of the bones. As phosphate is lost by excretion through the kidneys and bowels, the uptake of labelled phosphate by the bones will cease, and finally, when the labelled phosphate in the blood is nearly removed, there will be a slow exchange of labelled phosphate back from the bones to the blood. Thus it may be possible to detect labelled phosphate several weeks after administration. Examination of the separate organs of the body enables the passage of the phosphate to be followed, and similarly examination of sections of bone shows that the rate of exchange of phosphate varies with different bones. In the case of growing bone such as the teeth of rats, the greater activity is found in the tooth formed after the addition of the labelled phosphate; the part formed before the addition, however, also shows some activity, indicating that exchange of phosphate takes place at the same time as new growth.

714

在非放射性元素中,已经分离出了纯净状态的氘 (^{2}H);而重氧 (^{18}O) 和重氮 (^{15}N) 则已被浓缩到足以作为示踪剂的浓度。

将重氢和氧转化成水,经过仔细提纯后,测定后者密度,以对其进行精确的估计。重氮是用质谱仪来测定的。这种方法也可以用来测定重氧:它的优势在于不需要进行极为细致的纯化。

也许有反对意见认为,将同位素作为示踪剂引入活体可能会扰乱正常的生理状态。实际上,必须使用的放射性磷所占比例极小,其放射性可能与体内存在的钾的放射性相当,而且在衰变过程中它会转变成为硫原子。另外,体内原本就存在少量的非放射性元素的重同位素,因而使用浓度稍高一点的重同位素作为示踪剂应当不会引起排斥。对于氘来说,最好还是避免使用浓缩制剂。

2 月 9 日,哥本哈根理论物理研究所的赫维西教授组织召开了化学学会与生理学学会的联席会议,会议最后讨论了同位素在生物学中的应用。

在放射性磷的应用实例中,赫维西教授描述了静脉注射标记磷酸钠的实验结果;间隔性采集的定量血液样品显示,标记磷酸钠先快速减少,接着缓慢减少。机体骨骼及器官均有摄取,但以前者为主。将固态磷酸钙与同位素标记的磷酸盐溶液摇匀,可以看出,磷酸根离子与固态中的磷酸盐之间存在快速交换,交换速率取决于离子的浓度以及固体的表面积。在体内,骨骼中固态磷酸盐的量远大于血液中无机磷酸盐的量,因此大部分同位素标记的磷酸盐与骨骼表面的普通磷酸盐进行交换。随着磷酸盐经由肾和肠被排出体外,骨骼对同位素标记的磷酸盐的吸收也会停止,最后,当血液中同位素标记的磷酸盐被基本清除后,同位素标记的磷酸盐缓慢交换而后从骨骼返回血液中。因此有可能在注药几周后也能检测到同位素标记的磷酸盐。检测机体内的各个器官可以追踪磷酸盐的通路,而对各部分骨骼进行的类似检测表明:不同骨骼中磷酸盐的交换速率也不同。对于诸如鼠齿那样的处在生长中的骨骼,在添加同位素标记的磷酸盐后长出的牙齿中发现了更强的放射性;然而,在添加之前就已经形成的那部分牙齿也表现出一些放射性,这说明伴随着新的生长过程同时也发生着磷酸盐的交换。

Phosphorus atoms in an organic molecule such as lecithin do not exchange with labelled sodium phosphate. Thus, if active lecithin is found, it indicates that the lecithin molecule has been synthesized after the administration of the labelled sodium phosphate (in the presence of suitable enzymes) and thus a distinction can be made between old and new molecules. This method may be used, for example, to find the place of formation of the phosphatides in the yolk of hens' eggs. By killing a hen five hours after an injection of labelled phosphate, it was found that the liver and plasma phosphatide were very active (the former more than the latter) compared with that in the ovary and yolk; thus the phosphatide molecules formed during the last five hours do not originate in the ovary, but are taken by the ovary from the plasma and used in the building up of the yolk. It was also found that after the egg had left the ovary no labelled phosphatide was formed in it. This was shown by examination of eggs laid less than 20 hours after administration of labelled phosphate (the egg, after leaving the ovary, remains in the oviduct about 20 hours).

The use of radioactive sodium phosphate have also been of great help in a study of the formation of (goat) milk. Samples of blood and milk taken at intervals after administration of labelled phosphate were examined for the activity of the phosphate in the blood and the various phosphorus compounds in the milk. It was found that after three to four hours the milk inorganic phosphate was replaced by the active phosphate of the plasma; while if heavy water was injected at the same time as the active phosphate, after one hour water samples prepared from blood and milk had the same density, the difference in the rates being due to the fact that water molecules diffuse through the membranes at a greater rate than phosphate ions. From the rate at which the casein phosphorus becomes active, compared with the active inorganic phosphorus in the milk, it was estimated that the time of formation of casein in the gland cells was about one hour.

The fact that a few hours after addition of the labelled phosphate the milk phosphatides are only slightly active compared with the inorganic phosphate, indicates that the latter cannot be produced from the former, thus contradicting the view that the fats and inorganic phosphate are produced by the breaking up, in the milk gland, of the phosphatides of the blood. The investigation of problems of milk secretion was also referred to by Dr. S. K. Kon.

Labelled sodium phosphate has also been used in a study of the movements of phosphate in plants. Maize and sunflowers grown first in a culture solution were then transferred to a second solution containing labelled phosphate. The leaves which grew while the plant was in the second solution were examined for activity and compared with the leaves which grew in the first (inactive) solution. It was found that, after four days, the lower leaves had 80 percent of the activity of the upper leaves, showing that a rapid replacement of phosphate took place. Similar experiments have been carried out using heavy nitrogen as indicator.

The use of deuterium as an indicator was discussed by Prof H. S. Raper and Dr. W. E. van Heyningen. The former referred to the work of Cavanagh and Raper, in which fats

716

诸如卵磷脂等有机分子中的磷原子与同位素标记的磷酸钠之间不发生交换。因此，如果发现有放射性的卵磷脂，就意味着该卵磷脂分子是在施用同位素标记的磷酸钠之后合成的（在合适的酶存在的条件下），因此便可以区分以前合成的分子和最新合成的分子。例如，利用这种方法可以找到鸡蛋卵黄中的磷脂的形成部位。给一只母鸡注射同位素标记的磷酸盐，5 小时后将其杀死，发现相对于卵巢和卵黄中的磷脂来说，肝和血浆中的磷脂（前者尤甚于后者）具有更强的放射性；因此在最后5 小时内合成的磷脂分子并非生成于卵巢，而是由卵巢从血浆中获得并用于制造卵黄。此外还发现离开卵巢的鸡蛋中不会再形成同位素标记的磷脂。这可以通过检测注射标记磷酸盐后 20 小时内（鸡蛋在离开卵巢后，在输卵管中保存约 20 小时）生产的鸡蛋得以证明。

放射性磷酸钠的使用也为研究（山羊）奶的形成提供了很大的帮助。在注射同位素标记的磷酸盐后，间隔性采集血液和奶液样品，检测血液中磷酸盐的放射性以及奶液中的各种含磷化合物。3 到 4 小时之后，发现奶液中的无机磷酸盐被血浆中的放射性磷酸盐取代；然而如果在注射放射性磷酸盐的同时注射重水，那么 1 小时后由血液和奶液制得的水溶液样品就具有相同的放射强度，速率方面的差异可归结为水分子扩散通过膜的速率比磷酸根离子更快这一事实。根据酪蛋白变得具有放射性的速率，对比奶液中的放射性无机磷，可以估算出腺细胞中酪蛋白的形成时间大约为一小时。

在加入同位素标记的磷酸盐几个小时后，与无机磷酸盐相比，奶液中的磷脂仅具有轻微的放射性，这一事实意味着后者不是由前者生成的，从而与认为脂肪和无机磷酸盐是在乳腺中通过血液中磷脂的断裂而生成的观点相矛盾。科恩博士也谈到了关于泌乳问题的研究。

同位素标记的磷酸钠还被用于研究磷酸盐在植物体内的运动。首先将玉米和向日葵在培养液中进行培养，然后再将其转入另一份含同位素标记的磷酸盐溶液中。当植物体处于第二份溶液中时检测其生长叶片的放射性，并与在第一份溶液（无放射性）中生长的叶片进行比较。四天后，发现下部叶片的放射性是上部叶片的 80%，这表明发生了磷酸盐的快速置换。有人将重氮作为示踪剂进行了类似的实验。

雷珀教授和海宁根博士论述了将氚作为一种示踪剂的应用。雷珀教授提到了卡

labelled with deuterium were fed to animals and a study made of the rate of formation of the deuterium labelled lipins in the liver and kidney; the latter gave an account of the work of Schoenheimer and Rittenberg *et al.* on deuterium as an indicator in the study of intermediary metabolism, and referred especially to the uses and limitations of the method.

The use of heavy oxygen as an indicator was discussed by Dr. J. N. E. Day. He referred to the work of Aten and Hevesy, who examined the possibility of exchange of oxygen in sulphate, with other oxygen atoms present in the body, by injecting heavy sodium sulphate into rabbits, and concluded that there was little or no exchange. Reference was also made to the work of Day and Sheel on the use of heavy oxygen in animal respiration, in which the heaviness of the expired carbon dioxide was determined.

Radioactive sodium was dealt with by Dr. B. G. Maegraith, who mentioned experiments in which active sodium chloride had been injected into rabbits, and the distribution of the active sodium investigated. He suggested that this might be used to estimate the extracellular fluid content of the rabbit.

Dr. W. D. Armstrong described experiments dealing with the exchange of phosphorus of the enamel of teeth and the blood using radioactive phosphorus. A very slow exchange was noticed, indicating, not the formation of new molecules, but exchange of phosphate between enamel and blood. Mr. C. H. Collie referred to work of Collie and Morgan showing that radioactive sulphur can be used as an indicator.

Dr. D. Roaf described a method of producing radioactive phosphorus by irradiation of tricresyl phosphate with slow neutrons.

(**143**, 709-711; 1939)

瓦纳和他本人的工作，即给动物喂食用氕标记的脂肪并研究其肝和肾中氕标记脂类的形成速率；海宁根博士描述了舍恩海默与里滕伯格等人所作的将氕作为示踪剂用于研究中间代谢的工作，特别是谈到了方法的使用和限制。

戴博士论述了将重氧作为一种示踪剂的应用。他提到了阿滕和赫维西的工作，他们通过给兔子注射重硫酸钠研究了硫酸盐中的氧与机体内存在的其他氧原子之间发生交换的可能性，其结论是两者之间很少或完全不交换。此外戴博士还提到了他自己和席尔应用重氧研究动物呼吸作用的工作，在此研究中他们测定了呼出二氧化碳的重量。

梅格雷思使用了放射性钠，他提到了对兔子注射放射性氯化钠后研究放射性钠在体内的分布的实验。他暗示该方法可能可以用于估计兔子细胞外液的量。

阿姆斯特朗博士描述了使用放射性磷研究牙齿珐琅质中的磷和血液中的磷交换的实验。实验中可以观察到极为缓慢的交换，这意味着在珐琅质和血液之间发生了磷酸盐的交换，而不是新分子的形成。科利先生提到了他自己和摩根的工作，证明放射性硫可以作为示踪剂。

娄夫博士描述了一种用慢中子激发磷酸三甲苯酯产生放射性磷的方法。

（王耀杨 翻译；杨志 审稿）

The Living Coelacanthid Fish from South Africa

J. L. B. Smith

Editor's Note

James Smith's announcement of the discovery of the coelacanth on 18 March 1938 caused a sensation. This brief report on the coelacanth's anatomy, less than two months later, was intended to be a stopgap, pending a projected monographic treatment. A problem was that Smith worked in what was then a distant part of the world in which resources were few. Some of the letters he received "contained very harsh criticism about the loss of the carcass of this fish". "Few persons outside South Africa have any knowledge of our conditions", he complained. Without the "energy and determination" of a Miss Latimer of East London, the fish might have gone undescribed. "The genus *Latimeria* stands as my tribute", Smith wrote.

THE recent discovery of the Coelacanthid fish (*Latimeria chalumnoe* J. L. B. Smith) near East London, as described in *Nature* of March 18, p. 455, has aroused great interest. This has partly found expression in numerous requests from all parts of the world for the earliest possible publication of a detailed description. I am able to pursue my investigations only in very limited spare time. The preparation of an adequately illustrated detailed description, under present conditions, will occupy several months. I have therefore decided upon the somewhat unusual procedure of issuing a synopsis of the more important results of my investigations to date. This is in the form of brief outlines without discussion.

Few scientific workers are as fortunate as those who have concentrated upon Coelacanthid remains. The present specimen is a living tribute to the accuracy of their interpretations and reconstructions.

The specimen was somewhat damaged in the trawl, the skin having been broken in several places. Repairs were skilfully executed by the taxidermist.

Skull and head. The skull is unfortunately not quite intact. The basisphenoid and part of the structures around the foramen magnum were removed (and discarded) in mounting. The soft parts of the head appear to have been removed rather roughly. The remaining tissues are in poor condition. Only the structures left intact will be described as fact.

Air-bladder. According to the fairly definite evidence of the taxidermist, this organ was at the very most but feebly ossified.

The scales are cycloid and but little ossified. The proportion of residue after ignition of

720

在南非发现的现生腔棘鱼

史密斯

编者按

1938 年 3 月 18 日，詹姆斯·史密斯宣布发现了腔棘鱼，这引起了一场轰动。不到两个月后，在一篇精心构思的学术论文完成之前，发表了这篇简短的用作临时补缺的腔棘鱼解剖报告。那时史密斯在非常偏远的地方工作，那里的资源非常稀少。他收到的一些信件"对这个鱼的躯体的缺损进行了非常尖锐的批评"。对此他抱怨道，"南非之外几乎没有人了解我们的条件"。如果没有东伦敦拉蒂迈小姐的"能力和决心"，这个鱼可能根本不会得到记载。史密斯写道，"我以拉蒂迈命名这个属就是为了表达对她的感谢"。

最近，在东伦敦附近发现腔棘鱼（拉蒂迈鱼，史密斯）的消息引起了人们极大的关注，3 月 18 日出版的《自然》第 455 页曾对这一发现进行了描述。从世界各地发来的众多信件希望我们可以尽快发表一篇详尽的报告，人们对此的关注程度由此也可见一斑。但是我只能在十分有限的业余时间里从事研究。在当前条件下，若要准备一份图解充分、描述详细的记载，要花费几个月的时间。因此我决定采用一个不太常用的办法，就是发表一些至今为止较为重要的研究结果的摘要。这些摘要将以简明概要的形式给出，而不作详细讨论。

很少有科学工作者能够像专注于腔棘鱼遗迹研究的科学家那样幸运。当前的标本为他们准确地解释和重构腔棘鱼提供了一个鲜活的参照。

标本在拖网中有所损坏，鱼皮有几个地方已经破损。标本剥制师对此进行了很巧妙的修复。

头骨与头部。遗憾的是头骨并不十分完整。在安装固定过程中，基蝶骨与枕骨大孔周围的部分结构被移除（并且被丢弃了）。头部的软组织部分似乎被粗暴地移除了。保留下来的组织状况不佳。因此，下文将仅对保留完整的组织结构进行如实的描述。

鱼鳔。依照标本剥制师提供的较为确切的证据，这个器官至多只是轻微骨化。

鱼鳞为圆鳞，但轻微骨化。对身体中部无修饰部分的鱼鳞进行灼烧处理，其剩

the unornamented portion of a mid-body scale is very much less than, but qualitatively identical with, that obtained by similar treatment of a teleostean scale. The exposed surface of a scale varies from one fourth to one sixth of the total area. The ornamentation on the caudal scales is in the form of spines, on the rest of the body as tubercles. The tubercles are superficial only; each is set in a thin oval basal plate with corrugated surface. The plates are attached to the scales by tissue which is softened by alkali. The tubercles are of simple structure with a central cavity (see Fig. 1). The lateral line tubes are posteriorly widely bifurcate.

Fig. 1. Section of a scale (×50) of *Latimeria chalumnoe* J. L. B. Smith.

Fins. All the rays of the first dorsal, of the principal and of the supplementary caudal are spinate. A few of the rays of the second dorsal and of the anal are basally feebly spinate. The outer face of the pectoral is spinate. The pelvics only are quite smooth. The bony rays of the dorsal and caudal are articulated. The soft rays are finely articulated to their bases. All rays are composed of two fused lateral segments.

Dermal bones. The cheek-bones are a postorbital, a squamosal, a preopercular and a suborbital (lacrimojugal). On the middle of the lower surface of each cheek is a small bony stud which may be an obsolescent quadratojugal. The opercular is moderate in size. There are two dermal structures in a stage of arrested metamorphosis from scale to dermal bone, which are regarded as subopercular and interopercular respectively.

The intertemporals and the supratemporals appear to be fused. The exposed portion of each frontal is small, oval and flat. These bones are all feebly ornamented. The splenial and angular alone show externally on the lower jaw. The gulars are large and heavy.

Fronto-rostrals. Just beneath the skin there are nineteen bones in this series on each side. One large frontal and a smaller "pre"-frontal; nine "rostro-nasals", including a canal-bearing bone (No. 18) often named "premaxilla" in fossils.

722

余物远远少于对硬骨鱼类鱼鳞进行类似处理后得到的剩余物，但性质是相同的。每片鱼鳞外露的表面积占总面积的 1/4 到 1/6 不等。尾部鱼鳞上的装饰为皮刺形式，而在身体的其他部分是结节形式。结节仅在表面存在；每一个结节位于一个有波纹的薄的椭圆形基板上。基板通过已被碱软化过的组织固定在鱼鳞表面。结节是带有中央空腔的简单结构 (见图 1)。侧线管后端分叉较宽。

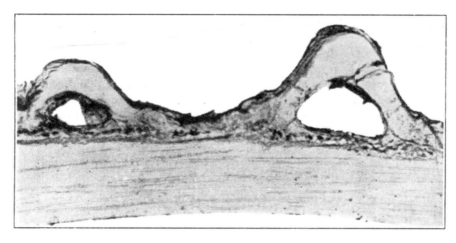

图 1. 拉蒂迈鱼（史密斯）的鳞片截面图（放大 50 倍）

鳍。第一背鳍、主鳍及副尾鳍的所有辐射线都是棘状的。第二背鳍和臀鳍的一些鳍条基部略微呈棘状。胸鳍的外表面也呈棘状。只有腹鳍完全平滑。背鳍和尾鳍的骨质鳍条是分节的。软鳍条精巧地分节到其基部。所有鳍条都由两个愈合的侧节构成。

真皮骨。颊骨包括眼窝骨、鳞状骨、前鳃盖骨与眶下骨（泪 – 轭骨）。每个颊的下表面中间是一个小的骨质柱，可能是退化的方轭骨。鳃盖骨中等大小。有两个真皮结构，处于从鳞到真皮骨的停止变形阶段，它们分别被当作是下鳃盖骨和间鳃盖骨。

间颞骨与上颞骨看起来像是愈合了。每个额骨的外露部分很小，呈椭圆扁平状。这些骨全都略有修饰。夹骨和隅骨在下颌上单独外露出来。咽喉板大而厚重。

额部 – 吻突。位于皮肤之下，这个系列每一边均有 19 块骨头。一块大的额骨与一块较小的"前"额骨；9 块"吻突 – 鼻"骨，包括一块含有管道的骨 (第 18 号)，在化石中这块骨通常被命名为"前颌骨"。

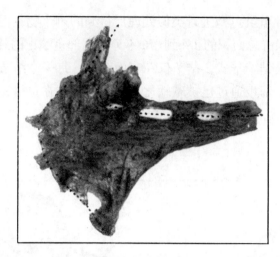

Fig. 2. Rostro-nasal, No. 18. Nat. size. Inner margin to the left. A piece of white paper has been inserted into the canal. The dotted lines show the course of sensory canals. The right-hand limb anastomoses with the suborbital.

There are eight bones in the "parafrontal" series, the anterior expanded bones having been named "antorbitals" in fossils. Besides these nineteen there are the small dentigerous rostral plates. Most of these bones are small and laminate.

Sensory canals. The main canals run much as have been shown in reconstructions. From the lateral line the canal passes through the supratemporals and intertemporals, thence "parafrontal" to the snout. There it has a small superior median branch. Just below that it gives off an inferior branch which is the infraorbital canal (running through the lateral limb of bone No. 18, which anastomoses with the suborbital). Below this junction is a commissure across the snout. The canal then runs downwards and curves outwards to end on the outer edge of the rostrum around the inner face of bone No. 18. Just behind the frontal is the junction of the infraorbital canal running downwards through the postorbital, which continues through the suborbital to the snout. The anterior limb of the squamosal anastomoses with the postorbital-suborbital anastomosis and carries the jugal canal which continues through the preopercular, thence as a tube in the skin obliquely over the lower outer face of the quadrate. It enters the angular very obliquely and thence runs forward on the lower margin just below the surface to the symphysis. There is a posterior branch on the lower surface of the angular.

Olfactory organs. There are on each side three "narial" apertures, two conventionally on the side of the snout before the eye, the third on the front of the rostrum. Each opening is the end of a simple tube which leads from a median capsule situated beneath a thin layer of mesethmoidal cartilage. The tubes and the capsule were apparently lined with fine rugose tissue. The capsule is depressed biconical in shape. There is also on each side of the rostrum an infero-lateral nasal tentacle, apparently imperforate. On each side of the snout, inferior and anterior to the median nasal cavity, lies a sac, typically covered below, laterally, and partly behind, by the prevomer, in front by bone No. 18. That bone lies against the

724

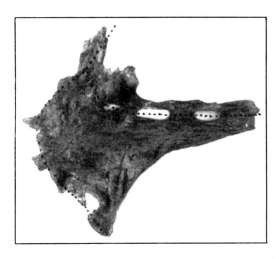

图 2. 吻–鼻，第18号。原始大小。左边为内缘。将一片白纸插入了管道中。虚线表示感觉管的线路。右侧分支与眶下骨吻合。

"侧额骨"系列有 8 块骨，在化石中前端膨大的骨通常被命名为"眶前骨"。除了这 19 块，还有小的有齿的吻骨板。这些骨多数较小且呈薄片状。

感觉管。如复原图中所示，总管延伸很长。从侧线开始，管道通过上颞骨与间颞骨，然后从"侧额骨"到口鼻部。在那里有一个小的上中分支。就在那之下分出一个下支，即眶下管 (贯穿第 18 号骨的侧支，与眶下骨汇合)。在这个连接点下面是一个横过口鼻部的接合处。然后管道下行并向外弯曲至第 18 号骨的内表面周围，于嘴外缘终止。眶下管下行直通后眶骨，穿过眶下骨延伸到口鼻部，眶下管的连接点恰好在额骨后面。鳞状骨前肢与后眶骨 – 眶下骨连接点吻合，其上的眶下管延伸至前鳃盖骨，继而斜开口于方骨下部外面的皮肤上。其十分倾斜地进入隅骨，然后沿表面下缘向前延伸至联合处。在隅骨下表面上有一后支。

嗅觉器官。在每边均有 3 个"鼻孔"孔隙，照例两个在眼睛前口鼻部的一边，第三个在吻的前端。每个孔都是一个单管的末端，管子从位于中筛软骨薄膜层之下的中囊导出。管子和囊明显与细皱褶组织在一条线上。囊呈扁平双锥形。嘴的每一边也有一条下侧鼻触须，明显无孔。在吻部的每一边，中央鼻腔的下部和前部，都生有一个囊，很有代表性的是，它的下面、侧面甚至部分后面都被前犁骨覆盖着，前面则被第 18 号骨覆盖。那块骨位于前犁骨后柱的外表面。这些囊与普通的"嗅觉

outer surface of the hind column of the prevomer. These sacs correspond with the usual "olfactory capsules", but do not appear to have any external narial opening. The nerve supply to these paired capsules appears to be from the olfactory lobe, and to enter at the upper inner portion of the surface. The nerve supply of the mesethmoidal (pineal?) cavity appears to come from farther back.

Respiratory organs. The branchial arches and appendages were lost. The arches are stated to have been strongly spinate. Remaining is a spinate epihyal and ceratohyal. Also a superficially ossified tuberculate copula ("tongue"), to which appear to have been attached four branchial arches. There is apparently a hyoidean gill-slit behind the free margin of the "preoperculum". The spiracles are small and probably functionless.

The palato-pterygo-quadrate apparatus is massive and typically coelacanthid. There is no hyomandibular. The prevomer is an unexpectedly solid structure, with the anterior edge of the autopalatine bearing against the hind surface of its outer columnar vertical process. The ectopterygoid overlaps the autopalatine and the pterygoid above.

Upper jaw. Maxillae are absent, probably also premaxillae. The upper jaw bears paired dermal plates attached to rostrals, prevomers, palatines and ectopterygoids. These plates are of fused small teeth, and each bears one or two large conical tusks. On the pterygoids and parasphenoid are conical granular teeth. There are two feeble granulate "epi"-pterygoid areas.

Lower jaw. There is a series of small "labial" dentate plates on the outer surface of the dentary. Superiorly are four dentate plates on each dentary. The anterior coronoid is small and bears granular teeth as well as several large tusks. The posterior coronoid is large and feebly granulate at the base. The articular-prearticular plate is very long and is granulate anteriorly only.

Several letters from overseas have contained very harsh criticism about the loss of the carcass of this fish. Few persons outside South Africa have any knowledge of our conditions. In the coastal belt only the South African Museum at Cape Town has a staff of scientific workers among whom is an ichthyologist. The other six small museums serving the coastal area are in extremely poor circumstances, and generally have only a director or curator, who cannot possibly be an expert in all branches of natural history. There are not uncommon fishes in the sea which to any of the latter would appear as strange as, if not stranger than, a coelacanthid. It was the energy and determination of Miss Latimer which saved so much, and scientific workers have good cause to be grateful. The genus *Latimeria* stands as my tribute.

(**143**, 748-750; 1939)

J. L. B. Smith: Rhodes University College, Grahamstown.

囊"相似，但似乎没有任何外部鼻孔开口。这些成对的囊的神经分布似乎来自嗅叶，并于表面的内上部进入。中筛（松果体？）腔的神经分布似乎来自更后方。

呼吸器官。鳃弓与附肢缺失。鳃弓被描述为强有力的棘状。棘状上舌骨与角舌骨得以保存。另外还有一个表面骨化的有结节的舌联桁（"舌状物"），似乎有 4 个鳃弓附在其上。在"前鳃盖骨"活动缘的后面有一个明显的 U 字形腮裂。呼吸孔小，而且可能已经丧失了功能。

腭骨 – 翼骨 – 方骨系列厚重，属于典型的腔棘鱼结构。没有舌颌骨。出人意料的是，前犁骨为实心结构，原腭骨的前缘靠在其外部柱状垂直突起的后表面。外翼骨重叠在原腭骨和翼骨上面。

上颌。上颌骨缺失，前上颌骨可能也缺失了。上颌有成对的真皮骨板附在吻骨突、前犁骨、腭骨及外翼骨上。这些骨板是小牙齿融合而成的，而且每个骨板都有一个或两个大的圆锥形长牙。在翼骨与副蝶骨上是圆锥形粒状牙齿。有两个略微成粒状的"上"翼骨区域。

下颌。齿骨外表面上有一系列小的"唇"齿状板。每个齿骨上是 4 个齿状板。前冠状骨小，具颗粒状牙齿，并且还有几个大的长牙。后冠状骨大，并在基部略微成粒状。关节 – 前关节骨板很长，并且仅前部呈粒状。

来自国外的一些信件对这个鱼的躯体的缺损进行了非常尖锐的批评。南非之外几乎没有人了解我们的条件。在海岸地带，仅在开普敦的南非博物馆有科学工作者，他们之中只有一位是鱼类学家。海岸区的其他 6 个小博物馆条件非常艰苦，一般只有一个主管或馆员，他们不可能精通博物学所有的分支学科。他们中的任何人对海中比腔棘鱼更加陌生或至少一样陌生的鱼早已见怪不怪。正是拉蒂迈小姐的能力和决心挽回了很多损失，科学工作者有足够的理由对其表示感谢。我以拉蒂迈命名这个属就是为了表达对她的感谢。

（田晓阳 翻译；陈平富 审稿）

Social Biology and Population Improvement

F. A. E. Crew *et al.*

Editor's Note

This document is remarkable in many ways, not least the auspicious timing. It is a response from many leading biologists to a question posed in the United States of how the world's population might be improved genetically. In the 1930s this question, seemingly anachronistic today, was generally viewed through a Darwinian lens, and the response in the US and Nazi Germany had tended to favour eugenics, partly through enforced sterilizations. That solution might have been expected to find favour with some of these signatories, such as Julian Huxley. But in fact the note is remarkably progressive, placing emphasis on the need for equality of opportunity before comparisons can be made, and challenging the notion that dominant classes are genetically superior.

IN response to a request from Science Service, of Washington, D.C., for a reply to the question "How could the world's population be improved most effectively genetically?", addressed to a number of scientific workers, the subjoined statement was prepared, and signed by those whose names appear at the end.

The question "How could the world's population be improved most effectively genetically?" raises far broader problems than the purely biological ones, problems which the biologist unavoidably encounters as soon as he tries to get the principles of his own special field put into practice. For the effective genetic improvement of mankind is dependent upon major changes in social conditions, and correlative changes in human attitudes. In the first place, there can be no valid basis for estimating and comparing the intrinsic worth of different individuals, without economic and social conditions which provide approximately equal opportunities for all members of society instead of stratifying them from birth into classes with widely different privileges.

The second major hindrance to genetic improvement lies in the economic and political conditions which foster antagonism between different peoples, nations and "races". The removal of race prejudices and of the unscientific doctrine that good or bad genes are the monopoly of particular peoples or of persons with features of a given kind will not be possible, however, before the conditions which make for war and economic exploitation have been eliminated. This requires some effective sort of federation of the whole world, based on the common interests of all its peoples.

Thirdly, it cannot be expected that the raising of children will be influenced actively by considerations of the worth of future generations unless parents in general have a very

728

社会生物学与人种改良

编者按

这篇文章从许多方面来说都是值得注意的，特别是本文发表的良好时机。它是许多杰出的生物学家对美国提出的关于世界人种的遗传改良问题的回应。尽管如今看来这个问题似乎已经过时，但是在 20 世纪 30 年代，科学家们从达尔文理论的视角出发对这个问题进行了综合的考虑，美国和纳粹德国的倾向于支持部分地通过强制性绝育来实现优生。原先预期这一解决方案可能会得到本文的部分签署者的支持，比如朱利安·赫胥黎，然而事实并非如此。实际上这篇短文是非常进步的，它强调了在作出比较之前首先需要均等的机会，并且质疑了统治阶层具有优等遗传基因的观点。

"怎样可以使世界人种得到最有效的遗传改良？"，华盛顿特区科学服务社希望科学工作者能够就这一问题给予回应，于是我们起草了这份补充声明，并在文章末尾联合署名。

与单纯的生物学问题相比，"怎样可以使世界人种得到最有效的遗传改良？"这一问题会引出更加宽泛的难题，一旦生物学家试图将自己专业领域的法则付诸实践，他就不得不面对这些问题。人类有效的遗传改良依赖于社会条件的重大变化，以及在人类态度方面与此相关的变化。首先，如果经济和社会条件不能为所有的社会成员提供大致均等的机会来取代社会成员从出生时就被划分为权力相差非常大的不同阶级，那就不存在可以用于估计和比较不同个体的内在价值的有效依据。

遗传改良的第二个主要障碍在于经济和政治条件，它催生了不同国家、不同民族和不同"种族"间的对立。然而，在消除导致战争和经济剥削的条件之前，要消除种族偏见以及认为优良或劣质基因专属于某一民族或具有特定特征的人这种不科学的信条是不可能的。这需要在全世界人民的共同利益的基础上，建立全世界的某种有效的联合。

第三，通常来说，除非父母有可观的经济保障，而且在生育和抚养每个孩子的过程中都能为他们提供充分的经济、医疗、教育和其他方面的帮助，从而保证拥有

considerable economic security and unless they are extended such adequate economic, medical, educational and other aids in the bearing and rearing of each additional child that the having of more children does not overburden either of them. As the woman is more especially affected by childbearing and rearing, she must be given special protection to ensure that her reproductive duties do not interfere too greatly with her opportunities to participate in the life and work of the community at large. These objects cannot be achieved unless there is an organization of production primarily for the benefit of consumer and worker, unless the conditions of employment are adapted to the needs of parents and especially of mothers, and unless dwellings, towns and community services generally are reshaped with the good of children as one of their main objectives.

A fourth prerequisite for effective genetic improvement is the legalization, the universal dissemination, and the further development through scientific investigation, of ever more efficacious means of birth control, both negative and positive, that can be put into effect at all stages of the reproductive process — as by voluntary temporary or permanent sterilization, contraception, abortion (as a third line of defence), control of fertility and of the sexual cycle, artificial insemination, etc. Along with all this the development of social consciousness and responsibility in regard to the production of children is required, and this cannot be expected to be operative unless the above-mentioned economic and social conditions for its fulfilment are present, and unless the superstitious attitude towards sex and reproduction now prevalent has been replaced by a scientific and social attitude. This will result in its being regarded as an honour and a privilege, if not a duty, for a mother, married or unmarried, or for a couple, to have the best children possible, both in respect of their upbringing and of their genetic endowment, even where the latter would mean an artificial—though always voluntary—control over the process of parenthood.

Before people in general, or the State which is supposed to represent them, can be relied upon to adopt rational policies for the guidance of their reproduction, there will have to be, fifthly, a far wider spread of knowledge of biological principles and of recognition of the truth that both environment and heredity constitute dominating and inescapable complementary factors in human wellbeing, but factors both of which are under the potential control of man and admit of unlimited but interdependent progress. Betterment of environmental conditions enhances the opportunities for genetic betterment in the ways above indicated. But it must also be understood that the effect of the bettered environment is not a direct one on the germ cells and that the Lamarckian doctrine is fallacious, according to which the children of parents who have had better opportunities for physical and mental development inherit these improvements biologically, and according to which, in consequence, the dominant classes and peoples would have become genetically superior to the underprivileged ones. The intrinsic (genetic) characteristics of any generation can be better than those of the preceding generation only as a result of some kind of *selection*, that is, by those persons of the preceding generation who had a better genetic equipment having produced more offspring, on the whole, than the rest, either through conscious choice, or as an automatic result of the way in which they lived. Under modern civilized conditions such selection is far less likely to be automatic than under primitive conditions,

更多的孩子不会成为他们过重的负担，否则考虑到抚养后代的代价，就不要指望生养小孩能带来积极主动的影响。因为女性会更多地受到分娩和养育孩子的影响，所以必须给予特别的保护以确保她的生育责任不会过分地干扰其参与一般的社会生活和工作的机会。除非有生产机构来保证消费者和工人的利益，工作条件适合父母尤其是母亲的需要，而且住宅、城镇及社区服务都基本上被改造成把对孩子有益作为它们的主要目标之一，这些目标才有可能实现。

有效遗传改良的第四个必要条件就是曾经奏效的生育控制手段的合法化、广泛传播以及通过科学研究使其进一步发展，不论消极的还是积极的，这些措施都能在生育过程的所有阶段实施——如自愿进行的暂时性或永久性绝育、避孕、流产（作为第三道防线）、控制生育力和生殖周期、人工授精等。与此同时，还需要与生育孩子相关的社会意识和社会责任的发展，并且只有上述所需的经济和社会条件都具备，以及现在盛行的关于性别和生育的迷信态度都已被科学的社会态度所取代的时候，这一发展才可能是切实有效的。对一位无论是已婚还是未婚的母亲或者一对夫妻而言，这将使得即便不会视生育为一项责任，也会视其为使拥有最好的孩子成为可能的一项荣誉和权利，这包括对他们的养育和遗传赋予，后者甚至意味着一种人为的——虽然通常是自愿的——对亲子关系的控制。

第五，在依靠广大民众或者代表他们的国家采取合理的政策来指导人们的生育之前，必须要广泛普及有关生物学原理的知识，广泛普及环境和遗传是共同构成人类福祉的主要的、不可或缺的互补因素的认识，但这两个因素都受到人类潜能的控制并且允许无限的但相互依存的进步。环境条件的改善能以上面提到的方式增加遗传改良的机会。但是也必须明白改善的环境并非直接影响生殖细胞，必须明白拉马克学说是错误的，根据该学说，拥有更好的身体和智力发育机会的父母生育的孩子能遗传这些生物学改良，而且按照这个学说可以得出，处于统治阶层的人比下层穷苦的人具有遗传上的优越性。仅仅作为某种**选择**的结果，任何一代的内在（遗传）特征就会比上一代的更好，即，总体而言，上一代中具有较好遗传资质的人可以通过有意识的选择或者作为其生活方式的无意识的结果从而产生比其他人更多的后代。与原始条件下相比，现代文明条件下的这种选择已经远不可能是无意识的了，因此要求对选择进行某些有意识的指导。然而，要想使这种选择成为可能，人们必

hence some kind of conscious guidance of selection is called for. To make this possible, however, the population must first appreciate the force of the above principles, and the social value which a wisely guided selection would have.

Sixthly, conscious selection requires, in addition, an agreed direction or directions for selection to take, and these directions cannot be social ones, that is, for the good of mankind at large, unless social motives predominate in society. This in turn implies its socialized organization. The most important genetic objectives, from a social point of view, are the improvement of those genetic characteristics which make (*a*) for health, (*b*) for the complex called intelligence, and (*c*) for those temperamental qualities which favour fellow-feeling and social behaviour rather than those (today most esteemed by many) which make for personal "success", as success is usually understood at present.

A more widespread understanding of biological principles will bring with it the realization that much more than the prevention of genetic deterioration is to be sought for, and that the raising of the level of the average of the population nearly to that of the highest now existing in isolated individuals, in regard to physical wellbeing, intelligence and temperamental qualities, is an achievement that would—so far as purely genetic considerations are concerned—be physically possible within a comparatively small number of generations. Thus everyone might look upon "genius", combined of course with stability, as his birthright. As the course of evolution shows, this would represent no final stage at all, but only an earnest of still further progress in the future.

The effectiveness of such progress, however, would demand increasingly extensive and intensive research in human genetics and in the numerous fields of investigation correlated therewith. This would involve the co-operation of specialists in various branches of medicine, psychology, chemistry and, not least, the social sciences, with the improvement of the inner constitution of man himself as their central theme. The organization of the human body is marvellously intricate, and the study of its genetics is beset with special difficulties which require the prosecution of research in this field to be on a much vaster scale, as well as more exact and analytical, than hitherto contemplated. This can, however, come about when men's minds are turned from war and hate and the struggle for the elementary means of subsistence to larger aims, pursued in common.

The day when economic reconstruction will reach the stage where such human forces will be released is not yet, but it is the task of this generation to prepare for it, and all steps along the way will represent a gain, not only for the possibilities of the ultimate genetic improvement of man, to a degree seldom dreamed of hitherto, but at the same time, more directly, for human mastery over those more immediate evils which are so threatening our modern civilization.

须首先意识到上述法则的重要性以及一种具有明智的指导性的选择所拥有的社会价值。

第六，有意识的选择还要求一个一致的方向或者多个可供选择的方向，并且这些方向不能是社会方面的，即，为了尽可能对人类有益，除非社会动机成为社会的主流。反过来看，这也暗示着它的社会化组织。从社会观点来看，最重要的遗传目标是使有助于（a）健康、（b）智力和（c）建立促成同情心和社会行为的性情素质的那些遗传特征得到改良，而非那些（如今被许多人最为看重的）有助于个人"成功"的素质，正如当前大家通常所理解的成功一样。

对生物学原理更广泛的理解将会使人们认识到有更多方面比预防遗传退化更值得被追求，另外就是关于身体健康程度、智力和性情素质的人种平均水平提高至接近现存的单独个体的最高水平，仅从遗传因素考虑，这在数量相对少的一代中是可能达到的。因而，当然与稳定性相结合，每个人都可以把"天赋"看作自己生来就有的权利。正如进化过程所显示的，这根本不代表最终阶段，而只是预示着将来还会进一步发展。

然而，这种进展的有效性要求对人类遗传学和其他与之相关的多个领域进行更加广泛和深入的研究。这将需要医学、心理学、化学和社会科学等多个学科的专家们以改善人类自身的内在构成作为中心主题而进行合作。人体组织的精细程度令人称奇，对其遗传特征的研究一直被特殊的困难所困扰，这些困难要求该领域的研究具有比迄今所预期的更大的规模以及更强的准确性和可分析性。然而，当人类的思想从战争、仇恨和为基本的生存方式而斗争转向共同追求更大目标的时候，它就能实现。

经济重建达到能够释放人类力量的阶段的那一天尚未来临，但是我们这一代人要随时为这一天做好准备并以此为己任。在沿着这条路奋进的途中，我们所经历的每一步都代表着收获，这种收获不仅是迄今为止还很难想象的人类最终遗传改良的可能性，而且同时，更直接的是，使人类能够更好地制服那些随时出现的威胁我们现代文明的邪恶力量。

F. A. E. Crew

C. D. Darlington

J. B. S. Haldane

S. C. Harland

L. T. Hogben

J. S. Huxley

H. J. Muller

J. Needham

G. P. Child

P. R. David

G. Dahlberg

Th. Dobzhansky

R. A. Emerson

C. Gordon

J. Hammond

C. L. Huskins

P. C. Koller

W. Landauer

H. H. Plough

B. Price

J. Schultz

A. G. Steinberg

C. H. Waddington

(**144**, 521-522; 1939)

克鲁	蔡尔德	科勒
达林顿	戴维	兰道尔
霍尔丹	达尔伯格	普劳
哈兰德	多布赞斯基	普赖斯
霍格本	爱默生	舒尔茨
赫胥黎	戈登	斯坦伯格
马勒	哈蒙德	沃丁顿
尼达姆	赫斯金斯	

（刘皓芳 翻译；刘京国 审稿）

Myosine and Adenosinetriphosphatase

W. A. Engelhardt and M. N. Ljubimowa

Editor's Note

Here biochemist Wladimir Engelhardt and his wife and former postgraduate student Militza Ljubimowa, both from Moscow's Academy of Sciences, show that myosine (now myosin), one of the contractile proteins of muscle, catalyses the breakdown of adenosinetriphosphate (ATP), and that ATP provides the energy for muscular contraction. Myosine, they suggest, splits ATP into adenosinediphosphate (ADP) and free phosphate ions, liberating energy along the way that can be used in muscle action. This led to the concept of ATP as a universal biochemical energy source that can be used by cells for many different functions. The molecule's breakdown liberates energy which enzymes harness to drive other chemical reactions.

ORDINARY aqueous or potassium chloride extracts of muscle exhibit but a slight capacity to mineralize adenosinetriphosphate. Even this slight liberation of phosphate is mainly due, not to direct hydrolysis of adenosinetriphosphate, but to a process of secondary, indirect mineralization, accompanying the transfer of phosphate from the adenylic system to creatine, the corresponding enzymes (for which the name "phosphopherases" is suggested) being readily soluble.

In contrast to this lack of adenosinetriphosphatase in the soluble fraction, a high adenosinetriphosphatase activity is associated with the water-insoluble proteins of muscle. This enzymatic activity is easily brought into solution by all the buffer and concentrated salt solutions usually employed for the extraction of myosine. On precipitation of myosine from such extracts, the adenosinetriphosphatase activity is always found in the myosine fraction, whichever mode of precipitation be used: dialysis, dilution, cautious acidification, salting out. On repeated reprecipitations of myosine, the activity per mgm. nitrogen attains a fairly constant level, unless denaturation of myosine takes place. Under the conditions of our experiments (optimal conditions have not been determined) the activity of myosine preparations ranged in different experiments from 350 to 600 microgram phosphorus liberated per mgm. nitrogen in 5 min. at 37°. Expressed as

$$Q_p \left(= \frac{\mu\text{gm. P}/31 \times 22.4}{\text{mgm. N} \times 6.25 \times \text{hour}} \right),$$

this gives values of 500–850.

Acidification to pH below 4, which is known to bring about the denaturation of myosine[1], rapidly destroys the adenosinetriphosphatase activity. Most remarkable is the extreme thermolability of the adenosinetriphosphatase of muscle: the enzymatic activity shown

肌球蛋白与三磷酸腺苷酶

恩格尔哈特，卢比莫娃

编者按

在这篇文章中，生物化学家弗拉基米尔·恩格尔哈特和他的妻子米丽莎·卢比莫娃（卢比莫娃之前是恩格尔哈特的研究生，他们都来自莫斯科科学院）指出，肌肉中一种可以收缩的蛋白——肌球蛋白（myosine，现在的拼法是myosin）催化了三磷酸腺苷（ATP）的分解，而ATP提供了肌肉收缩所需的能量。他们认为肌球蛋白将ATP分解为二磷酸腺苷（ADP）和自由的磷酸根阴离子，这一过程中释放的能量可被用于肌肉的运动。这引出了这样一种观点，即：ATP作为一种生化能源可以被细胞广泛地用在许多功能的实现中。这种分子断裂释放出来的能量可以被酶利用以驱动其他的化学反应。

通常肌肉中的水或氯化钾提取物仅具有微弱的矿化三磷酸腺苷的能力。即便是这种磷酸盐轻微的析出也不是主要由三磷酸腺苷（ATP）的直接水解引起的，而是由一个间接的矿化作用过程造成。这个过程通常伴随着磷酸盐从腺苷酸系统到肌氨酸的转移，并且会引起相应的酶（有人建议命名为"磷酸水解酶"）的可溶性增强。

三磷酸腺苷酶在肌肉可溶性组分中含量很少，而相比之下，很高的三磷酸腺苷酶活性却与肌肉中的水不溶性蛋白质相结合。所有用来提取肌球蛋白的缓冲液及高浓度盐溶液，都可以很容易地使这种酶的活性溶于其中。这使得人们在从肌肉抽提物中分离肌球蛋白时，无论是采用透析、稀释、谨慎酸化还是盐析的方法，往往都可以在分离得到的肌球蛋白组分中检测到三磷酸腺苷酶的活性。除非肌球蛋白在纯化过程中发生变性，否则即使反复析出肌球蛋白，以每毫克含氮量表示的比活性几乎都维持在一个恒定水平上。在我们的实验中（尚未确定最佳反应条件），37℃下5分钟内肌球蛋白抽提物的三磷酸腺苷酶活性在不同的实验中介于 350~600 微克磷 / 毫克氮。用下式表示得到的结果在 500~850 之间：

$$酶比活力 = \frac{每微克磷含量 \div 31 \times 22.4}{每毫克样品含氮量 \times 6.25 \times 小时}$$

人们已经知道，把溶液酸化到 pH 值在 4 以下会引起肌球蛋白的变性[1]，也会迅速地破坏三磷酸腺苷酶的活性。然而最值得注意的是肌肉中的三磷酸腺苷酶的极

737

by myosine solutions is completely lost after 10 min. exposure to 37°. This corresponds with the well-known thermolability of myosine[2]. In respect of its high thermolability adenosinetriphosphatase resembles the protein of the yellow enzyme, which when separated from its prosthetic group is also rapidly inactivated at 38° (Theorell[3]). Evidently in the intact tissue of the warm-blooded animal (all experiments were performed on rabbit muscles), some conditions must exist which stabilize the myosine against the action of temperature. A marked stabilizing effect on the adenosinetriphosphatase activity seems to be produced by the adenylic nucleotide itself. As can be seen from the accompanying graph, in the presence of adenosinetriphosphate the liberation of phosphate proceeds at 37° over a considerable period (Curves I, Ia and Ib), whereas the same myosine solution warmed alone to 37° for 10–15 min. shows on subsequent addition of adenosinetriphosphate an insignificant or no mineralization whatever.

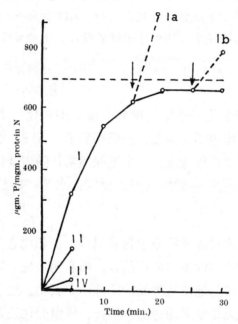

Crude buffer extracts accomplish a quantitative hydrolysis of the labile phosphate groups of adenosinetriphosphate; myosine, reprecipitated three times, liberates but 50 percent of the theoretical amount of phosphorus (see figure). It acts as true adenosine-*tri*-phosphatase and yields adenosinediphosphate, which is not further dephosphorylated and has been isolated in substance. This may serve as a convenient way of preparing adenosinediphosphate, instead of using crayfish muscle[4]. The adenosinediphosphatase is thus associated with the more soluble proteins, occupying an intermediate position between adenosinetriphosphatase and the most readily soluble phosphopherases.

Under no conditions tested could we obtain a separation of adenosinetriphosphatase from myosine. Either the activity was found in the myosine precipitate or else it was absent from the precipitates and from the remaining solution. This disappearance of the enzymatic activity we regard as the result of the start of denaturation of the very unstable myosine.

端不耐热性：肌球蛋白溶液在 37℃ 下放置 10 分钟，其中的酶的活性就会被完全破坏。这与公认的肌球蛋白的不耐热性是一致的 [2]。三磷酸腺苷酶的这种不耐热性与黄酶的蛋白质类似，黄酶与其辅基分离后在 38℃ 下也会很快失活（特奥雷尔 [3]）。显然，在温血动物（所有实验均在兔肌肉中进行）的完整组织中必定存在某种条件可以使肌球蛋白在温度作用下保持稳定。似乎腺苷酸自身就可以显著地使三磷酸腺苷酶活性保持稳定。从附图中可以看出，ATP 存在的情况下，在 37℃ 下相当长的时间内都有磷酸盐的生成（曲线 I、Ia 及 Ib），但单独加热同样的肌球蛋白溶液，使其在 37℃ 保持 10~15 分钟，再加入 ATP，无论如何其作用都是微乎其微的，或者根本就不存在矿化作用。

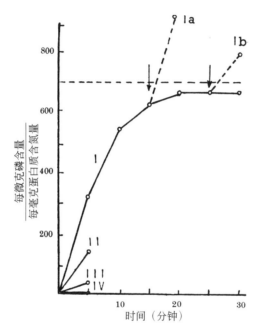

缓冲液原始提取物可以实现不稳定的 ATP 磷酸基团的定量水解。经过 3 次反复沉淀后，肌球蛋白析出的磷只有理论量的一半（见图）。它以纯的三磷酸腺苷形式存在并又生成二磷酸腺苷（ADP）。ADP 没有进一步脱去磷酸，并已从本体中分离出来。这也为制备 ADP 提供了一条新的途径，替代了使用小龙虾肌肉组织的方法 [4]。这也表明二磷酸腺苷酶结合可溶性更高的蛋白质，其溶解性介于三磷酸腺苷酶和那些极易溶解的磷酸水解酶之间。

在我们所尝试过的实验条件中，没有一种可以将三磷酸腺苷酶从肌球蛋白中分离出来。这种酶活性要么出现在肌球蛋白沉淀中，要么在沉淀及剩余溶液中都不出现。我们认为这种酶活性的消失可能是由于非常不稳定的肌球蛋白发生了变性。

We are led to conclude that the adenosinetriphosphatase activity is to be ascribed to myosine or, at least, to a protein very closely related to and at present not distinguishable from myosine. Thus the mineralization of adenosinetriphosphate, often regarded as the primary exothermic reaction in muscle contraction, proceeds under the influence and with the direct participation of the protein considered to form the main basis of the contractile mechanism of the muscle fibre.

(**144**, 668-669; 1939)

W. A. Engelhardt and M. N. Ljubimowa: Institute of Biochemistry, Academy of Sciences of the U.S.S.R., Moscow, August 7.

References:
1. v. Muralt, A., and Edsall, J. T., *J. Biol. Chem.*, **89**, 351 (1930).
2. Bate Smith, E. C., *Proc. Roy. Soc.*, B, **124**, 136 (1937).
3. Theorell, H., *Biochem. Z.*, **272**, 155 (1934); **278**, 263 (1935).
4. Lohmann, K., *Biochem. Z.*, **282**, 109 (1935).

　　总之，我们认为三磷酸腺苷酶活性应该来自肌球蛋白，或者，至少来自一个与肌球蛋白密切相关而目前我们尚不能将其分离出来的蛋白。因此，通常看作是肌肉收缩的主要放热反应的 ATP 矿化，在蛋白质的影响及直接参与下，可以认为是肌肉纤维收缩机理的主要基础。

（张锦彬 翻译；刘京国 审稿）

Action Potentials Recorded from Inside a Nerve Fibre

A. L. Hodgkin and A. F. Huxley

Editor's Note

Before this paper, no one had ever managed to record the electrical impulse, or "action potential", inside a nerve fibre. Here physiologists Alan Lloyd Hodgkin and Andrew Fielding Huxley (grandson of Thomas Henry Huxley) from the Laboratory of the Marine Biological Association, Plymouth, do just that, taking advantage of the considerable size of the nerve fibres (axon) of the giant squid. Their study proved that action potentials—self-propagating waves of electrochemical activity—arise at the cell surface, and for the first time measured their strength. The two scientists later won the 1963 Nobel Prize in Physiology or Medicine for their work on how action potentials arise and spread in nerve cells.

NERVOUS messages are invariably associated with an electrical change known as the action potential. This potential is generally believed to arise at a membrane which is situated between the axoplasm and the external medium. If this theory is correct, it should be possible to record the action potential between an electrode inside a nerve fibre and the conducting fluid outside it. Most nerve fibres are too small for this to be tested directly, but we have recently succeeded in inserting micro-electrodes into the giant axons of squids (*Loligo forbesi*)[1].

The following method was used. A 500 μ axon was partially dissected from the first stellar nerve and cut half through with sharp scissors. A fine cannula was pushed through the cut and tied into the axon with a thread of silk. The cannula was mounted with the axon hanging from it in sea water. The upper part of the axon was illuminated from behind and could be observed from the front and side by means of a system of mirrors and a microscope; the lower part was insulated by oil and could be stimulated electrically. Action potentials were recorded by connecting one amplifier lead to the sea water outside the axon and the other to a micro-electrode which was lowered through the cannula into the intact nerve beneath it. The micro-electrode consisted of a glass tube about 100 μ in diameter and 10–20 mm. in length; the end of the tube was filled with sea water, and electrical contact with this was made by a 20 μ silver wire which was coated with sliver chloride at the tip. Fig. 1 is a photograph of an electrode inside the living axon. The giant axon shows as a clear space and is surrounded by the small fibres and connective tissue which make up the rest of the nerve trunk. The silver wire can be seen inside the electrode and about 1 mm. from its tip. A small action potential was recorded from the upper end of the axon and this gradually increased as the electrode was lowered, until it reached

从神经纤维内记录到的动作电位

霍奇金，赫胥黎

编者按

在这篇论文之前，从未有人成功地测得神经纤维内的电脉冲或者"动作电位"。在此，来自普利茅斯英国海洋生物协会实验室的生理学家艾伦·劳埃德·霍奇金与安德鲁·菲尔丁·赫胥黎（托马斯·亨利·赫胥黎的孙子）利用巨型乌贼相当大的神经纤维（轴突）完成了这一壮举。他们的研究证明了动作电位——可自行传导的电化学波——产生于细胞表面，并且第一次测量了它们的强度。后来，这两位科学家因为他们在神经细胞中动作电位的产生和传播方面的研究工作而获得了 1963 年诺贝尔生理学或医学奖。

神经信号常与被称为动作电位的电学变化相联系。人们普遍认为，这种电位出现在位于轴突细胞质与外部介质之间的隔膜处。如果这一理论是正确的，那么我们就应该可以记录到置于神经纤维内部的电极与外部导电液之间的动作电位。由于大多数神经纤维都太小了，所以一直以来都难以直接测定这一电位，不过，最近我们成功地将微电极植入到了乌贼（福布斯式枪乌贼）的巨轴突中 [1]。

电极植入的方法记述如下。从第一星状神经中部分地剥离出一根 500 微米的神经轴突，并用锋利的剪刀划开一半。将一根微型套管植入开口，并用一根丝线将其缚于轴突内。套管和悬挂于其上的轴突被安置在海水中。利用一组镜面和显微镜从后面将轴突上部照亮，从而可以从前面和侧面对其进行观测；用油将轴突下部隔离，使其可以接受电刺激。把微电极穿过微型套管下放到套管下方没有触动过的神经上。把两个放大器分别通向轴突外部的海水和神经纤维内的微电极，通过这两个相连接的放大器即可记录动作电位。微电极由一个直径约 100 微米，长约 10~20 毫米的玻璃管构成；管的末端充入海水，用顶部覆盖有氯化银的一根 20 微米长的银导线与其进行电接触。图 1 是活体轴突内部的电极的照片。巨轴突呈现为明显的空白区，四周包围着小纤维和构成神经干其余部分的结缔组织。可以看到电极内部的银导线及其顶部约 1 毫米的部分。在轴突上端可以记录到小的动作电位，随着电极下降动作电位逐渐增大，到距离套管约 10 毫米时达到恒定数值，80~95 毫伏。在这个范围内，

a constant amplitude of 80–95 mv. at a distance of about 10 mm. from the cannula. In this region the axon appeared to be in a completely normal condition, for it survived and transmitted impulses for several hours. Experiments with external electrodes showed that the action potential was conducted for at least a centimetre past the tip of the micro-electrode.

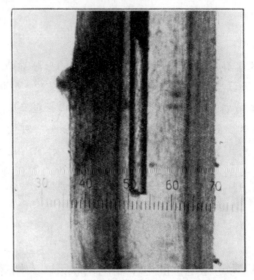

Fig. 1. Photomicrograph of electrode inside giant axon. 1 scale division = 33 μ.

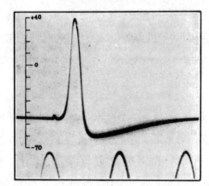

Fig. 2. Action potential recorded between inside and outside of axon. Time marker, 500 cycles/sec. The vertical scale indicates the potential of the internal electrode in millivolts, the sea water outside being taken at zero potential.

These results are important for two reasons. In the first place they prove that the action potential arises at the surface, and in the second, they give the absolute magnitude of the action potential as about 90 mv. at 20°C. Previous measurements have always been made with external electrodes and give values which are reduced by the short-circuiting effect of the fluid outside the nerve fibre.

The potential difference recorded between the interior and exterior of the resting fibre is about 50 mv. The potential difference across the membrane may be greater than this, because there may be a junction potential between the axoplasm and the sea water in

轴突看起来处于完全正常的状态，因为它可以存活并传输脉冲达几小时之久。外部电极实验表明，动作电位的传导至少越过微电极顶端 1 厘米。

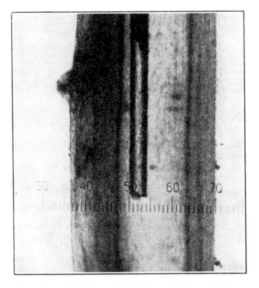

图 1. 植入巨轴突的电极的照片。1 刻度 = 33 微米。

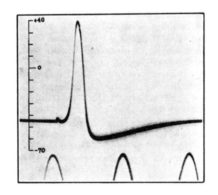

图 2. 记录到的轴突内部与外部之间的动作电位。计时器，500 周 / 秒。纵坐标以毫伏为单位显示内部电极的电势，以外部海水作为电势零点。

这些结果是很重要的，原因有两点。首先，它们证明动作电位出现于表面，其次，它们给出了动作电位的绝对数值，在 20℃ 时约为 90 毫伏。以前的观测都是用外部电极进行的，所以得到的数值由于神经纤维外部流体的短路效应而减小。

我们记录到的静息纤维内部与外部之间的电势差约为 50 毫伏。膜两边的电势差可能比这个数值更大些，因为在轴突细胞质与电极顶部的海水之间可能存在接界电

the tip of the electrode. This potential cannot be estimated, because the anions inside the nerve fibre have not been identified.

We wish to express our indebtedness to Mr. J. Z. Young, whose discovery of the giant axon in *Loligo* made this work possible.

(**144**, 710-711; 1939)

A. L. Hodgkin and A. F. Huxley: Laboratory of the Marine Biological Association, Plymouth, August 26.

Reference:
1. Young, J. Z., *Proc. Roy. Soc.*, B, **121**, 319 (1936).

势。由于尚未确定神经纤维内的负电荷，所以无法估计这一电势。

我们向扬先生表示感谢，正是他发现了枪乌贼的巨轴突才使得本项研究成为可能。

（王耀杨 翻译；金侠 审稿）

Gene and Chromosome Theory*

H. J. Muller

Editor's Note

Hermann Joseph Muller, an American geneticist, here reports on a recent meeting to discuss the current understanding of genes and chromosomes. Clearly the outbreak of war had severely disrupted proceedings, forcing Muller to rely in some cases on written abstracts. As his comments reveal, it was then still unclear whether the critical component of chromosomes, on which genes reside, was protein or nucleic acid—the genetic role of DNA was still unknown. That question was further complicated by studies on viruses, which, like chromosomes, are intimate mixtures of both substances. Much of the work Muller discusses examines gene mutation caused by ionizing radiation such as X-rays—the topic for which Muller himself was awarded a Nobel Prize seven years later.

THE treatment of gene and chromosome theory was on a far more analytical plane than ever before. Thus, the multiform variations of behaviour of chromosomes at meiosis, and the rules governing these variations, were shown by Darlington to trace back to variations in two primary factors: the region (centric or telic) in which pairing begins, and the time limit set to the pairing process; normally, these must be adapted to each other, so that a mixing of two systems tends to disturb the balance.

Regarding the mechanism whereby structural changes in chromosomes come about, a series of experiments by the present author and co-workers was reported, substantiating and extending, for Drosophila, the earlier conclusions of Stadler, of McClintock and of Sax on plant material, and of Muller and Belgovsky on Drosophila, to the effect that the manifold kinds of structural changes capable of surviving indefinitely are all caused by two distinct primary processes, succeeding one another. These are: (1) breakage of the chromonema at two or more points, followed (2) (though not until after the spermatozoon stage is passed) by two-by-two junction between the adhesive broken ends, giving a new linear order. Distant breakages, giving gross changes, were shown to result from separate individual ionizations, but nearby ones, giving minute rearrangements, to result from one and the same ionization by a spreading of its effect. The reports and demonstrations of a large series of separate investigators—Bauer, Fabergé, Demerec, Camara, Catcheside, Oliver and Belgovsky—in one way or another agreed with or led to one or more of the

---

* Although the Russian geneticists had withdrawn and most of the Germans had left before the time for their addresses, their abstracts had already been submitted, and in this Section the policy was followed of reading all such papers *in absentia*. They are accordingly included in this report, although only the papers of official members will be printed in the *Proceedings*. Also included here are papers given in joint meetings of this and the next Section.

748

基因与染色体理论*

分以往align的right

马勒

编者按

美国遗传学家赫尔曼·约瑟夫·马勒在此对最近一次有关基因和染色体现状的会议进行了综述。很显然，战争的爆发严重阻碍了会议进程，这使得马勒有些时候不得不依赖于一些书面的摘要。就像他的文章揭示的那样，当时仍然不清楚携带基因的染色体主要是由蛋白质组成还是由核酸组成——DNA 的遗传角色仍然是未知的。对病毒的研究使该问题更加复杂，和染色体类似，病毒也是由蛋白质和核酸混合组成的。马勒探讨的大部分工作都与电离辐射（例如 X 射线）诱导的基因突变有关，他自己也因在该领域的研究成果而在 7 年后获得了诺贝尔奖。

在对待基因与染色体理论方面现在正处于一个比以往任何时候都更具分析性的平台之上。因此，达林顿指出，在减数分裂阶段染色体行为多种形式的变化以及控制这些变化的规则，可以回溯到两个主要因素的变化：配对开始时的区域（着丝点区域或目的区域），以及配对过程受到的时间限制；通常情况下，这些都是必须彼此适应的，以至于混合两个体系便会倾向于扰乱平衡。

关于染色体中产生结构变化的机制，笔者与合作者们报道了用果蝇进行的一系列实验，证实并推广了斯塔德勒、麦克林托克和萨克斯之前基于植物材料得出的结论以及马勒和别利戈夫斯基之前基于果蝇得出的结论，大意是，能够保存下来的多种类型的结构变化可能是由两个截然不同的相继发生的重要过程引起的。它们是：(1) 染色丝在两个或更多个位点断裂，接着 (2) 黏性的断裂末端之间通过两两接合形成新的线性序列（不过要等精子阶段过去之后）。导致总体改变的远距离断裂看来是由个别独立的电离作用造成的，而导致微小重排的邻近距离的断裂则是由同一次电离及其效果的传播造成的。许多彼此独立的研究者——鲍尔、法贝热、德梅雷茨、卡马拉、卡奇赛德、奥利弗和别利戈夫斯基——的报道和论述以不同的方式同意或得到了一条或几条相同的结论，尽管有些研究者直到那时仍持有相反的观点。

尽管俄国遗传学家已经撤离，而且大多数德国学者在他们的演讲时间之前就已离去，但由于他们都已经提交了论文摘要，依会议条例，所有这类的论文也给予缺席宣读。因此它们也包含于这份报道之中，但只有正式与会成员的论文将会印在《会刊》中。此外，这里还包括在本部分和下一部分的联席会议中提交的论文。

same conclusions, although some investigators had until then held contrary views. Further light was thrown on chromosome structure and on the changes to which it is subject, in a special evening lecture by Metz, recounting his notable findings in Sciara.

A considerable group of papers analysed the special, though not absolutely distinctive, properties of heterochromatic regions of chromosomes. In this connexion, further illustrations were given of the high breakability of these regions by Kaufmann, Camara, Prokofyeva and Sidorov, and of their somatic variability (correlated in related cells) in respect to manner of chromatin staining, manner of aggregation of chromonemata and gene functioning. It was noted that all these properties extend, although to a degree diminishing with distance, beyond the originally heterochromatic regions into regions lying near to them in the chromonema, by a kind of "position effect" (Schultz, Prokofyeva, Panshin, Khvostova), and the important new point was brought out that the variations in staining—which, as proved by Caspersson and Schultz's studies of ultra-violet absorption spectra, fairly represent nucleic acid distribution—and the variations in gene activity are correlated with one another. Evidence was also adduced, by Prokofyeva and by Kaufmann, that small interstitial regions having some degree of heterochromaticity are scattered rather widely throughout the chromatin, and often coincide with regions that apparently originated relatively recently as duplications (which suggests that genes may become heterochromatic by a kind of denaturizing degeneration).

Analysing the mechanism of gene mutation on the basis of a great series of experiments, Timoféeff-Ressovsky brought out its causation (1) by individual atomic activations, apparently resulting from the accidental peaks of kinetic energy of thermal origin, as well as (2) by individual ionizations, resulting from radiation. The dependence on single ionizations was further strengthened by Rai-Choudhuri's finding that even radiation of intensities so low as 0.01 r./min. (a hundred times lower than the lowest previously used in such work) is so effective, ion for ion, as radiation of higher intensities. At the same time, the generality of the gene mutation effect of radiation was strengthened in another important way, by its definite extension to mammals (mice), in experiments of P. Hertwig.

In earlier work, no dependence of the frequency of the gene mutations upon the closeness of spacing of the ions within the radiation-paths could be detected, but newer work by Timoféeff-Ressovsky and his collaborators, reported at the meeting, suggested that a perceptible influence of this kind might be found by the use of the extremely closely spaced ions resulting from some neutron radiation. If so, it should be possible to estimate the "sensitive volume" for a gene mutation, that is, the amount of contiguous space occupied by material so constituted that one ionization, occurring anywhere within it, is capable of producing some one or more of a given series of alleles. Similar work, utilizing the frequency of chromosome breakage instead of that of gene mutations, was reported by Marshak. In this connexion, however, it should be noted that if, as seems likely, only a small proportion of the ionizations occurring within the region in question actually resulted in the effect looked for, this method would tend to lose its efficacy. Moreover, we have no basis for identifying the volume or area in question with that of the gene or

750

梅茨在一次特殊的晚间演讲中进一步阐述了染色体结构及其发生的变化，详细说明了他研究蕈蚊时取得的重要发现。

有相当数量的论文分析了染色体异染色质区域的一些虽非绝对独有但十分特殊的性质。关于这一点，考夫曼、卡马拉、普罗科菲耶娃和西多罗夫对这些区域的高度易断裂性及与染色质着色方式和染色丝凝聚及基因作用方式有关的细胞体易变性（在相关细胞中）进行了更深入的阐释。值得注意的是，所有这些性质都通过一种"位置效应"（舒尔茨、普罗科菲耶娃、潘申、赫沃斯托娃）延伸到最初的异染质区域之外——尽管在一定程度上随着距离而减少——进入染色丝中与其邻近的区域中。因而有人提出下面这一重要的新观点，着色的变化与基因活性的变化彼此关联——这已被卡斯帕森和舒尔茨通过紫外吸收光谱研究证明。紫外吸收光谱可以清楚地反映出核酸的分布。普罗科菲耶娃和考夫曼还发现，具有某种程度异染色质性的小间隙区域相当广泛地分散于染色质中并且经常与那些显然是新生成的重复段区域相一致（这意味着基因可以通过裂解方式而表现出异染色质性）。

在大量实验的基础上，季莫费耶夫–列索夫斯基对基因突变的机制进行了分析并提出其起因：（1）通过孤立的原子活化作用，它们显然是来自热动能的突然增值；（2）通过来自辐射的单一电离化作用。拉伊–乔杜里发现，在单一离子水平，即使是强度低达 0.01 r/min（比以前进行的这类研究中所用的最低强度还低 100 倍）的辐射也能像较高强度的辐射一样有效，这进一步证实了对单一电离化作用的依赖性。同时，赫特维希在哺乳动物（老鼠）中的实验通过另一种重要的方式证实了辐射对于基因突变的影响的普遍性。

在早期研究中，没能检测到基因突变频率是否依赖于放射线的离子间隔密度，但是在这次会议上报道的季莫费耶夫–列索夫斯基及其同事的近期研究表明，利用某种中子辐射产生的高密度离子，可以观测到这类影响。如果确实如此，就应该有可能估计出基因突变的"灵敏范围"，即这样形成的物质所占据的毗邻空间的总和，其中任何位置发生的一次电离化都可以产生一个或多个等位基因的特定序列。马沙克报道了利用染色体断裂频率而非基因突变频率进行的类似研究。不过就此而言，应该注意的是，若在可疑区域中仅仅发生小频率的电离化就导致预期结果的话，这种方法就会丧失有效性。此外，我们尚无法将可疑范围或区域与基因或染色丝自身的范围或区域对应起来。

chromonema itself.

In addition to the seemingly simple thermal effect of van't Hoff type, above-mentioned, there were shown—by Plough, Timoféeff-Ressovsky, Kerkis and Zuitin—to be decided increases in mutation frequency attending the abnormal physiological states of organisms subjected to temperature changes too rapid or too extreme for the organisms to adjust to them. This makes the search for special chemical influences affecting gene mutations seem more promising, despite certain negative results reported by Auerbach with carcinogenic substances. The sensitivity, as well as the intricacy, of the chemical complexes conditioning mutation was further evidenced by the strong dependence of the general mutation frequency upon the genetic complex present, as reported by Plough and by Tiniakov, and more especially by observations of Rhoades and of Harland showing certain enormous and highly specific mutational effects on particular genes, not previously known as "mutable genes", by other particular genes and genecombinations.

The series of papers dealing with the production of mutations by ultra-violet light showed the notable progress made in this field since the last congress. It was shown by Stadler that the curve representing the effectiveness of different wave-lengths in producing gene mutations and deficiencies in maize pollen begins at about 313 mμ and rises to a peak at 254 mμ, declining thereafter. This is suggestively similar to the absorption spectrum of nucleic acid and quite different from that of protein. Both Hollaender, working on fungus spores, and Knapp and Schreiber, on spermatozoids of Sphaerocarpus, reported results substantially similar to this, although their peaks of mutational effect (as well as of directly lethal effect) were at 265 mμ, which corresponds more exactly to the absorption peak for thymonucleic acid. In Hollaender's work, however, there was also a secondary peak, at 238 mμ; and another peculiarity in his results (one suggesting differential sensitivity of different spore stages) was a falling off in the mutation frequency of surviving individuals at very high doses.

In Stadler's work, a basis was found for drawing qualitative distinctions between several different classes of radiation effects. Thus, the frequency of abortive embryos at different wave-lengths, unlike that of gene mutations and deficiencies, failed to follow the nucleic acid curve, as it showed too high frequencies for the shorter wave-lengths. While these abortive embryos may after all represent some kind of non-genetic effect, the same cannot be said of sectional rearrangements of chromosomes (translocations). Stadler found that the latter were not produced by ultra-violet light, or were produced with a markedly lower frequency than by X-rays of the same gene mutation-producing strength.

This result, which was corroborated for Drosophila sperm by experiments carried out by Muller and Mackenzie, gives some ground for supposing that ultra-violet does not act by breaking the chromosomes, and that therefore gene mutations may not consist merely of linear rearrangements of ultra-small size, involving "intragenic" breakage and reunion. The latter idea, which is not yet actually refuted, would have tended to make the concept of the segmentation of the chromosomes into discrete "genes" a mere matter of verbal

除以上提到的范特霍夫类型的看似简单的热效应之外，普劳、季莫费耶夫-列索夫斯基、克尔基斯和祖伊京等人还指出，由于温度变化过快或过于极端以至于有机体无法适应而处于异常生理状态时，突变频率明显增加。这使得对于影响基因突变的特殊化学效应的研究看来更具前景，尽管奥尔巴赫在关于致癌物质的报道中存在某些阴性结果。普劳和京亚科夫的报道显示，基因突变的频率依赖于遗传体系，这进一步证实了化学复合物诱导突变发生的敏感性和复杂性。罗兹和哈兰德的观测也提供了进一步的证据。他们发现，某些特定的基因或基因组能够诱发另外的某些特定基因（之前并不知道这些特定基因就是"易突变基因"）发生大规模的高度特异性的突变。

一系列探讨因紫外线作用而产生突变的论文，展示了自上次会议以来该领域取得的重要进展。斯塔德勒展示了不同波长的光线引起玉米花粉的基因发生突变和缺陷的效果随波长变化的曲线。它始于约 313 纳米，于 254 纳米处达到峰值，此后逐渐下落。发人深省的是，这与核酸的吸收光谱类似，而与蛋白质的截然不同。研究真菌孢子的霍兰德和研究球葫苔属的游动精子的纳普与施赖伯都报道了在本质上与此类似的结果，只不过他们的结果中突变效应峰值（以及直接致死效应）出现在 265 纳米处，这与胸腺核酸的吸收峰对应得更为精确。不过，在霍兰德的研究中还出现了 238 纳米处的次级峰；他的结果中的另一个特异之处（暗示不同孢子时期的灵敏度不同）是在极高剂量时存活个体的突变频率下降。

斯塔德勒在研究中发现了定量区分不同类型辐射效应的基础。例如，不同波长下败育胚的频率——与基因突变和缺陷的频率不同——与核酸曲线并不一致，因为在较短波长处败育胚出现的频率非常高。虽然这些败育胚终究有可能代表某种非遗传效应，染色体的部分重排（易位）则不同。斯塔德勒发现后者不是由紫外线引发的，或者说相对于 X 射线而言，由紫外线引发相同的基因突变强度的概率极低。

被马勒和麦肯齐用果蝇精子进行的实验所确认的这一结果为下列假定提供了某种基础，即紫外线并不是通过使染色体断裂而起作用，因而基因突变可能不仅仅是由诸如"基因内"断裂和重接等极小尺度的线性重排组成的。后一观点——目前还没有被实质性地驳倒——将会倾向于将染色体区段的概念简称为离散的"基因"。斯

convenience. Possibly connected with the same series of problems was Stadler's further finding that the gene mutations produced by ultra-violet are far oftener "fractionals"—that is, confined to one of the chromatids derived from a given treated chromosome—than are those produced by X-rays.

For the first time in the history of genetics congresses a session was included on virus and protein studies in relation to the problem of the gene. A peculiar case of non-chromosomal, probably virus, "inheritance" in Drosophila, was reported by L'Héritier and Teissier. As was brought out by Mckinney, by Gowen and by Kausche, viruses, now known to be crystallizable nucleoproteins, have the distinctive combination of properties characteristic of genes, namely, mutation and (despite mutation) self-duplication, thus substantiating the concept (Muller, 1921, 1926) that viruses represent relatively free genes, and that the gene constitutes the basis of life.

An illuminating account was given by Astbury of his and other modern studies of the chemical structure of viruses and other proteins, with especial consideration of those features which might help to explain the gene's property of mutable self-duplication. He, as well as Caspersson and Schultz (who reported an increase of nucleic acid during periods of growth, both for chromosomes and for cytoplasm), directed attention to the role which nucleic acid may have in this process. The significant fact was reported by Astbury that the nucleic acid spacings are of the same magnitude as those within the protein (polypeptide) chain, a feature which would allow the nucleic acid to unite in parallel with the protein and so perhaps to serve as an intermediary in its synthesis. The paper of Mazia also was of interest in this connexion, since it showed that in salivary chromosomes the framework is not disintegrated by digestion either by pepsin or by nuclease, though it is by trypsin; hence it probably consists of protamine or histone chains, bound together laterally in some other way than by nucleic acid cross-connexions. Other chemical studies of nuclear material were reported by Gulick, which among other things cast doubt on the presence of iron in chromosomes, thus bringing their composition closer to that of the virus.

Our present knowledge of the internal structure of the tobacco mosaic virus particles, as disclosed by the pioneer X-ray diffraction studies of Bernal, Crowfoot et al., was described by D. Crowfoot as well as by Astbury. It was shown that in these rods, which are 15 mμ thick and at least ten times as long, the smallest possible chemical unit—that associated with one nucleotide—must contain about fifty-four amino-acid radicals, although there may be a geometrical sub-unit as small as one eighth of this volume (that is, about 1 mμ each way). But these units (or sub-units) are grouped according to a regular pattern into larger aggregates, about 7 mμ long, and the latter in turn are grouped in a regular way to form the aggregate of high order—the virus rod itself. The globular protein insulin, as well as the fibrous proteins, all show elementary units of about the same size, but the mode of aggregation varies both with the protein and with the conditions (pH, amount of water, etc.) under which it is being kept. Thus this type of analysis is already bridging the gap between the structures of the chemist and those of the microbiologist and geneticist.

754

塔德勒的进一步发现可能与同样的一系列问题有关联，即由紫外线产生的基因突变比由 X 射线产生的基因突变更为频繁地"断片化"——即突变局限于经过特定处理的染色体中的一条染色单体上。

在遗传学会议的历史上，这是第一次举办关于与基因问题相关的病毒和蛋白质研究的分会。莱里捷和泰西耶报道了果蝇中一种特殊的非染色体式的"遗传"，可能是病毒。如同麦金尼、高恩和考施所提出的那样，病毒，现在已经知道就是结晶核蛋白，具有基因性状性质的独特组合，即突变和（不考虑是否突变）自我复制，从而证实了病毒代表相对自由的基因以及基因构成生命基础的观点（马勒，1921 年，1926 年）。

阿斯特伯里就其本人的研究和另外一些关于病毒及其他蛋白质化学结构的现代研究进行了精彩的描述，并对那些可能会有助于解释基因的可突变自我复制的特征给予了特别的考虑。他，还有卡斯帕森和舒尔茨（他们报道了处于生长期时染色体和细胞质中核酸的增加），将注意力集中于核酸在这一过程中担当的角色。阿斯特伯里报道的一个重要事实是核酸间隔与蛋白质（多肽）链内的间隔具有同等的数量级，此特点将核酸和蛋白质联系起来，因此有可能在其合成过程中充当媒介。就此而言梅齐亚的论文也是值得注意的，因为该论文指出用胃蛋白酶或核酸酶消化并不能使唾液染色体的骨架分解，但用胰酶消化则能使其分解；因此它可能是由精蛋白或组蛋白链构成的，以核酸交叉连接以外的某种方式横向连接在一起。古利克报道了关于核物质的另外一些化学研究，这些研究和其他一些研究都对染色体中是否存在铁提出了疑问，从而使其组成与病毒的组成又接近了一些。

如同贝尔纳和克劳福特等人进行的先驱性的 X 射线衍射研究所提示的那样，克劳福特和阿斯特伯里阐述了现在我们关于烟草花叶病毒颗粒内部结构的知识。在这种粗为 15 纳米、长则至少 10 倍于此的杆状物中，最小的可能的化学单元——与一种核酸有关——必定包含约 54 个氨基酸残基，尽管也可能还存在一种大小仅为该体积 1/8（即每个维度方向上约 1 纳米）的几何亚单元，但是这些单元（或亚单元）依据一个规则模式聚合形成长约 7 纳米的较大的团聚体，后者再以某种规则方式聚合形成高级团聚体——那就是杆状病毒。球状蛋白胰岛素和纤维蛋白都呈现出同等尺寸的基本单元，但是在蛋白质的团聚方式以及保存所需条件（pH、水量等）方面却各不相同。于是，这类分析在化学家得到的结构与微生物学家和遗传学家得到的结构之间的裂谷上架起了桥梁。

A series of special conferences were held on problems of the gene, presided over chiefly by J. B. S. Haldane. These were very well attended. At these "gene conferences" many of the reports of those unable to attend were read, and many of the above and related questions concerning gene and virus structure and gene mutation were subjected to animated and searching discussion.

(**144**, 814-816; 1939)

近来召开了一系列关于基因问题的专门会议，主要是由霍尔丹主持的。这些会议举办得都很成功。在这些"基因会议"上宣读了很多无法亲自与会人员的报告，并且对上述关于基因、病毒结构和基因突变的很多相关问题都进行了活跃而深入的讨论。

（王耀杨 翻译；金侠 审稿）

Physiological Genetics

C. H. Waddington

Editor's Note

The English biologist Conrad Hal Waddington was one of the pioneers of modern developmental biology, in particular in building links between embryology and development. In this report he gives an overview of the developmental genetics of his time, describing how researchers were then seeking to understand how genes, cellular process and the external environment combine to determine the development and form of the organism. The advent of techniques for sequencing and manipulating DNA has now revolutionized the understanding of development, yet these questions are still at the heart of developmental biology and its interplay with evolution (called "evo devo"). In particular, the relative importance of genes and environment in development is still hotly debated.

ONE of the most active branches of genetics at the present time is the study of the ways in which genes affect developmental processes, and the section devoted to physiological genetics had a full and interesting programme. The problem of genic action is so complex and many sided that very many different methods of approach are possible, and examples of most of these can be found in the papers presented at the Congress.

The embryological approach was well exemplified in a paper by Landauer, in which he reviewed the correlated effects on different organs which are found both in fortuitous teratological specimens and also in abnormalities which are known to be dependent on genes. He suggested that many of these phenomena can be explained by the hypothesis of a general deleterious effect, often on the growth-rate, acting at a period which is critical for a certain set of developmental processes. A general embryological approach of a rather different kind was presented by Waddington, who discussed the relation of genes to developmental processes of the kind exemplified by the organizer reaction.

The importance of nuclear factors in particular steps of differentiation was analysed in more detail by Baltzer, who gave a summary of his well-known and important work on bastard merogons, in which an enucleated egg of one species of newt is fertilized by a sperm of a different species. He showed that some tissues are able to develop to late embryonic stages, while others die presumably owing to disharmonies of nucleus and cytoplasm during particular processes of differentiation; still other tissues, such as some of the anterior mesoderm, while able to live, lose their normal power of inducing other organs—in this instance the balancers and gills. Similar studies of the importance of particular elements in the nucleus were reported by Poulson in his studies of the abnormalities produced by total absence of certain genes (homozygous deficiencies) on the

758

发育遗传学

沃丁顿

编者按

英国生物学家康拉德·哈尔·沃丁顿是现代发育生物学的先驱之一，特别是在建立胚胎学和发育学之间的联系方面有着重要的贡献。在这份报告中他对其所处时代的发育遗传学进行了综述，描述了当时研究者不断探索试图弄清楚基因、细胞过程以及外部环境是如何共同决定生物的发育和类型的。如今DNA测序技术和操纵技术的进步已彻底改变了对发育的理解，然而这些问题仍然是发育生物学及其与进化的相互作用（即"进化发育生物学"）的核心部分。特别地，基因与环境在发育中的相对重要性仍然被热烈地讨论着。

目前遗传学最活跃的分支之一就是研究基因影响发育过程的方式，而且发育遗传学这一部分有一项完整又有意义的计划。基因行为的问题如此的复杂和多面，以至于可能有多种不同的研究途径，其中大部分的实例都可以在大会上提交的论文中找到。

兰道尔的论文从胚胎学途径进行了很好的例证。在论文中他综述了在偶发的畸形生物和与基因有关的异常物种中发现的不同器官的相关效应。他指出这些现象中的许多都能用一般有害效应假说来解释，它作用于某一发育过程至关重要的阶段，常常影响生长速率。沃丁顿则提出了另一种完全不同的胚胎学途径，他讨论了基因与由组织者效应例证的一类发育过程之间的关系。

巴尔策更加详细地分析了细胞核因素在分化的特定步骤中的重要性，他对自己著名且重要的关于无核卵杂交的工作进行了总结，实验中用不同物种的精子来使一种蝾螈的去除了细胞核的卵子受精。他的研究显示一些组织能够发育到晚期胚胎阶段，而其他的组织则死亡了，可能是由于在分化的特定过程中细胞核与细胞质的不协调造成的；还有一些其他的组织，例如一些前中胚层组织，能够生存，但是丧失了诱导其他器官——在这个例子中主要是平衡器和鳃——的正常功能。波尔森在他的关于果蝇的早期发育阶段由于某些基因的完全缺失（纯合缺失）而产生变异的研究中报道了证实细胞核内特定成分重要性的类似研究。

759

early development of Drosophila.

Several reports dealt with gene effects on processes which are chemically more or less defined. Thus we had further chapters in the important and rapidly developing researches of Price, Lawrence and others on flower pigments, and Beadle, Ephrussi and others on Drosophila eye colours. Perhaps the most important new contribution in this field was a fresh study by Sewall Wright of guinea pig coat colours, which he dealt with in a preliminary way many years ago. He was now able to suggest a scheme for the relations of the numerous genes which are known, and to present a mathematical theory of their quantitative interactions.

Quite a different method of approach to the same problem was reported by Schultz and Caspersson, who, by studying the chromosome itself, obtained data which allow of some speculation as to the chemical changes occurring under the immediate influence of genes. Owing to their strong absorption in the middle ultra-violet, the distribution and changes of the substances belonging to the nucleotide group can be followed in the living cell by spectrophotometric methods elaborated by Caspersson. It is found that in the nucleus the nucleic acids seem to the indispensable for the development of chromosomes, and that they are synthesized mainly in earliest prophase just before the chromosomes split. Schultz studied translocations of parts of the heterochromatic regions of the chromosomes, which are rich in nucleic acid, into normal euchromatic regions in Drosophila. In certain of these translocations the adult flies show variegation for characters affected by genes near the position of the break in the chromosome, as though frequent mutations of the genes had occurred in late stages of development. Schultz showed that this variegation is correlated with cytological effects in the salivary chromosome bands corresponding to the variegated genes; these effects ranged from appearance of excess nucleic acid to the assumption of a heterochromatic character and finally apparent disappearance. It is suggested that the appearance of excess nucleic acid is correlated with an inactivation of the gene as a developmental agent. The nucleic acid metabolism can be followed somewhat further, since it is found that the presence of extra heterochromatic material (for example, supernumerary Y-chromosomes) affects both the degree of variegation, the cytological correlates of variegation mentioned above, and also the nucleic acid content of the cytoplasm. The different types of nucleic acid occurring in chromosomes, nucleolus and cytoplasm can be determined spectroscopically and suggestions made as to their functional relations with the fundamental processes of gene action and gene reduplication.

It is possible, as Schultz suggests, that we may in this way obtain some insight into the changes of gene activity in different chromosome regions in the different tissues, which may provide a mechanism for the primary differentiation of the nuclei in development.

(**144**, 817-818; 1939)

一些报告讨论了基因对某些生物学过程的作用，从化学角度来看，这些生物学过程基本上是明确的。因此，我们还用另外的章节讨论了某些重要且迅速发展的研究方向，其中包括普赖斯、劳伦斯等对花色素的研究以及比德尔、伊弗鲁西等对果蝇眼睛颜色的研究。在这个领域中最重要的新发现可能是休厄尔·赖特对豚鼠皮毛颜色的新研究，数年前他就进行过初步的实验。现在他已经能够给出已知的为数众多的基因之间关系的结构图，并且提出它们之间定量关系的数学理论。

针对相同的问题，舒尔茨和卡斯帕森提出了一种截然不同的研究途径，他们通过研究染色体本身获得了一些数据，这些数据引发了人们对受基因直接影响而发生的化学改变的思考。由于其在中紫外区有强烈的吸收，卡斯帕森提出用分光光度计法检测活细胞内核苷酸类物质的分布和变化。他们发现，在细胞核中核酸对于染色体的发育似乎是必不可少的，而且它们主要是在染色体分离之前的最早期合成的。舒尔茨对果蝇中富含核酸的染色体上异染色质区域的一部分转移到正常常染色质区域的易位现象进行了研究。一些发生易位的成体果蝇由于受到染色体中断裂位点附近基因的影响而表现出性状的多样性，犹如发育后期这些基因发生的频繁突变。舒尔茨发现这种多样性与对应于唾液腺染色体条带中多样化的基因的细胞学效应有关；这些效应的作用从过量核苷酸的出现一直延伸到异染色质特征的假设，以及最终明显的消失。研究表明，过量核苷酸的出现与发育相关基因的失活有关。对核苷酸代谢的研究可以进一步深入，因为人们发现额外异染质物质的存在（例如超数 Y 染色体）能够影响多样性的程度、上述多样性的细胞学相关性以及细胞质中的核酸含量。可以借助分光光度计检测染色体、核仁和细胞质中不同类型的核酸，进而就能提出它们与基因行为和基因复制的基本过程之间的功能关系。

正如舒尔茨所提出的，我们很有可能能够通过这种方法深刻了解不同组织的不同染色体区域中基因行为的变化，从而了解发育过程中核初始分化的机制。

（毛晨晖 翻译；王晓晨 审稿）

The Relationship between Pithecanthropus and Sinanthropus

Editor's Note

In 1931, not long after his discovery of "Peking Man" (*Sinanthropus*), Davidson Black—perhaps inevitably—remarked on the similarity between this form and that of *Pithecanthropus* from Java. Relatively little cranial material of *Pithecanthropus* was available at the time. However, by the end of the 1930s, enough had been discovered to allow Gustav von Koenigswald and Franz Weidenreich to attempt this comparison. Their conclusion was that the two forms were no more different than people from two different modern human races. Indeed, both forms are now referred to *Homo erectus*. Neither scientist was to know it, but this was the last occasion any direct comparison could be made—the original Peking Man fossils disappeared during World War II.

DAVIDSON Black had remarked on the great similarity between the first skull of Sinanthropus to be found and the Pithecanthropus skull of Trinil, a condition which induced him to see in Pithecanthropus a Hominid form closely related to Sinanthropus (1931). The additional finds of the latter, unearthed in the interval, have confirmed Black's interpretation in every respect. But on the other hand, since the Pithecanthropus finds remained restricted to that rather incomplete specimen of Trinil, absolute evidence for his true Hominid character was lacking. In such circumstances, there was no other way open but to await the discovery of additional Pithecanthropus material before definitely solving this problem.

These discoveries materialized. Following the recovery of a rather primitive infantile skull (*Homo modjokertensis*) in 1936 and of a lower jaw of an adult individual, one of us (G. H. R. von K.), during 1937, discovered a skull in the undoubted Trinil deposits of Sangiran. This skull, preserved up to the basal region, conforms in every respect as to size, shape, and details to Dubois's Trinil specimen. Dubois, however, opposed the attribution of this skull to Pithecanthropus. Nevertheless, the details of the interior, as well as exterior surfaces of the skull, and also the skiagrams, delineating the otherwise indistinct sutures and breakage lines, show not the slightest trace of irregularity or deformation, such as would be unavoidable if the assembly of the fragments had been artificially adapted to a particular form. To this skull of Sangiran was added another skull fragment derived from the same deposits and of the same site during the summer of 1938, briefly described in *Nature* of October 15, 1938, p. 715, by us. We are now in a position to report on an additional Pithecanthropus find made this year. It concerns the lower part of an upper jaw of unusually large dimensions, comprising the processus alveolares of both sides with completely preserved nasal floor and palate, the complete left dental arch with all the teeth leading from the canine backward, and a part of the right dental arch up to the first molar.

爪哇猿人与中国猿人的关系

编者按

1931 年，戴维森·布莱克（步达生）在发现"北京人"（中国猿人）后不久，又对中国猿人与在爪哇发现的爪哇猿人之间的相似性进行了探讨，这项工作似乎是自然而然顺理成章的。当时爪哇猿人的头骨材料相对匮乏。然而到了 20 世纪 30 年代末，已经有了充足的材料，这使得古斯塔夫·冯·柯尼希斯瓦尔德（孔尼华）和弗朗茨·魏登赖希（魏登瑞）可以尝试进行比较研究。他们认为这两类猿人之间的差别并不比两个现代人种之间的差别大。确实，现在这两个类型都被归为直立人。然而两位科学家都没有预料到，这是他们最后一次有机会对这两类猿人进行直接的比较——北京人化石的原件在第二次世界大战期间遗失了。

步达生对已发现的第一件中国猿人头骨和特里尼尔爪哇猿人头骨之间极大的相似性进行了探讨，基于对比结果，他认为爪哇猿人是一种与中国猿人关系紧密的原始人类的新类型（1931 年）。后来发掘出的中国猿人的其他化石在各个方面都证实了步达生的观点。但另一方面，对于爪哇猿人化石的研究仅限于一件取自特里尼尔的相当不完整的标本，因此缺乏有关其真正人类形态的确实证据。在此情形下，彻底解决这个问题之前除了继续等待发现其他爪哇猿人化石资料之外，别无他法。

化石标本终于被发现了。1936 年在修复了一件相当原始的婴儿头骨（莫佐托克人）和一件成年个体的下颌之后，笔者之一（孔尼华博士）于 1937 年在桑吉兰无疑属于特里尼尔层的沉积中发现了一件头骨。这件头骨位于基底区上方，其大小、形状及细节等各方面都与杜布瓦的特里尼尔标本一致。然而，杜布瓦反对把这件头骨归为爪哇猿人。不过，这块头骨的内部细节与外部表面，以及能从中看出不同的模糊骨缝线与破裂线的 X 射线照片，都没有显示出丝毫不规则或变形的迹象。如果把这些碎片按照某种特别形式人为地拼接组装起来，就难免会造成一些不规则或变形迹象。我们在 1938 夏天于同一遗址的同一沉积层又发现了桑吉兰头骨的一件碎片，1938 年 10 月 15 日的《自然》第 715 页曾对其进行过简要的描述。现在我们要报道在这一年里所得到的有关爪哇猿人的另外一个发现。这是一件尺寸非常大的上颌的下部，它包括两侧的齿槽突，保存有完整的鼻腔底和腭骨；有从犬齿向后的全部牙齿的齐全左齿弓以及到第一白齿的部分右齿弓。门齿缺失，但是它们的齿槽被保存

The incisors were lost, but their alveoli are preserved (see accompanying reproductions, *a* and *c*). Afterwards the skull belonging to this jaw was also found. It comprises the posterior third of the brain case, including the entire basis.

Morphologically and geologically, we believe we are justified in attributing all these finds to Pithecanthropus. This type is thus represented by the following finds:

(1) Skull of Trinil (Dubois, 1891): Pithecanthropus skull I.

(2) Mandible of Kedung Brubus (Dubois, 1891): Pithecanthropus mandible A.

(3) Juvenile skull of *Homo modjokertensis* (Geol. Survey 1936)[*].

(4) Mandible of Sangiran (v. Koenigswald, 1936): Pithecanthropus mandible B.

(5) Skull of Sangiran (v. Koenigswald, 1937): Pithecanthropus skull II.

(6) Skull fragment of Sangiran (v. Koenigswald, 1938): Pithecanthropus skull III.

(7) (*a*) Maxilla of Sangiran (v. Koenigswald, 1939); (*b*) Skull fragment of Sangiran (v. Koenigswald, 1939): Pithecanthropus skull IV.

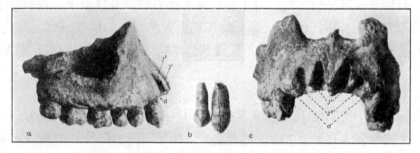

(*a*) Upper jaw of a male Pithecanthropus (Sangiran, January 1939), viewed from the right side. *d*, diastema; J^1, alveolus of J^1; J^2, alveolus of J^2. $\frac{2}{3}$ nat. size.

(*b*) Upper right canine and first premolar of a male individual of Sinanthropus (*F*IV), so as to demonstrate the protrusion of the canine. $\frac{2}{3}$ nat. size.

(*c*) The same as (*a*) but viewed from in front. $\frac{2}{3}$ nat. size.

The Sangiran skull (Pithecanthropus skull II) resembles the Trinil skull as closely as one egg another. The former (skull II) is only slightly smaller than skull I—its capacity being 835 c.c. as compared with 914 c.c. of the Trinil skull—but its parietal and occipital parts are relatively broader. The Sangiran fragment (Pithecanthropus skull III) is, in its preserved

---

[*] On the basis of my study of the original, I have now come to the conclusion that this infantile skull really represents a Pithecanthropus child. I shall report on this elsewhere.—F. W.

了下来（见所附的复制品，*a* 与 *c*）。后来这个下颌所属的头骨也被发现了。这件头骨包括脑壳的后 1/3，其中包含整个颅底部。

从形态学与地质学的角度分析，我们相信将所有这些化石归为爪哇猿人是合理的。因而爪哇猿人类型由以下化石来代表：

（1）特里尼尔头骨（杜布瓦，1891 年）：爪哇猿人头骨 I。

（2）凯登布鲁伯斯下颌骨（杜布瓦，1891 年）：爪哇猿人下颌骨 A。

（3）莫佐托克托人少年头骨，（《地质调查》，1936 年）[*]。

（4）桑吉兰下颌骨（孔尼华，1936 年）：爪哇猿人下颌骨 B。

（5）桑吉兰头骨（孔尼华，1937 年）：爪哇猿人头骨 II。

（6）桑吉兰头骨碎片（孔尼华，1938 年）：爪哇猿人头骨 III。

（7）（*a*）桑吉兰颌骨（孔尼华，1939 年）；（*b*）桑吉兰头骨碎片（孔尼华，1939 年）：爪哇猿人头骨 IV。

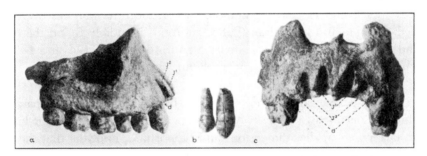

（*a*）男性爪哇猿人（桑吉兰，1939 年 1 月）上颌，右侧观察。*d*= 齿隙；$I^1=I^1$ 齿槽；$I^2=I^2$ 齿槽。实际尺寸的 $\frac{2}{3}$。

（*b*）中国猿人男性个体的右上犬齿和第一前臼齿（FIV），以此来显示凸出的犬齿。实际尺寸的 $\frac{2}{3}$。

（*c*）与（*a*）一样，但是是从前面观察。实际尺寸的 $\frac{2}{3}$。

桑吉兰头骨（爪哇猿人头骨 II）与特里尼尔头骨极为相似。只是前者（头骨 II）稍小于头骨 I（其颅容量为 835 毫升，特里尼尔头骨颅容量为 914 毫升），但其顶骨与枕骨相对较宽。桑吉兰头骨碎片（爪哇猿人头骨 III）的保存部分，不像其他两个（头骨 I 和 II）那样扁，但是在其他方面均与它们相似。另外，它的矢状

---

[*] 这篇文章是基于我最初的研究成果，现在我得出的结论是：这件婴儿头骨确实属于一个幼年爪哇猿人。我将在别处对此进行报道。——魏登瑞

part, not so flat as the other two (skulls I and II), but otherwise resembles them in every detail. In addition, it bears a distinctly pronounced crista sagittalis.

Of the Sinanthropus cranial material, skulls E and II of Locus L are most suitable for a comparison, having capacities of approximately 915 c.c. and 1,015 c.c., respectively. These skulls are slightly larger than the Pithecanthropus skulls, but they are the same in general form and particularly in height.

The main differences so far as the skull cap is concerned consist in that in Pithecanthropus (skulls I and II) the supraorbital tori pass directly over to the extraordinarily flattened forehead, whereas in Sinanthropus the supraorbital ridges are much more demarcated from the tuber-like vaulted but otherwise also strongly receding forehead. On the other hand, the obelion region in Sinanthropus is flat, while in Pithecanthropus it is rounded off. The greatest similarity is seen in the general form and structure of the temporal and occipital bones, and there is absolute conformity in some special details of these bones. The Pithecanthropus fragment III and Sinanthropus skull E show identical features even in apparently unessential structures. Beside the first mentioned sagittal crest, there is in the obelion region on each side of the sagittal suture a short groove which Black described and illustrated in Sinanthropus skull E.

With regard to the lower jaws, that from Kedung Brubus is characterized as representing Pithecanthropus by the exclusively basal location of a broad digastric fossa—the only usable criterion. This mandible corresponds in size and proportions to the small female Sinanthropus jaws A and H, and the mandible M II, more recently discovered, and not yet described. The jaw from Sangiran (Pithecanthropus mandible B), on the other hand, is large, and corresponds to the large male Sinanthropus jaws G I and K I, with the exception that the frontal section is considerably thicker than in the latter.

The upper jaw from Sangiran (see reproductions) has as yet nothing comparable among the Sinanthropus specimens, for the two upper jaw fragments of the latter known up to the present have much smaller dimensions and proportions, implying that they belong to female individuals, while the upper jaw from Sangiran must be ascribed to a male.

The differences in size and proportions of the upper and lower jaws of both Pithecanthropus and Sinanthropus, apparently chiefly due to sexual differences, also serve as a criterion of the cranial conditions in this respect. The lower and upper jaws from Sangiran are much too large for the small Pithecanthropus skulls I and II, whereas the lower jaw fragment of Kedung Brubus would seem to fit them better. It may be concluded, therefore, that the two Pithecanthropus skulls, regardless of their minor differences in size and thickness, must have belonged to female individuals, whereas the rather heavy Pithecanthropus skull IV represents undoubtedly an old male individual.

With respect to the dentition, the Pithecanthropus molars are larger than those of Sinanthropus available so far. But the lower incisors—so far as the size of the crowns can be estimated from that of the alveoli—and especially the lower canines of

脊非常明显。

在中国猿人头盖骨化石中，头骨 E 和发现于 L 地点的 II 号头骨最适于进行比较研究，其颅容量分别约为 915 毫升和 1,015 毫升。这些头骨稍大于爪哇猿人头骨，但它们的形状大体一样，特别是高度。

头盖骨的主要差别在于：爪哇猿人（头骨 I 和 II）的眶上圆枕直接过渡到非常扁平的前额，而中国猿人的眶上脊（眉脊）与拱状隆起且同样呈低平状的前额之间的界限更为明显。另一方面，中国猿人的顶孔区扁平，而爪哇猿人的顶孔区圆隆。最大的相似性在于颞骨和枕骨的一般形式和结构，而且这些骨头的某些特别的细节也完全相似。爪哇猿人 III 号头骨碎片与中国猿人头骨 E 呈现出完全相似的特征，即便在一些非重要的结构上也是如此。除首次提到的矢状脊外，在矢状缝每一边的顶孔点区都有一条短沟，步达生在中国猿人头骨 E 中对此作了描述与图示。

凯登·布鲁伯斯发现的下颌骨因其宽阔的二腹肌窝位置靠近底部——这是唯一可用的判断标准——被视为爪哇猿人的代表。在大小和比例上，这件下颌骨与尺寸较小的女性中国猿人下颌骨 A 与 H 以及最近发现的尚未被描述过的下颌骨 M II 皆相似。另一方面，桑吉兰下颌骨（爪哇猿人下颌骨 B）尺寸较大，相当于大型中国猿人男性下颌骨 G I 和 K I。只是桑吉兰下颌骨的前部切面明显厚于中国猿人颌骨。

发现于桑吉兰的上颌骨（见复制品）与中国猿人标本没有任何可比性，因为迄今发现的两件中国猿人上颌骨碎片的尺寸和比例都很小，这表明它属于女性个体，而出自桑吉兰的上颌骨应该属于男性。

对于爪哇猿人和中国猿人，其上下颌骨在大小和比例上的差异主要是由于性别差异引起的，这也是判定头骨情况的一个标准。与尺寸较小的爪哇猿人 I 号和 II 号头骨相比，发现于桑吉兰的下颌骨和上颌骨要大得多，而凯登·布鲁伯斯下颌碎片似乎与它们更加符合。因此，尽管这两件爪哇猿人头骨在大小和头骨厚度方面存在微小差异，我们仍然可以推断，它们一定是属于女性个体，而粗重的爪哇猿人 IV 号头骨则无疑应属于老年男性个体。

牙齿方面，爪哇猿人的臼齿大于迄今为止发现的所有中国猿人的臼齿。但是爪哇猿人的下门齿（目前可根据齿槽大小估测出齿冠大小）尤其是下犬齿，无疑都比

Pithecanthropus, are decidedly smaller than those of Sinanthropus. The canines of the Pithecanthropus upper jaw (see reproduction *a*) protrude considerably beyond the premolars, despite the fact that both are much worn. They conform in this respect to the Sinanthropus canines (see *b*) so far as male individuals are concerned. The pattern of the Pithecanthropus canine resembles that of Sinanthropus, but is less complicated by lacking the cingulum so characteristic of the latter. These differences are also true for the premolars and molars, in so far as no one of these teeth in Pithecanthropus shows such primitive characteristics as are found in Sinanthropus. Pithecanthropus, therefore, undoubtedly stands in this respect at the upper limit of the range of variation approaching the Neanderthal types. On the other hand, in respect to other features, Pithecanthropus is of a more primitive nature than Sinanthropus: for example, the second molar of both upper and lower jaw of Pithecanthropus is distinctly larger than the first, and the third lower molar the longest of all three. In addition, it is evident that—the first example of a fossil Hominid known hitherto—the upper canines of both sides are separated from the lateral incisors by a broad diastema, the width of which amounts to 6.2 mm. on the right side (*d*, in reproductions *a* and *c*). This width comes close to the average width known for male gorillas and corresponds to that of the male orang (average width, according to Remane, 6.8 mm. and 6.2 mm. respectively).

The dental arch of the Pithecanthropus upper jaw is long and relatively narrow. The front teeth, according to the alveoli, were ranged within a curved line and directed forward, whereas the molars form two straight and backwardly diverging rows. Thus, all the skeletal remains and teeth of Pithecanthropus and Sinanthropus so far available prove the close general relationship between the two types.

With respect to the affinity of the Pithecanthropus femora—that is to say, the so-called Trinil femur, and the five femora afterwards recovered by Dubois and also attributed to Pithecanthropus—it must be taken into consideration that the seven femora of Sinanthropus, most of them represented only by shafts, show significant differences when compared with those femora. All of the Sinanthropus femora display, among other characteristics, a marked degree of platymeria, and at the same time a very low pilaster index; while the supposed Pithecanthropus femora show no indication of this kind, and are in all respects identical with those of modern man. All this points against the probability of their belonging to Pithecanthropus.

Pithecanthropus and Sinanthropus undoubtedly represent the most primitive Hominid forms known hitherto, which, according to Boule, may be ranged collectively under the name Prehominids. Which of the two types must be taken as the more primitive cannot be decided with absolute certainty for the present. Fragments of Sinanthropus skulls suggest that this type includes also specimens the capacities of which did not exceed that of Pithecanthropus II—as, for example, Sinanthropus skull J—and, on the other hand, those with a very long and rather low cranium, as the Sinanthropus skull fragment H III. Nevertheless, it is certain that Pithecanthropus shows some significant characteristics which must be considered more primitive than those evident in Sinanthropus, especially

中国猿人的小。爪哇猿人上颌（见复制品 *a*）犬齿明显突出于前臼齿，尽管它们二者都磨损严重。就男性个体而言，在这方面它们与中国猿人的犬齿（见 *b*）一致。虽然爪哇猿人犬齿与中国猿人犬齿有相似之处，但由于缺少中国猿人特有的齿带，因而并不怎么复杂。前臼齿与臼齿也有这样的差异，也就是说，没有一个爪哇猿人的牙齿能够显示出在中国猿人中发现的这种原始的特征。因此，毋庸置疑，爪哇猿人在这方面是处在变化范围的上限，接近尼安德特人。另一方面，就其他特征而论，爪哇猿人比中国猿人更为原始：例如爪哇猿人上颌与下颌的第二臼齿均明显大于第一臼齿，而下颌第三臼齿是所有三个下颌臼齿中最长的。另外，两侧上犬齿明显与侧门齿分开，其间裂隙很宽，在右侧（*d*，见复制品 *a* 与 *c*）其宽度达 6.2 毫米，而这一特征在迄今为止发现的化石人类中还是第一次出现。这个宽度与雄性大猩猩的已知平均宽度接近，并与雄性猩猩的平均宽度一致（根据雷马内的测量，平均宽度分别为 6.8 毫米和 6.2 毫米）。

爪哇猿人上颌的齿弓较长且相对较窄。从齿槽来看，前齿沿曲线排列且朝向前方，然而臼齿形成直线向后分叉的两排。这样，迄今为止所有可供研究的爪哇猿人与中国猿人的骨骼遗迹和牙齿都证明这两种类型之间总体接近。

至于爪哇猿人股骨（即所谓的特里尼尔股骨，以及后来杜布瓦发现的同样属于爪哇猿人的 5 根股骨）的亲缘性，一定要考虑到，中国猿人 7 件股骨（其中大部分仅保留骨干部分）与这些股骨进行比较时呈现出了重大差别。其他特征还包括，所有的中国猿人股骨扁平程度明显，而且同时股骨的脊指数十分低；推测属于爪哇猿人的股骨没有显示出这种迹象，并且在各个方面都与现代人类的相同。所有这些特点都不支持将它们归为爪哇猿人。

毫无疑问，爪哇猿人与中国猿人代表了迄今为止所知道的最原始的化石人类类型，按照布勒的意见，它们可以一并划归为前人。目前还不能确定这两个类型中哪个类型更为原始。中国猿人头骨碎片表明这个类型也包括脑量不超过爪哇猿人 II 号头骨的标本（例如中国猿人头骨 J），此外还包括具有很长且相当低平的头盖骨的标本，如中国猿人 H III 号头骨碎片。不过，可以肯定爪哇猿人显示出的某些重要特征明显比中国猿人的更原始，特别是具有上颌齿隙。

the presence of a diastema in the upper jaw.

Considered from the general point of view of human evolution, Pithecanthropus and Sinanthropus, the two representatives of the Prehominid stage, are related to each other in the same way as two different races of present mankind, which may also display certain variations in the degree of their advancement.

The Prehominids are separated from the Neanderthal group by a considerable gap. On the other hand, an apparently close relationship exists between Pithecanthropus and *Homo soloensis*, the skulls of the latter appearing like an enlarged form of the former. Certain peculiarities of Pithecanthropus reappear in exactly the same form in *Homo soloensis*. Those traits which suggest an already more advanced type, like the greater cranial capacity, and several other structural features, can be derived directly from Pithecanthropus, and correspond to the condition in the Neanderthal stage already attained by *Homo soloensis*. The two available fragments of the tibia of *Homo soloensis* show no special peculiarities, with the exception of a pronounced platymeria, exhibiting only recent human characters in their general form and in details.

The finds reported herein show that Java has become the most important centre for the study of Prehominid forms. Not only Prehominids, but also the following evolutionary stage, *Homo soloensis*, are represented there. Furthermore, we know that the Wadjak man of Java represents another early form of recent man, whose upper jaw (Wadjak II) displays in some respects a most surprising resemblance to the Pithecanthropus upper jaw.

In conclusion, we wish to express our gratitude to the officers of the Government of the Netherlands East Indies, and the Carnegie Institution in Washington for their generous support, which made possible not only the more recent investigations in Java itself, but also our joint study, conducted in the Cenozoic Research Laboratory, Peiping Union Medical College, Peking, of recently obtained Pithecanthropus material.

G.H.R. von Koenigswald and Franz Weidenreich

(**144**, 926-929; 1939)

*   *   *

The article by Dr. G. H. R. von Koenigswald and Prof. F. Weidenreich on the relationship between Pithecanthropus and Sinanthropus in *Nature* of December 2 is eminently satisfactory to those anatomists who have not been able to understand why these two hominids should ever have been separated generically. Probably my colleague, Dr. S. Zuckerman, was the first to express doubt on the justification for this distinction, in an essay on Sinanthropus[1], and I have on more than one occasion urged that the Peking hominid is but a Chinese variant of Pithecanthropus[2].

从人类进化的一般观点来考虑，爪哇猿人和中国猿人，作为前人阶段的两个典型代表，就如同现代人类的两个不同人种一样，它们可能在进化程度上显示出某些差异，但彼此仍然是密切关联的。

前人与尼安德特人之间存在着相当大的时间间隔。另外，爪哇猿人与梭罗人明显存在着密切的关系，梭罗人头骨似乎是爪哇猿人头骨的放大版。爪哇猿人的某些特征在梭罗人化石上以完全一样的形式再现。那些象征着相对进步的特征，如较大的颅容量以及其他一些结构特征，可能是直接从爪哇猿人演化而来，并且与梭罗人已达到的尼安德特人阶段的情况相符合。现有的两件梭罗人胫骨碎片除呈现明显的扁平外，在一般形状和细节上没有显示出特别现代人类的特点。

在这里所报道的发现表明，爪哇已经成为最重要的研究前人类型的中心。因为那里不仅有被归为前人的爪哇猿人，还有作为其后续进化阶段的梭罗人。此外，我们还知道爪哇的瓦贾克代表着现代人类的另一种早期类型，其上颌（瓦贾克 II）在某些方面与爪哇猿人的上颌惊人地相似。

最后，我们对荷属东印度群岛政府的官员和华盛顿卡内基研究院表示感谢，感谢他们慷慨的支持，他们的支持不仅使得近期在爪哇当地的研究得以进行，也使得我们在北平协和医学院新生代研究室对最近获得的爪哇猿人材料进行共同研究成为可能。

孔尼华，魏登瑞

（田晓阳 翻译；刘武 审稿）

*　　*　　*

孔尼华博士和魏登瑞教授在 12 月 2 日《自然》杂志上发表的有关爪哇猿人和中国猿人之间关系的文章，为那些不能理解为什么这两种原始人类群体属于不同的种属的解剖学家提供了非常令人满意的解释。我的同事朱克曼博士可能是第一位对这一区分的合理性表示怀疑的人，他在一篇关于中国猿人的文章 [1] 中表达了这一观点，我也曾多次极力主张北京人只是爪哇猿人的一个中国变种而已 [2]。

While for some years Prof. Weidenreich, with his first-hand knowledge of the Chinese fossils, has been insisting on their supposed distinctive characters, and has put forward the thesis that they represent an early type of man still more primitive than Pithecanthropus, he now agrees not only that Pithecanthropus and Sinanthropus "are related to each other in the same way as two different races of present mankind", but also that in some significant characteristics Pithecanthropus is the more primitive of the two. If these conclusions are accepted (as they certainly should be) it now becomes necessary finally to discard the generic term Sinanthropus. The Chinese fossils should logically be referred to the species *Pithecanthropus erectus*, or, if it should be thought more desirable as a temporary convenience for purposes of reference, they might be conceded a specific distinction with the name *Pithecanthropus pekinensis*.

The article of Dr. von Koenigswald and Prof. Weidenreich raises again the whole question of the validity of the morphological evidence upon which physical anthropologists often seem to depend for their taxonomic conclusions. For example, the authors use the remarkable argument that, because the femora ascribed to Pithecanthropus of Java show a marked degree of platymeria such as is not found in the femora of "Sinanthropus", therefore they probably do not belong to Pithecanthropus. This statement would provide a pretty exercise in logical analysis. A similar process of reasoning would also lead to the conclusion that, because a femur excavated from a Neolithic barrow shows more platymeria than the femur of a modern Englishman, therefore it cannot belong to *Homo sapiens*! In any event, an acquaintance with recent literature might have informed the authors of a short, but not unimportant, paper by the late Dr. Dudley Buxton on platymeria and platycnemia[3] in which evidence is adduced in support of the thesis that these characters may have a nutritional basis, and no racial significance.

W. E. Le Gros Clark

(**145**, 70-71; 1940)

G. H. R. von Koenigswald: Bandoeng.

Franz Weidenreich: Peking Cenozoic Research Laboratory, Peiping Union Medical College, Peking.

W. E. Le Gros Clark: Department of Human Anatomy, University Museum, Oxford, Dec. 17.

References:
1. *Eugenics Rev.*, **24** (1931).
2. *Man*, **60** (April 1937); *Modern Quarterly*, 115 (April 1939); Presidential Address, Section H, British Association, 1939.
3. *J. Anat.*, **73**, 31 (1938-1939).

然而魏登瑞教授凭借自己多年来对中国化石的第一手了解，一直都坚持认为中国化石人类具有不同的特征，并且进一步提出中国猿人可以作为比爪哇猿人更原始的早期人类型的代表的观点。现在他不仅认可爪哇猿人和中国猿人"就如同现代人类的两个不同人种一样"，而且同意在某些主要特征上，二者相比，爪哇猿人要更原始一些。如果这些结论被认可了（事实上理应如此），那么现在就有必要废弃中国猿人这个属名。逻辑上应该将中国化石归为爪哇猿人直立种，或者，如果希望它更便于被临时引用，那么可以给它一个具有鲜明特征的名字，北京直立猿人。

孔尼华博士和魏登瑞教授的文章再次提出了形态学证据的有效性的问题，人类学家似乎常常将这些形态学证据作为分类学结论的依据。例如，作者下述值得注意的论证，即由于爪哇猿人的股骨具有显著的股骨扁平特点，而"中国猿人"的股骨却非如此，因此中国猿人很可能不属于爪哇猿人属。这样的阐述非常漂亮地应用了逻辑分析。那么同理也可得出如下结论：因为从新石器时代的古墓中发掘到的股骨显示出比现代英国人股骨更扁平的特点，所以不能将其归属于智人！但不管怎样，已故的达德利·巴克斯顿博士关于股骨和胫骨扁平度的文章[3]虽然短小却很重要，它可能使作者们对最近的文献有所认识，在这篇论文中，他举出了相关证据来支持如下观点：这些特征可能是由于营养问题造成的，而并不能反映种族的特征。

勒格罗·克拉克

（刘皓芳 翻译；刘武 审稿）

Some Biological Applications of Neutrons and Artificial Radioactivity*

J. H. Lawrence

Editor's Note

The discovery of nuclear transmutation and artificial means for inducing it stimulated a new era of fundamental physics. But as John Lawrence of the University of California here reports, it also triggered a revolution in biology and medicine. The use of radioactive tracers had already led to many new discoveries about the movement of sodium, potassium, calcium, iodine and other elements in the human body and in plants. Iodine, for example, quickly became concentrated in the thyroid gland. This effect also suggested the potential use of radioactive agents in the treatment of cancers. These studies were racing forward, in part because the new 220-ton cyclotron at Berkeley produced copious amounts of various radioactive elements, as well as powerful neutron beams.

T HE biological and medical sciences are being stimulated and benefited by the recent discoveries of the nuclear physicist in a manner similar to that following the discovery of the naturally occurring radioactive elements and the production of X-rays. The nuclear physicist can now induce radioactivity in practically all of the elements, and he can harness a beam of neutrons of intense biological activity. This new wonderland for the biologist has been brought about by such events as the first successful experiments of Joliot and Curie in artificial radioactivity, the discovery of the neutron by Chadwick, the discovery of heavy hydrogen by Urey, and the development of the cyclotron by E. O. Lawrence and his associates.

During the past four years, workers at the University of California have been intensely interested in the biological applications of these products of the physicist, and recently a new medical-biological laboratory dedicated to this study and housing the new 220-ton cyclotron—the William H. Crocker Radiation Laboratory—has been completed in Berkeley. Because of its ability to produce large quantities of the various radio-elements and neutron rays, the cyclotron is the nucleus of this unit. The Laboratory is, however, staffed not only by physicists, but also by chemists, biologists, cytologists, bacteriologists, physicians and radiologists—all of whom are interested in both the fundamental and practical problems concerned with the interaction of radiation and matter. My purpose here is to discuss briefly some of the investigations carried on in this laboratory or in conjunction with it. Unfortunately, there is not sufficient space to discuss the extensive and

---

* Based in part on a paper given before Section A (Mathematical and Physical Sciences) of the British Association, given on August 31 and September 1, 1939, at Dundee.

774

中子与人工放射性的若干生物学应用*

劳伦斯

编者按

核嬗变以及用于诱导核嬗变的人工方法的发现开创了基础物理学的一个新纪元。但是正如加利福尼亚大学的约翰·劳伦斯在这篇报告中所指出的，上述成果还引发了一场生物学和医学的革命。放射性示踪剂的使用使人们对钠、钾、钙、碘和其他元素在人体和植物体内的运动有了新认识。例如，钠在甲状腺中很快富集。这种效应也表明了放射剂在治疗癌症中的潜在用途。这些研究得到了飞速发展，这在一定程度上是因为伯克利的新的220吨回旋加速器可以产生大量不同的放射性元素和强中子束。

生物学和医学正在从核物理学家最近的发现中得到启发并且受益，在某种程度上类似于发现天然存在放射性元素与X射线的产生之后所带来的影响。现在，核物理学家几乎可以诱发所有元素的放射性，并且能对一束具有强烈生物活性的中子进行操控。生物学家涉足这一新奇领域是因为受到了下列事件的直接影响，这包括约里奥和居里第一次成功地进行了人工放射性实验，查德威克发现了中子，尤里发现了重氢，以及劳伦斯及其合作者研制出了回旋加速器。

在过去的四年中，加利福尼亚大学的研究人员一直对物理学家取得的这些成果在生物学中的应用具有强烈的兴趣，而且最近一座致力于这项研究的新医学 – 生物学实验室已经在伯克利建成，即威廉·克罗克辐射实验室，这个实验室拥有一台新的220吨回旋加速器。由于这台回旋加速器能够产生大量不同的放射性元素和中子射线，因此它是这个单位的核心。不过该实验室配备的工作人员并非只包括物理学家，此外还有化学家、生物学家、细胞学家、细菌学家、医师和放射学家，所有这些人都非常关注与辐射和物质之间相互作用有关的基础性和实用性问题。这里，我的目的是简要地讨论在这个实验室进行的或与其相关联的一些研究工作。遗憾的是，

---

* 部分依据1939年8月31日及9月1日在邓迪举行的英国科学促进会A分会(数学和物理科学)上宣读的一篇论文。

important work being done in this field in other laboratories.

When Hevesy first used radium D, an isotope of lead, as a "tracer" of lead movement in plants, the potential value of similar isotopes of elements which are important in physiological processes, such as phosphorus, sodium, iron and iodine, became apparent. These and other radioactive isotopes are now available; and, since they are chemically like their inactive relatives, their radiations simply label or tag them and enable the investigator (with the aid of a Geiger counter) to study their average exchange and distribution in biological and chemical processes, in health and disease. The effect of irradiation on the reaction is avoided by the use of sufficiently small "tracer" amounts. On the other hand, many of these isotopes may be used as potent sources of radiation if the metabolism of the element in question is not being studied. In Table 1 are listed some of the isotopes that are used in this University. A brief discussion of studies in which these isotopes are employed follows.

Table 1

Atomic number	Radio-element	Radiation	Half-life
1	Hydrogen (3)	Beta	150–170 days
6	Carbon (11)	Positron and gamma	20.5 minutes
11	Sodium (24)	Beta and gamma	14.8 hours
15	Phosphorus (32)	Beta	14.3 days
16	Sulphur (35)	Beta	88 days
17	Chlorine (34)	Positron and gamma	33 minutes
19	Potassium (42)	Beta and gamma	12.4 hours
20	Calcium (45)	Beta and gamma	180 days
26	Iron (59)	Beta and gamma	47 days
35	Bromine (82)	Beta and gamma	34 hours
53	Iodine (126)	Beta and gamma	13 days

Radioactive hydrogen is written $_1^3\text{H}$, in contrast to ordinary hydrogen ($_1^1\text{H}$) and heavy hydrogen ($_1^2\text{H}$), and has only recently been discovered by Alvarez and Cornog of this Laboratory. This newly labelled form is used in biological work and promises to become a valuable adjunct to heavy hydrogen, which has been so extensively used in "tracer" studies. Although radio-carbon has a very short half-life (20.5 minutes), Ruben, Hassid and Kamen are successfully using it in the study of photosynthesis. Leaves of barley plants grown in an atmosphere of C*O_2 form radioactive carbohydrates*, even when the plants are kept in darkness prior to exposure to C*O_2. The bulk of the radioactive material found in the plant is water-soluble and does not contain carbohydrate, carbonate, keto acids or pigments. Jenny and Overstreet of the College of Agriculture grew barley in the presence of K* and showed that the intake of ions is not a uni-directional process, but that ions of the same species may move into the root and out of the root at the same time. Stout and Hoagland of the same department, using the radioactive isotopes of potassium, sodium, phosphorus, and bromine, studied their upward movement after absorption by

---

*The radioactive form of an element is denoted by an asterisk.

没有足够的篇幅来讨论其他实验室在这一领域中所做的广泛而重要的工作。

当赫维西最早使用镭 D（铅的一种同位素）作为铅在植物体内运动的"示踪剂"时，那些在生理过程中十分重要的元素，例如磷、钠、铁和碘，它们的类似同位素的潜在价值变得明朗起来。这些以及其他一些放射性同位素现在都可以得到了；而且，既然它们的化学性质与它们相对应的非放射性元素的类似，那么它们的放射性只是给其加上标记或将其标示出来，并且使研究者（借助于盖革计数器）得以研究其在生物学和化学过程中、在健康和疾病状态间的平均交换和分布情况。通过使用剂量足够小的"示踪剂"，可以防止辐射对反应造成影响。另一方面，如果不研究涉及的元素的代谢，那么很多这样的同位素也许可以作为潜在的放射源使用。表 1列出了加利福尼亚大学使用的一些同位素。下面会对利用这些同位素进行的研究作简要论述。

表 1

原子序数	放射性元素	辐射	半衰期
1	氢（3）	β辐射	150～170天
6	碳（11）	正电子和γ辐射	20.5分钟
11	钠（24）	β和γ辐射	14.8小时
15	磷（32）	β辐射	14.3天
16	硫（35）	β辐射	88天
17	氯（34）	正电子和γ辐射	33分钟
19	钾（42）	β和γ辐射	12.4小时
20	钙（45）	β和γ辐射	180天
26	铁（59）	β和γ辐射	47天
35	溴（82）	β和γ辐射	34小时
53	碘（126）	β和γ辐射	13天

为了与普通氢（$_1^1H$）和重氢（$_1^2H$）相对照，放射性氢被写作 $_1^3H$，直到前不久它才被本实验室的阿尔瓦雷茨和考诺格发现。这种新的被标记的形式在生物学研究中得到了应用，并且有望成为重氢的一个有价值的补充，而重氢已被极为广泛地应用于"示踪剂"的研究。尽管放射性碳的半衰期很短（20.5 分钟），鲁宾、哈西德和卡门还是成功地将其用于对光合作用的研究之中。在 C^*O_2 环境中生长的大麦类植物的叶子[*]，会形成放射性的碳水化合物，即使这种植物在暴露于 C^*O_2 之前保存在黑暗中，结果也会如此。在植物中发现的大量放射性物质是水溶性的，而且并不包含碳水化合物、碳酸盐、酮酸或色素。农学院的珍妮和奥弗斯特里特使大麦生长在有 K*存在的条件下，结果发现离子的摄入并不是一个单向过程，实际上相同种类的离子也可能在移入根部的同时移出根部。同一院系的斯托特和霍格兰研究了钾、钠、磷和溴的放射性同位素被生长旺盛的柳树和天竺葵植物的根部吸收后的向上运动。他

---

[*] 某一元素的放射性形式用星号（*）标记。

the roots in actively growing willow and geranium plants. They found that when salts are absorbed by roots, some portion enters the xylem within very short periods of time, and is carried rapidly towards the leaves under the influence of transpiration. That the content of radioactivity in the bark as compared with the root is slight indicates that movement of salts in the former is very slow.

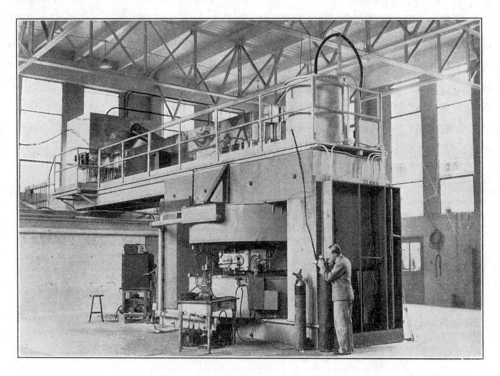

Fig. 1. The 200-ton 60-inch cyclotron in the Crocker Radiation Laboratory on the Berkeley campus of the University of California. Standing by the corner of the magnet and adjusting the helium flow to the target chamber is the late Dr. Harold Walke, of the University of Liverpool. On the table in the foreground is an experimental deuterium generator. The target chamber is behind, marked by the two port holes. Above and to the rear is seen the aluminium oscillator house or radio-frequency power supply. Photograph taken by Dr. Donald Cooksey, assistant director of the Radiation Laboratory.

Mullins, of the Department of Zoology, has investigated the effect of increasing activities of radio-sodium on the penetration of Na^+ into the single-celled alga Nitella. His results show that, below certain activity concentrations, there is no "radiation" effect, and point out the importance of using low activities in "tracer" work. Hamilton is making an extensive study of the physiology of the isotopes of sodium, potassium, iodine and other elements, in health and disease. Their rapid absorption and distribution is evidenced by their appearance in the hand a few minutes after oral administration of small amounts. The isotope of iodine is proving to be valuable in the study of thyroid physiology. Hamilton has demonstrated the marked concentration of ingested iodine in

们发现，当盐类被根吸收后，其中一部分会在很短的时间周期内进入木质部，并在蒸腾作用的影响下被迅速运往叶片。树皮中放射性成分的含量比根部的少，这意味着盐类在树皮中的运动是非常缓慢的。

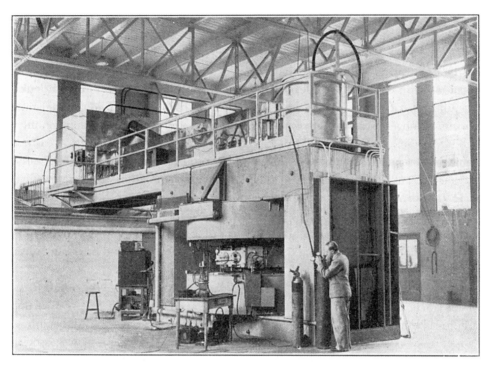

图 1. 位于加利福尼亚大学伯克利校区克罗克辐射实验室中的 200 吨重 60 英寸高的回旋加速器。站在磁铁边缘部、正调节通往靶室的氦气流的是利物浦大学已故的哈罗德·沃克博士。前部的桌子上放置的就是用于实验的氘发生器。靶室在其后面，以两个通道孔作为标志标示出来。上方向后的地方可以看到铝振荡器柜或射频电源。照片由辐射实验室助理主管唐纳德·库克西博士拍摄。

动物学系的马林斯曾研究过放射性钠的活性的增加对 Na^+ 穿透单细胞丽藻的能力的影响。他的研究结果表明，在一定的放射性浓度下不存在"辐射"效应，并且指出在"示踪剂"研究中使用低活性的重要性。汉密尔顿对健康状态和疾病状态下钠、钾、碘及其他元素的同位素进行了广泛的生理学研究。在口服少量试剂几分钟后它们就在手上出现，这证明了它们是被快速吸收并传播出去的。事实证明碘的同位素在甲状腺生理学的研究中是很有价值的。汉密尔顿已说明了被摄取的碘明显聚集在甲状腺中，并且已经能够通过将切除的腺体置于感光胶片上而对碘的分布进行拍照。

the thyroid gland and has been able to photograph the distribution of iodine by placing the excised gland on a photographic film. Anderson and her associates of the Institute of Experimental Biology have found that in adrenalectomized rats a single dose of "tagged" sodium is rapidly lost, while potassium tends to be retained.

Radio-phosphorus (P^{32}) has been the most extensively used radioactive isotope. The ease of manufacture in the cyclotron and its relatively long half-life (14.3 days) make it ideal for biological studies. Chaikoff and his associates in their studies of phospholipid metabolism have shown that various kinds of neoplasms in animals have individual rates of phospholipid turnover, and that cell type does not seem to be the determining factor. In association with Scott and Tuttle, the metabolism of labelled phosphorus in leukaemia in both animals and man is being investigated. The finding that in leukaemic mice the phosphorus turnover is apparently proportional to the degree of leukaemic infiltration suggested the use of radiophosphorus as a source of radiation in the treatment of leukaemia in humans. The concentration of the radio-phosphorus in the areas infiltrated with leukaemic cells tends to localize the therapeutic irradiation (beta-rays). Given by mouth in the form of sodium phosphate, phosphorus is well absorbed (75 percent) and slowly excreted (2 percent per day). The turnover of phosphorus in red cells is rapid, whereas the white cells retain it for longer periods of time. Numerous patients suffering from chronic leukaemia are being treated with this material, and in many instances remissions in the disease are obtained.

Tarver and Schmidt, of the Department of Biochemistry, have synthesized methionine from radio-sulphur and, after feeding it to rats, have shown that the radio-sulphur may later appear as cystine extracted from the tissues. Finally, Whipple and his associates at the University of Rochester, in their studies on normal and anaemic dogs, report the following: radio-iron is poorly absorbed by normal animals; anaemic animals absorb iron in proportion to their need for it; plasma is the medium for the transport of iron; the rapid appearance of iron in the red blood cells is spectacular.

The intense beam of neutrons produced by the cyclotron has made it possible to investigate their biological effects on various objects such as bacteria, plants, Drosophila eggs, animal tumours and normal mammals. This new penetrating form of radiation has intense biological effects, even greater than X-rays or gamma rays, on normal and tumour tissue, but when compared with X-rays, selectively affects some tissues more than others. Experiments on animals indicating that neutrons are more destructive to neoplastic tissue than to normal tissue suggested their trial in cancer therapy. In association with R. S. Stone, of the Department of Roentgenology, patients suffering from cancer are now regularly being treated with neutrons from the new 60-inch medical cyclotron. The recent experiments of Kruger in this Laboratory have opened up another possible application of neutrons to cancer therapy. He has demonstrated that cancers from mice placed in non-toxic concentrations of boric acid and irradiated with slow neutrons can be killed with doses of irradiation harmless to tissues not in contact with boron. The slow neutron is

安德森和她在实验生物学研究所的同事们曾发现，在肾上腺被切除的老鼠体内，单一剂量的"被标示"的钠迅速流失，而钾则往往得以保留。

放射性磷（^{32}P）已经成为应用最为广泛的放射性同位素。在回旋加速器中易于被制备且相对较长的半衰期（14.3 天）使它十分适用于生物学研究。柴可夫和他的同事们在他们对磷脂代谢的研究中已经表明，动物体内各种类型的肿瘤都有各自的磷脂转换率，而且细胞类型似乎并不是决定因素。在与斯科特和塔特尔的合作中，被标记的磷元素在患白血病的动物和人的体内的代谢正在被研究。结果发现，在患白血病的老鼠体内，磷的转换与白血病的浸润程度明显成正比，这一发现使人们想到在人类白血病治疗中把放射性磷作为辐射源来使用。放射性磷聚集在白血病细胞浸润的区域，这使得局部的放射治疗（β 射线）得以实现。将磷以磷酸钠的形式通过口服摄取，磷会被很好地吸收（75%）并被缓慢地排出（每天 2%）。磷在红细胞中的转换是很快的，而在白细胞中则会保持相对较长的一段时间。利用这种物质对患有慢性白血病的大量患者进行了治疗，并且在很多病例中病症都有所缓解。

生物化学系的塔弗和施密特用放射性硫合成了蛋氨酸，并且在把它喂给老鼠之后发现放射性硫随后可能作为从组织中提取出的胱氨酸而出现。最终，惠普尔及其在罗切斯特大学的合作者们在对正常的和患贫血症的狗进行了研究之后，给出了下面的结论：正常动物对于放射性铁的吸收效果很差；患贫血症的动物吸收的铁与它们所需铁元素的量成正比；血浆是输送铁的媒介；铁在红血球中的迅速出现是非常惊人的。

回旋加速器产生的强中子束使得研究其对各种不同对象的生物学效应成为可能，这些对象包括细菌、植物、果蝇卵、动物肿瘤和正常的哺乳动物等。这种新的贯穿辐射具有强烈的生物学效应，对于正常组织和肿瘤组织，它甚至比 X 射线或 γ 射线还要强，但是在与 X 射线进行比较时，它对某些组织的选择性影响要超过其他组织。动物实验表明，中子对肿瘤组织的破坏性比对正常组织的大，这就启发人们开始尝试将其用于癌症治疗。在与伦琴射线学系的斯通的合作研究中，癌症患者目前正在有规律地接受新的 60 英寸医用回旋加速器发出的中子的治疗。该实验室的克鲁格进行的最新实验研发出了另一种利用中子治疗癌症的可行性方法。他已经证实，将老鼠置于不会导致中毒浓度的硼酸之中并用慢中子进行辐射，在辐射剂量对未与硼接触的组织不构成伤害的条件下，便能杀死它体内的肿瘤。慢中子被硼核捕获。该结

captured by the boron nucleus. The combination emits two heavy ionizing particles in opposite directions—an alpha particle and a lithium nucleus—which traverse a distance of about 7 μ in tissue and thus approximate an explosion within the cell.

Fig. 2. The path of a 16-million electron volt deuteron beam traversing the air for a distance of nearly five feet. These particles emerge from the vacuum of the target chamber with a velocity of approximately 18,000 miles per second. In slowing down over the course of their path, they give up their energy to the air molecules, causing them to glow with a violet light. However, in practice, a target to be made radioactive is placed at the emergent point of the beam and thus bombarded. Where neutrons are desired the target used is beryllium. Photograph taken by Dr. Donald Cooksey, assistant director of the Radiation Laboratory.

Although the great contribution of the new nuclear physics to the problems of biology and medicine is certainly the "labelled" or "tagged" isotope, nevertheless it seems important to pursue the possibilities of artificial radioactivity and neutron rays in cancer therapy until a more satisfactory answer to this problem has been reached.

(**145**, 125-127; 1940)

John H. Lawrence: Radiation Laboratory, University of California, Berkeley.

合会在相反方向上释放出两个重离子，一个 α 粒子和一个锂核，它们在组织中的穿透距离约 7 微米，这近似于细胞内部的一个爆炸。

图2. 一束16兆电子伏特的氘核束在空气中穿行约5英尺距离产生的轨迹。这些粒子以近似每秒18,000英里的速度从靶室的真空中发出。在沿其轨迹行进的过程中它们不断减速，将能量转移给空气分子，使它们发出紫色的光。然而实际上，是因为将一个具有放射性的靶置于粒子束出现的位置上，所以受到了轰击。在希望得到中子的地方用铍来作靶。照片由辐射实验室的助理主管唐纳德·库克西博士拍摄。

　　毫无疑问"标记的"或"标示的"同位素是新核物理学对于生物学和医学问题所作出的一个巨大贡献，但是在我们对于癌症治疗问题得到更为令人满意的答案之前，继续探寻人工放射性和中子射线在癌症治疗中的可能似乎仍然是十分重要的。

（王耀杨 翻译；王乃彦 审稿）

Radium Treatment

Editor's Note

Here two scientists write to defend cancer radiotherapy using radium in the face of criticisms expressed in *Nature* by Leonard Hill. Sidney Russ, a renowned English physician, rejects Hill's accusation that radium therapy served the vested interests of the medical community, and argues that the risks are already understood and responsibly observed. Arthur Eve, a former collaborator with Rutherford in nuclear chemistry, also accuses Hill of using gossip and anecdote to attack a strategy that is used in general with care and caution, even if some mistakes have been made through ignorance. Radiotherapy to combat cancer is uncontroversial today, its debilitating side-effects accepted as a necessary evil. It serves a reminder of Paracelsus's famous dictum that the poison is in the dose.

REFERENCE is made in *Nature* of December 9, 1939, p. 973, to a paper (*J. Roy. Soc. Arts*, Dec. 8, 1939) entitled "The Penetration of Rays through the Skin, and Radiant Energy for the Treatment of Wounds", in which I express the view that we would be little the worse off if all the radium now buried in deep holes for security from bombing remained there, and states that the Cancer Act is Great Britain's reply to the question whether monetary influence determines the practice of radium therapy in Great Britain.

Radium, a destroyer of living cells in active division, cannot be used to attack cancer without also damaging living normal cells of the circulating blood, etc. It can only be used for accessible cancers in the skin or surfaces of the body, which can be removed in nearly all cases by the knife or the diathermy needle of the surgeon; and these last do not cause necrosis and incurable neuralgias, which have often followed the use of radium. Deaths due to leucopaenia have resulted from vain attempts to cure deep cancers by the use of radium in bombs. I can instance the damage done by radium by a case now attending the St. John Clinic and Institute of Physical Medicine; radium treatment of a small epithelioma in the skin on the side of the skull resulted in necrosis of the bone, probably incurable, in an area the size of the top of a sherry glass.

Loss of well-being and an incurable neuralgia were recorded by Mr. Furnival, the late distinguished surgeon, who died not long after radium treatment, in his case of cancer of the throat. Such an intolerable neuralgia was suffered by a relative of mine through treatment of a cancer of the root of the tongue, and he also died of a recurrence; and I have heard of many other such cases. There are good clinicians who hold the view that the use of radium favours the spread of metastases. Radium is popular because it can be used instead of the knife, which people dread, and therefore doctors use it.

镭疗

编者按

针对伦纳德·希尔在《自然》上的批评，两位科学家写了这篇文章为使用镭对癌症进行放射治疗作了辩护。悉尼·拉斯是英国著名医师，他反对希尔提出的关于镭疗法为医疗界提供既得利益的指责，并认为人们已经了解并负责任地注意到了镭疗法的危险。阿瑟·伊夫是卢瑟福在核化学领域的早期合作者，他也指责希尔用流言和奇闻来攻击一个总体来说被小心谨慎使用的方法，尽管曾经因为知识不够全面而犯过一些错误。现如今用放射疗法治疗癌症是没有争议的，其使人虚弱的副作用也被公认为是不可避免的。它使人想起了帕拉切尔苏斯著名的格言——毒药在于剂量。

1939 年 12 月 9 日《自然》杂志第 973 页提到了我的一篇题为《射线对皮肤的穿透性，以及用于伤口治疗的辐射能》的文章（《皇家艺术学会会刊》，1939 年 12 月 8 日），其中我表达了如下观点：如果我们基于防爆安全考虑而让储藏于深洞之中的镭继续待在那里，就可以使局面不再继续恶化。我认为，肿瘤法案就是英国对于是否由金融势力决定镭疗法在英国的应用这一问题的答复。

镭可以破坏分裂旺盛的活细胞，但是在攻击癌细胞的同时，必然也会攻击血液循环中的正常活细胞等。它只能被用于杀死那些位于皮肤或体表的易接近的肿瘤，但是所有这些病例几乎都可以用切除或透热治疗针的外科方法来治疗，而且后面这两种治疗方法还不会引起镭疗后的患者经常出现的组织坏死和难以治愈的神经痛。由于白血球减少症而导致的死亡，都是徒劳地试图用炸弹中的镭来治愈深部肿瘤而造成的。我能举出一个使用镭而造成伤害的实例，该患者目前正在圣约翰物理医学研究与治疗中心接受治疗；正是由于用镭治疗头颅侧面皮肤上的一小块上皮瘤，结果导致了可能终身无法治愈的颅骨坏死，其面积足有一个高脚酒杯的杯口那么大。

已故的杰出外科医生弗尔尼沃先生罹患喉癌，在使用镭疗后不久即去世。他在自己的病历中留下了健康状况下降和不可治愈的神经痛的记录。我的一位亲戚因舌根部癌症而接受镭疗后，也遭受到这种无法忍受的神经痛，并且死于癌症复发；我还听到过许多类似的病例。很多优秀的临床医师都认为，镭的使用容易导致肿瘤转移。镭之所以流行，是因为它可以代替人们畏惧的手术刀，因此医生们才使用它。

X-ray apparatus is now available which operates at a million volts or more, so that radiations approaching those of radium are produced. The dosage of radium is controlled by time and filtration; that of X-rays can be further controlled in wave-length and intensity. Radium may be chosen as more convenient for application in such a place as the larynx.

Both radium and X-rays can produce, not only necrosis, but also cancer, and Sir Norman Walker considers that the use of X-rays for lupus should be abandoned. I know of an excellent laboratory servant who, only by continued observation and treatment, is kept, so far, free from cancer resulting from scars due to X-ray treatment of lupus of the face.

We know that death has resulted in several workers who licked the paint off brushes when applying luminous, radioactive paint to watch dials, and cancer has resulted from a radium tube being left in the body. Those who mine radium-bearing ore die generally from cancer of the lung.

The *Lancet* (Dec. 23, 1939) says, "to ensure that radiation treatment is in charge of really competent workers, and to give opportunities for training therapists, a high degree of centralisation will be more effective than the creation of individual treatments units". "To use both radium and X-rays to the best advantage some surgical training and a good grasp of radiation physics are needed, and for the purpose of the Act clinical knowledge of cancer in all its manifestations as well".

While vast sums have been spent on radium, numbers of poor people have to die, unrelieved, of cancer, in their homes. The pressing needs are to spread knowledge, secure prevention where possible, and every early diagnosis, allowing a hopeful removal by the knife or diathermy needle, which can be used by surgeons everywhere, the use of radium or X-rays being reserved for one or two places in the body where surgical operation is very difficult, and to be used by specialists as the *Lancet* suggests. There is no trustworthy evidence that radium in weak doses has any beneficial action.

Leonard Hill

(**145**, 151; 1940)

*   *   *

Sir Leonard Hill returns in *Nature* of January 27, p. 151, to his statement "that we would be little the worse off if all the radium now buried in deep holes for security from bombing remained there", and seeks to make good this assertion by reminding us of some of the casualties of radiological practice. He has allowed these examples to stay in the forefront of his mind, instead of fitting them into the groundwork of experience, which every practising radiologist must do.

现在有了在一百万伏特或者更高电压下操作的 X 射线装置，因此可以产生出与镭接近的射线。镭的剂量可以通过时间和滤光作用得到控制，而 X 射线的剂量则可以通过波长和强度进一步控制。对于像咽喉这样的部位来说，选用镭疗似乎更为便利。

镭和 X 射线不仅会导致组织坏死，还可能会诱发癌症。诺曼·沃克先生甚至认为，应该放弃使用 X 射线来治疗狼疮的方法。我知道有一位优秀的实验室工作人员，仅仅依靠持续观察和治疗，到目前为止，避免了由于使用 X 射线治疗面部狼疮留下的伤疤而引发的癌症。

我们知道，有几个工人在使用发光放射性涂料涂抹表盘时因为舔食了刷子上的涂料而致死，另外还有因镭管遗留在体内而诱发癌症的病例。那些开采含镭矿石的工人通常会死于肺癌。

1939 年 12 月 23 日的《柳叶刀》杂志中提到，"要保证放射治疗由真正能够胜任的人来负责，并且要为放射治疗工作者提供培训的机会，高度的集中化治疗要比创建个体治疗单位更为有效"。"要想最出色地使用镭和 X 射线，需要一些专门的外科训练，并充分掌握辐射物理学，此外，按照肿瘤法案的要求，还需要掌握肿瘤的临床知识及其所有表现形式"。

虽然已经有大量资金投入镭疗，但仍有大批穷人得不到救治，只能在家中死于癌症。当前最紧迫的是传播相关知识、尽可能地保障预防以及力争对所有癌症都能做到早期诊断，以便增加通过手术或透热针将其切除的希望，这是任何地方的外科医生都能使用的方法，应该将镭或者 X 射线的使用限制在外科手术难以治疗的一两个身体部位，并且如《柳叶刀》杂志所建议的那样由专家们来施用。还没有可靠的证据证明小剂量的镭能带来什么好的效果。

伦纳德·希尔

*　　*　　*

伦纳德·希尔爵士在 1 月 27 日的《自然》杂志第 151 页中重申了他的观点，"如果我们基于防爆安全考虑而让储藏于深洞之中的镭继续待在那里，就可以使局面不再继续恶化"，他还试图通过提醒我们有一些放射治疗导致伤亡的事例来证实他的论断。他过于强调这些事例，却没有将它们与经验的基础结合起来考虑，而这一点恰恰是每一个放射科医生必须要做的。

The dangers attending the use of radium and X-rays have been the concern of the X-ray and Radium Protection Committee for many years. No Committee can possibly safeguard a patient against an unskilful application of rays, but radiologists have striven to limit such dangers by making a real specialty of their subject; and there are at present five universities or kindred bodies in Great Britain which grant medical diplomas in this subject. The subject has, indeed, reached a status where its exponents can afford to ignore the rather baser charges in question, but the dis-service of Sir Leonard is to the public, who pay undue attention to his *ex cathedra* statements.

It is not true that the 15,000 patients (mostly cancer patients) who received radium treatment in Great Britain during the year 1938 had such treatment because of the vested interests of the medical public; the vast majority had radium treatment because it was considered the best available for them. It should be remembered that more than 90 percent of the country's radium is held by big organizations, such as the Radium Commission, the King's Fund and the Medical Research Council; this is some guarantee that it is used by people of responsibility.

Sir Leonard would like to see "the use of radium or X-rays being reserved for one or two places in the body where surgical operation is very difficult". The publication "Medical Uses of Radium" has been issued yearly since 1922 by the Medical Research Council, and the Radium Commission has in recent years made annual reports on the results of treatment; from them the pertinent fact emerges that the medical profession continues year by year to treat various forms of cancer at many sites of the body, and there is a disposition to widen rather than restrict the field. As for radium in the boreholes, it is only right that the public should know that much of the radium put away for safety has now been brought into use for their treatment.

Sidney Russ

(**145**, 347; 1940)

*   *   *

Sir Leonard Hill writes, in *Nature* of January 27, with reference to radium treatment and expresses a belief that "we would be little the worse off if all the radium now buried in deep holes for security from bombing remained there. ..." This pessimism contrasts notably with the enlightened optimism of the article, in the same issue, written by Dr. John H. Lawrence, of the University of California, who is working with great opportunities at present lacking in Great Britain. Lawrence and his co-workers are pursuing "the possibilities of artificial radioactivity and neutron rays in cancer therapy until a more satisfactory answer to this problem has been reached". What we require in Great Britain is a well-organized radiological institute where the various possibilities of radiation can be developed and extended. It had always been my hope and ambition that such an institute

使用镭和 X 射线治疗的危险，多年以来一直受到 X 射线与镭防护委员会的关注。没有委员会能够向病人保证对射线的不熟练的应用不会导致事故，但是，放射科医生一直在努力打造真正的学科专长，从而降低射线治疗产生危险的概率；到目前为止，英国已经有五所大学和科研机构获得了该学科领域的医学许可证。实际上，该学科领域已经发展到了相当的程度，其倡导者完全可以忽视那些没有价值的质疑，但是伦纳德爵士的论述是对公众有害的，他对他的那些**权威**陈述过于关注了。

1938 年，英国有 15,000 位患者（大多数是癌症患者）接受了镭疗。但是他们接受镭疗的原因并不是由于医疗界的既得利益；绝大多数人接受镭疗是因为他们相信这是他们能选择的最好的治疗方法。必须指出的是，我们国家超过 90% 的镭是由大型团体组织掌控的，例如镭管理委员会、国王基金和医学研究理事会；这在一定程度上保证了镭是由负责任的人使用的。

伦纳德爵士希望看到"将镭或者 X 射线的使用限制在外科手术难以治疗的一两个身体部位"。自 1922 年以来，医学研究理事会每年都会出版发行《镭的医学应用》，近年来镭管理委员会也有发布关于镭疗效果的年度报告；报告中的有关事实表明，医学界每年都在使用镭治疗身体各个部位各种形式的癌症，镭疗使用的范围在扩大，而不是受到限制。至于矿井中的镭，公众应该知道大量基于安全考虑而被储存起来的镭现已被用于医疗才是对的。

悉尼·拉斯

* * *

伦纳德·希尔爵士在 1 月 27 日的《自然》杂志中谈到了镭疗，并表达了如下信念，"如果我们基于防爆安全考虑而让储藏于深洞之中的镭继续待在那里，就可以使局面不再继续恶化……"。这种悲观论调与同一期杂志中另一篇由约翰·劳伦斯博士撰写的文章中对该问题的开明乐观态度，形成了鲜明的对比；劳伦斯博士来自加州大学，他在极好的便利条件下工作，而这些条件正是目前英国所缺乏的。劳伦斯和他的同事们正在研究"癌症治疗中应用人工放射性与中子射线的可能性，直至对该问题得到一个更令人满意的答案"。现在英国急需的，就是一个组织良好并能使射线的各种可能应用得以开发和扩展的放射学中心。一直以来，我的期望和理想就是建立这样

would be founded as a memorial to Lord Rutherford, who was always wide awake as to the possibilities of radiotherapy and the proper means by which they could be furthered or achieved.

Sir Leonard Hill has collected a certain amount of gossip about radium and narrates a few cases of failure due to the misuse of radium which have come to his personal notice. To counterbalance his citation of deplorable failure, I could quote instances in my own experience where men with cancer of the throat, and elsewhere, have been treated by radium and returned in full health and happiness to their useful work and daily life.

It is admitted by all that the cause, prevention and cure of cancer have not yet been attained. In some cases, when early treatment has been given, there has been cure or palliation by three chief means, surgery, radium, X-rays. All three methods can be and have been grossly abused in some cases, but wisely applied in a vast number of instances. It is not proposed to abolish railway signals because a signalman has wrecked a train by pulling the wrong switch; the effort is made to improve the arrangement, to make if fool-proof, to have an efficient block system. Surgery is not condemned wholesale because of occasional deaths by incapacity, ignorance or carelessness. There have been deaths from dressings or swabs left in the wound. The wrong gas has been administered as anaesthetic. Overdoses of morphia have been given. A man can cut his throat with a safety razor. All such mistakes are no justification for complete disuse of the means employed.

It is, however, necessary to answer Sir Leonard Hill's ill-timed statement in a more positive sense, always remembering that while surgery has been under the guidance of men of high skill and intelligence for centuries—of men provided with every facility—on the other hand, both radium and X-rays are recent discoveries and naturally their applications to therapy are yet in their infancy.

A system of properly controlled and measured dosage, which can be repeated at will with exactitude, has scarcely yet been fully evolved. Certainly, in the past haphazard applications have produced deplorable results, but these are new avoidable in consequence of the research work already done on the proper direction of the radiation and on the determination of the magnitude of the dosage delivered to the growth and to the surrounding tissues.

Sir Leonard raises the question of the relative merits of radium and of X-rays and settles the matter to his own satisfaction with a positive assertion in favour of X-rays. In no part of the world has this difficult question yet been answered with sufficient scientific evidence to admit of certainty. The same uncertainty prevails as to the rival merits of X-rays of various voltages and wave-lengths. We may conjecture, but we cannot assert. Indeed these two important questions are forming part of an investigation by the Radium Beam Therapy Research under the Medical Research Council, and it is a matter for deep regret that this important work should be temporarily suspended by the exigencies of war.

一个中心来纪念卢瑟福勋爵，他对于放射治疗的可能性以及促使其进一步发展和完善的正确方法一直有着非常清醒的认识。

伦纳德·希尔爵士收集了一些关于镭的流言，又讲述了几个引起他关注的由于错误使用镭而导致失败的病例。为了与他所引用的悲惨的失败例证相对比，我也可以引用我亲身经历过的患喉癌或其他部位癌症的患者实例，他们接受了镭疗并完全康复，快乐地回到了正常的工作和生活中。

众所周知，人们对癌症的起因、预防和治疗还知之甚少。在癌症的早期治疗中，主要有三种方法来治愈或缓解病情：手术、镭和 X 射线。这三种方法都有可能并且已经在某些情况下被粗暴地滥用，但这只是极少数情况。没有人会因为曾有一个信号工人按错开关毁掉了一辆火车就提议废除铁路信号；人们会努力改进管理，确保系统万无一失，建立更高效的闭锁系统。外科手术并没有因为由无能、无知或疏忽导致的偶然死亡事故而遭到广泛的禁止。曾经出现过将敷料或棉签遗留在伤口中而导致死亡的事故，还曾发生过使用错误的气体作为麻醉剂以及施用了过量吗啡的事故。即使使用安全的剃须刀也有可能会割伤自己的喉咙。所有这些失误，并不能作为完全禁用相应方法和手段的正当理由。

不过，有必要对于伦纳德·希尔爵士不合时宜的陈述给予更积极的回应，我们必须始终记得，几个世纪以来，外科手术治疗一直是在具有高超技能和智慧且能获得各种所需设备的专业人士的指导下进行的，相比之下，镭和 X 射线都是最近发现的，很自然地，它们在治疗中的应用还处于初始阶段。

目前还没有完全开发出一个可随意并精确重复操作的、能合理控制和测量剂量的系统。无疑，过去对射线的肆意滥用已经造成了恶果，但在今天，这些恶果都是可以避免的，因为我们已经就实施辐射的正确操作和施用于肿瘤及周围组织的放射物剂量大小的确定进行了研究。

伦纳德爵士提出了关于镭与 X 射线相对价值的问题，他积极主张并赞成使用X 射线从而得到令自己满意的答案。世界上还没有任何一个地方对这个难题给出过足以令人信服的科学解答。就像不同电压和波长的 X 射线哪种效果更好依然没有被确定一样。我们可以猜测，却无法断言。实际上，这两个问题是医学研究理事会下属的镭射线治疗研究所正在进行的研究中的一部分，而令人深感遗憾的是，这一重要的研究将会由于紧急的战事而暂时中断。

Some of my well-informed friends point out to me that, in the case of cancer of the uterus, Wertheim's operation for the removal of the whole organ has been given up by surgeons throughout the world and replaced by the use of radium and subsequent wider irradiation with X-rays. This change was largely due to the influence of pioneer work done at the Curie Institute in Paris and Radiumhemmet of Stockholm. In fact, no sooner was the radium placed underground in September than gynaecologists implored that some of it should be made immediately available to save life and relieve distress.

In the case of cancer of the breast, both surgery and radiation are available, and it is a matter of expert advice to decide which is the better in a given case; always insisting on the importance of early and *correct* diagnosis. In carcinoma of the throat, treatment by the radium beam can be used without mutilation or loss of speech, and in cases too advanced for surgery. While surgery may be of some avail with cancer of the rectum and prostate, we have to admit with regret that all methods fail when the oesophagus or stomach is concerned. In less serious cases—such as skin and lip cancer—either surgery or radiation is effective, but most patients would prefer to avoid mutilation and scar by the simple and perfectly safe application of a few milligrams of radium, or its equivalent, for a few hours.

It is scarcely necessary to reply to Sir Leonard Hill's reference to those unfortunate girls, who licking radium paint from their brushes, accumulated radium in their system. Is it suggested that this is in the remotest degree connected with radium therapy? As to the miners' phthisis in the Joachimthal mines, it can unfortunately be matched with closely similar results in the gold mines of South Africa—a matter requiring the closest attention and medical research with a view to prevention and cure.

A. S. Eve

(145, 347-348; 1940)

Leonard Hill: St. John Clinic and Institute of Physical Medicine, Ranelagh Road, London, S. W. 1, Jan. 5.

Sidney Russ: Barnato Joel Laboratories, Middlesex Hospital, London, W. 1, Feb. 6.

A. S. Eve: Overponds Cottage, Shackleford, Surrey, Feb. 5.

一些博学的朋友告诉我，在治疗子宫癌时，全世界的外科医生们都已经放弃了韦特海姆提出的切除整个器官的手术，而代之以镭疗和随后的大范围 X 射线照射治疗。这一转变，很大程度上是受到巴黎的居里研究所以及斯德哥尔摩镭治疗医院所做的开创性研究的影响。事实上，那些镭刚刚于九月被置于地下后不久，就立即有妇科医生请求使用其中一部分来拯救生命和缓解病痛。

对于乳腺癌来说，手术和放射疗法都是可行的，在具体病例中哪种方法更好，是由专家建议来决定的事情；但早期发现和**正确**诊断总是很重要的。对于喉癌来说，使用镭射线照射可以避免残疾或者语言功能的丧失，比手术治疗更好。手术治疗对于直肠癌和前列腺癌可能会有帮助，但我们不得不遗憾地承认，所有的方法对于食道癌和胃癌都无效。在不是很严重的病例中——例如皮肤癌和唇癌——手术或者放射疗法都是有效的，但是大多数患者为了避免躯体受损或留下疤痕而更倾向于用几毫克镭（或者其等效物）进行几小时的简单而又绝对安全的放射治疗。

对于伦纳德·希尔爵士提到的那些因为舔食了刷子上的含镭涂料而在体内积累了镭的不幸的女孩们，则根本没有必要作出回答。这是与镭疗法毫不相干的事情，难道不是吗？至于约阿希姆斯塔尔矿场工人的肺结核，与南非金矿中发生的不幸事件非常类似——这是一件需要给予密切关注并从预防和治疗的角度进行医学研究的事情。

伊夫

（王耀杨 翻译；杨志 审稿）

Cancer-producing Chemical Compounds

J. W. Cook

Editor's Note

That some substances are carcinogenic was recognized at least since the eighteenth century, when some component of soot was deemed responsible for the high incidence of scrotal cancer in chimney sweeps. This led to a recognition that so-called aromatic hydrocarbons, common in tars and oils, are often carcinogens. Such compounds were also widely used in the dyestuffs industry, where they also posed a threat to industrial workers. By the time of this review by eminent chemist James Cook at Glasgow, the notion that carcinogenic compounds might feature not just in specialized petrochemical products but in human foodstuffs was starting to appear. Cook confesses that the variety of cancer-inducing agents made it hard to identify any generic chemical features among them.

IN the last resort, the degree of importance which is attached to the carcinogenic substances depends upon whether such compounds are concerned in the etiology of "spontaneous" human cancer. Perhaps closely bound up with this question is another unsolved problem of outstanding importance, namely, the manner in which these compounds bring about a transformation of normal cells into malignant cells. At least until answers are forthcoming to these questions, the carcinogenic compounds will continue to furnish useful material for the experimental study of cancer. Industrial cancer, in its various forms, has stimulated the researches which have brought to light the cancer-producing properties of the various carcinogenic agents, and in the preparation of the present brief survey of these agents regard has been paid to the correlation of the various forms of industrial cancer with their causative compounds.

In the earlier work on the carcinogenic properties of substances the skin of the mouse was usually employed as the test object. This was due to a number of reasons. Results could be expected comparatively rapidly; the ear of the rabbit, which had been first used, was less satisfactory in this respect. Moreover, the modes of application of the substances under examination were considerably restricted by the toxic and inflammatory properties of the crude mixtures which it was necessary to use. Many of these difficulties have been resolved by the availability of pure chemical compounds of high carcinogenic potency, and in recent years new techniques of administration have been developed, so that malignant tumours have been induced in a large number of different tissues, and in several different species. One outcome of these and other studies has been the revelation that, in certain strains of animal, tumours of a particular organ are apt to occur spontaneously. Thus, some strains of mice show a high incidence of mammary carcinoma; other strains show a high incidence of lung cancer; and there is at least one strain of mice in which liver-cell cancer (hepatoma) is apt to arise spontaneously. These findings indicate the caution

794

诱发癌症的化合物

早在 18 世纪，人们便认识到有些物质具有致癌性，当时发现煤烟中的某种组分与烟囱清洁工人的阴囊癌的高发病率有关，这使人们注意到在焦油和石油中普遍存在的芳香烃物质是致癌的。这类物质也广泛应用于染料工业，因而它们也会威胁产业工人的健康。当格拉斯哥的著名化学家詹姆斯·库克撰写这篇论文的时候，人们就开始认识到，致癌化合物可能不仅存在于特殊的石化产品中，也存在于人类的饮食中。库克坦言，引发癌症的物质种类繁多，以至于难以确定它们在化学特征上的共性。

判定化合物致癌程度最重要的依据是看这种化合物是否与人类"自发"癌症的发病原因有关。与此紧密相关的另一个十分重要但尚未解决的问题是这些化合物将正常细胞转变为恶性细胞的作用方式。至少在这些问题得到解决之前，致癌化合物将一直是癌症实验研究的有用材料。工业化引发的各种癌症促使人们对此展开研究，从而带来了揭示出致癌物质诱发癌症机制的希望。此外，在当前对这些物质进行简要描述的准备阶段，人们已经注意到工业引发的癌症与相应致癌物之间存在关联。

早期研究致癌物的特性时，经常用小鼠的皮肤作为测试对象。这是有多种原因的。首先，这样可以较快地得到结果，而最早使用兔子耳朵作为测试对象时这方面就不太令人满意。另外，因为研究中必须使用的天然混合物有毒性并能引起炎症反应，因此在实验中待检物质的给药方式受到很大程度的制约。现在，人们已经可以提纯出具有高致癌性的单一化合物，许多这样的问题已经得到了解决。近几年来，一些新的给药技术已经得到了发展，因而已经可以在好几种动物的多种组织中诱导出恶性肿瘤。这些以及其他一些相关研究的结果表明，对于某些品系的动物，肿瘤容易在某些特定器官中自发产生。因而，一些品系的小鼠表现出很高的乳腺癌发生率；另外一些品系表现出很高的肺癌发生率；此外，至少有一个品系的小鼠容易自发产生肝细胞癌（肝细胞瘤）。这些结果提示我们，当实验动物的一些器官发生了癌

that must be used in interpreting the results when cancers of such organs are found in experimental animals, especially when the tumours arise at sites other than that of application of the carcinogenic agent. Yet even so, tumours clearly attributable to the treatment have been found, usually at the site of application, in a variety of tissues of animals treated with carcinogenic compounds. In this respect the most versatile substances so far found are contained in the group of polycyclic hydrocarbons, mostly related to 1:2-benzanthracene (I), in which substituents are present at certain well-defined positions in the molecule. With these compounds malignant tumours have been obtained, usually in mice and rats, in such tissues as the skin, the subcutaneous tissues, the peritoneal cavity, the liver, the prostate, the forestomach, the brain, and the spleen; and this list is not exhaustive. Less widespread in their effect are members of other classes of compounds, where usually carcinogenic action has not been shown except in a single organ. In this connexion it needs to be borne in mind that these substances have not usually been so widely investigated as the polycyclic hydrocarbon class.

(I)   (II)

The earliest form of industrial cancer, recognized as such in the latter part of the eighteenth century, was the cancer of the scrotum to which chimney sweeps were specially liable. This was caused by soot, and the pursuit of the clue so provided culminated eventually in the isolation from coal tar of the individual compound responsible. This is 3:4-benzpyrene (II), an aromatic hydrocarbon, the relationship of which to 1:2-benzanthracene (I) is apparent from the formulae. 3:4-Benzpyrene is undoubtedly the principal cancer-producing constituent of coal tar. It has a high boiling point, and hence is present to an appreciable extent only in the highest boiling fractions of the tar. There are grounds for inferring that this or a similar compound is responsible for the carcinogenic properties shown to varying degrees by some of the mineral lubricating oils. Prolonged contact with industrial products of these types is now recognized as being fraught with danger, and the use of suitable precautions should lead to diminution if not to eradication of the form of industrial cancer which they are liable to cause.

The widespread use of tar in road surfaces, and the publication of statistics which appear to show that cancer of the lung is increasing at an alarming rate, have led to the suggestion that tarred road dust may be partly responsible for this increase. This suggestion has been tested experimentally; but although an increase in lung cancer was found in mice breathing air impregnated with road dust, this increase was not wholly related to the presence of tar in the dust, and the results of the experiments do not directly implicate such an agent in the increase of the human disease. Furthermore, it is considered by many

变，特别是肿瘤并不是发生在施用致癌剂的部位时，对结果的解释一定要小心谨慎。尽管如此，在用致癌物处理的动物的各种组织中，还是经常能观察到肿瘤出现在给药部位。迄今为止，人们发现的最常见的致癌物都属于多环烃类，其中大多数与1,2-苯并蒽（I）有关，一般是在该分子的某些特定位置上发生取代而形成的。通过施用这些化合物，可以在小鼠或大鼠的某些组织中成功诱导出恶性肿瘤，如皮肤、皮下组织、腹腔、肝脏、前列腺、前胃、脑、脾脏等等。还有一些其他类型的致癌物，其作用效果不是很广泛，一般只在单一的器官中显示出致癌活性。所以，这些物质并不像多环烃类那样受到广泛的关注。

(I)　　　　　　　　　　(II)

　　最早发现的由工业引发的癌症是烟囱清洁工们特别易患的阴囊癌，这种癌症在18世纪后半叶才被人们认识到。它是由煤烟引起的，对其发病原因的追踪使人们最终从煤焦油中分离出了单一的致癌化合物。这就是3,4-苯并芘（II），它是一种芳香烃，从分子式可以很容易地看出它与1,2-苯并蒽（I）的关系。3,4-苯并芘的确是煤焦油中主要的致癌组分。它有很高的沸点，因此只有在焦油最高沸点的馏分中，它的含量才能达到可测量的程度。有理由推断，某些矿物润滑油表现出的不同程度的致癌特性正是由于含有这种化合物或类似化合物。现在人们已经认识到，长时间接触这类工业产品是危险的，进行适当的防范即使不能完全消除至少也能够降低由这些物质诱发癌症的概率。

　　焦油被广泛应用于公路表面，而公布的统计数据也显示肺癌的发病率正在以惊人的速度增长，据此我们推测，导致肺癌发病率升高的部分原因可能是焦油路面的粉尘。人们已经通过实验检验了这种推测；让小鼠呼吸充满公路尘埃的空气，人们发现其肺癌发病率会升高，但是尽管如此，这种升高并不完全是由尘埃中存在焦油引起的，并且实验结果没有直接证明焦油就是导致人类肺癌病例增加的原因。许多

authorities that the recorded increase in lung cancer is largely accounted for by improved methods of diagnosis. Unconvincing attempts have also been made to implicate pollution of town air by soot, exhaust fumes, etc., and also tobacco smoking in the increase of lung cancer. However, the knowledge that the agencies in question may be, and sometimes are, associated with carcinogenic substances, does not allow such speculations to be too lightly dismissed.

The carcinogenic activity of 3:4-benzpyrene is of a high order, inasmuch as tumours arise in a large proportion of the treated animals, in a relatively short time. A somewhat greater potency is shown by 20-methylcholanthrene (III), a hydrocarbon first obtained by chemical transformation of the bile acids, and later indirectly from cholesterol. Other hydrocarbons of similar structure have similar high activity. An altogether higher order of activity, judged by the criterion of shortness of the latent period in the induction of skin tumours in mice, has recently been found in a small group of hydrocarbons typified by 9.10-dimethyl-1.2-benzanthracene (IV). These compounds, which are characterized by the presence of methyl groups in the positions shown, have in mice skin given tumours which frequently made their appearance within a month of the first application.

(III)   (IV)

It will be observed that the carcinogenic hydrocarbons thus far mentioned are all derived from 1:2-benzanthracene (I). A very considerable number of other carcinogenic derivatives of 1:2-benzanthracene is now known. These are purely synthetic compounds, not known to be associated with either industrial or naturally occurring products. Their chief interest lies in the large number of closely related compounds which have been shown to have such biological activity, and in the generalizations which it has been possible to arrive at regarding the correlation of carcinogenic activity with molecular structure and with other properties.

The benzanthracene group is not the only group of polycyclic hydrocarbons with carcinogenic properties. Feeble activity is shown by 3:4-benzphenanthrene (V), and systematic examination of homologues and derivatives now in progress is pointing to the conclusion that much enhanced activity is shown when suitable substituents are introduced into positions 1 and 2, but not into other positions of the molecule. Before 3:4-benzpyrene had been isolated from coal tar, it had been claimed erroneously that chrysene, also a coal tar constituent, had carcinogenic properties. This error appeared to be due to the

专家甚至认为有记录的肺癌病例的增加主要是因为诊断方法的进步。还有一些其他的实验尝试证明肺癌病例的增加与煤烟、废气等对城市空气的污染以及吸烟有关，但结果都不令人信服。尽管如此，以上这些物质还是有可能（或者是在某些情况下）与致癌作用有关，因此前面的推测不能完全被舍弃。

3,4-苯并芘的致癌活性非常高，因为用它处理过的动物很大一部分在相对较短的时间内就产生了肿瘤。20-甲基胆蒽（III）的致癌能力还要更强一点，这种烃类物质最先由胆汁酸通过化学转化得到，后来也可以通过胆固醇间接得到。其他一些具有类似结构的烃类也具有相似的高致癌活性。通过以小鼠皮肤癌诱导过程中潜伏期的长短作为判断标准，研究人员最近发现以 9,10-二甲基 -1,2 苯并蒽（IV）为代表的一小类烃具有更高的致癌活性。这些化合物的特点是在附图所示位置上存在甲基，它们可以诱发小鼠皮肤产生肿瘤。往往在首次施药后一个月内肿瘤就会出现。

我们可以看到，迄今为止有记载的致癌烃类都是由 1,2-苯并蒽（I）衍生而来的。现在已经知道了相当多的其他 1,2-苯并蒽的致癌衍生物。这些化合物完全都是人工合成的，与工业产物或纯天然生成的物质无关。这些化合物在研究领域的主要意义在于：这些结构类似并具有致癌活性的化合物种类非常多；通过研究其共性人们可能可以揭示致癌物活性与致癌物分子结构及其他特性之间的关系。

多环烃类化合物中并非只有苯并蒽家族才具有致癌活性。3,4-苯并菲（V）也具有微弱的致癌活性，根据目前正在进行的对其同系物及相应衍生物的系统检测，可以初步断定当在 1 位和 2 位上引入合适的取代基后其致癌活性会大大增加，而在分子结构中其他位置上引入取代基则无此作用。在从煤焦油中分离纯化出 3,4-苯并芘之前，人们一直错误地认为煤焦油中的另一种成分菌具有致癌活性。出现这种错误的原因应该是最初从煤焦油中获得的的菌由于提纯不彻底而含有杂质。诸如此类

incomplete purification of chrysene of coal tar origin. These and other circumstances have caused some attention to be devoted to chrysene derivatives, and a number of chrysene homologues, selected in a haphazard way, have been synthesized and found inactive when tested biologically. More recently a consideration of the structural relationship among the carcinogenic derivatives of 1:2-benzanthracene and 3:4-benzphenanthrene led C. L. Hewett (*J. Chem. Soc.,* in the press) to synthesize 1:2-dimethylchrysene (VI), and this hydrocarbon has been found to have definite carcinogenic activity when tested by application to the skin of mice.

(v)   (vi)

For many years it has been recognized that the operatives engaged in certain sections of the chemical industry, and especially in the manufacture of dyestuffs, are more liable to cancer of the urinary bladder than is the general population. This form of cancer was long known as "aniline cancer", and the prevailing opinion for many years has been that it is due to absorption of nitrogenous bases such as benzidine and the naphthylamines, especially β-naphthylamine. Until recently the evidence was purely circumstantial, and many unsuccessful attempts have been made to induce experimental tumours with these bases. Some two years ago, however, the production of bladder tumours in dogs given, subcutaneously and orally, large daily doses of a high grade of commercial β-naphthylamine was reported by American workers. It was doubtless the prevalence of this dye-workers' cancer, coupled with the known cell-proliferating properties of Biebrich Scarlet *R*, which led to Japanese researches which have shown that a number of relatively simple azo compounds have carcinogenic properties. The principal active compounds which have been revealed by this work are 4′-amino-2:3′-azotoluene (VII), which gives liver-cell tumours when fed to rats and mice, 2:3′-azotoluene (VIII), which gave malignant tumours of the urinary bladder in rats, and *p*-dimethylaminoazobenzene (IX), which is mainly carcinogenic towards the liver.

的相关情况使一些研究者将注意力投向了菎的衍生物。人们以随机选择的方式合成了大量菎的同系物，但通过生物学检测发现这些化合物并无致癌活性。最近，基于对 1,2-苯并蒽的各种致癌衍生物与 3,4-苯并菲的结构关系的考察，休伊特合成了（《英国化学会志》，即将出版）1,2-二甲基菎（VI），并且通过施用于小鼠皮肤进行测试发现这种烃类确实具有致癌活性。

多年来人们一直认为，在化学工业的某些部门特别是染料制造部门工作的工人比普通人群更容易患膀胱癌。长期以来，这种癌症被认为是"苯胺癌"，人们一直认为该病的起因是人体吸收了含氮碱基，例如联苯胺和萘胺，特别是 β-萘胺。但是直到最近仍然缺乏直接证据。人们进行了大量研究，试图通过实验的方法用这些碱基诱导肿瘤，但都没有成功。不过，美国的研究人员在两年多以前曾报道，每天给狗大剂量皮下注射和口服商品化的超纯 β-萘胺能够诱导出膀胱癌。毫无疑问，染料工人中膀胱癌的盛行与人们熟知的偶氮染料 R 所具有的使细胞增殖的特性有关。在此基础上，日本研究人员进一步研究发现许多相对比较简单的偶氮化合物也具有致癌活性。这项研究发现的具有致癌活性的化合物主要是 4'-氨基-2,3'-偶氮甲苯（VII），2,3'-偶氮甲苯（VIII）和对二甲氨基偶氮苯（IX）。4'-氨基-2,3'-偶氮甲苯能使大鼠和小鼠产生肝细胞瘤，2,3'-偶氮甲苯（VIII）能在大鼠中诱导出膀胱恶性肿瘤，对二甲氨基偶氮苯（IX）则主要对肝脏产生致癌作用。

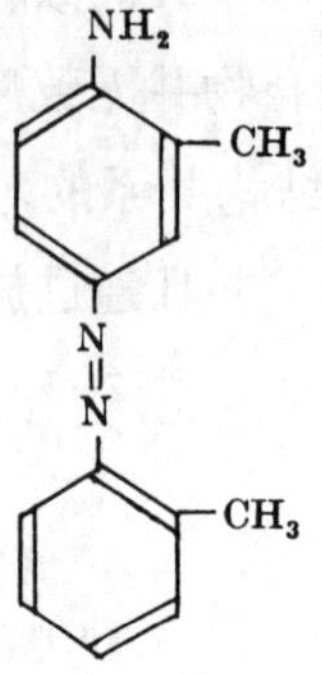

(VII) 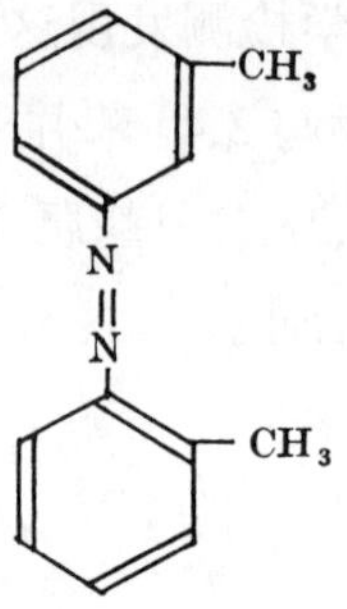   (VIII) 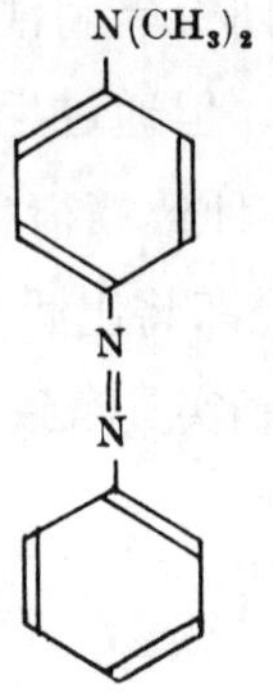   (IX)

In view of the possibility that contaminants of the naphthylamines might be responsible for the dye-workers' cancer, a number of possible transformation products have been administered to rats and mice in the research laboratories of the Royal Cancer Hospital, London. Mice treated with 2:2'-azonaphthalene (X) by application to the skin, or by subcutaneous injection, or by feeding, have developed many liver growths, some of them liver-cell carcinomas, but most were of a cholangiomatous type. Similar tumours were obtained with 2:2'-diamino-1:1'-dinaphthyl (XI), a product which arises easily by intramolecular change of the dihydride of (X), and also with 3:4:5:6-dibenzcarbazole (XII), which is formed by deamination of (XI). There is thus the possibility that the biological effects of this series of compounds are due to a common metabolite, and it is worth noting that the final product of the series (XII) has a structural resemblance to the carcinogenic polycyclic hydrocarbons.

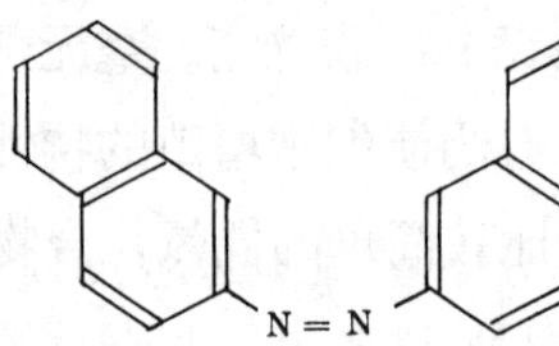

(X) 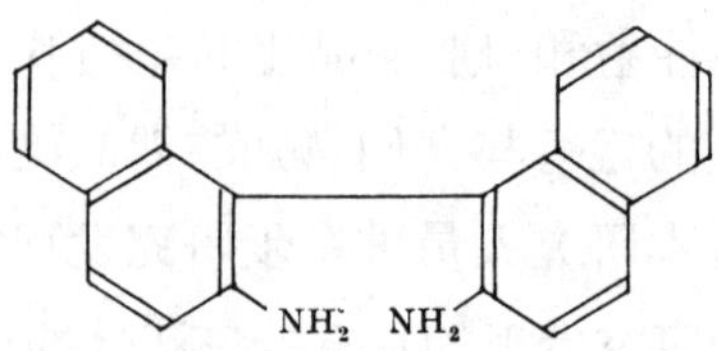   (XI)

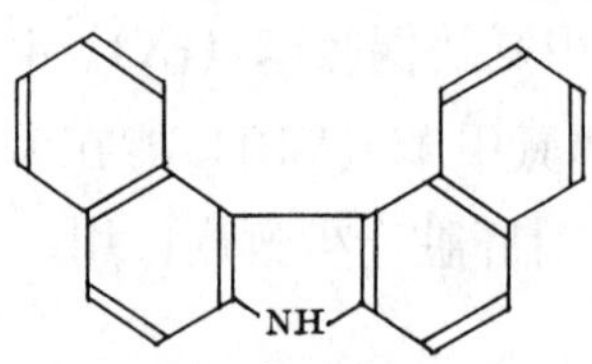

(XII)

One of the azo dyes found to be carcinogenic to the liver by the Japanese workers, namely, p-dimethylaminoazobenzene (IX), was formerly used as a food colouring matter under the name of "butter yellow" and has also been used in dyeing leather. Fortunately, its use in these respects appears now to be obsolete. In Great Britain the range of permitted food colouring matter is now very limited. With the co-operation of the Government

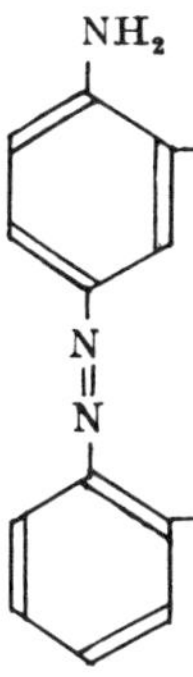

(VII) 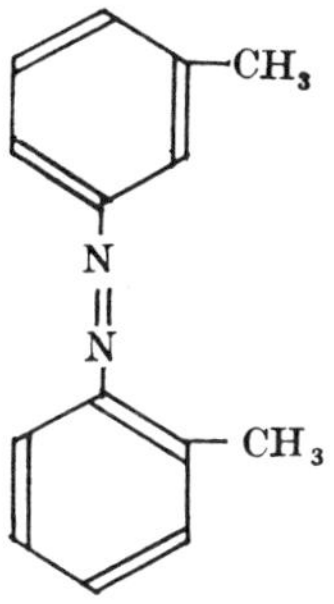　　　　(VIII) 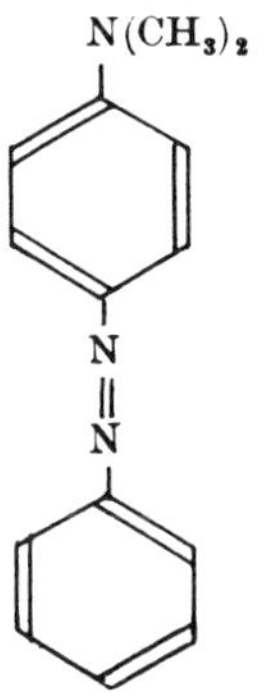　　　　(IX)

考虑到也有可能是萘胺中的杂质使染料工人患上癌症，因此，在伦敦皇家肿瘤医院的许多实验室中，研究人员对大鼠和小鼠施用了萘胺的多种可能的转化产物以进行研究。不管是通过皮肤表面用药，还是进行皮下注射，抑或通过饲喂给药，用 2,2′-偶氮萘（X）处理过的小鼠中都出现了肝脏增生，其中有一些小鼠出现了肝细胞癌，但大多数是胆管瘤类型的。对小鼠施用 2,2′-二氨基-1,1′-二萘（XI）或 3,4,5,6-二苯咔唑 (XII) 也可以得到相似的肿瘤。当 2,2′-偶氮萘（X）分子内发生二氢化可以很容易地得到 2,2′-二氨基-1,1′-二萘（XI）；而后者去氨基就能得到 3,4,5,6-二苯咔唑 (XII)。因此，这一生物学效应有可能是由同一代谢产物引起的，而且值得注意的是，这一系列化合物的代谢终产物（XII）具有与致癌多环烃类相似的分子结构。

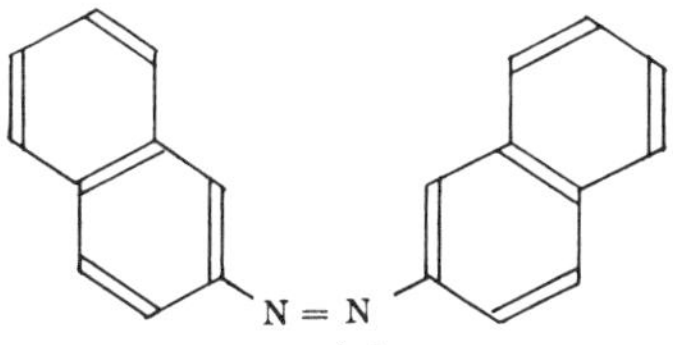

(x) 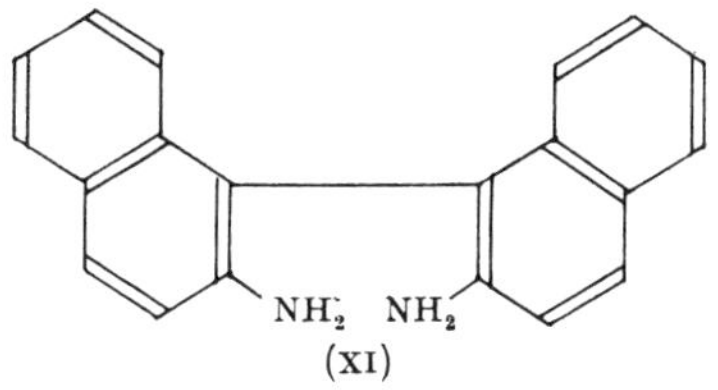　　　　(XI)

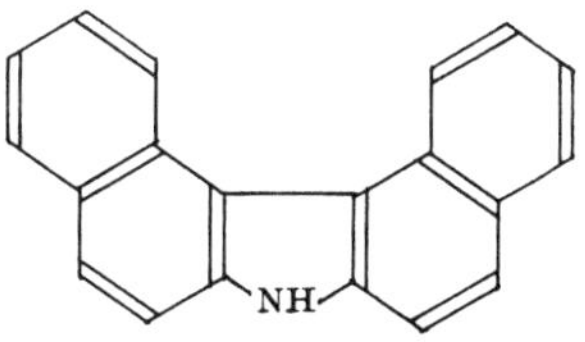

(XII)

日本的研究人员发现，有一种偶氮染料对二甲氨基偶氮苯（IX）对肝脏有致癌作用。这种先前被称为"甲基黄"的物质被用作食物色素，同时也用于皮革制品染色。幸运的是，现在这些行业已经很少使用这种物质了。如今在英国被允许用作食物色素的物质是非常有限的。在政府化学家的协助下，相关部门选取了一批被允许

Chemist, tests have been carried out in which relatively large amounts of a selection of these permitted dyes were regularly administered with the food to rats and mice. The compounds chosen were azo compounds bearing some structural resemblance to the azo compounds discussed in the present article. They are mostly water-soluble sulphonates, a circumstance which facilitates rapid elimination. In a few of the mice stomach tumours were obtained; but it is by no means certain that these were due to the dyes.

Existing knowledge of the structures of the various carcinogenic compounds and of the conditions under which they may be formed has led to various speculations regarding the possibility of such substances being present in human food. Some workers have recorded the production of skin tumours by heated fats and by tars prepared by heating coffee. It has been claimed also that wheat-germ oil prepared by a special extraction process produces sarcomatous tumours in rats. This claim has not thus far received independent confirmation, and at the present time there is no evidence that cancer of the internal organs is due to specific dietary constituents. However, it is evident that such lines of inquiry should be pursued.

A puzzling and in some ways disconcerting feature of the carcinogenic agents now known is their variety and their apparent lack of correlation. It may be recalled that cancer may be induced not only by the classes of compound reviewed in this article, but also by several other agencies. The malignant tumours which arise in consequence of exposure to ultra-violet light, X-rays, and radioactive substances may well be due to the production of carcinogenic compounds from normal constituents of the tissues.

There is, however, no evidence that the radiations exert their influence in this indirect manner. Cancer of the skin occurs in persons taking arsenic by mouth over long periods, and has also been found in workmen engaged in handling arsenical sheep dips. Teratoma of the testis in fowls can be induced by injection of zinc salts at the season of the year when the testis is actively secreting androgenic hormone. At other times also this type of growth may be produced if gonadotropic hormone is simultaneously administered, so that at least two factors seem to be involved.

Thus, in carcinogenesis we have a biological phenomenon which may be attributed to a variety of different substances and agencies. This is by no means unique, for the same is true of other biological phenomena; but it is a circumstance which adds to the difficulty of interpreting the biological properties of the carcinogenic compounds and in estimation their ultimate significance.

(**145**, 335-338; 1940)

J. W. Cook: University of Glasgow.

使用的染料，并通过定期地让小鼠和大鼠在进食时大量服用这些染料的方法对它们进行了测试。测试中选取的化合物都是与本文中讨论的化合物有一些类似结构的偶氮化合物。它们主要以水溶性的磺酸盐形式存在，这使得它们容易快速降解。实验中，部分小鼠出现了胃癌，但是不能完全肯定这就是由染料引起的。

目前，通过依据现有的对各种致癌化合物的结构以及产生条件的认识，人们提出了各种推测，这些推测都提到了这些致癌物质存在于人类食物中的可能性。一些研究者发现热油脂和加热咖啡时产生的焦油都可以诱发皮肤癌。还有研究表明，经某种特殊提取方法制备的麦胚油可以诱发大鼠产生肉瘤样肿瘤。这一报道迄今还没有得到确切的证实，而且目前也没有证据表明体内器官的肿瘤是由特殊的饮食成分造成的。不过，很明显这类研究还需要进一步深入。

目前已知的致癌物质有一个令人困惑的、在某些方面甚至是令人担忧的特点，这就是它们的多样性以及彼此之间缺乏明显关联。值得一提的是，不仅是本文所提到的这一系列化合物能够诱发癌症，很多其他物质也能诱发癌症。暴露于紫外线、X射线或放射性物质后诱发的恶性肿瘤，也可能是由于组织中正常成分产生的致癌物质引起的。

然而，还没有证据证明放射物以这种间接的方式发挥作用。长期口服含砷物质的人容易患皮肤癌，另外还发现那些处理含砷的羊用防腐浸液的工人也容易患皮肤癌。在家禽睾丸活跃分泌雄性激素期间给它们注射锌盐会诱导其产生睾丸畸胎瘤。在其他时间，如果同时施用促性腺激素也会诱发家禽产生这类肿瘤。这样看来，在肿瘤诱发过程中至少涉及这两个因素。

综上所述，在癌症发生过程中，我们看到了一种可能由多种不同的物质或因素引发的生物学现象。无独有偶，其他一些生物现象也是如此。但是，这种情况为阐明致癌化合物的生物学特性以及评价它们的最终意义增加了不少难度。

（吴彦 翻译；秦志海 审稿）

Molecular Structure of the Collagen Fibres

W. T. Astbury and F. O. Bell

Editor's Note

Collagen is the fibrous protein that constitutes connective tissue—tendons and such structures in living things. William T. Astbury from the University of Leeds had built a reputation for himself during the 1930s by attempting the X-ray analysis of long polymer molecules, even including DNA. Here he describes the atomic structure of the collagen molecule. Unfortunately, Astbury had no means of knowing that collagen consists of a triple helix formed by protein polymer molecules. The problem that Astbury set himself was solved only in the 1980s.

X-RAY studies of the fibrous proteins indicate that they fall almost exclusively into one or other of two main configurational groups, the keratinmyosin group and the collagen group[1]. The interpretation of the structure and properties of the former group is now well advanced and has frequently been reported on in *Nature* and elsewhere, but the structure of the latter, in spite of many investigations, has hitherto remained unexplained. It was suggested several years ago that the amino-acid residues in gelatin (which also gives the typical collagen diffraction pattern) are somehow grouped in threes with probably every third a glycine residue and every ninth a hydroxyproline residue, that the strong meridian arc of spacing about 2.86 A. is associated with the average length of a residue in the direction of the fire axis, and that such an average length could very well arise from an alternate *cis*- and *trans*-configuration[2]; but further progress was not possible for lack of experimental data. More recent chemical and X-ray evidence points now to a solution that is both simple and convincing.

(1) Bergmann[3] concludes that the average residue weight in gelatin is about 94, and that the chief residues are present in the proportions set out in the accompanying table:

Amino-acid Frequencies in Gelatin

Amino-acid	Wt.%	Mol. Wt.	Gm. Mol.	Frequency
Glycine ··	25.5	75	0.34	3 ($2^0 \cdot 3^1$)
Proline ··	19.7	115	0.17	6 ($2^1 \cdot 3^1$)
Hydroxyproline ··	14.4	131	0.11	9 ($2^0 \cdot 3^2$)
Alanine ··	8.7	89	0.098	9 ($2^0 \cdot 3^2$)
Arginine ··	9.1	174	0.052	18 ($2^1 \cdot 3^2$)
Leucine-*isoleucine* ··	7.1	131	0.054	18 ($2^1 \cdot 3^2$)
Lysine ··	5.9	146	0.040	24 ($2^3 \cdot 3^1$)

胶原纤维的分子结构

阿斯特伯里，贝尔

编者按

胶原纤维是构成生物体内肌腱等结缔组织的纤维状蛋白质。在 20 世纪 30 年代，利兹大学的威廉·阿斯特伯里对包括 DNA 在内的多种长形聚合物分子进行了 X 射线分析并由此闻名于世。他在这篇文章中描述了胶原纤维分子的原子结构。遗憾的是，阿斯特伯里当时还无法了解到胶原纤维是由三股蛋白质聚合物分子形成的螺旋构成的。他提出的问题直到 20 世纪 80 年代才得到解决。

对纤维状蛋白质的 X 射线研究表明，它们几乎可以毫无例外地被归入两种主要构型族中的任意一种，即角蛋白－肌球蛋白族和胶原蛋白族 [1]。对于第一种构型族的结构和性质的解析目前进展顺利，其相关报道常在《自然》或其他一些杂志上发表。对于第二种构型族的结构，尽管已进行了很多研究，但迄今为止仍没有得到解析。几年前有人提出，在结构上，明胶（它也能产生典型的胶原蛋白衍射图样）中的氨基酸残基以三个为一组，并且每到第三个就很可能是一个甘氨酸残基，而第九个则很可能是羟基脯氨酸残基，这样就把弧长约为 2.86 Å 的强子午线与沿纤维轴方向上一个残基的平均长度关联起来了，而这样的平均长度又正好可以由顺式与反式构型 [2] 相互交替产生；不过，由于缺乏实验数据一直无法取得进一步的进展。如今，最新的化学证据与 X 射线数据都指向了一个更为简单和令人信服的结果。

（1）伯格曼 [3] 得到的结论是，明胶中氨基酸残基的平均分子量约为 94，其中最主要的几种氨基酸残基出现的比例如下表所示：

明胶中氨基酸出现的频率

氨基酸	质量百分比	分子量	摩尔质量	频率
甘氨酸	25.5	75	0.34	3 $(2^0 \cdot 3^1)$
脯氨酸	19.7	115	0.17	6 $(2^1 \cdot 3^1)$
羟基脯氨酸	14.4	131	0.11	9 $(2^0 \cdot 3^2)$
丙氨酸	8.7	89	0.098	9 $(2^0 \cdot 3^2)$
精氨酸	9.1	174	0.052	18 $(2^1 \cdot 3^2)$
亮氨酸和异亮氨酸	7.1	131	0.054	18 $(2^1 \cdot 3^2)$
赖氨酸	5.9	146	0.040	24 $(2^3 \cdot 3^1)$

Thus not only are one third of the residues glycine residues, but also, except for one residue in eighteen, another third are either proline or hydroxyproline residues; that is to say, are of the form:

$$
\begin{array}{ccc}
\text{CH}_2 && \text{CH.OH} \\
\text{CH}_2 \quad \text{CH}_2 && \text{CH}_2 \quad \text{CH}_2 \\
\text{N}——\text{CH} \qquad \text{or} & & \text{N}——\text{CH} \\
\qquad \text{CO}—— && \qquad \text{CO}—
\end{array}
$$

The table shows too that there cannot be fewer than 72 residues in the gelatin "molecule" (there will not be any definite molecule of gelatin itself, but only large, and possibly somewhat modified, fragments of the original collagen pattern): the true number must be a fairly high multiple of this, possibly 576, to judge by the histidine content for example.

(2) The side-chain and backbone spacings in dry gelatin are about 10.4 A. and 4.4 A., respectively, while the density is about 1.32 gm./c.c. Suppose these two spacings to be inclined at an angle β, and the average length of a residue in the direction of the fibre axis to be L A., then

$$94 \times 1.65 = \frac{10.4 \times 4.4 \times L \times 1.32}{\sin\beta}$$

that is, $\qquad\qquad L = 2.6 \sin\beta \text{ (approx.)}$

and therefore L cannot be greater than about 2.6 A. This is only an approximate calculation, but it is sufficiently accurate to confirm that the strong meridian arc of spacing 2.86 A. is almost certainly associated with the average length of a residue.

(3) If it is actually equal to it, as seems most probable, and the residues follow one another in a row, then from the table the minimum length of the intramolecular pattern along the fibre axis is about 72×2.86 A. This length is not only too small to include the residues of other acids, such as histidine, omitted from the table, but also it is too small to account for the meridian spacings reported by Wyckoff and Corey[4] and by Clark and co-workers[5]. Their data are best explained by a sequence of 4×72 residues in a row, grouped in approximate sets of 12, 24 and 36. This gives a molecular weight of about 27,000, or a multiple thereof, corresponding to Svedberg's gliadin class[6].

(4) The proposed partial *cis*-configuration[2] is readily accounted for by the preponderance of imino residues. When we allow for this and the glycine content, there seems to be only one reasonable solution, represented by the scale model shown in the accompanying illustration. The basic sequence is $-P-G-R-$, where (with the exception of one residue

这样看来，不仅有 1/3 的残基是甘氨酸残基，而且，另外 1/3 的残基基本上不是脯氨酸残基就是羟脯氨酸残基（每 18 个氨基酸残基中有 1 个例外）；也就是说，它具有如下结构：

表中数据还显示，明胶"分子"（这里并非指明胶本身的某种确切分子，而仅仅是指那些大的并且可能受到某些修饰的具有胶原蛋白特性的片段）中的氨基酸残基数不可能少于 72 个，真实的残基数必定是 72 的很多倍，比如，根据组氨酸的含量推断出的残基数可能是 576。

（2）在干明胶中，氨基酸的侧链与主干的长度分别约为 10.4 Å 和 4.4 Å，而密度则约为 1.32 g/c.c.。假定侧链与主干之间的倾角为 β，而一个氨基酸残基沿纤维轴方向的平均长度为 L Å，那么：

$$94 \times 1.65 = \frac{10.4 \times 4.4 \times L \times 1.32}{\sin\beta}$$

即：
$$L = 2.6 \sin\beta \text{（近似值）}$$

因此 L 大约不会超过 2.6 Å。虽然只是粗略估算，不过这已经足够准确地证实弧长约为 2.86 Å 的强子午线与一个残基的平均长度是有关联的。

（3）如果实际长度确实如此（目前看来这是很有可能的），并且残基一个接一个地连成一行，那么根据表中的数据，分子内模块在沿着纤维轴方向上的最小长度约为 72×2.86 Å。这个长度太小了，不但根本无法包括上述表格中被忽略的其他氨基酸残基，如组氨酸，而且也不足以解释威科夫与科里 [4]、克拉克及其合作者们 [5] 所报道的子午线长度。对他们的数据所能作出的最佳解释是一条由 4×72 个氨基酸残基排成一行的序列，其中大约 12、24 或 36 个组成一组。这样得到的分子量大约是 27,000 或其若干倍，这与斯韦德贝里的醇溶蛋白类 [6] 是一致的。

（4）亚氨基残基 [2] 在数量上占优势地位的事实能很好地解释之前有人提出过的部分顺式构型。当我们考虑到这一点和甘氨酸的含量之后，看来似乎就只有一个合理的答案了，即如附图中展示的比例模型所代表的结构。基本序列是 –P–G–R–，其

in eighteen) *P* stands for either proline or hydroxyproline, *G* for glycine, and *R* for one or other of the remaining residues. The full-length pattern, and also variations within the collagen group as a whole, must arise by suitably modifying this simple theme.

Using the interatomic distances found in silk fibroin, the average length per residue in the pattern shown in the illustration works out to be 2.85 A., almost exactly the spacing of the strong meridian arc. Other points in favour of the model are: (*a*) there is no steric interference between the side-chains, the longer side-chains all lying on the side of the main-chain remote from the rings, leaving only the unobtrusive glycine side-chain (−*H*) on the same side as the rings; and (*b*) the polypeptide chain proceeds in a straight line, and any attempt to stretch it results in the side-chain (−*R*) swinging over towards the rings and the system coiling back upon itself: thus we have an explanation of the paradox that though the collagen configuration is shorter than that of the β-proteins, it is nevertheless practically inextensible.

The above solution of the collagen problem permits now of the broad generalization that all the extended forms of the fibrous proteins fall into either of two classes: they are built from polypeptide chains in either the *cis*- or the *trans*-configuration.

A fuller account of this investigation may be found in the first Procter Memorial Lecture[7], and a more detailed discussion still will be published elsewhere.

(**145**, 421-422; 1940)

中 P 代表脯氨酸或羟基脯氨酸（除去每 18 个中的一个例外），G 代表甘氨酸，R 则代表其他残基中的某一个。明胶分子内模块的全长以及明胶与胶原纤维组这个整体的差异，应该可以通过适当调整这个简单序列的形式而得到。

利用研究丝素蛋白时得到的原子间距离，可以推算出插图所示的模块中每个氨基酸残基的平均长度为 2.85 Å，基本上完全等同于强子午线弧长。还有一些支持这种模型的证据：(a) 侧链之间不存在位阻影响，较长的侧链全都位于主链上远离环的一侧，只有小得不起眼的甘氨酸侧链（$-H$）位于环的同一侧；(b) 多肽链沿着直线延伸，任何拉伸链的尝试都会使侧链（$-R$）转向环，从而使整个体系转个弯折回来。因而，我们便可以对胶原蛋白的构型比 β–蛋白短但几乎不能伸展这一看似矛盾的现象作出解释了。

上述关于胶原纤维问题的解决经过推广可以得到如下结论：纤维状蛋白质的所有伸展形式都可以归入两类中的一种，要么是由顺式构型的多肽链构成，要么是由反式构型的多肽链构成。

关于这项研究的更为完整的叙述可以在第一届普罗克特纪念讲演 [7] 中找到，我们还将在其他地方发表一份更详细的论述。

（王耀杨 翻译；刘京国 审稿）

W. T. Astbury and Florence O. Bell: Textile Physics Laboratory, University of Leeds, Jan. 31.

References:

1. Astbury, W. T., *C.R. Lab. Carlsberg*, **22**, 45 (1938) (Sørensen Jubilee Vol.); *Trans. Faraday Soc.*, **34**, 377 (1938); *Ann. Rev. Biochem.*, **8**,113(1939); *Ann. Rep. Chem. Soic.*, **35**, 198 (1939).

2. Astbury, W. T., *Trans. Faraday Soc.*, **29**, 193 (1933); *Cold Spring Harbor Symposia on Quantitative Biology*, **2**, 15 (1934); *Chem. Weekbl.*, **33**, 778 (1936). Astbury, W. T., and Atkin, W. R., *Nature*, **132**, 348 (1933).

3. Bergmann M., *J. Biol. Chem.*, **110**, 471 (1935). Bergmann, M., and Niemann, C., *ibid.*, **115**, 77 (1936).

4. Wyckoff, R. W. G., Corey, R. B., and Biscoe, J., *Science*, **82**, 175 (1935). Corey, R. B., and Wyckoff, R. W. G., *J. Biol, Chem.*, **114**, 407 (1936). Wyckoff, R. W. G., and Corey, R. B., *Proc. Soc. Expt. Biol. and Med.*, **34**, 285 (1936).

5. Clark, G. L., Parker, E. A., Schaad, J. A., and Warren, W. J., *J. Amer. Chem. Soc.*, **57**, 1509 (1935).

6. Svedberg, T., *Proc. Roy. Soc.*, B, **127**, 1 (1939).

7. Astbury, W. T., *J. Int. Soc. Leather Trades' Chemists* (in the press).

True and False Teleology

C. H. Waddington

Editor's Note

Teleology is the notion that a change is induced not solely by the action of physical forces but because it is being drawn toward some ultimate goal. While it normally has no place in physical science, the embryologist Conrad Waddington here argues, the notion does find a sensible use in biology. The existence of separate sexes in many organisms, he points out, allows for cross-breeding and rapid natural selection. Given this context, it then becomes natural to say that developmental mechanisms lead to the differentiation of the sexes for precisely this purpose: a teleological explanation. Today evolutionary biologists tend to be much more wary of invoking teleology, given how it has been abused by religious groups to distort evolutionary theory.

THERE has recently been a considerable revival of interest, largely due to the work of Darlington[1], in the teleology of different systems of reproduction. The logical status of teleological arguments is very different in this connexion from that in other spheres, since the "purpose" which is brought forward is the fulfilment of the conditions for rapid evolutionary advance under the influence of natural selection. That is to say, a genetic system which achieves its "purpose" provides in so doing the mechanism for its survival. The considerations which have led to the rejection of teleological arguments in other connexions therefore do not apply; though one might still question whether the teleological phraseology is the most convincing in which the arguments can be framed.

There is, however, a danger that the teleological method of argument will be carried over, by association, into regions in which it cannot be sustained. This seems to have occurred, to some extent, in the valuable article by Mather[2] in which he discusses the evolutionary significance of the formation of two different sexes in the diploid phase. He is not content to point out that the separation of the sexes is a mechanism for encouraging cross-breeding, but he contrasts this statement with some sentences, taken from a recent work of mine[3], on the developmental mechanisms involved, from which he deduces that "the sexes are separated supposedly in order to ensure that the gametes are differentiated". Such a view, he states later, must be rejected.

But such teleological statements should never arise in a discussion of developmental mechanisms. It is not sufficient to recognize that the development of two distinct sexes may be an evolutionary advantage; we have still to find out how it is done, and the "developmental–genetical idea" cannot be "dismissed". At the same time, this does not invalidate the arguments which Mather brings forward as to the evolutionary consequences of such a differentiation; in fact, he will find a statement of his main point, that the evolutionary advantage of having two distinct sexes is that it ensures cross-breeding, in the

正确的和错误的目的论

沃丁顿

编者按

目的论认为：任何变化都不完全是由物理力的作用引起的，它还取决于这种变化的某种终极目标。虽然这一观点在物理科学中没有得到认可，但是胚胎学家康拉德·沃丁顿在这篇文章中指出，这一观点在生物学中确实能得到合理的应用。他指出，很多有机体存在两性分离的状况，从而允许了杂交育种和快速的自然选择。在这种背景下，可以很自然地说，发育机制正是出于这种目的而导致了性别的分化：这是一种目的论的解释。考虑到一些宗教团体利用目的论来歪曲进化论，现在的进化生物学家往往更加谨慎地引述目的论。

最近，人们对许多不同繁殖体系的目的论重新燃起了相当大的兴趣，这在很大程度上是由达林顿[1]的工作引起的。目的论论点的逻辑状态在这种关联性上与其他领域是有很大区别的，因为，在自然选择的影响下，为了加快进化，被提出的"目的"就是条件的满足。也就是说，一个达到其"目的"的遗传体系就为自身的存活提供了途径。这些分析在其他的一些关联上会导致有悖于目的论的论点，因此不能被应用；尽管有人仍然存在疑问：在已有的各种观点中，目的论的表述是否是最令人信服的？

然而，这里存在一个危险：由于与一些它在其中不能成立的领域有着某种联系，目的论的论证方法将无法继续进行下去。从某种程度上来说，这种情况在马瑟[2]撰写的那篇极有价值的论文中似乎已经出现过，在文章中，他讨论了在二倍期中两性形成的进化意义。他指出，两性的分离是激发杂交育种的一个机制，他对这一结论并不满意，但是他将这种表述同我[3]最近的一篇论文中的一些有关发育机制的句子进行了比较，并推断出，"两性的分离也许是为了有意地确保配子的独特性"。他后来又说，这样的观点一定会被否定。

但是，在关于发育机制的讨论中，这种目的论的陈述应该不会出现。因为目前还不足以确认，发育出两个不同的性别就具有了进化优势；我们仍然不得不去找出它是怎么进行的，并且"发育-遗传的思想"也不能被"抛弃"。同时，这并没有使马瑟提出的关于这种性别分化的进化后果的观点失效；实际上，他将会在所引句子紧挨着的前面找到一句其主要观点的陈述，即拥有两种不同性别的进化优势在于它

sentence immediately preceding the ones he chooses to quote. But if the new teleology is to be received with the respect which is its due, it is of the greatest importance that it should not stray outside its own legitimate fields.

(**145**, 705; 1940)

C. H. Waddington: Department of Zoology, Cambridge, April 1.

References:

1. Darlington, C. D., "The Evolution of Genetic Systems". (Cambridge: University Press, 1939.)

2. Mather, K., *Nature*, 145, 484 (1940).

3. Waddinton, C. H., "An Introduction to Modern Genetics". (London: Allen & Unwin, 1939.)

确保了杂交育种。但是如果这种新的目的论将被人们怀着它应得的尊重加以接受，那么最重要的就是，它不应该游离在自己的合理范围之外。

（刘霞 翻译；赵见高 审稿）

True and False Teleology

K. Mather

Editor's Note

The geneticist Kenneth Mather responds to Waddington's criticisms (in the previous paper) of his views on teleology. While he was taken to task for using teleological expressions, he says he had only quoted from Waddington's own work. In any event, their difference on this point is trivial, as both would agree that evolution by natural selection is not in a true sense purposeful, even if it is often simpler to speak that way. However, Mather points to a true disagreement between himself and Waddington regarding the link between sexual separation in development and cross-fertilization in a species. Mather thus highlights the nub of a persistent problem: evolution indeed has no "goal", but it is often hard not to speak as though it does.

DR. Waddington's criticisms seem to be two. In the first place, I am taken to task for the unwarrantable use of teleological expressions, particularly in the specific case of my paraphrase of his own discussion of sex separation. Inasmuch, however, as the discussion was originally Waddington's and not mine, I can scarcely be called to account for its nature, whether teleological or otherwise. In any event, the point is trivial, as I feel confident that Darlington and Waddington would agree with me in regarding adaptation as the outcome of selection and in denying that it was purposeful, whether the discussion concerned genetical or morphological questions.

Secondly, I am criticized for wishing to "dismiss" the "developmental–genetical idea". This I have no desire to do in general as, clearly, developmental studies can contribute much to our understanding of genetics. But I do disagree with the specific idea, apparently held by Waddington in common with many others, that separation of the sexes in the diploid phase is essentially a reflection of gametic differentiation. They may be related developmentally in dioecious organisms, but this should not blind us to their wholly dissimilar genetical consequences. Gametic differentiation cannot of itself lead to regular outbreeding, as is well shown by Triticum and Pisum, where the two kinds of gametes are strikingly different but where self-fertilization is the rule. On the other hand, crossing between different individuals must always follow from dioecism. As Waddington, in the section of his book which I quoted, and in the previous sentence which I did not quote, relates outbreeding primarily to gametic differentiation and thence secondarily to separation of the sexes, I must disagree with him. Differentiation of the gametes has no place in the genetical relation between outbreeding and unisexuality.

(**145**, 705-706; 1940)

K. Mather: John Innes Horticultural Institution, Mostyn Road, Merton Park, London, S.W.19.

正确的和错误的目的论

马瑟

编者按

遗传学家肯尼思·马瑟回应了沃丁顿对于他的目的论观点的批评（见前一篇论文）。他说当他在工作中使用目的论的表述时，他只是援引了沃丁顿本人的工作。实际上，他们在这一观点上的区别是微小的，例如他们都同意通过自然选择的进化并不是真正意义上有目的的，尽管人们常常采用这种简单的表述。不过，马瑟指出他和沃丁顿的观点之间一个真正的区别在于物种发育中的两性分离和异体受精之间的关联。因此马瑟点明了一个长期存在的问题的症结所在：进化本身的确没有"目的"，但也不能否认，它看起来的确像是有目的的。

沃丁顿博士的批评似乎有两点。第一，我在工作中使用的目的论表述是没有依据的，尤其是我在解释他本人关于两性分离的讨论中使用了这种表达。然而，因为这个讨论最初是沃丁顿引发的，而不是我，所以很少有人来找我证明其性质是目的论的还是其他的。实际上，这一点是微不足道的，因为我相信，不管涉及的讨论是关于遗传学的还是关于形态学的，达林顿先生和沃丁顿先生都会和我一样认同适应是选择的结果，并且也和我一样否认它是有目的的。

第二，我想"抛弃""发育－遗传的思想"的想法受到了批评。一般情况下，我并不希望这样，因为很明显发育学的研究非常有助于我们理解遗传学。尽管如此，我确实不同意这个特殊的观点，即在二倍期中，两性的分离实际上是配子分化的结果，显然，沃丁顿和其他很多人都很认同这一观点。在雌雄异体的生物体内，两性的分离和配子的分化在发育上可能是相关的，但是，这不应该使我们忽视它们完全不同的遗传结果。配子的分化自身不能形成正常的杂交繁殖，这一点已经在小麦和豌豆的例子中得到了证实，这两种植物的配子极为不同，但是它们都是进行自体受精的。另一方面，雌雄异体的生物间都会伴随着不同个体之间的杂交。正如沃丁顿在我引用的他书中的句子中以及我没有引用的他以前的一些句子中所表述的那样，他首先将杂交繁殖与配子的分化相关联，继而与两性的分离相关联，我肯定不同意这一点。配子的分化与杂交繁殖和雌雄异体之间毫无遗传联系。

（刘霞 翻译；赵见高 审稿）

Dextran as a Substitute for Plasma

A. Grönwall and B. Ingelman

Editor's Note

The First World War prompted searches for artificial blood substitutes, and various colloids (liquid suspensions) were tested. Here two chemists in Sweden revisit the question towards the end of the Second World War. The qualifications for blood substitutes are stringent, in particular the need to evade the body's immune response and tendency to break down foreign substances. The paper reports some promise in the use of solutions of dextran, a sugar made by bacterial fermentation, as blood plasma (a carrier fluid for blood cells). Today's candidates are more sophisticated, mimicking the oxygen-carrying action of real blood—but such blood substitutes are still highly imperfect, and still urgently needed.

AS is well known, we have in blood, plasma and serum adequate media for the treatment of shock, for example, in cases of serious loss of blood or contusions. During the present War, however, it has proved impossible completely to supply the large requirements of these materials. It is therefore natural that physiologists and chemists are seeking for substances the aqueous solutions of which can replace the expensive and delicate blood or plasma.

In the course of the War of 1914–18, Bayliss[1] attempted to employ solutions of gum arabic for purposes of infusion. Later, other substances such as gelatin, polyvinyl alcohol, pectin, polyvinylpyrrolidone and others were tested to this end. The infusion of these colloids has, however, been attended by certain difficulties. Some of the substances tested have antigenic properties, whereas others cannot be broken down by the organism, for which reason they are stored in the organs, especially in the liver.

The conditions to be fulfilled by a foreign colloid in order that it may exercise a therapeutic effect in cases of shock are, in brief, as follows:

In all cases of shock, both in bleeding and in contusions and burns, it is essential to increase the volume of the circulating blood by the infusion of a liquid. This cannot be done satisfactorily with solutions of crystalloids. The infused liquids must instead contain colloids that exert the same colloidal osmotic pressure as the plasma proteins, or 300–400 mm. water. A condition for the exertion of this pressure by the colloids is that they must be of such a molecular size that they cannot pass through the walls of the capillaries.

The colloid must be suited to repeated intravenous injection in large quantities. It must also be completely atoxic and devoid of antigenic properties.

820

作为血浆替代物的葡聚糖

格伦瓦尔，英厄尔曼

编者按

第一次世界大战促使人们展开了对人工血液替代物的探寻，并且检验了各种胶体悬浮液。这篇文章中两位瑞典的化学家在第二次世界大战接近尾声的时候重提了这一课题。血液替代物应满足严格的条件限制，特别是能够避免引发身体的免疫反应以及分解外源物质的倾向。这篇文章报道了使用细菌发酵制得的葡聚糖溶液作为血浆（血细胞的液体环境）的可能性。现如今的血浆替代物要复杂得多，能够模拟真实血液的携氧功能，但是这样的血液替代物仍然很不完善，需求依然迫切。

众所周知，构成血液的血浆和血清可以用来治疗由严重失血或挫伤等原因造成的休克。然而，事实证明在当前战争期间对这些原料的大量需求完全不可能得到满足。因此很自然地，生理学家和化学家开始寻找其水溶液可以替代价格高昂且难以保存的血液或血浆的物质。

1914~1918 年战争期间，贝利斯[1]为达到输注的目的曾尝试使用阿拉伯树胶溶液。之后，为此目的还测试了其他物质，例如明胶、聚乙烯醇、果胶和聚乙烯吡咯烷酮等。不过，注射上述胶体总会伴随某些特定的困难。有些测试物质具有抗原性，而另一些则无法被机体降解，因而会在某些器官中累积，尤其是肝脏。

下面简要列出使异源性胶体发挥治疗休克的效果所应满足的条件：

在所有发生休克的情况下，包括失血、挫伤和烧伤，通过液体灌输的方式增加循环血量是必要的。用晶体溶液是不能令人满意地实现这一点的。输入的液体必须含有胶体，且能产生与血浆蛋白一样的、相当于 300~400 毫米水的胶体渗透压。使胶体产生渗透压的一个条件是它们必须具有不能透过毛细管壁的分子尺寸。

这种胶体必须适合进行重复的大剂量静脉注射。它还必须是完全无毒的，并且没有抗原性。

The solutions must not have a high viscosity. The viscosity should preferably be of the same order as that of the blood.

Finally, the substance should be of such a nature that the body can gradually rid itself thereof, so that it does not remain long in the blood and is not stored in the organs.

A substance not previously tested for this purpose and apparently fulfilling the requirements listed above is the neutral polysaccharide dextran. Dextran is a water-soluble high-molecular carbohydrate which is formed in solutions of sugar infected with the bacterium *Leuconostoc mesenteroides*. It has been possible to show that the dextran molecule is built up of glucose units, linked together in long, more or less branched chains[2]. The molecular weight of dextran may be very high, of the order of magnitude of many millions[3,4,5]. By partial hydrolysis dextran preparations of lower molecular weight, for example, of the order of 100,000–200,000, can be made[6]. The partially hydrolysed dextran, like the original substance, is inhomogeneous with respect to molecular weight.

By well-controlled partial hydrolysis it is possible to prepare dextran solutions for purposes of infusion in which the solute has a suitable molecular weight and which do not give rise to injuries or reactions even after repeated large infusions. The sedimentation reaction, however, is increased after infusion (which has also been observed after infusion of, for example, gum arabic). The viscosity and colloidal osmotic pressure of the 6 percent solutions employed (with 1–3 percent sodium chloride) are of the same order as those of blood[6,7].

The solutions can be autoclaved and the preparation distributed in concentrated solutions or in the form of dry powder.

If a normal infusion dose is injected intravenously into a dog, the dextran concentration in the blood falls to zero in the course of three to four days. During the whole of this period dextran can be detected in the urine. The dextran ejected with the urine has a lower molecular weight than that originally injected. Even after repeated large infusions, no storage in the organs can be demonstrated histologically[6,7].

As dextran is broken down by the organism, glucose and relatively low-molecular fragments of dextran are presumably formed, which can pass the kidney filter and be expelled with the urine.

The therapeutic effect was investigated experimentally in cases of shock from bleeding, histamine shock and contusion shock developed artificially in rabbits and cats. Rapid and lasting effects on the blood pressure, heart action and respiration were always registered[6].

The experiments on animals giving favourable results, a clinical investigation was therefore commenced, at first on a limited scale. As the first clinical tests also gave promising results, and as there is reason for supposing that dextran is better suited as a plasma substitute

这种胶体溶液的黏度绝对不能太高，最好与血液在同一水平。

最后，该物质还应该具有能被机体自身逐渐清除的特性，这样它才不会长期停留在血液中，也就不会在器官中累积。

有一种物质，之前并未出于这种目的进行检验，但是明显可以满足以上列出的条件，它就是中性的多糖，葡聚糖。葡聚糖是一种水溶性的高分子碳水化合物，可在肠系膜明串珠菌感染的蔗糖水溶液中形成。已经有可能表明葡聚糖分子是由葡萄糖单元构成的，它们彼此连接成长链，并具有或多或少的支链[2]。葡聚糖的分子量可以非常高，达到百万以上的数量级[3,4,5]。通过部分水解可以制备出较低分子量的葡聚糖[6]，例如分子量数量级为 100,000~200,000 的葡聚糖。与初始的葡聚糖一样，部分水解的葡聚糖的分子量也是不均一的。

通过精确控制部分水解，有可能可以制备出满足注射要求的葡聚糖溶液，其溶质具有适当的分子量，并且即使在重复的大剂量注射之后也不会引起损伤或不良反应。不过，注射后沉降反应有所增加（在注射诸如阿拉伯树胶等物质后也观测到这种情况）。所使用的 6%溶液（含有 1%~3% 的氯化钠）的黏度和胶体溶液渗透压[6,7]水平与血液一致。

这种葡聚糖溶液可以进行高压灭菌，产品以浓缩溶液或者干粉的形式进行保存分装。

如果给一只狗进行一次常规剂量的静脉注射，那么其血液中的葡聚糖浓度在三四天之内就会下降到零。整个周期中可以在尿液中检测到葡聚糖。通过尿液排出的葡聚糖比初始注入的葡聚糖的分子量低。组织学检验证实，即使经过重复的大剂量注射，葡聚糖在器官中也不发生累积[6,7]。

随着葡聚糖被机体降解，估计会有葡萄糖和分子量相对较低的葡聚糖片段形成，它们可以通过肾过滤，并随着尿液排出。

通过人为引发兔子和猫的失血休克、组胺休克和挫伤休克，人们对葡聚糖的疗效进行了实验研究。已有报道表明葡聚糖对血压、心搏和呼吸都有快速且持续的影响[6]。

动物实验给出了有利的结果，于是一项临床研究随之开始，一开始规模很小。由于最初的临床实验也给出了前景乐观的结果，并且有理由推测葡聚糖作为血浆替

than, for example, gum arabic, polyvinylpyrrolidone or pectin, it was considered justified to set in train a more thorough clinical investigation. An account of this will be submitted at a later stage.

We wish to thank Prof. Arne Tiselius for helpful advice and Profs. T. Svedberg and A. Westerlund for the provision of laboratory facilities. The research has been carried out with grants from A. B. Pharmacia, Stockholm, and Svenska Sockerfabriks A. B., Malmö.

(**155**, 45; 1945)

Anders Grönwall and Björn Ingelman: Institute of Physical Chemistry, University of Uppsala.

References:

1. Bayliss, W. M., *J. Pharm. Exp. Therap.*, **15**, 29 (1920).
2. Levi, J., Hawkins, L., and Hibbert, H., *J. Amer, Chem. Soc.*, **64**, 1959 (1942).
3. Grönwall, A., and Ingelman, B., *Acta Physiol. Scand.*, **7**, 97 (1944).
4. Ingelman, B., and Siegbahn, K., *Ark. Kem. Min. Geol.*,**18B**, No.1 (1944).
5. Ingelman, B., and Siegbahn, K., *Nature*, **154**, 237 (1944).
6. Grönwall, A., and Ingelman, B., *Acta Physiol. Scand.*, in the press.
7. Grönwall, A., and Ingelman, B., *Nordisk Medicin*, **21**, 247 (1944).

代物会比诸如阿拉伯树胶、聚乙烯吡咯烷酮或果胶更合适，因此着手实施一项更加彻底的临床研究是合理的。一份关于这方面的报告将于下一阶段提交。

我们要感谢阿尔内·蒂塞利乌斯教授提供了有益的建议，感谢斯韦德贝里和韦斯特隆德两位教授提供了实验设备。这项研究是在斯德哥尔摩法玛西亚公司和马尔摩斯文斯卡·索克法布里克斯公司的资助下进行的。

（王耀杨 翻译；莫韫 审稿）

A Labour-saving Method of Sampling

J. B. S. Haldane

Editor's Note

Some of the most important advances in statistical analysis during the early twentieth century came from geneticists and population biologists. Here J. B. S. Haldane describes a method for extracting statistical estimates from small samples. The problem he addresses is how to estimate what fraction of a population p have an attribute A—without knowing roughly how big p is, one can't be sure how big a representative sample must be. He shows that one can estimate p with a fixed standard error by counting up just the number of items in the sample that have the minority attribute A, which takes less time and labour.

IF a fraction p of a population have the attribute A, then it is well known that if m members out of a sample of N have this attribute, the best estimate of p is $\frac{m}{N}$, and its standard error is $\sqrt{\frac{m(N-m)}{N^3}}$ or $p\sqrt{\frac{1-p}{m}}$. Supposing, therefore, that we want our estimate of p to be correct within a standard error of 10 percent of its value, we must count a sample containing $100(1-p)$ members with the attribute A. If we do not know p roughly beforehand we do not know how large to take our sample. For example, if we wish to estimate the frequency of a type of blood corpuscle, and count 1,000 blood corpuscles in all, we should get such values as 20 ± 1.3 percent, or 1 ± 0.31 percent. The former value would be needlessly precise for many purposes. The latter would not differ significantly from an estimate of 2 percent.

The standard error is almost proportional to the estimated frequency if we continue sampling until a fixed number m of the minority with attribute A have been counted, and then stop. Supposing the total number in the sample is now N, we cannot use $\frac{m}{N}$ as an estimate of p. It can, however, be shown that $\frac{m-1}{N-1}$ is an unbiased estimate of p, with standard error very approximately $\frac{1}{N}\sqrt{\frac{m(N-m)}{N-1}}$, or $p\sqrt{\frac{1-p}{m-2}}$, which is nearly proportional to p when this is small. Thus to get a standard error of about 10 percent of the estimate we should have to count until we had observed a number m of the rarer type A, which only varies from 102 when p is very small, to 72 when it reaches 30 percent. If we were content with a standard error of $0.2p$ we could take a quarter of this value, and so on.

一种省力的抽样方法

霍尔丹

编者按

20世纪初，统计分析方面的一些重要进展均来自遗传学家和种群生物学家。这篇文章中霍尔丹描述了一种从少量样本中取样进行统计估算的方法。他将这个问题归结为如何估计一个种群 p 中有多少比例具有 A 特性——如果不知道 p 大概有多大，我们就不能预知需要多大的代表性样本。他指出，可以通过只对样本中少数具有 A 特性的样本的数目进行计数来得到具有固定标准误差的 p 的估计值，这么做既省时又省力。

如果种群中的一个部分 p 具有 A 特性，那么很明显在 N 个样本中有 m 个具有这种特性时，对 p 的最佳估计值为 $\frac{m}{N}$，且其标准误差是 $\sqrt{\frac{m(N-m)}{N^3}}$ 或 $p\sqrt{\frac{1-p}{m}}$。因此，假设我们希望对 p 的估计值的标准误差在10%的范围内，那么我们要计数的样本中就必须有 $100(1-p)$ 个具有 A 特性。如果预先不知道 p 的大概数值，我们就不知道该选取多大的样本。例如，如果我们在估计某种血细胞的出现频率时总共计数了1,000个血细胞，我们会得到像 20%±1.3% 或 1%±0.31% 这样的数值。多数情况下不必像前一个数那么精确，而后一个数值与2%的估计值相比并没有显著差别。

如果我们持续抽样，直到少数具有 A 特性的数量达到固定数字 m 后停止，那么标准误差几乎是与出现频率的估计值成比例。假如样本总量为 N，我们不能用 $\frac{m}{N}$ 作为对 p 的估计值。但是可以将 $\frac{m-1}{N-1}$ 作为对 p 的无偏估计，而且其标准误差非常近似于 $\frac{1}{N}\sqrt{\frac{m(N-m)}{N-1}}$ 或 $p\sqrt{\frac{1-p}{m-2}}$，在数值较小的情况下，这个标准误差值就几乎与 p 成比例。这样，为了使估计值的标准误差是10%左右，我们就应该连续计数样品直到具有 A 特性的数量达到 m，m 只在一定范围内变化，当 p 很小时 m 是102，当 p 接近30%时 m 是72。如果我们可以接受 $0.2p$ 的标准误差，那么 m 取这个值的 1/4 即可，以此类推。

My friend, Dr. R. A. M. Case, has for some time employed a method substantially equivalent to the above in his haematological work, and found it to result in a considerable saving of labour.

Full details will be published elsewhere.

(**155**, 49-50; 1945)

J. B. S. Haldane: Department of Biometry, University College, London, Dec. 7.

一段时间以来，我的好友凯斯博士在他的血液病学工作中使用了一种与此非常类似的方法，他发现这种方法确实省力。

全部细节将在别处发表。

（吴彦 翻译；赵见高 审稿）

Plant Viruses and Virus Diseases[*]

F. C. Bawden

Editor's Note

Towards the end of the Second World War, the scientific community's knowledge of the nature and mechanism of viruses was only rudimentary. This article by the director of the Rothamsted Experimental Station, Frederick C. Bawden, is a succinct summary of the state of knowledge at the time. The reality of viral infections was adequately supported by experimental work with plant viruses. Little was known of animal viruses such as those responsible for infectious diseases of humans. This naivety was removed by two imminent steps—the development of electron microscopes that made viruses visible for the first time and the discovery of the role of nucleic acids in biology by Watson and Crick in 1953.

THE existence of viruses was first deduced from work done in 1892 on tobacco plants suffering from mosaic, and much of what we now know of these elusive entities has come from further work on this and a few other plant diseases. It is far from certain that this knowledge can safely be applied to the causes of the many diseases, affecting all kinds of animals, higher plants and bacteria, that are now attributed to viruses. These cover a wide range of clinical conditions, and we know for certain of only two features that they have in common; their causes have neither been seen nor cultivated *in vitro*. If we wish, we can turn these negative features into what looks like a positive statement, by defining viruses as obligately parasitic pathogens too small to be resolved by microscopes using visible light. Indefinite as this is, it may still prove to be more precise than the facts warrant, for obligate parasitism is always postulated rather than proved, and serious attempts at cultivation have actually been made with very few viruses. Thus, when we speak of a virus disease, we usually mean merely an infectious disease with an invisible cause. Unless the resolving power of the microscope has some unsuspected significance in defining biological types, this obviously tells us nothing specific about the nature of viruses and might well cover a range of different entities.

This possibility seems increasingly likely when we try to generalize about plant virus diseases, for we find that no statements can be made about such features as symptoms, methods of infection, or distribution of virus in the host, to which there are no exception. This is far from conclusive, however, for what a virus does to a plant is as much a property of the plant as of the virus, and the same virus may produce very different effects in different plants. Also, although complete generalizations are impossible, there are some features shared by a number of different virus diseases, especially those met commonly in Nature.

---

[*] Substance of two lectures at the Royal Institution delivered on November 21 and 28.

植物病毒和病毒性疾病*

鲍登

编者按

第二次世界大战快结束的时候，科学界对病毒本质和机理的认识还非常初步。罗森斯得实验站的主任弗雷德里克·鲍登先生撰写的这篇文章对当时的认识状况进行了精炼的概括。病毒感染的事实得到了利用植物病毒所做的实验工作的充分支持，但是对于诸如引发人类感染性疾病的动物病毒，当时人们仍是知之甚少。这种无知被两种即将获得的重要进展——使得病毒首次能被直接观察到的电子显微镜技术的发展以及 1953 年沃森和克里克对核酸生物学功能的发现——所消除。

早在 1892 年，人们便通过对患有花叶病的烟草的研究而推断出了病毒的存在，我们今天对这些难以捉摸的实体的认识大多是通过对这种病毒以及其它几种植物病毒的进一步研究而获得的。这些知识能否被可靠地用于理解许多疾病的起因还不能确定，这些疾病影响着各类动物、高等植物以及细菌，现在我们把它们都归因于病毒的侵染。这些疾病涵盖了广泛的临床症状。我们只确切地知道它们具有两个相同的特点：这些致病因子既没有被观察到，也没有在体外条件下被成功培养过。如果愿意的话，我们可以将这些否定的特征转换成一种看上去像是肯定性的陈述：病毒可以被定义为一种小到利用光学显微镜无法分辨的专性寄生的病原体。尽管这样的定义很模糊，但它可能仍然比事实所表现出来的更加精确，因为专性寄生状态一直都只是被假设而从未得到证实，而且实际上只对少数几种病毒进行过体外培养的真正尝试。所以，当我们提到一种病毒性疾病的时候，我们通常仅仅指由一种不明病因导致的感染性疾病。除非显微镜的分辨率在定义生物类型方面的某种重要性不容置疑，否则这样的定义显然并没有告诉我们任何关于病毒本质的内容，这样的不明病因可能涵盖的是一个相当范围的不同实体。

当我们试图概括植物病毒性疾病的时候，这种可能性似乎更为明显，因为我们发现无法针对症状、感染方式或者病毒在寄主中的分布等特征作出任何不存在例外情形的论断。但这样的说法也远非结论性的，因为一种病毒对一种植株所作的，与病毒的特性和植株的特性都同样地相关，同一种病毒对不同的植株会产生完全不同的效应。另外，尽管目前还不可能提出一种全面的概括，但很多不同的病毒性疾病也存在一些共同的特征，特别是对那些自然界中常见的病毒性疾病而言。

---

*11 月 21 日和 28 日在皇家研究院所作的两场报告的内容。

The effects most frequently caused by viruses are a dwarfing of the host plants and an alteration of the colour and shape of the leaves. Instead of being uniformly dark green, the leaves may bear spots, rings or patches of light green, yellow or white, or they may be generally chlorotic without definite mottling. Deformation may show only as an alteration in the leaf outline, or the laminae may be so reduced that the leaves consist of little but the main veins; it may take the form of local hyperplasia, to give unusual outgrowths from the leaves or gall-like proliferations in stems. Symptoms tend to occur more generally over a whole plant than with most fungal or bacterial diseases, for in natural infections it is usual for viruses to spread through the vegetative parts of affected plants. In plants infected experimentally, however, symptoms are often restricted to local lesions, produced by the death of tissues around the entry point. Diagnosis from symptoms is by no means easy, for different viruses may cause almost identical symptoms in the same host, whereas the same virus may produce totally different clinical conditions in different hosts. A further complication is that many viruses are unstable and frequently change to give forms that produce different symptoms from those produced by the parent virus. To be recognized, a virus must cause changes in the appearance of some plants; but it need not necessarily cause changes in all susceptible hosts. Indeed, the phenomenon of the carrier—an infected individual showing no symptoms—is common in plants, and such carriers can be of considerable importance as unsuspected sources of infection for intolerant species.

In many virus diseases, three distinct phases can be identified. As a result of virus multiplication at sites of infection, lesions first appear on inoculated leaves. After a few days, the virus passes to the phloem, through which it travels rapidly to distant parts of the plant. It seems to have no autonomous movement, but to travel along with the translocation stream of elaborated food materials, away from tissues actively engaged in photosynthesis and towards regions of active growth. It is because of this, and not because they resist infection, that leaves already fully developed at the time the plant becomes infected rarely show symptoms. Thus results of systemic infection appear on the young, actively growing leaves; the later symptoms of this systemic phase often differ from those first produced, as the disease passes from an acute to a chronic stage. Often both stages are serious diseases; in many potato varieties, for example, leaf-drop-streak is succeeded by severe mosaic. Occasionally, however, the chronic stage is extremely mild, such as in tobacco plants with ring-spot, which recover from an acute necrotic disease and afterwards show few or no symptoms. The virus is present in such plants, but in smaller quantities than during the acute stage. The sequence of three phases is common, but by no means general, and the same virus may give different sequences in different hosts. Potato virus Y, for example, gives local lesions only in one host, local lesions followed by systemic symptoms of two kinds in a second host, whereas in a third it gives no local lesions and systemic symptoms of only one kind.

In addition to altering the external appearance of plants, viruses also produce internal changes. Some of these are simply modifications of normal structures or tissues, such as reduction of the chloroplasts or necrosis of the phloem; but the most characteristic involve the production of new kinds of intracellular inclusion bodies. These are not found in all

　　由病毒引起的最常见的影响包括：宿主植株的矮化、叶子颜色和形状的改变等。原本呈均匀暗绿色的叶片可能出现浅绿色、黄色或白色的斑点、斑环或斑块，或者叶片上虽没有明显的花斑但却普遍萎黄。变形可能只表示叶片外形的一种变化，或者叶片纹层减少只剩下少量的主要叶脉；变形可能以局部增生的形式展现，导致叶片异常地往外生长或者树干上产生类似瘿瘤那样的肿起。与大多数真菌性或细菌性疾病相比，这些症状往往更多发生于整棵植株，因为在自然的感染过程中，病毒通常扩散于植株正在生长的部分。然而，在用实验方法感染的植株中，由于进入点附近组织的死亡，损伤症状通常被限定在局部位置。通过症状来做出诊断并不容易，因为不同的病毒在同一种宿主中可能产生几乎相同的症状，而同一种病毒在不同的宿主中可能产生完全不同的症状。另一个使事情复杂化的因素是，许多病毒不稳定，频繁地变化成能引发与亲代病毒不同的病症的形式。如果要被识别的话，一种病毒必须要引起某些植株的外形上的改变，但问题是，病毒并不一定在所有易感宿主中都会引起这种改变。的确，携带现象——被感染的个体不表现出任何症状的情形——在植物中很常见。这种携带者作为敏感型植物的不被觉察的感染源具有相当重要的意义。

　　许多病毒性疾病的发病过程都可以分成三个明显不同的阶段。由于病毒在侵染部位的增殖，被接种的叶片上首先出现损伤。几天后，病毒会传递到韧皮部，通过韧皮部它就会迅速移动到植株的远端。病毒似乎并不存在自主运动，而只是随着精细营养物质的运输流而移动，避开光合作用活跃的组织，移向活跃生长的区域。那些在植株被病毒感染时已经完全发育的叶片很少表现出症状，正是因为这个原因，而不是因为它们可以抵御病毒感染。因此，全身性感染的结果大多出现在早期还处于活跃生长的叶片中；这种全身性感染的晚期症状经常与最初产生的症状不同，因为此时疾病从急性阶段进入到慢性阶段。这两个阶段经常都会产生严重的病症，例如，在多个马铃薯品种中，条纹落叶病后紧接着会出现严重的花叶病。然而，慢性阶段的症状偶尔也会非常轻微，例如患环斑病的烟草植株，开始表现为急性坏死病，然后就很少或者不表现出症状。病毒存在于这样的植株中，但数量比急性阶段少。这三个阶段的顺序通常就是这样的，但绝非全都如此，同一种病毒在不同的宿主中可能表现为不同的顺序。例如，马铃薯病毒 Y，在一种宿主中只表现为局部病变，在第二种宿主中是局部病变后表现出两种全身性症状，在第三种宿主中不表现出局部病变而仅仅出现一种全身性症状。

　　除了改变植株的外部特征，病毒也能使植株内部产生变化。这些内部变化中，有的只是对正常结构或组织的修饰，例如叶绿体的减少或者韧皮部的坏死，但最典型的是涉及新的类型的细胞内包涵体的产生。包涵体并非发现于所有病毒性疾病中，

virus diseases, but their formation appears to be specific to viruses, for similar bodies have not been found either in healthy plants or in those suffering from other kinds of disease. Different viruses give rise to different kinds of inclusion body, and produce them in varying numbers and in different tissues. The most general type is a vacuolar, amoeboid-like body found in the cytoplasm, but crystalline and fibrous inclusions also occur in infections with a number of different viruses. At least two viruses give rise to crystalline inclusions in the nuclei. The precise nature of these bodies is still uncertain; but we know that they contain virus, and their production can in part be simulated *in vitro*. It seems most likely that they are insoluble complexes produced by the viruses combining with some metabolic product of the diseased plants.

Symptomatology without proof of transmissibility is insufficient to assign a particular disease to the virus group, for similar kinds of symptoms can be caused by toxins, deficiencies of mineral nutrients and aberrant genes. With hosts that are easily grafted, transmission by grafting is usually the first method tried; for once organic union is established, all viruses that cause systemic symptoms readily pass from infected scions into healthy stocks. Indeed, grafting is the only method of transmission known for many virus diseases, and it has almost become the critical test of a plant virus disease.

Infection occurs only through wounds, but wounds that permit one virus to enter may not permit another. Many viruses are readily transmitted by rubbing healthy leaves with sap from diseased plants, but others are not; some of both these types are transmitted by insects. Several different explanations can be offered for the failure of inoculation to transmit viruses that are readily transmitted by insects. First, some viruses may be able to establish themselves only in deep-seated tissues, such as the phloem, which are not penetrated by ordinary inoculation methods. Secondly, conditions in the expressed sap of some hosts may be such that the viruses are rapidly destroyed or rendered non-infective. Thirdly, the virus content of sap from some diseased plants may be below that required for infection. Thus, although failure to transmit by inoculation is often used as a specific character of a virus, clearly it may equally well be a reflexion of some property of the host.

Insects do not seem to act simply as mechanical carriers of viruses, for no insect vectors are known for the two viruses most easily transmitted by inoculation; and there appear to be specific relationships between insects and the viruses they transmit. Individual viruses are usually transmitted by only a few related species of insect and not by others, though these may have similar feeding habits and be vectors of other viruses. Vectors are usually insects with sucking mouth-parts; the most important are aphides, leaf-hoppers, white-fly and thrips. Two main types of behaviour in the insect have been distinguished. Vectors of one type of virus can infect healthy plants immediately after feeding for a short time on a diseased plant, and these usually cease to be infective within a few hour. After feeding on diseased plants, vectors of the other type cannot infect healthy plants for some time, which varies from minutes to days with different viruses, and such vectors remain infective for long periods, often for their whole lives. Some workers believe that viruses of the

834

但是它们的形成似乎是病毒特有的，因为在健康植株或者那些患其他疾病的植株中还从来没有发现过类似的小体。不同的病毒产生不同类型的包涵体，以不同的数目产生于不同的组织中。最常见的包涵体类型是一种发现于细胞质内的与变形虫类似的空泡，但多种不同的病毒感染后也会产生晶体状和纤维状的包涵体。至少有两种已知的病毒可以导致细胞核中产生晶体状包涵体。这些小体的确切本质仍然不确定，但我们知道它们包含有病毒，在体外条件下可以部分模拟这些小体的产生过程。看上去最有可能的是，包涵体是由病毒与受感染植株的一些代谢产物形成的不溶性复合物。

那些没有证据可以确认其可遗传性的症状还不足以将一种特定的疾病归入病毒类中，因为相似的症状也可能由毒素、矿质营养素的缺乏或基因缺陷引起。对于容易嫁接的宿主植株，人们通常首先会尝试用嫁接传染的方法。一旦宿主与接穗之间建立了有机联系，引起全身性症状的所有病毒就很容易从受感染的接穗转移到健康的宿主植物中。实际上，嫁接是许多病毒性疾病已知的唯一传播方式，它几乎已经成为植物病毒性疾病的决定性检测手段。

病毒只能通过伤口造成感染，但允许一种病毒侵入的伤口不一定会允许另一种病毒侵入。通过取自受感染植株的汁液来涂抹健康叶片这种方式就可以使许多病毒得以传播，但另外一些病毒却不行。这两种类型的病毒中都有一些可以通过昆虫传播。对于通过昆虫很容易传播却不能通过接种传播的现象，我们可以提供几种不同的解释。第一，有些病毒只能在像韧皮部这样的深层组织中存活，而普通的接种方法并不能渗透到这么深的部位。第二，一些宿主挤压汁的条件可能使病毒在其中很快被破坏或丧失感染性。第三，来自受感染植株的树汁中的病毒含量可能低于成功感染所需的量。因此，虽然接种传播的失败经常被视为病毒特异性的表征，但很明显的是，它也完全可能只是宿主植株的某些特性的反映。

昆虫似乎并不仅仅是病毒的被动型携带者，因为对于两种很容易通过接种方式传播的病毒，我们到目前还未发现任何昆虫媒介，因此在昆虫和它们所传播的病毒之间似乎存在专一性关联。特定的病毒通常只能通过几种亲缘关系接近的昆虫来传播，而不能通过其他可能有着相似取食习性并且也可以作为其他病毒媒介的昆虫来传播。病毒传播者通常都是那些具有刺吸式口器的昆虫，最重要的代表包括蚜虫、叶蝉、飞虱和牧草虫等。在昆虫中已经发现了两种主要的行为方式。一类病毒的传播昆虫在摄食受感染植株后的很短一段时间内就能够马上侵染健康植株，并且它们的感染性通常在几小时后就会消失。另一类病毒的传播昆虫在摄食受感染植株后有一段时间是不能感染健康植株的，针对不同的病毒而言，这段时间可以从几分钟到几天不等，但这种携带者能在较长时期间内保持感染性，经常可以保持一生。一些

second type multiply in the insects. There is no obvious reason why they should not, and the theory would explain some of the now puzzling features of the behaviour of these viruses; but there is no conclusive evidence that insects ever contain more virus than they acquire while feeding on infected plants. Studies on the virus causing dwarf disease of rice supplies the best circumstantial evidence for multiplication. This virus is unique in being the only one known to pass from infective adults through eggs to their progeny. Progeny up to the seventh generation have once been found to be infective and from on infected egg the progeny have infected more than 1,000 plants. This is regarded by some workers as "overwhelming" evidence for multiplication, as they consider that the quantity of virus in the original eggs could not have been enough to give all the infections. But is this so? If the virus multiplied in the insects to anything like the extent it does in plants, then there would be no reason why the progeny should not continue to be infective indefinitely, and infect as many plants as they feed on. We know nothing of the size of this virus, but if it is of the same order as other plant viruses the sizes of which are approximately known, then 1,000 particles would weigh less than 10^{-14} gm., and many times this quantity could surely be contained in a leaf-hopper's egg without difficulty.

Transmission of some viruses has been achieved by linking diseased and healthy plants with the parasite dodder (*Cuscuta* sp.). This novel method of transmission promises to be valuable in extending the host ranges of some viruses to plants more favourable for study than those in which the viruses occur naturally. One of the greatest differences between individual viruses lies in the numbers of different plants they can attack. Some are known to infect hundreds of plant species, belonging to many different families and orders; others have been transmitted to only a few closely related species. This difference may be apparent rather than real, for viruses transmissible only by grafting or by insects will normally have host ranges restricted to plants which can be intergrafted or which can act as food plants for the insect vector.

For more than forty years, work on plant viruses was largely concerned with symptoms, transmission and host ranges. It showed that viruses could multiply and alter, and produced few results conflicting with the generally accepted conclusion that they were small organisms, essentially similar to bacteria. There were opposers of this, usually from among those studying tobacco mosaic virus, but they could offer nothing definite to support their alternative views. The intensive study during the last ten years of the viruses *in vitro* has led to results that necessitate considerable modification of the earlier views. They do not, however, justify the sweeping conclusions implied by such facile phrases as "lifeless molecules", which are increasingly applied to viruses.

What has been achieved is the successful application of the techniques of protein chemistry to the purification of a dozen or so viruses. This has shown us that the particles of these viruses are not organized cellularly like organisms, and that in many ways they resemble constituent parts of organisms rather than whole organisms. They can be obtained in forms chemically much simpler than bacteria, free from diffusible components, and with a much greater regularity of internal structure than is usual with

研究人员认为第二类病毒会在昆虫体内增殖。没有明显的证据表明为什么病毒不能在昆虫体内增殖。此外，这个理论也可以用来解释目前这些病毒行为的很令人费解的一些特征；但并没有确切的证据表明昆虫体内的病毒含量比它摄食受感染植株时所获得的更多。针对引起稻谷矮化病的病毒开展的实验研究为这样的增殖理论提供了最好的旁证。这种病毒很独特，它是人们知道的唯一一种能从具有感染性的成虫通过其卵细胞而传递给后代的病毒，甚至到第七代的昆虫都仍有传染性。在一次实验中，一个被感染的卵最后竟然感染了 1,000 多株植物！这被一些研究人员看作是增殖理论"无可辩驳"的证据，因为他们认为最初的卵里病毒的数量是不足以感染这么多植株的。确实如此吗？如果病毒在昆虫里增殖的程度与病毒在植株中增殖的程度是一样的话，那便无法解释为什么昆虫后代不继续具有无穷的感染性进而感染它们摄食的所有植株。我们目前对这种病毒的大小还一无所知，但是如果它的大小与我们大约知道大小的其他植物病毒处于同一数量级，那么 1,000 个病毒颗粒的重量应该低于 10^{-14} 克，一个叶蝉的卵可以轻易装下比这一数量多许多倍的病毒。

一些病毒的传播是通过寄生植物菟丝子（菟丝子属）将受感染植株与健康植株连接在一起而实现的。这种全新的传播方式在将一些病毒的宿主范围扩展到比自然发生的病毒感染的植株更适于开展研究的植株方面很有价值。不同病毒之间最大的差异之一在于侵染的植株种类范围不同。我们知道，有一些病毒可以感染上百种属于不同科和不同目的植物，而另外一些只能感染少数几种紧密相关的物种。这种差异可能仅仅是表面的，而非真实存在的，因为对于只依靠嫁接或者昆虫传播的病毒，它们正常的宿主范围仅局限于能进行相互嫁接的植株或能被昆虫携带者当作食物的植株。

四十多年来，对植物病毒的研究大多关注的是症状、传播和宿主范围。这些研究的结果表明，病毒能够增殖和变异，几乎没有结果与病毒是小的有机体且与细菌基本类似这一被人们普遍接受的结论相冲突。也有人反对这一观点，他们通常都是那些烟草花叶病毒的研究者，但他们不能提供明确的证据来支持他们的不同看法。近十年来对病毒进行的比较集中的体外研究结果表明需要对早期的观点做相当的修正。然而，这样的观察结果也并没有对"无生命的分子"这类草率术语所隐含的全新结论提供支持，尽管这种术语越来越多地被应用到病毒上。

人们已经成功地将蛋白质化学技术应用于对大约十几种病毒的纯化工作。这提示我们，这些病毒颗粒并非像生物有机体那样是由细胞组织构成的，并且它们在许多方面类似于有机体的组成部分而非整个有机体。研究人员得到的病毒的化学形式比细菌简单得多，不包含可扩散的成分，与生物有机体通常的情形相比其内部结构更加规则。目前纯化得到的病毒都与核蛋白具有相同的化学形式。它们都含有核

organisms. The viruses so far purified have all been obtained in the same chemical form, as nucleoproteins. They all contain nucleic acid of the ribose type, but the proportion of nucleic acid to protein varies with the individual viruses. It is far too early to conclude that all plant viruses are essentially nucleoproteins; but we can say that it will be a major discovery if one is found to be anything else, for those already purified cover a diversity of types, some known to be insect-transmitted and others not. They range from potato virus Y, which denatures and loses infectivity within a few days, to tobacco mosaic virus, which remains stable for years. Stability as a native protein, however, is not always the same thing as stability as a virus; the infectivity of preparations of any of these viruses can be destroyed by some treatments that have no appreciable effects on the physical, chemical and serological properties.

The shape of the particle is responsible for some of the most striking differences between the properties of preparations of different viruses. Solutions of purified tobacco mosaic virus, and of potato viruses X and Y, show phenomena characteristic of greatly elongated particles; they are anomalous in all their physical properties and are polydisperse. No true crystals have been prepared from these, but dilute solutions show anisotropy of flow strongly, and concentrated solutions are liquid crystalline. X-ray studies of solutions of tobacco mosaic virus have demonstrated a regularity of structure previously unsuspected in fluids, for the particles are arranged equidistant from one another so that the available space is filled uniformly. When mixed with their antisera, these rod-shaped virus particles precipitate almost immediately, giving bulky, fluffy precipitates resembling those produced by bacterial flagellar antigens.

Solutions of bushy stunt and tobacco necrosis viruses behave very differently and show none of the anomalous properties characteristic of elongated particles. By suitable treatments they can be induced to crystallize in forms characteristic of the individual virus. When mixed with their antisera, they precipitate more slowly than the rod-shaped viruses and, as might be expected with spherical particles, pack more closely to give dense, granular precipitates resembling those produced by somatic antigens.

What is the relationship between these isolated nucleoproteins, which in laboratory work behave much like preparations of other proteins, and the viruses as they occur in the plant? There is enough evidence now to show that these proteins are the viruses in the sense that they can initiate infection. Nevertheless, it would be premature to assume that, while active in the host plant, the viruses are chemically so simple as analysis of the purified preparations suggests. During the course of isolation, many materials are discarded as impurities; most of these are certainly constituents of the normal host, but some may well be specific products of virus activity. Any such are clearly not essential for infectivity; but if the virus were organized cellularly, they would be retained within a cell wall and would be accepted as integral parts of the virus, which would immediately look a much more complex body than does our naked protein particle.

糖型的核酸，只是核酸与蛋白质的比例随病毒种类的不同而不同。所有的植物病毒基本上都是核蛋白这样的说法还言之过早。但是，如果发现由其他物质组成的病毒的话，那肯定将是一个重大发现，因为已经被纯化的那些病毒涵盖了多种类型，已知其中一部分是靠昆虫传播的，另外一些却并非如此。它们涵盖了从马铃薯病毒 Y——在几天之内就会变性并失去感染性的一种病毒，到烟草花叶病毒——多年之后仍旧可以保持其感染性稳定不变的一种病毒。然而，天然蛋白质的稳定性和病毒的稳定性并不完全相同。这些病毒制备物的感染性可能会被某些对其物理、化学、血清学特性没有明显影响的处理过程破坏。

病毒颗粒的形态不同是造成不同病毒制备物特性之间具有惊人差异的原因。纯化出的烟草花叶病毒、马铃薯病毒 X 和 Y 的溶液都表现出了高度伸长颗粒的特征，它们的所有物理特性都反常，并处于多分散性的不均一状态。从这些溶液中人们没有得到过真正的晶体，但是它们的稀释溶液表现出很强的流动各向异性，而浓缩溶液是液晶态的。对烟草花叶病毒溶液的 X 射线研究揭示了一种先前未知的在液体中的规则结构，因为颗粒之间的排列是等距离的以至于那些可被利用的空间都被均一地填充了。当与它们的抗血清混合后，这些棒状病毒颗粒几乎立刻就会形成大块的蓬松沉淀，这类似于由细菌鞭毛抗原产生的沉淀。

浓密矮化病毒和烟草坏死病毒溶液的表现与上述情况极其不同，并不显示出任何伸长颗粒那样的特征性异常行为。经过适当的处理，它们能被诱导形成晶体，形态因病毒而异。当与它们的抗血清混合后，这些病毒沉淀的速度比那些棒状病毒慢。正如人们对球状颗粒所期望的那样，它们排列得更加密集，产生出粒状沉淀物，类似于菌体抗原产生的沉淀。

这些分离到的在实验室研究中行为非常类似于其他蛋白质制备物的核蛋白与植物中出现的病毒之间具有什么关系呢？就它们能够引发感染来说，现在已经有足够的证据表明，这些蛋白质就是病毒。然而，虽然对宿主植物有感染活性，但是现在就假定病毒在化学组成上就如我们对纯化的制备样品进行分析得到的结果所暗示的那么简单的话肯定还为时尚早。在分离过程中，许多物质都被当作杂质而除掉了，毫无疑问，这其中的大部分都是正常宿主的组成成分，但有一部分很可能就是植物中因为病毒的感染而产生的特异产物。无疑，这些杂质中的任何组分都不是病毒感染特性所必需的；但如果病毒是在细胞内组装而成的话，这些特异性杂质就会被保留在细胞壁以内，并成为病毒整体的一部分，这样一来，病毒颗粒立刻就变得比我们得到的裸露的蛋白质颗粒更加复杂了。

In the absence of specific tests for any product of virus activity, we have no positive evidence for their occurrence in plants, but evidence from various sources suggests that purification may be altering the viruses. Purified preparations of tobacco mosaic virus, for example, contain particles about 15 mμ wide but varying in length from less than 100 mμ to more than 1,000 mμ. There is nothing to show that the greatly elongated particles occur in the plant, and much to suggest that they are produced by the linear aggregation of small particles during the course of preparation. By taking suitable precautions, solutions of tobacco mosaic virus can be made that show little or no anisotropy of flow and behave serologically more like somatic antigens; but these are unstable and readily change into anisotropic solutions with serological behaviour characteristic of flagellar antigens. This change seems to be connected with the removal of other material from the small nucleoprotein particles, which then join together end-to-end. The change in size and shape may explain the failure to produce true crystals of this virus *in vitro*, though they occur abundantly in infected plants.

We know also that the purified virus readily combines with other proteins such as trypsin and ribonuclease, and that these can be removed again without affecting infectivity. May not similar combinations occur within the host, and be responsible for converting this nucleoprotein into a functioning system capable of multiplication and of the activities of which the results are so obvious?

In addition to the changes produced by purification, there is other evidence that virus does occur in the plant in forms with different properties from those of the purified nucleoproteins. Until recently, all laboratory work on plant viruses was done with the sap that is expressed from macerated infected leaves. This was thought to contain all the virus in the plant, for washing the fibrous residues gives little extra virus. However, these residues actually contain as much virus as does the sap; but normally this is insoluble, probably because it is combined with other substances, and special treatments are needed to get it into solution. It is possible that this insoluble virus is the biologically active system, whereas that free in the sap may be merely excess virus functioning as a mobile source of infection for other cells. We know so little about the multiplication of viruses, and of their activities within the host, that at present we must suspend judgment. But it is probably safest to regard the nucleoproteins as the chemical minima—equivalent to reproductive organs or embryonic viruses—which develop into working entities only when placed in an environment containing the materials or enzyme systems they lack in their purified state.

(**155**, 156-158; 1945)

F. C. Bawden: Rothamsted Experimental Station.

在对病毒感染导致的产物缺乏特异检测方法的情况下，它们在植物中是否的确存在，我们对此还缺乏肯定性的证据。但是源自各方面的证据都表明，纯化过程可能会改变病毒结构。例如，烟草花叶病毒的纯化制备物中就含有宽度在 15 纳米，长度从小于 100 纳米到大于 1,000 纳米不等的颗粒。没有证据表明这些极其伸长的颗粒是在植物中产生的，多方面证据暗示它们是在样品制备过程中由小颗粒通过线性聚集而成。通过采取适当的预防措施，我们能制备出很少出现甚至不出现流动各向异性的烟草花叶病毒溶液，其血清学行为更加类似于菌体抗原，但这样的溶液并不稳定，很容易转变成血清学行为具有鞭毛抗原特征的各向异性溶液。这种变化似乎与从小的核蛋白质颗粒中清除其他物质的过程有关，小的核蛋白质颗粒就是在这一过程中首尾相连的。这种大小和形态的变化也许可以解释为什么这种病毒在体外未能形成真正的晶体，尽管这样的晶体在受感染的植株中大量出现。

我们也知道纯化出的病毒很容易与胰岛素、核糖核酸酶这样的其他蛋白质结合，并且这些蛋白质也可以再次被去除而不影响病毒的感染性。在宿主植物中难道不会发生类似的结合现象而使得这种病毒核蛋白转化为研究结果明显表明的那样能够增殖并进行感染的功能系统吗？

除了纯化过程带来的改变之外，还有其他证据表明病毒的确能以某些特性有异于纯化核蛋白那样的形式存在于宿主植株中。直到最近，与植物病毒有关的所有实验室研究都是利用从被浸泡的受感染植株的叶子中得到的榨汁而开展的。人们认为这样的汁液中包含了植株中的所有病毒，因为洗涤剩下的纤维性残渣很少能得到更多的病毒。然而，这些残渣中实际上包含着和汁液中一样多的病毒，但通常这些病毒处于不溶状态，可能是因为这些病毒结合了其他物质，所以需要经过特殊的处理才能使它们进入溶液中。很可能这种不溶性病毒才是具有生物学活性的系统，而那些存在于汁液中的自由病毒可能仅仅是作为其他细胞的一种移动性感染源而发挥作用的额外病毒。我们对病毒在宿主内的增殖和感染机制知道得如此之少，以至于目前我们只能暂时不予评判。但是，将核蛋白看作是病毒的最基本化学结构——相当于生殖器官或胚胎病毒——也许是最稳妥的，只有把这些基本的化学结构放到它们处于纯化状态时所缺乏的某些物质或酶系统的环境中后，它们才能成为一个能够发挥其功能的完整实体。

（韩玲俐 翻译；昌增益 审稿）

Action of Penicillin on the Rate of Fall in Numbers of Bacteria *in vivo*

A. B. MacGregor and D. A. Long

Editor's Note

By 1945, penicillin's antibacterial activity was well known, but the drug was thought to have its effects by stopping growing cells from dividing. Alexander B. MacGregor and David A. Long show in this paper that the biggest drop in bacterial number happens within 15 minutes of penicillin being given, too quickly for the "bacteriostatic" theory to hold. They suggest that penicillin may sometimes kill bacteria.

DURING work on the use of penicillin pastilles in oral infections, of which a preliminary report has been published[1], numerous experiments have been undertaken in order to determine the total and differential fall in numbers of different species of bacteria in the mouth and the rate of this fall, under the influence of penicillin. This rate of fall is of particular interest, and as it appears that certain deductions on the mode of action of penicillin can be drawn from the experiments, it is considered that it is worth directing attention to them in a separate publication. The technique of determining the rate of fall was as follows:

0.1 c.c. of saliva was added to a measured quantity of normal saline. 1/50 c.c. of this mixture was then inoculated on to a blood agar plate. This was incubated at 37°C. for twenty-four hours, when the number of colonies was counted. From this figure the approximate number of bacteria per c.c. of saliva was calculated.

Immediately the first specimen of saliva had been obtained, a 500-unit penicillin pastille was placed in the buccal sulcus between the cheek and the teeth, and allowed to dissolve without sucking. A fresh pastille was inserted every thirty minutes. At fifteen-minute intervals further specimens were taken, and the number of bacteria later estimated in the manner described.

In these experiments saliva diluted with normal saline was used as the inoculum; it was therefore possible that the penicillin present in the saliva and transferred to the plate might be sufficient to inhibit growth: to guard against this action penicillinase was added to the medium.

The results of these experiments are shown in graph *A*, which represents the mean of three, though the findings have been constant in a far larger series of experiments using pastilles of different strengths.

在体内青霉素对于细菌数量减少速率的作用

麦格雷戈，朗

编者按

在 1945 年以前，青霉素的抗菌活性已是众所周知，不过普遍认为这种药物是通过阻止增殖期细菌的分裂而实现抑菌效果的。亚历山大·麦格雷戈和戴维·朗在这篇文章中指出，细菌数量在给药之后的 15 分钟之内发生最大幅度的下降，这个过程太快了，难以用"抑制细菌繁殖"的理论来解释。他们认为青霉素在一些情况下会杀死细菌。

在从事将青霉素锭剂用于口腔感染治疗的研究期间（初步报告已发表 [1]），已有大量实验被用来确定在青霉素作用下口腔中细菌总量的减少和不同菌种的减少量，以及减少的速率。此减少速率受到格外关注，并且似乎有可能从相关实验中对青霉素作用模式作出推断，因此我们认为有必要就这些实验单独发表一篇文章。测定细菌量减少速率的方法如下：

先向一定量的生理盐水中加入 0.1 毫升唾液，接着将 1/50 毫升此混合溶液接种到一块血琼脂平板上。在 37℃ 下培育 24 小时后进行菌落计数。根据这个数值可以估算出每毫升唾液中的细菌总数。

得到第一份唾液样品之后，立即将 500 单位的青霉素锭剂置入颊齿之间的口腔沟槽内，使其在不被吸入的前提下溶解。接着每隔 30 分钟置入新的锭剂。每隔 15 分钟采集一次样品，然后用上述方法估算细菌总数。

在上述实验中，使用经生理盐水稀释过的唾液作为接种物，因此存在于唾液中并被转移到血琼脂平板上的青霉素有可能就足以抑制细菌的生长。为排除这种情况我们向培养基中加入了青霉素酶。

曲线 A 显示了这些实验的结果，它代表了 3 次重复实验的平均值。不过，在用不同强度的青霉素锭剂进行的大量实验中结果一直保持恒定。

It will be seen from this graph that the maximum fall in the total number of organisms occurred within the first fifteen minutes after the application of the penicillin.

Increased salivation due to the presence of a pastille in the mouth could have been a factor in this rapid fall, and in order to exclude this possibility the experiments were repeated using pastilles made of base alone, without penicillin. Pastilles of this type were maintained in the mouth for one and a half hours, and estimations on the total number of organisms in the saliva carried out as in the previous experiments.

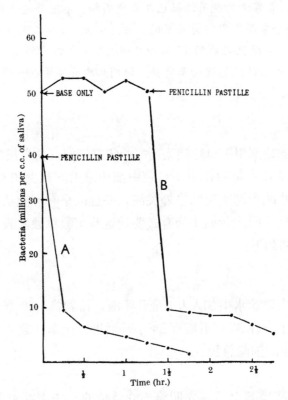

At one and a half hours, pastilles of the same base, but containing 500 units of penicillin each, were inserted and maintained for an equal period of time. The results expressed in graph *B* show that the pastille base alone produced no reduction in the total number of organisms, but substitution of the pastilles containing penicillin caused a fall in numbers comparable with the results shown in graph *A*. The possibility of the mechanical effect of salivation causing a reduction in numbers of organisms could therefore be excluded.

Consideration of the results shows that the most rapid fall in the total number of bacteria occurred in the first fifteen minutes after application of the penicillin. This rapidity of action is difficult to explain on the current hypothesis that penicillin is bacteriostatic, and would suggest that *in vivo*, when conditions of temperature, etc., are favourable, it may have a true bactericidal action.

(**155**, 201-202; 1945)

从图中可以看出，在使用青霉素后的最初 15 分钟内细菌总数出现最大幅度的减少。

口腔内存在锭剂导致的唾液分泌增多也可能造成细菌数量的快速下降。为了排除这种可能性，我们又用只含基质而不含青霉素的锭剂重复了这一实验。将这种不含青霉素的锭剂在口腔内放置一个半小时，用前面实验所采用的方法估计唾液中细菌的总数。

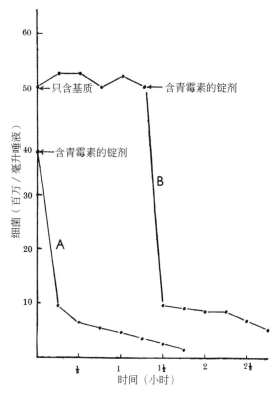

一个半小时之后，将既含有同样基质又含有 500 单位青霉素的锭剂置入口腔内，持续同样长的时间。结果如曲线 B 所示，只含基质的锭剂没有导致细菌总数的减少，而含青霉素的锭剂则导致了细菌总数的减少，并且这种减少程度与曲线 A 中的类似。由此可以排除唾液增多造成细菌数量减少的可能性。

以上结果显示细菌总数最大幅度的下降出现在使用青霉素后的最初 15 分钟内。这种作用的速度之快很难用现有的假说(即青霉素是抑菌剂)加以解释。这提示我们，当体内温度等条件都合适时，青霉素可能有直接的杀菌作用。

（王耀杨 翻译；金侠 审稿）

Alexander B. MacGregor and David A. Long: Hill End Hospital and Clinic, (St. Bartholomew's), St. Albans, Herts, Jan. 11.

Reference:

1. MacGregor and Long, *Brit. Med. J.*, ii, 686 (1944).

Biphasic Action of Penicillin and Other Sulphonamide Similarity

W. S. Miller *et al.*

Editor's Note

Here W. Sloan Miller, C. A. Green and H. Kitchen from the Royal Naval Medical School sum up the current state of knowledge about penicillin by comparing similarities between the antibiotic and antibacterial sulphonamide drugs. Both are selective and temperature-sensitive, and show two contrasting behaviours: stimulating growth at low concentrations and slowing it at high ones. They add that like sulphonamides, penicillin can also kill bacteria, a suggestion posited by MacGregor and Long in the previous paper. The authors also note, rather ominously, that "organisms can be trained to resist either penicillin or sulphonamides to a surprising degree."

SUBSTANCES generally acknowledged as being toxic to cells may have an opposite effect in higher dilution. This biphasic action—inhibition in high concentrations and stimulation in low concentrations—has been observed with a wide variety of substances, including narcotics, cyanide, pyrithiamine[1] and sulphonamides[2]. There is ample evidence that low concentrations of the last group stimulate bacterial growth; and it would appear that the period of active proliferation, which frequently precedes bacteriostasis by sulphonamides in higher concentrations, is a manifestation of the same phenomenon. We here report what appears to be an expression of the same effect occurring with penicillin.

The growth of sensitive bacteria in broth is quantitatively inhibited by suitably graded dilutions of penicillin, and the degree of inhibition can be measured turbidimetrically[3]. With the Oxford H staphylococcus (No. 6571 N.C.T.C.) as test organism, and measuring turbidity on the logarithmic scale of a Spekker photoelectric absorptiometer, we have obtained turbidity–penicillin concentration curves generally sigmoid in shape, with broth penicillin concentrations from 0.05 Oxford units per ml. to nil. Frequently, however, we have observed that tubes containing 0.005 U./ml., and sometimes 0.01 U./ml., have shown significantly more turbidity than those containing no penicillin. This effect appeared inconsistently when the incubation temperature was 37°C. At that temperature it has been noted after 4–24 hours of incubation, with a staphylococcal inoculum of between one and ten million per ml. broth, and with different samples of commercial sodium penicillin assaying from 84–820 units per mgm. It occurred in nutrient broth containing "Marmite" and 0.1 percent glucose, and in 10 percent horse serum broth without the addition of sugar. In a preliminary investigation of this phenomenon we have been unable to define the exact conditions necessary for its occurrence at 37°C., but the amount of bacterial inoculum and duration of subsequent incubation are certainly concerned; whereas the *p*H of the medium, age of inoculum, the order of mixing and initial temperatures of the various reagents in the test, within certain limits, are not critical.

青霉素与磺胺的类似性：两相行为及其他

米勒等

编者按

来自英国皇家海军医务学校的斯隆·米勒、格林和基钦在这篇文章中通过比较青霉素与磺胺类抗生素的相似之处，总结了目前对于青霉素的认识水平。它们都具有选择性和温度敏感性，而且都表现出了两种相反的作用：低浓度时促进细菌生长，而高浓度时抑制细菌生长。他们还认为青霉素和磺胺类药物一样也可以杀死细菌，这一观点由麦格雷戈和朗在之前的文章中提出。很不妙的是，作者们也注意到"有机体对青霉素以及磺胺类药物可产生惊人的耐药性"。

通常认为对细胞有毒性的物质在高度稀释状态下可能会有相反的效果。这种两相行为——高浓度时有抑制作用而低浓度时有促进作用——已经在很多种物质中被观测到，包括麻醉剂、氰化物、吡啶硫胺 [1] 和磺胺类药物 [2]。已有充分的证据表明，上文提到的最后一组物质在低浓度情况下能促进细菌增殖；而且，在高浓度磺胺类药物抑菌现象出现之前，经常出现的活性增殖期可能也是上述同一现象的表现形式之一。在这里，我们将阐述青霉素在此类效应中的表现。

适度逐级稀释的青霉素对肉汤中敏感细菌的增殖具有可量化的抑制作用，抑制作用的程度可以用比浊滴定法来测定 [3]。以牛津 H 葡萄球菌（No. 6571，N.C.T.C.）为待测有机体，把用斯佩克光电吸收计测定的浊度用对数的形式表示出来，我们可以得到基本上是 S 形的浊度 – 青霉素浓度曲线，其中，肉汤中青霉素浓度的范围为零到每毫升 0.05 牛津单位。但是，我们经常观测到含有浓度为 0.005 单位 / 毫升甚至 0.01 单位 / 毫升青霉素的试管比那些不含青霉素的试管呈现出明显更大的浊度。这种现象在 37℃ 的培育温度下时不时地出现。在 37℃ 的温度下，将浓度为每毫升 100 万 ~ 1,000 万个葡萄球菌菌种的肉汤培养 4 ~ 24 小时，并用浓度变化范围为 84 ~ 820 单位 / 毫克的不同商业青霉素钠盐进行实验分析，我们都观察到了上述现象。同时，这种现象在含马麦脱酸酵母和 0.1% 葡萄糖的营养肉汤以及含 10% 马血清但不添加糖类的肉汤中都会出现。在对这种现象的初步研究中，我们还无法确定在 37℃ 下出现该现象的准确条件，但是初始的菌种数量和随后的培养时间肯定是重要因素；在一定范围内，培养基的 pH 值、菌种的新老程度、测试中使用的各种试剂混合的顺序以及它们各自的初始温度都不是很关键。

We have been able to study this phenomenon more easily by incubation at temperatures below 37°C. It may be consistently reproduced by overnight (16 hr.) incubation at 24°C. of 20 ml. amounts of nutrient broth containing "Marmite" and dextrose (0.1 percent), with an inoculum of approximately 5,000,000 cocci per ml., from a 24-hour broth culture. Under these conditions a well-marked growth-stimulating effect has been repeatedly obtained with penicillin concentrations in the broth of about 0.01 Oxford units per ml., as shown in the accompanying graph from an actual test. The addition of *p*-amino-benzoic acid (5 mgm. per 100 ml.) to the medium makes no qualitative difference to the result, nor does the addition of 10 percent horse serum.

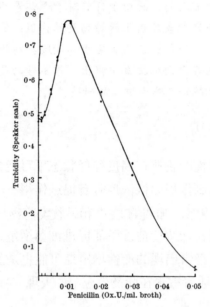

Turbidity-penicillin concentration curve.

Staphylococcal broth after 16 hr. incubation at 24°C. The concentrations lower than 0.01 U./ml. are 0.008, 0.006, 0.004, 0.002 and 0.001 U./ml. respectively. Test in duplicate.

The increased turbidity is not due to mere enlargement or distortion of the individual cocci[4]. The organisms from tubes at the peak of the curve (containing growth-stimulating dilutions of penicillin) are morphologically indistinguishable from those containing no penicillin. In fact, plate counts have provided unequivocal evidence that there may be twice as many viable bacteria in the penicillin growth stimulated cultures as in controls containing no penicillin. That this observation is caused by impurities seems unlikely in view of consistent reproducibility of the effect with pure crystalline penicillin; significant increases in turbidity over penicillinless controls can be obtained with as little as 0.0006 micrograms per ml.

Apart from recording the participation of penicillin in a rather general biological phenomenon, the object of this communication is to direct attention to the accumulating empirical evidence of the similarity penicillin and sulphonamide action. So far as we are aware, the only commonly held conception of the mode of action of penicillin is that it

850

在低于 37℃ 的温度下进行培养，研究这一现象就容易得多。取 20 毫升含有马麦脱酸酵母和右旋葡萄糖（0.1%）且经过 24 小时肉汤培养后的每毫升约 5,000,000 个球菌菌种的营养肉汤，在 24℃ 下经过一夜（16 小时）的培养，就可以重复观察到这一现象。在上述条件下，我们可以在青霉素浓度为每毫升约 0.01 牛津单位的肉汤中重复获得明显增强的促进增殖的效应，附图显示的是一次实际实验的结果。在培养基中添加对氨基苯甲酸（浓度为 5 毫克每 100 毫升）或添加 10% 的马血清对于结果都没有太大的影响。

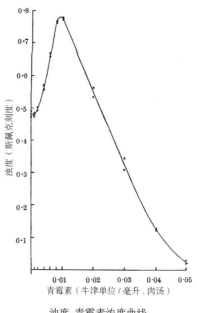

浊度-青霉素浓度曲线

葡萄球菌肉汤在24℃下培养16个小时。低于0.01单位/毫升的浓度值分别是0.008、0.006、0.004、0.002和0.001单位/毫升。检测进行两次。

浊度的增加并不仅仅是由于单个球菌的增大和变形[4]。位于曲线峰值处的试管中的有机体（含促进增殖的青霉素稀释液）与不含青霉素的试管中的有机体在形态上没有区别。事实上，平板计数结果已提供了明确的证据，在有青霉素促进增殖的培养液中可存活的细菌数量可以是不含青霉素的对照组中的两倍。由于使用纯的结晶青霉素进行了多次实验都得到了同一现象，所以这一现象看来不太可能是由杂质造成的；在青霉素浓度低到每毫升 0.0006 微克的对照组中也能看到浊度的明显上升。

除了报道青霉素也存在这种相当普遍的生物学现象以外，本文的目的还在于希望将大家的注意力引向越来越多的关于青霉素与磺胺类药物存在相似行为的经验证据上。就我们所知，关于青霉素作用模式的唯一共识是，它通过阻止正在生长的细胞的分裂而起到抑制细菌的作用。这是在加德纳的观察基础上得出的，加德纳观察

acts bacteriostatically by preventing division of growing cells. This is based on Gardner's observation that bacteria subjected to concentrations of penicillin too small to inhibit growth completely undergo distortion and enlargement[4]. Analogous morphological changes are frequently associated with sulphonamide action. It has been further proposed that penicillin acts only on dividing bacteria[5], and that this action is bactericidal[6]. One of the final conclusions reached by Henry, in a very comprehensive review of the mode of action of sulphonamides, is that they achieve their effect by stopping cell division[7]. In general, the antibacterial action of both penicillin and sulphonamides, *in vitro* and *in vivo*, appears to be primarily "bacteriostatic". Under certain experimental conditions, however, penicillin[6,8], like sulphonamides, may exert a "bactericidal" effect. Confusion is caused by drawing too fine a distinction between these terms.

The fundamental antibacterial action of penicillin and sulphonamides is inhibition of cell multiplication. Sulphonamides inhibit the growth of almost every variety of cell besides bacteria, although in widely varying concentration; there is as yet little evidence that penicillin will have such a general effect, but Cornman has reported survival of normal cells in penicillin solutions lethal to malignant cells[9]. Sulphonamide action is usually biphasic; our observations suggest that this may be true of penicillin. Primary bacterial proliferation preceding bacteriostasis, as occurs with sulphonamides, has recently been noted with penicillin and *Leptospira icterohoemorrhagioe*[10]. (As mentioned previously, this may well be another aspect of the biphasic phenomenon.) Penicillin growth-inhibition, in conformity with that of sulphonamides, appears to obey the law of mass action, in that (*a*) the inhibition is reversible, by removing the bacteria from contact with penicillin or destroying the penicillin with penicillinase, and (*b*) the inhibition is directly related to the penicillin concentration[3]. Both penicillin[6,8] and sulphonamide activity are directly related to the temperature. In the presence of a constant amount of sulphonamide, antibacterial activity is inversely related to the number of organisms present; this is a phenomenon which awaits satisfactory explanation; and the explanation is also required in respect of penicillin[6]. A feature of sulphonamide activity is that it varies from one bacterial species to another, from strain to strain, and even perhaps from organism to organism. Penicillin exemplifies this selectivity *par excellence*. Antibacterial effect is greatly influenced by the sulphonamide chemical structure, and scientific progress with penicillin must be seriously impeded until its structure is made known. By analogy, however, there is every reason to expect that substances chemically related to penicillin will have different bacterial "spectra". Chemical information is also lacking for a comparison of the effect of *p*H changes on penicillin action; at present this factor appears more important to sulphonamide action. Organisms can be trained to resist either penicillin or sulphonamides to a surprising degree, and with almost equal ease. Penicillin shares with sulphonamides synergism of antibacterial effect by antibodies and cellular defence mechanisms.

The commonly accepted theory of sulphonamide action, that of Woods and Fildes, casts *p*-amino-benzoic acid in an essential role[11]. Since this substance plays no similar part in penicillin action, some fundamental difference in mode of action might be presumed.

到当青霉素浓度小到完全不能抑制细菌生长时，细菌发生了变形和增大 [4]。类似的形态方面的变化经常被认为与磺胺类药物的作用有关。有人进一步提出，青霉素只对分裂中的细菌起作用 [5]，并且这种作用是杀菌作用 [6]。亨利在一篇关于磺胺类药物作用模式的非常全面的评述中给出的最终结论之一是，磺胺类药物通过终止细胞分裂而起作用 [7]。一般来说，青霉素和磺胺类药物无论是在体外还是在体内，抗菌作用看来主要都是"抑制性的"。不过，在某些特定的实验条件下，与磺胺类药物相似，青霉素 [6,8] 可以产生"杀菌性的"效果。对这些术语的意思区分得太细会造成困惑。

青霉素和磺胺类药物的基础抗菌作用是抑制细胞的增殖。除了细菌之外，磺胺类药物能抑制几乎所有类型细胞的增殖，当然对不同细胞起作用的浓度不同，变化范围很大；目前几乎没有证据表明青霉素也有如此普遍的效果，但是康曼已经报道了在对恶性肿瘤细胞有破坏作用的青霉素溶液中正常细胞能够正常存活 [9]。磺胺类药物的作用通常是两相的；我们的观测结果表明青霉素的作用可能也是两相的。如同在磺胺类药物中所发生的那样，最近我们注意到在青霉素和出血性黄疸钩端螺旋体体系中也发生了细菌抑制前的初期细菌增殖 [10]。（如同之前曾提到的那样，这很可能是两相现象的另一个方面。）青霉素抑制细菌增殖的作用与磺胺类药物的作用是一致的，似乎也遵循质量作用定律，具体表现在：(a) 在消除细菌与青霉素的接触或用青霉素酶破坏青霉素之后，抑制作用也会被消除。(b) 青霉素对细菌的抑制作用与青霉素浓度有直接的关系 [3]。无论是青霉素 [6,8] 还是磺胺类药物，其活性都与温度直接相关。在有恒定量的磺胺类药物存在时，抗菌活性与有机体的数量存在负相关的关系；这一现象仍有待人们作出合理的解释；对青霉素的研究也同样需要对这种现象给出合理解释 [6]。磺胺类药物活性的一个特征是：其对不同菌种、不同菌株甚至不同个体的活性都有所不同。青霉素为这一**超乎寻常的**选择性提供了例子。由于磺胺类药物的化学结构极大地影响了其抗菌效应，所以在搞清楚青霉素的结构之前，青霉素的科学应用进展必将遭遇巨大的阻碍。不过，通过类比，我们有足够的理由期待与青霉素具有化学相关性的物质会有不同的细菌"谱"。在比较不同 pH 值对青霉素作用的影响时还缺乏足够的化学信息；目前看来，pH 值对于磺胺类药物的作用更为重要。有机体对青霉素以及磺胺类药物可产生惊人的耐药性，不同有机体产生这种耐药性的难易程度相当。青霉素与磺胺类药物都具有通过抗体和细胞防御机制发挥抗菌效应的协同作用。

关于磺胺类药物的作用机制，被普遍接受的理论是由伍兹和法尔兹提出的，这个理论认为对氨基苯甲酸具有关键性的作用 [11]。既然对氨基苯甲酸这种物质在青霉

But Henry's conclusions throw considerable doubt on the Woods-Fildes explanation, and reconcile this apparent anomaly in sulphonamide-penicillin similarity[7]. There appears to be general agreement that sulphonamide bacteriostasis is achieved by direct inhibition of one or more enzymes. The profound biological activity shown by penicillin in trace concentrations would appear to be eminently explicable in terms of enzymic phenomena. It is not our purpose, however, to speculate on the mode of action of penicillin; but to suggest, on the basis of empirical observations available now, that it is not likely to be fundamentally unique. The differences that do exist between sulphonamides and penicillin, and which place the latter in its preeminent therapeutic position, appear to be differences of degree so far as mode of action is concerned.

Much of the technical work on which our observations are based was performed by laboratory assistants in the Royal Navy, to whom we are indebted. The crystalline penicillin was generously given by I. C. (P.), Ltd.

Addendum. Additional information has become available since this communication was written. It is now clear that at least three chemically different varieties of penicillin have already been identified and that their relative efficiencies for various bacteria are probably different[12]. The effect of pH on penicillin activity[13] is, in fact, in striking conformity with what has been reported for sulphonamides. The direct relationship of temperature to penicillin activity has been amplified[13,14]. Todd[15] has demonstrated the frequency with which primary multiplication occurs in cultures subjected to the influence of penicillin, and noted that Fleming originally reported this phenomenon in 1929.

(**155**, 210-211; 1945)

W. Sloan Miller, C. A. Green and H. Kitchen: Royal Naval Medical School.

References:

1. Woolley, D. W., and White, A. G. C., *J. Expt. Med.*, **78**, 489 (1943).

2. Finklestone-Sayliss *et al.*, *Lancet*, ii, 792 (1937). Green, H. N., *Brit. J. Expt. Path.*, **21**, 38 (1940). Green, H. N., and Bielschowsky, F., *Brit. J. Expt. Path.*, **23**, 1 (1942). Lamanna, C., *Science*, **95**, 304 (1942). Lamanna, C., and Shapiro, I. M., *J. Bact.*, **45**, 385 (1943). Colebrook *et al.*, *Lancet*, ii, 1323 (1936). McIntosh, J., and Whitby, E. H., *Lancet* (i), 431 (1939).

3. Foster, J. W., *J. Biol. Chem.*, **144**, 285 (1942). Foster, J. W., and Wilkers, B. L., *J. Bact.*, **46**, 377 (1943). Foster, J. W., and Woodruff, H. B., *J. Bact.*, **46**, 187 (1943). Joslyn, D. A., *Science*, **99**, 21 (1944). Lee *et al.*, *J. Biol. Chem.*, **152**, 485 (1944). McMahan, J. R., *J. Biol. Chem.*, **153**, 249 (1944).

4. Gardner, A. D., *Nature*, **146**, 837 (1940).

5. Miller, C. P., and Foster, A. Z., *Proc. Soc. Expt. Biol. N.Y.*, **56**, 205 (1944). Hobby, G. L., and Dawson, M. H., *Proc. Soc. Expt. Biol. N.Y.*, 181 (1944).

6. Bigger, J. W., *Lancet*, ii, 497 (1944).

7. Henry, R. J., *Bact. Rev.*, 7, 175 (1943).

8. Garrod, L. P., *Lancet*, ii, 673 (1944).

9. Cornman, I., *Science*, **99**, 247 (1944).

10. Alston, J. M., and Broom, J. C., *Brit. Med. J.*, **2**, 718 (1944).

11. Woods, D. D., and Fildes, P., *Chem. and Ind.*, **59**, 133 (1940).

12. See *Nature*, **154**, 725 (1944).

13. Garrod, *Brit. Med. J.*, i, 107 (1945).

14. Eagle and Musselman, *J. Exp. Med.*, **80**, 493 (1944).

15. *Lancet*, i, 74 (1945).

素作用机制中起的作用完全不同，我们就可以认为青霉素的作用模式与磺胺类药物存在某些根本的差异。但是亨利的结论对伍兹和法尔兹的解释提出了相当强的质疑，从而消除了磺胺类药物与青霉素类似性中的这个明显差异 [7]。看来一般性的共识是磺胺类药物的抑菌作用是通过直接抑制一种或多种酶来实现的。酶现象似乎可以很好地解释痕量浓度的青霉素表现出的奇妙的生物学活性。然而，我们的目的并不是要推测出青霉素的作用模式，而是在目前已有的经验观测的基础上提出这种作用模式从本质上看并不是单一的。就作用模式而言，磺胺类药物与青霉素之间的差异只是作用程度上的差异，这种差异使青霉素在治疗中显得非常优秀。

我们对皇家海军的实验室助理表示感谢，因为我们的观察所依赖的大量技术工作都是由他完成的。帝国化学（制药）有限公司慷慨地为我们提供了结晶青霉素。

附录：在写完这篇通讯之后我们又得到了一些新的信息。现在明确的是，至少已经鉴定出 3 种化学性质不同的青霉素变体，而且它们对于不同细菌的相对作用效果可能也是很不一样的 [12]。事实上，pH 值对于青霉素活性的影响 [13] 与之前报道的 pH 值对磺胺类药物的影响具有惊人的一致性。温度对青霉素活性产生的直接影响被夸大了 [13,14]。托德 [15] 展示了受青霉素影响的培养中初始增殖发生的频率，并且注意到弗莱明最早于 1929 年报道了这种现象。

（王耀杨 翻译；莫韫 审稿）

Mode of Action of Penicillin

G. Lapage

Editor's Note

Here Geoffrey Lapage summarises the ongoing debate about how penicillin works. We now know that penicillin blocks cell wall synthesis, killing bacteria by making them burst. Alexander Fleming's colleague Edgar W. Todd veered towards this conclusion in a 1945 paper demonstrating the ability of penicillin to lyse bacteria, but he also thought the antibiotic could kill without lysis. Irish bacteriologist Joseph Bigger doubted the ability of penicillin to kill non-dividing bacteria, suggesting the drug be given intermittently to increase efficacy. Others came much closer to the truth, suggesting that penicillin is bacteriostatic, bactericidal and bacteriolytic in that order. Lapage notes that certain bacteria can be made much more resistant to penicillin, and notes that a unified theory of resistance is still lacking.

IN an article on penicillin treatment in *Nature* (677, Nov. 25, 1944) reference was made to the work of Lieut.-Colonel J. W. Bigger (*Lancet*, 497, Oct. 14, 1944), who concluded that penicillin actually kills *Staphylococcus pyogenes*. He suggested that it kills them at the time of division and has no effect upon individual cocci which are not dividing. These, therefore, persist in broth cultures, which penicillin frequently fails to sterilize, and are the explanation of that failure. Bigger proposed to give penicillin intermittently, in the hope that these "persisters" would begin dividing in the intervals of the penicillin doses and so would be killed by the next dose. Bigger refers to the work of C. D. Gardner (*Nature*, **146**, 837; 1940), who found that, in weak concentrations of penicillin, cocci swelled to three times their normal size without division, and bacilli showed similar changes.

E. W. Todd (*Lancet*, 74, Jan. 20, 1945) also refers to this and other work in his report on his experiments on the bacteriolytic action of penicillin. Working with Pneumococcus Types I, II and III and with *Streptococcus viridans*, haemolytic streptococci, staphylococci and *Clostridium welchii*, he found that all the strains of these organisms which he used were lysed by penicillin, but that such organisms as *Bact. coli* and *Pseudomonas pyocyanea*, which resist penicillin, were not lysed by it. But penicillin, he concluded, can kill organisms without lysis. When lysis occurs, its rate depends on the actual or potential rate of multiplication of the organisms. Their multiplication, as G. L. Hobby, K. Meyer and E. Chaffee (*Proc. Soc. Expt. Biol., N.Y.*, **50**, 281; 1942) also found, is essential for the action of penicillin. "It would appear that bacteriostasis, bactericidal action and bacteriolysis may be different stages of a single process proceeding in that order." The most rapid lysis occurs with organisms at the maximal rate of multiplication. This may be the real reason why penicillin is so effective, that is, because young actively multiplying cultures are more susceptible to bacteriolysis, so that organisms in the phase which enables them most readily to invade the human body are also then most susceptible to lysis.

青霉素的作用机制

杰弗里·拉帕吉在这篇文章中对尚有争议的青霉素的作用机制进行了综述。我们现在知道青霉素阻断细胞壁的合成，使细菌胀裂而死。亚历山大·弗莱明的同事埃德加·托德在1945年的一篇描述青霉素对细菌的溶解能力的文章中得出这一结论，但他同时认为这种抗生素可以不通过溶解作用而杀死细菌。爱尔兰的细菌学家约瑟夫·比格对于青霉素杀死不分裂的细菌的能力颇为质疑，他建议通过间歇性给药来提高药效。另一些人的观点则更接近事实，他们认为青霉素依次具有抑制细菌增殖、杀菌和溶菌的性能。拉帕吉注意到有些细菌会对青霉素产生较强的耐药性，他还指出目前仍缺乏一个关于耐药性的统一理论。

《自然》（第677页，1944年11月25日）中一篇关于青霉素治疗的文章引用了比格中校的工作（《柳叶刀》，第497页，1944年10月14日），他的结论是青霉素直接杀死化脓葡萄球菌。不过，他认为青霉素杀死了处于分裂期的球菌，而对不分裂的个体则没有影响。因此，这些球菌继续存活于肉汤培养基中，青霉素无法将其清除，从而导致青霉素失效。于是，比格提出间歇性地加入青霉素，希望这些"顽固者"能在加药间歇期开始分裂从而被下一剂青霉素杀死。比格引用了加德纳的工作（《自然》，第146卷，第837页，1940年），后者发现在低浓度的青霉素中，球菌膨胀到正常体积的三倍而不分裂，杆菌也表现出类似的变化。

托德（《柳叶刀》，第74页，1945年1月20日）在关于青霉素溶菌作用实验的报道中也引用了这一研究和其他研究结果。在用Ⅰ型、Ⅱ型和Ⅲ型肺炎双球菌、绿色链球菌、溶血链球菌、葡萄球菌和韦氏梭菌进行的研究中，他发现所用的这些菌株都被青霉素溶解了，然而像大肠杆菌和绿脓杆菌等对青霉素具有抗性的细菌则不被溶解。但是他认为青霉素不通过溶菌作用也能杀死细菌。当溶菌发生时，其速率取决于细菌增殖的实际速率或潜在速率。霍比、迈耶和查菲（《实验生物学会会刊》，纽约，第50卷，第281页，1942年）同样也发现细菌的增殖对青霉素是否有效起关键作用。"看来抑菌、杀菌和溶菌可能是同一过程中依次进行的不同阶段。"最快的溶菌速度出现在增殖速率最大的细菌中。这也许就是青霉素如此有效的真正原因，也就是说，培养时间不长的增殖活跃的细菌更易受到溶菌作用影响，虽然这时细菌最容易侵入人体，但是同时也最容易被溶解。

These conclusions may be compared with those of Prof. L. P. Garrod (*Brit. Med. J.*, 108, Jan 27, 1945), who agrees that penicillin actually kills susceptible bacteria. He quotes the further opinion of L. A. Rautz and W. M. M. Kirby (*J. Immunol.*, **48**, 335; 1944) that penicillin is actually bactericidal. Garrod gives, however, only qualified support to Bigger's hypothesis that penicillin is bactericidal only to organisms when they are about to divide, which was, he says, also put forward by G. L. Hobby and M. H. Dawson (*Proc. Soc. Expt. Biol., N.Y.*, **56**, 178; 1944) and by C. P. Miller and A. Z. Foster (*ibid.*, **56**, 205). Against this hypothesis, Garrod maintains, are (1) his experiments on the effects of temperature; like other disinfectants, penicillin is more active at higher temperatures, but is even more active at 42°C., when bacterial growth ceases, than at 37°C.; incidentally, Garrod finds that its action is impaired by increase of the acidity between pH 7.0 and 5.0; (2) the fact that bacteria from both old and very young cultures are almost uniformly susceptible. Garrod therefore thinks that there is no conclusive evidence in support of Bigger's proposal to give penicillin intermittently, and claims that clinical experience supports his view. Penicillin treatment fails because the organisms are inaccessible inside necrotic areas or in undetected abscesses.

Further important conclusions drawn by Garrod are that nothing is to be gained by using higher concentrations of penicillin (cf. Sir A. Fleming, *Lancet*, 621, Nov. 11, 1944; see also *Nature*, **155**, 341, March 17, 1945), especially in local treatment. The idea that higher doses will be more effective does not apply to penicillin. The reverse is truer. A concentration of 1 unit per c.c. is not only just as effective as one of 1,000 units, but is often more effective. The only good reason for using stronger solutions in local treatment is to ensure that the concentration does not fall below the minimum fully effective level of about 0.1 unit per c.c. Garrod further emphasizes the importance of the purity of the penicillin which is being used experimentally. He found that all commercial penicillins tested were less active in higher than in low concentrations. Presumably impurities were responsible for this, and they cause serious obstacles to the study of the action of penicillin. It will be necessary to find out whether penicillin is a single substance of unvarying composition and uniform action.

Discussing these results in a valuable leading article, the *British Medical Journal* (123, Jan. 27, 1945) directs attention to the enormous variation in the susceptibility of various bacteria to penicillin. Some species classed as totally resistant are affected by higher concentrations of penicillin; for example, the typhoid bacillus and the salmonellas. H. F. Helmholz and C. Sung (*Amer. J. Dis. Children*, **68**, 236; 1944) have found that some resistant bacteria in the urine are affected by high concentrations, for example, *Proteus* and some strains of *B. coli*. Only *Bact. aerogenes* and *Pseudomonas pyocyanea* remained unaffected. The treatment of some infections of the urinary tract with penicillin might thus be effective. E. W. Todd, G. S. Turner and L. G. W. Drew (*Brit. Med. J.*, 111, Jan. 27, 1945) have found that *Staphylococcus* strain Oxford H. can be trained by growth in increasing quantities of penicillin to become 3,000 times more resistant to penicillin than it originally was. Similar results were obtained

上述结论可以与加罗德教授（《英国医学杂志》，第 108 页，1945 年 1 月 27 日）的结论相比较，加罗德教授也认为青霉素确实可以杀死敏感细菌。他进一步引述了劳兹和柯比的看法（《免疫学杂志》，第 48 卷，第 335 页，1944 年），即青霉素实际上是杀菌的。不过，加罗德对于比格提出的青霉素只对即将分裂的细菌才有杀菌作用的假说只给予了部分肯定；他提到霍比和道森（《实验生物学会会刊》，纽约，第 56 卷，第 178 页，1944 年）以及米勒和福斯特（同前，第 56 卷，第 205 页）也提出过类似的观点。加罗德提出的反对观点包括：（1）他关于温度影响的实验，与其他杀菌剂相似，青霉素在较高温度时活性更强，在 42℃时格外地强，而细菌的增殖在超过 37℃时就停止了；另外，加罗德发现在 pH 7.0 到 5.0 之间时青霉素的作用随着酸性增加而减弱；（2）无论是在久置的还是全新的培养基中，细菌对药物的敏感程度基本一致。加罗德由此认为，并没有决定性的证据支持比格关于间断性施用青霉素的提议，并宣称临床经验支持他的看法。青霉素治疗失败是因为那些细菌处于青霉素无法接触到的坏死区域内部或者未检测到的脓肿之中。

加罗德得出的更为重要的结论是，使用高浓度的青霉素不会有什么效果（对比弗莱明爵士的文章，《柳叶刀》，第 621 页，1944 年 11 月 11 日；另见《自然》，第 155 卷，第 341 页，1945 年 3 月 17 日），尤其是在治疗局部感染时。剂量越大效果越好的观念并不适用于青霉素。事实恰好相反。每毫升 1 个单位的青霉素不仅可以与 1,000 个单位的药效一样，而且常常更为有效。在局部治疗中使用较浓药液的唯一恰当的理由是为了保证药物浓度不会减少到约每毫升 0.1 个单位的最小有效剂量之下。加罗德进一步强调了实验用青霉素纯度的重要性。他发现所有被检测的商品青霉素在高浓度时都比低浓度时活性更差。估计这是由杂质造成的，而杂质严重阻碍了对青霉素作用的研究。有必要确定青霉素是否是一种组成不变、作用一致的单一物质。

在《英国医学杂志》中，一篇重要的前沿文章对上述结果进行了讨论（第 123 页，1945 年 1 月 27 日），并引导大家关注各种细菌对青霉素敏感性的巨大差异。某些被列为具有完全耐药性的细菌会受到较高浓度青霉素的影响，例如伤寒杆菌和沙门氏菌。赫姆霍尔兹和宋（《美国儿童疾病杂志》，第 68 卷，第 236 页，1944 年）发现尿液中的某些耐药性细菌也受高浓度青霉素影响，例如变形杆菌属和大肠杆菌的某些菌株，只有产气杆菌和绿脓杆菌仍然不受影响。因此用青霉素治疗某些尿道感染可能是有效的。托德、特纳和德鲁（《英国医学杂志》，第 111 页，1945 年 1 月 27 日）发现葡萄球菌类的牛津 H 菌株可以通过在浓度逐渐增大的青霉素中培养而达到比最初大 3,000 倍的耐药性。对另一种葡萄球菌菌株的培养也得到了类似结果。不

with another strain of *Staphylococcus*. Unlike other organisms which become "drug-fast", however, *Staphylococcus* lost this property rapidly in media not containing penicillin. The authors refer to work which showed, on the other hand, that pneumococcus type III, made resistant to penicillin, either by culture in media containing penicillin (G. Rake *et al.*, *J. Immunol.*, **48**, 271; 1944) or by passage through mice treated with penicillin (L. H. Schmidt and C. L. Sesler, *Proc. Soc. Expt. Biol.*, *N.Y.*, **52**, 353; 1943), retained its resistance. The nature of these phenomena of resistance requires further investigation. Although some organisms can produce a penicillinase which destroys penicillin (see, for example, the penicillinase produced by *B. subtilis* reported by E. S. Duthie, *Brit. J. Expt. Path.*, **25**, 96; 1944), resistance to penicillin apparently does not always depend on the production by the resistant organism of penicillinase. W. M. Kirby (*Science*, 452, June 2, 1944) has extracted a substance which is not penicillinase from *Staphylococcus* resistant to penicillin.

(**155**, 403-404; 1945)

过，与产生"耐药性"的细菌不同，葡萄球菌在不含青霉素的培养基中迅速失去耐药性。作者也引用了与此相反的另外一些研究结果：无论是通过在含青霉素的培养基中进行病菌培养（雷克等人，《免疫学杂志》，第 48 卷，第 271 页，1994 年）还是借助用青霉素治疗过的老鼠进行病菌传代（施密特和塞斯勒，《实验生物学会会刊》，纽约，第 52 卷，第 353 页，1943 年），抗青霉素的 III 型肺炎双球菌的耐药性保持不变。这些耐药性现象的机制还有待于进一步研究。尽管某些细菌能够产生破坏青霉素的酶（例如，达西报道了枯草杆菌产生的青霉素酶，《英国实验病理学杂志》，第 25 卷，第 96 页，1944 年），不过青霉素耐药性显然并不总是取决于耐药菌能否产生青霉素酶。柯比（《科学》，第 452 页，1944 年 6 月 2 日）已经从有青霉素抗性的葡萄球菌中提取出了一种并非青霉素酶的物质。

（王耀杨 翻译；金侠 审稿）

Penicillin Treatment of Venereal Disease and Spirochaetal Infections

G. Lapage

Editor's Notes

Penicillin was widely used to treat infections during the Second World War, and here Geoffrey Lapage summarises the antibiotic's use to treat gonorrhoea and syphilis. United States Army medical men had already noted that penicillin yielded immediate effects in syphilis, outperforming arsenical treatments. But whilst subsequent papers backed this up, the twice-daily repeated injections were seen as a problem. In the United States, where around 200,000 mega units of penicillin were produced, many syphilis trials were ongoing. So Lapage suggests that "we are justified in expending a large proportion of even the limited British supplies of penicillin on the study of its effects on syphilis." He also notes that penicillin is effective against other syphilis-like bacteria, as well as gonorrhoea.

THE remarkably successful treatment of gonorrhoea with penicillin was recorded in an earlier note on penicillin treatment (*Nature*, 677, Nov. 25, 1944). In that note also the opinion of United States Army medical men that the immediate effects of penicillin in the treatment of syphilis are better than those of arsenical preparations was recorded. Leading articles in the *Lancet* (853, Dec. 30, 1944) and the *British Medical Journal* (821, Dec. 23, 1944) discuss the whole question of penicillin treatment of human syphilis, with references to the relevant literature.

In the United States the first experiments on this problem were done on rabbits infected with syphilis, and J. F. Mahoney, R. C. Arnold and A. Harris (*Ven. Dis. Inform.*, **24**, 355; 1943) were apparently the first to record penicillin treatment of human syphilis. In Britain, E. M. Lourie and H. O. J. Collier (*Ann. Trop. Med. and Parasitol.*, **37**, 200; 1943) showed that penicillin will cure infections of mice with *Treponema recurrentis* and *Spirillum minus*. In co-operation with A. O. F. Ross and R. B. Wilson (*Lancet*, 845, Dec. 30, 1944) they report on the treatment of five cases of human syphilis with penicillin. All these cases had well-marked secondary lesions, and the immediate response "could not have been bettered by any known form of treatment". The spirochaetes and lesions disappeared at least as rapidly as they do under treatment with arsenicals and bismuth. But all these cases were in the secondary stage of the disease, and later observations upon them showed that only one of the five cases was apparently cured. It was therefore doubtful whether penicillin was as beneficial as arsenicals and bismuth would have been. These authors concluded that penicillin will not become suitable for routine civilian practice until frequently repeated day- and night-injections can be avoided.

青霉素治疗性病和螺旋体感染

拉帕吉

编者按

第二次世界大战期间，青霉素被广泛用于治疗感染性疾病，杰弗里·拉帕吉对使用青霉素治疗淋病和梅毒进行了总结。美国军医已经注意到青霉素对梅毒的即时疗效比砷剂更好。虽然后续的文章支持这种观点，但每天两次重复注射被视为一个问题。在美国，大约已生产出 200,000 百万单位的青霉素，许多治疗梅毒的实验正在进行之中。因此拉帕吉认为"我们也应扩大研究青霉素对梅毒的影响，即使它会消耗英国有限青霉素产量的大部分。"他还注意到青霉素对其他类似梅毒的致病菌及淋病也是有效的。

关于青霉素疗法的一篇早期文章（《自然》，第 677 页，1944 年 11 月 25 日）中就记录了用青霉素成功治疗淋病的实例，其中也记录了美国军医的观点，他们认为青霉素对梅毒的即时疗效好于砷剂。在参考相关文献的基础上，《柳叶刀》（第 853 页，1944 年 12 月 30 日）和《英国医学杂志》（第 821 页，1944 年 12 月 23 日）上的重要文章讨论了青霉素治疗人类梅毒的全部问题。

在美国，针对这个问题实施的首例实验是利用感染了梅毒的兔子完成的。显然，马奥尼、阿诺德和哈里斯（《性病学通报》，第 24 卷，第 355 页，1943 年）是最早记录青霉素治疗人类梅毒的人。在英国,劳里和科利尔（《热带医学与寄生虫学纪事》，第 37 卷，第 200 页，1943 年）发现青霉素能够治愈感染了回归热密螺旋体和鼠咬热螺旋体的小鼠。在与罗斯和威尔逊的合作研究中（《柳叶刀》，第 845 页，1944 年 12 月 30 日），他们报道了 5 例用青霉素治疗人类梅毒的病例。所有这些病例都存在明显的二期损伤，而且其即时疗效是"其他任何已知的治疗方法不能超越的"。螺旋体和病变部位消失的速度至少和用砷剂以及铋剂治疗的一样快。但是所有这些病例都处于该病的二期阶段，随后的观察发现 5 例中只有一例是明显治愈的。因此，有人质疑青霉素是否和砷剂以及铋剂一样有效。这些作者最后得出的结论是，除非能够避免日夜频繁重复注射的缺点，否则青霉素将不适合普通常规使用。

The problem of dosage in the treatment of syphilis is discussed by both the *British Medical Journal* and the *Lancet* (*loc. cit.*). In the United States, where so much more penicillin is available, extensive trials of it for the treatment of syphilis have been going on at thirty-one centres, and the *Lancet* discusses the reports on these and the supply of penicillin, stating that, by April 1944, the tentative production programme of the United States and Canada was, according to R. D. Coghill (*Chem. Engineer. News*, **22**, 588; 1944), of the order of 200,000 mega units (1 mega unit is 1 million Oxford units). There will be general agreement that we are justified in expending a large proportion of even the limited British supplies of penicillin on the study of its effects on syphilis. Arsenical treatment is more toxic and is not infallible; it involves supervision of the patient for a year or longer, and J. Marshall (*Nature*, **153**, 187; 1944) has pointed out that less than half the patients get enough of such treatment to ensure a cure-rate of 80 percent, because they default. One danger of future penicillin treatment is emphasized by both the *Lancet* and the *British Medical Journal* (*loc. cit.*). A patient may have both gonorrhoea and syphilis at the same time. The gonococcus is more susceptible to penicillin than the spirochaete of syphilis. Treatment with doses of penicillin which are sufficient to cure the gonorrhoea may therefore suppress the early signs of the syphilis, without being sufficient to cure this disease, especially if the syphilis is at an early stage when the only sign of it may be a hidden chancre. The diagnosis of syphilis may therefore be only made later when the secondary signs appear. F. L. Lydon and W. R. S. Cowe (*Brit. Med. J.*, 110, Jan. 27, 1945) also discuss this subject, adding the point that battle casualties treated with penicillin for gonorrhoea, for which it is, they agree, the drug of choice, may by incubating syphilis as well, which would thus escape detection. They think that routine blood-tests should be enforced by law upon the whole population. Similar cases of coincident infections with these two venereal diseases are discussed by F. A. Ellis (*J. Amer. Med. Assoc.*, **126**, 80; 1944) and by C. R. Wise and D. M. Spillsbury (*Brit. J. Surg.*, **32**, 214; 1944).

Penicillin seems to be very effective also against other spirochaetes and their relatives. Brigadier G. M. Findlay, Major K. R. Hill and A. Macpherson (*Nature*, 795, Dec. 23, 1944) report some success in the treatment with penicillin of yaws, due to *Spirochoeta pertenue* and of tropical ulcers infected with spirochaetes, fusiform bacilli and other organisms. Ulcers have caused, during 1944, the loss of 30,000 men-days among West African troops. A. B. MacGregor and D. A. Long (*Brit. Med. J.*, 686, Nov. 25, 1944) report the rapid disappearance of *Treponema vincenti*, the cause of Vincent's gingivitis, under treatment with penicillin incorporated in pastilles. J. M. Alston and J. C. Broom (*Brit. Med. J.*, 718, Dec. 2, 1944) report on their experiments on its action on nine strains of *Leptospira icterohoemorrhagioe*, the cause of Weil's disease (six strains were human, two were from rats and one from a dog) and on one strain of *L. canicola*, the cause of another form of leptospiral infection of man and dogs. Penicillin killed all these strains in cultures and also inhibited their multiplication. It also cured infections of guinea pigs with leptospira virulent to them, provided that it was given early enough (eighteen hours after infection). It did not prevent the development in the guinea pigs of serum antibodies or resistance to re-infection. It was not toxic to the guinea pigs as others have reported it to be. In the same issue of the *British Medical Journal* (p. 720), V. Lloyd Hart reports upon the treatment of

《英国医学杂志》和《柳叶刀》（文献同前）都对治疗梅毒时青霉素的剂量问题进行了讨论。在美国，由于有大量的青霉素可供使用，关于其治疗梅毒的大量实验正在 31 个中心开展。《柳叶刀》讨论了这些研究报告以及青霉素的供给问题，并指出：依据科格希尔的文章（《化学化工新闻》，第 22 卷，第 588 页，1944 年），截止到 1944 年 4 月，美国和加拿大的总产量预期可达到 200,000 百万单位（1 个百万单位就是 100 万个牛津单位）。在此将达成一致共识，即使我们在研究青霉素对梅毒的影响时消耗了英国有限供给的大部分青霉素，那也是有意义的。砷剂治疗具有较强的毒性而且并非总是有效；另外，它需要对患者进行一年或者更长时间的随访，而且马歇尔（《自然》，第 153 卷，第 187 页，1944 年）指出只有不到一半的患者能够得到充分治疗以确保 80% 的治愈率，因为他们不能坚持。《英国医学杂志》和《柳叶刀》（文献同前）都强调了将来青霉素疗法的一个风险。患者可能同时患有淋病和梅毒。淋病球菌较之梅毒螺旋体对青霉素更敏感。用足以治愈淋病的青霉素剂量治疗时可能因此抑制了梅毒的早期表现，而不足以治愈该病，尤其如果梅毒正处于早期阶段，此时它唯一的表现可能是一个隐性硬下疳。因此，只有在二期症状出现后才能作出梅毒的诊断。莱登和科维（《英国医学杂志》，第 110 页，1945 年 1 月 27 日）也讨论了这个问题并进行了补充，他们一致认为，用青霉素治疗患淋病的战伤人员时该首选药物可能使梅毒潜伏而无法被检测到。他们认为应该有法律强制性地对整个人群进行血常规检验。埃利斯（《美国医学会志》，第 126 卷，第 80 页，1944 年）和怀斯、斯皮尔斯伯利（《英国外科学杂志》，第 32 卷，第 214 页，1944 年）也对同时感染这两种性病的类似病例进行了讨论。

青霉素对其他螺旋体及其亲缘微生物似乎也十分有效。芬德利准将、希尔少校和麦克弗森（《自然》，第 795 页，1944 年 12 月 23 日）报道了成功地用青霉素治疗由细弱螺旋体感染引起的雅司病以及由螺旋体、梭状杆菌及其他生物感染引起的热带溃疡。1944 年，溃疡引起西非军队丧失 30,000 人工作日。麦格雷戈和朗（《英国医学杂志》，第 686 页，1944 年 11 月 25 日）报道了用整合到锭剂中的青霉素进行治疗可以使引起奋森氏牙龈炎的奋森氏密螺旋体快速消失。奥尔斯顿和布鲁姆（《英国医学杂志》，第 718 页，1944 年 12 月 2 日）报道了青霉素对 9 个品系的黄疸出血型钩端螺旋体的效果的实验，这种钩端螺旋体可引起威尔氏病（6 个品系来自人类，2 个来自大鼠，1 个来自狗）。此外还对 1 个种系的犬钩端螺旋体展开实验，其可引起人类和狗另一种钩端螺旋体病。青霉素杀死了培养基中的所有这些品系，并抑制了它们的增殖。如果用药足够早的话（感染后的 18 小时内），它也能治愈感染了恶性钩端螺旋体的豚鼠。它并不阻止豚鼠血清抗体以及对再次感染的抵抗力的形成。就像之前报道的其他动物一样，它对豚鼠没有毒性。在同一期的《英国医学杂志》

one Italian male suffering from Weil's disease. The results suggest that even the very small doses, given at relatively long intervals, had some curative effect; but Hart also emphasizes the need for early administration. It is, however, difficult to diagnose Weil's disease in its early stages. The same necessity for early administration is emphasized by Brig. E. Bulmer (*Brit. Med. J.*, 113, Jan. 27, 1945) in his summary of the treatment by various medical officers of sixteen cases of the same disease in Normandy. It is thought that Weil's disease is spread by infected rats, which pass the spirochaetes in their urine. The spirochaetes can live for a time in stagnant water, wells and sewers, so that men infect themselves by drinking and bathing. Up to December 1944, cases had been notified between mid-July and the end of September, and only from Normandy. It is, Bulmer thinks, surprising that cases have not occurred in the Low Countries, where there is "plenty of water". There was great difficulty in assessing the results of the penicillin treatment. The liver and kidneys are rapidly damaged by the spirochaete, so that penicillin should ideally be given before the diagnosis can be made. Inadequate doses of penicillin appeared, however, to shorten the duration of the fever and to cause dramatic improvement, especially when high doses were given. It did not appear to influence the damage done to the liver and kidney. In the same issue of the *British Medical Journal* (p. 119), A. E. Carragher reports on the treatment of one other case, a soldier invalided from France. After only six injections of penicillin the *Leptospira* disappeared from the blood and there was rapid clinical improvement.

Among other organisms of the spirochaete type are *Streptobacillus moniliformis* and *Spirillum minus*, the causative organisms of the two rat-bite fevers. The reasons for the conclusions that two organisms are concerned in the etiology of this disease have been discussed (*Lancet*, 540, Oct. 21, 1944), together with the effect of penicillin on them. F. R. Heilman and W. E. Herrell (*Proc. Staff Meeting, Mayo Clinic*, **19**, 257; 1944) and H. Eagle and H. J. Magnuson (*Pub. Health Rep. Wash.*, **59**, 583; 1944) have confirmed the results obtained by Lourie and Collier mentioned above. Heilman and Herrell found that penicillin cured mouse infections with *Sp. minus* and *Strept. moniliformis*, so that both forms of rat-bite fever may prove susceptible to it. The former responds dramatically to organic arsenicals, but the latter resists arsenic, sulphonamides and gold treatment. F. F. Kane (*Lancet*, 548, Oct. 21, 1944) reports on the infection of an Ulster boy with *Strept. moniliformis* as the result of a rat-bite, which was successfully treated with penicillin after gold treatment had failed. Eagle and Magnuson obtained cures with penicillin of infections of rats and mice with *Spirochoeta recurrentis* (=*Treponema novyi*), so that it is possible that penicillin may prove better than arsenic for the treatment of relapsing fever of man, which is caused by this organism.

(**155**, 459-461; 1945)

（第 720 页）中，劳埃德·哈特报道了对一位患有威尔氏病的意大利男性患者的治疗。结果显示即使是非常小的剂量，给药间隔期相对较长，也有一些治疗效果；但是哈特同样强调了早期治疗的必要性。然而，在早期阶段很难诊断出威尔氏病。布尔默（《英国医学杂志》，第 113 页，1945 年 1 月 27 日）在总结诺曼底多个医疗官员治疗 16 个相同病例的经验中也强调了早期治疗的必要性。人们认为威尔氏病是由感染的老鼠传播的，它们从尿液中排出螺旋体。这些螺旋体能够在静水、水井和下水道中生存一段时间，人类在饮用这些水或者在这些水中洗澡时受到感染。直到 1944 年 12 月，报告的病例发生在 7 月中旬和 9 月底，而且只来自诺曼底。布尔默认为没有在低地国家发现病例是很奇怪的，因为这些国家有"大量的水"。评价青霉素治疗的结果是非常困难的。螺旋体能快速损伤肝脏和肾脏，因此，较为理想的情况是，青霉素应该在给出诊断前使用。即使不足量的青霉素看起来也能够缩短发烧持续的时间，并具有显著的疗效，大剂量使用时效果更加明显。但它似乎不能影响肝脏和肾脏已经发生的损伤。在同一期《英国医学杂志》（第 119 页）中，卡拉格报道了对另一个病例的治疗——一名法国伤兵。在仅仅注射 6 次青霉素后，其血液中的钩端螺旋体就消失了，临床症状也获得了快速的改善。

螺旋体家族的其他生物包括念珠状链杆菌和小螺旋菌，分别是两种鼠咬热的病原体。判定这两种生物是这些疾病的病原体的理由以及青霉素对它们的效果都已经被讨论过了（《柳叶刀》，第 540 页，1944 年 10 月 21 日）。海尔曼和赫里尔（《梅奥诊所记录》，第 19 卷，第 257 页，1944 年）以及伊格尔和马格努森（《公共卫生报告》，第 59 卷，第 583 页，1944 年）已经证实了劳里和科利尔在前面提及的结果。海尔曼和赫里尔发现青霉素能够治愈感染了小螺旋菌和念珠状链杆菌的小鼠，因此这两种形式的鼠咬热都对青霉素敏感。前者对有机砷剂的治疗反应十分显著，但是后者可以耐受砷剂、磺胺类药物和金制剂。凯恩（《柳叶刀》，第 548 页，1994 年 10 月 21 日）报道了一位被鼠咬后感染念珠状链杆菌的北爱尔兰男孩在金制剂治疗失败后用青霉素成功地治愈。伊格尔和马格努森用青霉素治愈了感染回归热螺旋体的大鼠和小鼠，这证明青霉素在治疗人类由于感染该病菌而引起的回归热时很可能比砷剂更加有效。

（毛晨晖 翻译；金侠 审稿）

Comments on Chromosome Structure

I. Manton

Editor's Note

British cytologist Irene Manton wrote these comments on chromosome structure after an "appreciative perusal" of Erwin Schrödinger's *What is Life?*, a book that speculated on how genetic storage might work. Although the structure of DNA was yet to be deciphered, Schrödinger mused on the size of genes, and Manton says what Schrödinger "really wishes to discuss is the fundamental molecular unit of chromosome structure." At the time there were two extreme views: chromosome threads were either single- or many-stranded. Manton says the single-strand theory fits well with the proposed helical structure of chromosomes. And she uses Schrödinger's calculations to help estimate that a chromosome could harbour between "300 and 12,000 duplicate versions of the genetical material."

WHEN a great physicist takes the trouble to explain in simple language some of his matured thoughts on topics of general interest outside his own subject, it is an event for which one cannot be too grateful. The following remarks have been aroused by the appreciative perusal of Prof. E. Schroedinger's delightful little book "What is Life?"[1].

Without attempting to summarize the whole of Prof. Schroedinger's argument, it is valuable to notice the great stress which is laid on the existence of two very different methods of obtaining orderly behaviour of matter in Nature. In the inanimate world "order from disorder" is said to be the rule; the behaviour of matter in bulk being in most cases the expression of a statistical average of the behaviour of vast numbers of particles (atoms, molecules or the like) which, individually, may be doing the most diverse things under the sole compulsion of a tendency towards increased randomness. In biological systems, on the other hand, "order from order" is met with. In such a system the most complex sequence of events may be determined and set in motion by the pattern of arrangement in space of a comparatively minute number of individual particles occupying relatively fixed positions with regard to one another. The paramount importance of the pattern of atomic arrangement in the particular case of the genetical material carried by the chromosomes is, in Prof. Schroedinger's view, the most interesting discovery of our time.

Few biologists will probably wish to dispute this in general terms. Cytological comment is, however, aroused by the details of its presentation from the circumstance that Prof. Schroedinger, at various points, is thinking in terms of certain assumptions regarding chromosome structure which are by no means universally held. It may therefore be of interest to inquire what change of view, if any, will be entailed if these assumptions are altered.

关于染色体结构的评论

曼顿

编者按

英国细胞学家艾琳·曼顿在阅读了埃尔温·薛定谔的《生命是什么》这本"令人欣喜的小册子"之后写下了对于染色体结构的评论，这本书着力于解释遗传信息是如何工作的。尽管当时并不知道DNA的结构，薛定谔假想了基因的大小，而曼顿认为薛定谔"真正想讨论的是染色体结构的最基本分子单元"。当时有两种极端的观点：染色体要么是单链的要么是多链的。曼顿认为单链的理论与假设的染色体的螺旋结构更为相符。她进一步用薛定谔的计算方法估算出一个染色体能够储存"300~12,000个副本的遗传物质"。

当一位伟大的物理学家费神地去关注他本专业领域以外的大众话题，并用简单的语言发表自己成熟的观点时，人们对此应该无限感激。以下是我怀着感激的心情拜读了薛定谔教授所著的《生命是什么?》[1]这本令人欣喜的小册子后作的一些评论。

我并不打算总结薛定谔教授的全部观点，我只关注他着重强调的一个观点：在自然界中，不同的物质获得有序行为的方法有两种。在无生命的世界中，物质遵循从无序到有序的规则；在大多数情况下，大量物质的行为可以用数量巨大的粒子（如原子、分子或者其他相似物）行为的统计平均值来描述，而每个粒子可能都是在增加随机性的单向强迫力下做不同的运动。相反在生物系统中，物质遵循从有序到有序的规则；在这个系统里，单个粒子在空间中相对于其他粒子有固定的位置，其中一些数量不多的粒子会按照在空间排列的模式进行运动，从而产生一系列非常复杂的行为。以薛定谔教授的观点，在染色体所携带的遗传物质这个特例中，原子的排列模式是极其重要的，这是我们这个时代里最有意思的发现。

很少有生物学家愿意用简单的词语来讨论这个问题。薛定谔教授从不同的方面考虑了这种有关染色体结构的假说，虽然这种假说还没有被普遍认可。在此情形下，有关这种假说的细节引发了细胞学家们的讨论。令人感兴趣的是，如果这些假说发生变化，那么会使观点变成什么样呢（如果会有变化的话）。

Prof. Schroedinger is much impressed by the singleness of the "code script" in inheritance, by code script meaning the sum of hereditary material carried by a haploid nucleus (in genetical parlance this would be referred to as a genome). That only one chromosome set or genome is actually necessary for development is clearly shown by the existence of haploid organisms, for example, many of the lower plants, or cases of parthenogenesis in both plants and animals; it is therefore quite legitimate to ignore diploidy and polyploidy. Difficulties, however, appear at once if "singleness" is interpreted literally in a molecular sense, and that this is Prof. Schroedinger's interpretation seems clear from his very interesting discussion of the size of a gene.

It might perhaps be questioned whether the "size of a gene" is a desirable or legitimate use of words. In its original sense a "gene" meant nothing more than the physical basis of an externally visible mutation, and if a mutation can be caused, as Prof. Schroedinger suggests, by a change of atomic arrangement, of the nature of a quantum jump, occurring within a molecule, then a gene, strictly speaking, is the changed part of that molecule and nothing else. It is an unfortunate biological practice which Prof. Schroedinger cannot be personally blamed for following, that the word is now often used in so many extended senses that it has little precise meaning left. In discussing the maximum size of a "gene" from genetical data, the word denotes either the smallest piece of a chromosome which can have a genetically detectable effect, or the shortest distance between two mutations which can be separated by crossing-over; a numerical estimate of either of these by existing genetical or cytological methods may be expected to yield purely subjective values expressing present crudities of technique. In discussing the minimum size of a "gene" as deduced from induction of mutations by ionizing radiations, the word is apparently equated with the range of influence of the minimum degree of ionization required to induce a mutation. The thing that Prof. Schroedinger really wishes to discuss is the fundamental molecular unit of chromosome structure, which is not necessarily identical with any of these concepts. The use of the word gene for this also may seem particularly unfortunate to a cytologist because some idea of structural discontinuity is almost inevitably implied. Mutations are discontinuous and arranged in linear sequence along a chromosome. Whether the fundamental genetical material in which the mutations occur is or is not also discontinuous (like beads on a string) is quite another question and one which neither cytology nor genetics can yet determine. This linguistic difficulty would perhaps best be met if the word gene were deleted from the vocabulary; one of the most pregnant of Prof. Schroedinger's sentences could then be paraphrased as—"We believe a mutation to be of the nature of a quantum jump and the fundamental unit of chromosome structure—or perhaps the whole chromosome fibre—to be an aperiodic solid".

At this point the sense in which the "whole chromosome fibre" can be regarded as a unit becomes of importance. There are at present two extreme views as to this, both supported by some positive evidence. According to the one which Prof. Schroedinger is using, the

　　遗传学上代码指令的单一性给薛定谔教授留下了很深刻的印象。代码指令是指单倍体细胞核中遗传物质的总和（按遗传学的说法应该是指基因组）。事实上，对于生物发育来说只有一套染色体组或者基因组是必要的。单倍体生物的存在，例如许多低等植物，以及一些动植物的单性繁殖，就能证明这一点。因此我们可以完全忽略对二倍体和多倍体的研究。然而，如果"单一性"这个名词在分子意义上仅仅是字面意思（很明显，这是薛定谔教授从基因大小的有趣讨论中得出的名词），那就会出现很多困惑。

　　使用"基因的大小"这种表述是否可取或是否合理，是值得怀疑的。"基因"的本来意义仅仅是一种外在可见突变的物质基础，而按照薛定谔教授的观点，突变的产生源自于原子排列的改变，本质为量子的跃迁。突变首先发生在一个分子内，接着是基因，严格地说，基因仅仅是指分子中发生了变化的那部分而已。令人遗憾的是，在生物学上，"基因"这个词现在经常被广泛应用到许多延伸的领域，已经不存在确切的定义了，所以这并不能单单怪罪薛定谔教授一人。在讨论"基因"的最大值（从遗传学资料中分析）时，"基因"所代表的含义是最小的染色体片段（这个片段在遗传学上可以产生可见的效果）或者两个突变位点之间最短的距离（这两个突变能够被杂交实验所分离）。通过现有遗传学或者细胞学方法对上述任一数值进行估算，由于目前技术粗糙，只能得到纯粹主观的数值。在讨论"基因"的最小值时（从离子辐射诱导突变的推断中分析），很明显，"基因"等同于引起突变所需的最小的离子辐射范围。其实薛定谔教授真正想讨论的是染色体结构的最基本分子单元，与上述的这些概念并不相符。基因这个词汇被如此滥用对于细胞学家来说是非常不幸的，因为这些解释几乎不可避免地暗示了一些基因结构不连续性的观点：突变不连续地分布在一条染色体的线性序列中。那些发生突变的基本遗传物质是否同样具有不连续性呢（就像一条串起的珠链），这是一个新的问题，而且这个问题也不是现在的遗传学和细胞学所能解决的。如果把"基因"这个词从词汇里删去，那么这语言上的困难就很容易解决了。如果这样的话，薛定谔教授最富有深意的一句话将会被解释成："我们认为突变的本质是量子跃迁，染色体结构的基本单元或者说整个染色体纤维是一个非周期性的实心体。"

　　在这一点上，"整个染色体纤维"被视为一个单元变得尤为重要。目前对于这样的描述有两种极端的观点，这两种观点都有一些正面证据来支持。按照其中的一种

whole chromosome thread is a single structure at all times, except for a limited period during prophase when it is doubled in preparation for a nuclear division. The strongest piece of evidence in support of this is the differential behaviour of certain nuclear stages to irradiation by X-rays at dosages sufficient to cause gross chromosome fracture. Assuming that the statements in the literature are correct (for example, Riley 1936[2]) the position appears to be that in some organisms such as *Tradescantia*[3], irradiation at prophase can cause fracture of half-chromosomes (chromatid breaks) but at other stages only breakage of whole chromosomes. In interpreting this as meaning that the chromosome thread is single at all stages other than prophase it is not always realized that this singleness could be conferred by spiral structure and is not necessarily based on singleness of the genetical material.

The spiral structure of chromosomes (or perhaps more correctly the helical structure, since the geometrical figure involved would, in mathematics, be termed a helix) is probably not as widely known to scientific workers in general as its importance deserves and it is still far from being fully understood even by cytologists. That the apparent diameter of a fully developed chromosome is that of a helical coil has been known since 1880[4] for meiotic chromosomes, though owing to technical difficulty the basic facts for mitotic chromosomes have only recently been elucidated. Examples of the structure, as revealed by special treatment or in rare instances spontaneously, are shown in Figs. 1–4 for the different sorts of division in the fern *Osmunda*. Fig. 1 shows an unpaired chromosome at the first meiotic division, with four large coils. At this stage all cytologists are agreed that the genetical material is double in preparation for the second meiotic division and that in a paired chromosome genetical crossing-over between chromatids (half-chromosomes) has taken place; both chromatids, however, share a common spiral path. At the second meiotic division, Fig. 2, the two chromatids diverge so widely that they are only in contact at the region known as the centromere (marked *c* in the figure). A new spiral, differing in diameter and in number of coils from that at the previous division, affects each chromatid. At a somatic division, Fig. 3, the number of coils is increased still further but the two chromatids lie close together. It is, however, certain that the spiral in each has been independently formed for, in the case figured, direction of coiling was determined in corresponding parts of the sister chromatids and found to be opposite[5]. There must therefore be two separate spirals during prophase and it is possibly this which is detected by the X-ray breakages. At anaphase one of the two spirals passes to each pole, and that the helical coil at anaphase is indeed single seems to follow from the unusually clear case of Fig. 4. This differs from the preceding in that it had been subjected to X-rays some hours previously and many types of fusion and fracture are displayed. The unusual clarity of the undoubtedly single spiral in certain of the chromosomes is perhaps also an after-effect of the irradiation, but it accords very fully with the other evidence.

观点（薛定谔教授所采用的），整个染色体纤维自始至终都保持着一个单一结构，细胞有丝分裂前期的一小段时间除外，因为在这个时候，染色体丝为核分裂作准备而复制翻倍。支持这个观点的最强有力的证据是，在细胞核的不同阶段用剂量足以使所有染色体破裂的 X 射线照射后会发生不同的情况。假定文献（例如，赖利，1936 年 [2]）中的这种观点是正确的，那么同样的情况一定会出现在一些生物中，比如紫露草 [3]。在有丝分裂前期 X 射线辐射可以导致半数染色体破裂(染色单体破裂)，而在有丝分裂的其他时期就会导致整个染色体的破裂。对于在任何时期(除前期以外)染色体都是单一结构这样的解释，他们并没有意识到这种单一结构可以被赋予螺旋结构的形式，而且这种单一结构也不必一定要建立在遗传物质的单一结构上。

染色体的螺旋结构（称作螺旋体结构也许更确切，因为在数学上这种几何形状应当被定义为螺旋体）可能还不像它值得重视的程度那样，为一般的科学工作者所知。即便是对于细胞学家而言，对它的了解也是远远不够的。自 1880 年 [4] 开始对减数分裂染色体进行研究以来，由于技术上的困难，关于有丝分裂染色体的一些基础证据和事实直到最近才被阐明，但我们已经知道了完全形成螺旋状卷曲的整个染色体的表观尺寸。图 1~4 展示的是蕨类植物薇菜处于不同分裂时期的染色体结构的一些例子，包括经过特殊处理或者在自然条件下自发形成的很少见的一些结构。图 1 显示的是减数第一次分裂中尚未配对的染色体，可以看到 4 个巨大的卷曲。所有的细胞学家都一致认为，在这个阶段遗传物质处于双倍的状态，为减数第二次分裂作准备；与此同时，配对的染色单体（半染色体）之间会发生遗传物质的交叉互换，这两条配对的染色单体共享同一个螺旋轨道。在减数第二次分裂中，如图 2 所示，两条染色单体彼此分开，通过我们称之为着丝点的区域相互连接（在图中标记为 c）。一条在直径和螺旋圈数上都与先前分裂产生的螺旋不同的新螺旋影响了每一条染色单体。在体细胞分裂中，如图 3 所示，螺旋圈数仍然在进一步增加，但是两条染色单体却相互靠近。然而，可以肯定的是每一条染色单体的螺旋都是独立形成的，在这个图例中，螺旋盘绕的方向是由两条姐妹染色单体对应的部分决定的，并且两者盘绕方向相反 [5]。因此，在细胞分裂前期，必然可以通过 X 射线的破损作用检测到两条分离的螺旋。在细胞分裂后期，两条螺旋分别分到两极，我们可以见到相当清晰的单个螺旋卷曲，如图 4 所示。这与先前实验的区别在于，样品处理前经过了数小时 X 射线的照射，产生了多种类型的融合和破裂。在图 4 中，某些染色体中的单个螺旋显得非常清楚，也许这是射线照射后的效应，但它与其他的证据完全符合。

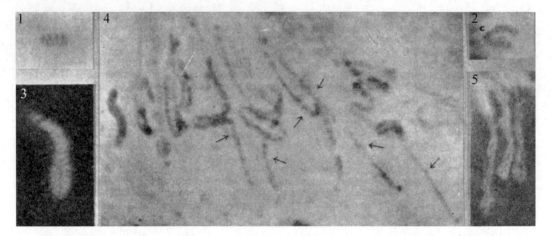

Fig. 1. Unpaired chromosome of *Osmunda* at the first meiotic division after ammonia treatment for spiral structure. Acetocarmine preparation photographed in clove oil by visual light. (×2,000.)

Fig. 2. Split chromosome at the second meiotic division in *Osmunda* after ammonia treatment for spiral structure, the two chromatids are attached only at the centromere (*c*). Acetocarmine preparation, ultra-violet photograph. (×2,000)

Fig. 3. Split chromosome at metaphase of the third spore division in *Osmunda* showing spiral structure without special treatment. Acetocarmine preparation, ultra-violet photograph (×4,000) negative print (the positive of this and others in Manton and Smiles, 1943[5]).

Fig. 4. Anaphase of the first spore division in *Osmunda* fixed 30 hours after irradiation of the uninucleate spore with X-rays at 2,500 r. showing fractures, fusions and abnormally clear spiral structure. Acetocarmine preparation, visual light photograph. (×1,000)

Fig. 5. Anaphase of the third spore division in *Todea* after ammonia treatment, showing lateral separation of component strands. Acetocarmine preparation, ultra-violet photograph (×4,000), negative print (the positive of this and other chromosomes in Manton, 1945[6]).

The other extreme view of chromosome structure, which is becoming increasingly accepted in the U.S.S.R. and the U.S.A. and (before the War) in Japan, though in Great Britain it has been somewhat opposed, is that a chromosome is fundamentally many-stranded at all stages. If the physical basis of unitary behaviour of chromosome or chromatid can in part at least be interpreted in terms of helical structure rather than molecular structure then perhaps the chief objection to this view has been removed. An example of the type of observational evidence on which the view is based is contained in Fig. 5[6]. The specimen here is at a later stage of anaphase than that of Fig. 4 and had not been subjected to X-rays. Instead it had been given the normal pretreatment with ammoniated alcohol which is generally necessary to make the spiral appear. The plant concerned is, however, not *Osmunda* but the closely related fern *Todea*, and the pretreatment and method of mounting acting together have in this case produced a remarkable lateral separation of longitudinally running strands. No less than four strands per chromosome are unmistakably present, most clearly countable, in the original print, in the leftmost chromosome. With the detection of quarter-chromosomes the limit of optical resolution has been reached, even with light of short wave-length; but it is by no means impossible that further subdivision would be found to be present if resolution could be extended.

874

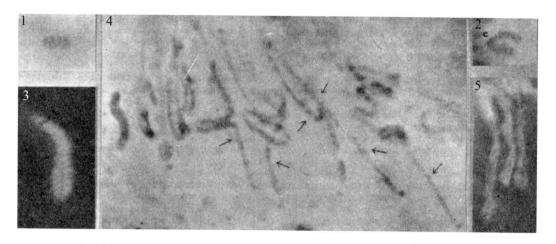

图 1. 薇菜减数第一次分裂中尚未配对的染色体。螺旋结构经过氨处理。在丁香油中加入醋酸洋红染色，可见光成像。（×2,000）

图 2. 薇菜减数第二次分裂中已经分开的染色体。螺旋结构经过氨处理。两条染色单体通过着丝点相互连接（c）。醋酸洋红染色，紫外光成像。（×2,000）

图 3. 薇菜孢子第三次分裂中期已经分开的染色体，出现了螺旋结构。没有经过特殊处理。醋酸洋红染色，紫外光成像（×4,000）负影像。（此实验的正影像和其他一些情况，在 1943 年曼顿和斯迈尔的文章中有介绍[5]。）

图 4. 薇菜孢子第一次分裂后期。用 2,500 r 的 X 射线照射单核孢子 30 小时，出现了破裂、融合和异常清晰的螺旋结构。醋酸洋红染色，可见光成像。（×1,000）

图 5. 块茎蕨中孢子第三次分裂后期。经过氨处理。显示了染色纤维组分的横向分离。醋酸洋红染色，紫外光成像（×4,000），负影像。（此实验的正影像和其他一些染色体的情况，在 1945 年曼顿的文章中有介绍[6]。）

　　关于染色体纤维的另外一种极端观点认为，染色体纤维在所有阶段都是多线状结构。苏联、美国以及日本（在战争爆发之前）的科学家正逐渐接受这种观点，尽管在英国还有略微的反对之声。如果染色体或者染色单体整体行为的物质基础可以（至少部分可以）被解释为螺旋结构而不是分子结构，那么针对这种观点的最主要的异议就不攻自破了。这种观点所依据的一个观察例证显示在图 5 中[6]。相对于图 4 而言，这个实验样品处于细胞分裂后期更靠后的一个阶段，没有经受过 X 射线的照射。相反，它采用氨化的乙醇进行正常的预处理过程，通常来讲这个过程对螺旋的解聚是必需的。实验所研究的植物不是薇菜，而是与之相近的蕨类植物块茎蕨。在这个例子中，样品经过预处理和固定后，在纵向排列的染色体上出现了显著的侧面横向分离。在每条染色体上至少存在 4 个条束，在最初发表的影像中最左边的那条染色体上，大部分条束都清楚可辨。我们能观察到四分染色体的存在，这已经达到了光学分辨率的极限，即使采用短波长的光也是如此；但是这绝不意味着进一步的细分是不可能的，只要分辨能力进一步提高，就一定会发现更进一步的细分。

It is not easy to determine the diameter of an object near the limit of visibility, but the fact that quarter-chromosomes are visible at all suggests that their thickness is likely to be of the same order as the wave-length of the light used (275 µµ). In round numbers and for the sake of argument this may be put as approximately 3,000 A. Now if Prof. Schroedinger's figures for the "size of a gene" be utilized, some simple arithmetic will provide a rough estimate of the possible limits of many-strandedness assuming this to exist. The higher limit for the "size of a gene" is given as 300 A. and there would therefore be room in a quarter chromosome of *Todea* for some 75 threads of this width. The lower limit of size is given as "ten atomic distances cubed" which, for a protein molecule, might perhaps be put as of the order of 50 A. cubed. There is room in a quarter chromosome of *Todea* for nearly three thousand threads of diameter 50 A. The real unit of chromosome structure is likely to lie in between these two extremes and it is probable that allowance must also be made for some empty spaces between the strands. Nevertheless the figures, rough as they are, indicate at least the possibility that in a whole chromosome, not one but between 300 and 12,000 duplicate versions of the genetical material may be present.

This is not quite the same as the unique phenomenon visualized by Prof. Schroedinger, though it is no doubt sufficiently close to it still to come under the general heading of "order from order" rather than "order from disorder". The importance attributed to atomic arrangement in the "aperiodic solid" of the unit fibre is almost certainly correct, and this property is shared by many other protoplasmic structures besides the chromosomes, notably by enzymes. If the view of chromosome structure put forward above be correct, however, a chromosome may be found to owe some of its peculiar powers not to the aperiodic fibre as such but to the fact that bundles of these are co-ordinated together in a manner recalling, though not necessarily exactly resembling, the periodic crystals. The "whole chromosome fibre" may in fact have to be visualized as an aperiodic solid in its longitudinal dimension but as periodic in its transverse dimension.

The recognition of an element of periodic structure in one dimension of the genetical material would perhaps be a minor emendation in the general philosophic view of a chromosome. The issues raised are, however, of immediate importance in cytology, and I trust that Prof. Schroedinger will forgive me if I have used his very interesting little book as an occasion for directing attention to them.

(**155**, 471-473; 1945)

I. Manton: University of Manchester.

References:

1. Schroedinger, E., "What is Life? The Physical Aspects of the Living Cell" (Cambridge, 1944).

2. Riley, H. P., *Cytologia*, 7, 139 (1936).

3. Catcheside, D. G., *Biol. Rev.*, **20**, 14 (1945) (a recent summary received since the above was written differs somewhat).

4. Baranetzky, J., *Bot. Zeit.*, **38**, 241 (1880).

5. Manton, I., and Smiles, J., *Ann. Bot.*, New Series, 7, 195 (1943).

6. Manton, I., *Amer. J.* Bot., in the press (1945).

在接近分辨率极限的情况下测定物体的直径是很困难的，但是能够观测到四分染色体，这从根本上说明它们的厚度可能和使用的光的波长（275 皮米）处于同一个数量级。为了便于讨论，以整数表示的染色体大小应该接近 3,000 Å。如果借用薛定谔教授对基因大小的描述，那也许用一些简单的算术就可以对这些多线状染色体结构的可能极限（假定是存在的）作一粗略的估计。如果基因大小的最高限是 300 Å，那么块茎蕨四分染色体大概可以容纳 75 条直径为 300 Å 的染色丝。如果基因大小的最低限为 10 倍原子间距的立方，比如对于一个蛋白分子来说大约就是 50 Å 的立方，那么块茎蕨四分染色体可以容纳大约 3,000 多条直径为 50 Å 的染色丝。染色体结构的真实大小很可能就介于这两个极限之间，而且很可能在线状结构之间留有一些空隙。尽管这些数据很粗略，但至少我们看到一种可能，就是在整个染色体中含有不止一个，而是 300~12,000 个副本的遗传物质。

尽管这些结果与薛定谔教授揭示的独特现象有些不同，但毫无疑问的是，把这种现象归入"从有序到有序"更为贴切，而非"从无序到有序"。在纤维单元"非周期性实心体"中原子的排列很重要，这种说法在某种程度上来说是正确的，许多原生质结构（除染色体外）都有这种性质，特别是酶。如果上述有关染色体结构的观点是正确的话，那么，染色体的一些特殊能力并不归功于非周期性纤维而归功于非周期性纤维束，这些纤维束在某种形式上相互协调，让人联想到周期性晶体，尽管两者并不完全相似。实际上"整个染色体纤维"也许在纵向维度上可以被看作是一个非周期性的实心体，但在横向维度上可以被看作是周期性的。

在遗传物质这样一个维度上探讨周期性结构的要素是对现有的染色体哲学观点的一点修正。这些争论在细胞学上是非常重要的。我相信薛定谔教授会原谅我引用他有趣的小册子作为引导这些争论的引子。

（刘振明 翻译；刘京国 审稿）

Artificial Protein Fibres: Their Conception and Preparation

author W. T. Astbury

Editor's Note

This review of the molecular structure of protein fibres by William Astbury begins with an observation commonly heard today of how apparently academic work can prove invaluable to applied science and industry. The study of natural fibrous proteins such as keratin using X-ray crystallography, which Astbury helped pioneer, suggested that it should be possible to make fibrous proteins from non-fibrous ones. Now industry was producing such artificial fibres for textiles from proteins in nuts, milk and beans. This involved "denaturing" the globular form of such proteins and aligning the strands into the parallel "β-sheet" form which, as demonstrated by silk, can be very strong. Such studies involve issues of protein folding and structural versatility that are still hot topics today.

A principal aim of science in relation to industry is to elucidate for the industrialist the nature of his working materials. All fundamental research into the structure and properties of things is therefore of potential value to industry; but this is a platitude to the man of science. To the nonscientific majority, however, it is not yet so obvious, and it is still regrettably necessary to make play with the more spectacular discoveries in order to attract proper support for research.

A recent development that has captured the imagination of the public is the discovery how to make artificial protein fibres from nuts, beans, milk, and other not visibly fibrous protein sources, and the official announcement during December 1944 by Imperial Chemical Industries, Ltd., of the successful production of "Ardil" from the protein of peanuts suggests an occasion for briefly re-telling the story, but this time more from the fundamental point of view than has been possible in the popular accounts that have appeared. I have no intention here of labouring the theme of what industry owes to science, or vice versa, or of emphasizing again that the great industries of the future must draw their sustenance from unremitting research; but I would certainly like to stress the case of the artificial protein fibres as being a most impressive example of the indivisibility of science. The discovery of the underlying principles was of purely academic origin, an outcome of X-ray and related studies of the molecular structure of biological tissues—studies that were neither supported by nor consciously dedicated to industry; the basic experiments were compounded of physics, chemistry, and biology, and were carried out in a university. Thereafter the development to commercial satisfaction was the work of industrial chemists and technologists.

It is not claimed, of course, that artificial protein fibres had never been produced

878

人造蛋白纤维：概念与制备

阿斯特伯里

编者按

威廉·阿斯特伯里在这篇关于蛋白纤维的分子结构的综述中，开篇便提出如今被广泛认可的学术研究对应用科学和工业具有非常显著的无法估量的作用。以阿斯特伯里为先驱，应用 X 射线晶体学对诸如角蛋白等天然纤维状蛋白质进行的研究表明，利用非纤维状蛋白质制备纤维状蛋白质应该是可能的。现在，工业上已经在利用来自坚果、牛奶以及豆类中的蛋白质生产这种人造纤维用于纺织了。这需要将蛋白质的球形结构"变性"，变成像丝绸中平行的"β 片层"一样非常强韧的线状结构。这类研究包括蛋白质折叠以及结构的多功能性等问题，这些在今天仍然是热点问题。

与工业有关的科学的一个主要目标就是为工业家阐明其工作原料的性质，因此对于物质的结构与性质的所有基础性研究对工业都具有潜在的价值；这对科学界人士而言可谓老生常谈。然而，对非科学界的大多数人而言显然不是这样的，并且令人遗憾的是，我们仍然必须要对那些重大发现加以强调才能吸引适当的研究支持。

最近有一项进展引起了公众的想象，这项进展便是发现了如何利用坚果、豆类、牛奶和其他一些看起来并非显然由纤维状蛋白质构成的物质来生产人造蛋白质纤维，并且帝国化学工业公司于 1944 年 12 月正式宣布成功地用花生蛋白生产出了"阿笛尔"纤维，这为我们简要地重新回顾人造蛋白纤维的产生历程提供了一个契机，不过这一次更多的是从基础研究的视角而非那些可能在通俗记述中已出现的方面进行回顾。我并非想要在这里细数工业受惠于科学之处或相反的看法，也不想再次强调未来的伟大工业必定从坚持不懈的研究中获得助力；但是我肯定乐于强调的是，人造蛋白纤维这个案例是一个最能给人留下科学具有不可或缺性这一深刻印象的实例。其基础性原理的发现有着纯粹的学术渊源，它正是 X 射线技术以及生物组织的分子结构的相关研究的结果——这些研究既未得到工业的支持也不曾有意识地致力于工业；基础性实验融合了物理学、化学和生物学等学科，并且在一所大学中实施。此后不断满足商业需求的发展就是工业化学家和技术专家的工作了。

当然，这并不是说以前从未生产过人造蛋白纤维——事实上，比如酪蛋白纤维

previously—indeed, it happens that the casein fibre, "Lanital", for example, was launched in Italy at almost the same time as the quite independent fundamental investigations about to be described were pointing the way to the general solution of the problem—but it is clear that in the absence of any structural picture of the protein molecule, and especially of the relation between the fibrous and the non-fibrous proteins, all such ventures were necessarily along empirical lines. The difference now is that what industry there was has been re-born as an inseparable part of protein science, with all the potentialities for advancement that this profoundest of molecular studies has to offer. It is in fact a logical *prediction* from the X-ray interpretation of protein denaturation that it should be possible to make fibrous from non-fibrous proteins: we can see now both what has to be done and the reason for it.

Since the beginning of the century, chemists have been increasingly convinced that all proteins are polypeptide chain systems, alone or in combination with various prosthetic groups; but there seemed to be a distinction between the fibrous and non-fibrous kinds in that the molecules of the latter are massive, rounded bodies that often aggregate to build orthodox, visible crystals; hence the name "corpuscular" proteins. With the growth of the concept of fibres as "molecular yarns" constructed from long chain-molecules, a concept to which many techniques have contributed but which first became "real" under the methods of X-ray analysis, there ceased to be any formal difficulty with regard to the protein fibres, as was demonstrated by Meyer and Mark when, in 1928, they interpreted the diffraction pattern given by natural silk (fibroin[1]); but the problem of the arrangement inside the corpuscular proteins remained, for sometimes the X-ray photographs showed sharp reflexions—characteristic to be sure of regular crystal lattices but not to be carried much beyond that on account of the large number of atomic parameters involved—but more often they showed simply two diffuse rings of spacing about $9\frac{1}{2}$ A. and $4\frac{1}{2}$ A., respectively. The first requirement was to explain these two rings, and this was done[2] on the basis of the X-ray data given by the *elastic* fibrous protein, keratin. Keratin did not fit in with the idea of extended polypeptide chains that had been found to suffice for fibroin; only the stretched form (β-keratin) could be interpreted on such a view, but the normal, unstretched form (α-keratin) demanded the postulate of a regularly folded configuration besides. The reversible intramolecular transformation between α- and β-keratin, corresponding to the transition between two distinct types of diffraction pattern, was recognized as providing the explanation of the well-known long-range elasticity of mammalian hairs and other keratinous tissues[3].

Among other things, then, the X-ray study of keratin brought out the two main points from which in due course the theory of artificial protein fibres followed naturally. The α-form revealed for the first time the existence of polypeptide chains that are normally in a folded state, while the β-form gave the average dimensions per amino-acid residue (and therewith an estimate of the order of density of proteins and the mass per unit area of protein monolayers[2,4]), and so bridged the gap to analytical chemistry and laid the foundations of a structural stoichiometry[5]. Thus was evolved the concept of the "polypeptide grid", and the two rings so common in protein diffraction patterns were

"拉尼塔"就是在意大利生产出来的，几乎在同一时间，就有人报道生产人造蛋白纤维这一问题的一般解决方法的独立的基础性研究——但是很明显，在对蛋白质分子的任何结构图像特别是对纤维状蛋白质与非纤维状蛋白质之间的关系一无所知的情况下，所有这些尝试必然是走经验路线。现在的不同之处在于，现代工业作为蛋白质科学不可分割的一部分而复兴，具有了最深刻的分子研究所赋予的发展潜能。事实上，根据蛋白质变性的 X 射线解释所作出的**预测**——可以利用非纤维状蛋白质制造纤维状蛋白质是符合逻辑的；现在我们既知道什么是必须要做的也知道这样做的理由。

自本世纪以来，化学家已经逐步确信所有的蛋白质都是独立的或与各种辅基相结合的多肽链体系；但是纤维状蛋白质和非纤维状蛋白质之间还存在着差别，后者的分子较大且为球形，经常聚集形成普遍的、可见的晶体，因此被称为"颗粒"蛋白。随着纤维就是由长链分子构成的"分子纱"这一概念的发展（很多技术都对此概念的发展有所贡献，而 X 射线分析方法首先使此概念成为"真的"），关于蛋白纤维不再存在任何结构上的困难，如同迈耶和马克在 1928 年解释天然丝（蚕丝蛋白[1]）的衍射图案时所阐明的那样；但是颗粒蛋白内部的排列问题仍然存在，因为有时 X 射线照片显示出明显的衍射——这固然是正方晶格的特征，但由于涉及大量的原子参量而进展不大——但是更经常的是它们仅仅呈现出两个间距分别约为 $9\frac{1}{2}$ 埃和 $4\frac{1}{2}$ 埃的散射环。首先要做的是解释这两个环，以角蛋白这种**弹性**纤维蛋白给出的 X 射线数据为基础已经做到了这一点[2]。角蛋白与之前发现的适用于蚕丝蛋白的伸展多肽链的想法并不符合；从这个角度只能解释其拉伸的结构（β–角蛋白），要解释正常的未拉伸结构（α–角蛋白）还需要假定一个有规则的折叠结构。α–角蛋白与β–角蛋白之间可逆的分子内转换对应着两种不同类型的衍射图案的转换，这被认为可以解释哺乳动物毛发和其他角蛋白组织的显而易见的较大幅度的弹性[3]。

此外，角蛋白的 X 射线研究给出了两个要点，它们适时而自然地引出了随后产生的人造蛋白纤维理论。α–结构第一次表明了在正常状态下处于折叠形式的多肽链的存在，而β–结构给出了每个氨基酸残基的平均尺度（从而可以估计蛋白质密度以及单层蛋白质单位面积质量的数量级[2,4]），因而弥补了分析化学的不足并为结构化学计量学奠定了基础[5]。由此发展出了"多肽网"的概念，并且认为蛋白质衍射图案中非常常见的两个环起源于相邻多肽链之间的两种主要连接模式，即侧链之间

identified as arising from the two principal modes of linkage between neighbouring polypeptide chains, namely, that between the side-chains and that between the backbones[2]. The corresponding spacings are now always referred to as the "side-chain" and "backbone" spacings.

The next advance came from an X-ray comparison of a number of protein preparations before and after wetting[6]. It was found that not only did the inner ring show the greater spacing variation from protein to protein, but also it generally showed a spacing increase on wetting. Both these properties would be expected if, as had been inferred from keratin, the reflexion represented the lateral separation of the main chains in the direction of the side-chains, and the evidence was in fact accepted as establishing the inference.

In the same investigation, among the preparations examined were egg and serum albumins that had been denatured (and coagulated) by heat. The obvious change brought about by this treatment was seen to be a marked sharpening of the backbone reflexion, and, less obvious, the appearance of at least one other outer ring of spacing about 3.6 A. (see Figs. 1a and 1b). In short, it became clear—what has been demonstrated since on many other protein preparations—that *the denaturation and aggregation of a corpuscular protein leads ultimately to a diffraction pattern like that given by disoriented* β-*keratin*. When keratin is stretched from the α- to the β-form, the side-chain reflexion remains and a strong backbone reflexion arises by the process of flattening the polypeptide grids—pulling out the folds, that is, that lie in planes transverse to the side-chains. This is a mechanical operation, but it appeared now that a similar sort of change could be brought about in the structure of the corpuscular proteins by thermal agitation, for example. (Later, it was shown[7] that the muscle protein, myosin, belongs to the same molecular family as keratin, and myosin can give either an oriented β-photograph by stretching or a disoriented β-photograph by heating.) In some way, therefore, the arrangement inside the corpuscular proteins was a generalization of the α-keratin idea; the situation was like α-keratin only more so. The polypeptide chains were there, but presumably they were folded and grouped in specific configurations from which in most cases they could be liberated fairly easily to produce a variety of nonspecific configurations. Subsequent aggregation (at least as regards the more organized regions that are responsible for the regular diffraction pattern) involved a building of "crystallites"—sometimes more perfect, sometimes less perfect, but at any rate of the type of the aggregates of polypeptide grids that constitute the crystallites of β-keratin. (Denaturation of a corpuscular protein as here described is the thoroughgoing irreversible phenomenon as it is usually understood in Great Britain. Reversible loss of solubility and specificity is sometimes described as "reversible denaturation", but such changes are conceivable without disorganization of the molecule as a whole. Sometimes, too, no clear distinction is recognized between denaturation and the aggregation of proteins that are already in an extended configuration[8]. Myosin "denatured" by simple drying is still in the folded α-form, but there is disorganization of the folds if it is heated.)

和骨架之间的连接模式 [2]。现在经常将其对应的间距称为"侧链"间距和"骨架"间距。

下一步进展来自对大量蛋白质制剂在浸湿前后进行的 X 射线比较 [6]。结果发现里面的环不仅随蛋白质的不同表现出较大的间距变化，而且在湿润条件下它通常表现出间距的增加。如果像已经从角蛋白推论得到的那样，衍射体现了主链在侧链方向上的横向间距，并且如果能够接受事实证据建立推论，那么上述这些性质就是可以预期的。

在同样的研究中，接受检验的制剂是已通过热变性（并且凝固）的卵清蛋白和血清蛋白。这种处理导致的显著变化是主链衍射明显锐化，另外不那么显著的则是出现了至少一个间隔约 3.6 埃的靠外的环（见图 1a 和图 1b）。简而言之，情况逐渐清晰——许多其他的蛋白质制剂已经证明——**颗粒蛋白的变性和聚集最终导致类似于无定向 β–角蛋白的衍射图案**。当角蛋白从 α–结构拉伸变成 β–结构时，侧链衍射维持不变，而一个强的主链衍射则由于多肽网拉平——将位于侧链横截平面内的折叠打开——而出现。这是一种机械作用，不过现在看来，可以通过诸如热搅动等方法使颗粒蛋白的结构发生这种类似的改变。（后面将会指出 [7]，肌肉中的蛋白——肌球蛋白与角蛋白属于同一分子家族，并且肌球蛋白既能通过拉伸而呈现定向的 β–照片也能通过加热而呈现无定向的 β–照片。）因此，在某种意义上，颗粒蛋白内部的排列是对 α–角蛋白情况的概括；从某种程度上来说，情况类似于角蛋白那样。根据推测，多肽是折叠的并且按特定构型形成肽基，这样在大多数情况下它们可以很容易地得到释放并产生多种非特异性结构。随后的聚合（至少对更有组织的区域而言这是产生规则衍射图案的原因）涉及"微晶"的构造——有时是完美的，有时则不那么完美，但无论如何都是构成 β–角蛋白微晶的多肽网的某种类型的聚合。（此处描述的颗粒蛋白的变性是完全不可逆的现象，就像在英国通常所理解的那样。在溶解性和特异性方面的可逆损失有时被称为"可逆变性"，但这类变化是在作为整体的分子没有解体的前提下发生的。有时候还不能清晰地分辨出蛋白质的变性和已经处于伸展构型的蛋白质的聚合 [8]。由简单的干燥引起"变性"的肌球蛋白仍具有折叠的 α–结构，但如果将它加热便会破坏折叠结构。）

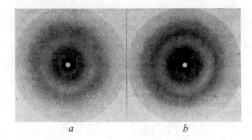

Fig. 1. X-ray powder photographs of (*a*) dried egg white, and (*b*) dried boiled egg white.

It followed at once from this line of argument that if, after unfolding the polypeptide chains, they could be drawn parallel, or approximately parallel, then artificial *fibres* would result; and the test would be the production of an *oriented* β-photograph. Decisive orientation effects were first obtained with denatured preparations of the seed globulins, edestin (from hemp seed) and excelsin (from Brazil nuts), and also with "poached" egg white[9]; while the first actual fibres were spun from strong urea solutions[9]. When edestin, for example, is dissolved in strong aqueous urea, the solution in time becomes very viscous, and elastic fibres may be produced either by drawing out the viscous mass or by squirting it through a capillary tube into water or dilute salt solution. (This sort of observation is not new, but it seems that X-rays first clearly exposed the reason for the rise in viscosity—the unfolding of round molecules to give polypeptide chains in extended configurations.) In general, though, such fibres have to be stretched farther in order to reveal definite orientation effects in the X-ray photographs. As a matter of interest, some of the early diffraction patterns are reproduced in Fig. 2. They illustrate the first demonstration by X-rays of the transformation of an originally crystalline protein into an elastic fibrous structure.

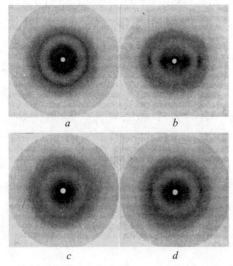

Fig. 2. (*a*) Disoriented β-keratin; (*b*) oriented β-keratin; (*c*) disoriented denatured edestin; (*d*) oriented (stretched) denatured edestin.

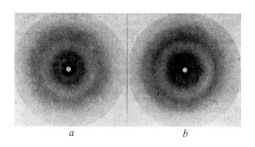

图 1. X 射线粉末照片。(a) 干燥的卵清蛋白，(b) 干燥的煮熟的卵清蛋白。

从这段论述中我们可以立刻得出，多肽链在打开折叠之后可能被拉伸成平行或近似于平行的状态，接着便会产生人造**纤维**；检测依据就是一张**定向** β–照片的产生。决定性的定向效应最初通过种球蛋白、麻仁球蛋白（来自大麻种子）和巴西果蛋白（来自巴西坚果）的变性制剂以及从"煮熟的"卵清蛋白获得 [9]，而最早实际出现的纤维从浓的尿素溶液中获得 [9]。例如，当把麻仁球蛋白溶解于浓的尿素溶液中时，溶液会立刻变得非常黏稠，通过拉出黏性物质或者通过一根毛细管将其喷入水或稀释的盐溶液中便可以产生弹性纤维。（这种现象并非是全新的，但似乎是 X 射线最先明确地揭示出其黏性增加的原因——球形分子中的折叠打开，形成伸展构型的多肽链。）不过一般地说，这种纤维还需进一步拉伸以便在 X 射线照片中呈现出明确的定向效应。根据其重要性，图 2 再现了一些早期的衍射图像。这些衍射图像是首次通过 X 射线方法证明从初始的结晶蛋白到弹性纤维构象转变的例证。

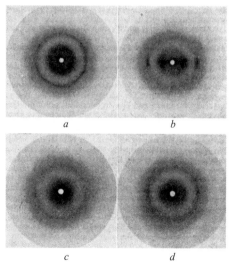

图 2. (a) 无定向 β–角蛋白；(b) 定向 β–角蛋白；(c) 无定向变性麻仁球蛋白；(d) 定向（拉伸的）变性麻仁球蛋白。

Experiments on continuous fibre production from urea solutions were then conducted in Prof. A. C. Chibnall's laboratory at the Imperial College of Science and Technology, London; but later, though considerable further advances were made, it was felt better to return to the more fundamental side of protein research. Thereafter, as indicated above, the production of "Ardil" (see Fig. 3) and its development to commercial satisfaction was the work of chemists of Imperial Chemical Industries, Ltd.[10]. "Ardil" is produced from the peanut globulin, arachin, and the original urea process has now been replaced by extraction and "maturing" with dilute alkali; also, striking new after-treatments have been evolved for improving the tensile properties of the fibres and their resistance to dyeing and finishing processes. Actually, the present stage of development was reached by the beginning of the War, but since then further work has been held up.

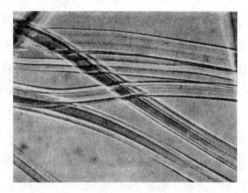

Fig. 3. Ardil fibres.

"Ardil" is a cream-coloured, crimped, elastic fibre that is soft and warm: it is a kind of artificial wool without surface scales. It greatly enhances the felting of wool, however, and dyes like wool, but it is not attacked by moths. Its elasticity arises from the circumstance that the unfolding of the original corpuscular molecules is imperfect, vestigial folds remaining that are an irregular counterpart of the regular α-folds to which the elasticity of keratin is due. As now produced, "Ardil" shows no orientation in its diffraction pattern, but an oriented β-pattern begins to appear on stretching. Fabrics have been made purely of "Ardil", but its best use is likely to be in combination with wool and other fibres. A by-product of its manufacture is, of course, arachis oil, of which peanuts contain 48–50 percent; furthermore, after the oil and protein have been extracted, the residue can be used for cattle food.

During the last few years a number of important papers have been published in America on the preparation of artificial protein fibres with the aid of detergents[11]. Products such as "Nacconol NRSF" (which is essentially dodecyl benzene sodium sulphonate) are found to act as excellent unfolding agents for the corpuscular proteins, and with the technique described it is possible to make fibres much stronger than anything reported previously. Using egg albumin, for example, the complex formed by mixing equal portions of 3 percent solutions of recrystallized egg albumin and detergent is first precipitated with saturated magnesium sulphate, the resulting "dough" is drawn out into fibres, which are

　　接着，伦敦帝国理工学院奇布诺尔教授的实验室进行了用尿素溶液持续生产纤维的实验研究；尽管已经取得了相当程度的进展，但后来他们还是觉得返回到蛋白质研究中更基础的方面比较好。从那以后，如同上面已指出的，"阿笛尔"的生产（见图3）及其针对商业需求的发展成为帝国化学工业公司化学家们的工作[10]。"阿笛尔"是由花生的球蛋白——花生球蛋白来生产的，并且最初的尿素工艺现在已经被稀碱抽提和"成熟"取代，而且还发展出令人惊奇的新的后期处理方法来改善纤维的拉伸性能以及它们对染色和修整工序的抗性。实际上，目前的发展水平在大战之初便已达到，但是从那以后进一步的研究便搁置起来。

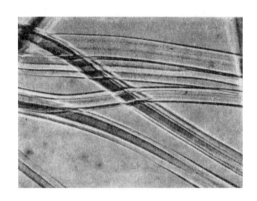

图 3. 阿笛尔纤维

　　"阿笛尔"是一种淡黄色的、有皱褶的弹性纤维，柔软而温暖：它是一种没有表面覆盖层的人造织物。这大大改善了织物的触感，尽管它能像羊毛一样被染色，却不会受到蛀虫的侵害。它的弹性来源于最初的颗粒状分子的不彻底伸展，保留下来的不完全折叠正是使得角蛋白具有弹性的规则 α–折叠的不规则对应物。现在生产出的"阿笛尔"在其衍射图案中不显示方向，但是在拉伸时开始呈现出定向的 β–图案。织物是完全由"阿笛尔"制成的，但其最佳用法可能是与毛线和其他纤维制品结合。当然，该产品的副产物是花生油——占花生的 48%~50%；此外，在提取出油和蛋白质之后，残余物可以用来作为家畜饲料。

　　最近几年，美国发表了大量关于借助去污剂制备人造蛋白纤维的重要论文[11]。已发现诸如"烷基芳基磺酸钠"（实质上就是十二烷基苯磺酸钠）等产品可以充当出色的颗粒蛋白展开剂，用前文描述的技术就有可能制造出比之前报道的任何纤维都更强的纤维。例如，以卵清蛋白为原料，将 3% 的重结晶卵清蛋白溶液与去污剂等量混合形成复合物，首先用饱和硫酸镁使其沉淀，将生成的"生面团"拉伸成纤维，接着将其用水洗涤，并用 60∶40 的丙酮－水溶液抽提，最后在流通蒸汽中将纤维拉

washed with water and extracted with 60:40 acetone–water solution, and then finally the fibres are stretched by about 400 percent in live steam. Before the stretching in steam, the orientation in the X-ray diffraction pattern is of the "crossed" type first observed with stretched films of poached egg white[9], that was interpreted as corresponding to chain-bundles broader than they are long; but after the stretching it is as in β-keratin with the chain-bundles lying along the direction of stretching. (The effect illustrated in Palmer and Galvin's paper[12] is slight and was at first thought not to be there; but it is now agreed that before stretching in steam there is no real discrepancy with the "poached egg effect". Private communication.) The oriented β-pattern given by these stretched egg albumin fibres is very good indeed: it is shown in Fig. 4, which is a reproduction of a photograph kindly sent me by Dr. Palmer. Egg albumin is one of the most typical and most studied of all the crystalline corpuscular proteins, and here in the end it is made to yield one of the best β-fibre photographs! The wheel has now come full circle.

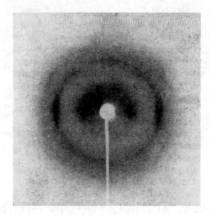

Fig. 4. Fibre pattern (β-keratin type) given by egg albumin fibres prepared by Lundgren and photographed by Palmer and Galvin.

Lundgren and O'Connell report that artificial fibres from egg albumin (and from chicken-feather keratin, which also responds to the technique just described) have been prepared with breaking strengths of more than 70×10^3 lb. per sq. in.: to appreciate what this means, it may be noted that in the same table they quote 72–100 for nylon, 46–74 for natural silk, and 17–25 for wool. Incidentally, lest it should seem somewhat indecent these days to talk about using egg white for making fibres, it should be added that the same authors point out that there are available annually in the United States more than 26,000,000 lb. of inedible technical egg white (much of which goes to waste) and more than 170,000,000 lb. of chicken feathers.

(**155**, 501-503; 1945)

W. T. Astbury: F.R.S., Textile Physics Laboratory, University of Leeds.

References:
1. Meyer, K. H., and Mark, H., *Ber.*, **61**, 1932 (1928).

888

长约 400%。在蒸汽中进行拉伸之前，X 衍射图案中的定向属于"交叉"类型，这种类型首先是在拉伸的熟蛋白薄膜中观察到的 [9]，这可以解释为对应的链束比其长度宽；但是在拉伸之后，图案与 β–角蛋白相似，其链束方向位于拉伸方向上。（帕尔默和高尔文的论文 [12] 中描述的效应是微弱的，最初认为它不存在；但是现在我们同意，在蒸汽中进行拉伸之前，该效应与"煮熟的鸡蛋效应"没有实质性的差别。私人交流。）由上述卵清蛋白纤维所给出的定向 β–图案实际上是很好的：图 4 显示了该图案，这是帕尔默博士友情赠送给我的一张照片的副本。卵清蛋白是所有结晶颗粒蛋白中最典型的、被研究得最多的一种，而这里终于用它得到了一张最好的 β–纤维照片！这样就圆满了。

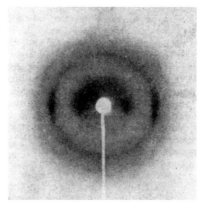

图 4. 由伦德格伦制备、帕尔默和高尔文照相的卵清蛋白纤维的纤维图案（β–角蛋白型）。

　　伦德格伦和奥康奈尔指出，现在已经能够由卵清蛋白来制备人造纤维了（还可以用鸡毛角蛋白，也适用于刚才描述的那种技术），其断裂强度超过 70×10^3 磅 / 平方英寸：要理解其中的含义，可以参考他们在同一个表格中提供的数据：尼龙为 72~100，天然丝为 46~74，羊毛为 17~25。顺便提一下，为了避免大家认为这些天来谈论的用卵清蛋白制造纤维这件事似乎有些不妥当，应当补充的是，这些作者指出，在美国每年都会有超过 26,000,000 磅的不可食用的用于工业技术的卵清蛋白（其中的大部分都被浪费了）和超过 170,000,000 磅的鸡毛。

（王耀杨 翻译；刘京国 审稿）

2. Astbury, W. T., *Trans. Faraday Soc.*, **29**, 193, 217 (1933).

3. For references see, for example, Astbury, W. T., *Nature*, **137**, 803 (1936).

4. See also Astbury, W. T., Bell, F. O., Gorter, E., and van Ormondt, J., *Nature*, **142**, 33 (1938).

5. Astbury, W. T., "Advances in Enzymology", **3**, 63 (1943).

6. Astbury, W. T., and Lomax, R., *J. Chem. Soc.*, 846 (1935).

7. Astbury, W. T., and Dickinson, S., *Nature*, **135**, 95 (1935); *Proc. Roy. Soc.*, B, **129**, 307 (1940).

8. See, for example, Coleman, D., and Howitt, F. O., *Nature*, **155**, 78 (1945).

9. Astbury, W. T., Dickinson, S., and Bailey, K., *Biochem. J.*, **29**, 2351(1935).

10. Traill, D., *Chem. and Ind.*, Feb. 24, 1945, p. 58.

11. Lundgren, H. P., *J. Amer. Chem. Soc.*, **63**, 2854 (1941); Lundgren, H. P., Elam, D. W., and O'Connell, R. A., *J. Biol. Chem.*, **149**, 183 (1943); Palmer, K. J., and Galvin, J. A., *J. Amer. Chem.* Soc., **65**, 2187 (1943); Palmer, K. J., *J. Phys. Chem.*, **48**, 12 (1944); Lundgren, H. P., and O'Connell, R. A., *Ind. Eng. Chem.*, **36**, 370 (1944).

12. Palmer, K. J., and Galvin, J. A., *J. Amer, Chem. Soc.*, **65**, 2187 (1943).

Significance of the Australopithecinae

W. E. L. G. Clark

Editor's Note

The Second World War inevitably put a stop to the previously avid search for evidence for the antiquity of human beings. As if to display the eagerness of palaeontologists to resume their previous work, Sir Wilfred Le Gros Clark, by then the doyen of British palaeontology, wrote the following article for *Nature* in which he expressed enthusiasm for the work of Raymond Dart and in particular for Dart's identification of the Taungs skull as a representative of the genus *Australopithecus*. He also urged further research in southern Africa for additional specimens.

IN 1924, the immature skull of a large ape-like primate was discovered in some lime workings at Taungs in the valley of the Harts River, South Africa. It was briefly described by Prof. R. A. Dart, who regarded it as representing an extinct race of apes intermediate between living anthropoid apes and man. To this extinct race he gave the name *Australopithecus africanus*. There followed a mild controversy on the interpretation of this fossil, but many anatomists quite properly preferred to wait before committing themselves to definite statements until a full and systematic report on the original remains should appear. Twelve years later, Dr. Robert Broom, who had decided to search for more remains of *Australopithecus*, paid a visit to a cave at Sterkfontein, near Krugersdorf. Here he found portions of skulls and jaws of a fossil primate similar to *Australopithecus* but (in his opinion) sufficiently distinct in some of its characters to be referred to a separate genus. He called it *Plesianthropus transvaalensis*. Then, in 1938, the remains of what were taken to represent still another type, called by Broom *Paranthropus robustus*, were brought to light at Kromdraai, two miles east of Sterkfontein. Thus there are now available for consideration three series of extinct ape-like primates from South Africa, which are believed to be representatives of one sub-family, the Australopithecinae. Excellent casts of the skull of *Australopithecus* have been available in Britain for many years now, and during the course of his excavations since 1936 Dr. Broom has been extremely generous in distributing casts of most of the valuable material which he has collected. Thus anatomists in Britain have for some time had this sort of evidence before them. Now there has appeared the long-awaited report on the Australopithecinae by Dr. Broom and Dr. Schepers[1]. In this monograph, which is abundantly illustrated and incorporates numerous comparative studies, Broom deals in considerable detail with the osteological material, while Schepers discusses the endocranial casts. Apart from the obvious fact that access to the original material is really necessary to complete the evidence on which to base a considered opinion, it is now possible, at least in general terms, to assess independently the significance of these remarkable fossils.

892

南方古猿亚科发现的意义

克拉克

编者按

第二次世界大战无疑使先前急切寻找古人类证据的工作停滞下来。似乎是为了表示古生物学家们对他们以前的工作又恢复了热情，时为英国首席古生物学家的威尔弗雷德·勒格罗·克拉克爵士在《自然》上发表了这篇文章，文中他对雷蒙德·达特的研究工作，尤其是对达特把汤恩头骨鉴定为南方古猿属的代表表现出极大的兴趣。为得到更多的标本他还强烈主张在南非进行更进一步的研究。

1924 年，在位于南非哈茨河流域的汤恩，人们在一些石灰岩矿区内发现了一种类似猿类的大型灵长类动物的幼年个体头骨。达特教授对其进行了简要描述，他认为这代表了一种已经灭绝的介于现存类人猿与人类之间的猿类。他将这种已经灭绝的猿类命名为南方古猿非洲种。随后产生了少许争论，但许多解剖学家在做出他们最终判断之前明智地选择了等待，等待一份关于这类化石详尽而系统的研究报告的发表。12 年后，致力于搜寻更多南方古猿化石的罗伯特·布鲁姆博士考察了克鲁格斯多普附近的斯泰克方丹的一个洞穴。在这里他发现了类似于南方古猿的灵长类动物的部分头骨和颌骨化石，但是（他认为）这些化石所具有的一些非常独特的特征使之能被划分为一个独立的属。他把这些化石命名为德兰士瓦迩人。随后在 1938年，在斯泰克方丹以东 2 英里的克罗姆德拉伊发掘出的化石又被用来代表另一个种类，布鲁姆将其命名为粗壮傍人。因此现在可以对来自南非的 3 组已经灭绝的类似猿类的灵长类动物化石进行研究，这些化石被认为属于同一个亚科，即南方古猿亚科。多年以来，在英国就可以得到极好的南方古猿头骨模型，而且自从 1936 年布鲁姆博士开始他的发掘工作以来，他一直都极为慷慨地分发出大部分他所收集到的极其珍贵的化石的模型。因此一段时间以来，英国的解剖学家们已经在使用这一类证据了。现在期盼已久的由布鲁姆博士和舍佩尔斯博士合著的关于南方古猿亚科的研究报告终于公之于世 [1]。这份专题论著图片丰富，并且结合图片进行了大量对比研究。在报告中，布鲁姆对骨骼材料做了非常详尽的论述，舍佩尔斯则主要对颅内模进行了论述。显而易见的事实是确实需要通过原始的化石标本来获取证据以便在此基础上建立一个成熟的学术观点，除此之外，至少现在有可能在大体上对这些著名化石的意义独立地做出评价。

Dr. Broom has demonstrated beyond any doubt at all that the Australopithecinae are extremely important for the study of human evolution, since they present an astonishing assemblage of simian and human characters. Such an assemblage, indeed, might well be postulated, entirely on indirect evidence, for hypothetical ancestors of the Hominidae. Thus it should be said at the outset of this review that Dart's original interpretation of the *Australopithecus* material has in several respects been completely vindicated. Some of the most outstandingly human features of the Australopithecinae are undoubtedly those of the teeth and jaws. In both the deciduous and permanent dentitions, the incisors and canines are of human rather than simian proportions and pattern. The deciduous premolars are quite similar to those of the human child, while the permanent premolars, though very large, have the distinctive human pattern. The permanent molar teeth, in spite of their size (which is exceeded only by male gorillas and certain large extinct apes such as *Sivapithecus giganteus*), also show some approach to man in the disposition of their cusps. The dental arcade forms a rounded curve as in man and not an elongated U-shape such as is characteristic of modern large apes. The nature of the wear of the teeth and the anatomy of the temporo-mandibular region show, also, that the teeth and jaws were used in human fashion. In contrast with the remarkably human features of the teeth, the skull as a whole resembles in its general proportions those of anthropoid apes; and in a number of details, for example, the great facial extension of the premaxilla, the contour of the mandibular symphysis, and the apparent absence of a foramen spinosum, it is entirely simian and departs widely from the human condition.

So far, then, the Australopithecinae might perhaps be regarded as a group of extinct apes, somewhat similar to the gorilla and chimpanzee, in which the characters of the dentition had developed (possibly independently) along lines almost identical with those of human evolution. But Dr. Broom has also, in his indefatigable search, brought to light some most important fragments of limb bones, which allow, and even make probable, a much more startling interpretation of these fossil remains. For example, of *Paranthropus* there are available the lower end of the humerus, the upper end of the ulna, and the talus. Judging from casts and Dr. Broom's illustrations, the humeral and ulnar fragments are entirely similar to those of *Homo sapiens*. Indeed, anatomists without the full evidence before them might well be excused if they expressed scepticism at their association with the skull of *Paranthropus*. But Dr. Broom states explicitly that the skull, jaw, humeral and ulnar fragments and the talus were all obtained from one mass of bone breccia less than a cubic foot in size, and that nowhere in any of the same deposits have remains of *Homo* come to light. Thus there seems no reason to doubt that this extinct ape-like creature had upper limbs of human proportions (at least so far as the elbow region is concerned), and which were evidently not used for brachiation. On the other hand, the talus is a remarkably small bone—in its dimensions, so far as these can be measured on a cast, it falls well short of the minimum recorded for modern races of mankind (cf. the data for Japanese women reported by B. Adachi[2]). Compared with the humeral fragment (which presumably belongs to the same individual) the size of the talus indicates a disproportionately small tarsus— more so, indeed, than would be expected if *Paranthropus* used its hind-limbs for the bipedal mode of progression characteristic of man. Further, the unusual medial extent of the

　　布鲁姆博士证明了南方古猿亚科对人类进化的研究是极为重要的，他的论证根本毋庸置疑，因为南方古猿亚科惊人地呈现出猿和人类的特征组合。事实上，完全依靠间接的证据很有可能去假定这样一种组合是假想的人科祖先所具有的。因此应该说在本文开头提到的达特关于南方古猿化石最初的解释有几个方面是完全正确的。毫无疑问，南方古猿亚科具有的那些最突出的人类特征为牙齿和颌骨的特征。在乳齿期和恒齿期，其门齿与犬齿的大小和形态与人类一致，而与猿的不同。其乳齿期的前臼齿与人类儿童的前臼齿非常相似，而其恒齿期的前臼齿尽管很大但却具有人类所特有的形态。恒齿期的臼齿，不管其尺寸大小（仅有雄性猩猩以及诸如巨型西瓦古猿之类的某些已灭绝的大型猿类的尺寸才比它大），其齿尖的排列同样与人类有些相似。它的齿弓形成了一个圆形曲线，这与人类相似，而与现代大型猿类所特有的一个拉长的 U 形不同。牙齿磨损的特征以及颞–下颌区的解剖学特征也表明，它的牙齿和颌骨的使用方式与人类一样。与其牙齿具有的显著人类特征形成鲜明对比的是，整体看来其头骨的比例大体上与类人猿的头骨相似；而且在很多细节上也是如此，例如，其面部前颌骨延伸很大，下颌联合部的外形以及棘孔的明显缺失，这完全是猿的特征，而与人类的特征相去甚远。

　　到目前为止，人们认为南方古猿亚科可能是一群已经灭绝的猿类，在某种程度上与大猩猩和黑猩猩相似，其齿系特征的演变（可能是独立地）与人类齿系的演化方式几乎相同。但是布鲁姆博士经过坚持不懈的搜寻，也发现了一些十分重要的肢骨破片，例如，傍人的肱骨下端、尺骨上端以及距骨，利用这些破片可对上述化石做出更加令人吃惊的解释。从模型以及布鲁姆博士的插图来判断，傍人的肱骨与尺骨破片与智人的非常相似。的确，在获得这些充分证据之前，如果解剖学家对其与傍人头骨存在联系表示怀疑，那是可以理解的。但是布鲁姆博士明确声明头骨、颌骨、肱骨破片与尺骨破片以及距骨都是从一堆小于一立方英尺的骨头角砾岩中获得的，而其他任何地方的相同堆积物中都没有发现人属的化石。因此似乎可以确定，该灭绝的类似猿类的动物的上肢与人类的上肢大小一致（至少就肘部而言），并且很明显它不是用来在树枝间游荡的。另一方面，距骨是一块尺寸非常小的骨，就对模型进行测量所得的尺寸而言，其远未达到现代人种的最小记录（参考安达报告的日本女性的数据[2]）。与肱骨破片（它与距骨大概属于同一个体）相比，距骨的尺寸表明跗骨的尺寸过小而不合比例——的确，尤其是加之此前推测傍人的后肢是用于人类所特有的两足行走，因此就更显得跗骨过小。而且，骨端关节面独特的中间部位似乎表明在距骨下关节处的灵活性非常好，而缩短的颈部显示其身体的大部分重量转移

articular surface of the head of the bone seems to indicate a very considerable mobility at the sub-talar joint, while the truncated neck suggests the transference of the major component of the weight of the body to the fore-part of the foot, a feature which Morton[3] has shown to be characteristic of the type of foot found in the great apes and therefore different from the trend shown in human evolution.

Of *Plesianthropus* there have been found the capitate bone of the carpus and the lower end of the femur. The former confirms the evidence of the arm bones of *Paranthropus*, that the Australopithecine upper limb closely corresponded to the modern human type, for the capitate bone seems to come well within the range of variation shown in the Bushman[4]. The lower end of the femur is perhaps the most important of all these limb-bone fragments, and Dr. Broom infers from it that *Plesianthropus* "walked, as does man, entirely or almost entirely on its hind feet". From the appearance of the femur as depicted in Dr. Broom's drawing of it, we are inclined to agree with his interpretation. But his description of this most important evidence is tantalizingly brief (it is confined to 34 lines!). Indeed, we could have wished that the author had dealt with all this limb material in much more detail. The illustrations, too, while they give a good general impression of the appearance of the bones, are not sufficiently accurate for comparative studies. For example, the text-figure of the *Paranthropus* talus, although stated to be natural size, actually represents the bone as somewhat larger than the cast. So much depends on this limb material for a proper assessment of the Australopithecinae. Thus, when Broom states at the conclusion of his section that "at Sterkfontein there are deposits of breccia probably as extensive as those of Choukoutien, and we may confidently assert that these are likely to yield dozens of skulls and probably fairly complete skeletons of the Sterkfontein ape-man", we can only implore all those concerned to see that facilities are provided forthwith in order that systematic excavations can be continued without interruption.

In addition to the skeletal material of the Australopithecinae, there were available for study the natural endocranial cast of *Australopithecus*, portions of endocranial casts (partly distorted) of *Plesianthropus*, and an endocranial cast of the left temporal region of *Paranthropus*. The study of endocranial casts of fossil man and apes always raises afresh the oft-debated question as to how far the sulcal pattern of the brain itself can really be inferred from them. Studies of endocranial casts of modern man and apes, and their direct comparison with the actual brains, have shown that in these forms attempts to delineate the sulcal pattern may be grossly misleading[5]. Dr. Schepers regards the conclusions drawn from such careful and objective studies as "pessimistic", and, of course, it may be true that, for some reason or other (and very conveniently for the palaeontologist!), the convolutions of the brain were more faithfully impressed on the bony walls of the cranial cavity in fossil than in living species. Nevertheless, neurological anatomists will be surprised to find that Dr. Schepers has apparently been able not only to map out with confidence the sulcal pattern over the whole cerebral hemisphere of *Australopithecus* and *Plesianthropus*, but also to delineate no less than twenty-six separate cyto-architectural areas and to compare them in their relative extent with those of modern apes and man. Cortical physiologists may likewise feel inclined to demur at some of the

到了足前部，这个特征与莫顿 [3] 发现的大型猿类的足部特征相同，因此也有别于人类足部进化的趋势。

迩人腕骨的头状骨与股骨的下端已被发现。前者进一步确认了傍人臂骨的证据，即南方古猿亚科的上肢与现代人类的类型几乎相同，因为头状骨似乎刚好在布须曼人 [4] 表现出的变化范围之内。股骨的下端可能是所有这些肢骨破片中最为重要的，并且布鲁姆博士据此推断迩人"像人一样完全或几乎完全依靠后足行走"。依据布鲁姆博士为其绘制的图中所示的股骨外观，我们愿意接受他的解释。但是他对于这个最为重要的证据的记述却简短到令人着急的程度（仅有 34 行！）。确实，我们希望作者更加详细地论述所有这些肢骨材料。尽管从整体上来看骨的外观画得还不错，但是对于比较研究来说这些插图也是不够精确的。例如，正文图中傍人距骨标明为实际大小，但实际上图中描绘的骨比模型稍微大一点。为了正确评价南方古猿亚科，要在很大程度上依赖于这些肢骨资料。因而，当布鲁姆在他这一部分的结论里说"在斯泰克方丹地区有角砾岩沉积，其分布范围可能像周口店的角砾岩沉积分布一样广泛，并且我们可以有把握地断定在这些角砾岩沉积中可能会出土许多头骨，还可能出土相当完整的斯泰克方丹猿人骨架"，我们仅能做的就是恳求所有相关方面都明白，要使系统发掘能够持续不间断地进行，他们需要立刻为此提供便利条件。

除南方古猿亚科的骨架材料之外，还可以利用天然的南方古猿颅内模、迩人的部分颅内模（有点变形）和傍人左颞骨区颅内模进行研究。关于化石人类和化石猿类的颅内模的研究总能再度引发那个一直存在争议的问题，即通过对这些内模的研究到底能够在多大程度上推断出大脑沟回的形态。对现代人类和现代猿类的颅内模进行的研究以及将它们与真实的大脑进行直接比较的研究表明，这些试图描绘颅内模沟回的研究方法可能引起很大的误导 [5]。如此谨慎而客观的研究所得出的结论在舍佩尔斯博士看来却是"悲观的"。并且，由于某种原因，在化石中大脑结构在颅腔内壁上留下的印模比在现存的种类中更为真实准确当然也可能是真的（这种解释给古生物学家的研究提供了便利！）。不过，如果神经解剖学家知道舍佩尔斯博士不仅有把握绘制出南方古猿与迩人的整个脑半球的沟回形态，而且还能描绘出 26 个以上单个的细胞结构区，并将其与现代猿类和现代人类的相应区域进行比较，他们一定会感到惊讶。有人认为仅仅通过对颅内模的考查便可推知大脑的功能（例如语言、抽象思维和运动技能），大脑皮层生理学家可能也会对此推论表示反对。但是在这部

inferences regarding functions such as speech, abstract thought and motor skill which, it is suggested, can be drawn simply from the examination of a cast of the inside of the skull. But Dr. Schepers, in the introduction to this section, makes his attitude perfectly plain, for he says: "In all attempts at representing these fossils in a reconstructed form, the method employed was that of claiming the maximum dimensions considered likely for the type. It is... much less satisfactory to hear that a new find of fossil material does *not* suggest any new hypothesis or does not corroborate evidence already at hand about a cognate theory, than to discover that at least someone is prepared to claim the maximum importance for his discovery. Such enthusiasm encourages criticism and comment, and is therefore to be recommended."

There may be some who doubt the propriety of preparing a scientific report with a bias of this kind, but at least it makes for lively reading. At the same time it does make it somewhat difficult for the reviewer not to appear in the role of the carping critic; and any appearance of carping criticism in the present instance is certainly to be avoided, if only because it might seem (unintentionally) to belittle the importance of this fossil material of which, in fact, the importance can scarcely be exaggerated. Also, the thoroughness with which Dr. Schepers has pursued his studies, and his evident eagerness to squeeze every drop of information from his material (even when this is defective) must evoke the greatest admiration for his assiduity. The cranial capacity of *Australopithecus* is estimated at 500 c.c., and of *Plesianthropus* at 435 c.c. The cranial capacity of *Paranthropus* is judged to be as high as 650 c.c.; but since this estimate is based on little more than the left temporal region of the skull, it is not easy to accept it without qualification. For the same reason, the contours of the reconstructed endocranial cast of *Paranthropus*, which are used for graphic comparison with casts of other primates, should probably be discounted as being too conjectural. Thus, so far as the absolute volume of the brain is concerned, the Australopithecinae appear to fall well within the limits of the anthropoid apes. On the other hand, if, as Dr. Schepers surmises might be the case, the endocranial casts hitherto obtained happen by chance to represent the lower limits of variation, the cranial capacity of the Australopithecinae may in its upper limits have transcended the range of variation found in modern large apes. Only the accession of further material can decide this point. But it remains a matter of very considerable interest that the volume of the Australopithecine brain *relative to the body size* (so far as this can be inferred in a very general way from fragments of the limb skeleton) does appear to have exceeded somewhat that of modern large apes.

Enough has been said in this review to indicate the intermediate position which many of the anatomical characters of the Australopithecinae in their combination occupy in relation to apes and men. Whether on the basis of this morphological evidence the South African representatives of this sub-family, which have so far been found, may be assumed to bear any direct relation to the line of human evolution, depends partly on the geological age of their remains. The evidence for this hitherto rests almost entirely on faunistic data. Dr. Broom is now of the opinion that the presence of a primitive hyaena, *Lycyaena*, and of two species of sabre-tooth tigers, in the deposits at Sterkfontein signifies a date not later than Upper Pliocene, and that the Taungs site is probably still older, perhaps

分的导言中，舍佩尔斯博士十分清晰地表明了自己的态度，他说："复原这些化石所能采取的所有方法中，这个方法声称最大限度地利用了颅内模提供的信息。……相对于得知新发现的化石材料**不能**支持任何新的假说，或者不能确证相关理论的已有证据，人们更乐于看到至少有人准备去声称他有了重大的发现。这种热情能够激发批评和评论，因此也受到欢迎。"

抱着这种偏见来准备科学报告至少是有助于活跃学术气氛的，尽管有人怀疑这种做法是否妥当。同时这也使那些吹毛求疵的评论家们想不挑剔都难；如果没有这么（无意地）强调化石的重要性，一定可以避免那些吹毛求疵的批评，但事实上，其重要性根本没有被夸大。此外，舍佩尔斯博士对研究的彻底投入，以及对从材料（即使有时候不完备）中尽量获取证据的强烈渴望，这些使人对他的敬意油然而生。据估计，南方古猿的颅容量为 500 毫升，而迩人的颅容量为 435 毫升，傍人的颅容量被认为高达 650 毫升；但这个估测主要依据头骨左颞骨区及其周边头骨的小部分信息，因此不宜无条件地接受。出于同样的原因，用来与其他灵长类动物颅内模作图样比较的复原傍人颅内模轮廓也因过于依赖推测而使其可信度大打折扣。因此，就脑的绝对容量而言，南方古猿亚科似乎正好在类人猿的范围之内。另一方面，因为舍佩尔斯博士的猜测可能就是事实，如果迄今获得的颅内模刚好代表了变化值的下限，那么南方古猿亚科颅容量的上限值就很可能超出了已发现的现代大型猿类的颅容量变化范围。只有获取更进一步的化石材料后才能确定这一点。但是还有一点相当重要，**相应于体型大小**（其体型大小大体上能从肢骨破片中推断出来），南方古猿亚科的脑量似乎确实比现代大型猿类的脑量大一些。

这篇评述充分论证了南方古猿亚科介于猿类和人类之间，其许多解剖学特征混合了猿类和人类的特点。以形态学证据为基础，迄今发现的南非该亚科的代表是否与人类进化有直接联系，这在一定程度上取决于其化石的地质年代。目前为止此类证据几乎完全依据动物群的资料。现在布鲁姆博士认为，斯泰克方丹堆积中发现的原始鬣狗（狼鬣狗属）及两种剑齿虎的化石表明其年代不晚于上新世晚期，而汤恩化石产地的年代可能还要早一些，可能是上新世中期。但是他明确指出，如同克罗姆德拉伊堆积一样，目前要测定此处确切的地质年代是不可能的。然而，南方古猿

Middle Pliocene. But he specifically states that, as with the deposits at Kromdraai, it is quite impossible at present to determine the geological age with certainty. However, the importance of the Australopithecinae is so great for the study of primate palaeontology, and particularly human palaeontology, that it becomes imperative to obtain more information about their antiquity. Even the skeletal remains, in spite of the magnificent work of Dr. Broom, still remain too scanty for firm conclusions regarding their significance. For Dr. Broom and Dr. Schepers they provide evidence for an assumption that the modern apes, and even the fossil apes of the *Dryopithecus-Sivapithecus* group, have no close relation to man (they are the result of a prolonged period of parallel development), and that the line of human evolution diverged from that of other primates so far back as Eocene times. On the other hand, the distinguished palaeontologist, Dr. W. K. Gregory (who has also had the opportunity of examining the original Australopithecine material) concludes[6] that "the evidence afforded by the morphology of the braincast and skull structure can hardly leave a well-founded doubt that the Australopithecine group were derived from the *Dryopithecus-Sivapithecus* stock*", and that "although it is too much to expect that the close structural approach of *Plesianthropus* towards *Sinanthropus* will discourage those who cling hopefully to the myth of Eocene man, all the facts... tend to confirm the conclusions of [those] who regard man as the result of a morphological revolution which took place during the later Tertiary period".

Such a divergence of opinion among recognized experts only serves to emphasize the need for still more fossil material. In this connexion we would quote a remarkable statement by Dr. Broom: "When some wealthy man or corporation undertakes the systematic exploration of our deposits... I think one can safely affirm that within three or four years we will discover more of the origin of man than has been revealed during the past hundred. Practically all the discoveries described in the present work, except the Taungs skull, were made by me in about two years working almost single-handed. Not only had I to find all the specimens, I had to develop them out of the matrix, give all the descriptions and make all the drawings."

Dr. Broom is a veteran palaeontologist unequalled today in experience and reputation. This monograph on the Australopithecinae is a fitting climax to a long life devoted to palaeontological research, and to a record of fossil-collecting which is quite unsurpassed for its rich discoveries. Moreover, with the preliminary work on *Australopithecus* contributed by Prof. Dart, he has very considerably enhanced the prestige of South African science in the eyes of the world generally. Surely one may express with some confidence the expectation that his magnificent contributions to the story of human evolution will be recognized by his fellow countrymen, to the extent that they will provide the funds and the facilities for the realization of his hopes for further discoveries at Sterkfontein and elsewhere. No greater service could be done to the study of human origins, and no more appropriate expression of gratitude could be made to Dr. Broom.

(157, 863-865; 1946)

亚科对于灵长类动物古生物学研究，特别是对人类古生物学研究，是非常重要的，因此迫切需要获得更多有关它们古代遗迹的资料。尽管布鲁姆博士做了大量的工作，但是考虑到结论的重要性，即便是骨化石也不足以证实这一结论。布鲁姆博士和舍佩尔斯博士对以下设想提供了证据：现代猿类，甚至是森林古猿–西瓦古猿群的化石猿类与人类都没有紧密联系（它们是长期平行进化的结果），人类进化路线早在始新世之后就已从其他灵长类动物的进化路线中分离出来了。另一方面，著名古生物学家格雷戈里博士（他也有机会观察到南方古猿亚科的化石原件）推断[6]："依据脑模形态和头骨结构提供的证据，还不足以对南方古猿亚科群的祖先来自森林古猿–西瓦古猿支干这个结论提出质疑"，而且"尽管难以期待迩人与中国猿人在结构上的相似会给信奉虚构出的始新世人的人泼冷水，但是所有的事实……却趋向于肯定那些人的结论，他们认为在第三纪晚期完成形态的进化后，人类诞生了"。

知名专家们在此观点上的分歧显示出他们仍然需要更多的化石材料。在此我们将引用布鲁姆博士的一段著名论述："当一些富有的人或者企业帮助我们对堆积物进行系统发掘的时候……我想人们可以坚信，在 3、4 年之内对于人类起源会有更多的发现，而这些发现会比在过去 100 年内已经发现的还要多。因为除汤恩头骨外，目前研究中所有的发现几乎都是依赖我一个人的力量在大约 2 年内完成的。我不仅发现了所有这些标本，还要把它们从围岩中挖掘出来，并对它们进行描述和绘图。"

古生物学家布鲁姆博士经验丰富，其经历与声望目前无人能及。这部论述南方古猿亚科的专著代表他长期致力于古生物学研究生涯的顶峰；同时，因其丰富的发现很难被超越，这也是化石收集记录的顶峰。达特教授为南方古猿的前期研究做出了很大的贡献，布鲁姆在此基础上大幅提高了南非的科学在世界范围内总体上的声望。当然，也许有人会满怀信心地期待，他的同胞能够认可他对人类进化史研究所做出的巨大贡献，甚至给他提供资金与便利使他得以实现自己的愿望：在斯泰克方丹及其他地方有进一步的发现。为人类起源研究做再多的工作也不为多，对布鲁姆博士表示再多感谢都不为过。

（田晓阳 翻译；董为 审稿）

References:

1. The South African Fossil Ape-Men: The Australopithecinae. By Dr. R. Broom and G. W. H. Schepers. (Transvaal Museum Memoir No. 2.) Pp. 272+18 plates. (Pretoria: Transvaal Museum, 1946.)

2. *Mitt. med. Fac. Kais.-Jap. Univ.*, **6**, 307 (1905).

3. Morton, D. J., *Amer. J. Phys. Anthrop.*, 7, 1 (1924).

4. See Kaufmann, H., and Sauter, M., *Arch. Suisses d'Anthrop. gén.*, **8**, 161 (1939).

5. See, for example, the most recent monograph on the subject, "Anthropoid and Human Endocranial Casts". By Pierre Hirschler. Pp. ix+150+11 plates. (Amsterdam: N. V. Noord-Hollandsche Uitgeversmij., 1942.) n.p.

6. *Ann. Transvaal Mus.*, **19**, 339 (1939).

Australopithecinae or Dartians

A. Keith

Editor's Note

Of all the anthropologists who objected to Raymond Dart's assertion that his *Australopithecus*, discovered at Taungs in Southern Africa in 1924, was an intermediate between apes and humans, Arthur Keith was perhaps the most vehement. This short note, published 23 years and many fossil discoveries later, was a plain and public admission of error: "I am now convinced that Prof. Dart was right and I was wrong". Keith suggested that australopithecines should be called "Dartians" by way of compensation. Keith's apology did not save his reputation. Less than five years later, the Piltdown Man remains, by which Keith had set such store in his own conception of human evolution, were exposed as forgeries.

WHEN Prof. Raymond Dart, of the University of the Witwatersrand, Johannesburg, announced in *Nature*[1] the discovery of a juvenile *Australopithecus* and claimed for it a human kinship, I was one of those who took the point of view that when the adult form was discovered it would prove to be near akin to the living African anthropoids—the gorilla and chimpanzee[2]. Like Prof. Le Gros Clark[3], I am now convinced, on the evidence submitted by Dr. Robert Broom[4], that Prof. Dart was right and that I was wrong; the Australopithecinae are in or near the line which culminated in the human form. My only complaint now is the length of the name which the extinct anthropoid of South Africa must for ever bear. Seeing that Prof. Dart not only discovered them but also rightly perceived their true nature, I have ventured, when writing of the Australopithecinae, to give them the colloquial name of "Dartians", thereby saving much expenditure of ink and of print. The Dartians are ground-living anthropoids, human in posture, gait and dentition, but still anthropoid in facial physiognomy and in size of brain. It is much easier to say there was a "Dartian" phase in man's evolution than to speak of one which was "australopithecine".

(**159**, 337; 1947)

Arthur Keith: Downe, Kent, Feb. 15.

References:
1. *Nature*, 115, 195 (1925).
2. *Nature*, 115, 234 (1925).
3. *Nature*, 159, 216 (1947).
4. "The South African Fossil Ape-Men: The Australopithecinae" (1946).

南方古猿亚科或达特猿

基思

编者按

雷蒙德·达特教授宣称他于 1924 年在南非汤恩发现的南方古猿是介于猿类和人类之间的中间类型，在所有反对他的这一结论的人类学家中，阿瑟·基思可能是反对最强烈的一个。23 年后，又有许多化石被发现，不久之后基思在这篇短文中公开坦率地承认错误："现在我深信达特教授是正确的，而我是错误的。"基思建议应该将南方古猿亚科称作"达特猿"作为对达特教授的补偿。但基思的道歉并没有挽回他的声誉。接下来不到 5 年的时间里，基思提出的作为人类进化理论重要证据的皮尔当人遗骸被曝光是伪造物。

南非约翰内斯堡威特沃特斯兰德大学的雷蒙德·达特教授在《自然》[1] 上宣布他发现了幼年南方古猿，并声称它同人类具有亲缘关系。当时我和其他一些人持有以下观点，即如若人们发现的是成年个体时，也许将能够证实它拥有和现存非洲类人猿——大猩猩和黑猩猩有更加接近的亲缘关系 [2]。和克拉克教授 [3] 一样，依据罗伯特·布鲁姆博士 [4] 提出的证据，现在我深信达特教授是正确的，而我是错误的；南方古猿亚科位于或者靠近这条最终进化为人类的线。现在我唯一不满的是这个已经灭绝的南非类人猿命名的长度。鉴于达特教授不仅发现了它们，而且正确地认识到了它们真正的本质，我冒险地提议在写南方古猿亚科时使用通俗的名字"达特猿"，这样能节省大量的油墨和印刷费用。达特猿是在地面生活的类人猿，在姿势、步态和齿系上与人类一致，但是在面部相貌和脑的大小上仍属类人猿。在人类进化过程中说存在一个"达特猿"时期比说一个"南方古猿亚科"时期要更容易。

（张玉光 张立召 翻译；赵凌霞 审稿）

Discovery of a New Skull of the South African Ape-man, *Plesianthropus*

R. Broom

Editor's Note

World War II had halted the search for fossil man in South Africa from 1941 until 1946, when General Smuts himself asked palaeontologist Robert Broom to continue his hunt for fossils of ape-men. Starting at the beginning of 1947, Broom searched the fruitful site at Kromdraai for three months, but found only one possible hominid bone. The hunt moved to Sterkfontein on 1 April and met success within a week: on 8 April the search was rewarded with remains of the ape-man *Plesianthropus* (now *Australopithecus*), including a partial skull of what Broom believed was an elderly female. In time, this fossil, named "Mrs Ples", became an iconic image of the South African ape-men.

BETWEEN 1924, when the Taungs ape-man was discovered, and 1941, when the jaws of the baby Kromdraai ape-man were found, remains of many individuals of this wonderful family, which some of us consider to be nearly related to man, were discovered; sufficient to show that in South Africa we may have the key which will solve the problem of the origin of man. But from 1941 until a few months ago, no further research was undertaken.

In 1946, a book was published on all we know of the South African ape-man; and the world awoke to the possibilities of the wonderful results that might be achieved by the further study of our caves. The United States has come to realize that South Africa is a more promising centre for the solution of the problem than even Java or China; and she seems determined to see that the problem must be solved, and solved soon.

At the beginning of this year, at the special request of General Smuts, I again started to hunt for more "missing links". My assistants and I commenced work at Kromdraai and continued there for three months. We found many interesting remains, including a very fine skull of a sabre-tooth tiger (*Meganthereon*) and the skull of a large type of "baboon", which belongs to a new genus; but only one bone that had possibly belonged to an ape-man.

一件新的南非猿人——迩人头骨的发现

布鲁姆

编者按

由于第二次世界大战，从 1941 年到 1946 年，在南非搜索人类化石的工作暂停了下来，后来斯穆茨将军亲自邀请古生物学家罗伯特·布鲁姆继续寻找猿人化石。从 1947 年年初开始，布鲁姆在曾有丰富收获的克罗姆德拉伊地区搜索了 3 个月，但仅发现了一件可能属于人科的骨。4 月 1 日起，搜索工作转移到了斯泰克方丹，并在 1 周内取得了重要发现：4 月 8 日发现了迩人（现称南方古猿）的化石，包括布鲁姆认为的属于老年女性的部分头骨。后来，这件化石被命名为"普莱斯夫人"，成为南非猿人的标志性形象。

1924 年，汤恩猿人的头骨化石被发现，接着在 1941 年，在克罗姆德拉伊地区发现了婴儿猿人的颌骨。在这一期间，发现了许多属于这个令人惊奇的猿人科的个体化石，有些人认为其与人类的关系较近；这些证据足以说明在南非我们可能找到解决人类起源问题的钥匙。但从 1941 年直到几个月前，再没有人进行过进一步的研究。

1946 年，一本关于所有已知南非猿人的书籍问世了；全世界都意识到通过对洞穴进行深入研究也许会取得非凡的成果。美国逐渐认识到南非是一个比爪哇或中国更能揭示人类起源问题的地方；她似乎坚信人类起源问题肯定能被解决，而且会很快得到解决。

今年年初，在斯穆茨将军的专程邀请下，我再次开始搜寻更多的"缺失环节"。我与助手们在克罗姆德拉伊开始工作，并在那里持续工作了 3 个月。我们发现了许多有趣的化石，包括一件保存完好的剑齿虎（巨颏虎）的头骨和一件属于一个新属的大型"狒狒"的头骨；但只发现了一件可能属于猿人的骨。

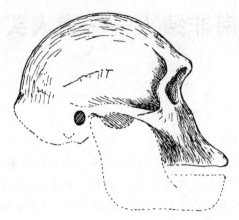

Fig. 1. Side view of skull of old female Sterkfontein ape-man *Plesianthropus transvaalensis* (Broom). $\frac{1}{3}$ natural size. The lower part of the occiput and part of the jugal arch are still embedded in matrix, and the lower jaw is not present.

Then on April 1 we started work at Sterkfontein, and almost immediately our labours were rewarded by sensational discoveries. On April 8 we found an isolated crushed snout of an adolescent *Plesianthropus* with some beautiful teeth; and a fragment of a snout of a child of possibly three years. This showed the perfect upper milk molars, and a note on the discovery was sent to *Nature* [see issue of May 3, p.602. Editors]. On April 11 two quite isolated teeth of Sterkfontein apes were found—one the beautiful upper canine of a male of perhaps thirty years, and a lower molar probably of a female of forty years.

But on April 18 a small blast cracked open a block of breccia, and there in the middle lay a perfect skull without the mandible of an adult *Plesianthropus*, with the brain case broken across. In ten days we had found the remains of five—possibly six—individuals of the Sterkfontein ape-man, *Plesianthropus*, and one of them represented by a complete skull.

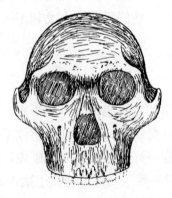

Fig. 2. Front view of skull of old female Sterkfontein ape-man, *Plesianthropus transvaalensis* (Broom). $\frac{1}{3}$ natural size.

The bones are very friable and the matrix not only rather hard lime, but also breccia with many large broken pieces of chert; it will take many weeks before the skull can be

908

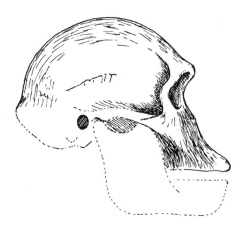

图 1. 老年女性斯泰克方丹猿人——德兰士瓦迩人头骨的侧面图（布鲁姆）。该图为真实尺寸的 $\frac{1}{3}$。枕骨的下部以及部分颧弓仍埋在基质中，没有发现下颌骨。

然后，我们于 4 月 1 日开始在斯泰克方丹工作，并且几乎在工作一开始就取得了重大发现。4 月 8 日，我们发现了一件青少年迩人被压碎的口鼻部骨化石，其上带有一些漂亮的牙齿；还发现了一件大约 3 岁孩子的口鼻部骨破片。这上面有保存完好的上乳臼齿，关于此发现的说明已发给《自然》杂志［见 5 月 3 日那一期，第 602 页，编者注］。4 月 11 日，我们发现了两颗单个的斯泰克方丹猿类的牙齿——一颗是约 30 岁男性的漂亮的上犬齿，另一颗可能属于 40 岁女性的下臼齿。

但在 4 月 18 日，一次小爆破炸开了一块角砾岩，在其中间发现一件保存完好的成年迩人头骨，只是没有下颌骨，头盖骨裂开了。10 天中我们发现了 5 个（也可能是 6 个）斯泰克方丹猿人（即迩人）个体的残骸，其中有一件是完整的头骨。

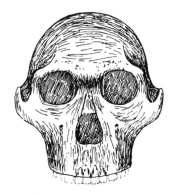

图 2. 老年女性斯泰克方丹猿人——德兰士瓦迩人头骨的正面图（布鲁姆）。该图为真实尺寸的 $\frac{1}{3}$。

这些骨十分易碎，而埋藏它们的基质不仅有颇为坚硬的石灰，而且还有很多大

completely developed from the matrix.

Enough, however, has now been done to reveal most of the more striking features of the skull. In the type skull the brain cavity was filled with matrix. In the newly discovered skull there is only a lining of lime crystals from about $\frac{1}{8}$ in. to nearly $\frac{1}{2}$ in. in thickness. When this layer is removed we will have every detail of the anatomy of the inner side of the base of the skull. The under surface of the base of the skull appears to be just as perfect; but it will take some further weeks of preparation to reveal it all.

So far, our labours have been concentrated on the details of the face, and in cleaning the cranial vault. The teeth are all lost, but many sockets remain, and we can say with much confidence that the skull is that of an elderly female.

The drawings I give will serve to show the general aspect of the face and the side view. I think there will be very general agreement that the being is not a chimpanzee or even closely allied to any of the living anthropoids, and that, though small, the skull has many resemblances to that of man.

The skull from glabella to opisthocranion is about 150 mm., and the greatest parietal width is about 100 mm. It is thus seen to have an index of about 66, and to be extremely dolichocephalic. As yet, we can only give a roughly approximate size for the brain cavity, but it seems probable that it will be about 500 c.c.

Of course, it will take many weeks before the skull is completely worked up and a full account can be published, but it seems well that the world should know that such a valuable skull has been discovered.

(**159**, 672; 1947)

R. Broom: Transvaal Museum, Pretoria, April 2.

块燧石的角砾岩，因此将头骨从这些基质中完好地取出需要几周时间。

不过现在这些发现完全可以揭示头骨中大部分比较显著的特征。在正型标本头骨的颅腔中充满了基质。在新发现的头骨中，仅内侧有一层厚度约 $\frac{1}{8}$ 英寸~$\frac{1}{2}$ 英寸的石灰晶体。当移除这层后，我们会观察到头骨底内侧的所有解剖学细节。头骨底的下表面看起来很完美；但是要揭示全部信息的话，还需要再做几个星期的准备工作。

到目前为止，我们的精力主要集中在研究脸部细节和清理头骨穹隆。虽然牙齿已经全部没有了，但是许多牙槽还在。我们可以确信地说，这是一件老年女性的头骨。

我提供的图画可用以展示其面部的轮廓及侧面情况。我认为大家会赞同这样的观点：这种生物不是黑猩猩，甚至不与任何现存类人猿有密切的亲缘关系，尽管头骨很小，但与人类头骨有很多相似之处。

该头骨从眉间点到颅后点约 150 毫米，在顶骨的最大颅宽约 100 毫米。因此可以看到这个头骨的颅指数约为 66，属极端的长颅型。到目前为止，我们对脑颅大小只能给出一个粗略的估计，可能约为 500 毫升。

当然，将这个头骨完全整理出来并发表一份完整的报告，还需要花费几个星期的时间，但是世界应该知道这样一件有价值的头骨已经被发现了，这一点看起来很好。

(刘皓芳 田晓阳 张立召 翻译；赵凌霞 审稿)

911

Jaw of the Male Sterkfontein Ape-man

R. Broom and J. T. Robinson

Editor's Note

Our extinct australopithecine cousins are known to have shown marked sexual dimorphism, with the males larger than the females. This was first shown with the discovery of a mandible (lower jaw) of a male *Plesianthropus* (now *Australopithecus*) at Sterkfontein in the Transvaal, to supplement the jaw of a female discovered earlier at the same site. The authors were impressed by how human the jaw looked, with signs of a chin, and few of the ape-like features they had expected. Their natural comparison was the Mauer mandible, the type specimen of the near-human Heidelberg Man (now *Homo heidelbergensis*). We now believe that all such signs of humanity were convergent, and probably not signs of direct descent from these australopithecines.

O N June 24, we blasted out at Sterkfontein, at a spot only about 8 ft. away from where we discovered the old female skull of *Plesianthropus*, the almost perfect lower jaw of a large male. The left mandible is complete except for the loss of the condyle and a little part of the margin of the angle. The whole symphysial region appears to be complete, while the right ramus is much broken and crushed. All the teeth of the left side are present, though worn.

The horizontal ramus is considerably larger than that of man but essentially similar. The ascending ramus is higher than in man, but otherwise not unlike that of the human jaw.

The front of the jaw is remarkably interesting. It does not slope rapidly backwards as in the living anthropoids but more downwards than backwards, giving an appearance not unlike that of the Heidelberg jaw. The front of the Sterkfontein jaw is narrower owing to the incisors being smaller. The symphysis, so far as can yet be seen, seems to agree fairly well with that of the Heidelberg jaw, and there appears to be no simian shelf.

The molars and premolars are much larger than in the Heidelberg jaw, and the canine very much larger. The molars and premolars are much worn. In the case of the 1st and 2nd molars, the whole outer sides of the crowns are worn off; but on the inner side parts of the enamel cusps still remain.

The canine is the most interesting tooth of the jaw. We know the unworn lower canine crown in the male *Plesianthropus*. It is much larger than in man and has a well-developed

斯泰克方丹男性猿人的颌骨

编者按

已经灭绝的人类近亲南方古猿亚科具有明显的两性异形，即男性个体大于女性。这一现象最早发现于南非德兰士瓦的斯泰克方丹，在此发现了一件男性迩人（现称南方古猿）的下颌骨，与先前在同一地点发现的女性颌骨的形态特征明显不同。通过与最接近现代人的海德堡人（人属海德堡种）的典型标本——毛尔下颌骨的比较，作者惊奇地发现，这件下颌骨的形态与人类的非常类似，已有下巴雏形的迹象，几乎没有类似猿类颌骨的特征。我们现在相信，人类的这些特征具有趋同性，现代人类所具有的特征可能并不是从南方古猿亚科直接演化而来的。

6月24日，我们炸开了位于斯泰克方丹的一处地点，这里距离我们先前发现那具老年女性迩人头骨的地方只有大概8英尺远，在这里我们发现了一件近乎完美的男性的下颌骨。此下颌骨的左侧基本完整，只缺失了髁突和下颌角边缘的一小部分。左右颌骨联合部看起来很完整，而右侧下颌支大部分已断裂并被压碎了。左侧的牙齿虽然有些磨损，但是全部保存了下来。

颌骨的水平支比人类的大很多，但是基本上与人类相似。上升支比人类的要高，但也与人类颌骨有着相似之处。

该颌骨的前部非常有趣，它并不像现存的类人猿的颌骨前部那样迅速向后倾斜，相比之下其向下倾斜的角度更大，这与海德堡下颌骨的外观有些类似。斯泰克方丹颌骨由于门齿比较小，前部显得比较窄。就我们目前所能看到的，其联合部与海德堡下颌骨颇有几分相似，已经看不出猿的框架了。

白齿和前白齿比海德堡下颌骨的大得多，犬齿也更大一些。白齿和前白齿磨损严重。虽然第一白齿和第二白齿的整个齿冠的表面都被磨损掉了，但是内侧齿尖的部分珐琅质还可见。

整个颌骨中最有趣的牙齿就是犬齿了。我们都知道男性迩人未被磨损的下犬齿齿冠比现代人类的大很多，并且具有一个发达、尖锐的主齿尖，以及一个小的后齿

and pointed main cusp and a small posterior cusp. In the female the upper canine we knew was worn down as in man, but in the male it was suspected that the lower canine passed in front of the upper as in the anthropoid apes. Now we find that the lower canine, while it may in the young animal pass up in front of the upper, is ground down, as age advances, by attrition with the upper, exactly as in man. The whole jaw is thus practically a human jaw.

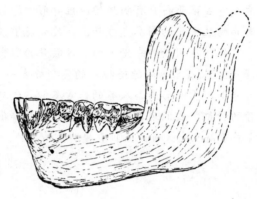

True side view of mandible of male *Plesianthropus transvaalensis* (Broom). $\frac{1}{2}$ natural size. The molars are badly worn along their outer sides. The canine has its crown ground down quite flat and in line with the other teeth. Parts of the outer sides of the premolars are in the counter slab and will later be replaced.

This jaw is much too large to have fitted the elderly female skull recently discovered; and the skull that belonged to this jaw must have been remarkably large. If a restoration of the skull is made from the known female skull, but large enough to have fitted this jaw, it is seen that we have a skull that is nearly human.

The brain of the female skull was only about 450 c.c., but the male skull that belonged to this jaw must have had a brain of 600 c.c., or perhaps even 700 c.c.

Another interesting point is that on the lower part of the front of the jaw there is a little bony thickening which might be regarded as an incipient chin.

This jaw seems to us to be of considerable importance on the question of man's origin.

(**160**, 153; 1947)

R. Broom and J. T. Robinson: Transvaal Museum, Pretoria, South Africa, July 3.

尖。而女性的上犬齿和男性的一样受到了磨损，但是有人怀疑男性的下犬齿像类人猿的一样长到上犬齿的前方。现在我们发现，正如人类一样，年轻动物随着年龄的增长，下犬齿可能向上生长至上犬齿的前方，由于上犬齿的磨擦，下犬齿便被碾磨掉了。因此整个颌骨实际上就是一个人类的颌骨。

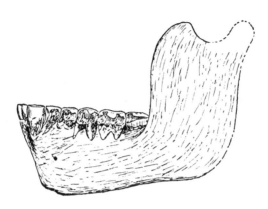

德兰士瓦迩人男性下颌骨侧面观（布鲁姆）。该图是真实尺寸的 $\frac{1}{2}$。臼齿颊侧面磨损严重。犬齿的齿冠被碾磨得很平整，与其他牙齿相平。前臼齿颊侧面部分位于相对面上，并且不久将会被替换。

由于这件颌骨太大，因此与最近发现的老年女性头骨无法匹配在一起；这一颌骨所属的头骨一定很大。如果根据已知的女性头骨的尺寸对该头骨进行复原，并使其足够大以匹配这个颌骨的话，那么我们就会得到一个接近现代人的头骨。

女性头骨的脑量只有450毫升左右，估计这具颌骨所属的男性头骨的脑量可达到600毫升，甚至700毫升。

另一个有趣的发现是，颌骨前方的下部有一点骨质增厚，被认为可能是下巴的雏形。

这个颌骨对我们研究人类起源问题可能具有相当重要的意义。

（刘皓芳 田晓阳 翻译；吴秀杰 审稿）

African Fossil Primates Discovered during 1947[*]

W. E. Le Gros Clark

Editor's Note

The 1930s and 1940s were a Golden Age for the discovery of fossil hominids, largely due to Robert Broom's work at Sterkfontein and Kromdraai in South Africa, and Louis Leakey's in Kenya. Here anatomist Le Gros Clark summarizes the work to date. He concentrated, naturally, on Broom's 1947 "ape-man" discoveries at Sterkfontein, showing how they put paid to any doubts arising after Dart's initial discovery of the Taung "baby" in 1924. But Le Gros Clark also looked at the much more ancient (and more apelike) *Proconsul*, remains of which had been emerging in Eastern Africa for some years, thanks to Louis Leakey and others. Leakey's own ventures into ape-man territory still lay more than a decade in the future.

THE elucidation of the fossil record of the group of mammals of which man himself is a member is clearly a matter of the most profound interest. But it is a study which in the past has progressed extremely slowly, for the reason that the remains of fossil Primates are usually found only at rare intervals. It is all the more remarkable, therefore, that the year 1947 was outstanding in the field of palaeontology for the great abundance of extinct representatives of the higher Primates, the Hominoidea, which have come to light. These discoveries were made in Africa, and it is particularly interesting to note that they followed so closely on the Pan-African Congress of Prehistory, held in Nairobi in January of that year, when the implications of earlier finds first gained a wide recognition. The discoveries are due primarily to two men of science, Dr. Robert Broom in South Africa and Dr. L. S. B. Leakey in Kenya.

The new discoveries made by Dr. Broom supplement those which he made between 1936 and 1941 at Sterkfontein and Kromdraai, and which he described in detail in a monograph published in 1946[1]. They are remarkable both in quantity and quality, for not only do they include considerable numbers of skulls, jaws, teeth and limb-bones of the Australopithecinae, but also some of this material is extraordinarily complete and well preserved.

It will be recalled that the first specimen of the Australopithecinae, an immature skull with an endocranial cast, was found at Taungs in 1924 and afterwards described by Prof. Raymond Dart[2]. Then in 1936 and later, Dr. Broom found portions of several adult skulls, jaws and teeth, as well as fragments of some limb-bones, at Sterkfontein

---

[*] Substance of a communication to the Linnean Society, April 22, 1948.

916

1947年间发现的非洲灵长类化石*

克拉克

编者按

20 世纪 30 和 40 年代是人科化石发现的黄金年代，这在很大程度上归功于罗伯特·布鲁姆在南非斯泰克方丹和克罗姆德拉伊的工作，以及路易斯·利基在肯尼亚的工作。本文中，解剖学家勒格罗·克拉克总结了迄今为止的工作。他自然地将注意力集中到 1947 年布鲁姆在斯泰克方丹发现的"猿人"，并展示了这些材料如何回应自 1924 年达特首次发现汤恩"小孩"头骨以来的各种质疑。然而，勒格罗·克拉克还将目光投向更为古老（以及更加类似猿类）的原康修尔猿，该物种的化石是数年前由路易斯·利基等人在东非发现的。而至少在未来十多年的时间内，利基仍将在猿人领域中进行他自己的冒险。

由于人类自身就是哺乳动物群体中的一员，所以对于这一群体化石记录的探究肯定是人类最感兴趣的事了。但是在过去，该方面的研究进展得极其缓慢，因为通常只偶尔才能发现灵长类的化石。因此 1947 年对古生物学领域来说是极不寻常的一年，因为发现了大量已灭绝高等灵长类的代表——人猿超科的化石，至此它们逐渐为人所知。这些发现是在非洲取得的，尤其有趣并要加以说明的是，这些发现恰巧发生在 1947 年 1 月在内罗毕举行的泛非史前学大会之前，而正是在这次会议上，该发现的意义首次得到广泛认可。这些发现主要归功于科学界的两个人，即南非的罗伯特·布鲁姆博士和肯尼亚的利基博士。

布鲁姆博士取得的新发现，是对他 1936 年 ~ 1941 年间在斯泰克方丹和克罗姆德拉伊发现的补充，他在 1946 年出版的一本专著中详细地描述了这些发现 [1]。这些发现在数量和质量上都是引人注目的，不仅因为它们包含了相当大量的南方古猿亚科的头骨、颌骨、牙齿和肢骨，也因为这些材料中有不少特别完整，保存得也特别好。

人们不会忘记南方古猿亚科的第一份标本，即一件具有颅内模的未成年头骨，它于 1924 年发现于汤恩，后来雷蒙德·达特教授对其进行了描述 [2]。1936 年及其后，布鲁姆博士在约翰内斯堡附近的斯泰克方丹和克罗姆德拉伊发现了几个成年个体的

---

* 向林奈学会所提交通讯的主要内容，1948 年 4 月 22 日。

917

and Kromdraai near Johannesburg. His report on this material made it clear that the Australopithecinae were ape-like creatures with brains of simian dimensions, but which at the same time showed such a remarkable assemblage of characters hitherto regarded as distinctive of the Hominidae that there could be little doubt of their importance for problems of human evolution. Among these hominid characters may be mentioned the morphology of the frontal region of the skull, the low level of the occipital torus, the construction of the tympanic region, the forward position of the foramen magnum, the palatal contour, the small canines, the bicuspid character of the anterior lower premolars, the morphology of the first deciduous molars, the flat wear of the molars, the details of the lower end of the femur, and certain features of the talus and the capitate bone. The evidence of the femur, combined with that of the foramen magnum, it should be noted, seemed to make it clear that the Australopithecinae were capable of assuming an erect posture closely approaching that of man.

This accumulation of evidence was sufficient to convince most anatomists that the importance of the Australopithecinae had not been exaggerated. A few critics, however, expressed scepticism, mainly, no doubt, because they had not had the opportunity of examining the original material for themselves. It is also a fact, however, that this material, although abundant, was in some respects fragmentary, and thus required a rather intimate acquaintance with the comparative osteology of the Hominoidea for a correct appreciation of its unusual characters. Early in 1947, Dr. Broom renewed his excavations at Sterkfontein, and in April announced the discovery of a practically complete adult Australopithecine skull, as well as the upper jaw of another adult, an upper jaw of an adolescent, and some isolated teeth[3]. The complete skull has been listed in the material from Sterkfontein as "Skull No. 5". Further excavations soon showed that many more remains of the Australopithecine fossils were embedded in the limestone matrix in the same area at Sterkfontein, and parts of at least five other skulls (some of which are fairly complete) have since come to hand. The preliminary notes on these discoveries so far published have now provided a remarkable vindication of the views previously expressed by those who had personally studied the earlier material described by Dart and Broom. For example, they appear to confirm the forward position of the foramen magnum and the human resemblances in the occipital and tympanic regions, while many other features such as the contour of the frontal region and the shape of the dental arcade conform closely with the evidence already supplied by less complete material. The discovery of a massive jaw in June 1947, with all the teeth in position, has also added further evidence that the canine was relatively small (compared with the premolars) and became worn down flat to the level of the adjacent teeth[4]. In all these skulls the endocranial capacity seems to vary about 500 c.c.

Dr. Broom, however, has not only found a number of skulls, but has also collected a considerable number of limb-bone fragments, including portions of a scapula, humerus and femur, a tibia, some ribs and vertebrae, and an almost complete os innominatum[5]. The last-named is without doubt the most remarkable find. In the shape of the ilium it corresponds very closely with that of man, and shows no resemblance at all to the os

918

头骨残块、颌骨和牙齿，以及一些肢骨破片。他对这些材料的报道表明，南方古猿亚科是一种脑量与猿相当的类似猿类的生物，但是该报道同时也展示了这些标本所具有的人科独有的一系列显著特征，以至于几乎无人怀疑其对解决人类进化问题所具有的重要性。在这些人科特征中，值得一提的包括头骨额区的形态、枕圆枕位置较低、鼓室区的构造、枕骨大孔位置前移、腭轮廓、犬齿较小、下前臼齿前部的双齿尖特征、第一乳臼齿的形态、臼齿平磨、股骨下端的细节以及距骨和头状骨的某些特征。需要说明的是，股骨以及枕骨大孔的证据似乎表明南方古猿亚科几乎和人类一样可以直立行走。

这些证据的积累足以使大多数解剖学家相信南方古猿亚科的重要性并未被夸大。然而，一些评论家表示了怀疑，毫无疑问，这主要是因为他们没有机会亲自查看原始材料。另一个事实是，尽管这些材料很丰富，但是从某些方面看，它们是破碎的，因此要想正确理解其不寻常的特征，就需要对人猿超科的比较骨骼学有相当缜密的认识。1947 年初，布鲁姆博士在斯泰克方丹重新开始了他的发掘工作，4 月他宣布发现了一件非常完整的成年南方古猿头骨、另一个成年南方古猿的上颌骨、一个未成年南方古猿的上颌骨以及一些单个的牙齿[3]。在斯泰克方丹化石材料中，该完整头骨被编为"5 号头骨"。随后进一步的发掘工作表明在斯泰克方丹同一区域的石灰岩基质中埋藏了更多的南方古猿亚科化石，而且在后续发掘中又获得了部分化石，包括至少代表 5 个其他头骨的化石（其中有些非常完整）。目前发表的关于这些发现的研究简报，为那些亲自研究过达特和布鲁姆以前描述的早期材料的那些人先前所持有的观点提供了非常有力的实证。例如，它们证实了枕骨大孔的前移、枕骨和鼓室区与人类的相像，而许多其他的特征，如额区的轮廓和齿弓的形状也与那些由不太完整的材料所提供的证据非常吻合。1947 年 6 月发现的粗大下颌骨，其所有牙齿原位保存，该标本进一步证明了犬齿（与前臼齿相比）相对较小，并且磨耗后与相邻牙齿处于相同水平[4]。所有这些头骨的颅容量好像都在 500 毫升左右变化。

然而，布鲁姆博士不仅发现了大量头骨，也搜集到了相当多的肢骨破片，包括肩胛骨、肱骨和股骨残块，一件胫骨、一些肋骨和椎骨以及一件几乎完整的髋骨[5]。最后一件无疑是最引人注目的发现。从髂骨的形状来看，它与人类的非常相符，而与类人猿的髋骨几乎没有任何相似之处。这的确为证明南方古猿亚科具有与人类一

innominatum of an anthropoid ape. Indeed, it provides the final proof (if further proof were needed) that the Australopithecinae stood and walked in approximately human fashion. So human in appearance is this fossil bone that the question may well be raised whether it really is part of an Australopithecine skeleton. But, apart from the fact that it is entirely consistent with the evidence previously provided by the femur and talus (and also the construction of the base of the skull), it was actually found by Dr. Broom embedded in a block of limestone matrix with portions of several other limb-bones, vertebrae and ribs and with a crushed Australopithecine skull. There can scarcely be any doubt, therefore, that all these bones are the remains of the skeleton of a single individual. It should be noted, also, that in certain features of the os innominatum, such as the details of the ischial tuberosity, it shows unusual characters which, it seems, are not paralleled either in the modern anthropoid apes or man.

It will naturally be some time before a detailed account of all this new material from Sterkfontein can be published, but private communications, together with unpublished photographs and drawings, make it clear that the information now available on the anatomy of the Australopithecinae is astonishingly complete. Stated briefly, they are ape-like creatures in respect of the general proportions of the braincase and jaws, but approach the Hominidae closely in the constructional details of the skull, in numerous features of the dentition, and particularly in the limb skeleton. There is no sign yet that the rich "lode" of Australopithecine material discovered by Dr. Broom is yet approaching exhaustion, and thus there is a reasonable possibility that further remains of the hand and foot skeleton may be found which will permit an even more precise definition of the status of these extinct hominoids. But, when all the evidence now available is published, systematists will inevitably be faced with the question whether the Australopithecinae should not be allocated to the Hominidae rather than the Pongidae.

Simultaneously with these remarkable discoveries of fossil hominoids in South Africa, discoveries of similar importance were being made in Early Miocene deposits of Kenya by Dr. L. S. B. Leakey and Dr. D. McInnes. In 1931, Dr. A. T. Hopwood described some fossil jaws and teeth recovered from Koru in Kenya and assigned them to three new genera of extinct anthropoid apes, *Proconsul*, *Xenopithecus* and *Limnopithecus*[6]. Afterwards, more Early Miocene material referable to these genera were obtained on Rusinga Island and adjacent areas by Leakey[7] and McInnes, and were described by the latter in 1943[8]. Following the Pan-African Congress on Prehistory, some of the delegates had the opportunity of visiting the sites of these discoveries and were impressed with the possibilities of further excavation. A British–Kenya Miocene Expedition was therefore organised with the aid of a grant from the Royal Society, with Dr. Leakey as field director. The expedition proved to be an outstanding success, for almost fifty specimens of fossil hominoids were collected[9]. This material is now being examined in the Department of Anatomy at Oxford, and the brief account which follows is entirely of a preliminary nature. Perhaps the most striking feature of the collection is the great variety of fossil apes which it represents. Before the first discoveries in Kenya, the only certain information available about fossil apes which existed earlier than the Middle Miocene was provided by two fragmentary jaws from the

920

样的直立行走方式提供了最终的证据（如果还需要进一步证据的话）。这件化石骨在形态上如此接近人类，以至于人们可能会怀疑其是否真的属于南方古猿亚科的部分骨。但是除了它与先前股骨和距骨（以及颅底的构造）提供的证据完全一致之外，实际上布鲁姆博士发现它时，它与其他的几块肢骨、椎骨和肋骨及一件压扁的南方古猿头骨一起埋在同一块石灰岩基质中。因此很难怀疑所有这些骨不是属于同一个个体的骨架残骸。另外应该说明的是，髋骨的某些特征，例如坐骨结节的细节，显示了它的确不同寻常，既不同于现代类人猿的，也不同于人类的。

对于在斯泰克方丹新发现的这批材料的详细研究成果，自然还需要一段时间才会发表，但是通过私人交流以及那些未发表的照片和绘图，我们可以确信现在获得的有关南方古猿亚科的解剖学信息已经达到了惊人的完整程度。简而言之，从脑颅和颌骨的总体比例来看，它们是类似猿类的生物；而从头骨的构造细节、齿系的诸多特征，尤其是从肢骨来看，它们又与人科很接近。然而还没有迹象表明布鲁姆博士发现的南方古猿亚科材料的丰富"矿脉"正在枯竭，因此仍然很有可能进一步发现手骨和脚骨的遗骸，这将有助于人们更准确地界定出这些已灭绝的人猿超科动物的身份。但是，当现在得到的所有证据得以发表时，分类学家们将不可避免地面对这样一个问题，即是否应该将南方古猿亚科归为猿科，而不是人科。

在南非发现这些引人注目的人猿超科动物化石的同时，利基博士和麦金尼斯博士在肯尼亚的中新世早期堆积物中也获得了类似的重要发现。1931 年，霍普伍德博士描述了从肯尼亚的科鲁挖掘的一些颌骨和牙齿化石，并将它们归入 3 个已灭绝的类人猿新属，即原康修尔猿、异猿和湖猿 [6]。后来，利基 [7] 和麦金尼斯在鲁辛加岛及其相邻地区得到了更多属于这些属的中新世早期材料，1943 年麦金尼斯对它们进行了描述 [8]。在泛非史前学大会之后，部分代表有幸参观了这些遗址，并对进一步发掘的可能性留下了深刻的印象。因此，在皇家学会资助下，一支以利基博士为野外领队的英国-肯尼亚中新世探险队组建起来了。事实证明这支探险队的组建是一项伟大的成功，因为他们收集了将近 50 件人猿超科动物化石标本 [9]。现在牛津大学解剖学系正在对这批材料进行研究，即将完成的简要记述仅仅是个初步认识。这批标本最显著的特征大概是它代表了各种各样的化石猿类。在肯尼亚的第一个发现之前，关于存在于中新世中期之前的猿化石唯一确切信息是从埃及渐新世的傍猴和原上猿的两块颌骨破片上获得的。现在，在肯尼亚得到的所有新材料使我们对中新世早期

Oligocene of Egypt, *Parapithecus* and *Propliopithecus*. Now, with the accession of all the new material from Kenya, we are faced with such a bewildering variety of Early Miocene apes that it is by no means an easy matter to sort them all out. Much of the material consists of jaws and teeth of an ape considerably larger than *Proconsul africanus* (Hopwood), and in which the relative sizes of the premolars and molars, as well as the cusp pattern of the latter, also show differences. There is little doubt, therefore, that they should be assigned to a different species. To this group belongs the almost complete mandible found on Rusinga Island in 1942, as has now been established by the discovery during 1947 of a portion of the maxilla of the same individual, with the cheek teeth in position. A still larger species is represented by portions of a huge mandible found by the British-Kenya Miocene Expedition at Songhor. This jaw is in some respects more massive than that of many adult male gorillas.

In all these specimens (which provisionally are regarded as constituting several species of the genus *Proconsul*), there are certain characteristic features. The incisors are relatively small in comparison with modern apes, and the symphysial region of the jaw in narrow, the axis of the symphysis tending to be more vertical; the canines are large and there is little definite evidence of any marked sexual difference; the first lower molar is proportionately rather small, the upper molars have a well-developed internal cingulum which in some cases is elaborately beaded, and the lower molars have very evident traces of an external cingulum. Some immature jaws of *Proconsul* which have been found are particularly instructive for the information which they give on the dental succession in these primitive apes. Of similar dimensions to *Proconsul*, but with a very different cusp pattern, are a number of upper teeth found at Rusinga and the surrounding area. These show some resemblance to the teeth of the Middle Miocene Indian genus, *Sivapithecus*, and in the absence of a strongly developed internal cingulum presumably represent a type which is more generalized (or perhaps more advanced) than *Proconsul*. But more material of this type is needed before a definite diagnosis can be made.

Portions of upper and lower jaws of *Xenopithecus* with the teeth in position have now made it possible to make a fairly complete study of the dentition of this genus, and to demonstrate that it differs from *Proconsul* mainly in its smaller size and in the elongation of the last lower molar. Still smaller than *Xenopithecus* is the gibbon-like *Limnopithecus*, of which a number of jaw fragments are now available and also a maxilla and palate with the upper premolars and molars. It now appears probable, from this material, that at least two species of *Limnopithecus* existed in East Africa in the Early Miocene.

Only a few limb-bones of the Early Miocene hominoids have so far been found, but these have provided information of unusual interest. The femur provisionally referred to *Proconsul* is a long, slender and straight bone, very similar to the Eppelsheim femur, which is probably that of *Dryopithecus*. The humerus is likewise of somewhat delicate construction and, though of about the same length as that of an adult chimpanzee, lacks the powerful muscular ridges which are associated with brachiating habits. The talus and calcaneus of *Proconsul* have already been described by McInnes. It is a striking fact that, in many

就有如此多样的猿类感到困惑，因此要想将它们全部进行分类，绝不是件易事。许多标本属于一种比非洲原康修尔猿（霍普伍德）大得多的猿类，包括它的颌骨和牙齿，其中前臼齿和臼齿的相对大小以及臼齿的齿尖式样都有差异。因此，毫无疑问，它们应该被归入另一物种。1942 年在鲁辛加岛发现的几乎完整的下颌骨也属于这类猿，1947 年间发现的上颌骨残块以及原位的颊齿也属于同一个体。英国－肯尼亚中新世探险队在松戈尔发现一个巨大下颌骨的残块，它代表了一种更大的物种，在某些方面这件颌骨比许多成年雄性大猩猩的颌骨还要粗大。

在所有这些标本中（我们暂时把它们当作原康修尔猿属下的几个不同物种），都存在几个典型特征。与现代猿类相比门齿较小；颌骨联合部狭窄，联合部轴线更加垂直；犬齿巨大，几乎没有明显的性别差异；相应地第一下臼齿有些小，上臼齿具有发育良好的内齿带，内齿带有时候呈现为精细的串珠状，下臼齿具有非常明显的外齿带痕迹。已发现的一些原康修尔猿的未成年颌骨对于了解这些原始猿类的牙齿更替情况很有帮助。鲁辛加岛及其周边地区发现的一些上牙与原康修尔猿的上牙大小相似，但是齿尖样式差异很大。这些表明它们与中新世中期的印度的西瓦古猿属的牙齿具有某些相像之处，并且缺少发达的内齿带，这些大概代表了它们是一种比原康修尔猿更广适（或可能是更高等）的物种。但这需要更多此类材料才可以做出准确的判断。

异猿的上、下颌骨残块及其原位保存的牙齿，使得现在对该属的齿系进行非常完整的研究成为可能，也可以证明它与原康修尔猿之间的差异主要在于其牙齿较小以及最后一颗下臼齿较长。比异猿更小的是一种类似长臂猿的湖猿，现在已经获得了一些它的颌骨破片，以及带有上前臼齿和臼齿的上颌骨和腭骨。现在从这些材料似乎可以判断，在中新世早期的东非很可能曾经存在至少两种湖猿。

目前为止只发现了少量中新世早期人猿超科动物的肢骨，但它们却提供了异常有趣的信息。暂时被归入原康修尔猿的股骨为一根又长又细又直的骨，它与发现于埃珀尔斯海姆的股骨非常相似，后者可能属于森林古猿。肱骨的构造同样有些精致，并且，尽管长度与成年黑猩猩的大致相同，但是缺少供肌肉附着的强壮脊，这些脊与臂行生活习性有关。麦金尼斯已经描述了原康修尔猿的距骨和跟骨。一个惊人的

of their details, they conform much more closely to the corresponding tarsal bones of the cercopithecoid monkeys than those of the modern large apes. It is probable, indeed, that *Proconsul* was not a brachiating specialist, but led a more cursorial type of existence. Combined with certain features of the femur, the tarsal bones indicate that these Early Miocene apes were lightly built and active creatures, capable of running and leaping with considerable agility.

These observations will certainly have an important bearing on the problem of the evolution of the higher Primates, for they suggest that the adoption of the brachiating habits which are characteristic of the modern apes may have been a relatively late acquisition. More limb-bone material of the Early Miocene apes of East Africa is needed in order to follow up the very interesting implications of the specimens already available, and we require also to know more about the structure of the skull (which we have already learned was much shorter in the facial region than in the recent large apes). There is a reasonable prospect that, in continuing their excavations during the coming season, Dr. Leakey and Dr. McInnes may be able to secure this important evidence. Meanwhile, it is not a little remarkable to find, as the result of the excavations in Kenya, that already in Early Miocene times East Africa was populated with numerous species of primitive apes ranging in close gradations of size from small creatures no larger than *Hylobates* to great apes of gorilloid dimensions.

(**161**, 667-669; 1948)

W. E. Le Gros Clark: F.R.S., Department of Human Anatomy, Oxford.

References:

1. Broom, R., and Schepers, G. W. H., *Transvaal Mus. Mem.*, No. 2 (1946).

2. Dart, R., *Nature*, **155**, 195 (1925).

3. Broom, R., *Nature*, **159**, 672 (1947).

4. Broom, R., and Robinson, J. T., *Nature*, **160**, 153 (1947).

5. Broom, R., and Robinson, J. T., *Nature*, **160**, 430 (1947).

6. Hopwood, A. T., *J. Linn. Soc.*, **38**, 437 (1933).

7. Leakey, L. S. B., *Nature*, **152**, 319 (1943).

8. McInnes, D. G., *J. East Africa and Uganda Nat. Hist. Soc.*, **17** (1943).

9. Le Gros Clark, W. E., *Nature*, **160**, 891 (1947).

924

事实是，与现代大型猿类相比，在许多细节上它们的跗骨与猴超科的猴类更加一致。原康修尔猿可能确实不适于臂行生活，而是更适于灵巧型行走的生存方式。这些跗骨与股骨的某些特征共同暗示了这些中新世早期猿类是身体结构轻盈并且很活跃的物种，它们能够非常敏捷地奔跑和跳跃。

上述观察对研究高等灵长类的进化问题具有相当重要的意义，因为它们揭示了作为现代猿类所特有的臂行生活习性可能是在较晚时候才获得的。要想进一步探寻这些现有标本所具有的科学意义，还需要在东非发现更多中新世早期猿类的肢骨材料，我们也需要对头骨的结构（我们现在已知道这些面骨区都比近代的大型猿类的要短）了解得更多。我们希望在接下来的野外发掘工作中，利基博士和麦金尼斯博士能够保证这一重要证据的安全。同时从肯尼亚的发掘结果不难看出，早在中新世早期，在东非就生活着许多大大小小的原始猿类，其小者比长臂猿还小，而大者却类似大猩猩的体形。

(刘皓芳 翻译；同号文 审稿)

A (?) Promethean *Australopithecus* from Makapansgat Valley

R. A. Dart

Editor's Note

While investigating stone-age localities in the Makapansgat Valley, Raymond Dart and his students picked over spoil-heaps from nearby lime-works, discovering fossils of a similar age to those from Sterkfontein. Among them was the occiput (rear skull) of a "robust" australopithecine, similar to *Paranthropus* discovered at Kromdraai. In this brief report, Dart notes that whereas none of the australopithecine localities had yielded stone tools, the wide variety of animal bones seemed to have been subjected to great heat. From this evidence, Dart conceived that australopithecines used fire (hence "promethean") and perhaps exploited the bones and horns of their prey as tools and weapons, what Dart called the "osteodontokeratic" culture. These ideas have since fallen from favour for lack of evidence.

DURING 1947 the Bernard Price Foundation for Palaeontological Research in the University of the Witwatersrand maintained an archaeological party in the Makapansgat Valley, twelve miles north-east of Potgietersrust. Their principal objective was to excavate the Cave of Hearths, a site of human habitation containing palaeoliths of Chell-Acheul facies embedded in limestone breccia.

About a mile lower down the same valley is the Limeworks site where no implements have been found and from the dumps of which students and members of the staff of the Anatomy Department have been recovering primate and other mammalian fossils for several years past. The presence of *Parapapio broomi*, Jones, in the lowest stratum of the Limeworks cavern breccia shows that this earliest stratum belongs to the same geological horizon as that of Sterkfontein. Consequently it was a happy confirmation of our expectations when Mr. James Kitching came across an australopithecine occiput in this type of breccia during September 1947, while searching among the Limeworks dumps.

The new occiput (see photograph) exhibits an *Australopithecus* comparable in brain size to *Paranthropus robustus*, Broom, from Kromdraai. Owing to partial synostosis of the sagittal and lambdoid sutures the cranial fragment comprises the major portion of the occipital bone including most of the right margin of the foramen magnum and the posterior third of each parietal bone. It displays several significant humanoid anatomical features, namely, an expanded planum occipitale; an inferior situation of the inion relative to the opisthocranion and a consequent downward deflexion of the palnum nuchale

在马卡潘斯盖河谷发现的一种普罗米修斯南方古猿（尚未确定）

达特

编者按

雷蒙德·达特和他的学生们在马卡潘斯盖河谷调查石器时代的遗址，在仔细检查石灰厂附近的废石堆时，发现了一些化石，这些化石和在斯泰克方丹发现的化石年代相近。其中有一块"粗壮型"南方古猿亚科的枕骨（后颅骨），与在克罗姆德拉伊发现的傍人非常类似。在这份简要的研究报告中，达特认为没有一个南方古猿亚科遗址中有石器工具，许多野生动物的骨似乎都被大火灼烧过。根据这一证据，达特认为南方古猿亚科能够使用火（因此称为"普罗米修斯"），并且也许能将他们捕获的猎物的骨和角加以利用，做成工具和武器。达特将其称之为"骨牙角"文化。由于缺乏证据，这些观点到现在还未被完全认同。

1947 年间，威特沃特斯兰德大学的伯纳德普莱斯古生物学研究基金会在距离波特希特斯勒斯东北方向 12 英里的马卡潘斯盖河谷设立了一个考古队。他们的主要目的是发掘哈斯洞穴，这是一处被埋入石灰石角砾岩的人类居住地遗址，发现有阿舍利期的旧石器。

马卡潘斯盖河谷下游约 1 英里处有一个石灰厂，在那里没有发现任何器具。在过去的几年里，解剖学系的学生和工作人员一直都在废墟中出土灵长类和其他哺乳动物的化石。在石灰厂洞穴的角砾岩最下层保留着琼斯命名的布鲁姆副狒的化石，这表明这一早期的角砾岩与斯泰克方丹的属于同一地质年代。随后在 1947 年 9 月间，詹姆斯·基钦先生在石灰厂废墟中进行搜寻时，在这种类型的角砾岩中偶然发现了一件南方古猿亚科的枕骨，这很好地验证了我们的猜测。

这件新枕骨（见图）所展现的南方古猿与布鲁姆在克罗姆德拉伊农场发现的粗壮傍人的脑部大小相当。由于矢状缝和人字缝部分骨联合，可以认为这些颅骨破片构成了绝大部分枕骨，包括枕骨大孔的大部分右侧区域和每一顶骨的后 1/3 部分。这块枕骨显示出几个显著的人类解剖特征，即延展的枕平面，枕外隆凸点相对于颅后点较低，向下偏斜的项平面（这反映了人类直立行走的姿势），中度发育的枕横圆

(thus proving the upright posture); the existence, but moderate development, of a torus transversus occipitalis (thus indicating the lack of a gorilloid "bullneck" in the type) and the presence of a sutura transversa occipitalis (or sutura mendosa) with subsidiary sulci forming a complex "Inca" bone (which has never been found hitherto in anthropoids).

Australopithecine occiput from the Makapansgat Valley

In all these features the Makapansgat *Australopithecus* distinguishes itself from its smaller-brained contemporary *Plesianthropus* at Sterkfontein. Further, if *Paranthropus* is geologically more recent (as Broom believes), the Makapansgat type probably differed considerably from *Paranthropus*. The final answer to this question must await the discovery of further remains.

But the most potent reason for separating the new fossil from its relatives is the difference between the bone breccia of Makapansgat Limeworks on one hand and the bone breccias of all the other manape sites on the other. The breccia shows first that the Makapansgat *Australopithecus* had advanced hunting habits and secondly that it was acquainted with the use of fire. Both of these inferences from the breccia were announced nine months before the discovery of the occiput in a communication to the First Pan-African Congress of Prehistory at Nairobi in January, 1947. They are briefly recapitulated here to explain why the name, *Australopithecus prometheus*, has been given to this specimen.

In addition to the Sterkfontein types of baboons such as *Parapapio broomi*, Jones, and *Parapapio jonesi*, Broom, the Makapansgat breccia has yielded fourteen species of Bovidae, eight of which appear to be new; three carnivores (namely, lion, hyena and jackal); two extinct pigs; a rhinoceros; a hippopotamus; and two giraffes (including the extinct *Griquatherium*). Thus apart from the elephant we have concrete evidence that these creatures drew upon every family of big game found in Southern Africa for their food supply. The dentitions of the big game specimens, however, show that these Australopithecines were more frequently successful in slaying juvenile and aged representatives of these various mammalian families. But the profusion of the bone breccia shows that they were capable and successful in their hunting of very big game.

枕（提示缺乏类似大猩猩的"粗壮的颈部"）以及一个枕横缝（或称之为假缝）和一些附带的沟槽，它们共同构成一个复合"印加"骨（目前为止在类人猿中尚未发现这种结构）。

马卡潘斯盖河谷的南方古猿亚科枕骨

马卡潘斯盖南方古猿所具有的这些特征使之与其同时代的脑量稍小的斯泰克方丹迩人区别开来了。此外，布鲁姆认为，如果从地质学角度来说傍人更接近于现代的话，那么马卡潘斯盖类型就可能与傍人存在相当大的差异了。这一问题的最终答案有待更多化石的发现才能解答。

但是将新发现的化石与其相关物种区分开来的最强有力的依据是马卡潘斯盖的石灰厂的骨角砾岩与所有其他人猿遗址的骨角砾岩之间的差异。角砾岩表明马卡潘斯盖南方古猿不仅已经改进了狩猎习性，而且已经熟悉了火的使用。在 1947 年 1 月于内罗毕举行的第一届泛非史前学大会上，已经宣布了我们从角砾岩得到的这两个推论，这比发现该枕骨的时间早 9 个月。在这里简要地概括一下以解释为什么将这一标本命名为普罗米修斯南方古猿。

除了斯泰克方丹类型的狒狒（如琼斯命名的布鲁姆副狒和布鲁姆命名的琼斯副狒）外，已经从马卡潘斯盖角砾岩出土了 14 种牛科动物（其中有 8 个似乎是新种），3 种肉食动物（即狮子、鬣狗和胡狼），2 种已灭绝的猪，1 种犀牛，1 种河马和 2 种长颈鹿（包括已灭绝的长颈鹿科中的 *Griquatherium* 属）。因此除了大象，我们现在有确实的证据表明这些南方古猿亚科是依靠在南非发现的每个科的大型猎物来满足自己的食物需求的。然而，大型猎物标本的齿系显示，这些南方古猿亚科通常更善于猎杀各种哺乳动物科中的幼年和老年动物。而丰富的骨角砾岩表明它们有能力成功猎取到很大的猎物。

In 1925 the presence of *Australopithecus* was unsuspected because of the great size of these bones and their charred appearance—so the specimens of this Makapansgat Valley Limeworks bone breccia were submitted to two chemists (Drs. Moir and Fox) for analysis to discover whether the bones had been subjected to fire. From the great size of the beasts represented in the breccia, the intentional fracturing or splitting of the bones, and corroborative reports of the chemists relative to the presence of carbon, the deposit was claimed at that time as a primitive human kitchen midden. Today it appears that all this "human" activity at Makapansgat Limeworks site is the handiwork of man's australopithecine predecessors. No traces of stone implements have been found in the Makapansgat Limeworks strata nor have they been found at Taungs, Sterkfontein and Kromdraai. But we have found at Makapansgat a number of large ungulate humeri the epicondylar ridges of which were battered and broken and apparently as the result of deliberate use prior to their fossilization. Most of the bones are fresh and firm, and display clean clear lines of gross fracture before fossilization, but others crack and crumble easily. These exhibit so fine a state of macroscopic and microscopic cracking as to preclude explaining their pre-fossilization fragmentation by the crude impact of stone or bone. The most likely force to have been responsible for their crumbling and altered condition is heat. That fire was the splintering agent is corroborated by their discoloration, disintegration and the chemical reports of free carbon made more than twenty years ago, and by the vesicular character of the adjacent glassy collophanites and clays recently discovered in the breccia by Dr. V. L. Bosazza and the comparative analysis he has made of ashy deposit from the breccia with ash from local Transvaal wood.

(**162**, 375-376; 1948)

Raymond A. Dart: Medical School, Hospital Street, Johannesburg, June 15.

　　1925年，南方古猿的存在尚未被意料到，因为这些骨太大并且它们的外观都是烧焦的，因此在马卡潘斯盖河谷石灰厂发现的骨角砾岩标本被送到了两位化学家（莫伊尔博士和福克斯博士）的手中，以分析这些骨是否曾受到过大火的灼烧。从角砾岩中发现的野兽个头之大、这些曾被故意折断或者拆分的骨，以及化学家们给出的关于碳存在的确证报告都表明，该沉积物在那时是作为原始人类的厨房内堆积物。现在，马卡潘斯盖河谷石灰厂遗址的所有这些"人类"的活动可能都是这些南方古猿亚科祖先们的行为。在马卡潘斯盖河谷石灰厂地层没有发现任何石器的踪迹，在汤恩、斯泰克方丹和克罗姆德拉伊也没有发现石器。但是我们在马卡潘斯盖发现了许多大型有蹄类动物的肱骨，其上髁的脊很多被击打破损，表明这些肱骨在石化前曾被使用过。大部分骨是新而结实的，表明石化前就有清晰的骨折纹路，其他的骨则易碎裂。这些骨的宏观裂缝和微观裂缝保存得如此之好，足以排除在石化前它们被石器或骨强烈击打而破碎的可能。导致这些裂缝和状态改变最可能的因素就是火。另外，这些骨的脱色现象、瓦解特点和含有的自由碳，在二十多年前的化学报告中就报道了，这些现象都支持裂缝是由火造成的。最近博萨扎博士发现当地角砾岩中含有多孔易燃的胶磷矿和黏土，以及他对角砾岩中的灰烬沉积物和德兰士瓦当地木材燃烧而成的灰烬所进行的比较分析，都进一步支持裂缝是由火造成的。

（刘皓芳 翻译；刘武 邢松 审稿）

Another New Type of Fossil Ape-man

R. Broom

Editor's Note

Robert Broom was perhaps the single most important figure in palaeoanthropology in South Africa in the early twentieth century. Without Broom's researches from the mid-1930s onwards, Dart's 1925 announcement of *Australopithecus africanus* from Taung would have remained a curiosity. By the late 1940s, Broom felt justified in saying that South Africa would yield most stages of human evolution: having unearthed slender australopithecines such as *Plesianthropus* from Sterkfontein and more robust forms such as *Paranthropus* from Kromdraai, he moved to a new site. Although only a mile from the main Sterkfontein quarry, Swartkrans quarry immediately yielded a very large form of robust australopithecine, which Broom called *Paranthropus crassidens* (now assigned to *P. robustus*, along with all other South African robust forms).

SOME years ago, I pointed out that in my opinion the cave deposits in the dolomite of the Transvaal when fully worked will give us the remains of most stages of early man and pre-man that have inhabited South Africa from probably Middle Pliocene to Recent. In the Sterkfontein area alone there are apparently in about ten square miles more than a hundred different cave deposits, and many of these are of quite different ages judging by the faunas. Almost all the animals at Kromdraai main deposit are quite different species from those at the main quarry at Sterkfontein. The jackals, the sabre-toothed tigers, the baboons, the dassies and the ape-men are quite different species.

At the beginning of November we started work at a new spot in association with the California University Expedition. Though the deposit is on the farm Swartkrans and only a mile from the main Sterkfontein quarry, the fauna so far as we have gone proves to be very different—whether older or younger we cannot yet say. Luckily we found teeth of a new type of ape-man within ten days, and a week later discovered much of a mandible with the complete lower premolars and molars.

The new mandible is not closely allied to that of the Sterkfontein ape-man *Plesianthropus*; the teeth are allied to those of the Kromdraai ape-man *Paranthropus*; but they are much larger and differ in a number of respects.

We have found two beautiful upper incisors and a perfect upper canine. These teeth are almost typically human, though a little larger than most human teeth. The canine has no deep infolding of the enamel on the lingual side as we have in the canines of *Plesianthropus*. It is also interesting to note that the canine and the 2nd incisor have been in contact, as each has been abraded by the other.

932

又一新型猿人化石

布鲁姆

编者按

20 世纪上半叶，罗伯特·布鲁姆可能是南非古人类学界最重要的人物。如果没有布鲁姆在 20 世纪 30 年代中期之后的研究，达特于 1925 年在汤恩发现的南方古猿非洲种至今仍然是个谜。到 20 世纪 40 年代末，布鲁姆提出人类进化的绝大部分阶段产生于南非大陆：他挖掘出纤细型南方古猿如斯泰克方丹的迩人、粗壮型南方古猿如克罗姆德拉伊的傍人，之后他转战到新的地点。在距斯泰克方丹采石场仅一英里的斯瓦特克朗斯采石场，布鲁姆随后发现了粗壮型南方古猿的又一巨型种类，并称之为傍人粗齿种（目前已与南非所有其他粗壮型物种一起归入粗壮傍人）。

多年前我曾指出，如果充分挖掘德兰士瓦的白云石洞穴沉积物，我们会发现居住在南非的早期人类和猿人绝大部分进化阶段的化石，时间大约在上新世中期到全新世。仅在斯泰克方丹地区，大约十平方英里的区域内就有一百多个不同的洞穴沉积物，我们从动物群来判断，其中许多洞穴沉积物都代表着迥然不同的年代。从克罗姆德拉伊主要沉积物中出土的几乎所有的动物都与在斯泰克方丹主采石场发现的物种有相当大的差异，如胡狼、剑齿虎、狒狒、蹄兔和猿人之间的差异非常大。

11 月初，我们和加利福尼亚大学探险队合作，在一新遗址处开始了发掘工作。尽管此处沉积物所在的斯瓦特克朗斯农场距离斯泰克方丹主采石场仅仅一英里，但是就我们发现的动物群而言，它们存在着很大差异，但是还不能断定二者谁的年代更久远。幸运的是，我们在 10 天之内就发现了一种新型猿人的牙齿，并在一星期后又发现了大半块下颌骨化石——具有完整的下前臼齿和臼齿。

这件新型下颌骨与斯泰克方丹猿人（迩人）的下颌骨并无紧密的亲缘关系；而牙齿与克罗姆德拉伊猿人（傍人）的牙齿具有亲缘关系，但它们的尺寸更大，并且在许多方面都有所不同。

我们发现了两颗完好的上门齿和一颗完美的上犬齿。这些牙齿几乎都是典型的人类牙齿，只不过比大多数人的牙齿稍大一点。犬齿并不像我们观察的迩人那样舌侧的珐琅质具有深层折叠现象。我们很感兴趣的一点是，其犬齿和第二门齿紧密接触，彼此间有相互磨损的痕迹。

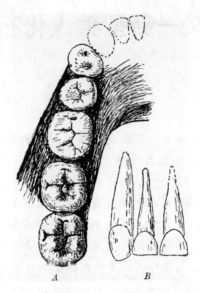

Teeth of *Paranthropus crassidens* Broom. $\frac{3}{4}$ Natural size. *A.* Left mandibular ramus. The 1st premolar is a little displaced in the specimen and has been restored to its natural position. The 3rd molar is drawn from the 3rd right molar reversed. *B.* Right upper incisors and canine.

The mandible was found in the same deposit but at a spot about 10 ft. from the isolated upper teeth, and it cannot belong to the same individual though it is clearly of the same species.

The mandible is very massive. The horizontal ramus is preserved from the 2nd premolar to the 2nd molar; the 1st premolar is also preserved but a little displaced. Most of the inner side of the symphysis resembles more closely that of the Heidelberg jaw than any other specimen of man or ape-man I know.

But the teeth are relatively huge. The drawing is of the occlusal view with the 1st premolar in position, and the left 3rd molar drawn from the right tooth reversed.

This new type of ape-man is not closely allied to either *Australopithecus* or *Plesianthropus*, but is allied to *Paranthropus*. When a skull is discovered it may prove to belong to a new genus; but provisionally we may call it *Paranthropus crassidens*.

As further evidence of the richness of our deposits, attention may be directed to the wonderful finds being made in the northern Transvaal at the Makapan caves. For about a couple of years the Bernard Price Institute has been working there, and Prof. R. A. Dart has recently announced the discovery of a remarkable ape-man occiput, and a few weeks later of a very fine mandible. These he has referred to a species of *Australopithecus*, and has called the animal *A. prometheus*. Though I am not convinced that he made fire, I am of opinion that the being belongs not only to a new species but also to a new genus.

934

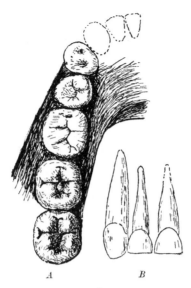

傍人粗齿种的牙齿（布鲁姆）。该图是真实尺寸的 $\frac{3}{4}$。A. 左下颌支。第一前臼齿在本标本中有点移位了，图中将其复原到了自然位置上。第三臼齿是根据右侧第三右臼齿镜像画出来的。B. 右上门齿和犬齿。

下颌骨和单颗的上牙发现于同一沉积物中，但它们被发现的地点相距约10英尺，因此尽管它们明显属于同一物种，但并不能将其归属于同一个体。

该下颌骨非常大。水平支从第二前臼齿到第二臼齿都被保存下来了；第一前臼齿虽然也保存下来了，但是有些移位。与我所知的人类或者猿人任何其他的标本相比，该颌骨联合部内侧面的大部分与海德堡下颌骨更为相像。

牙齿则相对较大。图中给出的是原位上第一前臼齿的咬合面，第三左臼齿是根据右侧牙齿镜像画出来的。

这种新型的猿人与南方古猿或者迩人都没有密切的亲缘关系，而与傍人具有亲缘关系。如果再发现一件头骨，就能证明它是否属于新属，目前我们可以暂时称其为傍人粗齿种。

随着我们的沉积物中证据进一步丰富，人们可能会把注意力转移到德兰士瓦北部的马卡潘洞穴中所取得的令人惊奇的发现。伯纳德普莱斯研究所花了大约两年时间对那里进行挖掘，达特教授最近宣称发现了一件引人注目的猿人枕骨，并在几周后又发现了一具非常完好的下颌骨。他将这些定为南方古猿的一个物种，并称之为普罗米修斯南方古猿。尽管我不相信这种猿人会生火，但是我认为它们不仅是一种新的物种，而且是一新属。达特教授还有其他一些非常重要的化石，对于这些化石

935

Prof. Dart has some other most important remains which he is describing. But I feel at liberty only to refer to what has been already announced.

(**163**, 57; 1949)

R. Broom: Transvaal Museum, Pretoria, Dec. 7.

他还在描述当中。但是我只想冒昧地谈论一下那些已经发表了的材料。

（刘皓芳 翻译；吴秀杰 审稿）

A New Type of Fossil Man

R. Broom and J. T. Robinson

Editor's Note

The cave site of Swartkrans, near Sterkfontein, yielded a very large form of robust australopithecine, which its discoverer, Robert Broom, named *Paranthropus crassidens*. But the same site also produced the jaw of another type of fossil man, unlike *Paranthropus* and more human. It was found in a lens of much more recent material than that which had yielded the australopithecines. Broom and Robinson compared the jaw with the famous Mauer mandible, exemplar of "Heidelberg Man" (now called *Homo heidelbergensis*), calling it *Telanthropus capensis*. This form was later reclassified as *Homo erectus*.

IN the cave at Swartkrans which has now yielded the jaws and skulls of the huge ape-man *Paranthropus crassidens*, there was found by Mr. J. T. Robinson, on April 29, 1949, the lower jaw of what is fairly manifestly a new type of man. Though this was discovered in the same cave as the large ape-man, it is clearly of considerably later date. In the main bone breccia of the cave deposit there has been a pocket excavated and refilled by a darker type of matrix. The pocket was of very limited extent, being only about 4 ft. by 3 ft. and about 2 ft. in thickness. The deposit was remarkably barren, there being no other bones in it except the human jaw and a few remains of very small mammals. We are thus at present unable to give the age of the deposit except to say that it must be considerably younger than the main deposit. If the main deposit is Upper Pliocene, not improbably the pocket may be Lower Pleistocene.

The jaw is smaller than many human jaws, though the 3rd molar is larger than in any known man. On the left ramus the three molars are preserved in good condition though a little worn, and the last two molars are well preserved in the right ramus. No other teeth are preserved, though we have sockets of all of them.

The jaw has been a little broken during fossilization, and slightly crushed; but otherwise it is nearly perfect except for the loss of most of the left condyle and the whole of the right. A very small part of the lower symphyseal region is lost. The symphysis runs downwards and slightly backwards, making an angle with the base of the ramus of about 75°. The depth of the symphysis is about 33 mm. The horizontal ramus is remarkably shallow. At the 1st molar it is only 29 mm. The base of the ramus is nearly level, and the angle is rounded and scarcely at all below the general level.

The ascending ramus has apparently been fairly broad, but very shallow. Fortunately the cast of the side of the one condyle is preserved, and the height of the back of the jaw is

一种新型人类化石

布鲁姆，鲁滨逊

编者按

邻近斯泰克方丹，在斯瓦特克朗斯的洞穴出土了一种大型的粗壮型南方古猿亚科化石，它的发现者罗伯特·布鲁姆将其命名为傍人粗齿种。虽然在同一个洞穴中还发现了另一种类型的古人类颌骨化石，但是这个颌骨与傍人的不同而更像人类。它所在的沉积物的年代比发现南方古猿亚科所在的沉积物的年代晚很多。布鲁姆和鲁滨逊将此颌骨与作为"海德堡人"（现在称为人属海德堡种）标本的著名的毛尔下颌骨化石进行比较，并将其命名为开普远人。这种类型后来被归为直立人。

在斯瓦特克朗斯一处洞穴中，已出土过一个巨型猿人傍人粗齿种的颌骨和头骨，1949 年 4 月 29 日，鲁滨逊先生在其中发现了一个下颌骨，很明显，它是一个新型古人类的下颌骨。尽管这个下颌骨与早先发现的大型猿人是在同一个洞穴里发现的，但是很明显这种新型古人类所处的年代要晚得多。在洞穴主要骨角砾岩的沉积物中有一个被挖掘过的凹坑的痕迹，凹坑曾被一种较暗类型的基质回填过。这个凹坑的大小约为 4 英尺×3 英尺，厚度约为 2 英尺。该处沉积物非常贫瘠，除了这一人类颌骨和少量小型哺乳动物的残骸外就没有其他骨了。因此我们现在除了能确定这些沉积物一定比主沉积物的年代晚很多外，并不能指出这些沉积物的年代。如果主沉积物是上新世晚期的，那么这处凹坑为下更新世的也并非不可能。

该颌骨比许多人类的颌骨小，但第三臼齿比现在任何已知人类的都要大。左下颌支的三颗臼齿除有一点磨损外都保存得很好，右下颌支最后两颗臼齿保存良好。其他牙齿都没有保存下来，但其牙槽都在。

由于石化作用，颌骨有一点破损和轻微的变形；但是除了缺少左侧大部分髁突和右侧全部髁突以外，其余部分的保存状况近乎完好。联合部的下面丢失了很小的一部分。联合部向下及稍向后方向进行，与下颌支基部形成约 75°角。联合部厚约 33 毫米。水平支非常浅，距第一臼齿处仅有 29 毫米。下颌支基部接近水平，角为圆形，几乎不低于总体水平线。

很明显，上升支相当宽阔但是很浅。幸运的是，一个髁突的侧面模型被保存下来了，而且颌骨背面的高度仅在颌骨水平线上方约 55 毫米处。最后一颗臼齿的外面

only about 55 mm. above the horizontal base of the jaw. Outside the last molar is a wide hollow as in the Heidelberg jaw, and the jaws of the ape-men. There is no simian shelf and the whole symphysis is not unlike that of Heidelberg man, but smaller. The mylohyoid groove runs up to the foramen as in typical human skulls. In *Paranthropus crassidens* the groove is, as in *Eoanthropus*, lower down.

The incisors and canines, so far as can be judged from the sockets, are human. The premolars have been a little larger than typically human premolars. The 1st molar is almost typically human in size and structure. It has five cusps and a trace of a sixth. The 2nd molar is also nearly human. It is larger than in *Homo*, and has a small sixth cusp. The 3rd molar is the largest of the three molars. It has five well-developed cusps and a small sixth.

The jaw in general structure comes nearest to that of Heidelberg man, but is smaller and has a lower horizontal ramus. The teeth differ markedly in the 3rd molar, being the largest of the series.

In the large size of the molars there is some resemblance to the condition seen in *Plesianthropus* and *Paranthropus*; but in this human jaw the molars are much smaller. In *Plesianthropus transvaalensis* the three molars measure in the male about 43 mm.; in *Paranthropus robustus* they measure 45 mm. In *Paranthropus crassidens* the three measure in the male about 51 mm., while in this new human jaw they only measure 38.4 mm. In the South African native the molars measure about 35 mm.

The new type of man represented by this fossil jaw we propose to call *Telanthropus capensis*. We regard him as somewhat allied to Heidelberg man, and intermediate between one of the ape-men and true man.

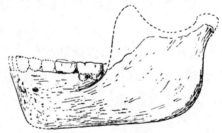

Fig. 1. Side view of lower jaw of *Telanthropus capensis* B. and R. (half-size)

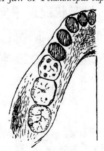

Fig. 2. Occlusal view of teeth of the left ramus of *Telanthropus capensis* B. and R. (half-size)

是一个大的空洞，这与海德堡下颌骨以及猿人颌骨的情况一样。它没有猿的框架，整个联合部与海德堡人的不同，但相对较小。下颌舌骨沟向上延伸至孔处，这与典型的人类头骨一样。傍人粗齿种的沟与曙人一样都是降低的。

目前根据牙槽可以判定其门齿和犬齿是属于人类的。前白齿比典型的人类前白齿大一点。第一臼齿在大小和结构方面都几乎与典型的人类牙齿一样。第一臼齿具有五个齿尖，并有第六齿尖的痕迹。第二臼齿也接近于人类的第二臼齿，但比人属的要大，并且具有一个小的第六齿尖。第三臼齿是三颗臼齿中最大的。该臼齿具有五个发育完全的齿尖和一个小的第六齿尖。

在整体结构上，该颌骨与海德堡人的最接近，只是稍小并且水平支较低。第三臼齿明显不同，是所有臼齿系列中最大的。

这种大尺寸的臼齿与在迩人和傍人中观察到的情况有些相似；但是在这具人类颌骨中，臼齿要小得多。德兰士瓦迩人男性的三颗臼齿的尺寸约为 43 毫米；粗壮傍人的尺寸为 45 毫米。傍人粗齿种男性的三颗臼齿的尺寸约为 51 毫米，而在这具新型人类颌骨中，只有 38.4 毫米。南非本土人的臼齿尺寸约为 35 毫米。

我们建议将这具颌骨化石代表的新型古人类命名为开普远人。我们认为它与海德堡人有些亲缘关系，介于猿人和真正的人类之间。

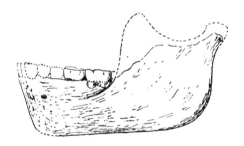

图 1. 开普远人下颌骨的侧面观（布鲁姆和鲁滨逊）（原尺寸的一半）

图 2. 开普远人的左侧下颌支牙齿咬合图（布鲁姆和鲁滨逊）（原尺寸的一半）

It might be thought that as *Plesianthropus transvaalensis*, of which we now know about a dozen skulls and about a hundred and fifty teeth, shows considerable variation, this supposed human jaw might be an extreme variant of *Paranthropus crassidens*. In man there are no doubt great variations, and the difference in size between the jaw of a small Bushman woman and the Wadjak and Heidelberg jaws is nearly as great as between our supposed human jaw and the huge *Paranthropus crassidens* jaw. We now have three good lower jaws and a number of isolated teeth of *P. crassidens*, and there is not much variation in either size or structure. It may be held that all these large jaws are male jaws, and the small jaw that of a female; but not only the size of the teeth but also the structure seems to rule out such a view. The 1st molar in the type of *P. crassidens* is about 16 mm. by 14.6 mm. In the supposed human jaw it is only 12 mm. by 11.5 mm. Further, the structure of the two teeth differ considerably. The typically human mylohyoid groove in our supposed man, and the certainly not typically human groove in *Paranthropus*, seem to make it certain that the two jaws belong to different genera. If we are right in believing that our new jaw is in structure intermediate between *P. crassidens* and *Homo*, it is but natural that there should be numerous resemblances to both.

(**164**, 322-323; 1949)

R. Broom and J. T. Robinson: Transvaal Museum, Pretoria, July 2.

　　从目前我们拥有的十多件头骨和150颗左右的牙齿来看，德兰士瓦迩人的变异相当大，我们可以推测人类的颌骨可能是傍人粗齿种的一种极端变异形式。人类的颌骨无疑也存在很大的变异，矮小的布须曼妇女颌骨、瓦贾克人颌骨以及海德堡下颌骨在大小上存在着较大的差异，这个差异相当于我们现在假定的人类颌骨与巨大的傍人粗齿种的颌骨之间的差异。我们现在有三件完好的下颌骨和傍人粗齿种的一些单个的牙齿，它们在大小或结构上都没有太大变异。也许有人认为所有这些大的颌骨都是男性的，小的颌骨则是女性的；但是，不论从牙齿的大小还是从结构上看，似乎都可以排除这种观点。傍人粗齿种类型的第一臼齿大约为 16 毫米 ×14.6 毫米。假定的人类颌骨只有 12 毫米 ×11.5 毫米。另外，两颗牙齿的结构差异很大。根据我们假定的人类的典型人类下颌舌骨沟以及傍人的显然非典型人类下颌舌骨沟，似乎可以确定这两件颌骨属于不同的属。我们认为这个新型颌骨在结构上介于傍人粗齿种和人属之间，如果我们的这一想法是正确的，那么二者之间具有很多的相像之处也就不足为奇了。

（刘皓芳 翻译；吴秀杰 审稿）